国外优秀数学著作
原 版 系 列

A Course in Analysis

—Vol. IV, Fourier Analysis, Ordinary Differential Equations, Calculus of Variations

分析学教程

——第4卷，傅里叶分析，常微分方程，变分法

[英] 尼尔斯·雅各布 (Niels Jacob)

[英] 克里斯蒂安·P. 埃文斯 (Kristian P. Evans) 著

（英文）

哈尔滨工业大学出版社
HARBIN INSTITUTE OF TECHNOLOGY PRESS

黑版贸审字 08 - 2020 - 190 号

图书在版编目(CIP)数据

分析学教程. 第 4 卷,傅里叶分析,常微分方程,变分法 = A Course in Analysis:Vol. IV,Fourier Analysis,Ordinary Differential Equations, Calculus of Variations:英文/(英)尼尔斯·雅各布(Niels Jacob),(英)克里斯蒂安·P. 埃文斯(Kristian P. Evans)著. —哈尔滨:哈尔滨工业大学出版社,2023.3

ISBN 978 - 7 - 5767 - 0614 - 7

I.①分⋯ II.①尼⋯ ②克⋯ III.①数学分析 – 英文 IV.①O17

中国国家版本馆 CIP 数据核字(2023)第 030372 号

FENXIXUE JIAOCHENG:DI-SI JUAN,FULIYE FENXI,CHANGWEIFEN FANGCHENG,BIANFENFA

World Scientific

策划编辑	刘培杰　杜莹雪
责任编辑	刘家琳
封面设计	孙茵艾
出版发行	哈尔滨工业大学出版社
社　　址	哈尔滨市南岗区复华四道街 10 号　邮编 150006
传　　真	0451 - 86414749
网　　址	http://hitpress. hit. edu. cn
印　　刷	哈尔滨博奇印刷有限公司
开　　本	720 mm×1 000 mm　1/16　印张 49.5　字数 860 千字
版　　次	2023 年 3 月第 1 版　2023 年 3 月第 1 次印刷
书　　号	ISBN 978 - 7 - 5767 - 0614 - 7
定　　价	118.00 元

(如因印装质量问题影响阅读,我社负责调换)

Preface

As in the previous volumes, in our preface we would like to thank those who have supported us in writing this volume. We are grateful to Saroj Limbu and in particular to James Harris and Laura Ladbrooke who have type-written the manuscript. Further thanks go to Huw Fry and Elian Rhind for proofreading parts of the text. We also want to thank the Department of Mathematics, the College of Science and Swansea University for providing us with funding for typewriting.

We would also like to highlight that figures 25.6-25.8, 25.11, 25.12, 25.14 and 25.15 were created using Mathematica and we would like to thank Wolfram for permission to do this.

Finally, we want to thank our publisher, in particular Tan Rok Ting and Ng Qi Wen, for a pleasant collaboration.

Niels Jacob
Kristian P. Evans
Swansea, April 2018

Introduction

The first three volumes of our Course cover material which must be considered as the fundamentals of any analysis training of undergraduates in mathematics. Starting with this volume, we treat material which is still basic for every serious student of analysis, however the general mathematics student might be choosing modules more selective. In this volume we discuss Fourier Analysis, Ordinary Differential Equations, and the basics of the Calculus of Variations (in the one-dimensional case) which includes some results on Analytical Dynamics, i.e. Hamiltonian Mechanics.

In Part 8, Fourier Analysis, we take for granted that the reader has a good understanding of Lebesgue's theory of integration as well as of complex analysis. We start by describing, in Chapter 1, the extraordinary role Fourier analysis has played (and still is playing) in the development of mathematics, not just analysis, a background knowledge of which shall help to better understand certain inner mathematical developments. Chapter 2 introduces trigonometric series and Fourier series and in Chapter 3 we discuss the basic properties of Fourier coefficients. We give a real-valued as well as complex-valued representation of the series and we discuss the importance of orthogonality. In particular we have a look at the decay properties of sequences of Fourier coefficients and their convergence behaviour.

The first central chapter is Chapter 4 where we handle pointwise convergence results and introduce the idea of summability. The Dirichlet kernel and the Fejer kernel are investigated and pointwise convergence resuts following Dirichlet's idea as well as Cesàro and Abel summability are discussed. We also have a look at the concept of "good kernels". Equally central as Chapter 4 is Chapter 5 in which we treat the L^2-theory of Fourier series. In this chapter we fully explore the ideas around orthonormality and series expansions as we prepare more general results on orthogonal series expansions. More specialized is Chapter 6, where we deal with further problems around the convergence of Fourier series. Dini's result is proved as are certain uniform convergence results discussed, for example for series with monotone sequences of Fourier coefficients. We also provide a proof (by following Fejer's modification) of Du-Bois Reymond's counter example of a continuous function with a Fourier series diverging at a point. Finally we show the Fejer-Lebesgue theorem about the almost everywhere Cesàro summability of the Fourier series of an L^1-function.

Chapter 7 highlights the relation between Fourier series and holomorphic functions in the unit disc, as well as their relations to harmonic functions. The Hardy space $\mathcal{H}^2$ is introduced and its basic properties are investigated. As a central tool we introduce the Blaschke product and we also discuss the Nevanlinna class as well as the Poisson-Jensen formula. We give extensions to the spaces $\mathcal{H}^p$ and discuss the boundary behaviour of Poisson integrals. In Chapter 8, we turn our attention first to Fourier series of measures and introduce basic notions such as Fourier-Stieltjes coefficients or positive definite functions. Assuming the Riesz representation theorem (which we will prove in Volume V), we prove the Herglotz-Bochner theorem and we discuss Poisson integrals of measures (on the circle). This led us to introduce Sobolev spaces on the circle and we give a first version of Sobolev's embedding theorem. The chapter closes with a discussion of the Gibbs phenomenon.

Chapter 9 takes up ideas from Chapter 5 and we start to discuss the expansion of L^2-functions into series with respect to certain complete systems of orthonormal functions such as Legendre or Hermite polynomials. This is a topic which we will follow up in Part 9 when dealing with certain second order linear differential equations (Sturm-Liouville operators), and some background material we also collect in Appendix III.

From Fourier series we move on to the Fourier transform. For this we first introduce the Schwartz space $\mathcal{S}(\mathbb{R}^n)$ which we will investigate further in Volume V as an example of a locally convex topological vector space. Of some independent importance for later purposes is the more detailed study of the Friedrichs mollifier. Now, in Chapter 11, we introduce the Fourier transform in $\mathcal{S}(\mathbb{R}^n)$ and give a thorough discussion of its property. The Fourier inversion theorem, Plancherel's theorem as well as the convolution theorem, the latter two we encounter already for Fourier series, are proved and their central role as tools is highlighted. We continue in Chapter 12 to discuss the Fourier transform in L^p-spaces, in particular for $p = 1$ and $p = 2$, and again we have a first look at Sobolev spaces and Bessel potential spaces. Chapter 13 looks at the Fourier transform of Borel measures on $\mathbb{R}^n$. Bochner's theorem is discussed and we have a first look at convolution semi-groups $(\mu_t)_{t \geq 0}$ of sub-probability measures. We close our investigations on the Fourier transform by treating four selected topics: the representation of the Fourier transform of rotational invariant functions with the help of Bessel functions, a first discussion of the Cauchy problem for the wave equation, the Poisson summation formula, and the uncertainty principle. In a quite extensive Appendix I we give a summary of how to embed the theory of Fourier series and the Fourier transform into commutative harmonic analysis, i.e. Fourier analysis on locally compact Abelian groups.

Before we turn in more detail to Part 9 "Ordinary Differential Equations" and Part 10 "Introduction to the Calculus of Variations" two general remarks are in order. Among all topics in more advanced analysis the teaching of ordinary differential equations is arguably most influenced by the emergence of advanced mathematical software. The classical theory has developed a certain systematic approach of how to deal with (systems of) ordinary differential equations and left us with many valuable formulae which, however, are often difficult to evaluate in concrete cases. Thus, analytical methods to deal with formulae have been developed too as have been approximation procedures invented. Nowadays, we may use software to evaluate these formulae, in fact for certain equations some software packages allow to find the solution directly. This does not reduce the theoretical importance of these formulae, however the teaching of and the practical work with ordinary differential equations will have to take the existence of software packages into account and make use of them. This cannot apply to an enterprise such as our Course in Analysis, it would single out one part and lead to quite a different presentation. We restrict ourselves here to derive the analytic theory, but we abstain from evaluating in examples complicated formulae by analytic tools (which can be replaced by proper software).

Differential equations should also be understood in a geometric context and many books are devoted to this idea. This however requires the reader to have some good knowledge in differential geometry and the theory of differentiable manifolds, topics we will discuss in Volume VII. Therefore we have reduced here the geometric aspects to a certain minimum and we will take up certain geometric aspects in the theory of ordinary differential equations and dynamical systems in Volume VII. Exceptions are Chapter 25 and the final chapter of Part 10. Solving first order partial differential equations requires a geometric approach, and we will encounter first order partial differential equations in the context of the calculus of variations and analytical dynamics, i.e. Hamiltonian systems. Hamiltonian systems will also be of great importance when discussing the propagation of singularities of solutions of certain higher order partial differential equations, e.g. the wave equation. We will, and we will have to, investigate such questions in Volume VI, thus before having properly introduced analysis on manifolds. But the calculus of variations and ordinary differential equations form the natural frame for handling Hamiltonian systems. We hope that our approach allows the reader to follow these ideas without having mastered advanced differential geometry.

Part 9 starts with an introductory chapter in which we discuss some first methods, e.g. separation of variables, or the variation of constants, and in which we fix some first terminology, e.g. linear system, homogeneous equation, initial value, etc. We also prove that every higher order equations can be transformed into a first order system such that the solutions correspond to each other. Chapter 16 provides us with the basic general existence theorems for initial value problems. Several variants of the Picard-Lindelöf theorem and the Peano theorem are proved. We show them with the help of fixed point theorems. While Banach's fixed point theorem is easily established, the Schander fixed point theorem needed for the proof of the Peano theorem requires the Brouwer fixed point theorem, a proof of which is discussed in Appendix IV assuming from algebraic topology the result that the sphere as boundary of a ball is no retract of the ball. The notion and the existence of a maximal solution is discussed. While Chapter 16 is devoted to a single equation, in Chapter 17 we extend these results to systems.

Due to their theoretical and practical importance we discuss in great detail linear systems of first order equations. In Chapter 18 we handle systems with constant coefficients. We show the relation to the eigenvalue problem for the system matrix and derive the standard formulae for the solutions. We assume the reader to have background knowledge of the Jordan normal form of a matrix. While for constant coefficient systems the formulation of results with the help of the exponential of the system matrix is a nice possibility to present results and opens the way to one-parameter operator semi-groups, we will see in Chapter 19 when dealing with systems with variable coefficients that matrix functions are in fact a very efficient tool. The basic results are obtained by investigating fundamental systems and the role of the Wronski determinant. As mentioned above, we do not give much emphasis on "evaluating" complicated formulae obtained for concrete examples, here is a place to use appropriate software.

Frobenius theory for second order differential equations with (real) analytic coefficients not only led to the deep results of L. Fuchs and H. Poincaré on differential equations nowadays called Fuchsian differential equations, but first of all it allows to handle, in a systematic way, some of the most important differential equations from mathematical physics obtained by separating partial differential equations, e.g. Laplace's equation, in curvilinear orthogonal coordinates. We prove Frobenius theorem in full generality and discuss several examples such as Airy's, Legendre's, Hermite's or Bessel's equation in detail. Then we turn to some discussion of qualitative properties of solutions leading to Sturm's oscillation theorem. Thus we introduce (regular) Sturm-Liouville problems. Regular Sturm-Liouville operators are the topic of Chapter 21 where we encounter for the first time boundary value and eigenvalue problems. We conduct a more detailed investigation of these eigenvalue problems, in particular we look at eigenvalues, eigenfunctions and eigenspaces and their relations when being associated with different eigenvalues. This also prepares our discussion of the spectral theory of (singular) Sturm-Liouville operators in Volume V. In addition we study Green's functions for regular Sturm-Liouville problems.

Many of the special functions we have encountered so far are related to the hypergeometric differential equation or the confluent hypergeometric differential equation. The central role of these equations in the theory of ordinary differential equations justifies, in fact demands, some treatment. Besides general existence considerations we look in more detail at transformations leaving the equations invariant and representations of well-known functions as hypergeometric or confluent hypergeometric functions. The topic of this chapter is a rather classical one, today maybe a bit neglected, but still of great importance.

Chapter 23 gives a first discussion of qualitative properties of solutions (or the solution manifold) of an ordinary differential equation. Several stability results are established and the concept of a Lyapunov function is introduced. We look at autonomous equations and we study the behaviour of solutions at equilibrium points. Eventually we turn to gradient systems and their properties.

Many textbooks on ordinary differential equations start early to discuss phase diagrams and flows. We believe that this should be done only when a proper geometric context is provided. For this reason we give a detailed introduction to the tangent space and the tangent bundle of a subset of $\mathbb{R}^n$ and then we introduce vector fields as sections of the tangent bundle. This allows us to understand the relation between a differential equation and an integral curve of a vector field in such a way that an extension to differentiable manifolds will become naturally. Once this is done in Chapter 24, in our final chapter on differential equations we look at phase diagrams and the flow associated with a differential equation. In particular we provide a first discussion of Liouville's theorem.

The Calculus of Variations from its beginning is closely related to mathematical physics, especially mechanics, as well as to differential geometry. It became of central importance in the study of holomorphic functions on Riemann surfaces, e.g. Riemann's use of Dirichlet's principle, which after the critics of Weierstrass ultimately led to the introduction of Hilbert spaces and the direct methods of the calculus of variations as tool for solving partial differential equations. Some of these topics we will take up in Volume V and in particular Volume VI. Here we intend to give a brief introduction to basic ideas and results. Chapter 26 sets the scene for one-dimensional results and we derive the Euler-Lagrange equations as necessary conditions for C^2-extremals. In Chapter 27 we investigate extremals, i.e. classical solutions to the Euler-Lagrange equations. We discuss regularity properties, first integrals and in particular a version of Noether's theorem linking symmetries to first integrals.

Sufficient conditions for extremals to be minimisers are more difficult to obtain. In Chapter 28 we study the second variation and derive the necessary conditions of Legendre. Our further discussion leads to the role of conjugate points and the Jacobi equation which we start to investigate.

It turns out that solving partial differential equations of first order is key to make further progress. For this reason we add a long chapter on these equations. We develop for linear equations first geometric ideas and then we look at quasi-linear equations. Although this leads away from our main goal, the connection of first order partial differential equations and the calculus of variations, quasi-linear partial differential equations allow best to understand regularity and existence problems as geometric problems (related to characteristics). We then give a full discussion of general non-linear first order partial differential equations not explicitly depending on the "time" parameter t. This discussion gives us the tools needed in Chapter 30 as well as in Volume VI when dealing with hyperbolic equations. We investigate the Hamilton-Jacobi differential equation as a major equation in analytical dynamics, we look at cyclic coordinates and constants of motion and we discuss many examples from mechanics.

As in the previous volumes, we have provided all (but the first) chapters with problems and detailed solutions. Of course, we now address a more matured readership which allows us to be more brief in some calculations.

Some of the topics we handle are quite standard, some are more advanced. This requires a different method of using references. For topics which we consider to be standard and which are treated in many books quite similarly we only give "global" references. However, we try to be very precise with references whenever dealing with more advanced or special material. We hope that this approach is appreciated by our readers and considered as fair by authors.

Contents

CONTENTS

List of Symbols

$\mathbb{N}$ natural numbers

$\mathbb{N}_0 = \mathbb{N} \cup \{0\}$,

$\mathbb{N}_0^n$ multi-indices (see more in Volume II)

$\mathbb{Z}$ integers

$\mathbb{Q}$ rational numbers

$\mathbb{R}$ real numbers

$\mathbb{R}_+ = \{x \in \mathbb{R} \mid x \geq 0\}$

$\overline{\mathbb{R}} = \mathbb{R} \cup \{-\infty, \infty\}$

$\mathbb{D}$ unit disc in $\mathbb{C}$

$\mathbb{T}$ torus

$\mathbb{T}^n$ n-dimensional torus

$B_r(x) = \{y \in \mathbb{R} \mid \|x - y\| < r\}$

$D_\rho(z_0) = \{z \in \mathbb{C} \mid |z - z_0| < \rho\}$

$S^{n-1} = \{x \in \mathbb{R}^n \mid \|x\| = 1\}$

$a \vee b = \max(a, b)$

$a \wedge b = \min(a, b)$

$\delta_{kl} = \begin{cases} 1, & k = l \\ 0, & k \neq l \end{cases}$ Kronecker delta

$\operatorname{sgn}(x)$ signum of $x \in \mathbb{R}$

$\operatorname{span} A$ span of a set of vectors

$\langle x, y \rangle$ scalar product in $\mathbb{R}^n$

$[A, B] = AB - BA$ commutator

$n!$ n factorial

$\binom{n}{k}$ binomial coefficient for $n, k \in \mathbb{N}$

$\binom{a}{k} = \frac{a(a-1)(a-2)\cdot\ldots\cdot(a-k+1)}{k!}$ general binomial coefficients

$(a)_n = a(a+1)(a+2) \cdot \ldots \cdot (a+n-1)$ Pochhammer symbol

$\alpha! = \alpha_1! \cdot \ldots \cdot \alpha_n!$ for $\alpha = (\alpha_1, \ldots, \alpha_n)$

$\binom{\alpha}{\beta} := \binom{\alpha_1}{\beta_1} \cdot \ldots \cdot \binom{\alpha_n}{\beta_n}$ for $\alpha = (\alpha_1, \ldots, \alpha_n)$

$\alpha \leq \beta \quad \alpha_j \leq \beta_j, \alpha, \beta \in \mathbb{N}_0^n$

$\alpha + \beta = (\alpha_1 + \beta_1, \ldots, \alpha_n + \beta_n)$ for $\alpha, \beta \in \mathbb{N}_0^n$

$x^\alpha = x_1^{\alpha_1} \cdot \ldots \cdot x_n^{\alpha_n}$ for $\alpha = (\alpha_1, \ldots, \alpha_n)$ and $x \in \mathbb{R}^n$

$M(m,n;\mathbb{R})$ vector space of all real $m \times n$ matrices

$M(n;\mathbb{R}) = M(n,n;\mathbb{R})$ $n \times n$-matrices with real elements

$GL(n;\mathbb{R})$ general linear group over $\mathbb{R}$ in dimension n

$SL(n;\mathbb{R})$ special linear group over $\mathbb{R}$ in dimension n

$O(n)$ orthogonal group in $\mathbb{R}^n$

$SO(n)$ special orthogonal group in $\mathbb{R}^n$

$S_p(2n)$ symplectic group in $\mathbb{R}^{2n}$

$U(n)$ unitary group in $\mathbb{C}^n$

$SU(n)$ special unitary group in $\mathbb{C}^n$

$\mathrm{Aut}(G)$ automorphism group of a domain

$f(A)$ image of A under f

$f^{-1}(A')$ pre-image of A' under f

$f^+ = f \vee 0$ positive part of f

$f^- = (-f) \vee 0$ negative part of f

$f * g$ convolution of f and g

J_f Jacobi matrix of f

$\det J_f$ Jacobi determinant of f

$\mathrm{Hess}(f)$ Hesse matrix of f

$\mathrm{supp}\, f$ support of f

ϵ_{w_0}	unit measure, Dirac measure at w_0
$\epsilon = \epsilon_0$	Dirac measure at $0 \in \mathbb{R}^n$
$\lambda^{(n)}$	Lebesgue-Borel and Lebesgue measure in $\mathbb{R}^n$
$\lambda_G^{(n)}$	restriction of the Lebesgue-Borel measure
$\nu * \mu$	convolution of measures
μ_G	Haar measure on the group G
$\overset{\circ}{G}$	interior of a set G
$\overline{G}$	closure of a set G
∂G	boundary of a set G
$\operatorname{diam}(A) = \sup\{d(x,y) \mid x,y \in A\}$	diameter of a set A
$\operatorname{dist}(A, B)$	distance of A and B
$\operatorname{dist}(A, x) = \operatorname{dist}(A, \{x\})$	
$\operatorname{dist}(x, A) = \operatorname{dist}(\{x\}, A)$	
$\dfrac{\mathrm{d}f}{\mathrm{d}x}(x) = f'(x)$	derivative of a function of a real variable
$\dfrac{\mathrm{d}^k f}{\mathrm{d}x^k} = f^{(k)}(x)$	higher order derivatives of a function of a real variable
$\dfrac{\mathrm{d}f}{\mathrm{d}z}(z) = f'(z)$	derivative of a function of a complex variable
$\dfrac{\mathrm{d}^k f(z)}{\mathrm{d}z^k} = f^{(k)}(z)$	higher order derivative of a function of a complex variable
$\partial^\alpha f, \partial_x^\alpha f$	partial derivates ($\alpha \in \mathbb{N}_0^n$)
$u_x = \dfrac{\partial u}{\partial x}$	
$\dot{u}(t) = \dfrac{\mathrm{d}u(t)}{\mathrm{d}t}$	

Δ_n Laplacian in $\mathbb{R}^n$

grad   gradient

$u(x+) = \lim\limits_{\substack{y \to x \\ y > x}} u(y)$

$u(x-) = \lim\limits_{\substack{y \to x \\ y < x}} u(y)$

$Fu, \hat{u}$ Fourier transform of the function u

$F^{-1}u$ inverse Fourier transform of
           the function u

$\hat{\mu}$ Fourier transform of the measure μ

$F_{\sin}, F_{\cos}$ Fourier sine/cosine transform

a_0, a_k, b_k real Fourier coefficients

c_k complex Fourier coefficients

$\hat{\mu}_k$ Fourier-Stieltjes coefficients

$S_N, S_N(f)$ partial sum of a Fourier series

$\mathcal{L}(u)$ Laplace transform of u

D_N N^{th} Dirichlet kernel

$D_N^{\sin}$ N^{th} $\sin$ $-$Dirichlet kernel

$D_N^{\cos}$ N^{th} $\cos$ $-$Dirichlet kernel

F_N N^{th} Fejer kernel

σ_N N^{th} Cesàro mean

$P_r(t) = \dfrac{1 - r^2}{1 - 2r\cos t + r^2}$ Poisson kernel (ball)

$A_r(f)(t)$ Abel mean

$M_p(f; r) = \left(\dfrac{1}{2\pi} \displaystyle\int_0^{2\pi} |f(re^{i\varphi})|^p \, d\varphi \right)^{\frac{1}{p}}$

$M(f)(x)$ Hardy-Littlewood maximal function

$J_\epsilon(f)$ Friedrichs mollifier

$(X, \lVert \cdot \rVert)$ or $(X, \lVert \cdot \rVert_X)$	normed space with norm $\lVert \cdot \rVert$ or $\lVert \cdot \rVert_X$
$(H, \langle \cdot, \cdot \rangle_H)$	Hilbert space with scalar product $\langle \cdot, \cdot \rangle_H$
$X \perp Y$	X is orthogonal to Y
$M^\perp$	orthogonal complement of H

$$\lVert f \rVert_{\infty, G} = \sup_{x \in G} |f(z)|$$

$$\lVert \cdot \rVert_{L^p} = \lVert \cdot \rVert_p \qquad \text{norm in } L^p(G)$$

$$\lVert u \rVert_{k, \infty} = \max_{0 \le l \le k} \lVert u^{(l)} \rVert_\infty$$

$$\lVert u \rVert_{\infty, K, L} \qquad \text{weighted sup-norm (23.25)}$$

$$\lVert (d_k)_{k \in \mathbb{Z}} \rVert_\infty = \sup_{k \in \mathbb{Z}} |d_k|$$

$$\lVert (d_k)_{k \in \mathbb{Z}} \rVert_p = \left(\sum_{k \in \mathbb{Z}} |d_k|^p \right)^{\frac{1}{p}}$$

$$p_{\alpha, \beta}(u) = \sup_{x \in \mathbb{R}^n} |x^\beta \partial^\alpha u(x)|$$

$$q_{m_1, m_2}(u) = \sup_{x \in \mathbb{R}^n} \left((1 + \lVert x \rVert^2)^{\frac{m_1}{2}} \sum_{|\alpha| \le m_2} |\partial^\alpha u(x)| \right)$$

Functions in the following spaces may be real- or complex-valued.

$C([a, b])$	continuous functions on $[a, b]$
$C^k([a, b])$	k times continuously differentiable functions on $[a, b]$
$C(X)$	continuous functions on X
$C_b(X)$	space of bounded functions on the metric space X
$C_0(X)$	space of all continuous functions with compact support

$C^k(G)$	k-times continuously differentiable functions on G
$C^\infty(G) = \cap_{k=1}^\infty C^k(G)$	
$C_0^\infty(G) = C^\infty(G) \cap C_0(G)$	
$C_\infty(\mathbb{R}^n)$	space of all continuous functions vanishing at infinity
C_{per}	space of all continuous 2π-periodic functions
$\mathcal{S}(\mathbb{R}^n)$	Schwartz space
$A(\mathbb{R}^n)$	Wiener algebra
$L^p(G)$	Lebesgue space of p−integrable functions
$L^1_{loc}(G)$	locally integrable functions
$H^s(\mathbb{R}^n), H^s(\mathbb{T})$	Bessel potential space
$W^{m,2}(\mathbb{R}^n)$	Sobolev space
$\mathcal{H}^2, \mathcal{H}^p$	Hardy spaces
$\mathcal{N}$	Nevanlinna class

$$l_p(\mathbb{Z}) = \{(d_k)_{k\in\mathbb{Z}} | d_k \in \mathbb{C}, \sum_{k\in\mathbb{Z}} |d_k|^p < \infty\}$$

$$l_\infty(\mathbb{Z}) = \{(d_k)_{k\in\mathbb{Z}} | d_k \in \mathbb{C}, \sup_{k\in\mathbb{Z}} |d_k|^p < \infty\}$$

$B(z)$	Blaschke product
$P(D) = \displaystyle\sum_{|\alpha|\le m} a_\alpha D^\alpha$	partial differential operator with constant coefficients
$W(t)$ or $W(u_1,\dots,u_n)(t)$	Wronski determinant
$G(x,y)$ or $G_{B^a,B^b}(x,y)$	Green's function
$D_{\mu_k} = \begin{pmatrix} \operatorname{Re}\mu_k & \operatorname{Im}\mu_k \\ -\operatorname{Im}\mu_k & \operatorname{Re}\mu_k \end{pmatrix}$	
B_r, F_l	special block matrices (18.44)
$\Lambda^s(D)u = F^{-1}((1+|\cdot|^2)^{\frac{s}{2}}\hat{u})$	

$\Gamma(z)$	gamma-function
$B(z, w)$	beta-function
$\zeta(z)$	zeta-function
$\vartheta(t)$	theta-function
J_k, J_ν	Bessel functions
Y_ν	Bessel function of second kind
K_ν, I_ν	modified Bessel function
Si	integral sine function
$_2F_1$	(Gauss) hypergeometric function
$_pF_q$	generalised hypergeometric function
$\Phi(\alpha, \gamma, x)$	Kummer function
$U(\alpha, \gamma; z)$	Kummer function of second kind
erf	error function
H_n	Hermite polynomial
L_n^α	Laguerre polynomial
P_n	Legendre polynomial
$P_n^{(\alpha, \beta)}$	Jacobi polynomial
C_n^λ	Gegenbauer polynomial
trig_ν	any of $\cos \nu, \sin \nu$ or $x \mapsto 1$

$$\mathcal{F}(u) = \int_a^b F(t, u(t), u'(t)) \, dt$$

$D_{A,B}(\mathcal{F})$	see (26.3)
$F_x(t, x, y) = \text{grad}_x F(t, x, y)$	
$F_x(t, x, y) = \text{grad}_y F(t, x, y)$	
$L(t, q, \dot{q})$	Lagrange function
$H(t, q, p), H(p, q)$	Hamilton function
H_q, H_p	$\text{grad}_q H(t, q, p), \text{grad}_p H(t, q, p)$

$$S = \int_a^b L(s, x(s), \dot{x}(s)) \, ds \qquad \text{action integral}$$

$(\delta \mathcal{F})(u; v)$ first variation

$(\delta^2 \mathcal{F})(u; v)$ second variation

Note in Chapter 24 and Appendix I quite a few special notations are introduced which, however, are essentially used in Chapter 24 and Appendix I, respectively. These notations are not listed in this index.

Part 8: Fourier Analysis

1 The Historical Place of Fourier Analysis in Mathematics

It is hard to give a precise description of what "Mathematical Analysis" constitutes. However, every reasonable definition will have to mention that it is, in particular, occupied with the investigation of (real-valued) functions. An attempt to write the history of mathematical analysis will have to start in the pre-Newton and pre-Leibniz times, it will discuss the introduction of "the Calculus" by Leibniz and Newton, and highlight the contributions of the Bernoulli family, as well as those of Euler, d'Alembert and Lagrange. But neither Euler, d'Alembert nor Lagrange worked with a precise notion of a function. In fact, only in the late 18th century and the early 19th century did mathematicians become aware of serious problems in the fundamental definitions and methods used in analysis and other mathematical disciplines up to that point of time. As an example we just want to mention the notion of "integral".

With the ancient method of exhaustion in mind, the integral as handled by Leibniz was a definite integral considered as an "infinite sum" of areas of rectangles approximating the area of a certain figure. The integral sign $\int$ is derived from "sum". Euler changed the emphasis (in light of the fundamental theorem of calculus) and put the notion of a primitive in the centre of the discussion. Hence he looked more at indefinite integrals and used the interpretation of the integral as an anti-derivative. This interpretation has dominated analysis for a long time.

In this short chapter, we do not want to give a historical account of the development of analysis and more general aspects of mathematics. We intend to explain why "Fourier Analysis" should be seen to be one, arguably the most important source for changing mathematical analysis, and indeed mathematics, from its state in the late 18th century to its modern approaches.

Jean-Baptiste Joseph Fourier (1768-1830) is one of the most impressive people of his times and we refer to [5], [24] and [54] for his biography. In particular, he excelled as a scientist, as the organizer of the French retreat from Egypt, as the coordinator of the publication of the famous "Description de l'Egypte, Antiquités" a multi-volume publication which founded modern

3

Egyptology (he was also the mentor of J. F. Champollion-Figeac who eventually could decipher the Stone of Rosetta), and as an administrator, namely as Préfet de Département l'Isère with Grenable as its capital. Later in his life he was elected Secrétaire Perpétual of the French Academy of Sciences. Shortly after his death, in France his scientific achievements were almost forgotten so that Victor Hugo in "Les Misérables" [65] could write "Il y avait à l'Académie des Science un Fourier célébré que la postérité a oublie et dans je ne sais quel grenier un Fourier obscur dont l'avenir souvenidra."[1] The other Fourier was of course Charles Fourier, the socialistic utopist who had no family relations to J.-B. J. Fourier.

However, outside France, in particular in Germany, his work was not forgotten, in fact it became extremely influential. This is due to P. G. J. (Lejeune) Dirichlet who as a young scholar visited Fourier shortly before his death. Dirichlet's research on the convergence of Fourier series triggered research which eventually was to change mathematics forever: the origin of Cantor's set theory lies in Cantor's efforts to solve some problems on Fourier series left open by Dirichlet. The origin of many modern notions and results in mathematical analysis goes back to attempts of Riemann, du Bois-Reymond, Cantor and many others to turn Fourier's idea into rigorous mathematics. To understand these developments, we start by looking at the definition of a function as it was used by Euler or Lagrange. Our main source in the following discussion is [90].

In Euler's "Introductio ad analysin infinitorium", published in 1748, we can read "A function of a variable is an analytic expression formed in some manner from the variable and numbers or constants" (cited after [90]). Of course the meaning of "analytic expression" is unclear but was well understood in its times as meaningful, and still in 1813 in the second edition of his "Théorie des fonctions analytiques" Lagrange stayed close to this idea and took it for granted that every function has a representation as a power series. However, mathematicians such as Euler, Daniel Bernoulli, d'Alembert and Lagrange were aware of problems related to their notion of a function as their bitter dispute about the solution of the problem of the vibrating string, i.e. the wave equation, shows.

---

[1] "There was a celebrated Fourier at the Academy of Science, whom posterity has forgotten and in some garret an obscure Fourier, whom the future will recall."

D'Alembert claimed that the general solution of the wave equation

$$\frac{\partial^2 u}{\partial t^2} - c^2 \frac{\partial^2 u}{\partial x^2} = 0 \tag{1.1}$$

is of the type

$$u(t, x) = \varphi(x + ct) + \psi(x - ct), \tag{1.2}$$

but he assumed that both φ and ψ are "analytic expressions", i.e. they need to have (second order) derivatives. The full problem of the vibrating string consists in solving the initial boundary value problem

$$\frac{\partial^2 u}{\partial t^2} - c^2 \frac{\partial^2 u}{\partial x^2} = 0 \quad \text{in } [0, T] \times [0, L], \tag{1.3a}$$

$$u(t, 0) = u(t, L) = 0 \quad \text{for } t \in [0, T], \tag{1.3b}$$

$$u(0, x) = g(x), \quad \frac{\partial u}{\partial t}(0, x) = h(x), \quad x \in [0, L]. \tag{1.3c}$$

Euler obtained, after d'Alembert, essentially a similar solution, but he insisted that (in our terminology) discontinuous solutions must be allowed. The discussion was turned into a different direction when Daniel Bernoulli suggested to obtain a solution as a superposition of "simple" vibrations which led him to consider series which we call now trigonometric series, i.e.

$$\sum_{\nu=0}^{\infty} A_\nu \cos \nu x \quad \text{or} \quad \sum_{\nu=1}^{\infty} B_\nu \sin \nu x. \tag{1.4}$$

However, the problem of convergence was essentially left open as the question of when a given function has a representation with the help of such series - recall that functions were not well-defined "analytic expressions". Thus, these mathematicians have been confronted with the problem to find reasons to justify

$$f(x) = \sum_{\nu=0}^{\infty} A_\nu \cos \nu x + \sum_{\nu=1}^{\infty} B_\nu \sin \nu x \tag{1.5}$$

for a given function f. Daniel Bernoulli pointed out that the orthogonality properties of trigonometric functions might be used to determine the coefficients A_ν and B_ν, and his results were published only posthumously in 1793.

In two long memoirs from 1759 and 1762, Lagrange derived a solution to (1.3a)-(1.3c). His result was

$$u(t) = \frac{2}{L} \int_0^L \left(\sum_{n=1}^\infty \sin \frac{n\pi x}{L} \cos \frac{n\pi ct}{L} \sin \frac{n\pi y}{L} \right) g(y) \, dy \qquad (1.6)$$

$$+ \frac{2}{\pi c} \int_0^L \left(\sum_{n=1}^\infty \frac{1}{n} \sin \frac{n\pi x}{L} \sin \frac{n\pi ct}{L} \sin \frac{n\pi y}{L} \right) h(y) \, dy.$$

His arguments are by no means rigorous by today's standards, but note that when interchanging in (1.6) integration and summation we obtain a Fourier series solution of (1.3a)-(1.3c). With these preparations we can turn to Fourier's work. The reader with more of an historical interest is referred to the work of I. Grattan-Guinness and J. R. Ravetz [49] as well as [48] in addition to the biographies already mentioned.

Fourier was interested in the mathematical laws that can be used to model the propagation of heat. He did a lot of experiments and measurements and eventually came up with an equation, more precisely, depending on the configuration (geometry), with different equations, we can nowadays easy summarize in the formula

$$\frac{\partial u(t, x)}{\partial t} = \sum_{k,l=1}^n \frac{\partial}{\partial x_k} \left(a_{kl}(x) \frac{\partial u(t, x)}{\partial x_l} \right), \quad n = 1, 2, \text{ or } 3. \qquad (1.7)$$

An initial distribution of the heat is assumed by the condition $u(x, 0) = h(x)$, and geometric constraints lead to boundary conditions. His attempt to solve (1.7) under the additional conditions eventually led him to seek solutions in form of trigonometric series. However he had to overcome several hurdles. As pointed out by Medveder [90], it was Fourier who introduced the notion of an arbitrary function and gave up the "analytic expressions". We quote from [90] p.49, the translation of parts of p.500 in [39]:

> "A general function $f(x)$ is a sequence of values or ordinates, each of which is arbitrary. . . It is by no means assumed that these ordinates are subject to any general law, they may follow one another in a completely arbitrary manner, and each of them is defined as if it were a unique quantity.

From the nature of the problem itself and the analysis applied
to it, it might seem that the transition from one ordinate to the
next must proceed in the continuous manner. But then we are
talking about special conditions, whereas the general equality (B)
consider by itself, is independent of these conditions. It is strictly
applicable to discontinuous functions also."

Thus we should not credit Dirichlet for a first modern definition of a function but Fourier (and Fourier Analysis). However, we know that Dirchlet
was mathematically close to Fourier and had intensive personal contacts. It
is also noteworthy that Fourier already allowed discontinuous functions.

The second insight of Fourier was that when trying to represent an arbitrary
function with the help of a trigonometric series, the coefficients should be
determined by term by term integration against $\cos kx$ and $\sin lx$ using the
orthonormality of the system of trigonometric functions. He then derived the
"Fourier" series for several concrete functions, some, considered as periodic
functions on $\mathbb{R}$, are discontinuous. However, he did not discuss the convergence of these series, in particular not the questions whether the series (if
convergent) represent the given functions.

Fourier also considered the case of the infinite strip which led him from
Fourier series to the cos- and sin-Fourier transform, hence to the Fourier
transform. He obtained the fundamental solution to the heat equation as
well as the inversion formula, but again, from our point of view there is a
lack of rigour.

The lack of rigour caused him several problems, in particular with Lagrange,
but eventually his ideas were accepted. Cauchy made a first attempt to
prove the convergence of a Fourier series to the function, but his proof has
gaps. It was Dirichlet in 1829 who gave for a first time a correct proof that
for certain functions the associated Fourier series converges to the function.
He had to require the function to be continuous and most interestingly, he
had to add the condition that the function has only a finite number of minima and maxima on the periodicity interval. Dirichlet stated that the latter
condition can be removed but never published a proof or came back to the
problem.

The problem of Cauchy's proof lies in the fact that (in our modern terminology) Cauchy has taken it for granted that the series converges absolutely, a fact he was used to by his work on holomorphic (analytic) functions and power series. Dirichlet realized that Fourier series need not have this property and his extra conditions allowed him to handle certain integrals key in his proof as integrals with non-negative integrands.

Thus, after the concept of an arbitrary function, the method to determine the Fourier coefficients by using the orthonormality of the trigonometric functions, Fourier series taught us to make a difference between convergence and absolute convergence and uniform convergence of series of functions.

In his "Habilitationsschrift" from 1854, Riemann pushed the topic further. He carefully analysed Dirichlet's result and related work. He pointed to the need of taking the different types of convergence into account, the need to handle integrable functions (and not every arbitrary function is integrable) as well as the role of Dirichlet's extra condition on maxima and minima. Riemann then concentrated himself first on the problem of integration and it was here where the Riemann integral, as a required tool to study the convergence of Fourier series was introduced. Riemann then continued the investigation of Dirichlet on the convvergence and properties of Fourier series.

The next steps forward are due to two mathematicians: Paul du Bois-Reymond and Gregor Cantor. The first one proved in 1873 by a counter example that Dirichlet's conjecture, i.e. the convergence of the Fourier series of a continuous function to the function is wrong. A further milestone was his result from 1874 on the coefficients of a trigonometric series. Du Bois-Reymond made several remarkable and lasting contributions to the theory of Fourier series (as well as to the other fields in analysis). His ideas on how to deal with "infinity", i.e. his "Infinitärcalcül" however were at the time not well-received and rarely understood. They had indirect influence, e.g. his PhD-student O. Hölder used in his thesis some of these ideas and came to a notion of continuity which is today called Hölder's continuity, but after Hardy [53] investigated these ideas in more detail they started to impact. Today, they are of interest in some aspects of theoretical computer science.

Of quite different nature was the impact of Cantor's studies. His starting point was the question of the uniqueness of a trigonometric series, more precisely in 1870 he discussed the question whether a given function which can be represented by a Fourier series converging at every point to the function may allow a further such representation by a different trigonometric series. He could prove the expected uniqueness result and extended the question in [20], published in 1872, to the situation where the series is allowed not to converge at certain points. This paper is the birth of set theory, in particular this theory of transfinite ordinal numbers. In order to allow infinite exceptional sets, he introduced what he has called the derivative of a point set (a subset of $\mathbb{R}$). Following up these notions led him to what Hilbert called a "paradise", i.e. set theory which is now the basis of all mathematics.

Cantor's insights and du Bois-Reymond's (counter)-example and other investigations enforced a revision of the notion of integrability (in the sense of Riemann) which was carried out essentially by H. Lebesgue (but was clearly influenced by the work of E. Borel).

Thus the study of Fourier's work ended in a complete revision of many concepts of analysis as well as the foundations of mathematics. There is, however, a further strand of developments that started by Fourier's work. We can reinterpret the development of a given function into a Fourier series as a development into eigenfunctions of a certain differential operator. These ideas, already indicated by d'Alembert in his work on the wave equation, were taken up by Liouville, who was a protegé of Fourier, and Sturm and culminated in what is now known as Sturm-Liouville theory for self-adjoint boundary value problems for second order differential operators. It opened the way to orthogonal expansions in particular with respect to systems of special functions of mathematical physics, e.g. Legendre, Hermite or Laguerre polynomials, or Bessel functions. In fact, it set the scene for modern spectral theory as well as analysis in (separable) Hilbert spaces. An historical account can be found in [87], we refer also to [16] and [17].

There are more developments in mathematics originating from Fourier analysis. However, we stop here not without mentioning that still Fourier analysis contributes with new and surprising results to the development of mathematics. Most of all our aim was to convince the reader that an early study of Fourier analysis will help to understand and follow the developments of other branches of analysis.

2 Trigonometric Series and Fourier Series

For two sequences $(a_k)_{k \geq 0}$ and $(b_k)_{k \geq 1}$ of real numbers we consider the formal term

$$\frac{a_0}{2} + \sum_{k=1}^{\infty}(a_k \cos kx + b_k \sin kx) \tag{2.1}$$

where $x \in \mathbb{R}$, and "formal" means that for the moment we do not require the series to converge. We call (2.1) a (formal) **trigonometric series**. The 2π-periodicity of the trigonometric functions cos and sin implies that in the case where a function is defined by the trigonometric series (2.1), then this function must be 2π-periodic too. We introduce

$$u_k(x) := a_k \cos kx + b_k \sin kx, \quad k \in \mathbb{N}, \tag{2.2}$$

and

$$u_0(x) = \frac{a_0}{2}. \tag{2.3}$$

Now we rewrite (2.1) as

$$\sum_{k=0}^{\infty} u_k(x). \tag{2.4}$$

Suppose that the two series $\sum_{k=1}^{\infty}|a_k|$ and $\sum_{k=1}^{\infty}|b_k|$ converge. It follows that

$$|u_k(x)| \leq |a_k| + |b_k|, \quad k \in \mathbb{N},$$

and therefore

$$\sum_{k=0}^{\infty}\|u_k\|_{\infty} \leq \frac{|a_0|}{2} + \sum_{k=1}^{\infty}|a_k| + \sum_{k=1}^{\infty}|b_k| < \infty. \tag{2.5}$$

The Weierstrass test, Theorem I.29.1, implies that the series $\sum_{k=0}^{\infty} u_k$ converges on $\mathbb{R}$ absolutely and uniformly, and since u_k, $k \in \mathbb{N}_0$, is continuous this series represents a 2π-periodic continuous function

$$u(x) := \sum_{k=0}^{\infty} u_k(x) := \lim_{N \to \infty} \sum_{k=0}^{N} u_k(x). \tag{2.6}$$

For the following it is convenient to write

$$S_N(x) := \sum_{k=0}^{N} u_k(x) = \frac{a_0}{2} + \sum_{k=1}^{N}(a_k \cos kx + b_k \sin kx) \tag{2.7}$$

11

for the $\mathbf{N}^{\text{th}}$ **partial sum** of $\sum_{k=0}^{\infty} u_k(x)$. As the function u defined by (2.6) is continuous and 2π-periodic it is Riemann as well as Lebesgue integrable over any interval of length 2π. We will need the following lemma which is a straightforward exercise in integration by parts and we refer to Problem 7 in Chapter I.13. We denote the **Kronecker symbol** δ_{kl} by

$$\delta_{kl} = \begin{cases} 1, & k = l \\ 0, & k \neq l. \end{cases} \tag{2.8}$$

Lemma 2.1. *For $k, l \in \mathbb{N}$ we have*

$$\frac{1}{\pi} \int_{-\pi}^{\pi} \cos kx \cos lx \, dx = \delta_{kl}; \tag{2.9}$$

$$\frac{1}{\pi} \int_{-\pi}^{\pi} \sin kx \sin lx \, dx = \delta_{kl}; \tag{2.10}$$

$$\frac{1}{\pi} \int_{-\pi}^{\pi} \sin kx \cos lx \, dx = 0; \tag{2.11}$$

$$\frac{1}{\pi} \int_{-\pi}^{\pi} 1 \cdot \sin kx \, dx = \frac{1}{\pi} \int_{-\pi}^{\pi} 1 \cdot \cos kx \, dx = 0. \tag{2.12}$$

Since (2.6) converges absolutely and uniformly we have for $k \in \mathbb{N}$

$$u(x) \cos kx = \sum_{n=0}^{\infty} u_n(x) \cos kx \tag{2.13}$$

as well as

$$u(x) \sin kx = \sum_{n=0}^{\infty} u_n(x) \sin kx \tag{2.14}$$

and again both series converge absolutely and uniformly. Hence we may integrate the series (2.6), (2.13) and (2.14) term by term, compare with Theorem I.25.27, to find

$$\frac{1}{\pi} \int_{-\pi}^{\pi} u(x) \, dx = \frac{1}{\pi} \int_{-\pi}^{\pi} \frac{a_0}{2} \, dx + \sum_{n=1}^{\infty} \left(\frac{1}{\pi} \int_{-\pi}^{\pi} a_n \cos nx \, dx + \frac{1}{\pi} \int_{-\pi}^{\pi} b_n \sin nx \, dx \right)$$

$$= a_0;$$

$$\frac{1}{\pi} \int_{-\pi}^{\pi} u(x) \cos kx \, dx = \frac{1}{\pi} \int_{-\pi}^{\pi} \frac{a_0}{2} \cos kx \, dx$$

$$+ \sum_{n=1}^{\infty} \left(\frac{1}{\pi} \int_{-\pi}^{\pi} a_n \cos nx \cos kx \, dx + \frac{1}{\pi} \int_{-\pi}^{\pi} b_n \sin nx \cos kx \, dx \right)$$

$$= a_k;$$

and

$$\frac{1}{\pi} \int_{-\pi}^{\pi} u(x) \sin kx \, dx = \frac{1}{\pi} \int_{-\pi}^{\pi} \frac{a_0}{2} \sin kx \, dx$$

$$+ \sum_{n=1}^{\infty} \left(\frac{1}{\pi} \int_{-\pi}^{\pi} a_n \cos nx \sin kx \, dx + \frac{1}{\pi} \int_{-\pi}^{\pi} b_n \sin nx \sin kx \, dx \right)$$

$$= b_k.$$

We now give

Definition 2.2. *Let* $u : \mathbb{R} \to \mathbb{R}$ *be a* 2π-*periodic function such that* $u|_{[-\pi,\pi]}$ *is integrable. We call*

$$a_0 := \frac{1}{\pi} \int_{-\pi}^{\pi} u(x) \, dx, \tag{2.15}$$

$$a_k := \frac{1}{\pi} \int_{-\pi}^{\pi} u(x) \cos kx \, dx, \quad k \in \mathbb{N}, \tag{2.16}$$

and

$$b_k := \frac{1}{\pi} \int_{-\pi}^{\pi} u(x) \sin kx \, dx, \quad k \in \mathbb{N}, \tag{2.17}$$

*the **Fourier coefficients** of the function* u.

Remark 2.3. In Definition 2.2 we left open whether integrability is Riemann or Lebesgue integrability. In general by integrability we mean Lebesgue integrability. By Theorem III.8.11 we know that if a function is Riemann integrable of $[a, b]$, it is Lebesgue integrable over $[a, b]$. Further we know that all continuous functions as well as all monotone functions on $[a, b]$ are Riemann integrable, hence when explicitly working with such functions we can use the Riemann integral and the "classical" fundamental theorem of calculus as well as the rules of integration being derived from the fundamental theorem.

So far we have proved

Theorem 2.4. *If a trigonometric series converges uniformly in $\mathbb{R}$ then it represents a continuous function u and the coefficients in (2.1) are the Fourier coefficients (2.15)-(2.17) of u.*

Definition 2.5. *Let $u : \mathbb{R} \to \mathbb{R}$ be a 2π-periodic function integrable over $[-\pi, \pi]$ with Fourier coefficients (2.15)-(2.17). We call the corresponding series (2.1) the* **Fourier series** *associated with u and we write*

$$u \sim \frac{a_0}{2} + \sum_{k=1}^{\infty}(a_k \cos kx + b_k \sin kx). \tag{2.18}$$

Note the important distinction: a trigonometric series is any series of the form (2.1), a Fourier series is a series of the type (2.1) where the coefficients are the Fourier coefficients of a 2π-periodic function integrable over $[-\pi, \pi]$. We will see later in Theorem 6.8 that the trigonometric series

$$\sum_{k=2}^{\infty} \frac{\sin kx}{\ln k} \tag{2.19}$$

converges for all $x \in \mathbb{R}$, but the limit is not an integrable function, i.e. the coefficients in (2.19) are not Fourier coefficients, and hence there is indeed a need for making the distinction between general trigonometric series and Fourier series. We will mainly investigate Fourier series, only a few results about general trigonometric series will be discussed.

Let $u : \mathbb{R} \to \mathbb{R}$ be a continuous 2π-periodic function. Hence u defines a Fourier series. We may ask whether this Fourier series converges pointwisely to the function u. As we will see in Theorem 6.12 the surprising answer due to P. du Bois-Reymond is: in general no.

We may weaken the question and look at 2π-periodic functions integrable over $[-\pi, \pi]$ and now we may ask whether the associated Fourier series converges almost everywhere (a.e.) with respect to the Lebesgue measure to this function. Again the answer is: in general no. We will discuss this result which is due to A. N. Kolmogorov in Chapter 6.

It is much easier to prove that the Fourier series of a square integrable function, i.e. a 2π-periodic function $u : \mathbb{R} \to \mathbb{R}$ square integrable over $[-\pi, \pi]$, converges in the quadratic mean, i.e. in the L^2-norm, to u. We will discuss this result in Chapter 5. One of the deepest results in analysis in the 20th century is the theorem of L. Carleson stating that the Fourier series of a square integrable function converges

14

almost everywhere to this function. The proof of this result is still too hard for a first course on Fourier analysis and we will not prove it, for a proof we refer to [47], [70] or [92], and of course to [21].

In this context it is worth mentioning that the series

$$\sum_{n=1}^{\infty} \frac{\sin nx}{n^\alpha} \tag{2.20}$$

converges for all $\alpha > 0$, compare with Theorem 6.8, however for $\alpha < \frac{1}{2}$ it cannot be the Fourier series of a square integrable function.

Thus we have set the programme: investigate the convergence of Fourier series. Before we start with this we want to give a representation of a trigonometric series using complex numbers and the complex exponential function (with purely imaginary argument). Using

$$e^{ikx} = \cos kx + i \sin kx \tag{2.21}$$

and

$$\cos kx = \frac{e^{ikx} + e^{-ikx}}{2}, \qquad \sin kx = \frac{e^{ikx} - e^{-ikx}}{2i}, \tag{2.22}$$

we find by substituting (2.22) into (2.1) that

$$\frac{a_0}{2} + \sum_{k=1}^{\infty}(a_k \cos kx + b_k \sin kx)$$

$$= \frac{a_0}{2} + \sum_{k=1}^{\infty}\left(a_k \left(\frac{e^{ikx} + e^{-ikx}}{2} \right) + b_k \left(\frac{e^{ikx} - e^{-ikx}}{2i} \right) \right)$$

$$= \frac{a_0}{2} + \sum_{k=1}^{\infty}\left(\left(\frac{a_k - ib_k}{2} \right) e^{ikx} + \left(\frac{a_k + ib_k}{2} \right) e^{-ikx} \right),$$

and with

$$c_k := \begin{cases} \frac{a_k - ib_k}{2}, & k \in \mathbb{N} \\ \frac{a_0}{2}, & k = 0 \\ \frac{a_{-k} + ib_{-k}}{2}, & -k \in \mathbb{N} \end{cases} \tag{2.23}$$

we get

$$\frac{a_0}{2} + \sum_{k=1}^{\infty}(a_k \cos kx + b_k \sin kx) = \sum_{k \in \mathbb{Z}} c_k e^{ikx}. \tag{2.24}$$

15

With

$$S_N^c(x) := \sum_{k=-N}^{N} c_k e^{ikx} \tag{2.25}$$

we can now study series of the type

$$v(x) := \lim_{N \to \infty} S_N^c(x) = \sum_{k \in \mathbb{Z}} c_k e^{ikx}, \tag{2.26}$$

where $c_k \in \mathbb{C}$. Assuming in (2.26) uniform convergence we find further for $l \in \mathbb{Z}$ that

$$\frac{1}{2\pi} \int_{-\pi}^{\pi} v(x) e^{-ilx} \, dx = \frac{1}{2\pi} \int_{-\pi}^{\pi} \sum_{k \in \mathbb{Z}} c_k e^{ikx} e^{-ilx} \, dx$$

$$= \sum_{k \in \mathbb{Z}} \frac{1}{2\pi} \int_{-\pi}^{\pi} c_k e^{i(k-l)x} \, dx.$$

However, for $k = l$ it follows that

$$\frac{1}{2\pi} \int_{-\pi}^{\pi} e^{i(k-l)x} \, dx = 1,$$

whereas for $k \neq l$ we know that

$$\frac{1}{2\pi} \int_{-\pi}^{\pi} e^{i(k-l)x} \, dx = 0,$$

thus

$$\frac{1}{2\pi} \int_{-\pi}^{\pi} e^{i(k-l)x} \, dx = \delta_{kl} \tag{2.27}$$

implying

$$c_k = \frac{1}{2\pi} \int_{-\pi}^{\pi} v(x) e^{-ikx} \, dx. \tag{2.28}$$

In order that the series (2.26) with $c_k \in \mathbb{C}$ corresponds to a series of type (2.1) it is necessary that

$$c_{-k} = \overline{c_k}, \quad k \in \mathbb{N}_0, \tag{2.29}$$

see Problem 1.

It is worth noting already the close relation of Fourier series to power series. Let

$$\sum_{k=0}^{\infty} c_k z^k, \quad z \in \mathbb{C}, \tag{2.30}$$

be a power series which we assume for a moment to converge on $S^1 \subset \mathbb{C}$. Every $z \in S^1$ can be written as $z = e^{it}$, $t \in [-\pi, \pi)$, and (2.30) becomes

$$\sum_{k=0}^{\infty} c_k e^{ikt} = \sum_{k=0}^{\infty} (\operatorname{Re} c_k + i \operatorname{Im} c_k)(\cos kt + i \sin kt)$$

and with $c_k = \frac{a_k - ib_k}{2}$, $k \in \mathbb{N}$, $c_0 = \frac{a_0}{2}$, we find

$$\operatorname{Re} \sum_{k=0}^{\infty} c_k e^{ikt} = \frac{a_0}{2} + \sum_{k=1}^{\infty} (a_k \cos kt + b_k \sin kt) \tag{2.31}$$

and

$$\operatorname{Im} \sum_{k=0}^{\infty} c_k e^{ikt} = \sum_{k=1}^{\infty} (-b_k \cos kt + a_k \sin kt). \tag{2.32}$$

The series (2.32) is called the **conjugate series** to the series (2.31), i.e. (2.1).

Instead of assuming that (2.30) converges on S^1 it is more helpful to assume that the radius of convergence of (2.30) is 1 which leads us to consider the two series

$$u(r, t) = \frac{a_0}{2} + \sum_{k=1}^{\infty} (a_k \cos kt + b_k \sin kt) r^k \tag{2.33}$$

and

$$v(r, t) = \sum_{k=1}^{\infty} (-b_k \cos kt + a_k \sin kt) r^k, \tag{2.34}$$

where $r \in [0, 1)$. The problem to investigate the convergence of (2.1) is now reduced to the study of $\lim_{r \to 1} u(r, t)$.

The following lemma is obvious:

Lemma 2.6. *If the series (2.1) represents a 2π-periodic function on $\mathbb{R}$ then this function is even if and only if $b_k = 0$ for all $k \in \mathbb{N}$, and this function is odd if and only if $a_k = 0$ for all $k \in \mathbb{N}_0$.*

Before we start to study the pointwise convergence of (2.1) we want to provide some examples of Fourier series. The starting point is always an integrable function $u : [-\pi, \pi) \to \mathbb{R}$ which we extend as a 2π-periodic function on $\mathbb{R}$, i.e. for $x = \mathbb{R} \backslash [-\pi, \pi)$ we first determine $n \in \mathbb{Z}$ such that $x = 2\pi n + y$, $y \in [-\pi, \pi)$ and then we define $u(x) := u(y)$. Since no confusion can arise we use for this extension the same symbol as for the original function. We also note that for any 2π-periodic function $v : \mathbb{R} \to \mathbb{R}$ we have with trig_ν being any of the functions $x \mapsto \cos \nu x$, $x \mapsto \sin \nu x$, $\nu \in \mathbb{N}$, or $x \mapsto 1$, that

$$\int_{a-\pi}^{a+\pi} v(x) \, \mathrm{trig}_\nu(x) \, \mathrm{d}x = \int_{-\pi}^{\pi} v(x) \, \mathrm{trig}_\nu(x) \, \mathrm{d}x, \qquad (2.35)$$

i.e. we may determine the Fourier coefficients associated with a 2π-periodic function ν by integrating over any interval of length 2π.

A further remark is in order. The periodic extension of a continuous function u on $[-\pi, \pi)$ is not necessarily continuous. This can hold only if $u(-\pi) = \lim_{x \nearrow \pi} u(x)$. Thus in light of (2.35) and its consequence for determining Fourier coefficients it follows that if $u(-\pi) \neq \lim_{x \nearrow \pi} u(x)$ we must consider the Fourier coefficients as being determined by a discontinuous function even when $u|_{[-\pi,\pi)}$ is continuous.

Example 2.7. We consider the 2π-periodic extension of the function $u_1 : [-\pi, \pi) \to \mathbb{R}$, $u_1(x) = x$. Since $u_1(-\pi) = -\pi \neq \lim_{x \nearrow \pi} u_1(x) = \pi$, the periodic extension is a discontinuous function on $\mathbb{R}$ with a jump discontinuity at the points $x = \pi + 2\pi n$, $n \in \mathbb{N}$, also see Figure 2.1 below. Since $x \mapsto x$ and $x \mapsto x \cos kx$, $k \in \mathbb{N}$, are odd functions we find for the Fourier coefficients

$$a_0 = \frac{1}{\pi} \int_{-\pi}^{\pi} x \, \mathrm{d}x = 0 \quad \text{and} \quad a_k = \frac{1}{\pi} \int_{-\pi}^{\pi} x \cos kx \, \mathrm{d}x = 0,$$

and further

$$b_k = \frac{1}{\pi} \int_{-\pi}^{\pi} x \sin kx \, \mathrm{d}x = \frac{2(-1)^{k+1}}{k}.$$

The Fourier series associated with u_1 is therefore given by

$$u_1(x) \sim \sum_{k=1}^{\infty} \frac{2(-1)^{k+1}}{k} \sin kx. \qquad (2.36)$$

We emphasise that at the moment we do not know whether this series represents u_1 at points other than $x = 2\pi n$, $n \in \mathbb{N}$. In Figure 2.1 we have plotted u_1 as well as the partial sums S_3 ($\cdots$) and S_5 (- - -) of the series in (2.36).

18

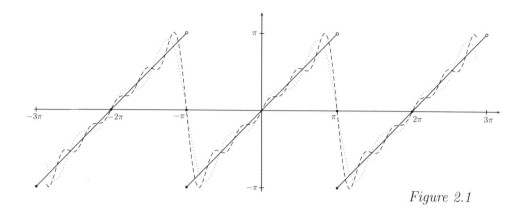

Figure 2.1

Example 2.8. A function similar to the one in Example 2.7 is the 2π-periodic extension of $u_2 : [-\pi, \pi) \to \mathbb{R}$ given by

$$u_2(x) = \begin{cases} \frac{-x - \pi}{2}, & x \in [-\pi, 0) \\ \frac{\pi - x}{2}, & x \in [0, \pi), \end{cases}$$

which is an odd function, see also Figure 2.2, thus $a_k = 0$ for $k \in \mathbb{N}_0$.

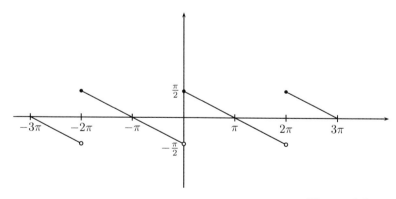

Figure 2.2

Since the 2π-periodic extension of u_2 is also the 2π-periodic extension of $\tilde{u}_2 : [0, 2\pi] \to \mathbb{R}$, $u_2(y) = \frac{\pi - y}{2}$, see Figure 2.2, we find for b_k, $k \in \mathbb{N}$, using integration by parts

$$b_k = \frac{1}{\pi} \int_0^{2\pi} \left(\frac{\pi - x}{2} \right) \sin kx \, dx$$

$$= -\frac{1}{2\pi k} (\pi - x) \cos kx \Big|_0^{2\pi} - \frac{1}{2\pi k} \int_0^{2\pi} \cos kx \, dx = \frac{1}{k}.$$

Hence we arrive at

$$u_2(x) \sim \sum_{k=1}^{\infty} \frac{\sin kx}{k}. \tag{2.37}$$

Example 2.9. Let u_3 be the 2π-periodic extension of $u_3 : [-\pi, \pi) \to \mathbb{R}$, $u_3(x) = x^2$, see Figure 2.3.

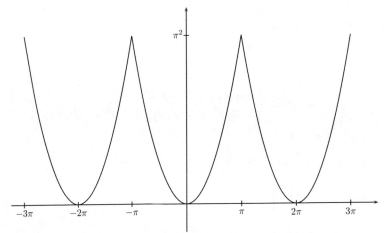

Figure 2.3

The functions $x \mapsto x^2 \sin kx$ are odd implying that

$$b_k = \frac{1}{\pi} \int_{-\pi}^{\pi} x^2 \sin kx \, dx = 0.$$

Further we have

$$a_0 = \frac{1}{\pi} \int_{-\pi}^{\pi} x^2 \, dx = \frac{2\pi^2}{3}$$

and

$$a_k = \frac{1}{\pi} \int_{-\pi}^{\pi} x^2 \cos kx \, dx = \frac{4(-1)^k}{k^2},$$

20

which yields

$$u_3(x) \sim \frac{\pi^2}{3} + \sum_{k=1}^{\infty} \frac{4(-1)^k}{k^2} \cos kx. \tag{2.38}$$

In this case we know more about the convergence of the series. Since

$$\left| \frac{4(-1)^k}{k^2} \cos kx \right| \le \frac{4}{k^2}$$

the series in (2.38) converges uniformly on $\mathbb{R}$ to a continuous function. However, at the moment we cannot decide whether the limit of this series is indeed u_3. A uniqueness result for Fourier series, i.e. Fourier coefficients, would be helpful at this place. In Figure 2.3 we have shown u_3 as well as the partial sums S_2 and S_4. Suppose that the series in (2.38) represents u_3 at $x = 0$. Then we will have

$$0 = \frac{\pi^2}{3} + \sum_{k=1}^{\infty} \frac{4(-1)^k}{k^2},$$

or

$$\sum_{k=1}^{\infty} \frac{(-1)^{k+1}}{k^2} = \frac{\pi^2}{12}. \tag{2.39}$$

If the series would represent the function at $x = \pi$ we would find

$$\pi^2 = \frac{\pi^2}{3} + \sum_{k=1}^{\infty} \frac{4(-1)^k}{k^2} \cos k\pi,$$

or

$$\frac{\pi^2}{6} = \sum_{k=1}^{\infty} \frac{1}{k^2} = \zeta(2), \tag{2.40}$$

where ζ denotes the Riemann zeta-function. Thus it seems that on some occasions we can use the Fourier series to find the values of series of real (or complex) numbers.

Remark 2.10. A. In the last few examples when calculating Fourier coefficients we have not always provided the details of the evaluation of the integrals. We assume that the reader of the Course is meanwhile familiar with finding integrals with the help of tables or web-based sources. As a rule, for standard integrals we will not show calculations, however when the evaluation of an integral is not standard or will lead to some additional insights we will give details.
B. Every partial sum of trigonometric or Fourier series is an arbitrarily often differentiable function on $\mathbb{R}$ with a holomorphic extension to $\mathbb{C}$. Considering the

Fourier series associated with a given integrable 2π-periodic function as a possible approximation of that function we must be prepared to encounter convergence problems for example at points of discontinuity. Worded differently, we shall expect at some points a rather weak notion of convergence or even substitutes for proper convergence, i.e. summability, see Chapter 4.

Example 2.11. Let us consider the 2π-periodic function $x \mapsto |\sin x|$. This is an even function and hence the Fourier coefficients b_k, $k \in \mathbb{N}$, must all vanish. The Fourier series associated with $x \mapsto |\sin x|$ will be a pure cosine series. In the case it represents the function $[0, \pi)$ we would have a representation of $x \mapsto \sin x$ by a cosine series. For the Fourier coefficients a_k we find

$$a_0 = \frac{1}{\pi} \int_{-\pi}^{\pi} |\sin x| \, dx = \frac{2}{\pi} \int_0^{\pi} \sin x \, dx = \frac{4}{\pi};$$

$$a_1 = \frac{1}{\pi} \int_{-\pi}^{\pi} |\sin x| \cos x \, dx = \frac{2}{\pi} \int_0^{\pi} \sin x \cos x \, dx$$

$$= \frac{1}{\pi} \int_0^{\pi} \sin 2x \, dx = 0;$$

and for $k \geq 2$ we get

$$a_k = \frac{1}{\pi} \int_{-\pi}^{\pi} |\sin x| \cos kx \, dx$$

$$= \frac{2}{\pi} \int_0^{\pi} \sin x \cos kx \, dx = -2 \frac{(-1)^k + 1}{\pi(k^2 - 1)}.$$

Therefore we have

$$|\sin x| \sim \frac{2}{\pi} - \frac{4}{\pi} \sum_{k=1}^{\infty} \frac{\cos 2kx}{4k^2 - 1}, \tag{2.41}$$

and as in Example 2.9 we conclude by noting that $\frac{1}{4k^2-1} \leq \frac{1}{k^2}$, $k \in \mathbb{N}$, that the Fourier series associated with $x \mapsto |\sin x|$ converges absolutely and uniformly to a continuous function. Supposing that this function is in fact $x \mapsto |\sin x|$ we find at $x = 0$ that

$$0 = \frac{2}{\pi} - \frac{4}{\pi} \sum_{k=1}^{\infty} \frac{1}{4k^2 - 1}$$

or

$$\sum_{k=1}^{\infty} \frac{1}{4k^2 - 1} = \frac{1}{2}. \tag{2.42}$$

Example 2.12. We start with the function $w : (0, \pi) \to \mathbb{R}$, $w(x) = 1$ for all $x \in (0, \pi)$. We may extend w to a constant $w(x) = 1$ for all $x \in \mathbb{R}$, this is clearly a 2π-periodic function, and the corresponding Fourier series is trivial, namely it is the constant 1, i.e. $a_0 = 2$ and $a_k = b_k = 0$ for $k \in \mathbb{N}$. However we can also extend w as an odd function to $[-\pi, \pi)$ by

$$w(x) := \begin{cases} 0, & x = -\pi \\ -1, & x \in (-\pi, 0) \\ 0, & x = 0 \\ 1, & x \in (0, \pi) \end{cases}$$

and then we consider its 2π-periodic extension, see Figure 2.4

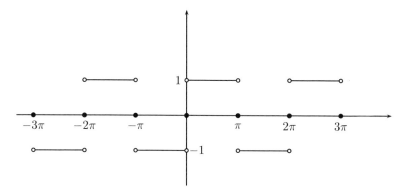

Figure 2.4

Now we find $a_0 = a_k = 0$ for $k \in \mathbb{N}$ and

$$b_k = \frac{1}{\pi} \int_{-\pi}^{\pi} w(x) \sin kx \, dx = \frac{2}{\pi} \int_0^{\pi} w(x) \sin kx \, dx$$
$$= \frac{2}{\pi} \int_0^{\pi} \sin kx \, dx = \frac{2}{\pi k} \left(1 - (-1)^k \right),$$

leading to

$$w(x) \sim \frac{4}{\pi} \sum_{k=1}^{\infty} \frac{\sin(2k-1)x}{2k-1}. \tag{2.43}$$

We close this introductory chapter with the remark that instead of 2π-periodic functions we may consider a function $f : \mathbb{R} \to \mathbb{R}$ with period $2L$, $L > 0$. With $x = \frac{Ly}{\pi}$ the function $u(y) = f\left(\frac{Ly}{\pi}\right)$ will have period 2π. The Fourier series associated with f is now

$$f(x) \sim \frac{a_0}{2} + \sum_{k=1}^{\infty} \left(a_k \cos\left(k\frac{\pi}{L}x\right) + b_k \sin\left(k\frac{\pi}{L}x\right) \right) \tag{2.44}$$

23

and for the Fourier coefficients we find

$$a_0 = \frac{1}{\pi} \int_{-\pi}^{\pi} f\left(\frac{Ly}{\pi}\right) dy = \frac{1}{L} \int_{-L}^{L} f(x)\, dx, \tag{2.45}$$

$$a_k = \frac{1}{\pi} \int_{-\pi}^{\pi} f\left(\frac{Ly}{\pi}\right) \cos ky\, dy = \frac{1}{L} \int_{-L}^{L} f(x) \cos\left(k\frac{\pi}{L}x\right) dx, \quad k \in \mathbb{N}, \tag{2.46}$$

and

$$b_k = \frac{1}{\pi} \int_{-\pi}^{\pi} f\left(\frac{Ly}{\pi}\right) \sin ky\, dy = \frac{1}{L} \int_{-L}^{L} f(x) \sin\left(k\frac{\pi}{L}x\right) dx, \quad k \in \mathbb{N}. \tag{2.47}$$

We leave the details to the reader, see Problem 5.

Our final example in this chapter is a simple observation. If $f : \mathbb{R} \to \mathbb{C}$ is a 2π-periodic function integrable on $[-\pi, \pi]$, then for every $x_0 \in \mathbb{R}$ the function f_{x_0}, $f_{x_0}(x) := f(x + x_0)$ has the same properties. For the Fourier coefficients of f_{x_0} we find

$$\frac{1}{2\pi} \int_{-\pi}^{\pi} f_{x_0}(x) e^{-ikx}\, dx = \frac{1}{2\pi} \int_{-\pi}^{\pi} f(x + x_0) e^{-ikx}\, dx$$

$$= \frac{1}{2\pi} \int_{-\pi}^{\pi} f(y) e^{-ik(y-x_0)}\, dy = e^{ik_0 x} \frac{1}{2\pi} \int_{-\pi}^{\pi} f(y) e^{-iky}\, dy.$$

Thus if we denote the Fourier coefficients of g by $(Fg)(k)$ we have

$$(Ff_{x_0})(k) = e^{ikx_0}(Ff)(k). \tag{2.48}$$

Problems

1. Let $u : \mathbb{R} \to \mathbb{R}$ be a 2π-periodic function such that $u|_{[-\pi,\pi]}$ is integrable and consider the complex Fourier coefficients c_k, $k \in \mathbb{Z}$, where $c_k = \frac{1}{2\pi} \int_{-\pi}^{\pi} u(x) e^{-ikx} dx$. Prove that representation (2.1) for the Fourier series associated with u only satisfies (2.23) if $c_{-k} = \overline{c_k}$.

2. Prove (2.35).

3.   a) Find the Fourier series (2.1) associated with the 2π-periodic extension of $u : [-\pi, \pi) \to \mathbb{R}$, $u(x) = |x|$. Sketch the graph of the extension in $[-3\pi, 3\pi]$.

   b) For $\gamma \in \mathbb{R} \setminus \mathbb{Z}$ consider the 2π-periodic extension of $v(x) = \cos \gamma x$, $x \in [-\pi, \pi)$. Find its Fourier series in the form of (2.1).

4. Let $\alpha \in \mathbb{R}$, $\alpha \neq 0$. We use the function $x \mapsto e^{\alpha x}$ to construct the following three 2π-periodic functions:

   a) f is the 2π-periodic extension of $x \mapsto e^{\alpha x}$, $x \in [-\pi, \pi)$;

   b) g is the 2π-periodic even extension of $x \mapsto e^{\alpha x}$, $x \in [0, \pi]$;

   c) h is the 2π-periodic odd extension of $x \mapsto e^{\alpha x}$, $x \in (0, \pi]$.

   (Note that we must set $h(0) := 0$.)

   In each case find the corresponding Fourier series in the form of (2.1).

5. Let $f : \mathbb{R} \to \mathbb{R}$ be a $2L$-periodic function, $L > 0$, which is integrable over $[-L, L]$ and has a uniformly convergent representation (2.44). By a direct calculation verify (2.45)-(2.47).

6. For $L > 0$ we define the function

$$
\tilde{g}(y) = \begin{cases} \cos \frac{\pi y}{L}, & 0 \leq y \leq \frac{L}{2} \\ 0, & \frac{L}{2} < y < L \end{cases}
$$

   and we consider its even $2L$-periodic extension g to $\mathbb{R}$. Sketch $g|_{[-L,L]}$ and find its associated Fourier series in the form of (2.44).

7. Assume that

$$
\sum_{k=1}^{\infty} \frac{\cos kx}{k^2} = \left(\frac{x - \pi}{2}\right)^2 - \frac{\pi^2}{12}
$$

   holds, see Lemma 5.8. Note that the series is uniformly convergent. Prove that

$$
\sum_{k=1}^{\infty} \frac{\sin kx}{k^3} = \frac{x(x - \pi)(x - 2\pi)}{12}
$$

   and deduce that

$$
\sum_{l=0}^{\infty} \frac{(-1)^l}{(2l + 1)^3} = \frac{\pi^3}{32}.
$$

8. Prove that

$$
\sum_{k=0}^{\infty} \frac{\cos kt}{3^k} = \frac{9 - 3\cos t}{10 - 6\cos t}.
$$

   Hint: consider the series $\sum_{k=0}^{\infty} \left(\frac{e^{it}}{3}\right)^k$.

9. Consider the series $\sum_{k=0}^{\infty} c_k e^{ik\varphi}$, $\varphi \in [-\pi, \pi)$, with $|c_k| \le M$ for all $k \in \mathbb{N}_0$. Note that this is a trigonometric series in complex form with $c_k = 0$ for $k \in -\mathbb{N}$. Prove that for $|z| \le r < 1$ a holomorphic function in $B_1(0) \subset \mathbb{C}$ is given by

$$u(z) = \tilde{u}(r, \varphi) = \sum_{k=0}^{\infty} c_k r^k e^{ik\varphi}, \quad z = re^{i\varphi}.$$

Now suppose that $c_k = \gamma^k$, $0 < \gamma < 1$, $k \in \mathbb{N}_0$. Find u and the largest disc $B_R(0) \subset \mathbb{C}$ in which u is holomorphic.

3 Fourier Series and Their Coefficients

We want to look more closely at the relations between 2π-periodic functions defined on $\mathbb{R}$ and functions defined on the unit circle S^1. We now allow all functions to be complex-valued. A point $p \in S^1$ is given by $p = e^{it}$, $t \in \mathbb{R}$, i.e. we consider S^1 as a subset of $\mathbb{C}$. Given $p \in S^1$, t is only uniquely determined up to an additive multiple of 2π, i.e. for $k \in \mathbb{Z}$ we have $p = e^{it} = e^{i(t+2\pi k)}$. We can enforce uniqueness in this representation by restricting t to any interval of length 2π, e.g. $t \in [-\pi, \pi)$ or $t \in [0, 2\pi)$. There is another way to clarify the situation by introducing on $\mathbb{R}$ the equivalence relation $x \sim_{2\pi} y$ if and only if $y = x + 2\pi k$ for some $k \in \mathbb{Z}$. From the algebraic point of view $2\pi\mathbb{Z}$ is a normal subgroup of $\mathbb{R}$, adding topology to our considerations, we can identify $2\pi\mathbb{Z}$ with a discrete, hence closed subgroup of $\mathbb{R}$ and now we may pass to the quotient $\mathbb{R}/\sim_{2\pi}$ as a group and topological space, thus we get a new Abelian topological group. This group is the one-dimensional **torus** and is denoted by $\mathbb{T}$ (or $\mathbb{T}^1$). We can identify $\mathbb{T}$ algebraically and topologically with S^1 and we will do this in Appendix I. Since periodicity in n (real) variables is linked to functions defined on the n-dimensional torus $\mathbb{T}^n = \mathbb{T}^1 \times \cdots \mathbb{T}^1$ (n copies), in the context of trigonometric series we prefer to use $\mathbb{T}$ instead of S^1 as notation for the unit circle, but still we identify $\mathbb{T}$ with a subset of $\mathbb{C}$.

The objects we are interested in are 2π-periodic functions on $\mathbb{R}$. For a 2π-periodic function $u : \mathbb{R} \to \mathbb{R}$ we can introduce the function $f : \mathbb{T} \to \mathbb{C}$ with the help of the relation

$$f(x) = f\left(e^{it}\right) = u(t), \quad x = e^{it}. \tag{3.1}$$

While t is not uniquely determined by e^{it}, the value of f is, which implies that complex-valued functions defined on $\mathbb{T}$ and 2π-periodic functions on $\mathbb{R}$ have a one-to-one correspondence. We use this one-to-one correspondence to introduce notions such as continuity, differentiability or integrability for functions defined on $\mathbb{T}$.

Definition 3.1. *Let $f : \mathbb{T} \to \mathbb{C}$ be a function. We call f **continuous** if the function u defined by (3.1) is continuous on $\mathbb{R}$. If u is m-times continuously differentiable on $\mathbb{R}$ we call f **m-times continuously differentiable**. If u is integrable over $[-\pi, \pi)$ we call f **integrable** on $\mathbb{T}$.*

We introduce

$$C(\mathbb{T}) := \left\{ f : \mathbb{T} \to \mathbb{C} \,\middle|\, u(t) = f\left(e^{it}\right), \, u \in C(\mathbb{R}) \right\}, \tag{3.2}$$

$$C^m(\mathbb{T}) := \left\{ f : \mathbb{T} \to \mathbb{C} \,\middle|\, u(t) = f\left(e^{it}\right), \, u \in C^m(\mathbb{R}) \right\}, \tag{3.3}$$

27

and

$$L^p(\mathbb{T}) := \left\{ f : \mathbb{T} \to \mathbb{C} \,\middle|\, u(t) = f\left(e^{it}\right), \, u \in L^p([-\pi, \pi)) \right\}. \tag{3.4}$$

The norm on $C(\mathbb{T})$ is the supremum norm $\|f\|_{\infty,\mathbb{T}}$ and the norm on $L^p(\mathbb{T})$ is denoted by $\|.\|_{L^p}$.

Remark 3.2. A. Definition 3.1 extends to the definition of pointwise continuity and differentiability.
B. Integrability means Lebesgue integrability and when dealing with L^p-spaces we identify functions being almost everywhere equal, i.e. $L^p(\mathbb{T})$ consists of equivalence classes of functions. However, when we want to work with Riemann integration theory, functions are assumed to be bounded as they are Riemann integrable functions, and the domain of integration will be the closed interval $[-\pi, \pi]$.
C. If f is integrable, the integral we consider is $\int_{-\pi}^{\pi} u(t)\, \mathrm{d}t$ where $f\left(e^{it}\right) = u(t)$, of course we may take any other interval of length 2π, or Borel subsets of $\mathbb{R}$ contained in an interval of length 2π.
D. In Appendix I we will have a closer look at continuity and integrability on $\mathbb{T}$, in particular we will introduce the Haar measure.

In order to simplify notations we will not write $f\left(e^{it}\right) = u(t)$ anymore with interpretation (3.1), but we just write $f(x)$, $x \in \mathbb{T}$, or $f(t)$, $t \in \mathbb{R}$. In the first case we consider f to be defined on $\mathbb{T}$, and in the second case we consider f as a 2π-periodic function on $\mathbb{R}$. When dealing with integrals we find for example

$$\int_{\mathbb{T}} f(x)\, \mathrm{d}x = \int_{-\pi}^{\pi} f(t)\, \mathrm{d}t. \tag{3.5}$$

In practice this slight abuse of notation will be very helpful and easy to cope with.

Given $f \in L^1(\mathbb{T})$ we can calculate its Fourier coefficients

$$c_k = \frac{1}{2\pi} \int_{-\pi}^{\pi} f(t) e^{-ikt}\, \mathrm{d}t, \qquad k \in \mathbb{Z}. \tag{3.6}$$

When switching from the system $\left(e^{-ik\cdot}\right)_{k \in \mathbb{Z}}$ to the system $1, \cos k\cdot, \sin k\cdot, k \in \mathbb{N}$, we find the relations

$$\begin{cases} c_0 = \frac{a_0}{2}, \\ c_k = \frac{a_k - ib_k}{2}, & k \in \mathbb{N}, \\ c_{-k} = \frac{a_k + ib_k}{2}, & k \in \mathbb{N}, \end{cases} \tag{3.7}$$

which yields

$$\begin{cases} a_0 = 2c_0, \\ a_k = c_k + c_{-k}, & k \in \mathbb{N}, \\ b_k = i(c_k - c_{-k}), & k \in \mathbb{N}, \end{cases} \tag{3.8}$$

28

however the coefficients a_k and b_k may now be complex numbers since we allow complex-valued functions. The condition

$$c_{-k} = \overline{c_k} \tag{3.9}$$

assures that the coefficients a_k and b_k are real numbers, also see (2.29). As before, for $f \in L^1(\mathbb{T})$ we write

$$f(t) \sim \sum_{k \in \mathbb{Z}} c_k e^{ikt} \tag{3.10}$$

for the Fourier series associated with f. We call

$$-i \sum_{k \in \mathbb{Z}} \operatorname{sgn}(k) c_k e^{ikx} \tag{3.11}$$

the **conjugate Fourier series** associated with f. Note that in the case where (3.9) holds, i.e. $c_{-k} = \overline{c_k}$, then (3.11) equals (2.32), see Problem 1. The partial sums we consider are the symmetric ones and we denote them now just by S_N, i.e.

$$S_N(t) = \sum_{k=-N}^{N} c_k e^{ikt}. \tag{3.12}$$

The central problem is in which sense does $\sum_{k \in \mathbb{Z}} c_k e^{ikt}$ represent f, in particular when do we have in some sense

$$\lim_{N \to \infty} S_N = f. \tag{3.13}$$

Let us change our point of view for a moment. Given $f \in L^1(\mathbb{T})$. By $(c_k)_{k \in \mathbb{Z}}$ a sequence of complex numbers is defined and since $f \in L^1(\mathbb{T})$ we have

$$|c_k| = \frac{1}{2\pi} \left| \int_{-\pi}^{\pi} f(t) e^{-ikt} \, dt \right| \leq \frac{1}{2\pi} \|f\|_{L^1}, \tag{3.14}$$

i.e. this sequence is bounded. Note that a sequence $(d_k)_{k \in \mathbb{Z}}$, $d_k \in \mathbb{C}$, is a mapping $d : \mathbb{Z} \to \mathbb{C}$, $d(k) = d_k$, the bounded sequences $(d_k)_{k \in \mathbb{Z}}$ are just the bounded mappings $d : \mathbb{Z} \to \mathbb{C}$. We introduce

$$l_\infty(\mathbb{Z}) := \left\{ (d_k)_{k \in \mathbb{Z}} \,\middle|\, d_k \in \mathbb{C} \text{ and } |d_k| \leq \gamma \text{ for some } \gamma \geq 0 \text{ and all } k \in \mathbb{Z} \right\} \tag{3.15}$$

as well as for $1 \leq p < \infty$

$$l_p(\mathbb{Z}) := \left\{ (d_k)_{k \in \mathbb{Z}} \,\middle|\, d_k \in \mathbb{C} \text{ and } \sum_{k \in \mathbb{Z}} |d_k|^p < \infty \right\}. \tag{3.16}$$

By

$$\|(d_k)_{k\in\mathbb{Z}}\|_\infty := \sup_{k\in\mathbb{Z}} |d_k| \tag{3.17}$$

a norm is given on $l_\infty(\mathbb{Z})$, see Problem 2, and from the Minkowski inequality in the form of Problem 8 of Chapter I.18 it also follows that

$$\|(d_k)_{k\in\mathbb{Z}}\|_2 := \left(\sum_{k\in\mathbb{Z}} |d_k|^2\right)^{\frac{1}{2}} \tag{3.18}$$

is a norm on $l_2(\mathbb{Z})$. We will soon prove below that

$$\|(d_k)_{k\in\mathbb{Z}}\|_p := \left(\sum_{k\in\mathbb{Z}} |d_k|^p\right)^{\frac{1}{p}}, \quad 1 \le p < \infty, \tag{3.19}$$

is a norm on $l_p(\mathbb{Z})$ and that the spaces $l_p(\mathbb{Z})$, $1 \le p < \infty$, are Banach spaces, see Theorem 3.12.

Thus we can define a mapping

$$F : L^1(\mathbb{T}) \to l_\infty(\mathbb{Z}) \tag{3.20}$$
$$f \mapsto ((Ff)(k))_{k\in\mathbb{Z}}, \quad (Ff)(k) := c_k.$$

The linearity of the integral immediately implies

Lemma 3.3. *The mapping F as defined in* (3.20) *is linear.*

Proof. For $f, g \in L^1(\mathbb{T})$ and $k \in \mathbb{Z}$ we have for $\lambda, \mu \in \mathbb{C}$ that

$$(F(\lambda f + \mu g))(k) = \frac{1}{2\pi} \int_{-\pi}^{\pi} (\lambda f + \mu g)(t)e^{-ikt}\, dt$$
$$= \frac{\lambda}{2\pi} \int_{-\pi}^{\pi} f(t)e^{-ikt}\, dt + \frac{\mu}{2\pi} \int_{-\pi}^{\pi} g(t)e^{-ikt}\, dt$$
$$= \lambda(Ff)(k) + \mu(Fg)(k).$$

$\square$

Using (3.20), i.e. the mapping F, we can reformulate some of the problems related to Fourier series associated with functions $f \in L^1(\mathbb{T})$, for example

- Is F injective, i.e. do the Fourier coefficients determine f?

- What is the image of $L^1(\mathbb{T})$ under F, i.e. which bounded sequences $(d_k)_{k\in\mathbb{Z}}$ are the Fourier coefficients of an $L^1(\mathbb{T})$-function?

- Given some subspace of $L^1(\mathbb{T})$, for example $L^2(\mathbb{T})$, $C(\mathbb{T})$ or $C^m(\mathbb{T})$. Can we find its image under F?

- Can we determine "nice" subspaces of $L^1(\mathbb{T})$ by looking at the behaviour of the Fourier coefficients, e.g. their decay?

$$\vdots$$

The last question raised above is linked to one of our earliest observations on Fourier series. Suppose a sequence $(c_k)_{k \in \mathbb{Z}}$ is given with the property that for some $\kappa > 0$ we have

$$|c_k| \leq \frac{\kappa}{|k|^2}, \quad k \in \mathbb{Z}. \tag{3.21}$$

Then the trigonometric series

$$\sum_{k \in \mathbb{Z}} c_k e^{ikt} = \lim_{N \to \infty} \sum_{|k| \leq N} c_k e^{ikt}$$

converges uniformly, hence it represents a continuous function $f \in C(\mathbb{T}) \subset L^1(\mathbb{T})$ and the Fourier coefficients of f are given by $(Ff)(k) = c_k$.
Suppose that $f \in C^2(\mathbb{T})$. For c_k, $k \in \mathbb{Z}$, it follows by integration by parts and from the 2π-periodicity of f that

$$
\begin{aligned}
c_k &= \frac{1}{2\pi} \int_{-\pi}^{\pi} f(t) e^{-ikt} \, \mathrm{d}t = \frac{1}{2\pi} \int_{-\pi}^{\pi} f(t) \left(\frac{1}{-ik} \right) \frac{\mathrm{d}}{\mathrm{d}t} e^{-ikt} \, \mathrm{d}t \\
&= -\frac{1}{2\pi ik} f(t) e^{-ikt} \Big|_{-\pi}^{\pi} + \frac{1}{2\pi ik} \int_{-\pi}^{\pi} \left(\frac{\mathrm{d}}{\mathrm{d}t} f(t) \right) e^{-ikt} \, \mathrm{d}t \\
&= \frac{1}{2\pi ik} \int_{-\pi}^{\pi} \left(\frac{\mathrm{d}f}{\mathrm{d}t}(t) \right) e^{-ikt} \, \mathrm{d}t.
\end{aligned}
$$

This calculation yields

$$ikc_k = ik(Ff)(k) = F\left(\frac{\mathrm{d}f}{\mathrm{d}t} \right)(k) \tag{3.22}$$

as well as the estimate

$$|c_k| \leq \frac{1}{2\pi|k|} \|f'\|_{L^1}. \tag{3.23}$$

From

$$c_k = \frac{1}{2\pi ik} \int_{-\pi}^{\pi} \left(\frac{\mathrm{d}f}{\mathrm{d}t}(t) \right) e^{-ikt} \, \mathrm{d}t$$

we obtain further

$$c_k = \frac{1}{2\pi i k} \int_{-\pi}^{\pi} \left(\frac{\mathrm{d}f}{\mathrm{d}t}(t) \right) \left(\frac{1}{-ik} \right) \frac{\mathrm{d}}{\mathrm{d}t} e^{-ikt} \, \mathrm{d}t$$

$$= \frac{1}{2\pi k^2} \int_{-\pi}^{\pi} \left(\frac{\mathrm{d}f}{\mathrm{d}t}(t) \right) \frac{\mathrm{d}}{\mathrm{d}t} e^{-ikt} \, \mathrm{d}t$$

$$= -\frac{1}{2\pi k^2} \int_{-\pi}^{\pi} \left(\frac{\mathrm{d}^2 f}{\mathrm{d}t^2}(t) \right) e^{-ikt} \, \mathrm{d}t$$

implying

$$|c_k| \leq \frac{1}{2\pi k^2} \|f^{(2)}\|_{L^1}. \tag{3.24}$$

Iterating this calculation gives

Lemma 3.4. *If $f \in C^m(\mathbb{T})$, $m \in \mathbb{N}_0$, then the Fourier coefficients c_k, $k \in \mathbb{Z}$, of f satisfy for $0 \leq l \leq m$*

$$(ik)^l c_k = (ik)^l (Ff)(k) = F\left(f^{(l)} \right)(k) \tag{3.25}$$

and we have the estimates

$$|c_k| \leq \frac{1}{2\pi |k|^m} \|f^{(m)}\|_{L^1}. \tag{3.26}$$

Corollary 3.5. *If $f \in C^2(\mathbb{T})$ then its associated Fourier series converges uniformly to f.*

Before we can proceed further we need some better understanding of the spaces $L^1(\mathbb{T})$, $L^p(\mathbb{T})$, $C(\mathbb{T})$ as well as $l_p(\mathbb{Z})$, $1 \leq p \leq \infty$.

We can reword Theorem III.8.21 as

Theorem 3.6. *The complex-valued step functions are dense in $L^p(\mathbb{T})$, $1 \leq p < \infty$.*

Furthermore using Theorem III.10.10 we can extend Theorem III.10.11 to

Theorem 3.7. *The space $C(\mathbb{T})$ is dense in $L^p(\mathbb{T})$, $1 \leq p < \infty$.*

Remark 3.8. The critical reader may object to the arguments given for Theorem 3.6 and Theorem 3.7 since we have not really (so far) considered the Borel σ-field on $\mathbb{T}$ and a measure on this σ-field as the underlying measure space for the construction of $L^p(\mathbb{T})$. In Appendix I we will see however that our introduction of $L^p(\mathbb{T})$ coincides with a more rigorous and formal one.

We further remind the reader of Corollary III.7.22 which we can reformulate in our context as

Proposition 3.9. *For* $1 \leq p < q \leq \infty$ *we have* $L^q(\mathbb{T}) \subset L^p(\mathbb{T})$ *and for* $1 \leq p < q < \infty$ *the estimate*

$$\|f\|_{L^p} \leq (2\pi)^{\frac{1}{p} - \frac{1}{q}} \|f\|_{L^q} \tag{3.27}$$

holds. In addition, for all $1 \leq p < \infty$ *we have*

$$\|f\|_{L^p} \leq (2\pi)^{\frac{1}{p}} \|f\|_{L^\infty}. \tag{3.28}$$

Now we turn to the spaces $l_p(\mathbb{Z})$.

Lemma 3.10. *The space* $(l_p(\mathbb{Z}), \|.\|_p)$, $1 \leq p \leq \infty$, *is a normed space.*

Proof. That $\|.\|_\infty$ is a norm on $l_\infty(\mathbb{Z})$ is proved in Problem 2. For $1 \leq p < \infty$ we find for $(d_k)_{k \in \mathbb{Z}} \in l_p(\mathbb{Z})$ that

$$\|(d_k)_{k \in \mathbb{Z}}\|_p = \left(\sum_{k \in \mathbb{Z}} |d_k|^p \right)^{\frac{1}{p}} = 0$$

if and only if $d_k = 0$ for all $k \in \mathbb{Z}$ and trivially we have $\|(d_k)_{k \in \mathbb{Z}}\|_p \geq 0$. For $\lambda \in \mathbb{C}$ it follows that

$$\|\lambda(d_k)_{k \in \mathbb{Z}}\|_p = \|(\lambda d_k)_{k \in \mathbb{Z}}\|_p$$

$$= \left(\sum_{k \in \mathbb{Z}} |\lambda d_k|^p \right)^{\frac{1}{p}} = \left(\sum_{k \in \mathbb{Z}} |\lambda|^p |d_k|^p \right)^{\frac{1}{p}}$$

$$= |\lambda| \left(\sum_{k \in \mathbb{Z}} |d_k|^p \right)^{\frac{1}{p}} = |\lambda| \|(d_k)_{k \in \mathbb{Z}}\|_p.$$

Finally we prove the triangle inequality. For $(a_k)_{k \in \mathbb{Z}}, (b_k)_{k \in \mathbb{Z}} \in l_p(\mathbb{Z})$ and $N \in \mathbb{N}$ we find using Minkowski's inequality in $\mathbb{R}^{2N+1}$, compare with Theorem I.23.17,

$$\left(\sum_{|k| \leq N} |a_k + b_k|^p \right)^{\frac{1}{p}} \leq \left(\sum_{|k| \leq N} |a_k|^p \right)^{\frac{1}{p}} + \left(\sum_{|k| \leq N} |b_k|^p \right)^{\frac{1}{p}}$$

$$\leq \left(\sum_{k \in \mathbb{Z}} |a_k|^p \right)^{\frac{1}{p}} + \left(\sum_{k \in \mathbb{Z}} |b_k|^p \right)^{\frac{1}{p}}$$

$$= \|(a_k)_{k \in \mathbb{Z}}\|_p + \|(b_k)_{k \in \mathbb{Z}}\|_p.$$

Now we may pass on the left hand side to the limit $N \to \infty$ to get

$$\|(a_k + b_k)_{k \in \mathbb{Z}}\|_p = \left(\sum_{k \in \mathbb{Z}} |a_k + b_k|^p\right)^{\frac{1}{p}} \leq \|(a_k)_{k \in \mathbb{Z}}\|_p + \|(b_k)_{k \in \mathbb{Z}}\|_p.$$

$\square$

Remark 3.11. Following the lines of the proof of the triangle inequality (**Minkowski's inequality**) in $l_p(\mathbb{Z})$, we can also prove **Hölder's inequality** stating for $1 < p, q < \infty$ and $\frac{1}{p} + \frac{1}{q} = 1$ if $(a_k)_{k \in \mathbb{Z}} \in l_p(\mathbb{Z})$ and $(b_k)_{k \in \mathbb{Z}} \in l_q(\mathbb{Z})$ then the sequence $(a_k b_k)_{k \in \mathbb{Z}}$ belongs to $l_1(\mathbb{Z})$ and we have

$$\sum_{k \in \mathbb{Z}} |a_k b_k| \leq \left(\sum_{k \in \mathbb{Z}} |a_k|^p\right)^{\frac{1}{p}} \left(\sum_{k \in \mathbb{Z}} |b_k|^q\right)^{\frac{1}{q}}. \tag{3.29}$$

Theorem 3.12. For $1 \leq p \leq \infty$ the normed spaces $(l_p(\mathbb{Z}), \|.\|_p)$ are complete, i.e. Banach spaces.

Proof. The case $p = \infty$ will be handled in Problem 2. For $n \in \mathbb{N}$ let $a^{(n)} = \left(a_k^{(n)}\right)_{k \in \mathbb{Z}}$ be a sequence in $l_p(\mathbb{Z})$ and assume that $\left(a^{(n)}\right)_{n \in \mathbb{N}}$ is a Cauchy sequence with respect to the norm $\|.\|_p$. Let $\epsilon > 0$ be given. Then there exists $N_0 = N_0(\epsilon) \in \mathbb{N}$ such that $m, l \geq N_0$ implies

$$\left\|a^{(m)} - a^{(l)}\right\|_p = \left(\sum_{k \in \mathbb{Z}} \left|a_k^{(m)} - a_k^{(l)}\right|^p\right)^{\frac{1}{p}} < \epsilon, \tag{3.30}$$

which yields for $m, l \geq N_0$ and $k \in \mathbb{Z}$ that

$$\left|a_k^{(m)} - a_k^{(l)}\right| < \epsilon. \tag{3.31}$$

Thus for every $k \in \mathbb{Z}$ a Cauchy sequence in $\mathbb{C}$ is given by $\left(a_k^{(n)}\right)_{n \in \mathbb{N}}$. By the completeness of $\mathbb{C}$ we find $a_k \in \mathbb{C}$, $k \in \mathbb{N}$, such that $\lim_{n \to \infty} a_k^{(n)} = a_k$. From (3.30) we deduce for every $M \in \mathbb{N}$ that $m, l \geq N_0$ implies

$$\left(\sum_{|k| \leq M} \left|a_k^{(m)} - a_k^{(l)}\right|^p\right)^{\frac{1}{p}} < \epsilon$$

which yields for $m \to \infty$ and $l \geq N_0$ that

$$\left(\sum_{|k| \leq M} \left|a_k - a_k^{(l)}\right|^p\right)^{\frac{1}{p}} \leq \epsilon,$$

and therefore for $l \geq N_0$

$$\left(\sum_{k \in \mathbb{Z}} \left| a_k - a_k^{(l)} \right|^p \right)^{\frac{1}{p}} \leq \epsilon. \tag{3.32}$$

Thus for $l \geq N_0$ the sequence $\left(a_k - a_k^{(l)} \right)_{k \in \mathbb{Z}}$ belongs to $l_p(\mathbb{Z})$ and therefore $a := (a_k)_{k \in \mathbb{Z}}$ must belong to $l_p(\mathbb{Z})$ too. But now (3.32) reads as

$$\left\| a - a^{(l)} \right\|_p \leq \epsilon$$

for $l \geq N_0$, i.e. $a^{(l)} \to a$ in $l_p(\mathbb{Z})$ and the theorem is proved. $\square$

The relations of the spaces $l_p(\mathbb{Z})$ and $l_q(\mathbb{Z})$ for $1 \leq p < q$ is given by

Proposition 3.13. *For $1 \leq p < q \leq \infty$ we have $l_p(\mathbb{Z}) \subset l_q(\mathbb{Z})$ and for $1 \leq p < q$ the estimate*

$$\left(\sum_{k \in \mathbb{Z}} |a_k|^q \right)^{\frac{1}{q}} \leq \left(\sum_{k \in \mathbb{Z}} |a_k|^p \right)^{\frac{1}{p}} \tag{3.33}$$

holds.

Proof. Since $a = (a_k)_{k \in \mathbb{Z}} \in l_p(\mathbb{Z})$ implies the convergence of the series $\sum_{k \in \mathbb{Z}} |a_k|^p$ it follows that the sequence $(a_k)_{k \in \mathbb{Z}}$ must be bounded, i.e. $l_p(\mathbb{Z}) \subset l_\infty(\mathbb{Z})$ for all $p \geq 1$. For $1 \leq p < q$ let $a = (a_k)_{k \in \mathbb{Z}} \in l_p(\mathbb{Z})$ and assume that $\left(\sum_{k \in \mathbb{Z}} |a_k|^p \right)^{\frac{1}{p}} = 1$. This implies that $|a_k| \leq 1$ and therefore $|a_k|^q \leq |a_k|^p$, consequently we have

$$\left(\sum_{k \in \mathbb{Z}} |a_k|^q \right)^{\frac{1}{q}} \leq \left(\sum_{k \in \mathbb{Z}} |a_k|^p \right)^{\frac{1}{q}} = 1;$$

i.e.

$$\left(\sum_{k \in \mathbb{Z}} |a_k|^q \right)^{\frac{1}{q}} \leq \left(\sum_{k \in \mathbb{Z}} |a_k|^p \right)^{\frac{1}{p}}.$$

Let $a = (a_k)_{k \in \mathbb{Z}} \in l_p(\mathbb{Z})$ be an arbitrary sequence and set $A := \left(\sum_{k \in \mathbb{Z}} |a_k|^p \right)^{\frac{1}{p}}$. It follows that $\left(\frac{a_k}{A} \right)_{k \in \mathbb{Z}} \in l_p(\mathbb{Z})$ and

$$\left(\sum_{k \in \mathbb{Z}} \frac{|a_k|^p}{A^p} \right)^{\frac{1}{p}} = \frac{1}{A} \left(\sum_{k \in \mathbb{Z}} |a_k|^p \right)^{\frac{1}{p}} = 1,$$

and therefore

$$\left(\sum_{k\in\mathbb{Z}}\frac{|a_k|^q}{A^q}\right)^{\frac{1}{q}} \leq 1 = \frac{1}{A}\left(\sum_{k\in\mathbb{Z}}|a_k|^p\right)^{\frac{1}{p}}$$

or

$$A\left(\sum_{k\in\mathbb{Z}}\frac{|a_k|^q}{A^q}\right)^{\frac{1}{q}} \leq \left(\sum_{k\in\mathbb{Z}}|a_k|^p\right)^{\frac{1}{p}},$$

proving the proposition. $\qquad\square$

Now we turn to the question whether $F : L^1(\mathbb{T}) \to l_\infty(\mathbb{Z})$ is injective or surjective. We start with the injectivity result by following [97].

Proposition 3.14. *If two continuous functions $f, g \in C(\mathbb{T})$ have the same Fourier coefficients then they are equal, i.e. $(Ff)(k) = (Fg)(k)$ for all $k \in \mathbb{Z}$ implies $f = g$.*

Proof. By the linearity of F it is sufficient to prove that if $(Ff)(k) = 0$ for all $k \in \mathbb{Z}$ then $f(t) = 0$ for all t. Let $f = u + iv$ for real-valued functions $u, v \in C(\mathbb{T})$. From

$$0 = 2\pi(Ff)(k) = \int_{-\pi}^{\pi} (u(t) + iv(t))e^{-ikt}\,\mathrm{d}t$$

we deduce

$$0 = \int_{-\pi}^{\pi} (u(t)\cos kt + v(t)\sin kt)\,\mathrm{d}t$$

and

$$0 = \int_{-\pi}^{\pi} (v(t)\cos kt - u(t)\sin kt)\,\mathrm{d}t,$$

which yields when applied for $m \in \mathbb{N}_0$ and $-m \in \mathbb{N}_0$

$$0 = \int_{-\pi}^{\pi} u(t)\cos mt\,\mathrm{d}t = \int_{-\pi}^{\pi} u(t)\sin mt\,\mathrm{d}t \qquad (3.34)$$

and

$$0 = \int_{-\pi}^{\pi} v(t)\cos mt\,\mathrm{d}t = \int_{-\pi}^{\pi} v(t)\sin mt\,\mathrm{d}t. \qquad (3.35)$$

Hence (3.34) and (3.35) hold for all $m \in \mathbb{N}_0$. Thus we have to prove: if for a continuous function $h \in C(\mathbb{T})$ we have for all $m \in \mathbb{N}_0$

$$0 = \int_{-\pi}^{\pi} h(t)\cos mt\,\mathrm{d}t = \int_{-\pi}^{\pi} h(t)\sin mt\,\mathrm{d}t \qquad (3.36)$$

then $h(t) = 0$ for all t. Suppose that (3.36) holds for $h \in C(\mathbb{T})$ and all $m \in \mathbb{N}_0$ and that for some $t_0 \in (-\pi, \pi)$ we have $h(t_0) \neq 0$. Using

$$\cos(\alpha + \beta) = \cos \alpha \cos \beta - \sin \alpha \sin \beta$$

and

$$\sin(\alpha + \beta) = \sin \alpha \cos \beta + \cos \alpha \sin \beta$$

we find that if (3.36) holds, then we also have

$$0 = \int_{-\pi}^{\pi} \frac{h(t + t_0)}{h(t_0)} \cos mt \, dt = \int_{-\pi}^{\pi} \frac{h(t + t_0)}{h(t_0)} \sin mt \, dt \qquad (3.37)$$

for all $m \in \mathbb{N}_0$, see Problem 4. If necessary, after switching from h to $-h$ we may assume $h(t_0) > 0$ and then there exists $\delta > 0$ such that $[-\delta, \delta] \subset (-\pi, \pi)$ and $\frac{h(t+t_0)}{h(t_0)} \geq \frac{1}{2}$ for all $t \in [-\delta, \delta]$. Consider $p \in C(\mathbb{T})$, $p(t) = 1 + \cos t - \cos \delta$. On $[-\delta, \delta]$ we find $p(t) \geq 1$ while for $t \in (-\pi, \pi) \backslash [-\delta, \delta]$ it follows that $|p(t)| < 1$. For $(p(t))^n = (1 + \cos t - \cos \delta)^n$ we have on $[-\delta, \delta]$ that $(p(t))^n \geq 1$ and for $t \in (-\pi, \pi) \backslash [-\delta, \delta]$ we find that $\lim_{n \to \infty} (p(t))^n = 0$. Expanding $(p(t))^n$ into functions $\cos lt$, $\sin lt$, $l \in \mathbb{N}_0$, $l \leq n$, we deduce now from (3.37) that

$$\int_{-\pi}^{\pi} \frac{h(t + t_0)}{h(t_0)} (p(t))^n \, dt = 0 \qquad \text{for all } n \in \mathbb{N}_0,$$

hence

$$\int_{-\delta}^{\delta} \frac{h(t + t_0)}{h(t_0)} (p(t))^n \, dt = \left(\int_{-\pi}^{-\delta} + \int_{\delta}^{\pi} \right) \left(\frac{h(t + t_0)}{h(t_0)} (p(t))^n \right) dt = 0.$$

By the dominated convergence theorem, Theorem III.7.33, we find that

$$\lim_{n \to \infty} \left(\int_{-\pi}^{-\delta} + \int_{\delta}^{\pi} \right) \left(\frac{h(t + t_0)}{h(t_0)} (p(t))^n \right) dt = 0$$

implying that

$$0 = \lim_{n \to \infty} \int_{-\delta}^{\delta} \frac{h(t + t_0)}{h(t_0)} (p(t))^n \, dt. \qquad (3.38)$$

However on $[-\delta, \delta]$ we have $\frac{h(t+t_0)}{h(t_0)} \geq \frac{1}{2}$ and $(p(t))^n \geq 1$, which contradicts (3.38) and the proposition is proved. $\qquad \square$

Theorem 3.15. *The mapping $F : L^1(\mathbb{T}) \to l_\infty(\mathbb{Z})$ as defined by (3.20) is injective, i.e. $(Ff)(k) = (Fg)(k)$ for all $k \in \mathbb{Z}$ implies $f = g$ $\lambda^{(1)}$-almost everywhere.*

Proof. Using again the linearity of F we have to show that $(Fh)(k) = 0$ for all $k \in \mathbb{Z}$ implies $h = 0$ $\lambda^{(1)}$-almost everywhere. From Corollary III.11.10 we know that $H(t) := \int_{-\pi}^{t} h(s)\,ds$ is an absolutely continuous, hence a continuous function and $\lambda^{(1)}$-almost everywhere differentiable. For $k \in \mathbb{Z}\backslash\{0\}$ we find

$$2\pi(FH)(k) = \int_{-\pi}^{\pi} H(t)e^{-ikt}\,dt = \int_{-\pi}^{\pi} e^{-ikt}\left(\int_{-\pi}^{t} h(s)\,ds\right)dt$$

$$= \int_{-\pi}^{\pi} h(s)\left(\int_{s}^{\pi} e^{-ikt}\,dt\right)ds$$

$$= \int_{-\pi}^{\pi} h(s)\left(\frac{e^{-ik\pi} - e^{-iks}}{-ik}\right)ds = 0$$

since for all $k \in \mathbb{Z}$ we have $(Fh)(k) = \frac{1}{2\pi}\int_{-\pi}^{\pi} h(s)e^{-iks}\,ds = 0$. We can now apply Proposition 3.14 to the continuous function G defined by

$$G(t) := H(t) - (FH)(0)$$

which yields that $H(t) - (FH)(0) = 0$. Moreover, by Proposition III.11.12 we deduce that $G'(t) = H'(t) = h(t)$ $\lambda^{(1)}$-almost everywhere, and hence $G'(t) = 0$ and the theorem follows. $\qquad\square$

Now we start to investigate the surjectivity of $F : L^1(\mathbb{T}) \to l_\infty(\mathbb{Z})$. The following result, called the Riemann-Lebesgue lemma implies that F is not surjective. It states that the Fourier coefficients of an L^1-function f must **vanish at infinity** in the sense that for every $\epsilon > 0$ there exists $N \in \mathbb{N}$ such that $|k| \geq N$ implies $|(Ff)(k)| < \epsilon$. As already indicated in Chapter 2, a proof will be provided in Example 6.9, the trigonometric series (2.19), i.e.

$$\sum_{k=2}^{\infty} \frac{\sin kx}{\ln k}$$

converges for all $x \in \mathbb{R}$ but is not to an integrable function. We calculate the sequence of the corresponding "complex" Fourier coefficients with the help of (3.7) to find

$$c_0 = 0, \quad c_k = \frac{-i}{2\ln k}, \quad k \in \mathbb{N}, \quad \text{and} \quad c_{-k} = \frac{i}{2\ln k}, \quad k \in \mathbb{N},$$

and they vanish at infinity. Hence F is not surjective even if we restrict its range to all sequences of complex numbers $(d_k)_{k \in \mathbb{Z}}$ vanishing at infinity. The Riemann-Lebesgue lemma will have further important applications when discussing the pointwise convergence of Fourier series. We state the result in a more general setting.

Theorem 3.16 (Lemma of Riemann-Lebesgue in $L^1(\mathbb{R})$**).** *Let $g \in L^1(\mathbb{R})$. Then the function*

$$(Fg)(\xi) := \frac{1}{\sqrt{2\pi}} \int_{\mathbb{R}} g(x)e^{-ix\xi} \, dx \qquad (3.39)$$

is a continuous function on $\mathbb{R}$ which vanishes at infinity, i.e. $Fg \in C_\infty(\mathbb{R})$.

Proof. The real and the imaginary parts of the function $(x, \xi) \mapsto g(x)e^{-ix\xi}$ satisfy all assumptions of Theorem III.8.1 and therefore $\xi \mapsto (Fg)(\xi)$ is a continuous function. Now let $\chi_{[a,b]}$ be the characteristic function of a bounded interval, $a < b$. For $\xi \neq 0$ we find

$$\left| \int_{\mathbb{R}} \chi_{[a,b]}(x)e^{-ix\xi} \, dx \right| = \left| \int_a^b e^{-ix\xi} \, dx \right| \qquad (3.40)$$

$$= \left| \frac{e^{-i\xi a} - e^{-i\xi b}}{i\xi} \right| \leq \frac{2}{|\xi|}$$

which implies the result for characteristic functions, hence for step functions. The real and the imaginary parts of g are real-valued L^1-functions and by Theorem III.8.21 the step functions are dense in (the real space) $L^1(\mathbb{R})$. Thus for $h \in \{\operatorname{Re} g, \operatorname{Im} g\}$ given, for $\epsilon > 0$ we can find a step function $s \in L^1(\mathbb{R})$ such that

$$\int_{\mathbb{R}} |h(x) - s(x)| \, dx < \frac{\epsilon}{2}. \qquad (3.41)$$

From (3.40) we deduce that for $\epsilon > 0$ we can also find $R > 0$ such that $|\xi| > R$ implies

$$\left| \int_{\mathbb{R}} s(x)e^{-ix\xi} \, dx \right| < \frac{\epsilon}{2}. \qquad (3.42)$$

Combining (3.42) with (3.41) we eventually obtain for $|\xi| > R$ that

$$\left| \int_{\mathbb{R}} h(x)e^{-ix\xi} \, dx \right| \leq \int_{\mathbb{R}} |h(x) - s(x)| \, dx + \left| \int_{\mathbb{R}} s(x)e^{-ix\xi} \, dx \right| < \epsilon$$

and the result follows. $\square$

Corollary 3.17 (Lemma of Riemann-Lebesgue in $L^1(\mathbb{T})$**).** *For $f \in L^1(\mathbb{T})$ its sequence of Fourier coefficients $((Ff)(k))_{k \in \mathbb{Z}}$ vanishes at infinity.*

Proof. We choose in Theorem 3.16 as g the function $f\chi_{[-\pi,\pi]}$ and $\xi = k \in \mathbb{Z}$. $\square$

Remark 3.18. A. The idea to prove first Theorem 3.16 which we will find useful later on is taken from [117].
B. From (3.8) we conclude also $\lim_{k\to\infty} a_k = \lim_{k\to\infty} b_k = 0$.

We want to determine the range of F when restricted to $L^2(\mathbb{T})$. Let $f \in L^2(\mathbb{T})$ and denote by $(c_k)_{k \in \mathbb{Z}}$ the sequence of its Fourier coefficients as defined by (3.6). For the Nth partial sum $S_N(t) = \sum_{|k| \leq N} c_k e^{ikt}$ we find

$$\|f - S_N\|_{L^2}^2 = \int_{-\pi}^{\pi} |f(t) - S_N(t)|^2 \, dt$$

$$= \int_{-\pi}^{\pi} (f(t) - S_N(t))\overline{(f(t) - S_N(t))} \, dt$$

$$= \int_{-\pi}^{\pi} f(t)\overline{f(t)} \, dt - \int_{-\pi}^{\pi} S_N(t)\overline{f(t)} \, dt$$

$$- \int_{-\pi}^{\pi} f(t)\overline{S_N(t)} \, dt + \int_{-\pi}^{\pi} S_N(t)\overline{S_N(t)} \, dt.$$

We observe that

$$\int_{-\pi}^{\pi} S_N(t)\overline{f(t)} \, dt = \sum_{k=-N}^{N} c_k \int_{-\pi}^{\pi} e^{ikt}\overline{f(t)} \, dt$$

$$= \sum_{k=-N}^{N} c_k \overline{\int_{-\pi}^{\pi} e^{-ikt} f(t) \, dt}$$

$$= 2\pi \sum_{k=-N}^{N} |c_k|^2,$$

and further

$$\int_{-\pi}^{\pi} f(t)\overline{S_N(t)} \, dt = \overline{\int_{-\pi}^{\pi} \overline{f(t)}S_N(t) \, dt} = 2\pi \sum_{k=-N}^{N} |c_k|^2,$$

as well as

$$\int_{-\pi}^{\pi} |S_N(t)|^2 \, dt = \int_{-\pi}^{\pi} S_N(t)\overline{S_N(t)} \, dt$$

$$= \sum_{|k| \leq N} \sum_{|l| \leq N} \int_{-\pi}^{\pi} c_k e^{ikt}\overline{c_l} e^{-ilt} \, dt$$

$$= 2\pi \sum_{|k| \leq N} \sum_{|l| \leq N} c_k \overline{c_l} \int_{-\pi}^{\pi} e^{i(k-l)t} \, dt$$

$$= 2\pi \sum_{|k| \leq N} |c_k|^2.$$

Thus we have

$$\|f - S_N\|_{L^2}^2 = \|f\|_{L^2}^2 - 2\pi \sum_{|k| \leq N} |c_k|^2 \qquad (3.43)$$

implying that

$$\sum_{|k| \leq N} |c_k|^2 \leq \frac{1}{2\pi} \|f\|_{L^2}^2$$

and therefore

$$\sum_{k \in \mathbb{Z}} |c_k|^2 \leq \frac{1}{2\pi} \|f\|_{L^2}^2, \qquad (3.44)$$

i.e. we have proved **Bessel's inequality** (3.44) and

Proposition 3.19. *The image of $L^2(\mathbb{T})$ under F is a subspace of $l_2(\mathbb{Z})$ and we have the estimate*

$$\|F(f)\|_2 \leq \frac{1}{\sqrt{2\pi}} \|f\|_{L^2}. \qquad (3.45)$$

Proof. It remains to note that $\|F(f)\|_2^2 = \sum_{k \in \mathbb{Z}} |c_k|^2$. $\qquad\square$

In Chapter 5 we will prove that $F(L^2(\mathbb{T})) = l_2(\mathbb{Z})$.

We close these considerations with a trivial remark. A **trigonometric polynomial** of degree m is by definition a function p of the form

$$p(t) = \sum_{|k| \leq m} c_k e^{ikt} \qquad (3.46)$$

with $|c_m| + |c_{-m}| > 0$, or when switching to real-valued functions

$$p(t) = \frac{a_0}{2} + \sum_{k=1}^{m} (a_k \cos kt + b_k \sin kt)$$

with $|a_m| + |b_m| > 0$. Clearly, trigonometric polynomials are their own Fourier series, which are finite sums, and they belong to $L^2(\mathbb{T})$. The calculation leading to (3.43) yields for a trigonometric polynomial $p(t)$ given by (3.46)

$$\|p\|_{L^2}^2 = \sum_{|k| \leq m} |c_k|^2. \qquad (3.47)$$

Extending (3.47) for $f \in L^2(\mathbb{T})$ to

$$\|f\|_{L^2}^2 = \sum_{k \in \mathbb{Z}} |c_k|^2$$

will be the key step to prove $F(L^2(\mathbb{T})) = l_2(\mathbb{Z})$.

We want to close this chapter by discussing the convolution of 2π-periodic functions $f : \mathbb{R} \to \mathbb{C}$ which are Riemann integrable over $[-\pi, \pi)$. Recall that Riemann integrable functions are by definition bounded and the product of two Riemann integrable functions is again Riemann integrable - note that this is an important difference to Lebesgue integration theory, Lebesgue integrable functions are not neccessarily bounded and in general $f, g \in L^1(\Omega)$ does not imply $f \cdot g \in L^1(\Omega)$. However , we know by Theorem III.8.11. that if f is Riemann integrable on $[-\pi, \pi)$ then f is Lebesgue integrable on $[-\pi, \pi)$.

Definition 3.20. *Let* $f, g : \mathbb{R} \to \mathbb{C}$ *be two* 2π*-periodic functions integrable over* $[-\pi, \pi)$. *We define their* **convolution** *by*

$$(f * g)(t) = \frac{1}{2\pi} \int_{-\pi}^{\pi} f(s)g(t - s) \, ds. \tag{3.48}$$

As f and g are 2π-periodic it follows that $f * g$ is 2π-periodic too and using a change of variable we find

$$(f * g)(t) = \frac{1}{2\pi} \int_{-\pi}^{\pi} f(t - s)g(s) \, ds. \tag{3.49}$$

Since f and g are Lebesgue integrable on $[-\pi, \pi)$ the function $(t, s) \mapsto f(s)g(t - s)$ is integrable over $[-\pi, \pi) \times [-\pi, \pi)$ and we may apply Fubini's theorem, Theorem III.9.17, to find

$$\int_{-\pi}^{\pi} (f * g)(t) \, dt = \frac{1}{2\pi} \int_{-\pi}^{\pi} \left(\int_{-\pi}^{\pi} f(s)g(t - s) \, ds \right) dt$$

$$= \frac{1}{2\pi} \int_{-\pi}^{\pi} f(s) \int_{-\pi}^{\pi} g(t - s) \, dt \, ds$$

$$= \frac{1}{2\pi} \int_{-\pi}^{\pi} f(s) \, ds \int_{-\pi}^{\pi} g(r) \, dr,$$

implying

$$\|f * g\|_{L^1} \le \frac{1}{2\pi} \|f\|_{L^1} \|g\|_{L^1}. \tag{3.50}$$

Moreover we find

$$\|f * g\|_{\infty,\mathbb{T}} \le \|f\|_{\infty,\mathbb{T}} \|g\|_{\infty,\mathbb{T}} \tag{3.51}$$

where we make explicit use of the fact that Riemann integrable functions are bounded and we use the notation $\|f\|_{\infty,\mathbb{T}} = \sup_{t \in [-\pi,\pi)} |f(t)|$.

The following proposition is proved by "inspection", see Problem 8, when making use of Lebesgue's integration theory which we are allowed to by Theorem III.8.11.

42

Proposition 3.21. *Let* $f, g, h : \mathbb{R} \to \mathbb{C}$ *be* 2π-*periodic functions Riemann integrable on* $[-\pi, \pi)$. *Then we have*

$$f * (g + h) = (f * g) + (f * h); \tag{3.52}$$

$$(\alpha f) * g = \alpha(f * g) = f * (\alpha g), \quad \alpha \in \mathbb{C}; \tag{3.53}$$

$$f * g = g * f; \tag{3.54}$$

and

$$(f * g) * h = f * (g * h). \tag{3.55}$$

Of central importance is the **convolution theorem.**

Theorem 3.22. *Let* $f, g : \mathbb{R} \to \mathbb{C}$ *be two* 2π-*periodic functions Riemann integrable on* $[-\pi, \pi)$ *with Fourier coefficients* $(Ff)(k)$ *and* $(Fg)(k)$, $k \in \mathbb{Z}$. *It follows that*

$$F(f * g)(k) = (Ff)(k)(Fg)(k) \tag{3.56}$$

holds for all $k \in \mathbb{Z}$.

Proof. Using Fubini's theorem, Theorem III.9.17, we find

$$
\begin{aligned}
F(f * g)(k) &= \frac{1}{2\pi} \int_{-\pi}^{\pi} (f * g)(t) e^{-ikt} \, \mathrm{d}t \\
&= \frac{1}{2\pi} \int_{-\pi}^{\pi} \frac{1}{2\pi} \left(\int_{-\pi}^{\pi} f(s) g(t - s) \, \mathrm{d}s \right) e^{-ikt} \, \mathrm{d}t \\
&= \frac{1}{2\pi} \int_{-\pi}^{\pi} f(s) e^{-iks} \left(\frac{1}{2\pi} \int_{-\pi}^{\pi} g(t - s) e^{-ik(t-s)} \, \mathrm{d}t \right) \mathrm{d}s \\
&= \frac{1}{2\pi} \int_{-\pi}^{\pi} f(s) e^{-iks} \left(\frac{1}{2\pi} \int_{-\pi}^{\pi} g(t) e^{-ikt} \, \mathrm{d}t \right) \mathrm{d}s \\
&= (Ff)(k)(Fg)(k).
\end{aligned}
$$

$\square$

Further we have

Theorem 3.23. *Let* $f, g : \mathbb{R} \to \mathbb{C}$ *be two* 2π-*periodic functions Riemann integrable on* $[-\pi, \pi)$. *Then* $f * g$ *is a continuous function.*

Proof. If one of the functions is continuous the result follows from (3.48) or (3.49) combined with Theorem III.8.1. Now let $(f_k)_{k \in \mathbb{N}}$ be a sequence of 2π-periodic continuous functions each Riemann integrable on $[-\pi, \pi)$ converging in $L^1(\mathbb{T})$, i.e. with respect to the norm $\|.\|_{L^1}$, to f, i.e. $\lim_{n \to \infty} \|f - f_n\|_{L^1} = 0$, see Theorem 3.7. It follows that

$$
\begin{aligned}
\|f * g - f_n * g\|_{\infty, \mathbb{T}} &= \|(f - f_n) * g\|_{\infty, \mathbb{T}} \\
&\leq \|f - f_n\|_{\infty, \mathbb{T}} \|g\|_{\infty, \mathbb{T}},
\end{aligned}
$$

implying that $f * g$ is the uniform limit of continuous functions, i.e. continuous. $\square$

Problems

1. Let $f \in L^2(\mathbb{T})$ and suppose that $c_{-k} = \overline{c_k}$. Prove that

$$-i \sum_{k \in \mathbb{Z}} \mathrm{sgn}(k) c_k e^{ikx} = \sum_{k=1}^{\infty} (-b_k \cos kx + a_k \sin kx).$$

2. Prove that on $l_\infty(\mathbb{Z})$ a norm is given by (3.17) and that $l_\infty(\mathbb{Z})$ with this norm is complete, i.e. a Banach space.

3. Prove Hölder's inequality (3.29).

4. Give the details of the argument leading to (3.37).

5.   a) Let $f : \mathbb{R} \to \mathbb{R}$ be a 2π-periodic function of the class C^1. Denote the Fourier coefficients of f by a_k and b_k and those of f' by α_k and β_k. Prove that $\alpha_k = kb_k$ and $\beta_k = -ka_k$, $k \in \mathbb{N}$.

    b) Let $g : \mathbb{R} \to \mathbb{C}$ be a continuous 2π-periodic function. Assume that $c_0 = 0$ and consider the primitive G of g, i.e. $G(t) = \int_0^t g(s)\,ds$, $t \in \mathbb{R}$. Prove that G is a 2π-periodic function and that its Fourier coefficients are given for $k \neq 0$ by $\gamma_k = \frac{c_k}{ik}$.

6.   a) Prove that $t \mapsto \sum_{k=0}^{\infty} \frac{\sin kt}{4^k}$ is a C^∞-function.

    b) Show that $t \mapsto \sum_{k=0}^{\infty} \frac{\cos 4^k t}{4^k}$ is a continuous function and use the Lemma of Riemann-Lebesgue to prove that this function is not differentiable on $\mathbb{R}$.

7. Let $u : \mathbb{R} \to \mathbb{C}$ be a 2π-periodic function of class C^1 with Fourier coefficients u_k, $k \in \mathbb{Z}$. Prove the estimate

$$\sum_{k \in \mathbb{Z}} |u_k| \leq |u_0| + \frac{\pi}{\sqrt{3}} \|u'\|_{L^2}.$$

Hint: you may use the fact that $\sum_{k=1}^{\infty} \frac{1}{k^2} = \frac{\pi^2}{6}$.

8. Provide the proof of Proposition 3.21.

9. Consider the function $g : \mathbb{R} \to \mathbb{C}$, $g(t) = \sum_{k \in \mathbb{Z}} c_k e^{ikt}$, and assume $|c_k| \leq \frac{M}{k^2}$, $k \neq 0$, and $|c_0| \leq M$. Hence g is a continuous function. Prove that $g * g * g$ is a C^4-function.

Hint: determine the decay of the Fourier coefficients of $g * g * g$.

10. Show with the help of the Lemma of Riemann-Lebesgue that if $h : [a, b] \to \mathbb{R}$ is a continuous function then

$$\lim_{\gamma \to \infty} \int_a^b h(t) \sin \gamma t \, dt = \lim_{\gamma \to \infty} \int_a^b h(t) \cos \gamma t \, dt = 0.$$

11. Prove that for $f \in L^2(\mathbb{T})$ the minimiser $d^{(\min)} = (d_{-N}^{\min}, \ldots, d_0^{\min}, \ldots, d_N^{\min}) \in \mathbb{C}^{2N+1}$ of the extremal problem

$$\inf \left\{ \left\| f - \sum_{|k| \le N} d_k \frac{e^{ik \cdot}}{\sqrt{2\pi}} \right\|_{L^2} \mid d_k \in \mathbb{C} \right\}$$

is given by $c = (c_{-N}, \ldots, c_0, \ldots, c_N)$, where c_k is the kth Fourier coefficient of f.

4 Pointwise Convergence and Summability

We want to investigate the question for which points $t_0 \in [-\pi, \pi)$ does a Fourier series of a given function f converge to $f(t_0)$? For this question to make sense f must be integrable and pointwisely defined, thus we cannot work in $L^1(\mathbb{T})$ since elements of $L^1(\mathbb{T})$ are equivalence classes of functions being $\lambda^{(1)}$-almost everywhere equal. For this reason, in this chapter we assume all our functions to be defined on $\mathbb{R}$, to be 2π-periodic and to be Riemann integrable on $[-\pi, \pi]$. By Theorem III.8.11 these functions are also Lebesgue integrable, hence they determine a unique element in $L^1(\mathbb{T})$ and we can apply all our previous results.

Let $f : \mathbb{R} \to \mathbb{C}$ be a 2π-periodic function Riemann integrable on $[-\pi, \pi]$ with Fourier coefficients $(c_k)_{k \in \mathbb{Z}}$ and associated Fourier series

$$f(t) \sim \sum_{k \in \mathbb{Z}} c_k e^{ikt}. \tag{4.1}$$

The Nth partial sum of $\sum_{k \in \mathbb{Z}} c_k e^{-ikt}$ is denoted by $S_N(t)$ or $S_N(f)(t)$ and we have

$$S_N(f)(t) = S_N(t) = \sum_{|k| \leq N} c_k e^{ikt} = \sum_{k=-N}^{N} \left(\frac{1}{2\pi} \int_{-\pi}^{\pi} f(s) e^{-iks} \, ds \right) e^{ikt}$$

$$= \sum_{k=-N}^{N} \frac{1}{2\pi} \int_{-\pi}^{\pi} f(s) e^{ik(t-s)} \, ds$$

$$= \frac{1}{2\pi} \int_{-\pi}^{\pi} f(s) \left(\sum_{k=-N}^{N} e^{ik(t-s)} \right) ds,$$

or with

$$D_N(t) := \sum_{|k| \leq N} e^{ikt} \tag{4.2}$$

we find

$$S_N(f)(t) = \frac{1}{2\pi} \int_{-\pi}^{\pi} f(s) D_N(t-s) \, ds = (f * D_N)(t). \tag{4.3}$$

The integral in (4.3) is a convolution and in order to investigate its behaviour for $N \to \infty$ we first need to understand D_N better.

Lemma 4.1. *For $N \in \mathbb{N}_0$ we have*

$$D_N(t) = \begin{cases} \frac{\sin(N+\frac{1}{2})t}{\sin \frac{t}{2}}, & t \in [-\pi, \pi] \setminus \{0\} \\ 2N + 1, & t = 0. \end{cases} \tag{4.4}$$

47

Proof. To find the value of $D_N(t)$ for $t = 0$, i.e. in order to prove consistency, we use the rules of de l'Hospital

$$\lim_{t \to 0} \frac{\sin \left(N + \frac{1}{2}\right) t}{\sin \frac{t}{2}} = \lim_{t \to 0} \frac{\left(N + \frac{1}{2}\right) \cos \left(N + \frac{1}{2}\right) t}{\frac{1}{2} \cos \frac{t}{2}}$$

$$= 2N + 1 = \sum_{|k| \leq N} e^{ik0}.$$

For $t \in [-\pi, \pi] \backslash \{0\}$ we note that

$$\sum_{|k| \leq N} e^{ikt} = \sum_{|k| \leq N} \left(e^{it}\right)^k = \sum_{k=0}^{N} \left(e^{it}\right)^k + \sum_{k=-N}^{-1} \left(e^{it}\right)^k$$

$$= \sum_{k=0}^{N} \left(e^{it}\right)^k + \sum_{k=1}^{N} \left(e^{-it}\right)^k$$

$$= \frac{1 - \left(e^{it}\right)^{N+1}}{1 - e^{it}} + \frac{1 - \left(e^{-it}\right)^{N+1}}{1 - e^{-it}} - 1$$

$$= \frac{1 - \left(e^{it}\right)^{N+1}}{1 - e^{it}} + \frac{\left(e^{-it}\right)^{N} - 1}{1 - e^{it}}$$

$$= \frac{e^{-iNt} - e^{i(N+1)t}}{1 - e^{it}} = \frac{e^{-i\left(N+\frac{1}{2}\right)t} - e^{i\left(N+\frac{1}{2}\right)t}}{e^{-i\frac{t}{2}} - e^{i\frac{t}{2}}}$$

$$= \frac{\sin \left(N + \frac{1}{2}\right) t}{\sin \frac{t}{2}}.$$

$\square$

Definition 4.2. *We call the function D_N defined by (4.4) (and periodically extended to $\mathbb{R}$) the Nth **Dirichlet kernel**.*

Note that D_N is a real-valued even function.

Corollary 4.3. *For all $N \in \mathbb{N}_0$*

$$\frac{1}{2\pi} \int_{-\pi}^{\pi} D_N(t) \, dt = 1 \tag{4.5}$$

holds.

Proof. From (4.3) we deduce for every $f : \mathbb{R} \to \mathbb{C}$ which is 2π-periodic and Riemann integrable over $[-\pi, \pi]$ that

$$S_N(f)(t) = \frac{1}{2\pi} \int_{-\pi}^{\pi} f(t - s) D_N(s) \, ds$$

and choosing f to be identically equal to 1 we obtain (4.5). $\square$

48

Lemma 4.4. *For a 2π-periodic function $f : \mathbb{R} \to \mathbb{C}$ Riemann integrable on $[-\pi, \pi]$ the Nth partial sum of the associated Fourier series has the representation*

$$S_N(f)(t) = \frac{2}{\pi} \int_0^{\frac{\pi}{2}} D_N(2s) \left(\frac{f(t + 2s) + f(t - 2s)}{2} \right) ds. \qquad (4.6)$$

Proof. From (4.3) we deduce

$$S_N(f)(t) = \frac{1}{2\pi} \int_{-\pi}^{\pi} D_N(s) f(t - s) \, ds$$

$$= \frac{1}{2\pi} \int_0^{\pi} D_N(s) f(t - s) \, ds + \frac{1}{2\pi} \int_{-\pi}^0 D_N(s) f(t - s) \, ds$$

and using the fact that D_N is an even function we get

$$\frac{1}{2\pi} \int_{-\pi}^0 D_N(s) f(t - s) \, ds = \frac{1}{2\pi} \int_0^{\pi} D_N(s) f(t + s) \, ds$$

which yields

$$S_N(f)(t) = \frac{1}{2\pi} \int_0^{\pi} D_N(s) \left(f(t - s) + f(t + s) \right) ds$$

$$= \frac{1}{\pi} \int_0^{\frac{\pi}{2}} D_N(2s) \left(f(t - 2s) + f(t + 2s) \right) ds$$

$$= \frac{2}{\pi} \int_0^{\frac{\pi}{2}} D_N(2s) \left(\frac{f(t + 2s) + f(t - 2s)}{2} \right) ds.$$

$\square$

Combining Lemma 4.4 with Corollary 4.3 we obtain

Theorem 4.5. *Let $u : \mathbb{R} \to \mathbb{C}$ be a 2π-periodic function Riemann integrable on $[-\pi, \pi]$ and let $t_0 \in (-\pi, \pi)$ be fixed. The Fourier series associated with f converges at t_0 to a value $S(t_0)$ if*

$$\lim_{N \to \infty} \frac{2}{\pi} \int_0^{\frac{\pi}{2}} D_N(2s) \left(\frac{f(t_0 + 2s) + f(t_0 - 2s)}{2} - S(t_0) \right) ds = 0. \qquad (4.7)$$

Proof. We note that by Corollary 4.3

$$S(t_0) = \frac{2}{\pi} \int_0^{\frac{\pi}{2}} S(t_0) D_N(2s) \, ds$$

49

and therefore

$$S_N(t_0) - S(t_0) = \frac{2}{\pi} \int_0^{\frac{\pi}{2}} D_N(2s) \left(\frac{f(t_0 + 2s) + f(t_0 - 2s)}{2} - S(t_0) \right) ds.$$

Now (4.7) yields

$$\lim_{N \to \infty} |S_N(t_0) - S(t_0)|$$

$$= \lim_{N \to \infty} \frac{2}{\pi} \left| \int_0^{\frac{\pi}{2}} D_N(2s) \left(\frac{f(t_0 + 2s) + f(t_0 - 2s)}{2} - S(t_0) \right) ds \right| = 0.$$

$\square$

For $f : \mathbb{R} \to \mathbb{C}$ a 2π-periodic function which is Riemann integrable on $[-\pi, \pi]$ we introduce the following notation, compare also with Chapter III.11. Let $t_0 \in (-\pi, \pi)$ be fixed. We set

$$f\left(t_0^+\right) := \lim_{\substack{t \to t_0 \\ t > t_0}} f(t) \quad \text{and} \quad f\left(t_0^-\right) := \lim_{\substack{t \to t_0 \\ t < t_0}} f(t) \tag{4.8}$$

as well as

$$f'_+(t_0) := \lim_{\substack{h \to 0 \\ h > 0}} \frac{f(t_0 + h) - f(t_0^+)}{h} \tag{4.9}$$

and

$$f'_-(t_0) := \lim_{\substack{h \to 0 \\ h > 0}} \frac{f(t_0 - h) - f(t_0^-)}{h}, \tag{4.10}$$

provided these limits exist. Now we can prove

Theorem 4.6. *Let $f : \mathbb{R} \to \mathbb{C}$ be a 2π-periodic function Riemann integrable on $[-\pi, \pi]$. Let $t_0 \in (-\pi, \pi)$ and suppose that $f(t_0^+)$, $f(t_0^-)$ and $f'_+(t_0)$, $f'_-(t_0)$ do exist. Then the Fourier series associated with f converges at t_0 to the mean value $\frac{1}{2}\left(f(t_0^+) + f(t_0^-)\right)$.*

Proof. We introduce the function

$$\varphi(t_0, s) := \frac{f(t_0 + 2s) + f(t_0 - 2s)}{2}$$

and consider

$$\psi(t_0, s) = \begin{cases} \frac{\varphi(t_0, s) - \frac{1}{2}(f(t_0^+) + f(t_0^-))}{\sin s}, & 0 < s < \frac{\pi}{2} \\ f'_+(t_0) + f'_-(t_0), & s = 0. \end{cases}$$

It follows that

$$\lim_{\substack{s \to 0 \\ s>0}} \psi(t_0, s) = \lim_{\substack{s \to 0 \\ s>0}} \left(\frac{f(t_0 + 2s) + f(t_0 - 2s) - f(t_0^+) - f(t_0^-)}{2 \sin s} \right)$$

$$= \lim_{\substack{s \to 0 \\ s>0}} \left(\frac{s}{\sin s} \right) \lim_{\substack{s \to 0 \\ s>0}} \left(\frac{f(t_0 + 2s) - f(t_0^+)}{2s} + \frac{f(t_0 - 2s) - f(t_0^-)}{2s} \right)$$

$$= f'_+(t_0) + f'_-(t_0),$$

i.e. $s \mapsto \psi(t_0, s)$ is continuous at $s = 0$, and since $s \mapsto \varphi(t_0, s)$ is integrable on $\left[0, \frac{\pi}{2}\right]$ the function $s \mapsto \psi(t_0, s)$ is also integrable on $\left[0, \frac{\pi}{2}\right]$. Now we find using the representation of D_N by (4.4) that

$$\int_0^{\frac{\pi}{2}} \left| \varphi(t_0, s) - \frac{1}{2} \left(f(t_0^+) + f(t_0^-) \right) \right| D_N(2s) \mathrm{d}s = \int_0^{\frac{\pi}{2}} \psi(t_0, s) \sin((2N+1)s) \, \mathrm{d}s$$

and the Lemma of Riemann-Lebesgue, Theorem 3.16, implies that

$$\lim_{N \to \infty} \int_0^{\frac{\pi}{2}} \psi(t_0, s) \sin((2N+1)s) \, \mathrm{d}s = 0.$$

Thus we have proved

$$\lim_{N \to \infty} \int_0^{\frac{\pi}{2}} \left(\frac{f(t_0 + 2s) + f(t_0 - 2s)}{2} - \frac{1}{2} \left(f(t_0^+) + f(t_0^-) \right) \right) D_N(2s) \, \mathrm{d}s = 0$$

and now Theorem 4.5 gives the result. $\qquad \square$

Corollary 4.7. *If in addition to the assumptions of Theorem 4.6 the function f is continuous at t_0, then the Fourier series associated with f converges at t_0 to $f(t_0)$, i.e.*

$$f(t_0) = \lim_{N \to \infty} S_N(f)(t_0) = \sum_{k \in \mathbb{Z}} c_k e^{ikt_0} \tag{4.11}$$

with $c_k = (Ff)(k)$.

Proof. By Theorem 4.6 we know that $(S_N(f)(t_0))_{N \in \mathbb{N}}$ converges to $\frac{f(t_0^+) + f(t_0^-)}{2}$ and the continuity of f at t_0 implies that $\frac{f(t_0^+) + f(t_0^-)}{2} = f(t_0)$. $\qquad \square$

Corollary 4.8. *For all $f \in C^1(\mathbb{T})$ the associated Fourier series converges at each point $t \in (-\pi, \pi)$ to $f(t)$.*

Proof. This follows from Theorem 4.6 since $f \in C^1(\mathbb{T})$ satisfies the assumptions of the theorem for all $t \in (-\pi, \pi)$. $\qquad \square$

Remark 4.9. Suppose that the function $f : \mathbb{R} \to \mathbb{C}$ is 2π-periodic and Riemann integrable on $[-\pi, \pi]$. Then the function $g_{\frac{\pi}{4}} : \mathbb{R} \to \mathbb{C}$ defined by $g_{\frac{\pi}{4}}(t) := f\left(t + \frac{\pi}{4}\right)$ is also 2π-periodic and Riemann integrable over $[-\pi, \pi]$, hence Theorem 4.6 and its corollaries apply to $g_{\frac{\pi}{4}}$, which in turn implies that these results also hold for f at the points π and $-\pi$.

Example 4.10. Consider the function u_2 from Example 2.8. It satisfies all assumptions of Theorem 4.6, the periodicity follows from the construction and as a piecewise differentiable function on $[-\pi, \pi]$ it is Riemann integrable. For all points $t \in [-\pi, \pi]\backslash\{0\}$ the Fourier series associated with u_2 converges at t to $u_2(t)$ while at $t = 0$ the Fourier series converges to 0, but $u_2(0) = \frac{\pi}{2}$.

Example 4.11. Let $u : [-\pi, \pi) \to \mathbb{R}$ be a bounded monotone function which we extend periodically to $\mathbb{R}$. This function satisfies again all the assumptions of Theorem 4.6. We assume in addition that on $[-\pi, \pi)$ the function u has at most finitely many jumps, say at $t_1, \ldots, t_M$. We know that then all points in $(-\pi, \pi)\backslash\{t_1, \ldots, t_M\}$ are points of continuity of u, see Chapter I.32, and therefore at these points t the Fourier series associated with u converges to $u(t)$. On the other hand, at points where u has a jump the value of the Fourier series is $\frac{u(t_j^+) + u(t_j^-)}{2}$, $j = 1, \ldots, M$.

In general for a continuous function it is not true that its Fourier series converges to the function, in fact as we will see in Chapter 6 the Fourier series of a continuous function may diverge. The question arises whether we can find a different way to represent a given function pointwisely with the help of its Fourier coefficients. In Problem 13 of Chapter I.15 we have proved for a sequence $(\alpha_n)_{n \in \mathbb{N}}$ of real numbers

$$\lim_{n \to \infty} \alpha_n = \alpha \quad \text{implies} \quad \lim_{n \to \infty} \frac{1}{n} \sum_{k=1}^{n} \alpha_k = \alpha, \tag{4.12}$$

i.e. if $(\alpha_n)_{n \in \mathbb{N}}$ converges to α then the sequence of the arithmetic means $\left(\frac{1}{n}\sum_{k=1}^{n} \alpha_k\right)_{n \in \mathbb{N}}$ converges to the same limit α. There are however sequences for which $\left(\frac{1}{n}\sum_{k=1}^{n} \alpha_k\right)_{n \in \mathbb{N}}$ converges while the sequence $(\alpha_n)_{n \in \mathbb{N}}$ does not converge. Take the sequence $\left((-1)^{n+1}\right)_{n \in \mathbb{N}}$ which has the accumulation points $+1$ and -1, but for the sequence of its arithmetic means we find

$$\frac{1}{n} \sum_{k=1}^{n} (-1)^{k+1} = \begin{cases} 0, & n \text{ is even} \\ \frac{1}{n}, & n \text{ is odd} \end{cases} \tag{4.13}$$

and therefore

$$\lim_{n \to \infty} \frac{1}{n} \sum_{k=1}^{n} (-1)^{k+1} = 0. \tag{4.14}$$

We may switch to sequences of functions $u_k : A \to \mathbb{C}$, $A \neq \emptyset$, $k \in \mathbb{N}$, with partial sums

$$T_N(x) = \sum_{k=1}^{N} u_k(x)$$

and consider instead of the sequence of partial sums $(T_N)_{N \in \mathbb{N}}$ the sequences

$$C_N(x) = \frac{1}{N} \sum_{k=1}^{N} T_k(x) = \sum_{k=1}^{N} \frac{1}{N} T_k(x).$$

More generally, we may look at weighted partial sums

$$B_N(x) = \sum_{k=1}^{N} b_{N,k} T_k(x), \qquad b_{N,k} \geq 0,$$

and ask whether for $x \in A$ the sequence $(B_N(x))_{N \in \mathbb{N}}$ has a limit. Such a summation procedure makes sense as a substitute for the convergence of the partial sums if

$$\lim_{N \to \infty} T_N(x) = T(x) \quad \text{implies} \quad \lim_{N \to \infty} B_N(x) = T(x).$$

It turns out that for Fourier series several summation procedures are available. We start with the most natural one, namely by looking at arithmetic means.

Definition 4.12. *Let $f : \mathbb{R} \to \mathbb{C}$ be a 2π-periodic function Riemann integrable on $[-\pi, \pi]$.*
A. *We call*

$$\sigma_N(f)(t) := \frac{1}{N} \sum_{k=0}^{N-1} S_k(f)(t), \qquad N \in \mathbb{N}, \tag{4.15}$$

*the Nth **Cesàro mean** of $(S_k(f))_{k \in \mathbb{N}}$ at t.*
B. *We say that the sequence $(S_k(f))_{k \in \mathbb{N}_0}$ is **Cesàro summable** at t if the sequence $(\sigma_N(f)(t))_{N \in \mathbb{N}}$ has a limit.*
C. *If $(\sigma_N(f)(t))_{N \in \mathbb{N}}$ converges to $f(t)$ we say that the Fourier series associated with f is **Cesàro summable to f** at the point t.*

From (4.3) we deduce

$$\sigma_N(f)(t) = \frac{1}{N} \sum_{k=0}^{N-1} S_k(f)(t) = \frac{1}{N} \sum_{k=0}^{N-1} (f * D_k)(t)$$

$$= \left(f * \left(\frac{1}{N} \sum_{k=0}^{N-1} D_k \right) \right)(t).$$

Definition 4.13. *For $N \in \mathbb{N}$ we call*

$$F_N(t) := \frac{1}{N} \sum_{k=0}^{N-1} D_k(t) \tag{4.16}$$

the Nth Fejer kernel.

Lemma 4.14. *For $N \in \mathbb{N}$ we have*

$$F_N(t) = \sum_{l=-N+1}^{N-1} \left(1 - \frac{|l|}{N}\right) e^{ilt} \tag{4.17}$$

as well as

$$F_N(t) = \frac{1}{N} \frac{\sin^2\left(\frac{Nt}{2}\right)}{\sin^2\left(\frac{t}{2}\right)}. \tag{4.18}$$

In particular $F_N(t) \geq 0$ for all t and

$$\frac{1}{2\pi} \int_{-\pi}^{\pi} F_N(t)\, \mathrm{d}t = 1. \tag{4.19}$$

Moreover, for every $\delta > 0$ we have

$$\lim_{N \to \infty} \int_{\delta \leq |t| \leq \pi} F_N(t)\, \mathrm{d}t = 0. \tag{4.20}$$

Proof. For $N \in \mathbb{N}$ we find

$$NF_N(t) = \sum_{k=0}^{N-1} \left(\sum_{l=-k}^{k} e^{ilt}\right) = \sum_{l=-N+1}^{N-1} \sum_{|k|=l} e^{ilt}$$

$$= \sum_{l=-N+1}^{N-1} (N - |l|)e^{ilt},$$

or

$$F_N(t) = \sum_{l=-N+1}^{N-1} \left(1 - \frac{|l|}{N}\right) e^{ilt},$$

and (4.17) is proved. Using the trigonometric identity $\sin \alpha \sin \beta = \frac{\cos(\alpha-\beta)-\cos(\alpha+\beta)}{2}$ or $\cos(\alpha - \beta) - \cos(\alpha + \beta) = 2 \sin \alpha \sin \beta$, we get with the help

54

of Lemma 4.1

$$N \sin^2 \left(\frac{t}{2}\right) F_N(t) = \sum_{k=0}^{N-1} \sin\left(\left(k+\frac{1}{2}\right)t\right) \sin\frac{t}{2}$$

$$= \frac{1}{2} \sum_{k=0}^{N-1} (\cos kt - \cos(k+1)t)$$

$$= \frac{1}{2}(1 - \cos Nt) = \sin^2\left(\frac{Nt}{2}\right)$$

which yields (4.18), i.e.

$$F_N(t) = \frac{1}{N} \frac{\sin^2\left(\frac{Nt}{2}\right)}{\sin^2\left(\frac{t}{2}\right)}.$$

Since we have $\frac{1}{2\pi}\int_{-\pi}^{\pi} D_N(t)\,dt = 1$ by Corollary 4.3 we deduce (4.19):

$$\frac{1}{2\pi}\int_{-\pi}^{\pi} F_N(t)\,dt = \frac{1}{N}\sum_{k=0}^{N-1}\frac{1}{2\pi}\int_{-\pi}^{\pi} D_k(t)\,dt = 1.$$

Finally, given $\delta > 0$ we can find some $\eta = \eta(\delta) > 0$ such that $\sin^2\frac{t}{2} \geq \eta > 0$ for $\delta \leq |t| \leq \pi$ which implies in light of (4.18) that

$$F_N(t) \leq \frac{1}{N\eta} \quad \text{for } \delta \leq |t| \leq \pi,$$

and now (4.20) follows. $\qquad\square$

Now we can prove

Theorem 4.15 (Fejér). *The Fourier series of every Riemann integrable 2π-periodic function $f : \mathbb{R} \to \mathbb{C}$ is Cesàro summable to $f(t_0)$ for every point t_0 of continuity of f.*

Proof. Let f be continuous at $t_0 \in [-\pi, \pi]$ and for $\epsilon > 0$ choose $\delta > 0$ such that $|t| < \delta$ implies $|f(t_0 - t) - f(t_0)| < \epsilon$. It follows that

$$\sigma_N(f)(t_0) - f(t_0) = (f * F_N)(t_0) - f(t_0)$$

$$= \frac{1}{2\pi}\int_{-\pi}^{\pi} F_N(t)f(t_0 - t)\,dt - f(t_0)$$

$$= \frac{1}{2\pi}\int_{-\pi}^{\pi} F_N(t)\left(f(t_0 - t) - f(t_0)\right)dt,$$

where we used (4.19). This implies that

$$|\sigma_N(f)(t_0) - f(t_0)| \leq \frac{1}{2\pi} \int_{-\pi}^{\pi} F_N(t) |f(t_0 - t) - f(t_0)| \, dt$$

$$\leq \frac{1}{2\pi} \int_{|t|<\delta} F_N(t) |f(t_0 - t) - f(t_0)| \, dt$$

$$+ \frac{1}{2\pi} \int_{\delta \leq |t| \leq \pi} F_N(t) |f(t_0 - t) - f(t_0)| \, dt$$

$$\leq \frac{\epsilon}{2\pi} \int_{-\pi}^{\pi} F_N(t) \, dt + \frac{2\|f\|_{\infty,\mathbb{T}}}{2\pi} \int_{\delta \leq |t| \leq \pi} F_N(t) \, dt.$$

From (4.20) we deduce the existence of $N_0 \in \mathbb{N}$ such that $N \geq N_0$ implies that $\int_{\delta \leq |t| \leq \pi} F_N \, dt < \epsilon$ and therefore we arrive by using once more (4.19) at

$$|\sigma_N(f)(t_0) - f(t_0)| \leq \epsilon + \frac{2\|f\|_{\infty,\mathbb{T}}}{2\pi} \epsilon = \kappa\epsilon,$$

and the theorem is proved. $\square$

Corollary 4.16. *For a continuous 2π-periodic function $f : \mathbb{R} \to \mathbb{C}$ the Fourier series is at every point t_0 Cesàro summable to $f(t_0)$.*

Remark 4.17. Calling $(S_N(f))_{N \in \mathbb{N}_0}$ **uniformly Cesàro summable** to f if $\lim_{N\to\infty} \|\sigma_N(f) - f\|_{\infty,\mathbb{T}} = 0$, then the proof of Theorem 4.15 yields that for a continuous function f the sequence $(S_N(f))_{N \in \mathbb{N}_0}$ is uniformly Cesàro summable to f. Moreover, since for $N \in \mathbb{N}_0$ the functions $\sigma_N(f)(\cdot)$ are trigonometric polynomials it follows that we can approximate every 2π-periodic continuous function uniformly by trigonometric polynomials, or in other words, the trigonometric polynomials are dense in $C(\mathbb{T})$ with respect to the supremum norm. This is the **Weierstrass approximation theorem** for continuous periodic functions.

In Problem 10 of Chapter I.29 we have proved Abel's convergence theorem, i.e. if the series $\sum_{k=0}^{\infty} a_k$ converges then we have $\lim_{\substack{x\to 1 \\ x<0}} \sum_{k=0}^{\infty} a_k x^k = \sum_{k=0}^{\infty} a_k$. This result stimulates a further way of summing series which is called Abel summation. As preparation we need a type of integration by parts formula for sums. Let $(u_k)_{k \in \mathbb{N}_0}$ and $(v_k)_{k \in \mathbb{N}_0}$ be two sequences of complex numbers and define $V_N := \sum_{l=0}^{N} v_l$, $V_{-1} = 0$. By inspection it follows that

$$\sum_{k=m}^{n} u_k v_k = \sum_{k=m}^{n-1} (u_k - u_{k-1}) V_k + u_n V_n - u_m V_{m-1}. \tag{4.21}$$

With $\Delta u_k = u_k - u_{k+1}$, $\Delta V_k = V_k - V_{k+1}$ formula (4.21) yields

$$\sum_{k=m}^{n-1} \Delta u_k V_k = -\sum_{k=m}^{n} u_k \Delta V_{k-1} + u_m V_{m-1} - u_n V_n. \qquad (4.22)$$

We can interpret (4.21) or (4.22) as an "integration by parts" formula for sums. Note that for $m = 0$ formula (4.21) reads as

$$\sum_{k=0}^{n} u_k v_k = u_n V_n + \sum_{k=0}^{n-1} (u_k - u_{k+1}) V_k, \qquad (4.23)$$

for more details we refer to Problem 7, see also [8].

Definition 4.18. *Let $(d_k)_{k \in \mathbb{N}_0}$ be a sequence of complex numbers and suppose that for all $0 \le r < 1$ the series*

$$A(r) := \sum_{k=0}^{\infty} d_k r^k \qquad (4.24)$$

*converges. We call $\sum_{k=0}^{\infty} d_k$ **Abel summable** to α if*

$$\lim_{r \to 1} A(r) = \lim_{r \to 1} \sum_{k=0}^{\infty} d_k r^k = \alpha. \qquad (4.25)$$

Theorem 4.19. *If the series $\sum_{k=0}^{\infty} d_k$ is Cesàro summable to α then it is Abel summable to α. In particular, if $\sum_{k=0}^{\infty} d_k$ converges to α then this series is also Abel summable to α.*

We prepare the proof of Theorem 4.19 by

Lemma 4.20. *Let $(d_k)_{k \in \mathbb{N}_0}$ be a sequence of complex numbers which is Abel summable to α. Then we have for $0 \le r < 1$ that*

$$\lim_{r \to 1} (1 - r) \sum_{k=0}^{\infty} S_k r^k = \alpha. \qquad (4.26)$$

Proof. Since $\lim_{r \to 1} \sum_{k=0}^{\infty} d_k r^k = \alpha$, i.e. the limit exists, we find for $0 \le r < 1$ that

$$\sum_{k=0}^{\infty} d_k r^k = (1 - r) \sum_{k=0}^{\infty} S_k r^k. \qquad (4.27)$$

Indeed, the convergence of $\sum_{k=0}^{\infty} d_k r^k$ implies $|d_k r^k| \leq C = C(r) < \infty$ and for $0 \leq \rho < r < 1$ it follows that

$$|S_n \rho^n| = \left| \sum_{k=0}^{n} d_k \rho^n \right| = \left| \rho^n \sum_{k=0}^{n} \frac{d_k r^k}{r^k} \right|$$

$$\leq C \rho^n \sum_{k=0}^{n} r^{-k} = \frac{C}{1-r} \rho^n \frac{1}{r^n},$$

which implies for $0 \leq \rho < r < 1$

$$\lim_{n \to \infty} S_n \rho^n = 0. \tag{4.28}$$

Using (4.23) we find

$$\sum_{k=0}^{n} d_k \rho^k = \sum_{k=0}^{n-1} S_k \left(\rho^k - \rho^{k+1} \right) + S_n \rho^n \tag{4.29}$$

$$= (1 - \rho) \sum_{k=0}^{n-1} S_k \rho^k + S_n \rho^n.$$

But now (4.28) implies (with ρ replaced by r) that

$$\lim_{r \to 1} (1 - r) \sum_{k=0}^{\infty} S_k r^k = \alpha. \tag{4.30}$$

$\square$

Proof of Theorem 4.19. As above we set $\sigma_N = \frac{1}{N} \sum_{k=0}^{N-1} S_k$ which gives

$$(N + 1) \sigma_{N+1} = \sum_{k=0}^{N} S_k.$$

We note that further by (4.23)

$$\sum_{k=0}^{M} S_k r^k = \sum_{k=0}^{M-1} \left(r^k - r^{k+1} \right) \sum_{l=0}^{k} S_l + r^M \sum_{k=0}^{M} S_k$$

and therefore, since $\sum_{k=0}^{\infty} d_k$ is Cesàro summable

$$\sum_{k=0}^{\infty} S_k r^k = (1 - r) \sum_{k=0}^{\infty} (k + 1) \sigma_{k+1} r^k, \tag{4.31}$$

where we used that for $0 \leq r < 1$

$$\lim_{M \to \infty} r^M \sum_{k=0}^{M} S_k = \lim_{M \to \infty} (M+1)\sigma_{M+1} r^M = 0,$$

since $\lim_{M \to \infty} \sigma_{M+1} = \alpha$. Combining (4.31) with (4.27) we find

$$\sum_{k=0}^{\infty} d_k r^k = (1-r)^2 \sum_{k=0}^{\infty} (k+1)\sigma_{k+1} r^k. \tag{4.32}$$

For $0 \leq r < 1$ we have

$$\frac{1}{(1-r)^2} = \sum_{k=0}^{\infty} (k+1) r^k$$

or

$$1 = (1-r)^2 \sum_{k=0}^{\infty} (k+1) r^k, \tag{4.33}$$

i.e.

$$\alpha = (1-r)^2 \sum_{k=0}^{\infty} (k+1) r^k \alpha. \tag{4.34}$$

Subtracting (4.34) from (4.32) yields

$$\sum_{k=0}^{\infty} d_k r^k - \alpha = (1-r)^2 \sum_{k=0}^{\infty} (k+1)\left(\sigma_{k+1} - \alpha\right) r^k$$

$$= (1-r)^2 \sum_{k=0}^{N} (k+1)(\sigma_{k+1} - \alpha) r^k$$

$$+ (1-r)^2 \sum_{k=N+1}^{\infty} (k+1)(\sigma_{k+1} - \alpha) r^k.$$

Now let $\epsilon > 0$ be given and choose $N = N(\epsilon)$ such that $k \geq N(\epsilon)$ implies $|\sigma_{k+1} - \alpha| < \epsilon$, which is possible since by assumption we have Cesàro summability of $\sum_{k=0}^{\infty} d_k$ to α. However, for every $N = N(\epsilon)$ fixed it follows for $r \to 1$ that

$$\lim_{r \to 1} (1-r)^2 \sum_{k=0}^{\infty} (k+1)(\sigma_{k+1} - \alpha) r^k = 0,$$

and the theorem is proved. $\qquad \square$

Example 4.21. Consider the sequence $d_k = (-1)^k(k+1)$, $k \in \mathbb{N}_0$. Since it is unbounded the series $\sum_{k=0}^{\infty} d_k = \sum_{k=0}^{\infty}(-1)^k(k+1)$ diverges. For the partial sums we find

$$\sum_{k=0}^{\infty}(-1)^k(k+1) = \begin{cases} \frac{N+2}{2}, & N \text{ even} \\ \frac{-(N+1)}{2}, & N \text{ odd} \end{cases}$$

implying for the arithmetic means

$$\frac{1}{N+1}\sum_{k=0}^{N}(-1)^k(k+1) = \begin{cases} \frac{1}{2} + \frac{1}{2(N+1)}, & N \text{ even} \\ -\frac{1}{2}, & N \text{ odd}, \end{cases}$$

and therefore this series is not Cesàro summable. However, since for $0 \le r < 1$ we have

$$\sum_{k=0}^{\infty}(-1)^k(k+1)r^k = \sum_{k=0}^{\infty}(-1)^k\left(\frac{d}{dr}r^{k+1}\right)$$

$$= \frac{d}{dr}\sum_{k=0}^{\infty}(-1)^k r^{k+1} = \frac{d}{dr}\left(r\sum_{k=0}^{\infty}(-r)^k\right)$$

$$= \frac{d}{dr}\left(\frac{r}{1+r}\right) = \frac{1}{(1+r)^2}$$

it follows that

$$\lim_{r \to 1}\sum_{k=0}^{\infty}(-1)^k(k+1)r^k = \lim_{r \to 1}\frac{1}{(1+r)^2} = \frac{1}{4},$$

i.e. the series $\sum_{k=0}^{\infty}(-1)^k(k+1)$ is Abel summable to $\frac{1}{4}$. This example shows that Cesàro summability and Abel summability are not equivalent, but we know already by Theorem 4.19 that Cesàro summability implies Abel summability.

Let $f : \mathbb{R} \to \mathbb{C}$ be a 2π-periodic function Riemann integrable over $[-\pi, \pi]$. Since the Fourier coefficients c_k, $k \in \mathbb{Z}$, of f are bounded it follows that for $0 \le r < 1$ the series

$$A_r(f)(t) := \sum_{k \in \mathbb{Z}} c_k e^{ikt} r^{|k|} \tag{4.35}$$

converges absolutely and uniformly on $[-\pi, \pi]$, in fact on $\mathbb{R}$, which implies that every rearrangement of the series converges to the same limit. Therefore we find with

$$d_k(t) := \begin{cases} c_0, & k = 0 \\ c_k e^{ikt} + c_{-k}e^{-ikt}, & k \in \mathbb{N} \end{cases} \tag{4.36}$$

60

that

$$A_r(f)(t) = \sum_{k=0}^{\infty} d_k(t) r^k. \tag{4.37}$$

Moreover, using the uniform convergence which allows us to interchange summation and integration we have

$$A_r(f)(t) = \sum_{k \in \mathbb{Z}} c_k e^{ikt} r^{|k|}$$

$$= \sum_{k \in \mathbb{Z}} r^{|k|} \left(\frac{1}{2\pi} \int_{-\pi}^{\pi} f(s) e^{-iks} \, ds \right) e^{ikt}$$

$$= \frac{1}{2\pi} \int_{-\pi}^{\pi} f(s) \left(\sum_{k \in \mathbb{Z}} r^{|k|} e^{-ik(s-t)} \right) ds.$$

With

$$P_r(t) := \sum_{k \in \mathbb{Z}} r^{|k|} e^{ikt} \tag{4.38}$$

we obtain

$$A_r(f)(t) = \frac{1}{2\pi} \int_{-\pi}^{\pi} f(s) P_r(s-t) \, ds = (f * P_r)(t). \tag{4.39}$$

Before proceeding further we prove

Lemma 4.22. *The following identity holds:*

$$P_r(t) = \frac{1 - r^2}{1 - 2r \cos t + r^2}, \quad 0 \le r < 1. \tag{4.40}$$

Moreover $P_r(t) \ge 0$ *for all t and*

$$\frac{1}{2\pi} \int_{-\pi}^{\pi} P_r(t) \, dt = 1. \tag{4.41}$$

Proof. From the definition of $P_r(t)$ we deduce

$$P_r(t) = \sum_{k \in \mathbb{Z}} r^{|k|} e^{ikt}$$

$$= \sum_{k=0}^{\infty} \left(r e^{it} \right)^k + \sum_{k=1}^{\infty} \left(r e^{-it} \right)^k$$

$$= \frac{1}{1 - r e^{it}} + \frac{1}{1 - r e^{-it}} - 1$$

$$= \frac{1 - re^{-it} + 1 - re^{it} - |1 - re^{it}|^2}{|1 - re^{it}|^2}$$

$$= \frac{1 - r^2}{1 - 2r\cos t + r^2} \geq 0$$

since by assumption $0 \leq r < 1$. The uniform convergence of (4.38) for $0 \leq r < 1$ implies further that

$$\int_{-\pi}^{\pi} \sum_{k \in \mathbb{Z}} r^{|k|} e^{ikt} \, dt = \sum_{k \in \mathbb{Z}} r^{|k|} \int_{-\pi}^{\pi} e^{ikt} \, dt$$

$$= 2\pi + \sum_{k \in \mathbb{Z} \setminus \{0\}} r^{|k|} \frac{1}{ik} e^{ikt} \Bigg|_{-\pi}^{\pi} = 2\pi$$

proving (4.41). □

Definition 4.23. *We call $P_r(t)$ the **Poisson kernel** at t for $r \in [0, 1)$.*

Thus (4.39) means that $A_r(f)(t)$ is obtained as convolution of f with the Poisson kernel at t. With our previous considerations on Abel summability in mind as well as (4.37) we introduce

Definition 4.24. *Let $f : \mathbb{R} \to \mathbb{C}$ be a 2π-periodic function Riemann integrable on $[-\pi, \pi]$. We say that the Fourier series associated with f is **Abel summable** to f at $t_0 \in [-\pi, \pi]$ if*

$$\lim_{r \to 1} A_r(f)(t_0) = f(t_0). \tag{4.42}$$

*If (4.42) holds uniformly for all $t_0 \in [-\pi, \pi]$ then we call the Fourier series associated with f **uniformly Abel summable** to f.*

Theorem 4.25. *Let $f : \mathbb{R} \to \mathbb{C}$ be a 2π-periodic function Riemann integrable over $[-\pi, \pi]$. The Fourier series associated with f is at every point t_0 of continuity of f Abel summable to $f(t_0)$. If f is continuous on $[-\pi, \pi]$ then the Fourier series associated with f is uniformly Abel summable to f.*

Proof. The proof is in its structure similar to that of Theorem 4.15 and we will soon understand why. We start with the observation that

$$1 - 2r\cos t + r^2 = (1 - r)^2 + 2r(1 - \cos t)$$

and therefore, for $\frac{1}{2} \leq r \leq 1$ and $0 < \delta \leq |t| \leq \pi$ we get

$$1 - 2r\cos t + r^2 \geq \kappa_\delta > 0$$

62

implying

$$P_r(t) \leq \frac{1 - r^2}{\kappa_\delta} \tag{4.43}$$

for $r \in \left[\frac{1}{2}, 1\right]$ and $0 < \delta \leq |t| \leq \pi$. From this we deduce

$$0 \leq \lim_{r \to 1} \int_{\delta \leq |t| \leq \pi} P_r(t) \, dt \leq \lim_{r \to 1} \frac{1 - r^2}{\kappa_\delta} \int_{\delta \leq |t| \leq \pi} 1 \, dt = 0. \tag{4.44}$$

Now let $t_0 \in (-\pi, \pi)$ be a point of continuity of f. Given $\epsilon > 0$ we can find $\delta > 0$ such that $|s| < \delta$ implies $|f(t_0 - s) - f(t_0)| < \epsilon$. It follows with (4.41) that

$$
\begin{aligned}
|A_r(f)(t_0) - f(t_0)| &= |(f * P_r)(t_0) - f(t_0)| \\
&= \left| \frac{1}{2\pi} \int_{-\pi}^{\pi} P_r(s) \left(f(t_0 - s) - f(t_0) \right) ds \right| \\
&\leq \frac{1}{2\pi} \int_{|s| < \delta} P_r(s) |f(t_0 - s) - f(t_0)| \, ds \\
&\quad + \frac{1}{2\pi} \int_{\delta \leq |s| \leq \pi} P_r(s) |f(t_0 - s) - f(t_0)| \, ds \\
&\leq \frac{\epsilon}{2\pi} \int_{|s| < \delta} P_r(s) \, ds + \frac{2\|f\|_{\infty, \mathbb{T}}}{2\pi} \int_{\delta \leq |s| \leq \pi} P_r(s) \, ds.
\end{aligned}
$$

From (4.44) it follows that we can find $\frac{1}{2} \leq R < 1$ such that $r \geq R$ implies $\int_{\delta \leq |s| \leq \pi} P_r(s) \, ds < \epsilon$ which yields

$$|A_r(f)(t_0) - f(t_0)| \leq \left(1 + \frac{\|f\|_{\infty, \mathbb{T}}}{\pi} \right) \epsilon, \tag{4.45}$$

i.e. at t_0 we have $\lim_{r \to 1} A_r(f)(t_0) = f(t_0)$. Furthermore, if (4.45) holds for every $t_0 \in (-\pi, \pi)$, which is in particular the case when f is continuous on $[-\pi, \pi]$, then the convergence of $A_r(f)(t_0)$ to $f(t_0)$ is uniformly in t. $\qquad\square$

When looking at the proofs of Theorem 4.15 and Theorem 4.25 it is obvious that they have the same structure: we are given a family of kernels $K_\nu(t)$ on $[-\pi, \pi]$ where ν could be either from a discrete set, e.g. $\mathbb{N}_0$, or from a continuous set of parameters, e.g. $\nu \in [0, 1)$, and $K_\nu(t)$ is an even function which has the following properties

$$K_\nu(t) \geq 0, \tag{4.46}$$

$$\frac{1}{2\pi} \int_{-\pi}^{\pi} K_\nu(t) \, dt = 1, \tag{4.47}$$

and for every $\delta > 0$ we have

$$\lim_{\nu \to \nu_0} \int_{\delta \leq |t| \leq \pi} K_\nu(t)\,dt = 0, \tag{4.48}$$

where ν_0 is a natural limit point of the parameter set, e.g. $\nu_0 = \infty$ in the case $\nu \in \mathbb{N}_0$ and $\nu_0 = 1$ for $\nu \in [0,1)$. If f is a 2π-periodic function Riemann integrable on $[-\pi, \pi]$ and t_0 a point of continuity of f then we have

$$\lim_{\nu \to \nu_0} (f * K_\nu)(t_0) = f(t_0). \tag{4.49}$$

The reader is asked to provide the proof of (4.49) in Problem 11. Kernels with the properties (4.46)-(4.48), or with (4.46) replaced by

$$\int_{-\pi}^{\pi} |K_\nu(t)|\,dt \leq \kappa, \tag{4.50}$$

are nowadays often called **good kernels**, see [114]. The Fejer kernel and the Poisson kernel are good kernels, the Dirichlet kernel is not a good kernel, see Problem 11.

Suppose (4.46)-(4.48) is satisfied and let f be continuous. For $t_0 = 0$ the equality (4.49) now reads

$$\lim_{\nu \to \nu_0} \frac{1}{2\pi} \int_{-\pi}^{\pi} f(s) K_\nu(s)\,ds = f(0) = \int_{-\pi}^{\pi} f(s)\epsilon_0(ds) \tag{4.51}$$

where ϵ_0 is the unit mass (or the Dirac measure) at 0. On the Borel sets of $[-\pi, \pi]$ we can define for every ν a probability measure by

$$m_\nu(A) := \frac{1}{2\pi} \int_{-\pi}^{\pi} \chi_A(s) K_\nu(s)\,ds,$$

and we can rewrite (4.51) as

$$\lim_{\nu \to \nu_0} \int_{-\pi}^{\pi} f(s) m_\nu(ds) = \int_{-\pi}^{\pi} f(s)\epsilon_0(ds), \tag{4.52}$$

which we can interpret as a type of (weak) convergence of the measures m_ν to ϵ_0. Families of kernels K_ν with these properties are called **approximation to the identity** or in the case of countable parameter sets **sequences of the type δ** or **Dirac sequences** and we will study them in more detail in Volume V.

We want to return to (4.35), or (4.39), and change our point of view. For $0 \le r < 1$ and $t \in [-\pi, \pi)$ we can interpret (r, t) as polar coordinates of a point (x, y) in the unit disk $B_1(0) \subset \mathbb{R}^2$, compare with Chapter II.12, and we note that changing the domain of t to $[0, 2\pi)$ does not have any serious effect. It is further convenient for the moment to switch the notation and use as in Chapter II.12 now φ instead of t. Thus (4.35) reads as

$$u(x(r, \varphi), y(r, \varphi)) := A_r(f)(\varphi) = \sum_{k \in \mathbb{Z}} c_k e^{ik\varphi} r^{|k|}. \tag{4.53}$$

Since this is a convergent power series with respect to r with radius of convergence 1, we have for $\varphi \in [-\pi, \pi]$ and $0 \le r < 1$

$$\frac{\partial}{\partial r} A_r(f)(\varphi) = \sum_{k \in \mathbb{Z}} c_k e^{ik\varphi} |k| r^{|k|-1}$$

and

$$\frac{\partial^2}{\partial r^2} A_r(f)(\varphi) = \sum_{k \in \mathbb{Z}} c_k e^{ik\varphi} |k| (|k| - 1) r^{|k|-2}.$$

The uniform convergence of $\sum_{k \in \mathbb{Z}} c_k (ik) e^{ik\varphi} r^{|k|}$ and of $\sum_{k \in \mathbb{Z}} c_k (ik)^2 e^{ik\varphi} r^{|k|}$ yield further that

$$\frac{\partial^2}{\partial \varphi^2} A_r(f)(\varphi) = -\sum_{k \in \mathbb{Z}} c_k k^2 e^{ik\varphi} r^{|k|},$$

and therefore we find with (II.12.1)

$$\Delta_2 u(x, y) = \frac{\partial^2}{\partial r^2} A_r(f)(\varphi) + \frac{1}{r} \frac{\partial}{\partial r} A_r(f)(\varphi) + \frac{1}{r^2} \frac{\partial^2}{\partial \varphi^2} A_r(f)(\varphi)$$

$$= \sum_{k \in \mathbb{Z}} c_k |k| (|k| - 1) e^{ik\varphi} r^{|k|-2} + \sum_{k \in \mathbb{Z}} c_k |k| e^{ik\varphi} r^{|k|-2}$$

$$- \sum_{k \in \mathbb{Z}} c_k |k|^2 e^{ik\varphi} r^{|k|-2}$$

$$= \sum_{k \in \mathbb{Z}} c_k \left(|k|^2 - |k| + |k| - |k|^2 \right) e^{ik\varphi} r^{|k|-2} = 0.$$

Thus, we find that in $B_1(0)$ the function $u(x(r, \varphi), y(r, \varphi)) := A_r(f)(\varphi)$ is harmonic and

$$\Delta_2 u(x, y) = \frac{\partial^2}{\partial r^2} (f * P_r)(\varphi) + \frac{1}{r} \frac{\partial}{\partial r} (f * P_r)(\varphi) + \frac{1}{r^2} \frac{\partial^2}{\partial \varphi^2} (f * P_r)(\varphi) = 0. \tag{4.54}$$

For $f \in C(\mathbb{T})$ it follows further

$$\lim_{r \to 1} (f * P_r)(\varphi) = f(\varphi). \tag{4.55}$$

We can summarize this result in the following

Theorem 4.26. *For a continuous function f on $\mathbb{T} = \partial B_1(0)$ there exists a function u harmonic in $B_1(0)$ with the representation*

$$u(x(r, \varphi), y(r, \varphi)) = (f * P_r)(\varphi) \tag{4.56}$$

and

$$\lim_{r \to 1} u(x(r, \varphi), y(r, \varphi)) = \lim_{r \to 1}(f * P_r)(\varphi) = f(\varphi). \tag{4.57}$$

We may call u a solution to the **radial Dirichlet problem** in the unique disc with given continuous boundary values $f(\varphi)$. The links of Fourier series and differential equations (ordinary or partial) will be studied in more detail in several places in these volumes.

Problems

1. Use the Lemma of Riemann-Lebesgue and the fact that $\int_0^\pi D_N(t)\, dt = \frac{\pi}{2}$ to prove that $\int_0^\infty \frac{\sin x}{x}\, dx = \frac{\pi}{2}$.
   Hint: first justify that for $\alpha > 0$

$$\int_0^\infty \frac{\sin x}{x}\, dx = \lim_{\gamma \to \infty} \int_0^\alpha \frac{\sin \gamma y}{y}\, dy$$

and further investigate the functions $g : (-2\pi, 2\pi) \to \mathbb{R}$ defined by

$$g(x) := \begin{cases} \frac{1}{x} - \frac{1}{2 \sin \frac{x}{2}}, & 0 < |x| < 2\pi \\ 0, & x = 0. \end{cases}$$

2.   a) Recall from Problem 3 a) of Chapter 3 the expansion

$$|x| \sim \frac{\pi}{2} - \frac{4}{\pi} \sum_{l=0}^\infty \frac{\cos(2l + 1)x}{(2l + 1)^2}$$

and deduce that $\sum_{k=1}^\infty \frac{1}{(2k-1)^2} = \frac{\pi^2}{8}$.

   b) Use Example 2.9 to find the value of $\sum_{k=1}^\infty \frac{(-1)^{k+1}}{k^2}$.

66

3. Let $(\gamma_k)_{k \in \mathbb{N}_0}$ be a sequence of complex numbers and denote by $S_{N+1} :=$ $\sum_{l=0}^{N} \gamma_l$ the corresponding partial sums. Further let $\sigma_{N+1} := \frac{1}{N+1} \sum_{l=0}^{N} S_l$. We call $\sum_{l=0}^{\infty} \gamma_l$ **Cesàro summable** to γ if $\lim_{N \to \infty} \sigma_{N+1} = \gamma$. Prove that if $\sum_{l=0}^{\infty} \gamma_l$ is Cesàro summable to γ then $\lim_{k \to \infty} \frac{\gamma_k}{k} = \lim_{N \to \infty} \frac{S_{N+1}}{N} = 0$.

4. Prove **Tauber's theorem:** if $\sum_{l=0}^{\infty} \gamma_l$ is Cesàro summable to γ and if $\lim_{k \to \infty} k\gamma_k = 0$ then the series $\sum_{l=0}^{\infty} \gamma_l$ converges to γ.

5. For which $x \in (-\pi, \pi)$ are the series $\frac{1}{2} + \sum_{k=1}^{\infty} \cos kx$ and $\sum_{k=1}^{\infty} \sin kx$ Cesàro summable?

6. Justify for the Fejer kernel F_{N+1} the estimates $F_{N+1}(t) \le N+1$ and $F_N(t) \le \frac{\pi^2}{(N+1)t^2}$, $0 < t \le \pi$.

7. Add some details to see that (4.21) holds and deduce with the help of these details (4.22).

8. Prove that $\sum_{k=0}^{\infty} \cos kt$ and $\sum_{k=1}^{\infty} \sin kt$ are for $t \in (-\pi, \pi) \setminus \{0\}$ Abel summable (by a direct application of the definition and not relying on the result of Problem 5).

9. Prove the following lemma: let $h(z) = \sum_{k=0}^{\infty} \alpha_k z^k$ be a power series with real coefficients and radius of convergence equal to 1. Suppose that for $t_0 \in [-\pi, \pi)$ the limit $r \to 1$ exists and $\lim_{r \to 1} h(re^{it_0}) = h(e^{it_0})$. Then the series $\sum_{k=0}^{\infty} \alpha_k \cos kt_0$ and $\sum_{k=1}^{\infty} \alpha_k \sin kt_0$ are Abel summable and their Abel limits are given by $\mathrm{Re}\, h(e^{it_0})$ and $\mathrm{Im}\, h(e^{it_0})$, respectively. Note that this result allows us to find Abel limits.

10. Use Problem 9 to investigate the Abel summability of $\sum_{k=0}^{\infty} \frac{\cos kt}{k}$.

11. Let $(K_n)_{n \in \mathbb{N}}$ be a family of good kernels, i.e. suppose that (4.48)-(4.50) hold. Further let $f : \mathbb{T} \to \mathbb{C}$ be a bounded integrable function. Prove: if f is continuous at t_0 then

$$\lim_{n \to \infty} (K_n * f)(t_0) = f(t_0).$$

5 Fourier Series of Square Integrable Functions

In Proposition 3.19 we have proved that the image of $L^2(\mathbb{T})$ under the mapping F as defined by (3.20) is a subspace of $l_2(\mathbb{Z})$ and for $f \in L^2(\mathbb{T})$ Bessel's inequality

$$\|F(f)\|_2 = \left(\sum_{k \in \mathbb{Z}} |c_k|^2\right)^{\frac{1}{2}} \leq \frac{1}{\sqrt{2\pi}} \left(\int_{-\pi}^{\pi} |f(t)|^2 \, dt\right)^{\frac{1}{2}} = \frac{1}{\sqrt{2\pi}} \|f\|_{L^2} \qquad (5.1)$$

holds. We want to prove that $F\left(L^2(\mathbb{T})\right) = l_2(\mathbb{Z})$ and that in (5.1) equality holds. In doing so we will have to introduce Hilbert spaces and the idea of orthogonality. We start with recollecting the notion of a scalar product space over the complex field, see Appendix I in Volume II. Let H be a vector space over $\mathbb{C}$, not necessarily finite dimensional. We call $\langle \cdot, \cdot \rangle_H : H \times H \to \mathbb{C}$ a (complex) **scalar product** on H if for all $x, y, z \in H$ and $\lambda, \mu \in \mathbb{C}$ we have

$$\langle \lambda x + \mu y, z \rangle_H = \lambda \langle x, z \rangle_H + \mu \langle y, z \rangle_H, \qquad (5.2)$$

$$\langle x, y \rangle_H = \overline{\langle y, x \rangle_H}, \qquad (5.3)$$

as well as

$$\langle x, x \rangle_H \geq 0 \quad \text{and} \quad \langle x, x \rangle_H = 0 \quad \text{if and only if } x = 0. \qquad (5.4)$$

If $\langle \cdot, \cdot \rangle$ is a scalar product on H we call $(H, \langle \cdot, \cdot \rangle_H)$ a **scalar product space** (over $\mathbb{C}$). For $x, y \in H$ the **Cauchy-Schwarz inequality**

$$|\langle x, y \rangle_H| \leq \langle x, x \rangle_H^{\frac{1}{2}} \langle y, y \rangle_H^{\frac{1}{2}} \qquad (5.5)$$

is satisfied which follows from

$$0 \leq \langle x + \lambda y, x + \lambda y \rangle_H = \langle x, x \rangle_H + \overline{\lambda} \langle x, y \rangle_H + \lambda \langle y, x \rangle_H + \lambda \overline{\lambda} \langle y, y \rangle_H$$

when choosing $\lambda = \frac{-\langle x, y \rangle_H}{\langle y, y \rangle_H}$ if $y \neq 0$, indeed this implies

$$0 \leq \langle x, x \rangle_H - \frac{|\langle x, y \rangle_H|^2}{\langle y, y \rangle_H} - \frac{|\langle x, y \rangle|^2}{\langle y, y \rangle_H} + \frac{|\langle x, y \rangle_H|^2}{\langle y, y \rangle_H}$$

or

$$|\langle x, y \rangle_H|^2 \leq \langle x, x \rangle_H \langle y, y \rangle_H.$$

In the case $y = 0$ but $x \neq 0$ we choose $\lambda = \frac{-\langle x, y \rangle_H}{\langle x, x \rangle_H}$, and if $x = y = 0$ the statement is trivial.

On $(H, \langle \cdot, \cdot \rangle_H)$ a norm is defined by

$$\|x\|_H := \langle x, x \rangle_H^{\frac{1}{2}}. \qquad (5.6)$$

It is trivial that $\|x\|_H \geq 0$ and $\|x\|_H = 0$ if and only if $x = 0$, as is the homogeneity $\|\lambda x\|_H = |\lambda| \|x\|_H$ a direct consequence of (5.2). The triangle inequality follows with the help of the Cauchy Schwarz inequality:

$$\|x + y\|_H^2 = \langle x + y, x + y \rangle_H = \langle x, x \rangle_H + \langle x, y \rangle_H + \langle y, x \rangle_H + \langle y, y \rangle_H$$
$$\leq \|x\|_H^2 + \|y\|_H^2 + 2|\langle x, y \rangle_H|$$
$$\leq \|x\|_H^2 + \|y\|_H^2 + 2\|x\|_H \|y\|_H = (\|x\|_H + \|y\|_H)^2.$$

With the help of $\| \cdot \|_H$ we can rewrite the Cauchy-Schwarz inequality as

$$|\langle x, y \rangle_H| \leq \|x\|_H \|y\|_H. \tag{5.7}$$

If the normed space $(H, \| \cdot \|_H)$, $\|x\|_H^2 = \langle x, x \rangle_H$, is a Banach space, i.e. **complete**, we call the scalar product space $(H, \langle \cdot, \cdot \rangle_H)$ a (complex) **Hilbert space**. Two elements $x, y \in H \backslash \{0\}$ are said to be **orthogonal** and we write $x \perp y$ if

$$\langle x, y \rangle = 0. \tag{5.8}$$

For two scalar product spaces $(H_1, \langle \cdot, \cdot \rangle_{H_1})$ and $(H_2, \langle \cdot, \cdot \rangle_{H_2})$ we call a linear mapping $T : H_1 \to H_2$ an isometry if for all $x, y \in H_1$ we have

$$\|Tx - Ty\|_{H_2} = \|x - y\|_{H_1} \tag{5.9}$$

and we will see in Problem 1 b) that this is indeed equivalent to

$$\langle Tx, Ty \rangle_{H_2} = \langle x, y \rangle_{H_1} \tag{5.10}$$

for all $x, y \in H_1$. In particular we have for all $x \in H_1$ that

$$\|Tx\|_{H_2} = \|x\|_{H_1} \tag{5.11}$$

which implies that an isometry is injective.

Example 5.1. A. The space $\mathbb{C}^n$, $n \in \mathbb{N}$, is a Hilbert space with respect to the scalar product

$$\langle x, y \rangle = \sum_{k=1}^{n} x_k \overline{y_k} \tag{5.12}$$

where $x = (x_1, \ldots, x_n)$, $y = (y_1, \ldots, y_n)$ and $x_l, y_l \in \mathbb{C}$.
B. On $l_2(\mathbb{Z})$ a scalar product is given by

$$\langle a, b \rangle_{l_2} = \sum_{k \in \mathbb{Z}} a_k \overline{b_k} \tag{5.13}$$

where $a = (a_k)_{k \in \mathbb{Z}}$ and $b = (b_k)_{k \in \mathbb{Z}}$ are sequences in $l_2(\mathbb{Z})$. From Theorem 3.12 we deduce that $l_2(\mathbb{Z})$ is a Hilbert space.

C. From Theorem III.8.20 it follows that for every measure space $(\Omega, \mathcal{A}, \mu)$ the space $L^2(\Omega)$ is a Hilbert space with respect to the scalar product

$$\langle u, v \rangle_{L^2} = \int_\Omega u(\omega)\overline{v(\omega)}\mu(d\omega). \tag{5.14}$$

But note that elements in $L^2(\Omega)$ are equivalence classes of pointwisely defined objects and we have to take the Lebesgue integral in (5.14). In most of our considerations we are dealing with functions $f : \mathbb{R} \to \mathbb{C}$ which are 2π-periodic and Riemann integrable on $[-\pi, \pi]$ and while a scalar product on this space is given by

$$\langle u, v \rangle_2 = \int_{-\pi}^{\pi} u(t)\overline{v(t)}\, dt \tag{5.15}$$

this space is neither complete nor does $\|u\|_2^2 = 0$ imply $u = 0$, which however will hold for continuous functions. However, $L^2(\mathbb{T})$ as we have defined it is a Hilbert space, compare with Appendix I for the definition of the appropriate measure on the Borel sets of $\mathbb{T}$.

Definition 5.2. *Let $(H, \langle \cdot, \cdot \rangle)$ be a Hilbert space. We call a family $(x_j)_{j \in I}$, $x_j \neq 0$, $I \neq \emptyset$, an **orthogonal system** in H if $j \neq l$ implies $x_j \perp x_l$. If in addition every vector x_j is a unit vector, i.e. $\|x_j\| = 1$, we call $(x_j)_{j \in I}$ an **orthonormal system** in H.*

Example 5.3. A. In $\mathbb{C}^n$ let $e_k = (0, \dots, 0, 1, 0, \dots, 0) \in \mathbb{C}^n$ where 1 stands in the kth position, $1 \le k \le n$. It follows that

$$\langle e_k, e_l \rangle = \delta_{kl}, \tag{5.16}$$

i.e. for $k \neq l$ these vectors are orthogonal and since they are unit vectors the set $\{e_1, \dots, e_n\} \subset \mathbb{C}^n$ is an orthonormal system in $\mathbb{C}^n$.

B. Now let $e_k := \left(e_k^{(n)} \right)_{n \in \mathbb{Z}}$, $k \in \mathbb{Z}$, be the sequence defined by $e_k^{(n)} = \delta_{nk}$, i.e. e_k is a sequence indexed by $\mathbb{Z}$ and the only non-zero element is 1 in the position $n \in \mathbb{Z}$. Clearly $\|e_k\|_{l_2} = 1$ and for $k, l \in \mathbb{Z}$, $k \neq l$, we find

$$\langle e_k, e_l \rangle_{l_2} = \sum_{n \in \mathbb{Z}} e_k^{(n)} \overline{e_l^{(n)}} = \sum_{n \in \mathbb{Z}} \delta_{nk}\delta_{nl} = 0, \tag{5.17}$$

i.e. $e_k \perp e_l$ in $l_2(\mathbb{Z})$ and $\{e_k \,|\, k \in \mathbb{Z}\}$ is an orthonormal system.

C. For $k \in \mathbb{Z}$ we define $e_k(t) = \frac{1}{\sqrt{2\pi}}e^{ikt}$ and note that

$$\int_{-\pi}^{\pi} |e_k(t)|^2\, dt = \frac{1}{2\pi} \int_{-\pi}^{\pi} 1\, dt = 1$$

as well as

$$\langle e_k, e_l \rangle_{L^2} = \int_{-\pi}^{\pi} e_k(t)\overline{e_l(t)}\, dt = \frac{1}{2\pi} \int e^{i(k-t)}\, dt = \delta_{kl}, \tag{5.18}$$

i.e. $e_k \perp e_l$ for $k \neq l$ and $\{e_k(\cdot)\,|\, k \in \mathbb{Z}\}$ is an orthonormal system in $L^2(\mathbb{T})$.

The scalar product $\langle \cdot, \cdot \rangle_{L^2}$ in $L^2(\mathbb{T})$ allows us to reinterpret the Fourier coefficients of $f \in L^2(\mathbb{T})$:

$$c_k = \frac{1}{2\pi} \int_{-\pi}^{\pi} f(t)e^{-ikt}\, dt = \frac{1}{\sqrt{2\pi}} \langle f, e_k \rangle_{L^2(\mathbb{T})} \tag{5.19}$$

with $e_k(t) = \frac{1}{\sqrt{2\pi}} e^{ikt}$. If we set

$$\gamma_k := \langle f, e_k \rangle_{L^2(\mathbb{T})} \tag{5.20}$$

we find

$$c_k = \frac{1}{\sqrt{2\pi}} \gamma_k \tag{5.21}$$

and for the Fourier series associated with $f \in L^2(\mathbb{T})$ it follows that

$$f(t) \sim \sum_{k \in \mathbb{Z}} c_k e^{ikt} = \frac{1}{\sqrt{2\pi}} \sum_{k \in \mathbb{Z}} \gamma_k e^{ikt}.$$

Now we have

$$\|(Ff)\|_2 = \left(\sum_{k \in \mathbb{Z}} |c_k|^2 \right)^{\frac{1}{2}} = \left(\sum_{k \in \mathbb{Z}} \left| \frac{1}{\sqrt{2\pi}} \gamma_k \right|^2 \right)^{\frac{1}{2}}$$

$$= \frac{1}{\sqrt{2\pi}} \left(\sum_{k \in \mathbb{Z}} |\gamma_k|^2 \right)^{\frac{1}{2}}$$

and Bessel's inequality (3.45) implies

$$\left(\sum_{k \in \mathbb{Z}} |\gamma_k|^2 \right)^{\frac{1}{2}} \leq \|f\|_{L^2}. \tag{5.22}$$

Definition 5.4. *Let $u_\nu, u : \mathbb{R} \to \mathbb{C}$ be 2π-periodic functions integrable on $[-\pi, \pi]$. We say that the sequence $(u_\nu)_{\nu \in \mathbb{N}}$ **converges in the quadratic mean** to u if*

$$\lim_{\nu \to \infty} \int_{-\pi}^{\pi} |u_\nu(t) - u(t)|^2\, dt = 0. \tag{5.23}$$

Remark 5.5. Convergence in the quadratic mean will imply convergence in $L^2(\mathbb{T})$ if we use our usual identification (3.1). While the limit in $L^2(\mathbb{T})$ is uniquely determined (as an equivalence class of Lebesgue almost everywhere equal functions), a limit in the quadratic mean is in general not uniquely defined, but it will determine a unique element in $L^2(\mathbb{T})$.

In our derivation of Bessel's inequality we have proved the equality

$$\|f - S_N\|_{L^2}^2 = \|f\|_{L^2}^2 - 2\pi \sum_{|k| \leq N} |c_k|^2$$

while we can rewrite with (5.21) as

$$\|f - S_N\|_{L^2}^2 = \|f\|_{L^2}^2 - \sum_{|k| \leq N} |\gamma_k|^2, \tag{5.24}$$

which has as immediate consequence

Corollary 5.6. *Let $f : \mathbb{R} \to \mathbb{C}$ be a 2π-periodic function Riemann integrable on $[-\pi, \pi]$. The sequence $(S_N(f))_{N \in \mathbb{N}_0}$ of its partial sums converges in the quadratic mean to f if and only if*

$$\lim_{N \to \infty} \sum_{|k| \leq N} |\gamma_k|^2 = \sum_{k \in \mathbb{Z}} |\gamma_k|^2 = \int_{-\pi}^{\pi} |f(t)|^2 \, \mathrm{d}t = \|f\|_{L^2}^2. \tag{5.25}$$

Let $f : \mathbb{R} \to \mathbb{C}$ be a 2π-periodic function Riemann integrable on $[-\pi, \pi]$ and denote by $(S_N)_{N \in \mathbb{N}_0}$ the sequence of the partial sums of the Fourier series associated with f. For every $M \in \mathbb{N}$ the partial sum S_M is a 2π-periodic function Riemann integrable on $[-\pi, \pi]$ and its associated Fourier series is of course S_M itself. Further, for $N < M$ the Nth partial sum of S_M is S_N and in addition we have

$$\|S_M\|_{L^2}^2 = \sum_{|k| \leq M} |\gamma_k|^2. \tag{5.26}$$

Combined with (5.24) the equality (5.26) implies for $M > N$

$$\|S_M - S_N\|_{L^2}^2 = \|S_M\|_{L^2}^2 - \sum_{|k| \leq N} |\gamma_k|^2 \tag{5.27}$$

$$= \sum_{N+1 \leq |k| \leq M} |\gamma_k|^2.$$

Bessel's inequality (3.44) implies the convergence of $\sum_{k\in\mathbb{Z}}|\gamma_k|^2$ and therefore the Cauchy criterion for series of real numbers yield that the sequence $(S_N)_{N\in\mathbb{N}_0}$ is a Cauchy sequence in $L^2(\mathbb{T})$, hence convergent to some limit $S \in L^2(\mathbb{T})$, i.e.

$$\lim_{N\to\infty} \|S_N - S\|_{L^2} = 0. \tag{5.28}$$

As a limit in $L^2(\mathbb{T})$ the element $S \in L^2(\mathbb{T})$ is an equivalent class of measurable functions which are square Lebesgue integrable and we cannot claim that $S = f$ or that $(S_N)_{N\in\mathbb{N}_0}$ converges in the quadratic mean to f. However, once we know that $(S_N)_{N\in\mathbb{N}_0}$ converges in the quadratic mean to f then we have of course $f = S$ almost everywhere.

Our aim is to prove

Theorem 5.7. *Let $f : \mathbb{R} \to \mathbb{C}$ be a 2π-periodic function Riemann integrable on $[-\pi, \pi]$. Then the sequence of the partial sums of its associated Fourier series converges in the quadratic mean to f.*

The proof is lengthy but quite standard and we follow the presentation of [37]. As preparation we first show two lemmata.

Lemma 5.8. *For $t \in [0, 2\pi]$ we have*

$$\sum_{k=1}^{\infty} \frac{\cos kt}{k^2} = \frac{(t-\pi)^2}{4} - \frac{\pi^2}{12}. \tag{5.29}$$

Proof. Since $\left|\frac{\cos kt}{k^2}\right| \le \frac{1}{k^2}$ this series converges uniformly to a continuous function $F(t) := \sum_{k=1}^{\infty} \frac{\cos kt}{k^2}$. By Example 2.8 we know that the Fourier series of the periodic extensions of $t \mapsto \frac{t-\pi}{2}$ is given by $-\sum_{k=1}^{\infty} \frac{\sin kt}{k} = \sum_{k=1}^{\infty} \frac{d}{dt}\left(\frac{\cos kt}{k^2}\right)$. In fact we know from Corollary 4.8 that on $(0, 2\pi)$ the series $-\sum_{k=1}^{\infty} \frac{\sin kt}{k}$ converges to $\frac{t-\pi}{2}$, however this convergence is not uniform. We will prove that for every $\delta \in (0, \pi)$ on $[\delta, 2\pi - \delta]$ the series $-\sum_{k=1}^{\infty} \frac{\sin kt}{k}$ converges uniformly. Once this is shown, integrating term by term yields

$$\sum_{k=1}^{\infty} \frac{\cos kt}{k^2} = \frac{(t-\pi)^2}{4} + c \tag{5.30}$$

and we may extend (5.30) by continuity to all $t \in [0, 2\pi]$. Integrating (5.30) gives

$$\int_0^{2\pi} \sum_{k=1}^{\infty} \frac{\cos kt}{k^2}\, dt = \int_0^{2\pi} \frac{(t-\pi)^2}{4}\, dt + 2\pi c = \frac{\pi^3}{6} + 2\pi c$$

74

and since

$$\int_0^{2\pi} \sum_{k=1}^{\infty} \frac{\cos kt}{k^2}\, dt = \sum_{k=1}^{\infty} \int_0^{2\pi} \frac{\cos kt}{k^2}\, dt = 0,$$

we find

$$c = -\frac{\pi^2}{12}, \tag{5.31}$$

i.e.

$$\sum_{k=1}^{\infty} \frac{\cos kt}{k^2} = \frac{(t-\pi)^2}{4} - \frac{\pi^2}{12}. \tag{5.32}$$

Now we prove the uniform convergence of $-\sum_{k=0}^{\infty} \frac{\sin kt}{k}$ on $[\delta, 2\pi - \delta]$. We note that

$$I_N(t) := \sum_{k=1}^{N} \sin kt = \operatorname{Im}\left(\sum_{k=1}^{N} e^{ikt}\right) = \operatorname{Im}\left(\sum_{k=1}^{N} (e^{it})^k\right)$$

and therefore

$$|I_N(t)| \le \left|\sum_{k=0}^{N} (e^{it})^k\right| = \left|\frac{1 - e^{i(N+1)t}}{1 - e^{it}}\right| \le \frac{2}{\left|e^{i\frac{t}{2}} - e^{-i\frac{t}{2}}\right|}$$

$$= \frac{1}{\sin\frac{t}{2}} \le \frac{1}{\sin\frac{\delta}{2}}.$$

Now we find for $M > N > 0$ that

$$\left|\sum_{k=N}^{M} \frac{\sin kt}{k}\right| = \left|\sum_{k=N}^{M} \frac{I_k(t) - I_{k-1}(t)}{k}\right|$$

$$= \left|\sum_{k=N}^{M} I_k(t)\left(\frac{1}{k} - \frac{1}{k+1}\right) + \frac{I_M(t)}{M+1} - \frac{I_{N-1}(t)}{N}\right|$$

$$\le \frac{1}{\sin\frac{\delta}{2}}\left(\frac{1}{N} - \frac{1}{M+1} + \frac{1}{M+1} + \frac{1}{N}\right) = \frac{2}{N\sin\frac{\delta}{2}},$$

which implies

$$\left|\sum_{k=N}^{\infty} \frac{\sin kt}{k}\right| \le \frac{2}{N\sin\frac{\delta}{2}}$$

from which the uniform convergence for $t \in [\delta, 2\pi - \delta]$ follows. $\qquad\square$

Lemma 5.9. *Let $f : \mathbb{R} \to \mathbb{R}$ be a 2π-periodic function such that $f|_{[-\pi,\pi]}$ is a step function. Then the Fourier series associated with f converges in the quadratic mean to f.*

Proof. For simplicity we will write f instead of $f|_{[-\pi,\pi]}$. First we consider the case

$$f(t) = \begin{cases} 1, & -\pi \le t < -\pi + a \\ 0, & -\pi + a \le t < \pi, \end{cases} \tag{5.33}$$

hence $0 \le a < 2\pi$. The (modified Fourier) coefficients γ_k, $k \in \mathbb{Z}$, are given by

$$\gamma_0 = \frac{1}{\sqrt{2\pi}} \int_{-\pi}^{\pi} f(t)\, dt = \frac{1}{\sqrt{2\pi}} \int_{-\pi}^{-\pi+a} 1\, dt = \frac{a}{\sqrt{2\pi}},$$

and for $k \ne 0$

$$\begin{aligned} \gamma_k &= \frac{1}{\sqrt{2\pi}} \int_{-\pi}^{-\pi+a} e^{-ikt}\, dt = \frac{1}{\sqrt{2\pi}} \frac{1}{(-ik)} e^{-ikt} \Big|_{-\pi}^{-\pi+a} \\ &= \frac{i(-1)^k}{k\sqrt{2\pi}} \left(e^{-ika} - 1 \right), \end{aligned}$$

and therefore, for $k \ne 0$ we find

$$|\gamma_k|^2 = \frac{1}{2\pi k^2} \left| e^{-ika} - 1 \right|^2 = \frac{1 - \cos ka}{\pi k^2}.$$

Since $x \mapsto \cos x$ is an even function and since $0 \le a \le 2\pi$ it folows by using Lemma 5.8 that

$$\begin{aligned} \sum_{k \in \mathbb{Z}} |\gamma_k|^2 &= \frac{a^2}{2\pi} + 2 \sum_{k=1}^{\infty} \frac{1 - \cos ka}{\pi k^2} \\ &= \frac{a^2}{2\pi} + \frac{2}{\pi} \sum_{k=1}^{\infty} \frac{1}{k^2} - \frac{2}{\pi} \sum_{k=1}^{\infty} \frac{\cos ka}{k^2} \\ &= \frac{a^2}{2\pi} + \frac{2}{\pi} \frac{\pi^2}{6} - \frac{2}{\pi} \left(\frac{(\pi - a)^2}{4} - \frac{\pi^2}{12} \right) = a. \end{aligned}$$

On the other hand we have

$$\|f\|_{L^2}^2 = \int_{-\pi}^{\pi} |f(t)|^2\, dt = \int_{-\pi}^{-\pi+a} 1\, dt = a,$$

i.e.

$$\|f\|_{L^2}^2 = \sum_{k \in \mathbb{Z}} |\gamma_k|^2.$$

Now we prove the general case. For a general step function $f|_{[-\pi,\pi]}$ satisfying our assumptions there exists finitely many step functions f_j of type (5.33) and coefficients $\alpha_j \in \mathbb{R}$, $1 \leq j \leq M$, such that

$$f(t) = \sum_{j=1}^{M} \alpha_j f_j(t)$$

and

$$S_N(f) = \sum_{j=1}^{M} \alpha_j S_N(f_j).$$

Now we find

$$\|f - S_N(f)\|_{L^2} = \left\| \sum_{j=1}^{M} \alpha_j(f_j - S_N(f_j)) \right\|_{L^2}$$

$$\leq \sum_{j=1}^{M} |\alpha_j| \|f_j - S_N(f_j)\|_{L^2}.$$

Since we know that $\lim_{N \to \infty} \|f_j - S_N(f_j)\|_{L^2} = 0$ it follows that

$$\lim_{N \to \infty} \|f - S_N(f)\|_{L^2} = 0$$

and the lemma is proved. □

Now we can prove the main result

Theorem 5.10. *Let $f : \mathbb{R} \to \mathbb{C}$ be a 2π-periodic function Riemann integrable over $[-\pi, \pi]$. Then its associated Fourier series converges in the quadratic mean to f.*

Proof. We may assume that f is real-valued and $|f|$ is bounded by 1, i.e. $|f(t)| \leq 1$ for all t. Otherwise we first split $f = \operatorname{Re} f + i \operatorname{Im} f$ and then we consider $\frac{f}{\|f\|_\infty}$ or $\frac{\operatorname{Re} f}{\|f\|_\infty}$ and $\frac{\operatorname{Im} f}{\|f\|_\infty}$, respectively. Given $\epsilon > 0$. The Riemann integrability of f on $[-\pi, \pi]$ implies the existence of two step functions φ and ψ on $[-\pi, \pi]$ with periodic extensions to $\mathbb{R}$ such that $-1 \leq \varphi \leq f \leq \psi \leq 1$ and

$$\int_{-\pi}^{\pi} |\psi(t) - \varphi(t)| \, \mathrm{d}t \leq \frac{\epsilon^2}{8}, \tag{5.34}$$

see Theorem I.25.15. With $v := f - \varphi$ we have

$$|v|^2 \leq |\psi - \varphi|^2 \leq 2(\psi - \varphi)$$

77

where we used that $|\varphi| \leq 1$ and $|\psi| \leq 1$. Therefore it follows by (5.34) that

$$\int_{-\pi}^{\pi} |v(t)|^2 \, dt \leq 2 \int_{-\pi}^{\pi} (\psi(t) - \varphi(t)) \, dt \leq \frac{\epsilon^2}{4}.$$

We denote by $S_N(f)$, $S_N(\varphi)$ and $S_N(v)$ the partial sums of the Fourier series corresponding to f, φ and v respectively, and we note that

$$S_N(f) = S_N(\varphi) + S_N(v).$$

By Lemma 5.9 we know that there exists $M \in \mathbb{N}$ such that for $N \geq M$ we have

$$\|\varphi - S_N(\varphi)\|_{L^2} < \frac{\epsilon}{2}$$

and (5.24) yields

$$\|v - S_N(v)\|_{L^2}^2 \leq \|v\|_{L^2}^2 \leq \frac{\epsilon^2}{4}.$$

Thus for $N \geq M$ we arrive at

$$\|f - S_N(f)\|_{L^2} \leq \|\varphi - S_N(\varphi)\|_{L^2} + \|v - S_N(v)\|_{L^2} \leq \frac{\epsilon}{2} + \frac{\epsilon}{2} = \epsilon$$

proving the theorem. $\qquad\square$

Remark 5.11. Theorem 5.10 states in particular that for a 2π-periodic function $f : \mathbb{R} \to \mathbb{C}$ which is Riemann integrable on $[-\pi, \pi]$ the Fourier coefficients c_k tend to zero for $|k|$ tending to infinity, i.e. $\lim_{|k| \to \infty} c_k = 0$. Thus we have given for these functions a further proof of the Lemma of Riemann-Lebesgue.

In light of Theorem 5.10 and the discussion preceding Corollary 5.6 we may ask whether for $f \in L^2(\mathbb{T})$ we can prove

$$\lim_{N \to \infty} \|f - S_N(f)\|_{L^2} = 0.$$

We need

Lemma 5.12. Let $\sum_{k \in \mathbb{Z}} c_k e^{ikt}$ be a trigonometric series with partial sums $S_N = \sum_{|k| \leq N} c_k e^{ikt}$. If for some $f \in L^2(\mathbb{T})$ and an infinite sub-sequence $(S_{N_j})_{j \in \mathbb{N}}$ we have

$$\lim_{j \to \infty} \|f - S_{N_j}\|_{L^2} = 0$$

then it follows that

$$c_k = \frac{1}{2\pi} \int_{-\pi}^{\pi} f(t) e^{-ikt} \, dt$$

for all $k \in \mathbb{Z}$, i.e. $\sum_{k \in \mathbb{Z}} c_k e^{ikt}$ is the Fourier series associated with f.

Proof. Let $k \in \mathbb{Z}$ and $j_0 \in \mathbb{N}$ such that $N_j > |k|$ for $j > j_0$. It follows that

$$\frac{1}{2\pi} \int_{-\pi}^{\pi} S_{N_j}(t) e^{-ikt} \, dt = \sum_{|l| \leq N_j} c_l \frac{1}{2\pi} \int_{-\pi}^{\pi} e^{i(l-k)t} \, dt = c_k$$

and therefore

$$\left| c_k - \frac{1}{2\pi} \int_{-\pi}^{\pi} f(t) e^{-ikt} \, dt \right| = \lim_{j \to \infty} \left| \frac{1}{2\pi} \int_{-\pi}^{\pi} (f(t) - S_{N_j}(t)) e^{-ikt} \, dt \right|$$

$$\leq \lim_{j \to \infty} \frac{1}{2\pi} \int_{-\pi}^{\pi} |f(t) - S_{N_j}(t)| \, dt$$

$$\leq \frac{1}{\sqrt{2\pi}} \lim_{j \to \infty} \left(\int_{-\pi}^{\pi} |f(t) - S_{N_j}(t)|^2 \, dt \right)^{\frac{1}{2}}$$

$$= \frac{1}{\sqrt{2\pi}} \lim_{j \to \infty} \|f - S_{N_j}\|_{L^2} = 0.$$

$\square$

As a consequence of Lemma 5.12 we get the **Fischer-Riesz** theorem:

Theorem 5.13. *For $f \in L^2(\mathbb{T})$ the partial sums of the Fourier series associated with f converge in $L^2(\mathbb{T})$ to f. Moreover, for every sequence $(c_k)_{k \in \mathbb{Z}}$ in $l_2(\mathbb{Z})$ there exists $g \in L^2(\mathbb{T})$ such that $\sum_{k \in \mathbb{Z}} c_k e^{ikt}$ is the Fourier series associated with g and the corresponding partial sums converge in $L^2(\mathbb{T})$ to g.*

Proof. We prove the second statement first. For the sequence $(S_N)_{N \in \mathbb{N}_0}$, $S_N = \sum_{|k| \leq N} c_k e^{ikt}$, we have with $M > N$ that

$$\|S_M - S_N\|_{L^2}^2 = \frac{1}{2\pi} \sum_{N+1 \leq |k| \leq M} |c_k|^2,$$

compare with (5.27). Since $\sum_{k \in \mathbb{Z}} |c_k|^2 < \infty$ it follows that $(S_N)_{N \in \mathbb{N}_0}$ is a Cauchy sequence in $L^2(\mathbb{T})$, hence it has a limit $g \in L^2(\mathbb{T})$, i.e.

$$\lim_{N \to \infty} \|S_N - g\|_{L^2} = 0.$$

Now Lemma 5.12 implies that

$$\frac{1}{2\pi} \int_{-\pi}^{\pi} g(t) e^{-ikt} \, dt = c_k$$

for all $k \in \mathbb{Z}$ and we have proved that $\sum_{k \in \mathbb{Z}} c_k e^{ikt}$ is the Fourier series associated with g. With these preparations the proof of the first part is easy. Given $f \in L^2(\mathbb{T})$.

We know from Bessel's inequality that the series $\sum_{k \in \mathbb{Z}} |c_k|^2$ converges and the part just proven implies that the trigonometric series $\sum_{k \in \mathbb{Z}} c_k e^{ikt}$ is the Fourier series associated with f and its partial sums converge in $L^2(\mathbb{T})$ to f. □

Corollary 5.14 (Parseval's equalities). *For $f, g \in L^2(\mathbb{T})$ we have*

$$\sum_{k \in \mathbb{Z}} |(Ff)(k)|^2 = \sum_{k \in \mathbb{Z}} |c_k|^2 = \frac{1}{2\pi} \int_{-\pi}^{\pi} |f(t)|^2 \, dt \tag{5.35}$$

and

$$\sum_{k \in \mathbb{Z}} (Ff)(k) \overline{(Fg)(k)} = \frac{1}{2\pi} \int_{-\pi}^{\pi} f(t) \overline{g(t)} \, dt, \tag{5.36}$$

where $c_k = (Ff)(k)$.

Proof. The equality (5.35) follows now from (3.43), i.e.

$$\|f - S_N\|_{L^2}^2 = \|f\|_{L^2}^2 - 2\pi \sum_{|k| \leq N} |c_k|^2,$$

when passing to the limit as N tends to infinity. The second equality follows from the polarization identity

$$\langle x, y \rangle_H = \frac{1}{4} \left(\|x + y\|_H - \|x - y\|_H + i\|x + iy\|_H^2 - i\|x - iy\|_H^2 \right) \tag{5.37}$$

which holds in every (complete) scalar product space, see Problem 1 a). □

For $f \in L^2(\mathbb{T})$ with associated Fourier series $\sum_{k \in \mathbb{Z}} c_k e^{ikt}$ and partial sums S_N, $N \in \mathbb{N}_0$, we can write (3.43) as

$$\|f - S_N\|_{L^2}^2 = 2\pi \sum_{|k| > N} |c_k|^2. \tag{5.38}$$

The equality (5.38) allows to discuss the rate of convergence of the L^2-approximation of f by the partial sums of the Fourier series associated with f. Suppose that we have the estimate

$$|c_k| \leq \kappa_0 |k|^{-\alpha} \tag{5.39}$$

for some $\alpha > 0$. For the remainder term $\sum_{|k| > N} |c_k|^2$ we find

$$2\pi \sum_{|k| > N} |c_k|^2 \leq 2\pi \kappa_0^2 \sum_{|k| > N} |k|^{-2\alpha} = 4\pi \kappa_0^2 \sum_{n = N+1}^{\infty} n^{-2\alpha}.$$

The integral comparison test for series yields

$$\sum_{n=N+1}^{\infty} n^{-2\alpha} \le \int_{N}^{\infty} x^{-2\alpha} \, dx = \frac{1}{2\alpha - 1} \frac{1}{N^{2\alpha-1}} < \infty$$

provided $\alpha > \frac{1}{2}$. Hence we have proved

Proposition 5.15. *Let $f \in L^2(\mathbb{T})$ and assume for the Fourier coefficients $c_k = (Ff)(k)$ of f the estimate (5.39) for all $k \in \mathbb{Z}$ and for some $\alpha > \frac{1}{2}$. Then we have*

$$\|f - S_N\|_{L^2}^2 \le \frac{4\pi\kappa_0}{2\alpha - 1} \frac{1}{N^{2\alpha-1}}. \tag{5.40}$$

For $C^m(\mathbb{T}) \subset L^2(\mathbb{T})$ we have by Lemma 3.4 the estimates

$$|c_k| \le \frac{1}{2\pi|k|^m} \|f^{(m)}\|_{L^1}$$

and now (5.40) implies

Corollary 5.16. *For $f \in C^m(\mathbb{T})$, $m \ge 1$, it follows*

$$\|f - S_N\|_{L^2}^2 \le \frac{2\pi\|f^{(m)}\|_{L^1}}{2m - 1} \frac{1}{N^{2m-1}}. \tag{5.41}$$

Remark 5.17. Further interesting considerations on the rate of convergence in (5.38) are made in [97].

We want to go a step further and to use the decay of the Fourier coefficients of $f \in L^2(\mathbb{T})$ in order to find differentiability properties of f. First let $p(t) = \sum_{|k| \le M} d_k e^{ikt}$, $d_k = (Fp)(k)$, be a trigonometric polynomial. It follows that

$$p(t) = \sum_{k \in \mathbb{Z}} (Fp)(k) e^{ikt}$$

$$= \sum_{k \in \mathbb{Z}} (Fp)(k)(1 + |k|^2)^{\frac{1}{2}} \frac{1}{(1 + |k|^2)^{\frac{1}{2}}} e^{ikt},$$

and by the Cauchy-Schwarz inequality we get

$$|p(t)| \le \sum_{k \in \mathbb{Z}} |(Fp)(k)|(1 + |k|^2)^{\frac{1}{2}} \frac{1}{(1 + |k|^2)^{\frac{1}{2}}}$$

$$\le \left(\sum_{k \in \mathbb{Z}} \frac{1}{1 + |k|^2} \right)^{\frac{1}{2}} \left(\sum_{k \in \mathbb{Z}} |(Fp)(k)|^2 (1 + |k|^2) \right)^{\frac{1}{2}}$$

or

$$\sup_{t\in[-\pi,\pi]} |p(t)| \le \left(\sum_{k\in\mathbb{Z}} \frac{1}{1+|k|^2} \right)^{\frac{1}{2}} \left(\sum_{k\in\mathbb{Z}} |(Fp)(k)|^2 (1+|k|^2) \right)^{\frac{1}{2}}.$$

We know that the trigonometric polynomials are dense in $L^2(\mathbb{T})$ and therefore we can derive a simple first version of **Sobolev's embedding theorem** for periodic functions.

Theorem 5.18. *Let $f \in L^2(\mathbb{T})$ and suppose that $\sum_{k\in\mathbb{Z}} |(Ff)(k)|^2 (1+|k|^2) < \infty$. Then f is equivalent to a continuous function $\tilde{f}$ and we have*

$$\|\tilde{f}\|_{\infty,\mathbb{T}} \le \left(\sum_{k\in\mathbb{Z}} \frac{1}{1+|k|^2} \right)^{\frac{1}{2}} \left(\sum_{k\in\mathbb{Z}} |(Ff)(k)|^2 (1+|k|^2) \right)^{\frac{1}{2}}. \qquad (5.42)$$

Proof. The estimate with $\tilde{f}$ replaced by f is clear from the preceding calculations and the density of the trigonometric polynomials in $L^2(\mathbb{T})$. The estimate also implies that the partial sums S_N, $N \in \mathbb{N}_0$, of the Fourier series associated with f form a Cauchy sequence with respect to the norm $\|\cdot\|_{\infty,\mathbb{T}}$. Indeed for $M > N > 0$ and with $\kappa_1 := \left(\sum_{k\in\mathbb{Z}} \frac{1}{1+|k|^2} \right)^{\frac{1}{2}}$ we find

$$\|S_N - S_M\|_{\infty,\mathbb{T}} \le \kappa_1 \sum_{N<|k|\le M} |(Ff)(k)|^2 (1+|k|^2) \qquad (5.43)$$

and the convergence of $\sum_{k\in\mathbb{Z}} |(Ff)(k)|^2 (1+|k|^2)$ implies the assertion. Since S_N is continuous and $C(\mathbb{T})$ is complete with respect to uniform convergence on $[-\pi, \pi]$ it follows that $(S_N)_{N\in\mathbb{N}}$ converges uniformly to a continuous function $\tilde{f} \in C(\mathbb{T})$. However $\tilde{f}$ has the same Fourier coefficients as f, so $f = \tilde{f}$ almost everywhere. $\square$

We can extend Theorem 5.18 further:

Theorem 5.19. *For $f \in L^2(\mathbb{T})$ suppose that for some $m \in \mathbb{N}$ we have $\sum_{k\in\mathbb{Z}} |(Ff)(k)|^2 (1+|k|^2)^m < \infty$. Then f is equivalent to a function $\tilde{f} \in C^{m-1}(\mathbb{T})$ and the estimates*

$$\|\tilde{f}^{(l)}\|_{\infty,\mathbb{T}} \le \left(\sum_{k\in\mathbb{Z}} \frac{1}{1+|k|^2} \right)^{\frac{1}{2}} \left(\sum_{k\in\mathbb{Z}} |(Ff)(k)|^2 \left(1+|k|^2\right)^{l+1} \right)^{\frac{1}{2}} \qquad (5.44)$$

hold for $0 \le l < m$.

Proof. For the difference $S_M - S_N$, $M > N \geq 0$, of two partial sums of the Fourier series associated with f we find

$$\frac{d^l}{dt^l}(S_M(t) - S_N(t)) = \frac{d^l}{dt^l} \sum_{N < |k| \leq M} (Ff)(k) e^{ikt}$$

$$= \sum_{N < |k| \leq M} (Ff)(k)(ik)^l e^{ikt},$$

and therefore

$$\left| \frac{d^l}{dt^l}(S_M(t) - S_N(t)) \right| \leq \sum_{N < |k| \leq M} |(Ff)(k)| \, (1 + |k|^2)^{\frac{l}{2}}$$

$$= \sum_{N < |k| \leq M} \frac{1}{(1 + |k|^2)^{\frac{1}{2}}} |(Ff)(k)| \, (1 + |k|^2)^{\frac{l+1}{2}}$$

$$\leq \left(\sum_{k \in \mathbb{Z}} \frac{1}{1 + |k|^2} \right)^{\frac{1}{2}} \left(\sum_{N < |k| \leq M} |(Ff)(k)|^2 \, (1 + |k|^2)^{l+1} \right)^{\frac{1}{2}},$$

i.e.

$$\left\| S_M^{(l)} - S_N^{(l)} \right\|_\infty \leq \kappa_1 \left(\sum_{N < |k| \leq M} |(Ff)(k)|^2 \, (1 + |k|^2)^{l+1} \right)^{\frac{1}{2}} \tag{5.45}$$

and we conclude as before that for $0 \leq l \leq m - 1$ the derivatives $S_N^{(l)}$ of the partial sums of the Fourier series associated with f form Cauchy sequences with respect to the norm $\|\cdot\|_{\infty, \mathbb{T}}$, hence they converge uniformly to continuous functions $\tilde{f}_l \in C(\mathbb{T})$. We know that $\tilde{f}_0$ and f have the same Fourier coefficients. In order to prove that $\tilde{f}^{(l)} = \tilde{f}_l$ we just note that both are continuous L^2-functions with the same Fourier coefficients. The estimate (5.44) now follows from the calculation leading to (5.45). $\square$

Definition 5.20. *Let $m \in \mathbb{N}_0$. The* **Bessel potential space** $H^m(\mathbb{T})$ *of periodic L^2-functions is defined by*

$$H^m(\mathbb{T}) := \left\{ u \in L^2(\mathbb{T}) \mid \sum_{k \in \mathbb{Z}} |(Fu)(k)|^2 (1 + |k|^2)^m < \infty \right\}, \tag{5.46}$$

and on $H^m(\mathbb{T})$ we introduce the scalar product

$$\langle u, v \rangle_{H^m(\mathbb{T})} := \sum_{k \in \mathbb{Z}} (Fu)(k) \overline{(Fv)(k)} (1 + |k|^2)^m \tag{5.47}$$

with corresponding norm

$$\|u\|_{H^m(\mathbb{T})} := \left(\sum_{k \in \mathbb{Z}} |(Fu)(k)|^2 (1 + |k|^2)^m \right)^{\frac{1}{2}}. \tag{5.48}$$

Theorem 5.21. *The space $\left(H^m(\mathbb{T}), \langle \cdot, \cdot \rangle_{H^m(\mathbb{T})} \right)$ is a Hilbert space.*

Exercise 5.22. *Prove Theorem 5.21.*

We want to put Theorem 5.18 into a different context. Let $f : \mathbb{R} \to \mathbb{C}$ be a 2π-periodic function which is absolutely continuous with derivative f' belonging to $L^2(\mathbb{T})$. For the Fourier coefficients c_k, $k \in \mathbb{Z}$, $k \neq 0$, we find (as is already known to us)

$$c_k = \frac{1}{2\pi} \int_{-\pi}^{\pi} f(t) e^{-ikt} \, \mathrm{d}t = -\frac{1}{2\pi} \int_{-\pi}^{\pi} \frac{1}{-ik} f'(t) e^{-ikt} \, \mathrm{d}t = \frac{c_k'}{ik},$$

where for the moment we denote by c_k', $k \neq 0$, the kth Fourier coefficient of f'. Since $f' \in L^2(\mathbb{T})$ by assumption, it follows that

$$\sum_{k \in \mathbb{Z}} |kc_k|^2 = \sum_{k \in \mathbb{Z}} |c_k'|^2 = \frac{1}{2\pi} \|f'\|_{L^2}^2$$

which however implies that $f \in H^1(\mathbb{T})$. Thus absolutely continuous 2π-periodic functions with derivatives belonging to $L^2(\mathbb{T})$ are elements of $H^1(\mathbb{T})$. On the other hand, for $f \in H^1(\mathbb{T})$ we get

$$\sum_{k \in \mathbb{Z}} |k(Ff)(k)|^2 \leq \|f\|_{H^1(\mathbb{T})}^2$$

and therefore $\sum_{k \in \mathbb{Z}} ikc_k e^{ikt}$ is the Fourier series associated with an L^2-function g which almost everywhere must coincide with f' since the partial sums corresponding to f' and g are the same. Thus we have proved

Corollary 5.23. *A function f belongs to the space $H^1(\mathbb{T})$ if and only if it is equivalent to an absolutely continuous 2π-periodic function $\tilde{f}$ with derivative $\tilde{f}' \in L^2(\mathbb{T})$.*

Problems

1.   a) Prove the polarisation identity (5.37).

       b) Let $(H_1, \langle \cdot, \cdot \rangle_{H_1})$ and $(H_2, \langle \cdot, \cdot \rangle_{H_2})$ be two Hilbert spaces over $\mathbb{C}$ and $T : H_1 \to H_2$ be a linear isometry. Prove that T is unitary, i.e. $\langle Tx, Ty \rangle_{H_2} = \langle x, y \rangle_{H_1}$ for all $x, y \in H_1$.

2. Let $\emptyset \neq M \subset H$ be a subset of the Hilbert space $(H, \langle \cdot, \cdot \rangle)$. The **orthogonal complement** $M^\perp$ of M is defined by

$$M^\perp := \{y \in H | \langle y, x \rangle = 0 \text{ for all } x \in M\}.$$

Prove:

   a) $M^\perp$ is a closed subspace of H;

   b) $M_1 \subset M_2$ implies $M_2^\perp \subset M_1^\perp$;

   c) $M \subset M^{\perp\perp} := (M^\perp)^\perp$;

   d) $M^\perp = (\overline{\text{span}M})^\perp$.

3. We call two subspaces G_1, G_2 of the Hilbert space $(H, \langle \cdot, \cdot \rangle)$ orthogonal and we write $G_1 \perp G_2$ if for every $x \in G_1$ and $y \in G_2$ the relation $\langle x, y \rangle = 0$ holds.

   a) Let $(e_k)_{k \in \mathbb{Z}}$ be the standard orthonormal system in $l_2(\mathbb{Z})$ as introduced in Example 5.3.B. Let $\emptyset \neq K_1, K_2 \subset \mathbb{Z}$ be a partition of $\mathbb{Z}$, i.e. $K_1 \cap K_2 = \emptyset$ and $K_1 \cup K_2 = \mathbb{Z}$. Prove that $\text{span}\{e_k | k \in K_1\} \perp \text{span}\{e_k | k \in K_2\}$ and deduce that $\overline{\text{span}\{e_k | k \in K_1\}} \perp \overline{\text{span}\{e_k | k \in K_2\}}$.

   b) Let $A_1, A_2 \subset \mathbb{R}$ be two disjoint Borel sets, i.e. $A_1 \cap A_2 = \emptyset$. Define

$$H_1 := \{u \in L^2(\mathbb{R}) | |u|_{A_1^\complement} = 0 \text{ a.e.}\}$$

and

$$H_2 := \{u \in L^2(\mathbb{R}) | |u|_{A_2^\complement} = 0 \text{ a.e.}\}.$$

Show that $H_1 \perp H_2$.

4. We call a sequence $(u_k)_{k \in \mathbb{N}}$, $u_k \in H$, in a Hilbert space $(H, \langle \cdot, \cdot \rangle)$ **weakly convergent** to $u \in H$ if for all $\varphi \in H$ we have $\lim_{k \to \infty} \langle u_k - u, \varphi \rangle = 0$.

   a) Prove that if $(u_k)_{k \in \mathbb{N}}$ converges in H to u, i.e. $\lim_{k \to \infty} \|u_k - u\| = 0$ then it converges weakly to u.

   b) Prove that the sequence $(e_k)_{k \in \mathbb{N}}$ considered as a subsequence of $(e_k)_{k \in \mathbb{Z}}$ as introduced in Example 5.3.B converges weakly to 0, but it does not converge in $l_2(\mathbb{Z})$.

   c) Prove that the sequence $\left(\frac{1}{\sqrt{2\pi}} e^{ik \cdot} \right)_{k \in \mathbb{N}}$ converges weakly in $L^2(\mathbb{T})$ to 0, but again it does not converge in $L^2(\mathbb{T})$.

5. Prove that the sequence $h_k : [-\pi, \pi] \to \mathbb{R}$, $k \in \mathbb{N}$, $h_k(x) = \left(\frac{x}{\pi} \right)^k$ converges in the quadratic mean to the function $h_0(x) = 0$ for all $x \in [-\pi, \pi]$. Does $(h_k)_{k \in \mathbb{N}}$ converge to h_0 pointwisely on $[-\pi, \pi]$?

6. We call $f \in L^2(\mathbb{T})$ an **L^2-Hölder continuous function** of order α, $0 < \alpha \leq 1$, if with $f_h(x) := f(x+h)$ we have $\|f_h - f\|_{L^2} \leq \kappa|h|^\alpha$. Prove **Bernstein's theorem**:

   If f is L^2-Hölder continuous of order $\alpha > \frac{1}{2}$ then the series of the Fourier coefficients of f converges absolutely and hence the Fourier series associated with f converges absolutely.

   Hint: first study the dyadic blocks $\sum_{2^m \leq |k| < 2^{m+1}} |c_k|$ and then apply Parseval's equality to $\|f_h - f\|_{L^2}$.

7. Suppose that $c_{-k} = \overline{c_k}$ and prove the following variant of Parseval's equality:

$$\frac{a_0^2}{2} + \sum_{k=1}^{\infty}(a_k^2 + b_k^2) = \frac{1}{\pi}\int_{-\pi}^{\pi} |f(t)|^2 \, dt.$$

   Now use the result of Example 2.9 to find the value of the series $\sum_{k=1}^{\infty} \frac{1}{k^4}$.

8. Prove that $H^m(\mathbb{T})$, $m \in \mathbb{N}$, is complete.

9. Let $f \in L^2(\mathbb{T})$, $f(t) = \sum_{k \in \mathbb{Z}} \gamma_k e_k(t)$, $e_k(t) = \frac{1}{\sqrt{2\pi}}e^{ikt}$. Show that $u := \sum_{k \in \mathbb{Z}} \frac{\gamma_k}{1+|k|^2} e_k(\cdot)$ belongs to $H^2(\mathbb{T})$ and solves the equation $-\Delta u + u = f$ in $L^2(\mathbb{T})$.

6 Further Topics on the Convergence of Fourier Series

Since their emergence Fourier series have stimulated a lot of research in mathematics: integration theory, functional analysis or most of all set theory, and many more subjects are off-springs of Fourier analysis. The reasons for this are the many surprising results showing up when investigating the convergence of Fourier series or more generally trigonometric series. In this chapter we will discuss some of these results which need more challenging proofs. Some readers may prefer to move immediately to the end of this chapter where we have summarized the results. We start by discussing the **integral sine** function

$$\text{Si}(x) := \frac{2}{\pi} \int_0^x \frac{\sin t}{t} \, dt, \quad x \geq 0. \tag{6.1}$$

Since $\lim_{t \to 0} \frac{\sin t}{t} = 1$ for $0 \leq x < \infty$ the integral is well defined. In Example I.28.19.B we have seen that also the improper integral

$$\frac{2}{\pi} \int_0^\infty \frac{\sin t}{t} \, dt = 1 \tag{6.2}$$

exists and in Example III.25.28 we have calculated the value. Thus we have

$$\text{Si}(0) = 0 \quad \text{and} \quad \lim_{x \to \infty} \text{Si}(x) = 1. \tag{6.3}$$

From the definition (6.1) it follows that

$$\text{Si}((n+1)\pi) - \text{Si}(n\pi) = \frac{2}{\pi} \int_{n\pi}^{(n+1)\pi} \frac{\sin t}{t} \, dt \tag{6.4}$$

and this term is non-negative for n even and non-positive for n odd. Moreover $|\text{Si}((n+1)\pi) - \text{Si}(n\pi)|$ is decreasing in n. Hence $\text{Si}(\cdot)$ has local maxima at $(2m-1)\pi$, $m \in \mathbb{N}$, and local minima at $2\pi m$, $m \in \mathbb{N}$. From these properties we deduce that $\text{Si}(\cdot)$ has a global maximum at π, and

$$0 \leq \text{Si}(x) \leq \text{Si}(\pi) \quad \text{for all } x \in [0, \infty). \tag{6.5}$$

Next we observe that we can extend (4.3) to elements of $L^1(\mathbb{T})$, i.e. for every $f \in L^1(\mathbb{T})$ we have

$$S_N(f)(t) = (f * D_N)(t) \tag{6.6}$$

where D_N is the Dirichlet kernel which we usually use in the representation (4.4).

Proposition 6.1. *Let $f \in L^1(\mathbb{T})$ and assume for some $0 < \delta < \pi$ that*

$$\int_{-\delta}^{\delta} \left| \frac{f(t)}{t} \right| dt < \infty. \tag{6.7}$$

In this case we have

$$\lim_{N \to \infty} S_N(f)(0) = 0. \tag{6.8}$$

Proof. Since $\lim_{s \to 0} \frac{\frac{s}{2}}{\sin \frac{s}{2}} = 1$ we find $M \geq 0$ such that for all $|s| \leq \pi$ it follows that $\left| \frac{\frac{s}{2}}{\sin \frac{s}{2}} \right| \leq M$. Therefore we have on $[-\pi, \pi]$

$$\left| \frac{f(s) \cos \frac{s}{2}}{\sin \frac{s}{2}} \right| \leq 2M \left| \frac{f(s)}{s} \right|$$

$$\leq 2M \left| \frac{f(s)}{s} \right| \chi_{[-\delta,\delta]}(s) + \frac{2M}{\delta} \chi_{[-\pi,\pi] \setminus [-\delta,\delta]}(s)$$

and therefore we can conclude that $s \mapsto \frac{f(s) \cos \frac{s}{2}}{\sin \frac{s}{2}}$ belongs to $L^1(\mathbb{T})$. Using (6.6) with D_N given by (4.4), noting that D_N is an even function and applying the formula $\sin(\alpha + \beta) = \sin \alpha \cos \beta + \cos \alpha \sin \beta$, we find further

$$(S_N f)(0) = \frac{1}{2\pi} \int_{-\pi}^{\pi} f(s) \frac{\sin \left(N + \frac{1}{2} \right) s}{\sin \frac{s}{2}} \, ds$$

$$= \frac{1}{2\pi} \int_{-\pi}^{\pi} f(s) \cos N s \, ds + \frac{1}{2\pi} \int_{-\pi}^{\pi} \frac{f(s) \cos \frac{s}{2}}{\sin \frac{s}{2}} \sin N s \, ds.$$

Since f and $s \mapsto \frac{f(s) \cos \frac{s}{2}}{\sin \frac{s}{2}}$ are L^1-functions, the Lemma of Riemann-Lebesgue, Theorem 3.16 in the form of Remark 3.18.B, now yields

$$\lim_{N \to \infty} (S_N f)(0) = 0.$$

$\square$

Let $f \in L^1(\mathbb{T})$ and define $f_{t_0}(s) := f(s + t_0)$ as in the beginning of Chapter III.10. We continue to write $f(\cdot + t_0)$ for $f_{t_0}(\cdot)$. Further note the linearity of $f \mapsto S_N(f)$ which implies

$$S_N(f(\cdot + t_0) - f(t_0)) = S_N(f(\cdot + t_0)) - f(t_0)$$

since $S_N(f(t_0)) = f(t_0)$.

Theorem 6.2 (U. Dini). *If $f \in L^1(\mathbb{T})$ satisfies for some $\delta > 0$*

$$\int_{-\delta}^{\delta} \left| \frac{f(s + t_0) - f(t_0)}{s} \right| ds < \infty \tag{6.9}$$

then we have

$$\lim_{N \to \infty} (S_N f)(t_0) = f(t_0). \tag{6.10}$$

Proof. We apply Proposition 6.1 to $g(s) = f(s + t_0) - f(t_0)$ and we note that by (2.48) we have $(S_N f)(t + t_0) = \sum_{|k| \leq N} (Ff)(k) e^{ik(t+t_0)}$. $\square$

Many authors prefer to replace (6.9) by

$$\int_0^{\delta} \left| \frac{f(t_0 + s) + f(t_0 - s) - 2f(t_0)}{2} \right| ds < \infty \tag{6.11}$$

which of course follows from (6.9) by the triangle inequality. In fact we can combine Theorem 4.5 with such a condition in the following way:

Theorem 6.3. *Let $S(t_0)$ be a real number such that for some $0 < \delta < \pi$ we have*

$$\int_0^{\delta} \left| \frac{f(t_0 + s) + f(t_0 - s) - 2S(t_0)}{s} \right| ds < \infty. \tag{6.12}$$

Then it follows that

$$\lim_{N \to \infty} (S_n f)(t_0) = S(t_0). \tag{6.13}$$

Proof. Using the representation

$$S_N(f)(t_0) = \frac{1}{2\pi} \int_0^{\pi} (f(t_0 + s) + f(t_0 - s)) \frac{\sin \left(N + \frac{1}{2} \right) s}{\sin \frac{s}{2}} ds,$$

compare with the proof of Lemma 4.4, we first find

$$\begin{aligned}
(S_N f)(t_0) - S(t_0) &= \frac{1}{2\pi} \int_0^{\pi} (f(t_0 + s) + f(t_0 - s) - 2S(t_0)) \frac{\sin \left(N + \frac{1}{2} \right) s}{\sin \frac{s}{2}} ds \\
&= \frac{1}{2\pi} \int_0^{\pi} (f(t_0 + s) + f(t_0 - s) - 2S(t_0)) \frac{\sin \left(N + \frac{1}{2} \right) s}{\frac{s}{2}} ds \\
&\quad + \frac{1}{2\pi} \int_0^{\pi} (f(t_0 + s) + f(t_0 - s) - 2S(t_0)) \left(\frac{1}{\sin \frac{s}{2}} - \frac{1}{\frac{s}{2}} \right) \sin \left(N + \frac{1}{2} \right) s \, ds \\
&= I_1(N) + I_2(N). \tag{6.14}
\end{aligned}$$

From (6.12) and the Lemma of Riemann-Lebesgue we first deduce that $I_1(N)$ tends to 0 for $N \to \infty$, i.e.

$$\lim_{N \to \infty} \frac{1}{\pi} \int_0^\pi \left(\frac{f(t_0 + s) + f(t_0 - s) - 2S(t_0)}{s} \right) \sin \left(N + \frac{1}{2} \right) s \, ds = 0.$$

We claim that $I_2(N)$ also tends to 0 as $N \to \infty$. This will follow again from the Lemma of Riemann-Lebesgue if we can prove that $\left(\frac{1}{\sin \frac{s}{2}} - \frac{1}{s} \right)$ is uniformly bounded on $[0, \pi]$ or $\left(\frac{1}{\sin s} - \frac{1}{s} \right)$ is uniformly bounded on $\left[0, \frac{\pi}{2} \right]$. For this we observe first that for all x we have $|x - \sin x| \le \frac{|x|^3}{6}$, which we may obtain from the Taylor expansion of sine, another way is shown in Problem 2. Now it follows that

$$\left| \frac{1}{\sin s} - \frac{1}{s} \right| = \left| \frac{s - \sin s}{s \sin s} \right| \le \frac{|s|^3}{6|s||\sin s|} \le \frac{|s|^2}{6} \cdot \frac{|s|}{|\sin s|},$$

and using that $\lim_{s \to \infty} \frac{s}{\sin s} = 1$ we obtain the boundedness of $\frac{|s|}{6} \frac{|s|}{|\sin s|}$ for $s \in \left[0, \frac{\pi}{2} \right]$ and the claim follows. $\qquad\square$

Corollary 6.4. *If for $f \in L^1(\mathbb{T})$ the symmetric Hölder condition*

$$|f(t_0 + s) + f(t_0 - s) - 2f(t_0)| \le \kappa |s|^\alpha \tag{6.15}$$

holds at t_0 for some $0 < \alpha < 1$ and all $0 \le s \le \delta$ with some $\kappa > 0$, then the Fourier series associated with f represents f at t_0, i.e.

$$f(t_0) = \lim_{N \to \infty} S_N(f)(t_0).$$

A further result we can derive from Theorem 6.3 is

Theorem 6.5. *If $f \in L^1(\mathbb{T})$ is absolutely continuous then the Fourier series associated with f converges for all $t \in [-\pi, \pi]$ to $f(t)$.*

Proof. We first recall Theorem III.11.17 which provided the integration by parts formula for absolutely continuous functions: Let $f, g \in L^1([a, b])$ and

$$F(x) = \int_a^x f(t) \, dt, \qquad G(x) = \int_a^x g(t) \, dt,$$

then we have

$$\int_a^b f(t) G(t) \, dt = F \cdot G \Big|_a^b - \int_a^b F(t) g(t) \, dt. \tag{6.16}$$

By (6.14) we have

$$S_N(f)(t_0) - f(t_0) = \frac{1}{2\pi} \int_0^\pi (f(t_0 + s) + f(t_0 - s) - 2f(t_0)) \frac{\sin \left(N + \frac{1}{2} \right) s}{\frac{s}{2}} \, ds + I_2(N),$$

and we know already that $\lim_{N\to\infty} I_2(N) = 0$ where

$$I_2(N) = \frac{1}{2\pi} \int_0^\pi (f(t_0 + s) + f(t_0 - s) - 2f(t_0)) \left(\frac{1}{\sin \frac{s}{2}} - \frac{1}{\frac{s}{2}} \right) \sin \left(N + \frac{1}{2} \right) s \, ds.$$

Note that

$$\frac{1}{2\pi} \int_0^\pi (f(t_0 + s) + f(t_0 - s) - 2f(t_0)) \frac{\sin \left(N + \frac{1}{2} \right) s}{\frac{s}{2}} \, ds$$

$$= \frac{2}{\pi} \int_0^\pi \left(\frac{f(t_0 + s) + f(t_0 - s)}{2} - f(t_0) \right) \frac{\sin \left(N + \frac{1}{2} \right) s}{s} \, ds.$$

Since f is absolutely continuous, so is the function G, $G(s) = \frac{f(t_0+s)+f(t_0-s)}{2} - f(t_0)$, and therefore there exists $g \in L^1(\mathbb{T})$ such that

$$G(s) = \int_0^s g(r) \, dr, \quad 0 \le s \le \pi.$$

Furthermore, since

$$\text{Si}(s) = \frac{2}{\pi} \int_0^s \frac{\sin r}{r} \, dr,$$

it follows that Si is absolutely continuous and we can rewrite the integral of interest as

$$\frac{2}{\pi} \int_0^\pi \left(\frac{f(t_0 + s) + f(t_0 - s)}{2} - f(t_0) \right) \frac{\sin \left(N + \frac{1}{2} \right) s}{s} \, ds$$

$$= \int_0^\pi G(s) \left(\text{Si} \left(N + \frac{1}{2} \right) s \right)' \, ds$$

$$= \left(G(\pi) \text{Si} \left(\left(N + \frac{1}{2} \right) \pi \right) - \int_0^\pi G'(s) \text{Si} \left(\left(N + \frac{1}{2} \right) s \right) \right) ds$$

where we have used that $\text{Si}(0) = 0$. Since $G' = g \in L^1(\mathbb{T})$ and $0 \le \text{Si}\left(\left(N + \frac{1}{2} \right) s \right) \le \text{Si}(\pi)$ by (6.5), and since $\lim_{N\to\infty} \text{Si}\left(\left(N + \frac{1}{2} \right) \pi \right) = 1$ we first observe

$$\lim_{N\to\infty} G(\pi) \text{Si} \left(\left(N + \frac{1}{2} \right) \pi \right) = G(\pi)$$

and by the dominated convergence theorem we find

$$\lim_{N\to\infty} \int_0^\pi G'(s) \text{Si} \left(\left(N + \frac{1}{2} \right) s \right) ds = \int_0^\pi G'(s) \, ds = G(\pi) - G(0),$$

but $G(0) = \frac{f(t_0)+f(t_0)}{2} - f(t_0) = 0$. Thus it follows that

$$\lim_{N \to \infty} \frac{1}{2\pi} \int_0^\pi (f(t_0 + s) + f(t_0 - s) - 2f(t_0)) \frac{\sin\left(N + \frac{1}{2}\right)s}{\frac{s}{2}} \, ds = 0$$

and the theorem is proved. $\square$

Since $f \in L^1(\mathbb{T})$ has an absolutely continuous primitive

$$G(t) = \int_{-\pi}^t f(t) \, dt \tag{6.17}$$

which belongs to $L^1(\mathbb{T})$ we find for

$$H(t) := \int_{-\pi}^t (f(s) - (Ff)(0)) \, ds \tag{6.18}$$

that $H(-\pi) = H(\pi) = 0$, $H'(t) = f(t) - (Ff)(0)$ and therefore

$$ik(F(H))(k) = (Ff)(k) - \delta_{k0}(Ff)(0)$$

or for $k \neq 0$

$$(F(H))(k) = \frac{(Ff)(k)}{ik}. \tag{6.19}$$

Now Theorem 6.5 implies the everywhere convergence of

$$H(t) = (F(H))(0) + \sum_{k \in \mathbb{Z} \setminus \{0\}} \frac{(Ff)(k)}{ik} e^{ikt}, \tag{6.20}$$

and we obtain

Corollary 6.6. *Let $f \in L^1(\mathbb{T})$. Then the series*

$$\sum_{k \in \mathbb{Z} \setminus \{0\}} \frac{(Ff)(k)}{k} \tag{6.21}$$

converges. Moreover, for every interval $(a,b) \subset [-\pi, \pi]$ we have

$$\lim_{N \to \infty} \int_a^b S_N(f)(t) \, dt = \int_a^b f(t) \, dt. \tag{6.22}$$

92

Proof. Clearly (6.21) follows from (6.20) and it remains to prove (6.22). Since by (6.17) and (6.20)

$$G(b) - G(a) = \sum_{k \in \mathbb{Z} \setminus \{0\}} \frac{(Ff)(k)}{ik} \left(e^{ikb} - e^{ika} \right)$$

$$= \sum_{k \in \mathbb{Z} \setminus \{0\}} (Ff)(k) \int_a^b e^{ikt} \, dt$$

it follows that

$$G(b) - G(a) = \int_a^b f(t) \, dt - (b-a)(Ff)(0)$$

$$= \lim_{N \to \infty} \int_a^b (S_N(t) - (Ff)(0)) \, dt,$$

and (6.22) follows. □

In order to exploit the convergence of $\sum_{k \in \mathbb{Z} \setminus \{0\}} \frac{(Ff)(k)}{k}$ further we need some preparation. In particular we want to consider series with monotone decreasing coefficients, hence they will be real-valued and therefore it makes sense to switch to series in the representation

$$\frac{a_0}{2} + \sum_{k \in \mathbb{N}} (a_k \cos kt + b_k \sin kt) \tag{6.23}$$

with $a_k, b_k \in \mathbb{R}$. For $t = 0$ the convergence of (6.23) will never reveal some information about the coefficients b_k, while (6.23) can only converge for $t = 0$ if $\sum_{k=1}^\infty a_k$ converges. Given the structure of (6.23) we also prefer to work with the "modified" Dirichlet kernels

$$D_N^{\cos}(t) = \frac{1}{2} + \sum_{k=1}^N \cos kt = \frac{\sin\left(N + \frac{1}{2}\right)t}{2 \sin \frac{t}{2}} \tag{6.24}$$

and

$$D_N^{\sin}(t) = \sum_{k=1}^N \sin kt = \frac{\cos \frac{t}{2} - \cos\left(N + \frac{1}{2}\right)t}{2 \sin \frac{t}{2}}, \tag{6.25}$$

see Problem 6 to find the relations to $D_N(t)$. From (6.24) and (6.25) it follows immediately

$$|D_N^{\cos}(t)| \le \frac{\pi}{2|t|}, \quad 0 < |t| \le \pi \tag{6.26}$$

and

$$\left|D_N^{\sin}(t)\right| \leq \frac{\pi}{|t|}, \quad 0 < |t| \leq \pi, \tag{6.27}$$

and further using bounds for $\sin\frac{x}{2}$ on $[\delta, 2\pi - \delta]$, $\delta \in (0, 2\pi)$, we find

$$\left|D_N^{\cos}(t)\right| \leq \frac{\pi}{2\delta} \tag{6.28}$$

and

$$\left|D_N^{\sin}(t)\right| \leq \frac{\pi}{8}. \tag{6.29}$$

As further preparation we use (4.23) to prove **Abel's Lemma**, see [8].

Theorem 6.7. *Let $(u_k)_{k \in \mathbb{N}_0}$ be a decreasing sequence of non-negative real numbers with $\lim_{k \to \infty} u_k = 0$. Further let $v_k : K \to \mathbb{C}$, $k \in \mathbb{N}_0$, be a sequence of functions such that $V_N(t) := \sum_{k=0}^{N} v_k(t)$ is a uniformly bounded family, i.e. $|V_N(t)| \leq M$ for all $t \in K$ and all $N \in \mathbb{N}_0$. Then the series $S(t) := \sum_{k=0}^{\infty} u_k v_k(t)$ converges on K uniformly and we have*

$$|S(t)| \leq M u_0 \quad \text{for all } t \in K. \tag{6.30}$$

Proof. We define

$$S_N(t) := \sum_{k=0}^{N} u_k v_k(t)$$

and by (4.23) we find

$$S_N(t) = u_N V_N(t) + \sum_{k=0}^{N-1} (u_k - u_{k-1}) V_k(t) \tag{6.31}$$

or

$$S_N(t) - u_N V_N(t) = \sum_{k=0}^{N-1} (u_k - u_{k-1}) V_k(t).$$

We observe that for $t \in K$

$$|(u_k - u_{k-1}) V_k(t)| \leq (u_k - u_{k-1}) M$$

and that the series $\sum_{k=0}^{\infty} M(u_k - u_{k+1}) = M u_0$ converges where we used that $u_k \geq u_{k+1}$ and $\lim_{k \to \infty} u_k = 0$. Now it follows that the series $\sum_{k=0}^{\infty} (u_k - u_{k+1}) V_k(t)$ converges uniformly and absolutely on K. Using again that $|V_N(t)| \leq M$ and $\lim_{N \to \infty} u_N = 0$ we deduce from (6.31) that $(S_N(t))_{N \in \mathbb{N}_0}$ must converge on K uniformly to some limit $S(t) = \sum_{k=0}^{\infty} u_k v_k(t)$ and that (6.30) holds. $\square$

Now we can prove, see again [8].

Theorem 6.8. A. *Let $(a_k)_{k \in \mathbb{N}_0}$ be a monotone decreasing sequence of non-negative real numbers converging to 0. Then the series $\frac{a_0}{2} + \sum_{k=1}^{\infty} a_k \cos kt$ converges everywhere on the set $\mathbb{R} \backslash \{s = 2\pi k \mid k \in \mathbb{Z}\}$ and for $0 < \delta < 2\pi$ and all $\delta \leq t \leq 2\pi - \delta$ the convergence is uniform.*
B. *Let $(b_k)_{k \in \mathbb{N}}$ be a monotone decreasing sequence of non-negative numbers converging to 0. Then the series $\sum_{k=1}^{\infty} b_k \sin kt$ converges for all t and for every $\delta \in (0, \pi)$ and $0 < \delta \leq t \leq 2\pi - \delta$ the convergence is uniform.*

Proof. **A.** We apply Abel's lemma, Theorem 6.7, with $u_k = a_k$, $k \in \mathbb{N}_0$, $v_0 := \frac{1}{2}$ and $v_k(t) := \cos kt$, $k \in \mathbb{N}$. It follows that $V_N(t) = D_N^{\cos}(t)$ and since $|D_N^{\cos}(t)|$ is by (6.28) uniformly bounded in $0 \leq \delta \leq t \leq 2\pi - \delta(t)$ we obtain uniform convergence on $[\delta, 2\pi - \delta]$. Now, for $t \in (0, 2\pi)$ we can find $\delta(t) \in (0, \pi)$ such that $t \in [\delta(t), 2\pi - \delta(t)]$ and the pointwise convergence of the series follows for every $t \in (0, 2\pi)$, and by periodicity for all $t \in \mathbb{R} \backslash \{s = 2\pi k \mid k \in \mathbb{Z}\}$.
B. We argue as in part A, but we replace u_0 by 0 and u_k by b_k, $k \in \mathbb{N}$, as well as v_0 by 0 and $v_k(t)$ by $\sin kt$, $k \in \mathbb{N}_0$. Now we apply (6.29) and deduce from Abel's lemma the uniform convergence in $[\delta, 2\pi - \delta]$ for every $\delta \in (0, \pi)$, hence the pointwise convergence follows in $\mathbb{R} \backslash \{s = 2\pi k \mid k \in \mathbb{Z}\}$, but for $s = 2\pi k$ we have $\sin 2\pi k = 0$ and the convergence is trivial. $\qquad\square$

Example 6.9. The trigonometric series $\sum_{k=2}^{\infty} \frac{\sin kt}{\ln k}$ converges for all $t \in \mathbb{R}$. Since the series $\sum_{k=2}^{\infty} \frac{1}{\ln k}$ diverges we conclude that there exists a convergent trigonometric series which is not absolutely convergent.

Corollary 6.10. *There exists a convergent trigonometric series which is not a Fourier series.*

Proof. The function $f(t) = \sum_{k=2}^{\infty} \frac{\sin kt}{\ln k}$ is well defined on $\mathbb{R}$ by Theorem 6.7. Suppose $f \in L^1(\mathbb{T})$. Then by Corollary 6.6 the series $\sum_{k=2}^{\infty} \frac{1}{k \ln k}$ must converge which is however not the case. $\qquad\square$

By Theorem 6.5 we know that the Fourier series associated with an absolutely continuous function converges for all $t \in [-\pi, \pi]$ to $f(t)$. We will now provide an example which is due to L. Fejer [30] showing that for a continuous function the associated Fourier series need not converge at all points to the value of the function, in fact it need not even converge. A first example of such a continuous function was given by P. Du Bois-Reymond [25], see also [26]. In our presentation we follow [28]. We need

Lemma 6.11. *The partial sums $S_N(t)$ of the trigonometric series $\sum_{k=1}^{\infty} \frac{\sin kt}{k}$ are uniformly bounded.*

Proof. First we note that by Theorem 6.8.B the series $\sum_{k=1}^{\infty} \frac{\sin kt}{k}$ converges for all $t \in \mathbb{R}$. In fact we know by Example 2.8 combined with the proof of Lemma 5.8 that this series represents on $(0, 2\pi)$ the function $t \mapsto \frac{\pi - t}{2}$ and it is the Fourier series associated with the function

$$f(t) = \begin{cases} \frac{-\pi - t}{2}, & -\pi \leq t < 0 \\ \frac{\pi - t}{2}, & 0 \leq t < \pi. \end{cases} \tag{6.32}$$

For the Cesàro sum $\sigma_N(f)(t)$ of f we find, see the remark following Definition 4.12,

$$\sigma_N(f)(t) = f * \left(\frac{1}{N} \sum_{k=0}^{N-1} D_k \right)(t) = (f * F_N)(t)$$

where F_N is the Fejér kernel. Since $\frac{1}{2\pi} \int_{-\pi}^{\pi} F_N(t)\, dt = 1$ and $F_N(t) \geq 0$ for all t we get

$$|\sigma_N(f)(t)| = \left| \int_{-\pi}^{\pi} f(t - s) F_N(s)\, ds \right| \leq \frac{\pi}{2}. \tag{6.33}$$

Moreover, noting that f is an odd function, it follows that

$$|S_N(f)(t) - \sigma_{N+1}(f)(t)| = \left| S_N(f)(t) - \frac{1}{N+1} \sum_{l=1}^{N} S_l(f)(t) \right|$$

$$= \left| \sum_{k=1}^{N} \frac{1}{k} \sin kt - \frac{1}{N+1} \sum_{l=1}^{N} \sum_{j=1}^{l} \frac{1}{j} \sin jt \right|$$

$$= \frac{1}{N+1} \left| \sum_{k=1}^{N} \sin kt \right| \leq \frac{N}{N+1} \leq 1.$$

Combining the estimate with (6.33) yields

$$|S_N(f)(t)| \leq |\sigma_{N+1}(f)(t)| + |S_N(f)(t) - \sigma_{N+1}(f)(t)|$$
$$\leq \frac{\pi}{2} + 1.$$

$\square$

For $m \in \mathbb{N}$ let us consider the trigonometric polynomial

$$g_m(t) := \sum_{k=1}^{m} \frac{1}{k} \left(\cos(m - k)t - \cos(m + k)t \right).$$

Using the formula $\cos\alpha - \cos\beta = 2\sin\frac{\beta-\alpha}{2}$ we find

$$g_m(t) = 2\sin mt \sum_{k=1}^{m} \frac{\sin kt}{k}. \tag{6.34}$$

We have

$$g_1(t) = 1 - \cos 2t,$$

$$g_2(t) = \frac{1}{2} + \cos t - \cos 3t - \frac{1}{2}\cos 4t,$$

and for general m

$$g_m(t) = \frac{1}{m} + \frac{1}{m-1}\cos t + \frac{1}{m-2}\cos 2t + \cdots + \cos(m-1)t \tag{6.35}$$
$$- \cos(m+1)t - \frac{1}{2}\cos(m+2)t - \cdots - \frac{1}{m}\cos 2mt.$$

Since by Proposition 6.1 the sums $\sum_{k=1}^{m}\frac{\sin kt}{k}$ are uniformly bounded it follows that for some $M \geq 0$ we have

$$|g_m(t)| \leq M \quad \text{for all } m \in \mathbb{N} \text{ and } t \in \mathbb{R}.$$

We deduce that for every subsequence $(m_k)_{k\in\mathbb{N}}$ the series

$$G(t) := \sum_{k=1}^{\infty} \frac{1}{k^2} g_{m_k}(t) \tag{6.36}$$

converges on $\mathbb{R}$ uniformly to a continuous 2π-periodic function. Note that the series in (6.36) is not given in the usual representation of a trigonometric series. We want to find the partial sums S_N of the Fourier series associated with the continuous function G. Our aim is to determine eventually the subsequence $(m_k)_{k\in\mathbb{N}}$ such that the sequence $(S_N(0))_{n\in\mathbb{N}}$ is unbounded implying that the Fourier series associated with G diverges at 0.

Each function g_m is a trigonometric polynomial and therefore a finite Fourier series. For the partial sum S_{mn} of g_m we derive for $n \geq 2m$ that $S_{mn} = g_m$ which follows most easily from (6.35). From (6.34) we deduce that $g_m(0) = 0$, i.e. $S_{mn}(0) = 0$ for $n \geq 2m$, while for $n < 2m$ we see from (6.35) that $S_{mn}(0) \geq 0$. The Fourier coefficients of G may be calculated by term by term integration in (6.36) and therefore we find

$$S_n(t) = \sum_{k=1}^{\infty} \frac{1}{k^2} S_{m_k n}(t). \tag{6.37}$$

Note that S_n is of course representable as a finite sum of trigonometric function, only terms of sine and cosine functions $\sin lt$ and $\cos lt$ enter into (6.37) for l in a finite range, however the corresponding coefficients must be calculated by an infinite series and (6.37) is the more convenient expression. From (6.37) we get

$$S_n(0) = \sum_{k=1}^{\infty} \frac{1}{k^2} S_{m_k n}(0), \tag{6.38}$$

and further we find for every m

$$S_{mm}(0) = 1 + \frac{1}{2} + \cdots + \frac{1}{m} > \int_1^{\infty} \frac{dt}{t} = \ln(m). \tag{6.39}$$

This now implies

$$S_{m_k}(0) = \sum_{j=1}^{\infty} \frac{1}{j^2} S_{m_j m_k}(0) \geq \frac{1}{k^2} S_{m_k m_k}(0) > \frac{1}{k^2} \ln(m_k).$$

Now we choose $m_k = 2^{(k^4)}$ to obtain

$$S_{m_k}(0) > \frac{1}{k^2} \ln\left(2^{(k^4)}\right) = k^2 \ln 2, \tag{6.40}$$

implying that the sequence $(S_n(0))_{n \in \mathbb{N}}$ is unbounded. Hence we have proved

Theorem 6.12 (P. Du Bois-Reymond, L. Fejer). *There exists a 2π-periodic continuous function the associated Fourier series of which diverges at a point.*

Eventually we want to discuss an almost everywhere summability result, more precisely we want to look at the Cesàro summability of the Fourier series associated with an L^1-function. For the one-dimensional case we may state Lebesgue's differentiation theorem, Theorem III.11.29, as follows

Theorem 6.13. *For $f \in L^1(\mathbb{R})$ we have*

$$\lim_{h \to 0} \frac{1}{2h} \int_{-h}^{h} (f(t + x_0) - f(x_0)) \, dt = 0 \tag{6.41}$$

for almost all $x_0 \in \mathbb{R}$.

A reformation of (6.41) is

$$\lim_{h \to 0} \frac{1}{h} \int_{x_0}^{x_0+h} f(t) \, dt = f(x_0) \tag{6.42}$$

for almost all $x_0 \in \mathbb{R}$. Moreover, there are obvious formulations for $f \in L^1(\mathbb{T})$ or 2π-periodic functions Lebesgue integrable over $[-\pi, \pi)$. Since

$$\frac{1}{2h} \int_{-h}^{h} (f(t + x_0) - f(x_0))\, dt = \frac{1}{h} \int_{0}^{h} \left(\frac{f(x_0 + t) + f(x_0 - t)}{2} - f(x_0) \right) dt \qquad (6.43)$$

we can replace (6.42) by

$$\lim_{h \to 0} \frac{1}{h} \int_{0}^{h} \left(\frac{f(x_0 + t) + f(x_0 - t)}{2} - f(x_0) \right) dt = 0 \qquad (6.44)$$

to hold for almost all $x_0 \in \mathbb{R}$. In addition we find

$$\lim_{h \to 0} \frac{1}{h} \int_{x_0}^{x_0 + h} |f(t) - c|\, dt = |f(x_0) - c| \qquad (6.45)$$

almost everywhere which follows from (6.42) by replacing f by $g_C(t) = |f(t) - c|$. Taking now $|f(x + t) - f(x - t) - 2f(x)|$ we arrive at

Corollary 6.14. *For $f \in L^1(\mathbb{T})$ there exists a set $E \subset [-\pi, \pi)$ of Lebesgue measure zero such that for all $x \in [-\pi, \pi) \backslash E$ we have*

$$\lim_{h \to 0} \frac{1}{h} \int_{0}^{h} |f(x + t) + f(x - t) - 2f(x)|\, dt = 0. \qquad (6.46)$$

Remark 6.15. Note that the above considerations also outline a proof for the fact that the complement of the Lebesgue set $\Lambda(f)$ of f has measure zero, see Remark III.11.30.

Now we can prove

Theorem 6.16. *(Fejer-Lebesgue) For $f \in L^1(\mathbb{T})$ there exists a set $E \subset [-\pi, \pi]$, $\lambda^{(1)}(E) = 0$, such that for all $x \in [-\pi, \pi] \backslash E$ the Fourier series associated with f is Cesàro summable to $f(x)$, i.e.*

$$\lim_{N \to \infty} \sigma_N(f)(x) = f(x) \quad \lambda^{(1)}\text{-almost everywhere.} \qquad (6.47)$$

Proof. (compare with K. Stromberg [117]) Given $f \in L^1(\mathbb{T})$ and determine $E \subset [-\pi, \pi]$, $\lambda^{(1)}(E) = 0$, according to Corollary 6.14. Now fix $x \in [-\pi, \pi] \backslash E$ and define

$$g(h) := \int_{0}^{h} |f(x + t) + f(x - t) - 2f(x)|\, dt. \qquad (6.48)$$

We set $g(\pi) = a$. Next we choose δ, $0 < \delta < \pi$, such that $0 < |h| \le \delta$ implies $\left| \frac{1}{h} g(h) \right| < \frac{\epsilon}{13}$. By (6.46) and the fact that $x \notin E$ such a δ exists. As before we denote by F_N the Nth Fejer kernel and from Lemma 4.14 we may deduce that there exists $N_0 \in \mathbb{N}$ such that $N \geq N_0 > \frac{1}{\delta}$ implies

$$0 \leq F_N(t) < \frac{\epsilon}{a + t} \qquad \text{for } \delta \leq t < \pi. \tag{6.49}$$

As in the proof of Theorem 4.15 we find

$$\left| \sigma_N(f)(x_0) - f(x_0) \right| \leq \frac{1}{2\pi} \int_0^\pi F_N(s) \left| f(x_0 + s) + f(x_0 - s) - 2f(x_0) \right| ds$$

$$= \frac{1}{2\pi} \left(\int_0^{\frac{1}{N}} + \int_{\frac{1}{N}}^\delta + \int_\delta^\pi \right) (F_N(s) | f(x_0 + s) + f(x_0 - s) - 2f(x_0) |) \, ds.$$

For the following calculation we use properties of F_N as stated in Lemma 4.14, the obvious bound $F_N(t) \leq N + 1$ following from its definition and the bound

$$F_N(t) \leq \frac{\pi^2}{(N+1)t}, \qquad 0 < t \leq \pi, \tag{6.50}$$

which follows from (4.18), also see Problem 6 of Chapter 4. First we find

$$\int_0^{\frac{1}{N}} F_N(s) \left| f(x_0 + s) + f(x_0 - s) - 2f(x_0) \right| ds$$

$$\leq (N+1) \int_0^{\frac{1}{N}} \left| f(x_0 + s) + f(x_0 - s) - 2f(x_0) \right| ds$$

$$\leq (N+1)g\left(\frac{1}{N}\right) \leq 2Ng\left(\frac{1}{N}\right) \leq \frac{2\epsilon}{13}, \tag{6.51}$$

where we also used that $N > \frac{1}{\delta}$. For the second integral we get by using (6.50) that

$$\int_{\frac{1}{N}}^\delta F_N(s) \left| f(x_0 + s) + f(x_0 - s) - 2f(x_0) \right| ds$$

$$\leq \int_{\frac{1}{N}}^\delta \left| f(x_0 + s) + f(x_0 - s) - 2f(x_0) \right| \frac{\pi^2}{(N+1)s^2} \, ds,$$

and integration by parts yields

$$\int_{\frac{1}{N}}^\delta \left| f(x_0 + s) + f(x_0 - s) - 2f(x_0) \right| \frac{\pi^2}{(N+1)s^2} \, ds$$

$$= \frac{\pi^2}{N+1} \left(\frac{1}{\delta^2} g(\delta) - N^2 g \left(\frac{1}{N} \right) + 2 \int_{\frac{1}{N}}^{\delta} g(s) s^{-3} \, \mathrm{d}s \right)$$

$$\leq \frac{\pi^2}{N+1} \frac{1}{\delta} \frac{\epsilon}{13} + \frac{2\pi^2}{(N+1)} \int_{\frac{1}{N}}^{\delta} \frac{\epsilon}{13} s^{-2} \, \mathrm{d}s$$

$$\leq \frac{\pi^2 \epsilon}{13} + \frac{2\pi^2 \epsilon}{13(N+1)} \left(N - \frac{1}{\delta} \right) < \frac{3\pi^2 \epsilon}{13}, \tag{6.52}$$

where we used that $N > \frac{1}{\delta}$, i.e. $\frac{1}{(N+1)\delta} < 1$, and that $g(h) \geq 0$. Finally we observe that by (6.49)

$$\int_{\delta}^{\pi} F_N(s) |f(x_0 + s) + f(x_0 - s) - 2f(x_0)| \, \mathrm{d}s$$

$$\leq \int_{\delta}^{\pi} |f(x_0 + s) - f(x_0 + s) - 2f(x_0)| \frac{\epsilon}{a+1} \, \mathrm{d}s$$

$$\leq \frac{\epsilon}{a+1} g(\pi) < \epsilon. \tag{6.53}$$

From (6.51), (6.52) and (6.53) we now deduce that

$$|\sigma_N(f)(x_0) - f(x_0)| \leq \frac{1}{2\pi} \left(\frac{2\epsilon}{13} + \frac{3\pi^2 \epsilon}{13} + \epsilon \right) < \epsilon.$$

$\square$

Since convergence implies Cesàro summability we derive from Theorem 6.16

Corollary 6.17. *If for* $f \in L^1(\mathbb{T})$ *the associated Fourier series converges almost everywhere to a function* g, *then* $f = g$ *almost everywhere.*

Further, with the help of Theorem 4.19, we obtain

Corollary 6.18. *For* $f \in L^1(\mathbb{T})$ *exists a set* $E \subset [-\pi, \pi)$ *of measure zero, i.e.* $\lambda^{(1)}(E) = 0$, *such that for all* $x \in [-\pi, \pi) \backslash E$ *the Fourier series associated with* f *is Abel summable.*

The result of Theorem 6.12 was extended to the statement that for every set $E \subset [-\pi, \pi]$ of measure zero, i.e. $\lambda^{(1)}(E) = 0$, there exists a 2π-periodic continuous function with the property that the sequence of the partial sums of the associated Fourier series diverges at every point E, we refer to Theorem 3.4 in [73]. A theorem due to A. N. Kolomogorov [76] states that there exists a Fourier series which is at all points divergent, see Theorem 3.6 in [73]. This result is an extension of Kolomogorov's earlier result [75] on the existence of an L^1-function

the associated Fourier series of which diverges $\lambda^{(1)}$-almost everywhere, we refer to
L. Grafakos [47], p.195-200 for a detailed proof. N. N Lusin conjectured that for an
L^2-function the associated Fourier series always converges $\lambda^{(1)}$-almost everywhere
in 1966 L. Carleson [21] could prove this conjecture.

His result was extended by R. Hunt [66] who showed that the Fourier series associated with an L^p-function for every $p > 1$ converges $\lambda^{(1)}$-almost everywhere to
the function. The proofs of these results are beyond the scope of a first discussion
of Fourier series as we intend in this Course in Analysis.

We will pick up some of the convergence and divergence problems related to Fourier
series in Volume V within our treatment of functional analysis and operator theory.
We then will consider partial sums not merely as linear combinations of trigonometric (exponential) functions, but we will view S_N, $N \in \mathbb{N}$, as a family of linear
operators between certain Banach spaces.

The following table gives an overview of some of the important convergence and
summability results for the Fourier series $S(f)$ associated with a 2π-periodic function $f : \mathbb{R} \to \mathbb{C}$.

Further properties of f	S(f)
Riemann integrable	may diverge at all points
Riemann integrable	converges in the quadratic mean
continuous	may diverge at some points (on a set of measure zero)
absolutely continuous	converges at all points
$f \in L^1(\mathbb{T})$	may diverge almost everywhere
$f \in L^p(\mathbb{T})$, $1 < p < \infty$,	converges almost everywhere
$f \in L^2(\mathbb{T})$	converges in the norm $\|.\|_{L^2}$
continuous	uniformly Cesàro summable
continuous	uniformly Abel summable
$f \in L^1(\mathbb{T})$	almost everywhere Cesàro summable

Problems

1.   a) Use Leibniz' criterion for alternating series to prove once again that
the integral $\int_0^\infty \frac{\sin t}{t}\, dt$ exists as an improper Riemann integral.

b) Show that $t \mapsto \frac{\sin t}{t}$ is not Lebesgue intergrable on $[0, \infty)$.

2. Prove that $|x - \sin x| \leq \frac{|x|^3}{6}$ for $0 \leq x \leq \frac{\pi}{2}$.

3. Let $u \in L^1(\mathbb{T})$ and $\delta > 0$. The **integral modulus of continuity** of u is defined by

$$w_1(u, \delta) := \sup_{u \leq |h| \leq \delta} \int_{-\pi}^{\pi} |u(x + h) - u(x)| \, dx.$$

a) Show that the Fourier coefficients c_k, $k \in \mathbb{Z} \setminus \{0\}$, of $u \in L^1(\mathbb{T})$ satisfy the estimates

$$(*) \qquad |c_k| \leq \frac{1}{4\pi} w_1 \left(u, \frac{\pi}{|k|} \right).$$

b) In the case where $c_{-k} = \overline{c_k}$ deduce from $(*)$ that

$$|a_k| \leq \frac{1}{2\pi} w_1 \left(u, \frac{\pi}{k} \right) \quad \text{and} \quad |b_k| \leq \frac{1}{2\pi} w_1 \left(u, \frac{\pi}{k} \right), k \in \mathbb{N}.$$

4. Denote by $V(u)$ the total variation of a function u. Suppose that $u : \mathbb{R} \to \mathbb{R}$ is 2π-periodic and of bounded variation $V(u)$ on $[-\pi, \pi]$, hence integrable. Prove the estimates $|a_k| \leq \frac{V(u)}{2k}$ and $|b_k| \leq \frac{V(u)}{2k}$, $k \in \mathbb{N}$.

5. Assume that the series $\sum_{k=2}^{\infty} \frac{\sin kx}{k \ln k}$ converges uniformly and prove that it does not converge absolutely. Compare with the result of Example 6.9.

6. Verify the formulae (6.24) and (6.25).

7. Let $f \in L^1(\mathbb{T})$ be essentially bounded, i.e. there exists $M \geq 0$ such that $|f(x)| \leq M$ a.e. Show that the Fejer sums $\sigma_N(f)$ are uniformly bounded by M, i.e. $|\sigma_N(f)(x)| \leq M$ for all $x \in \mathbb{T}$ and $N \in \mathbb{N}$.

7 Holomorphic Functions, Harmonic Functions, Hardy Spaces

In this chapter we will also work with holomorphic functions $f : G \to \mathbb{C}$, where $G \subset \mathbb{C}$ is a region, i.e. an open and connected set. In most cases G will be the unit disc $D := B_1(0) \subset \mathbb{C}$. Points in $\mathbb{C}$ are denoted by $z = x + iy$ and we freely switch from the notation $g(z)$ to $g(x, y)$ (instead of $g(x + iy)$). This applies in particular when decomposing f into its real and imaginary parts

$$f(z) = u(z) + iv(z) = u(x, y) + iv(x, y). \tag{7.1}$$

We know that if f is holomorphic then u and v are harmonic, i.e. they are arbitrary often differentiable and satisfy the Laplace equation

$$\Delta u(x, y) = \Delta v(x, y) = 0. \tag{7.2}$$

Recall that given a harmonic function u on G we call v the **conjugate harmonic function** to u if $f = u + iv$ is holomorphic in G.

For a holomorphic function $f = u + iv$ the mean-value equality

$$f(z_0) = \frac{1}{2\pi} \int_0^{2\pi} f\left(z_0 + re^{it}\right) dt \tag{7.3}$$

holds and this implies for $w \in \{u, v\}$ that

$$w(z_0) = \frac{1}{2\pi} \int_0^{2\pi} w\left(z_0 + re^{it}\right) dt \tag{7.4}$$

for every $\overline{B_r(z_0)} \subset G$. In Proposition II.9.20 we proved the maximum principle for harmonic functions: Let $G \subset \mathbb{C}$ be a region and suppose that $u \in C(\overline{G})$ is harmonic in G. Then u attains its maximum on ∂G, i.e.

$$\max_{x \in \overline{G}} u(x) = \max_{x \in \partial G} u(x). \tag{7.5}$$

Now let f be a holomorphic function of the form

$$f(z) = \sum_{k=0}^{\infty} A_k z^k \tag{7.6}$$

where we assume that the radius of convergence of the power series is at least 1. Using polar coordinates $z = re^{i\varphi}$, we find in $B_1(0)$

$$f\left(re^{i\varphi}\right) = \sum_{k=0}^{\infty} A_k r^k e^{ik\varphi}. \tag{7.7}$$

For $r \in (0, 1)$ fixed we can interpret the right hand side of (7.7) as the Fourier series associated with the function $\varphi \mapsto f_r\left(e^{i\varphi}\right)$. Recall $f_r\left(e^{it}\right) = f\left(re^{it}\right)$. Note that the uniform convergence of $\sum_{k=0}^{\infty} A_k z^k$ for $|z| < 1$ allows us to identify $\sum_{k=0}^{\infty} A_k r^k e^{ik\varphi}$ with the Fourier series associated with f_r and corresponding Fourier coefficients $c_k(r) = A_r r^k$, $k \in \mathbb{N}_0$, and $c_k(r) = 0$ for $-k \in \mathbb{N}$. With $A_k = \alpha_k + i\beta_k$ we find for the real and imaginary part of $f_r(\varphi) = \sum_{k=0}^{\infty} A_k r^k e^{ik\varphi}$

$$\sum_{k=0}^{\infty} (\alpha_k + i\beta_k) r^k \left(\cos k\varphi + i\sin k\varphi\right)$$

$$= \sum_{k=0}^{\infty} \left(r^k(\alpha_k \cos k\varphi - \beta_k \sin k\varphi) + ir^k(\beta_k \cos k\varphi + \alpha_k \sin k\varphi)\right)$$

$$= \sum_{k=0}^{\infty} r^k(\alpha_k \cos k\varphi - \beta_k \sin k\varphi) + i \sum_{k=0}^{\infty} r^k(\beta_k \cos k\varphi + \alpha_k \sin k\varphi).$$

Note that the series representing the imaginary part is the conjugate series of the one representing the real part, see (2.32). Of course we know that each of the series in the last line represents a harmonic function in $B_1(0)$. Indeed, for every $k \in \mathbb{N}_0$ a function $g_k(r, \varphi) = \gamma_k r^k \cos k\varphi + \delta_k r^k \sin k\varphi$ is harmonic in $\mathbb{R}$ as follows from a direct calculation when applying the Laplace operator in polar coordinates, compare with (II.12.1), to $g_k(r, \varphi)$:

$$\left(\frac{\partial^2}{\partial r^2} + \frac{1}{r}\frac{\partial}{\partial r} + \frac{1}{r^2}\frac{\partial}{\partial \varphi^2}\right)\left(\gamma_k r^k \cos k\varphi + \delta_k r^k \sin k\varphi\right)$$
$$= r^{k-2}\left(k^2\gamma_k - k\gamma_k + k\gamma_k - k^2\gamma_k\right)\cos k\varphi + r^{k-2}\left(k^2\delta_k - k\delta_k + k\delta_k - k^2\delta_k\right)\sin k\varphi$$
$$= 0.$$

Thus we may start with the harmonic function $u : B_1(0) \to \mathbb{R}$ defined by

$$u(r, \varphi) = \sum_{k=0}^{\infty} \left(a_k r^k \cos k\varphi + b_k r^k \sin k\varphi\right) \tag{7.8}$$

for, say bounded coefficients a_k and b_k, and define

$$v(r, \varphi) := \sum_{k=0}^{\infty} \left(-b_k r^k \cos k\varphi + a_k r^k \sin k\varphi\right) \tag{7.9}$$

106

which is a further harmonic function in $B_1(0)$ and with $A_k := a_k - ib_k$ it follows that

$$f(z) = \sum_{k=0}^{\infty} A_k r^k e^{ik\varphi} = \sum_{k=0}^{\infty} A_k z^k$$

is in D holomorphic with decomposition $f(z) = u(z) + iv(z)$. Thus v is the conjugate harmonic function to u given by (7.8), and as already remarked, its (Fourier) series is the conjugate (Fourier) series of the one associated with u. A natural question is now: given a harmonic function in $D = B_1(0)$, when does it admit representation (7.8)? We may change our point of view and consider (7.7) as Abel means of the series $\sum_{k=0}^{\infty} A_k e^{ik\varphi}$ which we can interpret as a trigonometric series with coefficients $c_k = A_k$ for $k \in \mathbb{N}_0$ and $c_k = 0$ for $-k \in \mathbb{N}$. Suppose that $\sum_{k=0}^{\infty} |A_k|^2 < \infty$, i.e. the series $\sum_{k=0}^{\infty} A_k e^{ik\varphi}$ is associated with some L^2-function. For $0 < r < 1$ we define

$$M_2(f;r) := \left(\frac{1}{2\pi} \int_0^{2\pi} \left| f\left(re^{i\varphi} \right) \right|^2 d\varphi \right)^{\frac{1}{2}}, \tag{7.10}$$

and Parseval's equation yields

$$M_2(f;r) = \left(\sum_{k=0}^{\infty} |A_k|^2 r^{2k} \right)^{\frac{1}{2}}. \tag{7.11}$$

In complex variable theory and in particular in the theory of Hardy spaces it is more common to let φ be in the range $[0, 2\pi]$ rather than $[-\pi, \pi]$ and for this reason we do the same in this chapter. Since by assumption we have $\sum_{k=0}^{\infty} |A_k|^2 < \infty$ it follows that

$$\|f\|_{\mathcal{H}^2} := \lim_{r \to 1} M_2(f;r) = \left(\sum_{k=0}^{\infty} |A_k|^2 \right)^{\frac{1}{2}}, \tag{7.12}$$

and clearly $\| \cdot \|_{\mathcal{H}^2}$ is a norm.

Definition 7.1. *The **Hardy space** $\mathcal{H}^2$ consists of all holomorphic functions $f : D \to \mathbb{C}$ with power series expansion $f(z) = \sum_{k=0}^{\infty} A_k z^k$ for which $\sum_{k=0}^{\infty} |A_k|^2 < \infty$.*

From our previous considerations we deduce

Corollary 7.2. *The statements $f \in \mathcal{H}^2$ and $\lim_{r \to 1} M_2(f;r) < \infty$ are equivalent.*

We can extend the sequence $(A_k)_{k \in \mathbb{N}_0}$ of Taylor coefficients of f to the sequence $(A_k)_{k \in \mathbb{Z}}$ by setting $A_k = 0$ for $-k \in \mathbb{N}$. The sequence $(A_k)_{k \in \mathbb{Z}}$ is an element of $l_2(\mathbb{Z})$ since

$$\sum_{k \in \mathbb{Z}} |A_k|^2 = \sum_{k=0}^{\infty} |A_k|^2 < \infty.$$

Since $f \in \mathcal{H}^2$, and therefore by Theorem 5.13 there exists a function $f^* \in L^2(\mathbb{T})$ such that the Fourier series associated with f^* is given by $\sum_{k=0}^{\infty} A_k e^{ik\varphi}$, and we have $\lim_{N \to \infty} \left\| \sum_{k=0}^{N} A_k e^{ik\cdot} - f^* \right\|_{L^2} = 0$. From Corollary III.7.37 we can deduce that a subsequence of $\left(\sum_{k=0}^{N} A_k e^{ik\cdot} \right)_{N \in \mathbb{N}_0}$ converges $\lambda^{(1)}$-almost everywhere to f^*, however this is not sufficient for our purposes. It is a natural question how f^* relates to f and a naive way of interpreting f^* is as the boundary values of f. However, so far we do not know whether or in which sense f has boundary values. Denote as before by $P_r(\varphi) = \sum_{k \in \mathbb{Z}} r^{|k|} e^{ik\varphi} = \frac{1-r^2}{1-2r\cos\varphi+r^2}$ the Poisson kernel. From Theorem III.8.1, see also Problem 11 of Chapter 4, it follows that for $g \in L^1(\mathbb{T})$ the function $P_r * g$ is continuous.

Theorem 7.3. *Let* $f \in \mathcal{H}^2$, $f(z) = \sum_{k=0}^{\infty} A_k z^k$, *and denote by* $f^* \in L^2(\mathbb{T})$ *the function associated with the Fourier series* $\sum_{k=0}^{\infty} A_k e^{ik\varphi}$. *For* $z = re^{i\varphi}$ *and* $0 \le r < 1$ *we have*

$$f\left(re^{i\varphi}\right) = (P_r * f^*)(\varphi) \tag{7.13}$$

and for almost all φ *the radial limit*

$$\lim_{r \to 1} (P_r * f^*)(\varphi) = f^*(\varphi) \tag{7.14}$$

exists.

Proof. Let $\rho \in (0, 1)$ and recall that $A_k = 0$ for all $k \in -\mathbb{N}$. With $f_\rho(\varphi) := f\left(\rho e^{i\varphi}\right)$ it follows that

$$(P_r * f_\rho)(\varphi) = \sum_{k=0}^{\infty} r^k \rho^k A_k e^{ik\varphi} = f_{r\rho}(\varphi) \tag{7.15}$$

and since

$$F(f^* - f_\rho)(k) = \left(1 - \rho^k\right) A_k,$$

Plancherel's theorem yields

$$\|f^* - f_\rho\|_{L^2}^2 = 2\pi \sum_{k=1}^{\infty} \left(1 - \rho^k\right)^2 |A_k|^2.$$

From (7.15) and the fact that $0 \le r < 1$ we deduce further

$$\|f_{r\rho} - P_r * f^*\|_{L^2}^2 = 2\pi \sum_{k=0}^{\infty} r^{2k}(1 - \rho^k)^2 |A_k|^2$$

which implies in $L^2(\mathbb{T})$ that

$$f_r = P_r * f^*.$$

Since for $r \in [0,1)$ $f_r(\cdot)$ is continuous as is $(P_r * f^*)(\cdot)$ it follows that (7.13) holds. Since $A_k = 0$ for $k \in -\mathbb{N}$ the term $(P_r * f^*)(\varphi)$ is the rth Abel mean of f^*, i.e.

$$(P_r * f^*)(\varphi) = \sum_{k=0}^{\infty} r^k A_k e^{ik\varphi} = \sum_{k \in \mathbb{Z}} r^k A_k e^{ik\varphi}$$

and by Corollary 6.18 we deduce that outside a set of measure zero we have

$$\lim_{r \to 1} (P_r * f^*)(\varphi) = f^*(\varphi).$$

$\square$

The next result clarifies the structure of $\mathcal{H}^2$.

Theorem 7.4. *The vector spaces* $\{g \in L^2(\mathbb{T}) \,|\, (Fg)(k) = 0 \,for\, k \in -\mathbb{N}\}$ *and* $\mathcal{H}^2$ *are isomorphic and* $\mathcal{H}^2$ *is complete with respect to* $\|\cdot\|_{\mathcal{H}^2}$.

Proof. Let $f \in \mathcal{H}^2$, $f(z) = \sum_{k=0}^{\infty} A_k z^k$, and set $A_k = 0$ for $-k \in \mathbb{N}$. We define $J : \mathcal{H}^2 \to \{g \in L^2(\mathbb{T}) \,|\, (Fg)(k) = 0 \,for\, k \in -\mathbb{N}\}$ by

$$J(f)(\varphi) = \sum_{k=0}^{\infty} A_k e^{ik\varphi}.$$

Since

$$\|J(f)\|_{L^2} = \sqrt{2\pi} \left(\sum_{k=0}^{\infty} |A_k|^2 \right)^{\frac{1}{2}}$$

it follows that J is well defined, linear and injective. Now, given $h \in \{g \in L^2(\mathbb{T}) \,|\, (Fg)(k) = 0 \,for\, k \in -\mathbb{N}\}$. We define $f(z) := \sum_{k=0}^{\infty}(Fh)(k)z^k$ which implies $f \in \mathcal{H}^2$ and $J(f) = h$. It remains to prove that $\mathcal{H}^2$ is complete and we follow closely [103]. For this let $(f_k)_{k \in \mathbb{N}}$ be a Cauchy sequence with respect to $\|\cdot\|_{\mathcal{H}^2}$. We apply the Cauchy integral formula to $f_k - f_l$ for $|z| \leq r < R$ where we integrate over the circle $|\zeta| = R$. It follows that

$$(R - r)|f_k(z) - f_l(z)| \leq \frac{1}{2\pi} \int_0^{2\pi} \left| f_k\left(Re^{i\varphi}\right) - f_l\left(Re^{i\varphi}\right) \right| d\varphi$$

$$\leq \left(\frac{1}{2\pi} \int_0^{2\pi} \left| f_k\left(Re^{i\varphi}\right) - f_k\left(Re^{i\varphi}\right) \right|^2 d\varphi \right)^{\frac{1}{2}}$$

$$\leq \|f_k - f_l\|_{\mathcal{H}^2},$$

which implies the uniform convergence of $(f_k)_{k \in \mathbb{N}}$ on every compact subset of D to some holomorphic function $f : D \to \mathbb{C}$. Now let $\epsilon > 0$ be given and choose $l > k$ such that $\|f_k - f_l\|_{\mathcal{H}^2} < \epsilon$. This implies for every $r < 1$ that

$$M_2(f - f_l; r) = \lim_{k \to \infty} M_2(f_k - f_l; r) \leq \epsilon$$

or $\lim_{l \to \infty} \|f - f_l\|_{\mathcal{H}^2} = 0$, and in particular $f \in \mathcal{H}^2$. $\qquad\square$

Since the Poisson kernel is real-valued we can take in (7.13) and (7.14) the real (and imaginary) part and find that

$$u(r, \varphi) = \operatorname{Re} f(re^{i\varphi}) = (P_r * \operatorname{Re} f^*)(\varphi) \tag{7.16}$$

is in $B_1(0) \subset \mathbb{R}^2$ a harmonic function with almost everywhere radial limit (boundary values) $\operatorname{Re} f^*$. This observation leads to many interesting new questions: can we use the Poisson kernel to construct in $B_1(0)$ harmonic functions for other classes of "boundary functions"? Is it possible to improve the boundary behaviour, i.e. to allow more and different approach directions of $z = re^{i\varphi}$ to $\partial B_1(0) (= \mathbb{T})$? We also may ask whether we can extend the definition of $\mathcal{H}^2$ and then Theorem 7.3 and 7.4 to some L^p-setting. Here we can only discuss some of these results, for more we refer to [27], [33], [73], [77] or [103], to mention a few classical texts. Before addressing the problems mentioned above we need some preparation in dealing with Blaschke products, the Poisson integral and sub-harmonic functions.

First let us note that for a holomorphic function $f : G \to \mathbb{C}$, $G \subset \mathbb{C}$ a region, with the property that $f(z) \neq 0$ for all $z \in G$ the function $z \mapsto \ln |f(z)|$ is harmonic, see Problem 1, a result which follows from a straightforward calculation using the Cauchy-Riemann differential equations and the fact that $\operatorname{Re} f$ and $\operatorname{Im} f$ are harmonic. The mean-value theorem for harmonic functions now implies for $B_r(z_0) \subset G$

$$\ln |f(z_0)| = \frac{1}{2\pi} \int_0^{2\pi} \ln \left| f\left(z_0 + re^{i\varphi}\right)\right| \, d\varphi \tag{7.17}$$

provided $f(z) \neq 0$ in a neighbourhood of $\overline{B_r(z_0)}$.

For $\zeta \in D \backslash \{0\}$ we consider the Möbius transformation

$$w(z; \zeta) = \frac{\bar{\zeta}(\zeta - z)}{|\zeta|(1 - z\bar{\zeta})} = \frac{-\bar{\zeta}z + |\zeta|^2}{-|\zeta|\bar{\zeta}z + |\zeta|} \tag{7.18}$$

which maps $\mathbb{C} \backslash \left\{\frac{1}{\bar{\zeta}}\right\}$ biholomorphic onto $\mathbb{C} \backslash \left\{\frac{1}{|\zeta|}\right\}$, the point ζ onto 0 and the point 0 onto $|\zeta|$. Moreover, for $|z| = 1$ it follows that $|w(z; \zeta)| = 1$, in fact $w(\cdot; \zeta)$ maps $\overline{D}$ biholomorphic into itself. We extend $w(z; \zeta)$ for $\zeta = 0$ by

$$w(z; 0) = z. \tag{7.19}$$

For $0 < |\zeta| < r$ it now follows

$$w\left(\frac{z}{r}; \frac{\zeta}{r}\right) = r\frac{\bar{\zeta}(\zeta - z)}{|\zeta|(r^2 - z\bar{\zeta})} \tag{7.20}$$

which is a holomorphic mapping in $|z| < r$, vanishes only at $z = \zeta$ and $w\left(\frac{z}{r}; \frac{\zeta}{r}\right) = 1$ for $|z| = r$. We want to use $w(\cdot; \zeta)$ to "control" the zeroes of f. Suppose that f is holomorphic in $\overline{B_r(0)}$ and have only one simple zero (multiplicity 1) at ζ. The function

$$f_1(z) := f(z)\frac{1}{w\left(\frac{z}{r}, \frac{\zeta}{r}\right)}$$

has no zero in $\overline{B_r(0)}$ and on $\partial B_r(0)$, i.e. for $|z| = r$, we have

$$|f_1(z)| = |f(z)|.$$

If f has M zeroes $\zeta_1, \ldots, \zeta_M$ in $\overline{B_r(0)}$ each listed accordingly to its multiplicity then

$$f_1(z) := f(z)\frac{1}{\prod_{j=1}^M w\left(\frac{z}{r}, \frac{\zeta_j}{r}\right)} \tag{7.21}$$

has no zero in $\overline{B_r(0)}$. Hence $\ln |f_1(z)|$ is harmonic and by the mean-value theorem we have

$$\ln |f_1(0)| = \frac{1}{2\pi} \int_0^{2\pi} \ln |f_1\left(re^{i\varphi}\right)| \, d\varphi. \tag{7.22}$$

If in addition $f(0) \neq 0$, i.e. $\zeta_j \neq 0$ for $j = 1, \ldots, M$ and $|\zeta_j| \neq r$ we find

$$\ln |f(0)| = \frac{1}{2\pi} \int_0^{2\pi} \ln |f\left(re^{i\varphi}\right)| \, d\varphi \tag{7.23}$$

as well as with any branch of the logarithmic function

$$\ln |f_1(z)| = \ln |f(z)| - \log \prod_{j=1}^M w\left(\frac{z}{r}, \frac{\zeta_j}{r}\right).$$

Using now $f(0) \neq 0$, $w\left(\frac{z}{r}, \frac{\zeta_j}{r}\right) = 1$ for $|z| = r$, and (7.23) we get the **Poisson-Jensen formula**

$$\ln |f(0)| + \ln \prod_{j=1}^M \frac{r}{|\zeta_j|} = \frac{1}{2\pi} \int_0^{2\pi} \ln |f\left(re^{i\varphi}\right)| \, d\varphi \tag{7.24}$$

or

$$\ln|f(0)| + \sum_{j=1}^{M} \ln \frac{r}{|\zeta_j|} = \frac{1}{2\pi} \int_0^{2\pi} \ln \left| f\left(re^{i\varphi}\right) \right| d\varphi, \tag{7.25}$$

which yields **Jensen's inequality** when recalling that $\frac{r}{|\zeta_j|} > 1$:

$$\ln|f(0)| \leq \frac{1}{2\pi} \int_0^{2\pi} \ln \left| f\left(re^{i\varphi}\right) \right| d\varphi. \tag{7.26}$$

When taking in (7.25) some terms out of the sum we obtain

$$\ln|f(0)| + \sum_{l=1}^{\tilde{M}} \ln \frac{r}{|\zeta_{j_l}|} \leq \frac{1}{2\pi} \int_0^{2\pi} \ln \left| f\left(re^{i\varphi}\right) \right| d\varphi \tag{7.27}$$

with $\tilde{M} \leq M$. In the case where f has a zero at 0 of order m we may first consider instead of f the function $z \mapsto z^{-m} f(z)$.

We want to move a step further and study the infinite product $\prod_{k=1}^{\infty} w(z, \zeta_k)$.

Proposition 7.5. *Let $(\zeta)_{k\in\mathbb{N}}$ be a sequence of complex numbers such that $|\zeta_k| < 1$ and $\sum_{k=1}^{\infty}(1 - |\zeta_k|) < \infty$. Suppose m of these numbers are equal to 0 and define*

$$B(z) := \prod_{k=1}^{\infty} w(z, \zeta_k) = z^m \prod_{\zeta_k \neq 0}^{\infty} w(z, \zeta_k) = z^m \prod_{\zeta_k \neq 0} \frac{\bar{\zeta}_k(\zeta_k - z)}{|\zeta_k|(1 - z\bar{\zeta}_k)}. \tag{7.28}$$

Then this infinite product converges absolutely and uniformly in $\overline{B_r(0)} \subset D$ for every $r < 1$. Moreover we have $|B(z)| < 1$ for $|z| < 1$.

Definition 7.6. *The product $B(z)$ in (7.28) is called the **Blaschke product** corresponding to the sequence $(\zeta_k)_{k\in\mathbb{N}}$. The assumption $\sum_{k=1}^{\infty}(1 - |\zeta_k|) < \infty$ is called the **Blaschke condition**.*

Proof of Proposition 7.5. We put $\alpha_k := \frac{\bar{\zeta}_k(\zeta_k - z)}{|\zeta_k|(1 - z\bar{\zeta}_k)}$ and we have to prove that $\sum_{k=1}^{\infty}|1 - \alpha_k|$ converges uniformly in $\overline{B_r(0)}$ to a non-zero limit where we pretend for the proof that $m = 0$. Once this result is proved it follows that $\sum_{k=1}^{\infty} \ln|1 - \alpha_k|$ will converge implying the uniform and absolute convergence of $B(z)$ by extending Proposition I.30.10 to the complex case, see also Problem 11 and 12 of Chapter

112

III.13. We note that

$$
\begin{aligned}
|1 - \alpha_k| &= \left| 1 - \frac{\overline{\zeta_k}(\zeta_k - z)}{|\eta_k|(1 - z\overline{\zeta}_k)} \right| \\
&= \frac{||\zeta_k| + z\overline{\zeta}_k|}{|\zeta_k||(1 - z\overline{\zeta}_k)|} |1 - |\zeta_k|| \\
&\leq \frac{1 + |z|}{|1 - z\overline{\zeta}_k|} |1 - |\zeta_k|| \leq \frac{1 + r}{1 - r} |1 - |\zeta_k||,
\end{aligned} \tag{7.29}
$$

where in the last step we used the triangle inequality and the converse triangle inequality as well as $|\zeta_k| < 1$ and $|z| < r$. Since by assumption $\sum_{k=1}^{\infty}(1 - |\zeta_k|) < \infty$ the convergence as claimed is proved. Further we note that for $|z| < 1$ it follows that each factor in (7.28) is less than 1 implying that $|B(z)| < 1$. $\qquad\square$

Remark 7.7. In our treatment of the Blaschke product we used much the discussion in [73].

The Blaschke product allows us on the one hand side to construct a holomorphic function with given zeroes. On the other hand it allows us to factorize some holomorphic functions into a product one factor of which has no zeroes. Suppose that $(\zeta_k)_{k\in\mathbb{N}}$ are the zeroes (multiplicity taken into account) of the holomorphic function $f : D \to \mathbb{C}$ and assume that $\sum_{k=1}^{\infty} |1 - |\zeta_k|| < \infty$, i.e. that the Blaschke condition is fulfilled. Denote by $B_f = B$ the Blaschke product constructed with the help of the sequence $(\zeta_k)_{k\in\mathbb{N}}$. The function

$$
g(z) := f(z)(B_f(z))^{-1} \tag{7.30}
$$

is holomorphic in D, has no zeroes in D and since $|B_f(z)| < 1$ it follows $|f(z)| < |g(z)|$ for $z \in D$. Thus we have derived the **canonical factorization** of f, i.e.

$$
f(z) = g(z)B_f(z), \tag{7.31}
$$

where the holomorphic function g has no zeroes in D. It remains to find a suitable class of holomorphic functions satisfying the Blaschke condition.

Definition 7.8. *The **Nevanlinna class** $\mathcal{N}$ consists of all holomorphic functions $f : D \to \mathbb{C}$ for which*

$$
q_{\mathcal{N}}(f) := \sup_{0 < r < 1} \frac{1}{2\pi} \int_0^{2\pi} \ln^+ \left| f\left(re^{i\varphi}\right) \right| d\varphi < \infty, \tag{7.32}
$$

where $\ln^+ \alpha = \begin{cases} 0, & 0 < \alpha < 1 \\ \ln \alpha, & \alpha \geq 1 \end{cases}$.

113

Proposition 7.9. *For $f \in \mathcal{N}$ the Blaschke condition is satisfied, i.e. if $(\zeta_k)_{k \in \mathbb{N}}$ denotes the non-zero zeroes of f in D counted according to their multiplicity then $\sum_{k=1}^{\infty}(1 - |\zeta_k|) < \infty$.*

Proof. We may assume $f(0) \neq 0$, i.e. $\zeta_k \neq 0$, otherwise we consider $z \mapsto z^m \tilde{f}(z)$. The convergence of $\sum_{k=1}^{\infty}(1-|\zeta_k|)$ is equivalent to the convergence of $\prod_{k=1}^{\infty} |\zeta_k|$, see Chapter I.30. Hence it is sufficient to show that the series $\sum_{k=1}^{\infty} \ln |\zeta_k|$ is bounded from below. For this we first note that

$$\frac{1}{2\pi} \int_0^{2\pi} \ln |f(re^{i\varphi})| \, d\varphi \leq q_{\mathcal{N}}(f)$$

and next we use Jensen's inequality in the variant of (7.26). We first fix $\tilde{M}$ and choose r sufficiently close to 1 so that in $B_r(0)$ we have at least $\tilde{M}$ zeroes of f. Now (7.26) yields

$$\ln |f(0)| + \sum_{l=1}^{\tilde{M}} \ln \frac{r}{|\zeta_{l_j}|} \leq \frac{1}{2\pi} \int_0^{2\pi} \ln |f(re^{i\varphi})| \, d\varphi \leq q_{\mathcal{N}}(f)$$

or

$$\ln |f(0)| - q_{\mathcal{N}}(f) \leq \sum_{l=1}^{\tilde{M}} \ln |\zeta_{l_j}| - M \ln r.$$

For $r \to 1$ we obtain

$$\ln |f(0)| - q_{\mathcal{N}}(f) \leq \sum_{l=1}^{\tilde{M}} \ln |\zeta_{l_j}|,$$

and since $\tilde{M}$ was arbitrarily chosen it follows that $\sum_{k=1}^{\infty} \ln |\zeta_k|$ converges. $\square$

For every $p \geq 1$ and all $t \geq 0$ it follows that $\ln^+ t \leq t^p$ and therefore we have

Lemma 7.10. *The inclusion $\mathcal{H}^2 \subset \mathcal{N}$ holds.*

Proof. We need only to note that

$$\frac{1}{2\pi} \int_0^{2\pi} \ln^+ |f(re^{i\varphi})| \, d\varphi \leq \frac{1}{2\pi} \int_0^{2\pi} |f(re^{i\varphi})|^2 \, d\varphi$$

and therefore

$$q_{\mathcal{N}}(f) \leq \|f\|_{\mathcal{H}^2}.$$

$\square$

Now we can prove

Theorem 7.11. *For $f \in \mathcal{H}^2$ we consider the canonical factorization $f = gB_f$. It follows that $g \in \mathcal{H}^2$ and*

$$\|g\|_{\mathcal{H}^2} = \|f\|_{\mathcal{H}^2}. \tag{7.33}$$

Proof. Clearly (7.33) implies $g \in \mathcal{H}^2$. Denote by $(\zeta_k)_{k \in \mathbb{N}}$ the zeroes of f with $\zeta_k \neq 0$ and let $B_f^N(z)$ be the (finite) Blaschke product constructed with the first N zeroes of f, as usual, multiplicity must be taken into account. Consider now $g_N := f \cdot \left(B_f^N\right)^{-1}$. Since for $r \to 1$ we have $B_f^N\left(re^{i\varphi}\right) \to 1$ uniformly we deduce that $\|g_N\|_{\mathcal{H}^2} = \|f\|_{\mathcal{H}^2}$, recall

$$\|h\|_{\mathcal{H}^2} = \left(\sum_{k=1}^{\infty} |A_k(h)|^2\right)^{\frac{1}{2}}$$

$$= \lim_{r \to 1} M_2(h; r) = \lim_{r \to 1} \left(\frac{1}{2\pi} \int_0^{2\pi} \left|h\left(re^{i\varphi}\right)\right|^2 d\varphi\right)^{\frac{1}{2}}.$$

Since $\left|B_f^N(z)\right| < 1$ the functions $|g_N|$ increase to $|g|$ and the monotone convergence theorem yields

$$M_2(g; r) = \lim_{N \to \infty} M_2(g_N; r), \quad 0 < r < 1.$$

By (7.11) the function $r \mapsto M_2(f; r)$ is increasing and this implies $\lim_{N \to \infty} M_2(g_N; r) \leq \|f\|_{\mathcal{H}^2}$, hence we have proved that $\|g\|_{\mathcal{H}^2} \leq \|f\|_{\mathcal{H}^2}$. On the other hand, since $|f(z)| \leq |g(z)|$ in D we find

$$\left(\frac{1}{2\pi} \int_0^{2\pi} \left|f\left(re^{i\varphi}\right)\right|^2 d\varphi\right)^{\frac{1}{2}} \leq \left(\frac{1}{2\pi} \int_0^{2\pi} \left|g\left(re^{i\varphi}\right)\right|^2 d\varphi\right)^{\frac{1}{2}}$$

which gives immediately $\|f\|_{\mathcal{H}^2} \leq \|g\|_{\mathcal{H}^2}$ and the theorem is proved. $\square$

For $f \in L^p(\mathbb{T})$, $1 \leq p < \infty$, we want to study

$$u\left(re^{i\varphi}\right) := (P_r * f)(\varphi) = \frac{1}{2\pi} \int_{-\pi}^{\pi} P_r(\varphi - \vartheta) f(\vartheta) \, d\vartheta, \quad 0 < r < 1. \tag{7.34}$$

For simplicity we write for $h : B_1(0) \to \mathbb{C}$ now $h(x, y) = h(z) = h\left(re^{i\varphi}\right)$ and we observe that for $R < 1$

$$\int_{B_R(0)} |h(x, y)|^p \, dx \, dy = \int_0^R \int_{-\pi}^{\pi} \left|h\left(re^{i\varphi}\right)\right|^p r \, d\varphi \, dr \tag{7.35}$$

115

holds. From (7.34) we deduce that for $\frac{1}{p} + \frac{1}{q} = 1$, $p > 1$,

$$\begin{aligned}
|u(x,y)|^p = \left|u\left(re^{i\varphi}\right)\right|^p &= \left|\frac{1}{2\pi}\int_{-\pi}^{\pi} P_r(\varphi - \vartheta)f(\vartheta)\,d\vartheta\right|^p \\
&\leq \left(\frac{1}{2\pi}\int_{-\pi}^{\pi} P_r(\varphi - \vartheta)|f(\vartheta)|\,d\vartheta\right)^p \\
&\leq \left(\frac{1}{2\pi}\int_{-\pi}^{\pi} P_r^{\frac{1}{q}}(\varphi - \vartheta)P_r^{\frac{1}{p}}(\varphi - \vartheta)|f(\vartheta)|\,d\vartheta\right)^p \\
&= \frac{1}{(2\pi)^p}\left(\int_{-\pi}^{\pi} P_r(\varphi - \vartheta)\,d\vartheta\right)^{\frac{p}{q}}\left(\int_{-\pi}^{\pi} P_r(\varphi - \vartheta)|f(\vartheta)|^p\,d\vartheta\right) \\
&= \frac{1}{2\pi}\int_{-\pi}^{\pi} P_r(\varphi - \vartheta)|f(\vartheta)|^p\,d\vartheta.
\end{aligned}$$

Now it follows for $R < 1$ that

$$\begin{aligned}
\int_{B_R(0)} |u(x,y)|^p\,dx\,dy &\leq \int_0^R \int_{-\pi}^{\pi} \frac{1}{2\pi}\int_{-\pi}^{\pi} P_r(\varphi - \vartheta)|f(\vartheta)|^p\,d\vartheta\,r\,d\varphi\,dr \\
&= \frac{1}{2\pi}\int_0^R \int_{-\pi}^{\pi}\int_{-\pi}^{\pi} P_r(\varphi - \vartheta)\,d\varphi|f(\vartheta)|^p\,d\vartheta\,r\,dr \\
&= \int_0^R \int_{-\pi}^{\pi} |f(\vartheta)|^p\,d\vartheta\,dr = \frac{R^2}{2}\|f\|^p_{L^p(\mathbb{T})},
\end{aligned}$$

and for $R \to 1$ we get

$$\int_{B_1(0)} |u(x,y)|^p\,dx\,dy \leq \frac{1}{2}\|f\|^p_{L^p(\mathbb{T})}$$

or

$$\|u\|_{L^p(B_1(0))} \leq \frac{1}{2^{\frac{1}{p}}}\|f\|_{L^p(\mathbb{T})}, \tag{7.36}$$

and this estimate also holds for $p = 1$.

Theorem 7.12. *For $f \in L^p(\mathbb{T})$, $1 \leq p < \infty$, the function $u : B_1(0) \to \mathbb{C}$, defined by (7.34) belongs to $L^p(B_1(0))$ and $\operatorname{Re} u$ as well as $\operatorname{Im} u$ are harmonic.*

Proof. From (7.36) we conclude that $u \in L^p(B_1(0))$, $1 \leq p < \infty$. If f is real-valued the calculation leading to (4.54) yields that u is in every open ball $B_R(0)$, $0 < R < 1$ harmonic, hence it is harmonic in $B_1(0)$. Since for complex-valued functions f we have $\operatorname{Re} u(x,y) = \frac{1}{2\pi}\int_{-\pi}^{\pi} P_r(\varphi - \vartheta)\operatorname{Re} f(\vartheta)\,d\vartheta$ and $\operatorname{Im} u(x,y) = \frac{1}{2\pi}\int_{-\pi}^{\pi} P_r(\varphi - \vartheta)\operatorname{Im} f(\vartheta)\,d\vartheta$ the theorem follows. $\square$

We now extend the definition of $\mathcal{H}^2$ to the Hardy spaces $\mathcal{H}^p$, $1 \leq p < \infty$. For this let $f : D \to \mathbb{C}$ be a holomorphic function with power series representation $f(z) = \sum_{k=0}^{\infty} A_k z^k$. We define for $1 \leq p < \infty$

$$M_p(f;r) := \left(\frac{1}{2\pi} \int_0^{2\pi} \left| f\left(re^{i\varphi} \right) \right|^p \, d\varphi \right)^{\frac{1}{p}}. \tag{7.37}$$

Definition 7.13. *Let $1 \leq p < \infty$. The* **Hardy space** $\mathcal{H}^p$ *consists of all holomorphic functions $f : D \to \mathbb{C}$ for which*

$$\|f\|_{\mathcal{H}^p} := \lim_{r \to 1} M_p(f;r) < \infty \tag{7.38}$$

holds.

Remark 7.14. A. For $p = 2$ we recover $\mathcal{H}^2$. **B.** For all $p \geq 1$ $\mathcal{H}^p \subset \mathcal{N}$ and for $p_1 < p_2$ it follows that $\mathcal{H}^{p_1} \subset \mathcal{H}^{p_2}$.

Exercise 7.15. *Prove that $\| \cdot \|_{\mathcal{H}^p}$ is a norm and that $\mathcal{H}^p$ equipped with this norm is a Banach space.*

Lemma 7.16. *The mapping $r \mapsto M_p(f;r)$ is increasing.*

Proof. We know by (7.11) that the result is correct for $p = 2$. First let us suppose that $1 \leq p < \infty$ and that f has no zeroes in D. We consider $h(z) = f^{\frac{p}{2}}(z)$ and since f has no zeroes we may choose any branch of $f^{\frac{p}{2}}(\cdot)$. It follows for $r_1 < r_2 < 1$ that

$$M_p^p(f;r) = \frac{1}{2\pi} \int_0^{2\pi} \left| f\left(r_1 e^{i\varphi} \right) \right|^p \, d\varphi = \frac{1}{2\pi} \int_0^{2\pi} \left| h\left(r_1 e^{i\varphi} \right) \right|^2 \, d\varphi$$

$$\leq \frac{1}{2\pi} \int_0^{2\pi} \left| h\left(r_2 e^{i\varphi} \right) \right|^2 \, d\varphi = \frac{1}{2\pi} \int_0^{2\pi} \left| f\left(r_2 e^{i\varphi} \right) \right|^p \, d\varphi.$$

Now if f has in $B_{r_2}(0)$ the zeroes $\zeta_1, \dots, \zeta_M$ (counted according to their multiplicity) we can write

$$f_1(z) = f(z) \left(\prod_{j=1}^{M} w\left(\frac{z}{r_2}; \frac{\zeta_j}{r_2} \right) \right)^{-1}$$

and for $|z| < r_2$ it follows that $|f(z)| \leq |f_1(z)|$ whereas on $|z| = r_2$ we have $|f(z)| = |f_1(z)|$. Moreover f_1 has no zeroes in $|z| \leq r_2$. Now we find for $r_1 < r_2 < 1$ that

$$M_p^p(f;r_1) \leq M_p^p(f_1;r_1) \leq M_p^p(f_1;r_2) = M_p^p(f;r_2).$$

$\square$

Next we transfer Theorem 7.11 to the space $\mathcal{H}^p$:

Theorem 7.17. *Let $f \in \mathcal{H}^p$, $p \geq 1$, with canonical factorization $f = gB_f$. Then $g \in \mathcal{H}^p$ and*

$$\|g\|_{\mathcal{H}^p} = \|f\|_{\mathcal{H}^p}. \tag{7.39}$$

Proof. Again we note that (7.39) implies $g \in \mathcal{H}^p$. With $(\zeta_k)_{k \in \mathbb{N}}$ as in the proof of Theorem 7.11 being the non-zero zeroes of f and with the corresponding notations we find that $g_N := f\left(B_f^N\right)^{-1}$ converges on every set $B_r(0)$, $0 \leq r < 1$, uniformly to g, and $\left|B_f^N(f)\right|$ converges uniformly to 1 as $|z| \to 1$. Now we use Lemma 7.16 and the monotone convergence theorem to conclude that $g_N \in \mathcal{H}^p$ and $\|g_N\|_{\mathcal{H}^p} = \|f\|_{\mathcal{H}^p}$. For $r < 1$ we deduce

$$M_p^p(g; r) = \lim_{N \to \infty} M_p^p(g_N; r) \leq \lim_{N \to \infty} \|g_N\|_{\mathcal{H}^p}^p = \|f\|_{\mathcal{H}^p}^p,$$

i.e. for all $r < 1$ we have $M_p(g; r) \leq \|f\|_{\mathcal{H}^p}$, implying that $\|g\|_{\mathcal{H}^p} \leq \|f\|_{\mathcal{H}^p}$. As in the proof of Theorem 7.11 the converse inequality $\|f\|_{\mathcal{H}^p} \leq \|g\|_{\mathcal{H}^p}$ follows from $|f(z)| \leq |g(z)|$. $\square$

Theorem 7.18. *Let $1 \leq p < \infty$ and $f \in \mathcal{H}^p$. For almost all $\varphi \in \mathbb{T}$ the radial limit*

$$f^*(\varphi) := \lim_{r \to 1} f\left(re^{i\varphi}\right) \tag{7.40}$$

exists, belongs to $L^p(\mathbb{T})$ and we have

$$\|f\|_{\mathcal{H}^p} = \frac{1}{(2\pi)^{\frac{1}{p}}} \|f^*\|_{L^p}. \tag{7.41}$$

Proof. By Theorem 7.3 we know that the result holds for $p = 2$. Since $\mathcal{H}^p \subset \mathcal{N}$ we can use the canonical factorization $f(z) = g(z)B_f(z)$ to define $G(z) := (g(z))^{\frac{p}{2}}$ (again we may choose any branch of $(g(\cdot))^{\frac{p}{2}}$). It follows that $G \in \mathcal{H}^2$ and for almost all φ we have

$$\lim_{r \to 1} G\left(re^{i\varphi}\right) = G^*(\varphi), \tag{7.42}$$

where we use a notation analogously to the proof of Theorem 7.3. We find that for those φ for which (7.42) holds we must have

$$\lim_{r \to \infty} g\left(re^{i\varphi}\right) = g^*\left(e^{i\varphi}\right)$$

and $\left|g^*\left(e^{i\varphi}\right)\right|^{\frac{p}{2}} = |G^*(\varphi)|$. Since $\lim_{r \to 1} B_f\left(re^{i\varphi}\right) = 1$ we deduce that

$$f^*(\varphi) = \lim_{r \to 1} f\left(re^{i\varphi}\right) \tag{7.43}$$

118

exists almost everywhere and $|f^*(\varphi)|^{\frac{p}{2}} = |G^*(\varphi)|$ almost everywhere which implies by Theorem 7.17 that

$$\|f\|_{\mathcal{H}^p}^p = \|g\|_{\mathcal{H}^p}^p = \|G\|_{\mathcal{H}^2}^2$$
$$= \frac{1}{2\pi}\|G^*\|_{L^2}^2 = \frac{1}{2\pi}\|f^*\|_{L^p}^p.$$

$\square$

Let $p \geq 2$ and $f \in \mathcal{H}^p$, in particular we have $f \in \mathcal{H}^2$ and therefore we can apply Theorem 7.3, i.e. $f^*(\varphi) = \sum_{k=0}^{\infty} A_k e^{ik\varphi}$ in $L^2(\mathbb{T})$ where $f(z) = \sum_{k=0}^{\infty} A_k z^k$. Thus for $r < 1$ we have $f(re^{i\varphi}) = (P_r * f^*)(\varphi)$. Since in addition $\lim_{r\to 1}(P_r * f^*)(\varphi) = f^*(\varphi)$ almost everywhere, the monotonicity of $r \mapsto M_p(f;r)$ combined with the dominated convergence theorem yields

Corollary 7.19. *For $p \geq 2$ and $f \in \mathcal{H}^p$ we have $f(re^{i\varphi}) = (P_r * f^*)(\varphi)$ and*

$$\lim_{r\to 1} \|P_r * f^* - f^*\|_{L^p(\mathbb{T})} = 0. \tag{7.44}$$

We now want to turn to $\mathcal{H}^1$ and we aim to get the result of Corollary 7.19 for $p = 1$ which will entail the result for the full range $1 \leq p < \infty$.

By Theorem 7.18 we know already the almost everywhere existence of the radial limits f^* of $f \in \mathcal{H}^p$, $1 \leq p < \infty$, as well as the fact that $f^* \in L^p(\mathbb{T})$ and the estimate $\|f\|_{\mathcal{H}^p} \leq \frac{1}{(2\pi)^{\frac{1}{p}}}\|f^*\|_{L^p}$. The following result will be helpful:

Proposition 7.20. *Every $f \in \mathcal{H}^1$ admits a factorization $f = h_1 \cdot h_2$ with $h_1, h_2 \in \mathcal{H}^2$.*

Proof. Let $f = gB_f$ be the canonical factorization of f. Since g has no zeroes in D the function $h_1(z) = (g(z))^{\frac{1}{2}}$ is well defined in D (and we may choose any branch). Clearly we have $|h_1(z)|^2 = |g(z)|$, i.e. $h_1 \in \mathcal{H}^2$ and $\|h_1\|_{\mathcal{H}^2}^2 = \|g\|_{\mathcal{H}^1}$. For $h_2(z) = g^{\frac{1}{2}}(z)B_f(z)$ we find again $|h_2(z)| = |g(z)|^{\frac{1}{2}}|B_f(z)|$ implying $h_2 \in \mathcal{H}^2$ and $\|h_2\|_{\mathcal{H}^2}^2 = \|g\|_{\mathcal{H}^1}$. Obviously we have $f = h_1 \cdot h_2$. $\square$

Theorem 7.21. *For $f \in \mathcal{H}^1$ with corresponding radial limit $f^* \in L^1(\mathbb{T})$ we have*

$$\lim_{r\to\infty} \int_0^{2\pi} \left|f(re^{i\varphi}) - f^*(e^{i\varphi})\right| d\varphi = 0. \tag{7.45}$$

119

Proof. As before, for $g : D \to \mathbb{C}$ the function g_r is defined by $g_r(\varphi) = g\left(re^{i\varphi}\right)$, $0 \leq r < 1$. Let $f = h_1 \cdot h_2$ with $h_1, h_2 \in \mathcal{H}^2$ as in Proposition 7.20. It follows that $f^* = h_1^* h_2^*$ almost everywhere and

$$
\begin{aligned}
f_r - f^* &= h_{1,r} h_{2,r} - h_1^* h_2^* \\
&= h_2^*(h_{1,r} - h_1^*) + h_{1,r}(h_{2,r} - h_2^*).
\end{aligned}
$$

This decomposition implies

$$
\int_0^{2\pi} \left| f\left(re^{i\varphi}\right) - f^*\left(e^{i\varphi}\right) \right| d\varphi
$$

$$
\leq \int_0^{2\pi} \left| h_2^*\left(e^{i\varphi}\right) \right| \left| h_{1,r}(\varphi) - h_1^*\left(e^{i\varphi}\right) \right| d\varphi
$$

$$
+ \int_0^{2\pi} \left| h_{1,r}(\varphi) \right| \left| h_{2,r}(\varphi) - h_2\left(e^{i\varphi}\right) \right| d\varphi
$$

$$
\leq \left(\int_0^{2\pi} \left| h_2^*(e^{i\varphi}) \right|^2 d\varphi \right)^{\frac{1}{2}} \left(\int_0^{2\pi} \left| h_{1,r}(\varphi) - h_1^*\left(e^{i\varphi}\right) \right|^2 d\varphi \right)^{\frac{1}{2}}
$$

$$
+ \left(\int_0^{2\pi} \left| h_{1,r}(\varphi) \right|^2 d\varphi \right)^{\frac{1}{2}} \left(\int_0^{2\pi} \left| h_{2,r}(\varphi) - h_2^*\left(e^{i\varphi}\right) \right|^2 d\varphi \right)^{\frac{1}{2}}.
$$

Since $h_2^* \in L^2(\mathbb{T})$ and $\left(\frac{1}{2\pi} \int_0^{2\pi} \left| h_1\left(re^{i\varphi}\right) \right|^2 d\varphi \right)^{\frac{1}{2}} \leq \|h_1\|_{\mathcal{H}^2}$ by Lemma 7.16, we may apply Corollary 7.19 for $p = 2$ to the functions h_1 and h_2 and our estimate gives

$$
\lim_{r \to 1} \int_0^{2\pi} \left| f\left(re^{i\varphi}\right) - f^*\left(e^{i\varphi}\right) \right| d\varphi = 0.
$$

$\square$

If $f \in \mathcal{H}^1$ and $f(z) = \sum_{k=0}^{\infty} A_k z^k$ we deduce from Theroem 7.21 that the Fourier coefficients of f^* must be equal to A_k for $k \geq 0$ and they must vanish for $-k \in \mathbb{N}$ and hence it follows

Corollary 7.22. *For $f \in \mathcal{H}^1$ we have $f_r(\varphi) = f\left(re^{i\varphi}\right) = (P_r * f^*)(\varphi)$ for some $f^* \in L^1(\mathbb{T})$.*

This corollary finally yields

Corollary 7.23. *If $f \in \mathcal{H}^p$, $1 \leq p < \infty$, then $f\left(re^{i\varphi}\right) = (P_r * f^*)(\varphi)$ for some $f^* \in L^p(\mathbb{T})$ and (7.44) holds.*

Remark 7.24. For $p > 1$ in Volume V we will provide a different proof to Corollary 7.23 using weak convergence.

We may interpret the combined results of Corollary 7.23 and Theorem 7.18 in different directions. We may start with $f^* \in L^p(\mathbb{T})$. Then, according to Theorem 7.12, we can extend f^* with the help of the Poisson integral to a function $f : D \to \mathbb{C}$ which belongs to L^p and for which $\operatorname{Re} f$ and $\operatorname{Im} f$ are harmonic in D, in particular they are C^∞-functions. The interesting question now is when is $\operatorname{Im} f$ the conjugate harmonic function to $\operatorname{Re} f$. In this case f would be holomorphic. The Fourier coefficients of f^* then will determine whether $f \in \mathcal{H}^p$. From (7.8) and (7.9) we may deduce sufficient conditions for $\operatorname{Re} f$ and $\operatorname{Im} f$ to be conjugate harmonic functions.

On the other hand we may start with $f \in \mathcal{H}^p$ and we will obtain Fourier series associated with radial boundary data of holomorphic functions belonging to $L^p(\mathbb{T})$.

Within our Course we just give an introduction to first ideas and results on (classical) Hardy spaces, and our main sources have been the monographs of Katznelson [73] and W. Rudin [103]. These spaces are of great importance in complex analysis and we refer to the standard references [27] and [77]. Hardy spaces had been extended to the n-dimensional Euclidean space by E. M. Stein and G. Weiss [115] and C. L. Fefferman and E. M. Stein [29], see for this also the monographs [116], [113] as well as [123] and [47] or [120].

Problems

1. Let $f : G \to \mathbb{C}$, $G \subset \mathbb{C}$ open, be a holomorphic function and assume for some open subset $\tilde{G} \subset G$ that $f(z) \neq 0$ for all $z \in \tilde{G}$. Prove that the function $(x, y) \mapsto \ln |f(x + iy)|$ is harmonic in $\tilde{G}$.

2.   a) Let $u_0 \in C(\mathbb{T})$ be a continuous function with absolutely convergent Fourier series $u_0(\varphi) = \sum_{k=1}^\infty a_k \cos k\varphi$, $\varphi \in \mathbb{T}$, representing u. Prove that $u(r, \varphi) = \sum_{k=1}^\infty a_k r^k \cos k\varphi$ is in $B_1(0)$ a harmonic function satisfying $\lim_{r \to 1} u(r, \varphi) = u_0(\varphi)$.

   b) Let $u_0 \in C(\mathbb{T})$ be a continuous function with absolutely convergent Fourier series $u_0(\varphi) = \sum_{k=1}^\infty b_k \sin k\varphi$ which we assume to represent u_0. Find a function u harmonic in $B_1(0)$ such that $\lim_{r \to \infty} \frac{\partial u(r, \varphi)}{\partial r} = u_0(\varphi)$.

3. Solve Exercise 7.15.

4. Let $f : B_1(0) \to \mathbb{C}$ be a bounded holomorphic function for which the radial limit $f^*(\varphi) = \lim_{r \to 1} f(re^{i\varphi})$ exists and is integrable. Suppose that f is not

identically zero and for $0 < r < 1$ define

$$\mu_r(f) := \frac{1}{2\pi} \int_0^{2\pi} \ln |f(re^{i\varphi})| d\varphi.$$

Show that $\mu_r(f) \le \mu_s(f), 0 < r < s < 1$.
Hint: use the Poisson-Jensen formula.

5. Let $f : B_1(0) \to \mathbb{C}$ be a non-constant bounded holomorphic function. Is it possible for this function to have zeroes at the points $z_k := \frac{k-1}{k}, k \in \mathbb{N}$?
Hint: note that $f \in \mathcal{N}$ and use Proposition 7.9.

6. Prove **Hardy's theorem:** if $f \in \mathcal{H}^1$ then $\sum_{k=1}^{\infty} \frac{|a_k|}{k} < \infty$.

8 Selected Topics on Fourier Series

In this chapter we discuss some topics, not necessarily related to each other, which go beyond a first introduction to Fourier series. Several of these topics will be picked up in later chapters or volumes. In some cases we will need some "auxiliary results" which are best discussed and proved in a different context, as for example the Riesz representation theorem which we prove in full detail in Volume V. In such a case we give a precise statement of the result, a reference to existing literature, but we will postpone the proof.

We start with investigating the Fourier series associated with a finite (positive) measure on $\mathbb{T} = S^1$. A Borel measure μ on $\mathbb{T}$ is finite or bounded if $\mu(\mathbb{T}) = \|\mu\| < \infty$ and we call $\|\mu\|$ the total mass of μ. If $\|\mu\| = 1$ we call μ a probability measure. The set of all (positive) bounded Borel measures on $\mathbb{T}$ is denoted by $\mathcal{M}_b^+(\mathbb{T})$. Since $\left|e^{ikt}\right| = 1$ for all $k \in \mathbb{Z}$ and $t \in \mathbb{R}$ we can define the **Fourier-Stieltjes coefficients of a measure** $\mu \in \mathcal{M}_b^+(\mathbb{T})$ by

$$\hat{\mu}_k := (F\mu)(k) := \frac{1}{2\pi} \int_{-\pi}^{\pi} e^{-ikt} \mu(\mathrm{d}t). \tag{8.1}$$

Since

$$|\hat{\mu}_k| \leq \frac{1}{2\pi} \|\mu\| \tag{8.2}$$

the sequence $(\hat{\mu}_k)_{k \in \mathbb{Z}}$ is clearly bounded. Thus we can associate with each bounded Borel measure $\mu \in \mathcal{M}_b^+(\mathbb{T})$ a sequence $(\hat{\mu}_k)_{k \in \mathbb{Z}} \in l_\infty(\mathbb{Z})$ and therefore a formal series

$$\mu \sim \sum_{k \in \mathbb{Z}} \hat{\mu}_k e^{ikt} \tag{8.3}$$

which we call the **Fourier-Stieltjes series** associated with μ. However, at the moment we do not know anything about the convergence or summability of (8.3). Our first question is whether we can characterize the sequences $(\hat{\mu}_k)_{k \in \mathbb{Z}}$ forming the Fourier-Stieltjes coefficients of bounded Borel measures. We need

Definition 8.1. *A sequence $(c_k)_{k \in \mathbb{Z}}$ of complex numbers is called **positive definite** if for every $N \in \mathbb{N}$ and every selection of integers $\xi_1, \ldots, \xi_N \in \mathbb{Z}$ the matrix $(c_{\xi_k - \xi_l})_{k,l=1,\ldots,N}$ is positive semidefinite, i.e. for all $\lambda_1, \ldots, \lambda_N \in \mathbb{C}$ we have*

$$\sum_{k,l=1}^{N} c_{\xi_k - \xi_l} \lambda_k \overline{\lambda}_l \geq 0. \tag{8.4}$$

Recall that a matrix $(a_{kl})_{k,l=1,\ldots,N}$, $a_{kl} \in \mathbb{C}$, is called positive semidefinite or **positive Hermitian** if $a_{kl} = \overline{a_{lk}}$ and for all $\lambda_1, \ldots, \lambda_N \in \mathbb{C}$

$$\sum_{k,l=1}^{N} a_{kl} \lambda_k \overline{\lambda}_l \geq 0 \tag{8.5}$$

holds.

Exercise 8.2. *Prove that $(c_k)_{k \in \mathbb{Z}}$ is positive definite if and only if for every $N \in \mathbb{N}$ and all choices of $\lambda_{-N}, \ldots, \lambda_N \in \mathbb{C}$ we have*

$$\sum_{k,l=-N}^{N} c_{k-l} \lambda_k \overline{\lambda}_l \geq 0. \tag{8.6}$$

It is clear that if $(c_k)_{k \in \mathbb{Z}}$ and $(d_k)_{k \in \mathbb{Z}}$ are two positive definite sequences that $(c_k + d_k)_{k \in \mathbb{Z}}$ is also positive definite as in $(\eta c_k)_{k \in \mathbb{Z}}$ for $\eta \geq 0$. Moreover, if for each $n \in \mathbb{N}$ a positive definite sequence $\left(c_k^{(n)}\right)_{k \in \mathbb{Z}}$ is given and $\lim_{n \to \infty} c_k^{(n)} = c_k$ for every $k \in \mathbb{Z}$ with some $c_k \in \mathbb{C}$, then $(c_k)_{k \in \mathbb{Z}}$ is also positive definite, i.e. the componentwise (pointwise) limit of a sequence of positive definite sequences is positive definite. This follows immediately from (8.5). We will need

Lemma 8.3. *Let $\alpha = (\alpha_{kl})_{k,l=1,\ldots,N}$ and $\beta = (\beta_{kl})_{k,l=1,\ldots,N}$ be two positive Hermitian matrices and define $\gamma = (\gamma_{kl})_{k,l=1,\ldots,N}$ by $\gamma_{kl} = \alpha_{kl}\beta_{kl}$. Then γ is a positive Hermitian matrix.*

Proof. For β there exists a positive Hermitian matrix $\delta = (\delta_{kl})_{k,l=1,\ldots,N}$ such that $\beta = \delta\delta^*$ where $\delta_{kl}^* = \overline{\delta}_{kl}$, i.e. $\beta_{kl} = \sum_{j=1}^{N} \delta_{kj}\overline{\delta}_{lj}$. Now we find with $\lambda_1, \ldots, \lambda_N \in \mathbb{C}$ that

$$\sum_{k,l=1}^{N} \alpha_{kl}\beta_{kl}\lambda_k\overline{\lambda}_l = \sum_{j=1}^{N} \left(\sum_{k,l=1}^{N} \alpha_{kl}(\delta_{kj}\lambda_k)\overline{(\delta_{lj}\lambda_l)} \right) \geq 0.$$

$\square$

Proposition 8.4. *For positive definite sequences $(c_k)_{k \in \mathbb{Z}}$ and $(d_k)_{k \in \mathbb{Z}}$ we have*

$$c_0 \geq 0 \quad and \quad c_{-k} = \overline{c}_k; \tag{8.7}$$

$$|c_k| \leq c_0; \tag{8.8}$$

$$(\overline{c}_k)_{k \in \mathbb{Z}} \ and \ (\mathrm{Re}\, c_k)_{k \in \mathbb{Z}} \ are \ positive \ definite, \tag{8.9}$$

$$(c_k \cdot d_k)_{k \in \mathbb{Z}} \ is \ positive \ definite. \tag{8.10}$$

Proof. The statement (8.7) is clear from the properties of positive Hermitian matrices. To see (8.8) we choose in (8.4) for $N = 2$ the values $\xi_1 = 0$ and $\xi_2 = k \in \mathbb{Z}$. It follows that the matrix $\begin{pmatrix} c_0 & c_{-k} \\ c_k & c_0 \end{pmatrix}$ must be positive Hermitian, hence $\det \begin{pmatrix} c_0 & c_{-k} \\ c_k & c_0 \end{pmatrix} \geq 0$, but $\det \begin{pmatrix} c_0 & c_{-k} \\ c_k & c_0 \end{pmatrix} = c_0^2 - |c_k|^2$ by (8.7), hence (8.8) follows. The fact that $(\bar{c}_k)_{k \in \mathbb{Z}}$ is positive definite we deduce immediately from (8.4) which in turn by $\operatorname{Re} c_k = \frac{c_k + \bar{c}_k}{2}$ implies that $(\operatorname{Re} c_k)_{k \in \mathbb{Z}}$ is positive definite. Finally we note that (8.10) is a direct consequence of Lemma 8.3. $\qquad \square$

We return to the Fourier-Stieltjes coefficients of a bounded Borel measure $\mu \in \mathcal{M}_b^+(\mathbb{T})$ and prove

Lemma 8.5. *The sequence $(\hat{\mu}_k)_{k \in \mathbb{Z}}$ of Fourier-Stieltjes coefficients of a bounded Borel measure is positive definite.*

Proof. We fix $N \in \mathbb{N}$ and choose $\xi_1, \ldots, \xi_N \in \mathbb{Z}$ as well as $\lambda_1, \ldots, \lambda_N \in \mathbb{C}$ to find that

$$
\sum_{k,l=1}^{N} c_{\xi_k - \xi_l} \lambda_k \overline{\lambda_l} = \sum_{k,l=1}^{N} \frac{1}{2\pi} \int_{-\pi}^{\pi} e^{-i(\xi_k - \xi_l)t} \mu(\mathrm{d}t) \lambda_k \overline{\lambda_l}
$$

$$
= \sum_{k,l=1}^{N} \frac{1}{2\pi} \int_{-\pi}^{\pi} \left(e^{-i\xi_k t} \lambda_k \right) \overline{\left(e^{-i\xi_l t} \lambda_l \right)} \mu(\mathrm{d}t)
$$

$$
= \frac{1}{2\pi} \int_{-\pi}^{\pi} \left| \sum_{k=1}^{N} e^{-i\xi_k t} \lambda_k \right|^2 \mu(\mathrm{d}t) \geq 0.
$$

$\qquad \square$

Example 8.6. Let $(\nu_m)_{m \in \mathbb{N}}$ be a sequence of non-negative real numbers such that $\sum_{m=1}^{\infty} \nu_m = \nu < \infty$ and denote for $t_0 \in [-\pi, \pi)$ by ϵ_{t_0} the unit mass at t_0, i.e. for $f \in C(\mathbb{T})$ we have $\int_{-\pi}^{\pi} f(t) \epsilon_{t_0}(\mathrm{d}t) = f(t_0)$. It follows that $\mu := \sum_{m=1}^{\infty} \nu_m \epsilon_{t_m}$, $t_m \in [-\pi, \pi)$, is a finite Borel measure on $\mathbb{T}$ with total mass $\|\mu\| = \nu$. For $k \in \mathbb{Z}$ we find

$$
\hat{\mu}_k = \frac{1}{2\pi} \int_{-\pi}^{\pi} e^{-ikt} \mu(\mathrm{d}t) = \frac{1}{2\pi} \int_{-\pi}^{\pi} e^{-ikt} \sum_{m=1}^{\infty} \nu_m \epsilon_{t_m}(\mathrm{d}t)
$$

$$
= \frac{1}{2\pi} \sum_{m=1}^{\infty} \nu_m \int_{-\pi}^{\pi} e^{-ikt} \epsilon_{t_m}(\mathrm{d}t) = \frac{1}{2\pi} \sum_{m=1}^{\infty} \nu_m e^{-ikt_m}.
$$

In particular, for $\nu_k = 0$ if $k > 1$ and $\nu_1 = 1$, we have for $t_1 = 0$ that

$$\hat{\epsilon}_k = \frac{1}{2\pi} \tag{8.11}$$

and for general $t_1 \in [-\pi, \pi)$ it follows

$$\hat{\mu}_k = (\hat{\epsilon_{t_1}})_k = \frac{1}{2\pi} e^{-ikt_1}. \tag{8.12}$$

Example 8.7. Let $f \in L^1(\mathbb{T})$, $f \geq 0$ almost everywhere. We can interpret f as a Radon-Nikodym density with respect to $\lambda_{\mathbb{T}}$ of a measure μ, i.e. $\mu = f\lambda_{\mathbb{T}}$, and the integrability of f ensures μ is a bounded Borel measure. For the Fourier-Stieltjes coefficients $\hat{\mu}_k$ of μ we find

$$\hat{\mu}_k = \frac{1}{2\pi} \int_{-\pi}^{\pi} e^{-ikt} \mu(\mathrm{d}t) = \frac{1}{2\pi} \int_{-\pi}^{\pi} e^{-ikt} f(t)\,\mathrm{d}t = c_k,$$

or

$$(F\mu)(k) = (Ff)(k), \tag{8.13}$$

in other words $\hat{\mu}_k$ is the kth Fourier coefficients c_k of f.

Let $\mu \in \mathcal{M}_b^+(\mathbb{T})$ and $f \in C(\mathbb{T})$. The mapping

$$l_\mu : C(\mathbb{T}) \to \mathbb{C}, \qquad l_\mu(f) = \int_{-\pi}^{\pi} f(t)\,\mathrm{d}t, \tag{8.14}$$

is linear and positivity preserving in the sense that $f \geq 0$ implies $l_\mu(f) \geq 0$. Moreover we have

$$|l_\mu(f)| \leq \|f\|_\infty \|\mu\| \tag{8.15}$$

which implies that for a sequence of functions $f_k \in C(\mathbb{T})$ converging uniformly to $f \in C(\mathbb{T})$ the sequence $(l_\mu(f_k))_{k \in \mathbb{N}}$ converges in $\mathbb{C}$ to $l_\mu(f)$. In this sense l_μ defines a continuous linear functional on the Banach space $(C(\mathbb{T}), \|\cdot\|_\infty)$ which is positivity preserving. The following **Riesz representation theorem** which we will prove in Volume V states that every continuous linear and positivity preserving functional on $C(\mathbb{T})$ is of the type (8.14).

Theorem 8.8. *Let $l : C(\mathbb{T}) \to \mathbb{R}$ be a linear, continuous and positivity preserving functional. Then there exists a unique Borel measure $\mu \in \mathcal{M}_b^+(\mathbb{T})$ such that*

$$l(f) = \int_{-\pi}^{\pi} f(t)\mu(\mathrm{d}t) \tag{8.16}$$

holds for all $f \in C(\mathbb{T})$.

With the help of the Riesz representation theorem we can now prove the converse of Lemma 8.5

Theorem 8.9 (Herglotz-Bochner). *For every positive definite sequence $(c_k)_{k \in \mathbb{Z}}$ exists a unique Borel measure $\mu \in \mathcal{M}_b^+(\mathbb{T})$ such that $\bar{c}_k = 2\pi \hat{\mu}_k$.*

Proof. First we note that for a sequence $\alpha = (\alpha_k)_{k \in \mathbb{Z}}$ we can define "partial sums" $S_N(\alpha)(t) = \sum_{|k| \leq N} \alpha_k e^{ikt}$ and corresponding Cesàro sums

$$\sigma_{N+1}(\alpha)(t) = \frac{1}{N+1} \sum_{k=0}^{N} S_k(\alpha)(t) = \sum_{|k| \leq N} \left(1 - \frac{|k|}{N+1}\right) \alpha_k e^{ikt}. \tag{8.17}$$

Given $(c_k)_{k \in \mathbb{Z}}$ we define for fixed $N \in \mathbb{N}_0$

$$\lambda_k := \begin{cases} e^{-ikt}, & |k| \leq N \\ 0, & |k| > N \end{cases}, \tag{8.18}$$

and since $(c_k)_{k \in \mathbb{Z}}$ is a positive definite sequence we find

$$\sum_{k,l=-N}^{N} c_{k-l} \lambda_k \bar{\lambda}_l = \sum_{k,l=-N}^{N} c_{k-l} e^{-i(k-l)t} = \sum_k \eta_k c_k e^{-ikt}$$

where η_k is the number of ways we can represent k as the difference between two integers m, n with $|m|, |n| \leq N$, i.e. $\eta_k = \max(2N+1-|k|, 0)$. This implies further that

$$0 \leq \sum_{k,l=-N}^{N} c_{k-l} \lambda_k \bar{\lambda}_l = \sum_k \max(2N+1-|k|, 0) c_k e^{-ikt} = (2N+1)\sigma_{2N}(c)(t). \tag{8.19}$$

From the definition of $\sigma_{2N}(c)$ it follows that

$$2\pi \int_{-\pi}^{\pi} \sigma_{2N}(c)(t)\, dt = \frac{2\pi}{2N+1} \sum_{k=0}^{2N} \sum_{|l| \leq k} \int_{-\pi}^{\pi} c_l e^{ilt}\, dt = c_0. \tag{8.20}$$

Let $q(t) = \sum_{|k| \leq M} \gamma_k e^{ikt}$ be a trigonometric polynomial and define

$$L_c(q) := \sum_{|k| \leq M} \bar{c}_k \gamma_k. \tag{8.21}$$

For $N \geq M$ we find

$$\int_{-\pi}^{\pi} q(t)\sigma_{2N}(\overline{c})(t)\, dt = \int_{-\pi}^{\pi} \sum_{|k| \leq M} \gamma_k e^{ikt} \sum_{l=0}^{2N} \frac{1}{2N+1} \sum_{|j| \leq l} \overline{c}_j e^{-ijt}\, dt$$

$$= \sum_{|k| \leq M} \sum_{l=0}^{2N} \sum_{|j| \leq l} \frac{1}{2N+1} \gamma_k \overline{c}_j \int_{-\pi}^{\pi} e^{i(k-j)t}\, dt$$

$$= 2\pi \sum_{|k| \leq M} \overline{c}_k \gamma_k$$

and we deduce that

$$L_c(q) = \frac{1}{2\pi} \lim_{N \to \infty} \int_{-\pi}^{\pi} q(t)\sigma_{2N}(\overline{c})(t)\, dt. \tag{8.22}$$

Since

$$\left| \frac{1}{2\pi} \int_{-\pi}^{\pi} q(t)\sigma_{2N}(\overline{c})(t)\, dt \right| \leq \sup_{t \in [-\pi,\pi]} |q(t)| c_0 = c_0 \|q\|_\infty$$

we can use the Weierstrass approximation theorem in form of Remark 4.17 to now extend L_c to $C(\mathbb{T})$: take $f \in C(\mathbb{T})$ and note that by Remark 4.17 the trigonometric polynomial $\sigma_{2N}(f)(t)$ converges uniformly to f. Define

$$L_c(f) := \frac{1}{2\pi} \lim_{N \to \infty} L_c(\sigma_{2N}(f)) \tag{8.23}$$

to find

$$|L_c(f)| \leq c_0 \|f\|_\infty. \tag{8.24}$$

Using

$$\sigma_{2N}(f)(t) = \sum_{l=0}^{2N} \frac{1}{2N+1} \sum_{|j| \leq l} (Ff)(j)e^{ijt}$$

and

$$L_c(\sigma_{2N}(f)) = \frac{1}{2N+1} \sum_{k \in \mathbb{Z}} \max(2N+1-|k|, 0)(Ff)(k)\overline{c}_k$$

$$= \frac{1}{2N+1} \sum_{|k| \leq 2N} (2N+1-|k|)(Ff)(k)\overline{c}_k$$

$$= \sum_{|k| \leq 2N} \left(1 - \frac{|k|}{2N+1}\right)(Ff)(k)\overline{c}_k$$

128

we arrive for $f \geq 0$ at

$$L_c(f) = \frac{1}{2\pi} \lim_{N \to \infty} \sum_{|k| \leq 2N} \left(1 - \frac{|k|}{2N+1}\right) \int_{-\pi}^{\pi} f(t)e^{-ikt}\, dt\, \bar{c}_k \qquad (8.25)$$

$$= \frac{1}{2\pi} \lim_{N \to \infty} \int_{-\pi}^{\pi} f(t)\sigma_{2N}(\bar{c})(t)\, dt \geq 0,$$

since by (8.9) and (8.19) we know that $\sigma_{2N}(\bar{c})(t) > 0$.
Thus L_c defines a linear continuous functional on $C(\mathbb{T})$ which is positivity preserving, hence there exists a unique bounded Borel measure $\mu \in \mathcal{M}_b^+(\mathbb{T})$ such that

$$L_c(f) = \int_{-\pi}^{\pi} f(t)\mu(\mathrm{d}t).$$

For $f(t) = e^{-ikt}$ we obtain

$$L_c\left(e^{-ik\cdot}\right) = \int_{-\pi}^{\pi} e^{-ikt}\mu(\mathrm{d}t) = 2\pi\hat{\mu}_k,$$

and by the very definition of L_c, i.e. (8.21), we have with $q(t) = e^{-ikt}$ that $L_c\left(e^{-ik\cdot}\right) = \bar{c}_k$. It remains to prove the uniqueness of μ. Suppose that for μ_1 and μ_2 we have $\hat{\mu}_{1,k} = \hat{\mu}_{2,k}$ for all $k \in \mathbb{Z}$. This implies that for every trigonometric polynomial q we have $\int_{\mathbb{T}} q(t)\mu_1(\mathrm{d}t) = \int_{\mathbb{T}} q(t)\mu_2(t)$ and since the trigonometric polynomials are dense in the Banach space $(C(\mathbb{T}), \|\cdot\|_\infty)$ it follows $\int_{\mathbb{T}} f(t)\mu_1(\mathrm{d}t) = \int_{\mathbb{T}} f(t)\mu_2(\mathrm{d}t)$ for all $f \in C(\mathbb{T})$ implying by the uniqueness part of the Riesz representation theorem (or by a direct argument using measure theory) that $\mu_1 = \mu_2$. ☐

Remark 8.10. This is a remark for readers who have already some background in functional analysis. It looks artificial to link $\hat{\mu}_k$ to $\bar{c}_k$ and not to c_k. The advantage comes when dealing with duality in Banach spaces over $\mathbb{C}$. It is often more convenient to consider the duality pairing $\langle \cdot, \cdot \rangle : X \times X^* \to \mathbb{C}$ for a Banach space X and its dual X^* to be conjugate linear in the second argument, i.e. $\langle x, \alpha y^* \rangle = \bar{\alpha}\langle x, y^* \rangle$, as we have it in the case of the scalar product in $L^2(\mathbb{T})$ over $\mathbb{C}$. We can now read (8.25) as

$$L_c(f) = \lim_{N \to \infty} \sum_{|k| \leq 2N} \left(1 - \frac{|k|}{2N+1}\right) (Ff)(k)\overline{\hat{\mu}}_k \qquad (8.26)$$

or with the duality pairing $C_b(\mathbb{T}) \times \mathcal{M}_b(\mathbb{T})$, where $\mathcal{M}_b(\mathbb{T})$ denotes the bounded complex measures on $\mathbb{T}$,

$$\langle f, \mu \rangle = \lim_{N \to \infty} \sum_{|k| \leq 2N} \left(1 - \frac{|k|}{2N+1}\right) (Ff)(k)\overline{\hat{\mu}}_k. \qquad (8.27)$$

We have $\langle f, \mu \rangle = \int_{\mathbb{T}} \overline{f(t)} \mu(\mathrm{d}t)$ and in particular if the series $\sum_{k \in \mathbb{Z}} (Ff)(k) \overline{\hat{\mu}}_k$ is Cesàro summable we find an extension of **Parseval's equality**

$$\langle f, \mu \rangle = \frac{1}{2\pi} \sum_{k \in \mathbb{Z}} (Ff)(k) \overline{\hat{\mu}}_k. \tag{8.28}$$

Further, for $f \in L^2(\mathbb{T})$ and $\mu = g\lambda_{\mathbb{T}}$, $g \in L^2(\mathbb{T})$ and $g \geq 0$ almost everywhere we find $\hat{\mu}_k = (Fg)(k)$ and

$$\langle f, g \rangle := \int_{\mathbb{T}} f(t) \overline{g(t)} \, \mathrm{d}t = \frac{1}{2\pi} \sum_{k \in \mathbb{Z}} (Ff)(k) \overline{(Fg)(k)}.$$

Let $\mu \in \mathcal{M}_b^+(\mathbb{T})$ with Fourier-Stieltjes coefficients $\hat{\mu}_k$. Since the sequence $(\hat{\mu}_k)_{k \in \mathbb{Z}}$ is a positive definite sequence we find by (2.29) for $0 \leq r < 1$ that

$$u(r, \varphi) = \sum_{k \in \mathbb{Z}} r^{|k|} \hat{\mu}_k e^{ik\varphi} = \hat{\mu}_0 + \sum_{k=1}^{\infty} r^k \hat{\mu}_k \left(a_k \cos k\varphi + b_k \sin k\varphi \right)$$

with a_k and b_k calculated from (3.8) with $c_k = \hat{\mu}_k$, where we used that $\hat{\mu}_k = \overline{\hat{\mu}_{-k}}$. It follows that u is harmonic in D. On the other hand the definition of the Poisson kernel P_r and the uniform convergence of $\sum_{k \in \mathbb{Z}} r^{|k|} e^{ik\varphi}$ on compact subsets of D yields

$$
\begin{aligned}
(P_r * \mu)(\varphi) &:= \frac{1}{2\pi} \int_{-\pi}^{\pi} \frac{1 - r^2}{1 - 2r \cos(\varphi - \vartheta) + r^2} \mu(\mathrm{d}\vartheta) \\
&= \frac{1}{2\pi} \int_{-\pi}^{\pi} \sum_{k \in \mathbb{Z}} r^{|k|} e^{i(\varphi - \vartheta)k} \mu(\mathrm{d}\vartheta) \\
&= \sum_{k \in \mathbb{Z}} r^{|k|} e^{i\varphi k} \frac{1}{2\pi} \int_{-\pi}^{\pi} e^{-ik\vartheta} \mu(\mathrm{d}\vartheta) \\
&= \sum_{k \in \mathbb{Z}} r^{|k|} \hat{\mu}_k e^{ik\varphi}.
\end{aligned}
$$

Thus for every positive Borel measure $\mu \in \mathcal{M}_b^+(\mathbb{T})$ a non-negative harmonic function in D is given by

$$u(r, \varphi) = (P_r * \mu)(\vartheta) = \frac{1}{2\pi} \int_{-\pi}^{\pi} \frac{1 - r^2}{1 - 2r \cos(\varphi - \vartheta) + r^2} \mu(\mathrm{d}\vartheta). \tag{8.29}$$

Note that the harmonicity of u can also be directly proved when differentiating in $P_r * \mu$ under the integral sign and noting that the Poisson kernel is a harmonic function itself.

We will return to the Herglotz-Bochner theorem and properties of non-negative harmonic function several times in this and the following volumes.

The next observation picks up (8.28) and combines it with Bessel potential spaces introduced in Definition 5.20. Let $\mu \in \mathcal{M}_b^+(\mathbb{T})$ with Fourier-Stieltjes coefficients $\hat{\mu}_k$. Since $|\hat{\mu}_k| \leq \hat{\mu}_0$ we find for every $s < -\frac{1}{2}$

$$\sum_{k \in \mathbb{Z}} |\hat{\mu}_k|^2 \left(1 + |k|^2\right)^{-s} \leq |\hat{\mu}_0|^2 \sum_{k \in \mathbb{Z}} \left(1 + |k|^2\right)^{-s}$$

$$\leq |\hat{\mu}_0|^2 \left(1 + 2 \sum_{k \in \mathbb{N}} |k|^{-2s}\right) < \infty.$$

If we introduce for $s \in \mathbb{R}$

$$\|u\|_{H^s(\mathbb{T})} = \left(\sum_{k \in \mathbb{Z}} |(Fu)(k)|^2 \left(1 + |k|^2\right)^s\right)^{\frac{1}{2}} \tag{8.30}$$

we find for $s > \frac{1}{2}$, $\mu \in \mathcal{M}_b(\mathbb{T})$ and $f \in H^s(\mathbb{T})$, where $f \in H^s(\mathbb{T})$ if and only if $\|f\|_{H^s(\mathbb{T})} < \infty$, that

$$|\langle f, \mu \rangle| = \frac{1}{2\pi} \left|\sum_{k \in \mathbb{Z}} (Ff)(k)\overline{\hat{\mu}}_k\right| \leq \frac{1}{2\pi} \|f\|_{H^s(\mathbb{T})} \|\mu\|_{H^{-s}(\mathbb{T})}. \tag{8.31}$$

There are many questions left open: for which objects can we define $\|\cdot\|_{H^s(\mathbb{T})}$ when $s < 0$, what are the properties of the set $H^s(\mathbb{T})$ for $s < 0$, why does the first equality in (8.31) hold, etc. Eventually we will resolve these problems and understand why these considerations allow us not only to extend the "objects" (functions, measures) defined on $\mathbb{T}$, but also to define derivatives of measures and more general objects, i.e. distributions in the sense of L. Schwartz, see in particular Volume V.

Measures are rather rough objects while holomorphic functions are quite smooth. We want to study the decay of the Fourier coefficients of holomorphic functions. Thus we consider 2π-periodic functions $f : \mathbb{R} \to \mathbb{C}$ which are restrictions of some holomorphic functions $h : G \to \mathbb{C}$ where G is a region in $\mathbb{C}$ and $[-\pi, \pi] \subset G$. Such a function f is integrable over $[-\pi, \pi]$, is arbitrarily often differentiable, and for $m \in \mathbb{N}_0$ we must have $f^{(m)}(-\pi) = f^{(m)}(\pi)$. Furthermore we find for every $m \in \mathbb{N}_0$ the estimate

$$|(Ff)(k)| \leq \frac{1}{|k|^m} \left\|f^{(m)}\right\|_\infty. \tag{8.32}$$

The Cauchy inequalities, Theorem III.21.15, imply the existence of some $\gamma > 0$ such that

$$\left\| f^{(m)} \right\|_\infty \leq \gamma^m m!, \tag{8.33}$$

and therefore we obtain for all $k \in \mathbb{Z}$ and all $m \in \mathbb{N}_0$

$$|(Ff)(k)| \leq \frac{\gamma^m m!}{|k|^m} \leq \left(\frac{\gamma m}{|k|} \right)^m. \tag{8.34}$$

Let $a := \frac{\gamma}{\rho}$ with $\rho > 0$ such that $\frac{\gamma}{\rho} < 1$. We choose m_0 such that $m_0 \leq \frac{|k|}{\rho} < m_0 + 1$ which yields

$$|(Ff)(k)| \leq \left(\frac{\gamma m_0}{|k|} \right)^{m_0} \leq \left(\frac{\gamma}{\rho} \right)^{m_0}$$
$$= a^{m_0} = \frac{1}{a} a^{m_0+1} < \frac{1}{a} a^{\frac{|k|}{\rho}}.$$

We now choose $\alpha > 0$ such that $a^{\frac{1}{\rho}} < e^{-\alpha} < 1$ and find with $\beta = \frac{1}{\alpha}$ that

$$|(Ff)(k)| \leq \beta e^{-\alpha|k|}, \quad \alpha > 0. \tag{8.35}$$

Thus we have proved one direction of

Theorem 8.11. *Let $f : \mathbb{R} \to \mathbb{C}$ be a 2π-periodic function which is the restriction of a holomorphic function $h : G \to \mathbb{C}$, $[-\pi, \pi] \subset G$ and G is a region in $\mathbb{C}$. Then estimate (8.35) holds for the Fourier coefficients of f. Conversely, if for a 2π-periodic function $f : \mathbb{R} \to \mathbb{C}$ integrable over $[-\pi, \pi]$ the estimates (8.35) hold for all $k \in \mathbb{Z}$ then there exists a holomorphic function h defined in an open (complex) neighbourhood of $[-\pi, \pi]$ such that $h|_{[-\pi,\pi]} = f$.*

Proof. It remains to prove the last statement. Since $[-\pi, \pi]$ is compact it is sufficient to show to that for every $t_0 \in [-\pi, \pi]$ there exists $R = R(t_0) > 0$ such that in $(-R + t_0, t_0 + R)$ we have

$$f(t) = \sum_{m=0}^{\infty} \frac{f^{(m)}(t_0)}{m!} (t - t_0)^m. \tag{8.36}$$

First we note due to (8.35) the series $\sum_{k \in \mathbb{Z}} (Ff)(k) e^{ikt}$ and the series $\sum_{k \in \mathbb{Z}} \frac{d^m}{dt^m} \left((Ff)(k) e^{ikt} \right)$ converge in $[-\pi, \pi]$ absolutely and uniformly implying that f is arbitrarily often differentiable and

$$\frac{d^m}{dt^m} f(t) = \sum_{k \in \mathbb{Z}} (Ff)(k) (ik)^m e^{ikt}.$$

132

From (8.35) we deduce

$$\left|f^{(m)}(t)\right| \leq \beta \sum_{k \in \mathbb{N}} |k|^m e^{-\alpha|k|} < \infty.$$

The integral test for series yields when looking at $\int_0^\infty s^m e^{-\alpha s}\, ds$ that

$$\left|f^{(m)}(t)\right| \leq 2\beta \int_0^\infty s^m e^{-\alpha s}\, ds \leq 2\frac{\beta}{\alpha}\frac{m!}{\alpha^m}, \tag{8.37}$$

where we used that

$$\int_0^\infty s^m e^{-\alpha s}\, ds = \frac{1}{\alpha}\int_0^\infty \left(\frac{t}{\alpha}\right)^m e^{-t}\, dt = \frac{1}{\alpha^{m+1}}\int_0^\infty t^m e^{-t}\, dt$$

$$= \frac{1}{\alpha}\frac{\Gamma(m+1)}{\alpha^m} = \frac{1}{\alpha}\frac{m!}{\alpha^m}.$$

Now we find for $t_0 \in [-\pi, \pi]$ that

$$\sum_{m=0}^\infty \frac{\left|f^{(m)}(t_0)\right|}{m!}|t-t_0|^m \leq \frac{2\beta}{\alpha}\sum_{m=0}^\infty \left|\frac{t-t_0}{\alpha}\right|^m$$

which converges for $|t-t_0| < \alpha$. Thus in the neighbourhood of every $t_0 \in [-\pi, \pi]$ we can expand f into a convergent power series. For the remainder term we find by (I.29.16)

$$\left|R_{f,t_0}^{(m+1)}(t)\right| \leq 2\frac{\beta}{\alpha}\frac{(m+1)!}{\alpha^{m+1}} \cdot \frac{|t-t_0|^{m+1}}{(m+1)!} = 2\frac{\beta}{\alpha}\left|\frac{t-t_0}{\alpha}\right|^{m+1}$$

which tends to 0 for $m \to \infty$ and $|t-t_0| < \alpha$. It follows that the series $\sum_{m=0}^\infty \frac{f^{(m)}(t_0)}{m!}(t-t_0)^m$ is indeed in $|t-t_0| < \alpha$ the Taylor series of f. □

We want to explain a general phenomenon, known as the **Gibbs phenomenon**, by discussing the Fourier series associated with the 2π-periodic extension of the function

$$u : [-\pi, \pi) \to \mathbb{R}, \quad u(x) = \begin{cases} \frac{-x-\pi}{2}, & x \in [-\pi, 0) \\ \frac{\pi-x}{2}, & x \in [0, \pi), \end{cases} \tag{8.38}$$

compare with Example 2.8 and Figure 2.2. The Fourier series associated with u is

$$\sum_{k=1}^\infty \frac{\sin kx}{k} \tag{8.39}$$

which for $x \in [-\pi, \pi) \setminus \{0\}$ converges pointwisely to $u(x)$, whereas for $x = 0$ it converges to $0 \neq u(0) = \frac{\pi}{2}$, note $0 = \frac{u(0^+) + u(0^-)}{2}$, see Theorem 4.6. Moreover, by Theorem 6.8 we know that for every δ, $0 < \delta < \pi$, the convergence of $\sum_{k=1}^{\infty} \frac{\sin kx}{k}$ in $(\delta, \pi - \delta)$ and in $(-\pi + \delta, -\delta)$ is uniformly.

Given $\epsilon > 0$ we can find $N_0 \in \mathbb{N}$ such that $N \geq N_0$ implies that $|S_N(0) - 0| < \epsilon$, however we cannot expect for $x \in (0, \pi)$ close to 0 the term $|S_N(x) - u(x)|$ to be small. We will prove that $\lim_{N \to \infty} S_N(u) \left(\frac{\pi}{N} \right) > u(0) = \frac{\pi}{2}$.

For the partial sum $S_N(u)(x) = \sum_{k=1}^{N} \frac{\sin kx}{k}$ and $x \in (0, \pi)$ we find

$$S_N(u)(x) = \int_0^x \sum_{k=1}^{N} \cos kt \, dt$$

$$= \frac{1}{2} \int_0^x (D_N(t) - 1) \, dt = -\frac{x}{2} + \frac{1}{2} \int_0^x D_N(t) \, dt.$$

We note, see Problem 6, that

$$\frac{1}{2} D_N(t) = \frac{\sin \left(N + \frac{1}{2} \right) t}{2 \sin \frac{t}{2}} = \frac{1}{2} \sin Nt \cot \frac{t}{2} + \frac{1}{2} \cos Nt$$

$$= \frac{\sin Nt}{t} + \tilde{g}(t) \sin Nt + \frac{1}{2} \cos Nt$$

where

$$\tilde{g}(t) = \frac{1}{2} \cot \frac{t}{2} - \frac{1}{t}, \quad t \in (0, \pi).$$

The rules of de l'Hospital give

$$\lim_{t \to 0} \left(\frac{1}{2} \cot \frac{t}{2} - \frac{1}{t} \right) = 0$$

and it follows that we can extend $\tilde{g}$ to 0 and π by setting $\tilde{g}(0) = 0$ as well as $\tilde{g}(\pi) = 0$. Now we may consider the 2π-periodic extensions g of $\tilde{g}$. Since $\cot \frac{\pi}{2} = 0$ the function $g|_{[0,\pi)} = \tilde{g}$ decreases from $\tilde{g}(0) = 0$ to $\tilde{g}(\pi) = -\frac{1}{\pi}$. Note that our extension is continuous at 0 but has a jump discontinuity at π. Thus for $x \in (0, \pi)$ we have

$$S_N(u)(x) = -\frac{x}{2} + \frac{1}{2} \int_0^x D_N(t) \, dt$$

$$= -\frac{x}{2} + \int_0^x \frac{\sin Nt}{t} \, dt + \int_0^x \tilde{g}(t) \sin Nt \, dt + \frac{1}{2} \int_0^x \cos Nt \, dt$$

or

$$\left| S_N(u)(x) - \int_0^x \frac{\sin Nt}{t}\, dt \right| = \left| -\frac{x}{2} + \int_0^x \tilde{g}(t) \sin Nt\, dt + \frac{1}{2} \int_0^x \cos Nt\, dt \right|$$

$$\leq \frac{x}{2} + \frac{|\sin Nx|}{2N} + \left| \int_0^x \tilde{g}(t) \sin Nt\, dt \right|.$$

For $x = \frac{\pi}{N}$ it follows that

$$\left| S_N(u)(x) - \int_0^x \frac{\sin Nt}{t}\, dt \right| \leq \frac{\pi}{2N} + \left| \int_0^{\frac{\pi}{N}} \tilde{g}(t) \sin Nt\, dt \right|$$

$$\leq \frac{1}{N} \left(\frac{\pi}{2} + \pi \right) = \frac{3\pi}{2N},$$

i.e.

$$\lim_{N \to \infty} S_N(u) \left(\frac{\pi}{N} \right) = \int_0^\pi \frac{\sin t}{t}\, dt = \frac{\pi}{2} \operatorname{Si}(\pi).$$

We want to find a lower bound for $\frac{\pi}{2} \operatorname{Si}(\pi)$. The function $f(t) = \sin t - t + \frac{t^2}{\pi}$ satisfies $f(0) = f(\pi) = 0$ and $f'(t) = \cos t - 1 + \frac{2t}{\pi}$, which implies that for $t \in \left(0, \frac{\pi}{2}\right)$ we have $f'(t) > 0$, i.e. f is increasing on $\left(0, \frac{\pi}{2}\right)$, and $f'(t) < 0$ for $t \in \left(\frac{\pi}{2}, \pi\right)$, i.e. f is decreasing on $\left(\frac{\pi}{2}, \pi\right)$. Thus $f(t)$ must be positive in $(0, \pi)$ or $\frac{\sin t}{t} > 1 - \frac{t}{\pi}$ for $0 < t < \pi$. This now implies

$$\int_0^\pi \frac{\sin t}{t}\, dt > \int_0^\pi \left(1 - \frac{t}{\pi} \right) dt = \frac{\pi}{2}.$$

Thus we come to the following conclusion: for N large the value of u is $u\left(\frac{\pi}{N}\right) = \frac{\pi}{2}\left(1 - \frac{1}{N}\right)$ and we have

$$\lim_{N \to \infty} \left(S_N(u) \left(\frac{\pi}{N} \right) - u \left(\frac{u}{N} \right) \right) = \int_0^\pi \frac{\sin t}{t}\, dt - \frac{\pi}{2} > 0. \tag{8.40}$$

A numerical approximation of $\frac{\pi}{2} \operatorname{Si}(\pi)$ is $1.8519\ldots$ which leads to $\frac{\pi}{2} \operatorname{Si}(\pi) \approx 1.1789 \cdot \frac{\pi}{2}$, see [28]. As mentioned in the beginning of these considerations, the existence of such an overshoot of $S_N(u)$ close to a jump discontinuity of a 2π-periodic function u can be proved in general, the following formulation of the result is related to the one in [19].

Theorem 8.12. *Let $u : \mathbb{R} \to \mathbb{R}$ be a 2π-periodic function, integrable on $[-\pi, \pi]$ and C^1 on $\mathbb{R} \backslash \{2\pi\mathbb{Z}\}$. Suppose u has a jump discontinuity at $x = 0$. Then we have*

$$\lim_{N \to \infty} \left(S_N(u) \left(\frac{\pi}{N} \right) - u \left(\frac{\pi}{N} \right) \right) = \left(\frac{u(0^+) - u(0^-)}{2} \right) \left(\frac{2}{\pi} \int_0^\infty \frac{\sin t}{t}\, dt - 1 \right). \tag{8.41}$$

Problems

1. Solve Exercise 8.2.

2. Let μ be a bounded measure on $\mathbb{T}$. Find the Fourier-Stieltjes coefficients of $\nu := (1 - \cos t)\mu$ in terms of $\hat{\mu}_k$.

3. Let $S := \{\varphi \in \mathbb{T} | \frac{\pi}{6} \leq \varphi \leq \frac{\pi}{4}\}$. Find the Fourier-Stieltjes coefficients of the measure $\mu := \chi_S \lambda_{\mathbb{T}}$ where $\chi_S(\varphi) = \begin{cases} 1, & \varphi \in S \\ 0, & \varphi \notin S \end{cases}$ and $\lambda_{\mathbb{T}}$ is the Lebesgue measure restricted to $[-\pi, \pi)$.

4. Let $(G, +)$ be an Abelian group. We call a function $u : G \to \mathbb{C}$ **positive definite** if for every $N \in \mathbb{N}$ and any choice $x_1, \ldots, x_N \in G$ the matrix $(u(x_k - x_l))_{k,l=1,\ldots,N}$ is positive Hermitian. Prove the estimate

$$|u(x) - u(y)|^2 \leq 2u(0)(u(0) - \operatorname{Re}u(x - y)).$$

Hint: for $x, y \in G$, $u(x) \neq u(y)$, consider the positive Hermitian matrix
$$A_u = \begin{pmatrix} u(0) & \overline{u(x)} & \overline{u(y)} \\ u(x) & u(0) & u(x-y) \\ u(y) & \overline{u(x-y)} & u(0) \end{pmatrix}$$
and discuss the corresponding quadratic form $\langle A_u c, c \rangle$ with $c_1 = \lambda$, $c_2 = \frac{\lambda|u(x)-u(y)|}{u(x)-u(y)}$, $c_3 = -c_2$, $\lambda \in \mathbb{R}$ arbitrary.

5. Let $(\mu^{(n)})_{n \in \mathbb{N}}$, $\mu^{(n)} \in M_b^+(\mathbb{T})$, be a sequence such that for all $f \in C(\mathbb{T})$ we have $\lim_{n \to \infty} \int_{\mathbb{T}} f \, d\mu^{(n)} = f(0)$. Prove that for the sequence $(\hat{\mu}_k^{(n)})_{n \in \mathbb{N}}$, $k \in \mathbb{Z}$, of Fourier-Stieltjes coefficients it follows for all $k \in \mathbb{Z}$

$$\lim_{n \to \infty} \hat{\mu}_k^{(n)} = \frac{1}{2\pi}.$$

6. Show the equality $\frac{1}{2}D_N(t) = \frac{1}{2}\sin Nt \cot \frac{t}{2} + \frac{1}{2}\cos Nt$.

7. Let f satisfy the assumptions of Theorem 8.11 and $g \in L^1(\mathbb{T})$.

   a) Show that the Fourier coefficients of $f|_{[-\pi,\pi]} * g$ decay exponentially, i.e. for some constants $\gamma_1, \gamma_2 > 0$ we have $|F(f|_{-\pi,\pi} * g)(k)| \leq \gamma_1 e^{-\gamma_2|k|}$. Deduce that $f|_{[-\pi,\pi]} * g$ is the restriction of a holomorphic function defined and holomorphic in an open complex neighbourhood of $[-\pi, \pi]$.

   b) Give an example of a function $f \in L^1(\mathbb{T})$ whose Fourier coefficients decay faster than any power, i.e. for $m \in \mathbb{N}_0$ there exists $\kappa(m) \geq 0$ such that for all $k \in \mathbb{Z}$
   $$|(Ff)(k)| \leq \kappa(m)|k|^{-m}$$

136

holds, but the estimate (8.35) does not hold.

Hint: take $h \in C_0^\infty(\mathbb{R})$, $\operatorname{supp} h \subset \left[-\frac{\pi}{2}, \frac{\pi}{2}\right]$ and consider the 2π-periodic extension g of $h|_{[-\pi,\pi]}$.

8. Suppose that f satisfies the assumptions of Theorem 8.11. Prove that $f|_{[-\pi,\pi]} \in \bigcap_{m \in \mathbb{N}} H^m(\mathbb{T})$.

9 Orthonormal Expansions and Special Functions

As indicated in Chapter 1, Fourier series arose from the attempt to obtain and represent "general" solutions of certain (partial) differential equations as a superposition of relatively simple and well understood solutions, i.e. trigonometric functions. A key step was to identify the coefficients of such an expansion as Fourier coefficients which are integrals, and the "orthogonality" of trigonometric functions is crucial in this approach. Recall that orthogonality is a notion depending on the integral. We have seen in Chapter 5 that the L^2-theory of Fourier series is in particular satisfactory which is due to the underlying Hilbert space structure.

In this chapter we will outline the concept of orthonormal expansions in separable Hilbert spaces, a topic which we will pick up and investigate in much more detail in Volume V. The trigonometric functions will be replaced by an orthonormal base and the L^2-theory of Fourier series will turn out to be just a special case of an orthonormal series expansion. Our principal goal in this chapter is to introduce some orthonormal bases formed by special functions in certain L^2-spaces, e.g. Legendre polynomials, etc. These functions are often related to orthogonal curvilinear coordinates as well as to corresponding ordinary differential equations of second order and they reflect certain symmetries. In Part 9 we will discuss these ordinary differential equations. The relation to spectral theory will be investigated in Volume V and the applications to partial differential equations in Volume VI.

Let $(H, \langle \cdot, \cdot \rangle)$ be a complex Hilbert space with scalar product $\langle \cdot, \cdot \rangle : H \times H \to \mathbb{C}$ satisfying (5.2)-(5.4) and denote by $\|x\| := \langle x, x \rangle^{\frac{1}{2}}$ the corresponding norm, see (5.6). For the scalar product we have the Cauchy-Schwarz inequality (5.7) as well as the **parallelogram law**

$$\|x + y\|^2 + \|x - y\|^2 = 2 \left(\|x\|^2 + \|y\|^2 \right) \tag{9.1}$$

and the **polarization identity**

$$\langle x, y \rangle = \frac{1}{4} \left(\|x + y\|^2 - \|x - y\|^2 + i\|x + iy\|^2 - i\|x - iy\|^2 \right), \tag{9.2}$$

both equalities are justified by a straightforward calculation. We call $x, y \in H$, $x \neq 0$ and $y \neq 0$, **orthogonal** if

$$\langle x, y \rangle = 0 \tag{9.3}$$

and we write for this

$$x \perp y. \tag{9.4}$$

A family $(x_j)_{j\in I}$, $x_j \in H$, $x_j \neq 0$, $I \neq \emptyset$, is called an **orthogonal** system in H if $x_j \perp x_l$ for all $j \neq l$. If in addition $\|x_j\| = 1$ for all $j \in I$ we call $(x_j)_{j\in I}$ and **orthonormal** system, compare with Definition 5.2.

Definition 9.1. *We call two non-empty subsets $M_1, M_2 \subset H$ **orthogonal** and write $M_1 \perp M_2$ if $x \perp y$ for all pairs $(x, y) \in M_1 \times M_2$. The **orthogonal complement** $M^\perp$ of $\emptyset \neq M \subset H$ is defined as*

$$M^\perp := \left\{ x \in H \,|\, \langle x, y \rangle = 0 \text{ for all } y \in M \right\}. \tag{9.5}$$

Obviously we have $\{0\}^\perp = H$ and $H^\perp = \{0\}$.

Lemma 9.2. *For every $M \neq \emptyset$ the set $M^\perp$ is a closed linear subspace of H and $M \cap M^\perp = \{0\}$. Moreover we have $M \subset (M^\perp)^\perp$.*

Proof. Let $x, y \in M^\perp$ and $\lambda, \mu \in \mathbb{C}$. For $z \in M$ we find

$$\langle \lambda x + \mu y, z \rangle = \lambda \langle x, z \rangle + \mu \langle y, z \rangle = 0.$$

Let $(x_\nu)_{\nu\in\mathbb{N}}$ be a sequence in $M^\perp$ converging in H to $x \in H$. For $y \in M$ we find

$$\langle x, y \rangle = \left\langle \lim_{\nu\to\infty} x_\nu, y \right\rangle = \lim_{\nu\to\infty} \langle x_\nu, y \rangle = 0,$$

which means that $x \in M^\perp$, thus we have proved that $M^\perp \subset H$ is a closed subspace. The identity $M \cap M^\perp = \{0\}$ is trivial. Let $y \in (M^\perp)^\perp$ mean $\langle y, z \rangle = 0$ for all $z \in M^\perp$ which is of course satisfied for $y \in M$, hence $M \subset (M^\perp)^\perp$. In general we cannot expect $M = (M^\perp)^\perp$ to hold since $(M^\perp)^\perp$ is always a closed linear subset space of H but this is not necessarily true for M. $\square$

Definition 9.3. *A Hilbert space is called separable if it contains a countable dense set, i.e. there exists a set $\{x_\nu \in H \,|\, \nu \in \mathbb{N}\}$ such for every $y \in H$ a subsequence of $(x_\nu)_{\nu\in\mathbb{N}}$ converges in H to y.*

Lemma 9.4. *In a separable Hilbert space H every orthogonal subset is at most countable.*

Proof. Let $M = (x_j)_{j\in I}$ be an orthogonal subset of H. We may assume that $\|x_j\| = 1$, otherwise we switch to the set $\tilde{M} = \left(\frac{x_j}{\|x_j\|} \right)_{j\in I}$. For $j, l \in I$, $j \neq l$, it follows

$$\|x_l - x_j\|^2 = \langle x_l, x_l \rangle - \langle x_l, x_j \rangle - \langle x_j, x_l \rangle + \langle x_j, x_j \rangle = 2.$$

This implies that the family of open balls $\left(B_{\frac{1}{2}}(x_j) \right)_{j\in I}$, $B_{\frac{1}{2}}(x_j) = \{x \in H \,|\, \|x - x_j\| < \frac{1}{2}\}$, are mutually disjoint. Now let $(y_k)_{k\in\mathbb{N}}$ be a countable dense set in H. Then for every ball $B_{\frac{1}{2}}(x_j)$, $j \in I$, at least one element y_{k_j} of $(y_k)_{k\in\mathbb{N}}$ belongs to this ball, implying that there are at most countable many of these balls. $\square$

Of importance for our purposes in this chapter is the fact that the **Gram-Schmidt orthonormalization procedure** also works in separable Hilbert spaces.

Theorem 9.5. *Let H be a separable Hilbert space and $(x_j)_{j \in \mathbb{N}}$ be a countable linearly independent subset, $x_j \neq 0$ for all $j \in \mathbb{N}$. Then there exists a countable orthonormal subset $(y_k)_{k \in \mathbb{N}}$ such that any element y_k is a linear combination of the elements $x_1, \ldots, x_k$, more precisely*

$$y_k = a_{k1}x_1 + \cdots + a_{kk}x_k, \qquad a_{kk} \neq 0. \tag{9.6}$$

Further, every x_j has the representation

$$x_j = b_{j1}y_1 + \cdots + b_{jj}y_j, \qquad b_{jj} \neq 0. \tag{9.7}$$

Up to a factor ± 1 the elements of $(y_k)_{k \in \mathbb{N}}$ are uniquely determined by those of $(x_j)_{j \in \mathbb{N}}$.

Proof. We start by defining

$$y_1 := a_{11}x_1$$

with a_{11} satisfying

$$\langle y_1, y_1 \rangle = a_{11}^2 \langle x_1, x_1 \rangle = 1,$$

implying

$$a_{11} = \frac{1}{b_{11}} = \frac{\pm 1}{\langle x_1, x_1 \rangle^{\frac{1}{2}}}.$$

Suppose now that $y_1, \ldots, y_{k-1}$ are already constructed as a linearly independent set satisfying (9.6) and (9.7). It follows that

$$x_k = b_{k1}y_1 + \cdots + b_{k\,k-1}y_{k-1} + h_k$$

where $\langle h_k, y_j \rangle = 0$ for $j = 1, \ldots, k-1$. Since for $j = 1, \ldots, k-1$

$$0 = \langle h_k, y_j \rangle = \langle x_k - b_{k1}y_1 - \ldots - b_{k\,k-1}y_{k-1}, y_j \rangle$$
$$= \langle x_k, y_j \rangle - b_{kj}\langle y_j, y_j \rangle,$$

we find

$$b_{kj} = \frac{\langle x_k, y_j \rangle}{\langle y_j, y_j \rangle} = \langle h_k, y_j \rangle.$$

Since $\langle h_k, h_k \rangle = 0$ would imply that $\{x_1, \ldots, x_n\}$ is linear dependent we deduce $\langle h_k, h_k \rangle > 0$. Now we define

$$y_k := \frac{h_k}{\langle h_k, h_k \rangle},$$

and by construction (9.7) holds with $a_{kk} = \dfrac{1}{\langle h_k, h_k \rangle^{\frac{1}{2}}}$ as well as $\langle y_k, y_j \rangle = 0$ for $k \neq j$ and $\langle y_k, y_k \rangle = 1$. Moreover (9.6) holds and the theorem is proved. $\qquad \square$

The above result becomes important when looking for representations of elements of a separable Hilbert space with respect to orthonormal sets.

Definition 9.6. *Let H be a separable Hilbert space. We call an orthonormal set $(y_j)_{j \in \mathbb{N}}$ an **orthonormal basis** of H if it is linearly independent, in particular $y_j \neq 0$ for all $j = 1, 2 \ldots$, and if the closure of the span of $(y_j)_{j \in \mathbb{N}}$ is equal to H, i.e.*

$$H = \overline{\text{span}\{y_j \,|\, j \in \mathbb{N}\}}.$$

Remark 9.7. We refer to Appendix II.A.I where linear combinations and the span of a subset of a vector space is discussed. We want to emphasise that a linear combination of vectors is always a finite sum, even in infinite dimensional spaces.

Corollary 9.8. *Every separable Hilbert space has an orthonormal basis.*

Proof. We just have to apply Gram-Schmidt orthonormalisation procedure to a countable dense subset of H. $\qquad\square$

Example 9.9. The space $l_2(\mathbb{Z})$ introduced in Chapter 3 is a separable Hilbert space. Clearly, by

$$\langle (c_k)_{k \in \mathbb{Z}}, (d_k)_{k \in \mathbb{Z}} \rangle := \sum_{k \in \mathbb{Z}} c_k \overline{d_k}$$

a scalar product generating $\|\cdot\|_2$ is given. Hence from Theorem 3.12 we deduce that $l_2(\mathbb{Z})$ is complete, i.e. a Hilbert space. A countable dense subset of $l_2(\mathbb{Z})$ is for example given by all sequences $(d_k)_{k \in \mathbb{Z}}$, $d_k = a_k + ib_k$, $a_k, b_k \in \mathbb{Q}$. This implies in combination with the Fischer-Riesz theorem, Theorem 5.13, that $L^2(\mathbb{T})$ is separable too. We may choose all convergent Fourier series with rational Fourier coefficients as a countable dense set in $L^2(\mathbb{T})$. We also know that the trigonometric polynomials are dense in $L^2(\mathbb{T})$ which follows from the fact that we can approximate every $u \in L^2(\mathbb{T})$ in the norm $\|\cdot\|_{L^2}$ by the partial sums of the Fourier series associated with u. This implies however that the trigonometric polynomials with rational coefficients are dense in $L^2(\mathbb{T})$ and they form a countable set, see also Problem 1.

Definition 9.10. *Let H be a separable Hilbert space and $(y_k)_{k \in \mathbb{N}}$ an orthonormal subset. We call $(y_k)_{k \in \mathbb{N}}$ **complete** if for every $x \in H$ the **completeness relation** or **Parseval's equality***

$$\sum_{k=1}^{\infty} |\langle x, y_k \rangle|^2 = \|x\|^2 \qquad (9.8)$$

holds.

Our aim is to prove that the completeness of $(y_k)_{k\in\mathbb{N}}$ is equivalent to being an orthonormal basis. Let H be a separable Hilbert space and $(y_k)_{k\in\mathbb{N}}$ an orthonormal basis. For $x \in H$ we define its **coordinates** or **generalized Fourier coefficients** with respect to $(y_k)_{k\in\mathbb{N}}$ by

$$c_k := \langle x, y_k \rangle. \tag{9.9}$$

The formal series

$$\sum_{k=1}^{\infty} c_k y_k = \sum_{k=1}^{\infty} \langle x, y_k \rangle y_k \tag{9.10}$$

is called the **generalized Fourier series** or **orthonormal expansion** of x with respect to $(y_k)_{k\in\mathbb{N}}$. The Nth **partial sum** of (9.10) is defined as usual by

$$S_N := \sum_{k=1}^{N} c_k y_k. \tag{9.11}$$

It follows that

$$\begin{aligned}
\|x - S_N\|^2 &= \langle x - S_N, x - S_N \rangle \\
&= \langle x, x \rangle - \langle x, S_N \rangle - \langle S_N, x \rangle + \langle S_N, S_N \rangle \\
&= \|x\|^2 - \sum_{k=1}^{N} \langle x, c_k y_k \rangle - \sum_{k=1}^{N} \langle c_k y_k, x \rangle + \sum_{k,l=1}^{N} c_k \overline{c_l} \langle y_k, y_l \rangle \\
&= \|x\|^2 - \sum_{k=1}^{N} |c_k|^2 - \sum_{k=1}^{N} |c_k|^2 + \sum_{k=1}^{N} |c_k|^2 \\
&= \|x\|^2 - \sum_{k=1}^{N} |c_k|^2,
\end{aligned}$$

thus we arrive at

$$\|x - S_N\|^2 = \|x\|^2 - \sum_{k=1}^{N} |c_k|^2 \tag{9.12}$$

and

$$\sum_{k=1}^{N} |c_k|^2 = \|x\|^2 - \|x - S_N\|^2 \leq \|x\|^2. \tag{9.13}$$

As $N \to \infty$ we obtain

Theorem 9.11. *Let H be a separable Hilbert space with orthonormal basis $(y_k)_{k\in\mathbb{N}}$. Then for every $x \in H$ **Bessel's inequality***

$$\sum_{k=1}^{\infty} |c_k|^2 \leq \|x\|^2$$

holds, compare with Proposition 3.19.

Theorem 9.12. *Let H be a separable Hilbert space and $(y_k)_{k\in\mathbb{N}}$ an orthonormal set. The set $(y_k)_{k\in\mathbb{N}}$ is an orthonormal basis if and only if the completeness relation holds. In this case every $x \in H$ has the representation*

$$x = \sum_{k=1}^{\infty} \langle x, y_k \rangle y_k := \lim_{N \to \infty} \sum_{k=1}^{N} \langle x, y_k \rangle y_k. \tag{9.14}$$

Proof. Let $(y_k)_{k\in\mathbb{N}}$ be an orthonormal basis. It follows that every $x \in H$ can be approximated in H by a sequence $(L_N)_{N\in\mathbb{N}}$, $L_N = L_N(x) = \sum_{k=1}^{N} \alpha_k y_k$ and for L_N we find

$$\|x - L_N\|^2 = \|x\|^2 - \sum_{k=1}^{N} \langle x, \alpha_k y_k \rangle - \sum_{k-1}^{N} \langle \alpha_k y_k, x \rangle + \sum_{k=1}^{N} |\alpha_k|^2$$

$$= \|x\|^2 - \sum_{k=1}^{N} \overline{\alpha}_k c_k - \sum_{k=1}^{N} \alpha_k \overline{c}_k + \sum_{k=1}^{N} |\alpha_k|^2$$

$$= \|x\|^2 - \sum_{k=1}^{N} |c_k|^2 + \sum_{k=1}^{N} |\alpha_k - c_k|^2. \tag{9.15}$$

Comparing (9.15) with (9.12) we find

$$\|x - S_N\|^2 \leq \|x - L_N\|^2,$$

i.e. it follows that $\lim_{N\to\infty} \|x - S_N\|^2 = 0$ for every $x \in H$ which gives (9.15) and equality in Bessel's inequality, i.e. the completion relation holds for every orthonormal basis of H. Now suppose that $(y_k)_{k\in\mathbb{N}}$ is an orthonormal set for which the completeness relation holds. Then it follows for every $x \in H$

$$\|x - S_N\|^2 = \|x\|^2 - \sum_{k=1}^{N} |c_k|^2 \to 0 \quad \text{as } N \to \infty,$$

implying that $x \in \overline{\text{span}\{y_k \,|\, k \in \mathbb{N}\}}$ and this proves the theorem. $\qquad\square$

Corollary 9.13. *An orthonormal system $(y_k)_{k\in\mathbb{N}}$ is complete in the separable Hilbert space H if and only if $\langle x, y_k \rangle = 0$ for all $k \in \mathbb{N}$ implies $x = 0$.*

Proof. Suppose $(y_k)_{k\in\mathbb{N}}$ is complete. By (9.8) we find $\|x\|^2 = \sum_{k=1}^{\infty} |\langle x, y_k \rangle|^2$ for all $x \in H$ and if $\langle x, y_k \rangle = 0$ for all $k \in \mathbb{N}$ it follows that $x = 0$. Now suppose $\langle z, y_k \rangle = 0$ for all $k \in \mathbb{N}$ implies $z = 0$. From Bessel's inequality we deduce that for all $x \in H$ the series $\sum_{k \in \mathbb{N}} \langle x, y_k \rangle y_k$ converges in H. Indeed, the sequence $(S_N)_{N \in \mathbb{N}}$, $S_N = \sum_{k=1}^{N} \langle x, y_k \rangle y_k$ is a Cauchy sequence since for $N < M$ we have

$$\|S_N - S_M\|^2 = \left\| \sum_{k=N+1}^{M} \langle x, y_k \rangle y_k \right\|^2 = \sum_{k=N+1}^{M} |\langle x, y_k \rangle|^2.$$

Consider $z := x - \sum_{k \in \mathbb{N}} \langle x, y_k \rangle y_k$. We find

$$\langle z, y_k \rangle = \langle x, y_k \rangle - \sum_{l \in \mathbb{N}} \langle x, y_l \rangle \langle y_l, y_k \rangle$$

$$= \langle x, y_k \rangle - \langle x, y_k \rangle = 0$$

and by our assumption we can conclude that $z = 0$ or $x = \sum_{k \in \mathbb{N}} \langle x, y_k \rangle y_k$, which implies

$$\|x\|^2 = \langle x, x \rangle = \sum_{k \in \mathbb{N}} \langle x, y_k \rangle \overline{\langle x, y_k \rangle} = \sum_{k \in \mathbb{N}} |\langle x, y_k \rangle|^2,$$

i.e. the completeness of $(y_k)_{k \in \mathbb{N}}$. □

Remark 9.14. While in general we prefer to work with Hilbert spaces over $\mathbb{C}$, in particular when handling operators and dealing with spectral theory, in some cases it is more natural to work with Hilbert spaces over $\mathbb{R}$. This is for example the case when working with special functions of mathematical physics or orthogonal polynomials, see below. All the results discussed so far remain valid, some as the polarization identity simplify in the obvious way.

Let $(a, b) \subset \mathbb{R}$, $-\infty \leq a < b \leq \infty$, be an open interval and $w : (a, b) \to \mathbb{R}_+$ a continuous function which vanishes at most on a set of measure zero. By $w\lambda = w\lambda^{(1)}|_{(a,b)}$, $\lambda^{(1)}$ being the Lebesgue measure, a measure on the Borel sets of (a, b) is defined which obviously has a density with respect to the Lebesgue measure. The space $L^2((a, b), w\lambda)$ is a real Hilbert space with respect to the scalar product

$$\langle u, v \rangle_w := \int_a^b u(x) v(x) w(x) \, dx \tag{9.16}$$

compare also with Theorem III.8.20. In the case where w is fixed and it is clear which function w is chosen we just write $\langle u, v \rangle$. We add the assumption that for every $k \in \mathbb{N}_0$ the functions $m_k : (a, b) \to \mathbb{R}$, $m_k(x) = x^k$, belong to $L^2((a, b), w\lambda)$, i.e. we assume that for $k \in \mathbb{N}_0$

$$\int_a^b x^{2k} w(x) \, dx < \infty. \tag{9.17}$$

By the fundamental theorem of algebra the set $\{m_k \,|\, k \in \mathbb{N}_0\}$ is a linearity independent set in $L^2((a,b), w\lambda)$, however in general it will be neither orthonormal nor orthogonal. Using the Gram-Schmidt procedure, for every w fulfilling the assumptions stated above we can now construct with the help of $\{m_k \,|\, k \in \mathbb{N}_0\}$ an orthonormal system in $L^2((a,b), w\lambda)$. Some of these orthonormal systems are quite important, they form an orthonormal basis and are linked to certain differential equations. We will discuss two of these orthonormal systems below and pick them up in Part 9 where we will deliver a more thorough discussion. Here we are just interested in identifying them as examples of orthonormal bases in certain Hilbert spaces which allows us to study concrete orthonormal expansions beyond Fourier series.

A helpful result is

Lemma 9.15. *Let $\{q_n \,|\, n \in \mathbb{N}_0\}$ be a family of polynomials such that the degree of q_n is n, i.e. $q_n(t) = \sum_{k=0}^{n} a_{nk} t^k$, $a_{nn} \neq 0$. If Q is a polynomial of degree m then $Q \in \operatorname{span} \{q_n \,|\, 0 \leq n \leq m\}$.*

Proof. Let Q be a polynomial of degree m. We can find a real number c_m such that $Q - c_m q_m$ is a polynomial of degree $m - 1$. Suppose that for $0 < k \leq m$ there exists constants $c_m, c_{m-1}, \ldots, c_{m-k}$ such that $Q - \sum_{l=m-k}^{m} c_l q_l$ is of degree $m - k - 1$. Then we can find a constant c_{m-k-1} such that $Q - \sum_{l=m-k-1}^{m} c_l q_l$ is of degree $m - l - 2$ and by mathematical induction the result follows. $\square$

Applying for a given w the Gram-Schmidt procedure directly to $\{m_k \,|\, k \in \mathbb{N}_0\}$, $m_k(x) = x^k$, will necessarily lead to rather long (but elementary) calculations. The same holds when using determinant theory to prove the orthogalization procedure (which was the original approach by Gram). However in some cases there are indirect ways which give even more insights into the structure of the system of polynomials we are dealing with after having applied the Gram-Schmidt procedure to $\{m_k \,|\, k \in \mathbb{N}_0\}$. We start by discussing in detail Legendre polynomials which refers to the case $(a, b) = (-1, 1)$ and $w(x) = 1$ for all $x \in (-1, 1)$.

Definition 9.16. *For $n \in \mathbb{N}_0$ we call*

$$P_n(t) := \frac{1}{2^n n!} \frac{d^n}{dt^n} (t^2 - 1)^n, \quad t \in (-1, 1), \tag{9.18}$$

*the nth **Legendre polynomial**.*

Using the binomial theorem we find

$$(t^2 - 1)^n = \sum_{j=0}^{n} \binom{n}{j} (-1)^j t^{2n-2j} \tag{9.19}$$

and it follows that $P_n(t)$ is a polynomial of degree n with leading coefficient $\frac{(2n)!}{2^n(n!)^2}$ which follows from differentiating the right hand side of (9.19). In fact we have

$$P_n(t) = \sum_{j=0}^{\left[\frac{n}{2}\right]} \frac{(-1)^j(2n-2j)!}{2^nj!(n-j)!(n-2j)!}t^{n-2j}. \tag{9.20}$$

Lemma 9.17. *The Legendre polynomials form an orthogonal system in* $L^2((-1,1))$. *Further, if q is a polynomial and $\langle P_n,q\rangle = 0$ for all $n \in \mathbb{N}_0$ then $q = 0$.*

Proof. First we show that for $k > l$ we have

$$\langle P_k, P_l\rangle = \int_{-1}^1 P_k(t)P_l(t)\,\mathrm{d}t = 0, \tag{9.21}$$

which is equivalent to proving that

$$\int_{-1}^1 \frac{\mathrm{d}^k}{\mathrm{d}t^k}(t^2-1)^k\frac{\mathrm{d}^l}{\mathrm{d}t^l}(t^2-1)^l\,\mathrm{d}t = 0. \tag{9.22}$$

Note that the derivatives up to order $n-1$ of $(t^2-1)^n$ vanish at 1 and -1 and therefore integration by parts yields

$$\int_{-1}^1 \frac{\mathrm{d}^k}{\mathrm{d}t^k}(t^2-1)^k\frac{\mathrm{d}^l}{\mathrm{d}t^l}(t^2-1)^l\,\mathrm{d}t$$

$$= (-1)^k\int_{-1}^1(t^2-1)^n\frac{\mathrm{d}^{k+l}}{\mathrm{d}t^{k+l}}(t^2-1)^l\,\mathrm{d}t = 0$$

since $k + l > 2l$. Now suppose q is a polynomial and $\langle q, P_n\rangle = 0$ for all $n \in \mathbb{N}_0$. The Legendre polynomials satisfy the assumptions of Lemma 9.15. Therefore, if q is of degree m then $q \in \mathrm{span}\{P_n \mid 0 \le n \le m\}$, i.e.

$$q = \sum_{k=0}^m c_k P_k.$$

For $0 \le l \le m$ we find further

$$0 = \langle q, P_l\rangle = \sum_{k=0}^m c_k\langle P_k, P_l\rangle = c_l\|P_l\|_{L^2}^2$$

implying $c_k = 0$ for $k = 0, \ldots, m$, i.e. $q = 0$. $\qquad\square$

147

Before preceding further we note that Theorem III.10.11 also holds for open sets $G \subset \mathbb{R}^n$: for an open set $G \subset \mathbb{R}^n$ the continuous functions with compact support, i.e. the space $C_0(G)$, is dense in $L^2(G)$. (In Volume V we will discuss such type of result again and in more detail.)

Theorem 9.18. *Let $(q_n)_{n \in \mathbb{N}_0}$ be an orthogonal system in $L^2((-1,1))$ with the property that it contains for every $k \in \mathbb{N}_0$ at least one polynomial of degree k. Then we can find constants c_l, $l \in \mathbb{N}_0$, such that $\{q_n \mid n \in \mathbb{N}_0\} = \{c_l P_l \mid l \in \mathbb{N}_0\}$. Moreover the system of the Legendre polynomials $\{P_n \mid n \in \mathbb{N}_0\}$ is complete in $L^2((-1,1))$.*

Proof. Let $(q_m)_{m \in \mathbb{N}_0}$ be an orthogonal system in $L^2((-1,1))$ containing for every $k \in \mathbb{N}_0$ at least one polynomial of degree k. Clearly, one of the q_m's must be equal to a constant $c_0 \neq 0$. We now proceed by induction. Suppose that for some constants $c_j \neq 0$ the polynomials $c_j P_j$, $j = 0, \ldots, n-1$ belong to $\{q_m \mid m \in \mathbb{N}_0\}$ and let $Q_n \in \{q_m \mid m \in \mathbb{N}_0\}$ have degree n. Since $c_0 P_0, \ldots, c_{n-1} P_{n-1}, Q_n \in \{q_m \mid m \in \mathbb{N}_0\}$ it follows for $j = 0, \ldots, n-1$ that $\langle P_j, Q_n \rangle = 0$ as well as $\langle P_j, P_n \rangle = 0$. Hence for $\alpha, \beta \in \mathbb{R}$ and for $j = 0, \ldots, n-1$ we have $\langle P_j, \alpha Q_n + \beta P_n \rangle = 0$. Since P_n and Q_n are of order n we can find α_0, β_0, $\alpha_0 \neq 0 \neq \beta_0$, such that $\alpha_0 Q_n + \beta_0 P_n$ is a polynomial of degree $n-1$. However, $\{P_j, \mid j = 0, \ldots, n-1\}$ is a linearly independent set and span $\{P_j \mid j = 0, \ldots, n-1\}$ is by Lemma 9.15 the set of all polynomials of degree less than or equal to $n-1$. This implies that

$$\alpha_0 Q_n + \beta_0 P_n = \sum_{j=0}^{n-1} \mu_j P_j$$

and further

$$\|\alpha_0 Q_n + \beta_0 P_n\|_{L^2}^2 = \left\langle \sum_{j=0}^{n-1} \mu_j P_j, \alpha_0 Q_n + \beta_0 P_n \right\rangle$$

$$= \sum_{j=0}^{n-1} \mu_j \langle P_j, \alpha_0 Q_n + \beta_0 P_n \rangle = 0,$$

or $\alpha_0 Q_n + \beta_0 P_n = 0$, i.e. $Q_n = -\frac{\beta_0}{\alpha_0} P_n$. Thus multiplying P_n with a suitable constant c_n we obtain $c_n P_n \in \{q_m \mid m \in \mathbb{N}_0\}$ and we have proved $\{c_n P_n \mid n \in \mathbb{N}_0\} \subset \{q_m \mid m \in \mathbb{N}_0\}$ for suitable constants $c_n \neq 0$.

Next we prove that the orthogonal system $\{P_n \mid n \in \mathbb{N}_0\}$ is complete. We note that $\{P_n \mid n \in \mathbb{N}_0\}$ is complete if and only if $\left\{ \tilde{P}_n = \frac{P_n}{\|P_n\|_{L^2}} \mid n \in \mathbb{N}_0 \right\}$ is complete and we are going to apply Corollary 9.13. Let $f \in L^2((-1,1))$ and $\langle f, P_n \rangle = 0$

for all $n \in \mathbb{N}_0$. We know that $C_0((-1,1)) \subset C([-1,1])$ is dense in $L^2((-1,1))$ and by Weierstrass' approximation theorem in the form of Remark II.14.4.B the restrictions of polynomials to $[-1,1]$ are dense in $C([-1,1])$ with respect to the sup-norm. Thus for $f \in L^2((-1,1))$ and $\epsilon > 0$ we can find first $g \in C_0((-1,1))$ such that $\|f - g\|_{L^2} < \epsilon$ and then for g we can find a polynomial q such that $\|g - q\|_\infty < \epsilon$. Now we get

$$
\begin{aligned}
\|f\|_{L^2}^2 &= \langle f, f \rangle = \langle f - g, f \rangle + \langle g - q, f \rangle + \langle q, f \rangle \\
&\leq \|f - g\|_{L^2} \|f\|_{L^2} + \|g - q\|_{L^2} \|f\|_{L^2} + |\langle q, f \rangle| \\
&\leq \|f - g\|_{L^2} \|f\|_{L^2} + \sqrt{2} \|g - q\|_\infty \|f\|_{L^2} + |\langle q, f \rangle| \\
&\leq \epsilon \|f\|_{L^2} + \sqrt{2}\epsilon \|f\|_{L^2} + |\langle q, f \rangle|.
\end{aligned}
$$

Since q is a polynomial, say of degree N it follows from Lemma 9.15 that $q = \sum_{k=0}^{N} \tilde{c}_k \tilde{P}_k = \sum_{k=0}^{N} c_k P_k$ and hence $\langle q, f \rangle = 0$. Thus we arrive at

$$
\|f\|_{L^2}^2 \leq \epsilon(1 + \sqrt{2}) \|f\|_{L^2}
$$

implying $f = 0$. Thus we conclude that $\left\{ \tilde{P}_n \mid n \in \mathbb{N}_0 \right\}$ and hence $\{P_n \mid k \in \mathbb{N}_0\}$ is complete. Now the inclusion $\{c_n P_n \mid n \in \mathbb{N}_0\} \subset \{q_m \mid m \in \mathbb{N}_0\}$ combined with the completeness of $\{P_n \mid n \in \mathbb{N}_0\}$ yields $\{q_m \mid m \in \mathbb{N}_0\} = \{c_n P_n \mid n \in \mathbb{N}_0\}$. $\square$

Corollary 9.19. *There exists constants γ_n, $n \in \mathbb{N}_0$, such that $\{\gamma_n P_n \mid n \in \mathbb{N}_0\}$ is an orthonormal basis of $L^2((-1,1))$ and the Gram-Schmidt procedure applied to $\{m_k \mid k \in \mathbb{N}_0\}$ will give $\{\gamma_n P_n \mid n \in \mathbb{N}_0\}$.*

Lemma 9.20. *For $n \in \mathbb{N}_0$ we have*

$$
\int_{-1}^{1} |P_n(t)|^2 \, dt = \|P_n\|_{L^2}^2 = \frac{2}{2n+1}. \tag{9.23}
$$

Proof. Integration by parts yields

$$
\begin{aligned}
\int_{-1}^{1} |P_n(t)|^2 \, dt &= \frac{1}{2^{2n}(n!)^2} \int_{-1}^{1} \frac{d^n}{dt^n}(t^2 - 1)^n \frac{d^n}{dt^n}(t^2 - 1)^n \, dt \\
&= \frac{(-1)^n}{2^{2n}(n!)^2} \int_{-1}^{1} (t^2 - 1)^n \frac{d^{2n}}{dt^{2n}}(t^2 - 1)^n \, dt \\
&= \frac{(2n)!}{2^{2n}(n!)^2} \int_{-1}^{1} (1 - t^2)^n \, dt.
\end{aligned}
$$

We now reduce the integral in the last line to $B(n+1, n+1)$ where $B(x,y)$ is the beta-function, see (I.31.31). Using the substitution $1 - t = 2s$ and (I.31.34), i.e.

$B(x, y) = \frac{\Gamma(x)\Gamma(y)}{\Gamma(x+y)}$, we find

$$\int_{-1}^{1} \left(1 - t^2\right)^n dt = \int_{-1}^{1} (1-t)^n (1+t)^n dt = 2 \int_{0}^{1} 2^n s^n 2^n (1-s)^n ds$$

$$= 2 \cdot 2^{2n} B(n+1, n+1) = 2 \cdot 2^{2n} \frac{\Gamma(n+1)\Gamma(n+1)}{\Gamma(2n+2)}$$

$$= 2 \cdot 2^{2n} \frac{(n!)^2}{(2n+1)!}.$$

Thus we find

$$\int_{-1}^{1} |P_n(t)|^2 dt = \frac{2}{2n+1}.$$

$\square$

Eventually we have proved

Theorem 9.21. *By* $\left\{ \sqrt{\frac{2n+1}{2}} P_n \,\middle|\, n \in \mathbb{N}_0 \right\}$ *an orthonormal basis of* $L^2((-1,1))$ *is given and for every* $f \in L^2((-1,1))$ *the orthonormal expansion*

$$f(t) = \sum_{k=0}^{\infty} \left\langle f, \sqrt{\frac{2k+1}{2}} P_k \right\rangle_{L^2((-1,1))} P_k(t) \qquad (9.24)$$

holds as equality in $L^2((-1,1))$.

Remark 9.22. Our discussion of the Legendre polynomials made use of the presentation in [121], [34] and [85].

The second family of orthogonal polynomials we want to investigate are the **Hermite polynomials** defined for $n \in \mathbb{N}_0$ by

$$H_n(t) = (-1)^n e^{t^2} \frac{d^n}{dt^n} e^{-t^2}. \qquad (9.25)$$

From (9.25) we derive

$$e^{-t^2} H_n(t) = (-1)^n \frac{d^n}{dt^n} e^{-t^2} = -\frac{d}{dt}\left(e^{-t^2} H_{n-1}(t)\right)$$

$$= e^{-t^2} \left(2t H_{n-1}(t) - H_{n-1}'(t)\right)$$

which yields the formula

$$H_n(t) = 2t H_{n-1}(t) - H_{n-1}'(t) \qquad (9.26)$$

and since $H_0(t) = 1$ we derive from (9.26) that $H_n(t)$ is of degree n. The natural space to consider the Hermite polynomials is in the space $L^2\left(\mathbb{R}, e^{-t^2}\lambda^{(1)}\right)$, i.e. we now choose $(a, b) = (-\infty, \infty)$ and $w(t) = e^{-t^2}$.

Proposition 9.23. *The Hermite polynomials form an orthogonal system in* $L^2\left(\mathbb{R}, e^{-t^2}\lambda\right)$ *and we have*

$$\|H_n\|^2_{L^2\left(\mathbb{R}, e^{-t^2}\lambda\right)} = 2^n n!\sqrt{\pi}. \tag{9.27}$$

Proof. We will write $\langle\cdot,\cdot\rangle$ for the scalar product in $L^2\left(\mathbb{R}, e^{-t^2}\lambda\right)$. For $n > m$ we find

$$\langle H_m, H_n\rangle = \int_{\mathbb{R}} H_m(t)H_n(t)e^{-t^2}\,\mathrm{d}t$$

$$= \int_{\mathbb{R}} H_m(t)(-1)^n\frac{\mathrm{d}^n}{\mathrm{d}t^n}e^{-t^2}\,\mathrm{d}t$$

$$= \int_{\mathbb{R}}\left(\frac{\mathrm{d}^n}{\mathrm{d}t^n}H_m(t)\right)e^{-t^2}\,\mathrm{d}t = 0,$$

since H_m is a polynomial of degree m and $m < n$. A remark is needed for the last step which is an integration by parts for an improper integral. Since for every polynomial $q(t)$ we have $\lim_{|t|\to\infty} q(t)e^{-t^2} = 0$ the boundary terms vanish. Thus the orthogonality of the Hermite polynomials in $L^2\left(\mathbb{R}, e^{-t^2}\lambda\right)$ is proved. To find the norm of H_n we use the calculation made above and note

$$\langle H_n, H_n\rangle = \int_{\mathbb{R}}\frac{\mathrm{d}^n}{\mathrm{d}t^n}H_n(t)e^{-t^2}\,\mathrm{d}t.$$

For the leading coefficient h_n of H_n we have $\frac{\mathrm{d}^n}{\mathrm{d}t^n}H_n(t) = n!h_n$, and (9.20) yields $h_n = 2^n$. Thus we arrive at

$$\langle H_n, H_n\rangle = 2^n n!\int_{\mathbb{R}} e^{-t^2}\,\mathrm{d}t = 2^n n!\sqrt{\pi}.$$

$\square$

We want to show that the Hermite polynomials form a complete orthogonal system in $L^2\left(\mathbb{R}, e^{-t^2}\lambda\right)$. Our proof depends on the following result which is implied by Corollary 12.10 which we will prove in Chapter 12.

Proposition 9.24. *If* $f \in L^1(\mathbb{R})$ *and if for all* $\xi \in \mathbb{R}$ *we have* $\int_{\mathbb{R}} e^{ix\xi}f(x)\,\mathrm{d}x = 0$, *then* $f = 0$ *almost everywhere.*

Using Proposition 9.24 we can prove (partly by following [34])

Theorem 9.25. *The Hermite polynomials form a complete orthogonal system in* $L^2\left(\mathbb{R}, e^{-t^2}\lambda\right)$.

Proof. For $f \in L^2\left(\mathbb{R}, e^{-t^2}\lambda\right)$ we find by the Cauchy-Schwarz inequality (used in $L^2\left(\mathbb{R}, e^{-t^2}\lambda\right)$)

$$\int_{\mathbb{R}} |f(t)|e^{|t\xi|}e^{-t^2}\,\mathrm{d}t \le \left(\int_{\mathbb{R}} |f(t)|^2 e^{-t^2}\,\mathrm{d}t\right)^{\frac{1}{2}} \left(\int_{\mathbb{R}} e^{2|t\xi|}e^{-t^2}\,\mathrm{d}t\right)^{\frac{1}{2}}$$

for all $\xi \in \mathbb{R}$, i.e. $t \mapsto |f(t)|e^{|t\xi|}$ is an element in $L^2\left(\mathbb{R}, e^{-t^2}\lambda\right)$. Now let $f \in L^2\left(\mathbb{R}, e^{-t^2}\lambda\right)$ such that $\langle f, H_n \rangle = 0$ for all $n \in \mathbb{N}_0$, i.e.

$$\int_{\mathbb{R}} f(t)H_n(t)e^{-t^2}\,\mathrm{d}t = 0 \quad \text{for all } n \in \mathbb{N}_0,$$

which is of course equivalent to

$$\int_{\mathbb{R}} f(t)\frac{H_n(t)}{\|H_n\|_{L^2(\mathbb{R}, e^{-t^2}\lambda)}}e^{-t^2}\,\mathrm{d}t = 0 \quad \text{for all } n \in \mathbb{N}_0.$$

We note that $e^{it\xi} = \sum_{\nu=0}^{N} \frac{(it\xi)^{\nu}}{\nu!}$ and that for all $N \in \mathbb{N}_0$

$$\left|\sum_{\nu=0}^{N} \frac{(it\xi)^{\nu}}{\nu!}\right| \le \sum_{\nu=0}^{N} \frac{|t\xi|^{\nu}}{\nu!} \le e^{|t\xi|}.$$

By the dominated convergence theorem we now find

$$\int_{\mathbb{R}} e^{it\xi}f(t)e^{-t^2}\,\mathrm{d}t = \sum_{\nu=0}^{\infty} \frac{(i\xi)^{\nu}}{\nu!}\int_{\mathbb{R}} t^{\nu}f(t)e^{-t^2}\,\mathrm{d}t.$$

Since t^{ν} is a polynomial of degree ν and the Hermite polynomials satisfy the assumptions of Lemma 9.15 we can write $t^{\nu} = \sum_{j=0}^{\nu} c_{j,\nu}H_j(t)$ implying

$$\int_{\mathbb{R}} e^{it\xi}f(t)e^{-t^2}\,\mathrm{d}t = \sum_{\nu=0}^{\infty} \frac{(i\xi)^{\nu}}{\nu!}\sum_{j=0}^{\nu} c_{j,\nu}\langle f, H_j \rangle = 0.$$

Now by Proposition 9.24 we deduce $f(t)e^{-t^2} = 0$ almost everywhere, hence $f = 0$ almost everywhere and we have proved the completeness of the Hermite polynomials in $L^2\left(\mathbb{R}, e^{-t^2}\lambda\right)$. $\square$

Eventually we have justified

Theorem 9.26. *The system* $\left\{ \dfrac{H_n}{\sqrt{2^n n!\sqrt{\pi}}} \,\middle|\, n \in \mathbb{N}_0 \right\}$ *is an othonormal basis in* $L^2\left(\mathbb{R}; e^{-t^2}\lambda\right)$ *and every* $f \in L^2\left(\mathbb{R}, e^{-t^2}\lambda\right)$ *admits the representation*

$$f(t) = \sum_{k=0}^{\infty} \left\langle f, \frac{H_k}{\sqrt{2^k k!\sqrt{\pi}}} \right\rangle \frac{H_k(t)}{\sqrt{2^k k!\sqrt{\pi}}}, \tag{9.28}$$

where this equality is an equality in $L^2\left(\mathbb{R}, e^{-t^2}\lambda\right)$.

Of course, it remains open to now find conditions of f in (9.24) and (9.28), respectively, which will ensure the convergence of these series almost everywhere or pointwisely or uniformly to the function.

The result stated in Proposition 9.24 is a statement about the Fourier transform and the Fourier transform will be the object in the following chapters.

Problems

1. Prove that $L^2(\mathbb{T})$ is separable.

2. Prove that the two subspaces $L^2_{\text{even}}(\mathbb{T})$ and $L^2_{\text{odd}}(\mathbb{T})$ of the real vector space $L^2(\mathbb{T})$ where
$$L^2_{\text{even}}(\mathbb{T}) = \{u \in L^2(\mathbb{T}) | u(-t) = u(t)\}$$
and
$$L^2_{\text{odd}}(\mathbb{T}) = \{u \in L^2(\mathbb{T}) | u(-t) = -u(-t)\}$$
form an orthogonal decomposition of $L^2(\mathbb{T})$, i.e. $L^2(\mathbb{T}) = L^2_{\text{even}}(\mathbb{T}) \oplus L^2_{\text{odd}}(\mathbb{T})$, where "$\oplus$" stands for the direct sum, and $L^2_{\text{even}}(\mathbb{T}) \perp L^2_{\text{odd}}(\mathbb{T})$.

3. Let H be a separable Hilbert space and assume for a denumerable independent set $M \subset H$ that span $M = H$. Prove that H must be finite dimensional.

4. Let H be a separable Hilbert space. Prove that there exists an isometric linear mapping $j : H \to l_2(\mathbb{N})$ where the space $l_2(\mathbb{N})$ is the closed subspace of $l_2(\mathbb{Z})$ consisting of all sequences $(c_k)_{k\in\mathbb{Z}}$ with $c_k = 0$ for $k < 1$.

5. For the Legrendre polynomials P_n prove the identity
$$\sum_{m=0}^{\infty} P_n(x)z^n = (1 - 2xz + z^2)^{-\frac{1}{2}}, \quad |z| < 1, x \in [-1, 1].$$

Hint: apply the Cauchy formula for derivatives to (9.18) to obtain for a suitable curve γ

$$P_n(x) = \frac{1}{2\pi i} \int_\gamma \frac{(\zeta^2 - 1)^n}{2^n (\zeta - x)^{n+1}} \, d\zeta.$$

6. Denote by g the function $g(x,t) = (1 - 2xt + t^2)^{-\frac{1}{2}}$ defined for $|t| < 1$ and $x \in [-1, 1]$.

   a) Verify for g the differential equations

   $$(1 - 2xt + t^2) \frac{\partial g}{\partial t}(x, t) + (t - x)g(x, t) = 0$$

   and

   $$(1 - 2xt + t^2) \frac{\partial g}{\partial x}(x, t) - tg(x, t) = 0.$$

   b) By inserting into the two differential equations of part a) the power series $g(x, t) = \sum_{n=0}^\infty P_n(x)t^n$ derive the formulae

   $$(n + 1)P_{n+1}(x) - (2n + 1)xP_n(x) + nP_{n-1}(x) = 0, \quad n \in \mathbb{N},$$

   and

   $$P'_{n+1}(x) - 2xP'_n(x) + P'_{n-1}(x) - P_n(x) = 0, \quad n \in \mathbb{N}.$$

   c) Use part b) to prove

   $$((1 - x^2)P'_n(x))' + n(n + 1)P_n(x) = 0.$$

7.   a) For the function $h(t) = t^3$ find the coefficients of the expansion into Legendre polynomials.

   b) Find the nth coefficient of the expansion into Legendre polynomials of $g : [-1, 1] \to \mathbb{R}$, $g(t) = \chi_{[-1,0]}(t)$.

8. For $g_n(x) = x^{2n}$, $x \in \mathbb{R}$, prove that

   $$c_{2k}(g_n) := \langle g_n, \tilde{H}_{2k} \rangle = \frac{(2n)!}{2^{2n}(2k)!(k-n)!}, \quad k \leq n,$$

   where $\tilde{H}_n(x) = \left(\frac{1}{2^n n! \sqrt{n}} \right)^{\frac{1}{2}} H_n(x)$.
   Hint: use Legendre's duplication formula for the Γ-function, see Theorem I.31.12.

9. For $\alpha > -1$ we define the **Laguerre polynomials** L_n^α, $n \in \mathbb{N}_0$, by

$$L_n^\alpha(t) = \frac{t^{-\alpha}e^t}{n!} \frac{\mathrm{d}^n}{\mathrm{d}t^n}(t^{\alpha+n}e^{-t}), \quad t > 0.$$

Use Leibniz' rule to show that

$$L_n^\alpha(t) = \sum_{k=0}^n (-1)^k \frac{(n+\alpha)(n-1+\alpha)\cdots(k+1+\alpha)}{k!(n-k)!}t^k.$$

Further prove that for $n \neq m$ the Laguerre polynomials L_n^α and L_m^α are orthogonal in the space $L^2((0, \infty), \mu_\alpha)$ where $\mu_\alpha = m_\alpha \lambda^{(1)}$, $m_\alpha(t) = \chi_{(0,\infty)}(t)t^\alpha e^{-t}$.

10   The Schwartz Space

As indicated in the Introduction and discussed in more detail in Chapter 1, when dealing with functions defined on $\mathbb{R}$ or more generally on $\mathbb{R}^n$ we have to replace Fourier series by Fourier integrals. It was L. Schwartz who in [108] and [109] pointed out that the space $\mathcal{S}(\mathbb{R}^n)$, nowadays called the Schwartz space, is best suited to handle the Fourier integral and the Fourier transform. In this chapter we want to introduce this space.

Before turning to $\mathcal{S}(\mathbb{R}^n)$ we recollect for the reader's convenience the multi-index notation as introduced in Chapter II.5 together with some results of multi-dimensional calculus. Elements $\alpha, \beta \in \mathbb{N}_0^n$ with components $\alpha = (\alpha_1, \dots, \alpha_n)$, $\beta = (\beta_1, \dots, \beta_n)$ where $\alpha_j, \beta_j \in \mathbb{N}_0$, are called **multi-indices** and we agree to the following conventions:

$$\alpha + \beta := (\alpha_1 + \beta_1, \dots, \alpha_n + \beta_n); \tag{10.1}$$

$$|\alpha| := \alpha_1 + \cdots + \alpha_n; \tag{10.2}$$

$$\alpha! := \alpha_1! \cdot \ldots \cdot \alpha_n!; \tag{10.3}$$

$$\alpha \leq \beta \text{ if } \alpha_j \leq \beta_j \text{ for all } 1 \leq j \leq n; \tag{10.4}$$

$$\alpha - \beta := (\alpha_1 - \beta_1, \dots, \alpha_n - \beta_n), \quad \beta \leq \alpha; \tag{10.5}$$

$$\binom{\alpha}{\beta} := \binom{\alpha_1}{\beta_1} \cdot \ldots \cdot \binom{\alpha_n}{\beta_n}. \tag{10.6}$$

For $x \in \mathbb{R}^n$ (or $z \in \mathbb{C}^n$) and $\alpha \in \mathbb{N}_0^n$ we set

$$x^\alpha := x_1^{\alpha_1} \cdot \ldots \cdot x_n^{\alpha_n} \quad (z^\alpha = z_1^{\alpha_1} \cdot \ldots \cdot z_n^{\alpha_n}). \tag{10.7}$$

It is helpful to note that the number of all multi-indices $\alpha \in \mathbb{N}_0^n$ with $|\alpha| \leq k$ is $\frac{(n+k)!}{n!k!}$ and the number of all multi-indices $\alpha \in \mathbb{N}_0^n$ with $|\alpha| = k$ is $\frac{(n+k-1)!}{(n-1)!k!}$. The **binomial theorem** now reads for $x, y \in \mathbb{R}^n$ and $\alpha \in \mathbb{N}_0^n$ as

$$(x+y)^\alpha = \sum_{\beta \leq \alpha} \binom{\alpha}{\beta} x^\beta y^{\alpha-\beta}, \tag{10.8}$$

see Lemma II.5.20. A polynomial in n variables $x = (x_1, \dots, x_n) \in \mathbb{R}^n$ of degree m can be written as

$$p(x) = \sum_{|\alpha| \leq m} a_\alpha x^\alpha, \quad a_\alpha \in \mathbb{R}, \tag{10.9}$$

where we assume that for at least one α_0 with $|\alpha_0| = m$ we have $a_{\alpha_0} \neq 0$. An extension to complex variables is obvious. For $u \in C^k(G)$, $G \subset \mathbb{R}^n$ open, we write

$$\partial^\alpha u = \partial_1^{\alpha_1} \cdot \ldots \cdot \partial_n^{\alpha_n} u = \frac{\partial^{|\alpha|} u}{\partial x_1^{\alpha_1} \cdot \ldots \cdot \partial x_n^{\alpha_n}}, \tag{10.10}$$

sometimes, where appropriate, we will write ∂_x^α or $\partial_{x_j}^{\alpha_j}$. With this notation we have Leibniz's rule

$$\partial^\alpha(u \cdot v) = \sum_{\beta \leq \alpha} \binom{\alpha}{\beta} \partial^\beta u \partial^{\alpha-\beta} v, \tag{10.11}$$

see Lemma II.5.21. For the Faà di Bruno formula for the partial derivative of composite functions we refer to (II.5.34). In general we will work with complex-valued functions $u : G \to \mathbb{C}$, $G \subset \mathbb{R}^n$, and partial derivatives of $u = f + ig$, $f, g : G \to \mathbb{R}$, are given by $\partial^\alpha u = \partial^\alpha f + i\partial^\alpha g$.

It turns out that the following alteration of the notation used in Volume II is extremely useful: we now define

$$D^\alpha u := (-i\partial)^\alpha u = (-i)^{|\alpha|} \partial^\alpha u. \tag{10.12}$$

Taylor's formula in several variables is given, see Theorem 11.9.4, by

$$u(x) = \sum_{|\alpha| \leq k} \frac{\partial^\alpha u(x_0)}{\alpha!} (x - x_0)^\alpha + \sum_{|\alpha| = k+1} \frac{(\partial^\alpha u)(x_0 + \vartheta(x - x_0))}{\alpha!} (x - x_0)^\alpha \tag{10.13}$$

for some $\vartheta \in [0, 1]$. This formula holds for C^{k+1}-functions defined on an open set $G \subset \mathbb{R}^n$ and for points $x, x_0 \in G$ with the property that the line segment connecting x and x_0 belongs to G. A different way to write the remainder term $R_k(x_0, x)$ in (10.13) is

$$R_k(x_0, x) = \int_0^1 (1 - t)^k \sum_{|\alpha| = k+1} \frac{k+1}{\alpha!} (x - x_0)^\alpha \partial^\alpha u(tx + (1 - t)x_0) \, dt, \tag{10.14}$$

see Problem 1.

We will often need the estimates

$$\sum_{|\alpha| \leq k} x^{2\alpha} \leq (1 + \|x\|^2)^k \leq \sum_{|\alpha| \leq k} \gamma_{\alpha,k} x^{2\alpha} \leq C_{k,n} \sum_{|\alpha| \leq k} x^{2\alpha} \tag{10.15}$$

which hold for $x \in \mathbb{R}^n$, $\alpha \in \mathbb{N}_0^n$ and $k \in \mathbb{N}_0$ where

$$\gamma_{\alpha,k} := \frac{k!}{\alpha!(k - |\alpha|)!}. \tag{10.16}$$

Here $\|\cdot\|$ is the Euclidean norm in $\mathbb{R}^n$. The scalar product in $\mathbb{R}^n$ is denoted either by $x \cdot y$ or when more convenient by $\langle x, y \rangle$. A further helpful result is **Peetre's inequality**

$$\frac{1 + \|x\|^2}{1 + \|y\|^2} \leq 2 \left(1 + \|x - y\|^2\right), \quad x, y \in \mathbb{R}^n, \tag{10.17}$$

compare with Problem 8 of Chapter I.23.

For $\alpha, \beta \in \mathbb{N}_0^n$ and $u \in C^\infty(\mathbb{R}^n)$ we define

$$p_{\alpha,\beta}(u) := \sup_{x \in \mathbb{R}^n} \left| x^\beta \partial^\alpha u(x) \right| \tag{10.18}$$

and for $m_1, m_2 \in \mathbb{N}_0$ we set

$$q_{m_1,m_2}(u) := \sup_{x \in \mathbb{R}^n} \left(\left(1 + \|x\|^2 \right)^{\frac{m_1}{2}} \sum_{|\alpha| \leq m_2} |\partial^\alpha u(x)| \right). \tag{10.19}$$

Lemma 10.1. *If for some $u \in C^\infty(\mathbb{R}^n)$ we have $p_{\alpha,\beta}(u) < \infty$ for all $\alpha, \beta \in \mathbb{N}_0^n$ then $q_{m_1,m_2}(u) < \infty$ for all $m_1, m_2 \in \mathbb{N}_0$ and vice versa.*

Proof. Let $u \in C^\infty(\mathbb{R}^n)$ such that for all $\alpha, \beta \in \mathbb{N}_0^n$ we have $p_{\alpha,\beta}(u) < \infty$ and choose $m_1, m_2 \in \mathbb{N}_0$. It follows that

$$\left(1 + \|x\|^2 \right)^{\frac{m_1}{2}} \sum_{|\alpha| \leq m_2} |\partial^\alpha u(x)| \leq \left(1 + \|x\|^2 \right)^{m_1} \sum_{|\alpha| \leq m_2} |\partial^\alpha u(x)|$$

$$\leq \sum_{|\beta| \leq m_1} \gamma_{\beta,m_1} x^{2\beta} \sum_{|\alpha| \leq m_2} |\partial^\alpha u(x)|$$

$$\leq C_{m_1,n} \sum_{|\beta| \leq m_1} \sum_{|\alpha| \leq m_2} \left| x^{2\beta} \partial^\alpha u(x) \right|$$

$$\leq C_{m_1,n} \sum_{|\beta| \leq m_1} \sum_{|\alpha| \leq m_2} p_{\alpha,2\beta}(u) < \infty,$$

hence $q_{m_1,m_2}(u) < \infty$. Conversely, suppose that for all $m_1, m_2 \in \mathbb{N}_0$ we have $q_{m_1,m_2}(u) < \infty$. Given $\alpha, \beta \in \mathbb{N}_n^0$ we choose $m_1 \geq |\beta|$ and $m_2 \geq |\alpha|$ to find

$$\left| x^\beta \partial^\alpha u(x) \right| \leq \left(1 + \|x\|^2 \right)^{\frac{m_1}{2}} \sum_{|\alpha| \leq m_2} |\partial^\alpha u(x)| \leq q_{m_1,m_2}(u) < \infty$$

implying that $p_{\alpha,\beta}(u) < \infty$. $\square$

Definition 10.2. *The **Schwartz space** $\mathcal{S}(\mathbb{R}^n)$ consists of all arbitrarily often differentiable functions $u : \mathbb{R}^n \to \mathbb{C}$ with the property that for all $\alpha, \beta \in \mathbb{N}_0^n$ we have $p_{\alpha,\beta}(u) < \infty$, i.e.*

$$\mathcal{S}(\mathbb{R}^n) := \left\{ u \in C^\infty(\mathbb{R}^n) \,\middle|\, p_{\alpha,\beta}(u) < \infty \text{ for all } \alpha, \beta \in \mathbb{N}_0^n \right\}. \tag{10.20}$$

Elements of $\mathcal{S}(\mathbb{R}^n)$ are also called **rapidly decreasing functions** and it is trivial that $C_0^\infty(\mathbb{R}^n)$, the space of all arbitrarily often differentiable functions with compact support, is a subset of $\mathcal{S}(\mathbb{R}^n)$. The function $x \mapsto e^{-\|x\|^2}$ belongs to $\mathcal{S}(\mathbb{R}^n)$ but not to $C_0^\infty(\mathbb{R}^n)$, see Problem 4b. From Lemma 10.1 we deduce immediately

Corollary 10.3. *We have*

$$\mathcal{S}(\mathbb{R}^n) = \left\{ u \in C^\infty(\mathbb{R}^n) \,\middle|\, q_{m_1,m_2}(u) < \infty \ \text{for all } m_1, m_2 \in \mathbb{N}_0 \right\}. \tag{10.21}$$

Proposition 10.4. *The Schwartz space $\mathcal{S}(\mathbb{R}^n)$ is an algebra over $\mathbb{C}$. Moreover, for $\gamma \in \mathbb{N}_0^n$ and $u \in \mathcal{S}(\mathbb{R}^n)$ we have $\partial^\gamma u \in \mathcal{S}(\mathbb{R}^n)$. If $p(x) = \sum_{|\gamma| \le m} a_\gamma x^\gamma$, $a_\gamma \in \mathbb{C}$, is a polynomial then $p \cdot u \in \mathcal{S}(\mathbb{R}^n)$, and if $w \in C^\infty(\mathbb{R}^n)$ such that $\|\partial^\gamma w\|_\infty < \infty$ for all $\gamma \in \mathbb{N}_0^n$ we also have $w \cdot u \in \mathcal{S}(\mathbb{R}^n)$ for all $u \in \mathcal{S}(\mathbb{R}^n)$. In addition for $\gamma \in \mathbb{N}_0^n$ and $u \in \mathcal{S}(\mathbb{R}^n)$ it follows that $\partial^\gamma u \in L^1(\mathbb{R}^n) \cap C_b(\mathbb{R}^n) \subset L^p(\mathbb{R}^n)$ for every $p \ge 1$.*

Proof. Let $u, v \in \mathcal{S}(\mathbb{R}^n)$, $\alpha, \beta, \gamma \in \mathbb{N}_0^n$ and $\lambda, \mu \in \mathbb{C}$. Since

$$\left| x^\beta \partial^\alpha \overline{u(x)} \right| = \left| x^\beta \partial^\alpha u(x) \right|$$

and

$$\left| x^\beta \partial^\alpha \left(\lambda u(x) + \mu v(x) \right) \right| \le |\lambda| \left| x^\beta \partial^\alpha u(x) \right| + |\mu| \left| x^\beta \partial^\alpha v(x) \right|,$$

it is clear that $\mathcal{S}(\mathbb{R}^n)$ is a $\mathbb{C}$-vector space and $u \in \mathcal{S}(\mathbb{R}^n)$ implies $\overline{u} \in \mathcal{S}(\mathbb{R}^n)$. Furthermore we have

$$\left| x^\beta \partial^\alpha (u \cdot v)(x) \right| = \left| x^\beta \sum_{\gamma \le \alpha} \binom{\alpha}{\gamma} \partial^{\alpha-\gamma} u(x) \partial^\gamma v(x) \right|$$

$$\le \sum_{\gamma \le \alpha} \binom{\alpha}{\gamma} \left| x^\beta \partial^{\alpha-\gamma} u(x) \right| |\partial^\gamma v(x)|$$

$$\le \sum_{\gamma \le \alpha} \binom{\alpha}{\gamma} p_{\alpha-\gamma,\beta}(u) p_{\gamma,0}(u)$$

implying $u \cdot v \in \mathcal{S}(\mathbb{R}^n)$, i.e. $\mathcal{S}(\mathbb{R}^n)$ is indeed a $\mathbb{C}$-algebra. For $\gamma \in \mathbb{N}_0^n$ we find

$$\left| x^\beta \partial^\alpha \left(\partial^\gamma u(x) \right) \right| = \left| x^\beta \partial^{\alpha+\gamma} u(x) \right| \le p_{\alpha+\gamma,\beta}(u),$$

thus $\partial^\gamma u \in \mathcal{S}(\mathbb{R}^n)$. For $\alpha, \beta \in \mathbb{N}_0^n$ we calculate

$$
\left| x^\beta \partial^\alpha \left(\left(\sum_{|\gamma| \le m} a_\gamma x^\gamma \right) u(x) \right) \right| = \left| x^\beta \sum_{|\gamma| \le m} a_\gamma \partial^\alpha \left(x^\gamma u(x) \right) \right|
$$

$$
= \left| x^\beta \sum_{|\gamma| \le m} a_\gamma \sum_{\delta \le \alpha} \binom{\alpha}{\delta} \left(\partial^{\alpha-\delta} x^\gamma \right) \left(\partial^\delta u(x) \right) \right|
$$

$$
\le M \sum_{|\gamma| \le m} \sum_{\substack{\delta \le \alpha \\ \gamma - \alpha + \delta \ge 0}} \left| x^{\beta + \gamma - \alpha + \delta} \partial^\delta u(x) \right|
$$

$$
\le M \sum_{|\gamma| \le m} \sum_{\substack{\delta \le \alpha \\ \gamma - \alpha + \delta \ge 0}} p_{\delta, \beta + \gamma - \alpha + \delta}(u),
$$

i.e. $p \cdot u \in \mathcal{S}(\mathbb{R}^n)$. Analogously we find

$$
\left| x^\beta \partial^\alpha (w \cdot u)(x) \right| = \left| x^\beta \sum_{\gamma \le \alpha} \binom{\alpha}{\gamma} \left(\partial^{\alpha - \gamma} w(x) \right) \left(\partial^\gamma u(x) \right) \right|
$$

$$
\le c \left(\max_{\gamma \le \alpha} \left\| \partial^{\alpha - \gamma} w \right\|_\infty \right) \sum_{\gamma \le \alpha} p_{\gamma, \beta}(u)
$$

which yields $w \cdot u \in \mathcal{S}(\mathbb{R}^n)$. In order to prove the integrability of $\partial^\alpha u$, $u \in \mathcal{S}(\mathbb{R}^n)$, we first note that with $\omega_n = \lambda^{(n)}(B_1(0)) = \frac{\pi^{\frac{n}{2}}}{\Gamma(\frac{n}{2}+1)}$ we have for all $k \in \mathbb{N}_0$

$$
\int_{\mathbb{R}^n} \frac{\|x\|^k}{(1 + \|x\|^2)^{\frac{n+k+1}{2}}} \, dx = n\omega_n \int_0^\infty \frac{r^{k+n-1}}{(1+r^2)^{\frac{n+k+1}{2}}} \, dr
$$

$$
\le n\omega_n \int_0^1 \frac{r^{k+n-1}}{(1+r^2)^{\frac{n+k+1}{2}}} \, dr + n\omega_n \int_1^\infty r^{-2} \, dr < \infty.
$$

Since

$$
\left(1 + \|x\|^2 \right)^{\frac{n+1}{2}} |\partial^\gamma u(x)| \le \left(1 + \|x\|^2 \right)^{\frac{n+1}{2}} \sum_{|\alpha| \le |\gamma|} |\partial^\alpha u(x)|
$$

$$
\le q_{n+1, |\gamma|}(u)
$$

we get

$$
|\partial^\gamma u(x)| \le \frac{q_{n+1, |\gamma|}(u)}{(1 + \|x\|^2)^{\frac{n+1}{2}}}
$$

161

which implies

$$\int_{\mathbb{R}^n} |\partial^\gamma u(x)| \, dx \leq \int_{\mathbb{R}^n} \left(1 + \|x\|^2\right)^{\frac{-n-1}{2}} q_{n+1,|\gamma|}(u) \, dx$$

$$= q_{n+1,|\gamma|}(u) \int_{\mathbb{R}^n} \left(1 + \|x\|^2\right)^{\frac{-n-1}{2}} \, dx < \infty,$$

i.e. $\partial^\gamma u \in L^1(\mathbb{R}^n)$, and since $p_{\gamma,0}(u) = p_{0,0}(\partial^\gamma u) < \infty$ we find that $\partial^\gamma u \in L^1(\mathbb{R}^n) \cap C_b(\mathbb{R}^n)$ which of course implies $\partial^\gamma u \in L^p(\mathbb{R}^n)$ for $p \geq 1$. $\qquad\square$

Remark 10.5. Using a calculation similar to that used in the proof of Proposition 10.4 we find

$$\left(1 + \|x\|^2\right)^{\frac{m}{2}} |\partial^\gamma u(x)| \leq \frac{1}{\left(1 + \|x\|^2\right)^{\frac{n+1}{2}}} \left(1 + \|x\|^2\right)^{\frac{m+n+1}{2}} \sum_{|\alpha| \leq |\gamma|} |\partial^\alpha u(x)|$$

$$\leq \frac{1}{\left(1 + \|x\|^2\right)^{\frac{n+1}{2}}} q_{m+n+1,|\gamma|}(u)$$

implying that if $u \in \mathcal{S}(\mathbb{R}^n)$ then for every $m \in \mathbb{N}_0$ and $\gamma \in \mathbb{N}_0^n$ the function $\left(1 + \|\cdot\|^2\right)^{\frac{m}{2}} |\partial^\gamma u|$ belongs to $L^1(\mathbb{R}^n) \cap C_b(\mathbb{R}^n) \subset L^p(\mathbb{R}^n)$, $p \geq 1$. This result can also be deduced from Proposition 10.4 when noting that for every polynomial p we have $p \cdot u \in \mathcal{S}(\mathbb{R}^n)$ if $u \in \mathcal{S}(\mathbb{R}^n)$.

Since $u, v \in \mathcal{S}(\mathbb{R}^n)$ implies that $u, v \in L^p(\mathbb{R}^n)$, $1 \leq p < \infty$, their convolution

$$(u * v)(x) = \int_{\mathbb{R}^n} u(x - y) v(y) \, dy = \int_{\mathbb{R}^n} u(y) v(x - y) \, dy \qquad (10.22)$$

is well defined and by Young's inequality, see Theorem III.10.13, we have

$$\|u * v\|_{L^p} \leq \|u\|_{L^1} \|v\|_{L^p}. \qquad (10.23)$$

From Corollary III.8.6 we deduce that for $u, v \in \mathcal{S}(\mathbb{R}^n)$ we have $u * v \in C^\infty(\mathbb{R}^n)$ and

$$\partial^\alpha (u * v)(x) = \int_{\mathbb{R}^n} \left(\partial_x^\alpha u(x - y)\right) v(y) \, dy = \int_{\mathbb{R}^n} u(y) \, \partial_x^\alpha v(x - y) \, dy. \qquad (10.24)$$

We claim

Proposition 10.6. *The convolution $u * v$ of $u, v \in \mathcal{S}(\mathbb{R}^n)$ belongs to $\mathcal{S}(\mathbb{R}^n)$.*

Proof. We know already that $u * v \in C^\infty(\mathbb{R}^n)$. For $m_1 \in \mathbb{N}_0$ we find by Peetre's inequality

$$\left(1 + \|x\|^2\right)^{\frac{m_1}{2}} = \frac{\left(1 + \|x\|^2\right)^{\frac{m_1}{2}}}{\left(1 + \|y\|^2\right)^{\frac{m_1}{2}}} \left(1 + \|y\|^2\right)^{\frac{m_1}{2}}$$

$$\leq 2^{\frac{m_1}{2}} \left(1 + \|x - y\|^2\right)^{\frac{m_1}{2}} \left(1 + \|y\|^2\right)^{\frac{m_1}{2}}$$

and therefore

$$\left(1 + \|x\|^2\right)^{\frac{m_1}{2}} \sum_{|\alpha| \leq m_2} |\partial^\alpha (u * v)(x)| = \sum_{|\alpha| \leq m_2} \left(1 + \|x\|^2\right)^{\frac{m_1}{2}} \left| \int_{\mathbb{R}^n} (\partial_x^\alpha u(x - y)) \, v(y) \, \mathrm{d}y \right|$$

$$\leq \sum_{|\alpha| \leq m_2} \int_{\mathbb{R}^n} \left(1 + \|x\|^2\right)^{\frac{m_1}{2}} |\partial_x^\alpha u(x - y)| \, |v(y)| \, \mathrm{d}y$$

$$\leq \sum_{|\alpha| \leq m_2} 2^{\frac{m_1}{2}} \int_{\mathbb{R}^n} \left| \left(1 + \|x - y\|^2\right)^{\frac{m_1}{2}} \partial_x^\alpha u(x - y) \right| \left| \left(1 + \|y\|^2\right)^{\frac{m_1}{2}} v(y) \right| \, \mathrm{d}y$$

$$\leq 2^{\frac{m_1}{2}} \sum_{|\alpha| \leq m_2} \left\| \left(1 + \| \cdot \|^2\right)^{\frac{m_1}{2}} \partial_x^\alpha u \right\|_{L^1} \left\| \left(1 + \| \cdot \|^{\frac{m_1}{2}}\right) v \right\|_{L^1}$$

and since for $w \in \mathcal{S}(\mathbb{R}^n)$, $\alpha \in \mathbb{N}_0^n$ and $m \in \mathbb{N}_0$ the function $\left(1 + \| \cdot \|^2\right)^{\frac{m}{2}} |\partial^\alpha w|$ belongs to $L^1(\mathbb{R}^n)$ the result follows. $\square$

Next we want to introduce for sequences in $\mathbb{S}(\mathbb{R}^n)$ a notion of convergence.

Definition 10.7. *We call a sequence $(u_k)_{k \in \mathbb{N}}$, $u_k \in \mathcal{S}(\mathbb{R}^n)$, **convergent in** $\mathcal{S}(\mathbb{R}^n)$ to $u \in \mathcal{S}(\mathbb{R}^n)$ if for all $\alpha, \beta \in \mathbb{N}_0^n$ we have*

$$\lim_{k \to \infty} p_{\alpha, \beta}(u_k - u) = 0. \tag{10.25}$$

Remark 10.8. A. Choosing $\beta = 0$ in (10.25) we arrive at

$$\lim_{k \to \infty} \|\partial^\alpha u_k - \partial^\alpha u\|_\infty = 0 \tag{10.26}$$

for every $\alpha \in \mathbb{N}_0^n$, i.e. convergence in $\mathcal{S}(\mathbb{R}^n)$ implies uniform convergence of all partial derivatives. In particular it follows that the limit u is uniquely determined. **B.** From Lemma 10.1 we deduce that $(u_k)_{k \in \mathbb{N}}$ converges in $\mathcal{S}(\mathbb{R}^n)$ to u if and only if for $m_1, m_2 \subset \mathbb{N}_0$ we have

$$\lim_{k \to \infty} q_{m_1, m_2}(u_k - u) = 0. \tag{10.27}$$

C. If the sequence $(u_k)_{k \in \mathbb{N}}$ converges in $\mathcal{S}(\mathbb{R}^n)$ to u then it also converges in $L^p(\mathbb{R}^n)$, $1 \le p < \infty$, to u. Since we already know that $\lim_{k \to \infty} \|u_k - u\|_\infty = 0$ it is sufficient to prove that $\|u_k - u\|_{L^1} \to 0$ as $k \to \infty$. To see this we note

$$|u_k(x) - u(x)| = \left(1 + \|x\|^2\right)^{\frac{-n-1}{2}} \left(1 + \|x\|^2\right)^{\frac{n+1}{2}} |u_k(x) - u(x)|$$
$$\le \left(1 + \|x\|^2\right)^{\frac{-n-1}{2}} q_{n+1,0}(u_k - u)$$

implying

$$\int_{\mathbb{R}^n} |u_k(x) - u(x)| \, dx \le c q_{n+1,0}(u_k - u)$$

and the claim follows.

Definition 10.9. *A sequence $(u_k)_{k \in \mathbb{N}}$, $u_k \in \mathcal{S}(\mathbb{R}^n)$, is called a **Cauchy sequence** in $\mathcal{S}(\mathbb{R}^n)$ if for each pair $\alpha, \beta \in \mathbb{N}_0^n$ we have that for every $\epsilon > 0$ there exists $N(\epsilon)$ such that $k, l \ge N(\epsilon)$ implies $p_{\alpha,\beta}(u_k - u_l) < \epsilon$.*

Remark 10.10. We can replace in Definition 10.9 the family $p_{\alpha,\beta}$, $\alpha, \beta \in \mathbb{N}_0$ by the family q_{m_1,m_2}, m_1, m_2. This follows from Lemma 10.1. Thus $(u_k)_{k \in \mathbb{N}}$ is a Cauchy sequence in $\mathcal{S}(\mathbb{R}^n)$ if for every pair $m_1, m_2 \in \mathbb{N}_0$ we have that for every $\epsilon > 0$ there exists $N(\epsilon)$ such that $k, l \ge N(\epsilon)$ implies $q_{m_1,m_2}(u_k - u_l) < \epsilon$.

Theorem 10.11. *The Schwartz space $\mathcal{S}(\mathbb{R}^n)$ is (sequentially) complete, i.e. if $(u_k)_{k \in \mathbb{N}}$ is a Cauchy sequence in $\mathcal{S}(\mathbb{R}^n)$ then it has a limit in $\mathcal{S}(\mathbb{R}^n)$.*

Proof. Let $(u_k)_{k \in \mathbb{N}}$ be a Cauchy sequence in $\mathcal{S}(\mathbb{R}^n)$. It follows that for every $\alpha \in \mathbb{N}_0^n$ the sequence $(\partial^\alpha u_k)_{k \in \mathbb{N}}$ is a Cauchy sequence with respect to the sup-norm, hence it has a uniform limit $v_\alpha \in C_b(\mathbb{R}^n)$, i.e. $\lim_{k \to \infty} \|\partial^\alpha u_k - v_\alpha\|_\infty = 0$. We set $u := v_0$ and claim $\partial^\alpha u = v_\alpha$. This follows from Remark 10.8.A by a short induction. Observe that if $\alpha - \epsilon_j \ge 0$, $\epsilon_j = (0, \ldots, 0, 1, 0, \ldots, 0) \in \mathbb{N}_0^n$ with the 1 in position j, then we have

$$v_\alpha = \lim_{k \to \infty} \partial^\alpha u_k = \lim_{k \to \infty} \partial^{\epsilon_j} \partial^{\alpha - \epsilon_j} u_k$$
$$= \partial^{\epsilon_j} \lim_{k \to \infty} \partial^{\alpha - \epsilon_j} u_k = \partial^{\epsilon_j} v_{\alpha - \epsilon_j}.$$

The induction hypothesis $v_{\alpha - \epsilon_j} = \partial^{\alpha - \epsilon_j} u$ now implies that $v_\alpha = \partial^\alpha u$. It remains to prove that for all $\alpha, \beta \in \mathbb{N}_0^n$ we have $\lim_{k \to \infty} p_{\alpha,\beta}(u_k - u) = 0$, or equivalently that for all $m_1, m_2 \in \mathbb{N}_0$ it follows $\lim_{k \to \infty} q_{m_1,m_2}(u_k - u) = 0$. Let $m_1, m_2 \in \mathbb{N}_0$ and let $\epsilon > 0$ be given. We choose $N(\epsilon) \in \mathbb{N}$ such that for $k, l \ge N(\epsilon)$ it follows that

$$\sup_{x \in \mathbb{R}^n} \left(1 + \|x\|^2\right)^{\frac{m_1}{2}} \sum_{|\alpha| \le m_2} |\partial^\alpha u_k(x) - \partial^\alpha u_l(x)| < \epsilon.$$

Then we find

$$\left(1+\|x\|^2\right)^{\frac{m_1}{2}} \sum_{|\alpha|\leq m_2} |\partial^\alpha u_k(x) - \partial^\alpha u(x)|$$

$$= \lim_{l\to\infty} \left(1+\|x\|^2\right)^{\frac{m_1}{2}} \sum_{|\alpha|\leq m_2} |\partial^\alpha u_k(x) - \partial^\alpha u_l(x)|$$

$$\leq \limsup_{l\to\infty} q_{m_1,m_2}(u_k - u_l) \leq \epsilon,$$

which yields

$$\lim_{k\to\infty} q_{m_1,m_2}(u_k - u) = 0.$$

$\square$

In Chapter III.10 we introduce the **Friedrichs mollifier**. Let a be defined by

$$a^{-1} := \int_{\mathbb{R}^n} \exp\left((\|x\|^2 - 1)^{-1}\right) \mathrm{d}x$$

and define

$$j(x) := \begin{cases} a\exp\left((\|x\|^2 - 1)^{-1}\right), & \|x\| < 1 \\ 0, & \|x\| \geq 1. \end{cases} \tag{10.28}$$

It follows that $j \in C_0^\infty(\mathbb{R}^n)$, $\int_{\mathbb{R}^n} j(x)\,\mathrm{d}x = 1$, $j(x) \geq 0$, and $\mathrm{supp}\, j = \overline{B_1(0)}$. For $\epsilon > 0$ we set $j_\epsilon(x) := \epsilon^{-n} j\left(\frac{x}{\epsilon}\right)$ implying $j_\epsilon(x) \geq 0$, $\int_{\mathbb{R}^n} j_\epsilon(x)\,\mathrm{d}x = 1$, and $\mathrm{supp}\, j_\epsilon = \overline{B_\epsilon(0)}$. For $u \in L^p(\mathbb{R}^n)$ we now define

$$J_\epsilon(u)(x) := (j_\epsilon * u)(x) = \int_{\mathbb{R}^n} j_\epsilon(x-y)u(y)\,\mathrm{d}y. \tag{10.29}$$

We have proved that $J_\epsilon(u) \in C^\infty(\mathbb{R}^n) \cap L^p(\mathbb{R}^n)$ and that

$$\mathrm{supp}\, J_\epsilon(u) \subset \mathrm{supp}\, u + \overline{B_\epsilon(0)}. \tag{10.30}$$

The function

$$\varphi(x) := J_1\left(\chi_{B_2(x_0)}\right)(x) = \int_{\mathbb{R}^n} j(x-y)\chi_{B_2(x_0)}(y)\,\mathrm{d}y \tag{10.31}$$

belongs to $C^\infty(\mathbb{R}^n)$, for $x \in B_1(x_0)$ we have $\varphi(x) = 1$, whereas for all $x \in \mathbb{R}^n$ we have $0 \leq \varphi \leq 1$ and $\mathrm{supp}\, \varphi \subset \overline{B_4(x_0)}$. With this preparation we can prove

Lemma 10.12. *The space $C_0^\infty(\mathbb{R}^n)$ is sequentially dense in $\mathcal{S}(\mathbb{R}^n)$, i.e. for every $u \in \mathcal{S}(\mathbb{R}^n)$ exists a sequence $(\varphi_k)_{k\in\mathbb{N}}$, $\varphi_k \in C_0^\infty(\mathbb{R}^n)$, converging in $\mathcal{S}(\mathbb{R}^n)$ to u. Furthermore, the Schwartz space $\mathcal{S}(\mathbb{R}^n)$ is dense in $C_\infty(\mathbb{R}^n)$ with respect to the sup-norm.*

Proof. Let φ be as in (10.31) with $x_0 = 0$, and for $k \in \mathbb{N}$ define $\varphi_k(x) := \varphi\left(\frac{x}{k}\right) u(x)$. Clearly $\varphi_k \in C_0^\infty(\mathbb{R}^n)$ and for $\|x\| \leq k$ we have $\varphi_k(x) = u(x)$. Now it follows for $\alpha, \beta \in \mathbb{N}_0^n$

$$p_{\alpha,\beta}(\varphi_k - u) \leq \sup_{x \in \mathbb{R}^n \backslash B_k(0)} \left| x^\beta \partial^\alpha \left(\varphi\left(\frac{x}{k}\right) u(x) \right) \right|$$

$$= \sup_{x \in \mathbb{R}^n \backslash B_k(0)} \left| x^\beta \sum_{\gamma \leq \alpha} \binom{\alpha}{\gamma} k^{-|\gamma|} (\partial^\gamma \varphi)\left(\frac{x}{k}\right) \partial^{\alpha-\gamma} u(x) \right|$$

$$\leq \sup_{x \in \mathbb{R}^n \backslash B_k(0)} \left| x^\beta \varphi\left(\frac{x}{k}\right) u(x) \right| + \sup_{x \in \mathbb{R}^n \backslash B_k(0)} \left| x^\beta \sum_{0 \neq \gamma \leq \alpha} \binom{\alpha}{\gamma} k^{-|\gamma|} (\partial^\gamma \varphi)\left(\frac{x}{k}\right) \partial^{\alpha-\gamma} u(x) \right|$$

$$\leq \sup_{x \in \mathbb{R}^n \backslash B_k(0)} \left| x^\beta u(x) \right| + \frac{C}{k} \max_{0 \neq \gamma \leq \alpha} \|\partial^\gamma \varphi\|_\infty \max_{0 \neq \gamma \leq \alpha} \sup_{x \in \mathbb{R}^n \backslash B_k(0)} \left| x^\beta \partial^{\alpha-\gamma} u(x) \right|$$

$$\leq \sup_{x \in \mathbb{R}^n \backslash B_k(0)} \left| x^\beta u(x) \right| + \frac{C}{k} \max_{0 \neq \gamma \leq \alpha} \|\partial^\gamma \varphi\|_\infty \max_{0 \neq \gamma \leq \alpha} p_{\alpha-\gamma,\beta}(u).$$

The first term tends to 0 as k tends to infinity since $x \mapsto x^\beta u(x)$ vanishes at infinity and since $\max_{0 \neq \gamma \leq \alpha} \|\partial^\alpha \varphi\|_\infty \max_{0 \neq \gamma \leq \alpha} p_{\alpha-\gamma,\beta}(u)$ is finite the second term tends to 0 too. Thus we have proved that for all $\alpha, \beta \in \mathbb{N}_0^n$ it follows $\lim_{k \to \infty} p_{\alpha,\beta}(\varphi_k - u) = 0$.

If we choose $\alpha = \beta = 0$, i.e. $p_{\alpha,\beta}(\cdot) = \|\cdot\|_\infty$, our calculation also yields that we can approximate every continuous function vanishing at infinity, $u \in C_\infty(\mathbb{R}^n)$, uniformly by a continuous function with compact support, compare with Theorem II.14.20. We will prove now that every continuous function with compact support can be approximated in the norm $\|\cdot\|_\infty$ by elements of $C_0^\infty(\mathbb{R}^n)$. Since $C_0^\infty(\mathbb{R}^n) \subset \mathcal{S}(\mathbb{R}^n) \subset C_\infty(\mathbb{R}^n)$ this will imply that $\mathcal{S}(\mathbb{R}^n)$ is dense in $C_\infty(\mathbb{R}^n)$ with respect to the sup-norm. We note first that $u \in C_0(\mathbb{R}^n)$ is uniformly continuous, also see Problem 2. Now we use the Friedrichs mollifier. For $\eta > 0$ and $u \in C_0(\mathbb{R}^n)$ we have $J_\eta(u) \in C_0^\infty(\mathbb{R}^n)$ with supp $J_\eta(u) \subset$ supp $u + \overline{B_\eta(0)}$. The uniform continuity of u implies that for every $\epsilon > 0$ there exists $\delta > 0$ such that $\|x - y\| < \delta$ implies $|u(x) - u(y)| < \epsilon$. This yields

$$\left| u(x) - J_\eta(u)(x) \right| = \left| \int_{\mathbb{R}^n} j_\eta(x - y)(u(x) - u(y)) \, dy \right|$$

$$\leq \int_{\|x-y\|<\delta} |u(x) - u(y)| j_\eta(x - y) \, dy + 2\|u\|_\infty \int_{\|x-y\|\geq\delta} j_\eta(x - y) \, dy$$

$$\leq \epsilon + 2\|u\|_\infty \int_{\|y\|\geq\delta} j_\eta(y) \, dy$$

$$\leq \epsilon + 2\|u\|_\infty \int_{\|z\|\geq\frac{\delta}{\eta}} j(z) \, dz,$$

and it follows

$$\|u - J_\eta(u)\|_\infty \le \epsilon + 2\|u\|_\infty \int_{\|z\| \ge \frac{\delta}{\eta}} j(z) \, dz.$$

For $\eta \to 0$ it follows that $\frac{\delta}{\eta} \to \infty$ and hence $\int_{\|z\| \ge \frac{\delta}{\eta}} j(z) \, dz \to 0$, implying for all $\epsilon > 0$ that

$$\limsup_{\eta \to 0} \|u - J_\eta(u)\|_\infty < \epsilon,$$

or

$$\lim_{\eta \to 0} \|u - J_\eta(u)\|_\infty = 0.$$

$\square$

Remark 10.13. In Volume V we will introduce on $\mathcal{S}(\mathbb{R}^n)$ a locally convex, metrizable topology and we will see that with this topology we can turn $\mathcal{S}(\mathbb{R}^n)$ into a complete metric space in which $C_0^\infty(\mathbb{R}^n)$ is a dense subset.

Since $\mathcal{S}(\mathbb{R}^n)$ is a $\mathbb{C}$-vector space we can consider linear operators or mappings $T : \mathcal{S}(\mathbb{R}^n) \to \mathcal{S}(\mathbb{R}^n)$. Note that $\mathcal{S}(\mathbb{R}^n)$ is infinite dimensional and we shall not expect for T that all statements can be proved for linear mappings from $\mathbb{R}^n \to \mathbb{R}^m$ or from $\mathbb{C}^n$ to $\mathbb{C}^m$. For two linear operators $T_j : \mathcal{S}(\mathbb{R}^n) \to \mathcal{S}(\mathbb{R}^n)$ we can consider $(\lambda T_1 + \mu T_2)(u) = \lambda T_1 u + \mu T_2 u$, $\lambda, \mu \in \mathbb{C}$, which defines a further linear operator from $\mathcal{S}(\mathbb{R}^n)$ into itself.

Definition 10.14. *We call a linear operator* $T : \mathcal{S}(\mathbb{R}^n) \to \mathcal{S}(\mathbb{R}^n)$ ***continuous*** *if for every sequence* $(u_k)_{k \in \mathbb{N}}$ *converging in* $\mathcal{S}(\mathbb{R}^n)$ *to some limit* $u \in \mathcal{S}(\mathbb{R}^n)$ *the sequence* $(Tu_k)_{k \in \mathbb{N}}$ *converges in* $\mathcal{S}(\mathbb{R}^n)$ *to* Tu, *i.e. if for all* $\alpha, \beta \in \mathbb{N}_0^n$ *we have* $\lim_{k \to \infty} p_{\alpha,\beta}(u_k - u) = 0$ *then it follows that* $\lim_{k \to \infty} p_{\alpha,\beta}(Tu_k - Tu) = 0$.

Suppose that $T : \mathcal{S}(\mathbb{R}^n) \to \mathcal{S}(\mathbb{R}^n)$ has the property that for every sequence $(v_k)_{k \in \mathbb{N}}$ converging in $\mathcal{S}(\mathbb{R}^n)$ to 0 (zero-function) it follows that $(Tv_k)_{k \in \mathbb{N}}$ converges to 0. In this case we call T **continuous at** $0 \in \mathcal{S}(\mathbb{R}^n)$. Let $(u_k)_{k \in \mathbb{N}}$ be a sequence converging in $\mathcal{S}(\mathbb{R}^n)$ to u. Now $(u_k - u)_{k \in \mathbb{N}}$ converges in $\mathcal{S}(\mathbb{R}^n)$ to 0 and therefore $(T(u_k - u))_{k \in \mathbb{N}}$ converges in $\mathcal{S}(\mathbb{R}^n)$ to 0. Since $p_{\alpha,\beta}(T(u_k - u)) = p_{\alpha,\beta}(Tu_k - Tu)$ we have proved

Corollary 10.15. *A linear operator* $T : \mathcal{S}(\mathbb{R}^n) \to \mathcal{S}(\mathbb{R}^n)$ *is continuous if and only if it is continuous at* 0.

Remark 10.16. A. It is obvious that if $T_1, T_2 : \mathbb{S}(\mathbb{R}^n) \to \mathcal{S}(\mathbb{R}^n)$ are continuous then for $\lambda, \mu \in \mathbb{C}$ the operator $\lambda T_1 + \mu T_2 : \mathcal{S}(\mathbb{R}^n) \to \mathcal{S}(\mathbb{R}^n)$ is continuous too, as is the composition $T_2 \circ T_1 : \mathcal{S}(\mathbb{R}^n) \to \mathcal{S}(\mathbb{R}^n)$.
B. From Lemma 10.1 we deduce that in Definition 10.14 we can replace the family $(p_{\alpha,\beta})_{\alpha,\beta \in \mathbb{N}_0^n}$ by the family $(q_{m_1,m_2})_{m_1,m_2 \in \mathbb{N}_0}$, see also Problem 7.

Lemma 10.17. *Suppose that $T : \mathcal{S}(\mathbb{R}^n) \to \mathcal{S}(\mathbb{R}^n)$ is a linear operator with the property that for every $\alpha, \beta \in \mathbb{N}_0^n$ there exists $\alpha_1, \beta_1, \alpha_2, \beta_2, \ldots, \alpha_M, \beta_M \in \mathbb{N}_0^n$, $M = M(\alpha, \beta)$, such that*

$$p_{\alpha,\beta}(Tu) \leq c \max_{1 \leq j \leq M} p_{\alpha_j, \beta_j}(u) \tag{10.32}$$

holds for all $u \in \mathcal{S}(\mathbb{R}^n)$, then T is continuous.

Proof. If $(u_k)_{k \in \mathbb{N}}$, $u_k \in \mathcal{S}(\mathbb{R}^n)$, converges in $\mathcal{S}(\mathbb{R}^n)$ to 0, then (10.32) says that for all $\alpha, \beta \in \mathbb{N}_0^n$ the terms $p_{\alpha,\beta}(Tu_k)$ converge to $0 \in \mathcal{S}(\mathbb{R})$, i.e. $(Tu_k)_{k \in \mathbb{N}}$ converges to $0 \in \mathcal{S}(\mathbb{R}^n)$ and the result follows. $\square$

Remark 10.18. A. In (10.32) we can replace on each side the family $(p_{\alpha,\beta})_{\alpha,\beta \in \mathbb{N}_0^n}$ by the family $(q_{m_1,m_2})_{m_1,m_2 \in \mathbb{N}_0}$ and the statement of Lemma 10.17 still holds, see Problem 7. Further, since for $a_k \geq 0$, $1 \leq k \leq N$, we have

$$\max_{1 \leq k \leq N} a_k \leq \sum_{k=1}^{N} a_k \leq N \max_{1 \leq k \leq N} a_k$$

we can replace (10.32) by

$$p_{\alpha,\beta}(Tu) \leq c' \sum_{j=1}^{M} p_{\alpha_j, \beta_j}(u). \tag{10.33}$$

B. In Volume V we will prove that the continuity of T at 0 implies (10.32), i.e. (10.32) is equivalent to the continuity of T.

Example 10.19. Every linear partial differential operator $P(D) = \sum_{|\gamma| \leq m} a_\gamma D^\gamma$ with constant coefficients $a_\gamma \in \mathbb{C}$ is continuous from $\mathcal{S}(\mathbb{R}^n)$ into $\mathcal{S}(\mathbb{R}^n)$. Indeed for $u \in \mathcal{S}(\mathbb{R}^n)$ we find with $\alpha, \beta \in \mathbb{N}_0^n$

$$p_{\alpha,\beta}(P(D)u) = p_{\alpha,\beta}\left(\sum_{|\gamma| \leq m} a_\gamma D^\gamma u \right)$$

$$= \sup_{x \in \mathbb{R}^n} \left| x^\beta \partial^\alpha \left(\sum_{|\gamma| \leq m} a_\gamma \partial^\gamma u \right)(x) \right|$$

$$\leq \sum_{|\gamma| \leq m} |a_\gamma| \sup_{x \in \mathbb{R}^n} \left| x^\beta \partial^{\gamma+\alpha} u(x) \right|$$

$$= \sum_{|\gamma| \leq m} |a_\gamma| p_{\alpha+\gamma, \beta}(u) \leq c \max_{|\gamma| \leq m} \sum_{|\gamma| \leq m} p_{\alpha+\gamma, \beta}(u)$$

and it follows $P(D)u \in \mathcal{S}(\mathbb{R}^n)$ as well as the continuous of $P(D)$.

Example 10.20. Let $w \in \mathcal{S}(\mathbb{R}^n)$ be fixed and define $T_w : \mathcal{S}(\mathbb{R}^n) \to \mathcal{S}(\mathbb{R}^n)$ by $T_w u := w * u$. We claim that T_w is continuous. We know already that $T_w u \in \mathcal{S}(\mathbb{R}^n)$ by Proposition 10.6. The proof of Proposition 10.6 yields further for $m_1, m_2 \in \mathbb{N}_0$ that

$$q_{m_1,m_2}(T_w u) = \sup_{x \in \mathbb{R}^n} \left(1 + \|x\|^2\right)^{\frac{m_1}{2}} \sum_{|\alpha| \leq m_2} |\partial^\alpha (w * u)(x)|$$

$$\leq 2^{\frac{m_1}{2}} \sum_{|\alpha| \leq m_2} \int_{\mathbb{R}^n} \left| \left(1 + \|x - y\|^2\right)^{\frac{m_1}{2}} \partial_x^\alpha w(x - y) \right| \left| \left(1 + \|y\|^2\right)^{\frac{m_1}{2}} u(y) \right| \, dy$$

$$\leq 2^{\frac{m_1}{2}} \sum_{|\alpha| \leq m_2} \int_{\mathbb{R}^n} \left| \left(1 + \|x - y\|^2\right)^{\frac{m_1}{2}} \partial_x^\alpha w(x - y) \right| \, dy \, q_{m_1,0}(u)$$

$$= 2^{\frac{m_1}{2}} \left(\sum_{|\alpha| \leq m_2} \left\| \left(1 + \| \cdot \|^2\right)^{\frac{m_1}{2}} \partial^\alpha w \right\|_{L^1} \right) q_{m_1,0}(u),$$

where we have used the translation invariance of the Lebesgue measure. Thus we have proved for every $m_1, m_2 \in \mathbb{N}_0$ that

$$q_{m_1,m_2}(T_w u) \leq C_w q_{m_1,0}(u)$$

and the continuity of T_w follows.

Problems

1. Give a short sketch of the proof of formula (10.14).

2. Prove that $u \in C_\infty(\mathbb{R}^n)$ is uniformly continuous.

3. Use the Cauchy-Schwarz inequality and Young's inequality to show for $h \in L^1(\mathbb{R}^n)$ and $u, v \in L^2(\mathbb{R}^n)$ the estimate

$$\left| \int_{\mathbb{R}^n} \left(\int_{\mathbb{R}^n} u(x) v(y) h(x - y) \, dy \right) dx \right| \leq \|h\|_{L^1} \|u\|_{L^2} \|v\|_{L^2}.$$

   Now prove that for every $M \in \mathbb{N}$ and all $u, v \in \mathcal{S}(\mathbb{R}^n)$ and $h \in L^1(\mathbb{R}^n)$ the integral $\int_{\mathbb{R}^n} \left(\int_{\mathbb{R}^n} (1 + \|x - y\|^2)^M h(x - y) u(x) v(y) \, dy \right) dx$ is finite.
   Hint: use Peetre's inequality.

4.   a) Prove that if $T : \mathbb{R}^n \to \mathbb{R}^n$ is a bijective linear mapping then $u \in \mathcal{S}(\mathbb{R}^n)$ implies $u \circ T \in \mathcal{S}(\mathbb{R}^n)$.

b) Show that for a symmetric positive definite $n \times n$-matrix A the function $x \mapsto f(x) = e^{-\langle Ax, x \rangle}$, $x \in \mathbb{R}^n$, belongs to $\mathcal{S}(\mathbb{R}^n)$.

c) Prove that the function $u : \mathbb{R} \to \mathbb{R}$, $u(x) = e^{-x^2} \sin(e^{x^2})$ is arbitrarily often differentiable and decays faster than every power, but that it does not belong to $\mathcal{S}(\mathbb{R}^n)$.

5. Denote by $O_M(\mathbb{R}^n)$ the set of all $g \in C^\infty(\mathbb{R}^n)$ such that for every $\alpha \in \mathbb{N}_0^n$ there exists a constant $c = c(\alpha, g) \geq 0$ and an integer $k = k(\alpha, g) \geq 0$ for which $|\partial^\alpha g(x)| \leq c(1 + \|x\|^2)^k$ holds. Prove that $u \in \mathcal{S}(\mathbb{R}^n)$ and $g \in O_M(\mathbb{R}^n)$ imply $g \cdot u \in \mathcal{S}(\mathbb{R}^n)$.

6. Prove that the composition $T_2 \circ T_1$ of two linear and continuous operators $T_j : \mathcal{S}(\mathbb{R}^n) \to \mathcal{S}(\mathbb{R}^n)$ is continuous.

7. Let $T : \mathcal{S}(\mathbb{R}^n) \to \mathcal{S}(\mathbb{R}^n)$ be a linear operator and $(u_k)_{k \in \mathbb{N}}$, $u_k \in \mathcal{S}(\mathbb{R}^n)$, a sequence as well as $u \in \mathcal{S}(\mathbb{R}^n)$. Prove that the following conditions are equivalent:

i) T is continuous;

ii) if for all $\alpha, \beta \in \mathbb{N}_0^n$ we have $\lim_{k \to \infty} p_{\alpha, \beta}(u_k - u) = 0$ then it follows that $\lim_{k \to \infty} q_{m_1, m_2}(Tu_k - Tu) = 0$ for all $m_1, m_2 \in \mathbb{N}_0$;

iii) if for all $m_1, m_2 \in \mathbb{N}_0$ we have $\lim_{k \to \infty} q_{m_1, m_2}(u_k - u) = 0$ then it follows that $\lim_{k \to \infty} p_{\alpha, \beta}(Tu_k - Tu) = 0$ for all $\alpha, \beta \in \mathbb{N}_0^n$;

iv) if for all $m_1, m_2 \in \mathbb{N}_0$ we have $\lim_{k \to \infty} q_{m_1, m_2}(u_k - u) = 0$ then it follows that $\lim_{k \to \infty} q_{n_1, n_2}(Tu_k - Tu) = 0$ for all $n_1, n_2 \in \mathbb{N}_0$.

8. Let $u \in \mathcal{S}(\mathbb{R})$ and $\alpha, \beta \in \mathbb{N}_0$. We set

$$\nu_{\alpha, \beta, L^2}(u) := \left(\int_{\mathbb{R}} |x^\beta \partial^\alpha u(x)|^2 dx \right)^{\frac{1}{2}}.$$

Prove that for every $\alpha, \beta \in \mathbb{N}_0$ there exist finitely many multi-indices $\alpha_1, \beta_1, \ldots, \alpha_m, \beta_m$ such that

$$p_{\alpha, \beta}(u) \leq C \max_{1 \leq k \leq m} \nu_{\alpha_k, \beta_k, L^2}(u).$$

Moreover for $\alpha, \beta \in \mathbb{N}_0$ there exist finitely many multi-indices $\gamma_1, \delta_1, \ldots, \gamma_N, \delta_N$ such that

$$\nu_{\alpha, \beta, L^2}(u) \leq C' \max_{1 \leq j \leq N} p_{\alpha_j, \beta_j}(u).$$

9.   a) Let $h \in L^p(\mathbb{R}^n)$, $1 \leq p < \infty$, and define the linear mapping $l_h :$ $\mathcal{S}(\mathbb{R}^n) \to \mathbb{C}$ by $l_h(\varphi) := \int_{\mathbb{R}^n} h(x)\varphi(x)\,\mathrm{d}x$. Prove that l_h is well defined and that for every sequence $(\varphi_k)_{k \in \mathbb{N}}$, $\varphi_k \in \mathcal{S}(\mathbb{R}^n)$, converging in $\mathcal{S}(\mathbb{R}^n)$ to $\varphi \in \mathcal{S}(\mathbb{R}^n)$ it follows that $\lim_{k \to \infty} l_h(\varphi_k) = l_h(\varphi)$, i.e. l_h is (sequentially) continuous from $\mathcal{S}(\mathbb{R}^n)$ to $\mathbb{C}$.

b) Let $\mu \in \mathcal{M}_b^+(\mathbb{R}^n)$ be a bounded measure and define $l_\mu : \mathcal{S}(\mathbb{R}^n) \to \mathbb{C}$ by $l_\mu(\varphi) = \int_{\mathbb{R}^n} \varphi\,\mathrm{d}\mu$. Prove that l_μ is well defined, linear and (sequentially) continuous from $\mathcal{S}(\mathbb{R}^n)$ to $\mathbb{C}$.

10. Let $P(x, D) = \sum_{|\alpha| \leq m} a_\alpha(x)\partial^\alpha$ be a linear differential operator with coefficients $a_\alpha \in O_M(\mathbb{R}^n)$. Prove that $P(x, D) : \mathcal{S}(\mathbb{R}^n) \to \mathcal{S}(\mathbb{R}^n)$ is continuous.

11 The Fourier Transform in $\mathcal{S}(\mathbb{R}^n)$

In this chapter we study the Fourier transform in Schwartz space. To a large extent we follow Section 3.1 of [69] which partially used [68]. We start with

Definition 11.1. *For $u \in \mathcal{S}(\mathbb{R}^n)$ we define its Fourier transform $\hat{u}$ by*

$$\hat{u}(\xi) := (2\pi)^{-\frac{n}{2}} \int_{\mathbb{R}^n} e^{-ix\cdot\xi} u(x)\, \mathrm{d}x. \tag{11.1}$$

Remark 11.2. A. Since for every $\xi \in \mathbb{R}^n$ the function $x \mapsto e^{-ix\cdot\xi}u(x)$ belongs to $L^1(\mathbb{R}^n)$ the integral in (11.1) is well defined, hence a function $\hat{u} : \mathbb{R}^n \to \mathbb{C}$ is defined by (11.1).
B. Sometimes we will write $F(u)(\xi)$ or $(F_{x\mapsto\xi}u)(\xi)$ for $\hat{u}(\xi)$.
C. The constant $(2\pi)^{-\frac{n}{2}}$ is chosen in order to have equality in Plancherel's theorem, see Theorem 11.12 below. Other choices of the constant in (11.1) are used and equally well justified, for example to have the constant 1 in the convolution theorem, see Theorem 11.14 below.

Theorem 11.3. *The Fourier transform F is a continuous linear mapping from $\mathcal{S}(\mathbb{R}^n)$ into itself.*

Proof. First we prove that for $u \in \mathcal{S}(\mathbb{R}^n)$ the function $\hat{u}$ defined by (11.1) belongs to $\mathcal{S}(\mathbb{R}^n)$. For this let $\alpha, \beta \in \mathbb{N}_0^n$ and note that

$$
\begin{aligned}
\xi^\beta \partial_\xi^\alpha \hat{u}(\xi) &= \xi^\beta \partial_\xi^\alpha \left((2\pi)^{-\frac{n}{2}} \int_{\mathbb{R}^n} e^{-ix\cdot\xi} u(x)\, \mathrm{d}x \right) \\
&= (2\pi)^{-\frac{n}{2}} \int_{\mathbb{R}^n} \xi^\beta \partial_\xi^\alpha \left(e^{-ix\cdot\xi} \right) u(x)\, \mathrm{d}x \\
&= (2\pi)^{-\frac{n}{2}} \int_{\mathbb{R}^n} \xi^\beta (-ix)^\alpha e^{-ix\cdot\xi} u(x)\, \mathrm{d}x \\
&= (2\pi)^{-\frac{n}{2}} \int_{\mathbb{R}^n} \partial_x^\beta \left(e^{-ix\cdot\xi} \right) i^{|\beta|} (-ix)^\alpha u(x)\, \mathrm{d}x \\
&= (2\pi)^{-\frac{n}{2}} (-i)^{|\beta|+|\alpha|} \int_{\mathbb{R}^n} e^{-ix\cdot\xi} \partial_x^\beta (x^\alpha u(x))\, \mathrm{d}x,
\end{aligned}
$$

where in the last step we integrated by parts taking into account that the boundary terms vanish. Since with u the function $x \mapsto \partial_x^\beta (x^\alpha u(x))$ also belongs to $\mathcal{S}(\mathbb{R}^n)$, hence to $L^1(\mathbb{R}^n)$, we conclude that for all $\alpha, \beta \in \mathbb{N}_0^n$ we have

$$p_{\alpha,\beta}(\hat{u}) = \sup_{\xi \in \mathbb{R}^n} \left| \xi^\beta \partial_\xi^\alpha \hat{u}(\xi) \right| \leq (2\pi)^{-\frac{n}{2}} \int_{\mathbb{R}^n} \left| \partial_x^\beta (x^\alpha u(x)) \right| \mathrm{d}x < \infty, \tag{11.2}$$

i.e. $\hat{u} \in S(\mathbb{R}^n)$. Now the linearity of the Fourier transform considered as a mapping $F : S(\mathbb{R}^n) \to S(\mathbb{R}^n)$ is obvious. Finally we prove the continuity of F. By Leibniz' rule we find

$$\partial_x^\beta (x^\alpha u(x)) = \sum_{\gamma \le \beta} \binom{\beta}{\gamma} (\partial_x^\gamma x^\alpha) \partial_x^{\beta-\gamma} u(x),$$

and since

$$\int_{\mathbb{R}^n} \left(1 + \|x\|^2\right)^{\frac{-n-1}{2}} \, \mathrm{d}x = c_n < \infty$$

we find

$$\int_{\mathbb{R}^n} \left|\partial_x^\beta (x^\alpha u(x))\right| \mathrm{d}x \le c_n \sum_{\gamma \le \beta} \binom{\beta}{\gamma} \sup_{x \in \mathbb{R}^n} \left(\left(1 + \|x\|^2\right)^{\frac{n+1}{2}} \left|\partial_x^\gamma (x^\alpha)\right| \left|\partial_x^{\beta-\gamma} u(x)\right| \right)$$

$$\le C \max_{\substack{\gamma \le \beta \\ |\sigma| \le n+1+|\alpha|}} p_{\beta-\gamma,\sigma}(u)$$

implying by (11.2) the continuity of $F : S(\mathbb{R}^n) \to S(\mathbb{R}^n)$. $\qquad\square$

Corollary 11.4. *For $\alpha, \beta \in \mathbb{N}_0^n$ and $u \in S(\mathbb{R}^n)$*

$$\xi^\beta D_\xi^\alpha \hat{u}(\xi) = (-1)^{|\alpha|} \left(D_x^\beta (x^\alpha u(\cdot)) \right)\hat{\ }(\xi) \tag{11.3}$$

holds, where $D_\xi^\alpha = (-i\partial_\xi)^\alpha$.

Proof. Using the calculation in the proof of Theorem 11.3 we deduce

$$\xi^\beta D_\xi^\alpha \hat{u}(\xi) = (2\pi)^{-\frac{n}{2}} (-1)^{|\alpha|} \int_{\mathbb{R}^n} (-i)^{-|\beta|} \partial_x^\beta \left(e^{-ix\cdot\xi} \right) x^\alpha u(x) \, \mathrm{d}x$$

$$= (2\pi)^{-\frac{n}{2}} (-1)^{|\alpha|} \int_{\mathbb{R}^n} (-i)^{|\beta|} e^{-ix\cdot\xi} \partial_x^\beta (x^\alpha u(x)) \, \mathrm{d}x$$

$$= (2\pi)^{-\frac{n}{2}} (-1)^{|\alpha|} \int_{\mathbb{R}^n} e^{-ix\cdot\xi} D_x^\beta (x^\alpha u(x)) \, \mathrm{d}x$$

$$= (-1)^{|\alpha|} F_{x \mapsto \xi} \left(D_x^\beta (x^\alpha u(\cdot)) \right)(\xi).$$

$\qquad\square$

Remark 11.5. A. In terms such as $F_{x \mapsto \xi} \left(D_x^\beta (x^\alpha u(\cdot)) \right)$ there is some abuse of notation which however is easy to interpret and makes the presentation easier and hence we, as many other authors, will make use of it.

B. For a differential operator $P(D) = \sum_{|\alpha| \le m} a_\alpha D^\alpha$, $a_\alpha \in \mathbb{C}$, formula (11.3) yields

$$(P(D)u)\hat{\ }(\xi) = P(\xi)\hat{u}(\xi), \quad u \in S(\mathbb{R}^n). \tag{11.4}$$

Lemma 11.6. *The function* $g(x) = e^{-\frac{\|x\|^2}{2}}$ *is a fixed point of* F, *i.e.*

$$\hat{g}(\xi) = e^{-\frac{\|\xi\|^2}{2}}. \tag{11.5}$$

Proof. Since

$$\int_{\mathbb{R}^n} e^{-ix\cdot\xi} e^{-\frac{\|x\|^2}{2}}\, \mathrm{d}x = \prod_{j=1}^n \int_{\mathbb{R}} e^{-ix_j\xi_j} e^{-\frac{x_j^2}{2}}\, \mathrm{d}x_j$$

it is sufficient to consider the case $n = 1$. So let $h(t) = e^{-\frac{t^2}{2}}$, $t \in \mathbb{R}$. We know that $h \in \mathcal{S}(\mathbb{R})$ and it satisfies the differential equation

$$iD_t h(t) + t h(t) = 0,$$

but now Corollary 11.4 implies that

$$\tau \hat{h}(\tau) + i D_\tau \hat{h}(\tau) = 0$$

which yields that $\hat{h}(\tau) = c e^{-\frac{\tau^2}{2}}$. The constant c is determined as

$$c = \hat{h}(0) = (2\pi)^{-\frac{1}{2}} \int_{\mathbb{R}} e^{-\frac{t^2}{2}}\, \mathrm{d}t = 1.$$

$\square$

Remark 11.7. A. The reader may compare Lemma 11.6 with Example II.22.19 where for the case $n = 2$ the calculations are given in much more detail, note that g is a real-valued function and h is an even function, hence

$$\int_{\mathbb{R}} e^{-i\tau t} h(t)\, \mathrm{d}t = \int_{\mathbb{R}} h(t) \cos \tau t\, \mathrm{d}t.$$

B. We can now interpret Problem 13 of Chapter III.25 differently: the function $x \mapsto \frac{1}{\cosh\left(\sqrt{\frac{\pi}{2}}x\right)}$ is a further fixed point of the Fourier transform, but this first requires to justify that the integral $(2\pi)^{-\frac{1}{2}} \int_{-\infty}^{\infty} e^{-ix\xi} \frac{1}{\cosh\left(\sqrt{\frac{\pi}{2}}x\right)}\, \mathrm{d}x$ is a Fourier transform.

We want to prove that $F : \mathcal{S}(\mathbb{R}^n) \to \mathcal{S}(\mathbb{R}^n)$ is a bijective mapping with a continuous inverse.

Theorem 11.8. *For* $u \in \mathcal{S}(\mathbb{R}^n)$ *the functions defined by*

$$\left(F^{-1}u\right)(\eta) := (2\pi)^{-\frac{n}{2}} \int_{\mathbb{R}^n} e^{i\eta\cdot y} u(y)\, \mathrm{d}y \tag{11.6}$$

belongs to $\mathcal{S}(\mathbb{R}^n)$ *and* F^{-1} *maps* $\mathcal{S}(\mathbb{R}^n)$ *linearly and continuously into itself. Furthermore the Fourier transform* $F : \mathcal{S}(\mathbb{R}^n) \to \mathcal{S}(\mathbb{R}^n)$ *is bijective with continuous inverse* F^{-1}.

Proof. That $F^{-1}u \in S(\mathbb{R}^n)$ and $F^{-1} : S(\mathbb{R}^n) \to S(\mathbb{R}^n)$ is continuous is proved along the lines of Theorem 11.3 and left to the reader. We are going to prove that $F \circ F^{-1} = F^{-1} \circ F = \text{id}$ where id denotes the identity on $S(\mathbb{R}^n)$. For $u, v \in S(\mathbb{R}^n)$ we find

$$\int_{\mathbb{R}^n} \hat{u}(\xi) v(\xi) e^{ix \cdot \xi} \, d\xi = (2\pi)^{-\frac{n}{2}} \int_{\mathbb{R}^n} v(\xi) e^{ix \cdot \xi} \int_{\mathbb{R}^n} e^{-iy \cdot \xi} u(y) \, dy \, d\xi$$

$$= \int_{\mathbb{R}^n} u(y) (2\pi)^{-\frac{n}{2}} \int_{\mathbb{R}^n} e^{-i(y-x) \cdot \xi} v(\xi) \, d\xi \, dy$$

$$= \int_{\mathbb{R}^n} u(y) (Fv)(y-x) \, dy$$

$$= \int_{\mathbb{R}^n} (Fv)(z) u(z+x) \, dz.$$

With $v(x) := e^{-\frac{\epsilon^2 \|x\|^2}{2}}$ we deduce from Lemma 11.6 that

$$\int_{\mathbb{R}^n} \hat{u}(\xi) e^{-\frac{\epsilon^2 \|x\|^2}{2}} e^{ix \cdot \xi} \, d\xi = \epsilon^{-n} \int_{\mathbb{R}^n} e^{-\frac{\|z\|^2}{2\epsilon^2}} u(x+z) \, dz \tag{11.7}$$

$$= \int_{\mathbb{R}^n} e^{-\frac{\|y\|^2}{2}} u(x+\epsilon y) \, dy,$$

where we used the formula

$$F(g_\epsilon)(\xi) = \epsilon^{-n} (Fg) \left(\frac{\xi}{\epsilon} \right), \tag{11.8}$$

which holds for all $g \in S(\mathbb{R}^n)$, $g_\epsilon(x) = g(\epsilon x)$. Indeed, (11.7) follows by a substitution:

$$\int_{\mathbb{R}^n} e^{-ix \cdot \xi} g(\epsilon x) \, dx = \int_{\mathbb{R}^n} e^{-i(\epsilon x) \cdot \frac{\xi}{\epsilon}} g(\epsilon x) \, dx = \epsilon^{-n} \int_{\mathbb{R}^n} e^{-iy \cdot \frac{\xi}{\epsilon}} g(y) \, dy.$$

Applying the dominated convergence theorem to (11.7) we get as $\epsilon \to 0$

$$\int_{\mathbb{R}^n} \hat{u}(\xi) e^{ix \cdot \xi} \, d\xi = \int_{\mathbb{R}^n} e^{-\frac{\|y\|^2}{2}} u(x) \, dy = (2\pi)^{\frac{n}{2}} u(x),$$

which gives

$$u(x) = (2\pi)^{-\frac{n}{2}} \int_{\mathbb{R}^n} e^{ix \cdot \xi} \hat{u}(\xi) \, d\xi = F^{-1}(\hat{u})(x).$$

Thus we find $F^{-1} \circ F = \text{id}$ implying that F is injective. Interchanging the role of F and F^{-1} in our calculation we arrive at $F \circ F^{-1} = \text{id}$ implying the surjectivity of F. So F is bijective with continuous inverse F^{-1}. $\qquad \square$

Definition 11.9. *We call*

$$\left(F^{-1}u\right)(\eta) := (2\pi)^{-\frac{n}{2}} \int_{\mathbb{R}^n} e^{i\eta \cdot y} u(y) \, \mathrm{d}y \tag{11.9}$$

*the **inverse Fourier transform** of $u \in \mathcal{S}(\mathbb{R}^n)$.*

Remark 11.10. A. We will also use the notation $F^{-1}_{y \mapsto \eta}(u)(\eta)$.
B. On $\mathcal{S}(\mathbb{R}^n)$ we find $F^4 = \mathrm{id}$ or $F^{-1} = F^3$ and moreover, since

$$(2\pi)^{-\frac{n}{2}} \int_{\mathbb{R}^n} \overline{e^{-ix \cdot \xi} u(x)} \, \mathrm{d}x = (2\pi)^{-\frac{n}{2}} \int_{\mathbb{R}^n} e^{ix \cdot \xi} \overline{u(x)} \, \mathrm{d}x$$

we have

$$\overline{(Fu)(\xi)} = F^{-1}(\overline{u})(\xi). \tag{11.10}$$

We now collect some useful properties of the Fourier transform.

Lemma 11.11. A. *Let $T_a : \mathbb{R}^n \to \mathbb{R}^n$, $T_a x = a + x$, be the translation operator. For $u \in \mathcal{S}(\mathbb{R}^n)$ we find*

$$(u \circ T_a)\hat{\ }(\xi) = e^{ia \cdot \xi} \hat{u}(\xi). \tag{11.11}$$

B. *If $T : \mathbb{R}^n \to \mathbb{R}^n$ is a bijective linear mapping and $u \in \mathcal{S}(\mathbb{R}^n)$ then $u \circ T \in \mathcal{S}(\mathbb{R}^n)$ and*

$$(u \circ T)\hat{\ }(\xi) = \frac{1}{|\det T|} \left(\hat{u} \circ \left(T^{-1}\right)^t\right)(\xi) \tag{11.12}$$

where T^t is the transposed mapping (matrix). In particular for the reflection $Sx = -x$ we have

$$(u \circ S)\hat{\ }(\xi) = \hat{u}(-\xi), \tag{11.13}$$

and

$$(u \circ H_\lambda)\hat{\ }(\xi) = \lambda^{-n} \hat{u}\left(\frac{\xi}{\lambda}\right), \qquad \lambda > 0, \tag{11.14}$$

for the homothetic mapping $H_\lambda(x) = \lambda x$.

Proof. **A.** For $a \in \mathbb{R}^n$ and $u \in \mathcal{S}(\mathbb{R}^n)$ we have

$$(u \circ T_a)\hat{\ }(\xi) = (2\pi)^{-\frac{n}{2}} \int_{\mathbb{R}^n} e^{-ix \cdot \xi} u(x + a) \, \mathrm{d}x$$

$$= (2\pi)^{-\frac{n}{2}} \int_{\mathbb{R}^n} e^{-i(y-a) \cdot \xi} u(y) \, \mathrm{d}y$$

$$= e^{ia \cdot \xi} \hat{u}(\xi).$$

B. With $y = Tx$ we get by the chain rule for $v(x) = u(y)$ that

$$(\partial^\alpha v)(x) = \left(\sum_{|\gamma| \le |\alpha|} c_{\alpha\gamma} (\partial^\alpha u) \circ T^{-1} \right)(y)$$

implying that $v = u \circ T \in \mathcal{S}(\mathbb{R}^n)$. Now we find

$$(u \circ T)\hat{}(\xi) = (2\pi)^{-\frac{n}{2}} \int_{\mathbb{R}^n} e^{-ix\cdot\xi} u(Tx)\, dx$$

$$= \frac{1}{|\det T|} (2\pi)^{-\frac{n}{2}} \int_{\mathbb{R}^n} e^{-i(T^{-1}y)\cdot\xi} u(y)\, dy$$

$$= \frac{1}{|\det T|} (2\pi)^{-\frac{n}{2}} \int_{\mathbb{R}^n} e^{-iy\cdot(T^{-1})^t\xi} u(y)\, dy$$

$$= \frac{1}{|\det T|} \left(\hat{u} \circ (T^{-1})^t \right)(\xi).$$

$\square$

The following two estimates for the Fourier transform have far reaching consequences.

Theorem 11.12. *For $u \in \mathcal{S}(\mathbb{R}^n)$*

$$\|\hat{u}\|_\infty \le (2\pi)^{-\frac{n}{2}} \|u\|_{L^1} \tag{11.15}$$

and

$$\|\hat{u}\|_{L^2} = \|u\|_{L^2} \tag{11.16}$$

hold.

Remark 11.13. The equality (11.16) is a first variant of **Plancherel's theorem**, see Theorem 12.17 below. Using the polarisation identity we derive from (11.16)

$$\langle u, v \rangle_{L^2} = \langle \hat{u}, \hat{v} \rangle_{L^2}, \quad u, v \in \mathcal{S}(\mathbb{R}^n), \tag{11.17}$$

where $\langle \cdot, \cdot \rangle_{L^2}$ denotes the scalar product in $L^2(\mathbb{R}^n)$ considered as a vector space over $\mathbb{C}$.

Proof of Theorem 11.12. For $u \in \mathcal{S}(\mathbb{R}^n)$ we find

$$\|\hat{u}\|_\infty = (2\pi)^{-\frac{n}{2}} \sup_{\xi \in \mathbb{R}^n} \left| \int_{\mathbb{R}^n} e^{-ix\cdot\xi} u(x)\, dx \right|$$

$$\le (2\pi)^{-\frac{n}{2}} \int_{\mathbb{R}^n} |u(x)|\, dx = (2\pi)^{-\frac{n}{2}} \|u\|_{L^1},$$

proving (11.15). In order to show (11.16) we start with $u, v \in \mathcal{S}(\mathbb{R}^n)$ and note that

$$\int_{\mathbb{R}^n} u(x)\hat{v}(x)\,dx = \int_{\mathbb{R}^n} u(x)(2\pi)^{-\frac{n}{2}} \int_{\mathbb{R}^n} e^{-ix\cdot\xi}v(\xi)\,d\xi\,dx$$

$$= (2\pi)^{-\frac{n}{2}} \int_{\mathbb{R}^n} \int_{\mathbb{R}^n} e^{-ix\cdot\xi}u(x)v(\xi)\,dx\,d\xi$$

$$= \int_{\mathbb{R}^n} \hat{u}(\xi)v(\xi)\,d\xi.$$

We now choose for v the function $v = \bar{\hat{u}}$ and since $F\left(\bar{\hat{u}}\right) = \overline{F \circ F^{-1}u} = \bar{u}$ we arrive at (11.16). □

Next we prove the **convolution theorem** in $\mathcal{S}(\mathbb{R}^n)$:

Theorem 11.14. *For $u, v \in \mathcal{S}(\mathbb{R}^n)$ it follows*

$$(u \cdot v)\hat{}\,(\xi) = (2\pi)^{-\frac{n}{2}}\,(\hat{u} * \hat{v})\,(\xi) \tag{11.18}$$

as well as

$$(u * v)\hat{}\,(\xi) = (2\pi)^{\frac{n}{2}}\hat{u}(\xi) \cdot \hat{v}(\xi). \tag{11.19}$$

Proof. We start by taking in (11.17) the functions $y \mapsto e^{-iy\cdot\xi}u(y)$ and $y \mapsto \overline{v(y)}$ and we find

$$\left\langle e^{-i\langle\cdot,\xi\rangle}u, \bar{v}\right\rangle_{L^2} = \int_{\mathbb{R}^n} e^{-iy\cdot\xi}u(y)v(y)\,dy = (2\pi)^{\frac{n}{2}}\,(u \cdot v)\hat{}\,(\xi)$$

for the left hand side in (11.17), whereas for the right hand side we get

$$\left\langle F\left(e^{-i\langle\cdot,\xi\rangle}u\right), F\bar{v}\right\rangle_{L^2} = \int_{\mathbb{R}^n}\left(F_{y\to\eta}\left(e^{-iy\cdot\xi}u(y)\right)\right)(\eta)\overline{\left(F_{y\to\eta}\left(\overline{v(y)}\right)\right)}(\eta)\,d\eta$$

$$= (2\pi)^{-\frac{n}{2}}\int_{\mathbb{R}^n}\int_{\mathbb{R}^n} e^{-iy\cdot\xi}e^{-iy\cdot\eta}u(y)\,dy\,\hat{v}(-\eta)\,d\eta$$

$$= \int_{\mathbb{R}^n}\hat{u}(\xi+\eta)\hat{v}(-\eta)\,d\eta = \int_{\mathbb{R}^n}\hat{u}(\xi-\eta)\hat{v}(\eta)\,d\eta,$$

where we have used Lemma 11.11 as well as (11.10). Thus we have shown (11.18). Now, in order to prove (11.19) we choose in (11.18) $\hat{u}$ and $\hat{v}$ instead of u and v and hence we obtain

$$(\hat{u} \cdot \hat{v})\hat{}\,(\xi) = (2\pi)^{-\frac{n}{2}}\left((\hat{u})\hat{} * (\hat{v})\hat{}\right)(\xi)$$

or

$$(\hat{u} \cdot \hat{v})\,(\xi) = (2\pi)^{-\frac{n}{2}}F^{-1}\left((\hat{u})\hat{} * (\hat{v})\hat{}\right)(\xi).$$

With $\tilde{w}(x) = w^{\sim}(x) = w(-x)$ we find $Fu = \left(F^{-1}(u)\right)^{\sim}$ or $(\hat{u})^{\hat{}} = \tilde{u}$. Hence we deduce

$$F^{-1}\left((\hat{u})^{\hat{}} * (\hat{v})^{\hat{}}\right)(\xi) = F^{-1}(\tilde{u} * \tilde{v})(\xi)$$

and since $\tilde{u} * \tilde{v} = (u * v)^{\sim}$ as well as $F^{-1}\tilde{w} = \hat{w}$ we arrive eventually at (11.19). $\square$

Example 11.15. In this example we start to discuss the heat equation $\frac{\partial w(t,x)}{\partial t} - \Delta_n w(t,x) = 0$ with the help of the Fourier transform. In particular we want to develop ideas which later on will lead to the theory of strongly continuous one-parameter operator semigroups. We start by calculating the Fourier transform with respect to $x \in \mathbb{R}^n$ of

$$h_t(x) := (4\pi t)^{-\frac{n}{2}} e^{-\frac{\|x\|^2}{4t}}. \tag{11.20}$$

The function $(t, x) \mapsto h_t(x)$, $t > 0$, $x \in \mathbb{R}^n$, is called the **heat kernel**. We rewrite $h_t(x)$ as

$$h_t(x) = (4\pi t)^{-\frac{n}{2}} \left(g \circ H_{\frac{1}{\sqrt{2t}}}\right)(x)$$

with $g(x) = e^{-\frac{\|x\|^2}{2}}$ and $H_\lambda(x) = \lambda x$. With the help of (11.14) we obtain

$$\hat{h}_t(\xi) = (4\pi t)^{-\frac{n}{2}} \left(\frac{1}{\sqrt{2t}}\right)^{-n} \hat{g}\left(\frac{\xi}{\frac{1}{\sqrt{2t}}}\right)$$

$$= (2\pi)^{-\frac{n}{2}} e^{-t\|\xi\|^2}. \tag{11.21}$$

For $u \in \mathcal{S}(\mathbb{R}^n)$ and $t > 0$ we consider

$$(T_t u)(x) = (h_t * u)(x) = \int_{\mathbb{R}^n} h_t(x - y) u(y) \, dy. \tag{11.22}$$

Since $\frac{\partial}{\partial t} h_t(x) - \Delta_n h_t(x) = 0$ we find that

$$\left(\frac{\partial}{\partial t} - \Delta_n\right)(T_t u)(x) = \int_{\mathbb{R}^n} \left(\frac{\partial}{\partial t} - \Delta_{n,x}\right) h_t(x - y) u(y) \, dy = 0. \tag{11.23}$$

This equality can be obtained by a different calculation using the Fourier transform:

$$
F_{x \mapsto \xi}\left(\left(\frac{\partial}{\partial t} - \Delta_n\right) T_t u\right) = \left(\left(\frac{\partial}{\partial t} - \Delta_n\right) T_t u\right)^{\widehat{}}(\xi)
$$

$$
= \left(\frac{\partial}{\partial t} + \|\xi\|^2\right) (T_t u)^{\widehat{}}(\xi) = \left(\frac{\partial}{\partial t} + \|\xi\|^2\right) (h_t * u)^{\widehat{}}(\xi)
$$

$$
= (2\pi)^{\frac{n}{2}} \left(\frac{\partial}{\partial t} + \|\xi\|^2\right) \hat{h}_t(\xi) \hat{u}(\xi)
$$

$$
= (2\pi)^{\frac{n}{2}} \left(\frac{\partial}{\partial t} + \|\xi\|^2\right) (2\pi)^{-\frac{n}{2}} e^{-t\|\xi\|^2} \hat{u}(\xi)
$$

$$
= \left(\frac{\partial}{\partial t} e^{-t\|\xi\|^2} + \|\xi\|^2 e^{-t\|\xi\|^2}\right) \hat{u}(\xi) = 0.
$$

Since $h_t, u \in \mathcal{S}(\mathbb{R}^n)$ an application of the inverse Fourier transform yields again

$$
\left(\frac{\partial}{\partial t} - \Delta_n\right) T_t u = 0.
$$

Next we consider for $t_1, t_2 > 0$ the composition $T_{t_2} \circ T_{t_1}$. Using the Fourier transform we get

$$
\left((T_{t_2} \circ T_{t_1}) u\right)^{\widehat{}}(\xi) = \left(T_{t_2} (T_{t_1} u)\right)^{\widehat{}}(\xi)
$$

$$
= \left(h_{t_2} * (T_1 u)\right)^{\widehat{}}(\xi) = (2\pi)^{\frac{n}{2}} \hat{h}_{t_2}(\xi) \cdot (T_1 u)^{\widehat{}}(\xi)
$$

$$
= (2\pi)^n \left(\hat{h}_{t_2} \cdot \hat{h}_{t_1} \cdot \hat{u}\right)(\xi) = e^{-(t_2+t_1)\|\xi\|^2} \hat{u}(\xi)
$$

$$
= (2\pi)^{\frac{n}{2}} \left((2\pi)^{-\frac{n}{2}} e^{-(t_2+t_1)\|\xi\|^2} \hat{u}(\xi)\right)
$$

$$
= \left(h_{t_2+t_1} * u\right)^{\widehat{}}(\xi) = \left(T_{t_2+t_1} u\right)^{\widehat{}}(\xi).
$$

Again we use that h_t, u and $h_t * u$ belong to $\mathcal{S}(\mathbb{R}^n)$ and by applying the inverse Fourier transform to $\left(T_{t_2} \circ T_{t_1}\right)^{\widehat{}} u = \left(T_{t_2+t_1} u\right)^{\widehat{}}$ we arrive at $\left(T_{t_2} \circ T_{t_1}\right) u = T_{t_2+t_1} u$ or the **semi-group** property

$$
T_{t_2} \circ T_{t_1} = T_{t_2+t_1} \tag{11.24}
$$

for the family of operators $(T_t)_{t>0}$, $T_t : \mathcal{S}(\mathbb{R}^n) \to \mathcal{S}(\mathbb{R}^n)$, where $T_t u = h_t * u$. Finally we claim

$$
\lim_{t \to 0} \|T_t u - u\|_\infty = 0. \tag{11.25}
$$

First we note for $u \in \mathcal{S}(\mathbb{R}^n)$ that

$$
u(x) = F^{-1}(\hat{u})(x) = (2\pi)^{-\frac{n}{2}} \int_{\mathbb{R}^n} e^{ix \cdot \xi} \hat{u}(\xi) \, d\xi
$$

which implies

$$\|u\|_\infty \le (2\pi)^{-\frac{n}{2}} \|\hat{u}\|_{L^1}. \tag{11.26}$$

This estimate is now applied to $T_t u - u$ to find

$$\|T_t u - u\|_\infty \le (2\pi)^{-\frac{n}{2}} \|(T_t u - u)^\wedge\|_{L^1}$$
$$= (2\pi)^{-\frac{n}{2}} \int_{\mathbb{R}^n} \left| e^{-t\|\xi\|^2} - 1 \right| |\hat{u}(\xi)| \, d\xi$$

and since $\left| e^{-t\|\xi\|^2} - 1 \right| \le 2$ and $\lim_{t\to 0} e^{-t\|\xi\|^2} = 1$ the dominated convergence theorem implies

$$\lim_{t\to 0} \|T_t u - u\|_\infty = 0.$$

Eventually we have proved that for $t > 0$, $x \in \mathbb{R}^n$ a solution to

$$\frac{\partial}{\partial t} w(t, x) - \Delta_n w(t, x) = 0$$

is given by $w(t, x) := T_t u(x)$, $u \in \mathcal{S}(\mathbb{R}^n)$, and

$$\lim_{\substack{t\to 0 \\ t>0}} w(t, x) = u(x)$$

uniformly on $\mathbb{R}^n$.

Our aim here was to see the Fourier transform at work, we will return to this example at several occasions and will provide a refined analysis.

The space $C_0^\infty(\mathbb{R}^n)$ is a subset of $\mathcal{S}(\mathbb{R}^n)$ and therefore for every $u \in C_0^\infty(\mathbb{R}^n)$ the Fourier transform is well defined. Since $u \in C_0^\infty(\mathbb{R}^n)$ and $\xi \in \mathbb{R}^n$

$$\int_{\mathbb{R}^n} e^{-ix\cdot\xi} u(x) \, dx = \int_{\text{supp } u} e^{-ix\cdot\xi} u(x) \, dx$$

and for every $z \in \mathbb{C}^n$ fixed the function $x \mapsto e^{-i(x_1 z_1 + \cdots + x_n z_n)} u(x)$ is integrable over supp u, since by assumption supp u is compact, we can extend $\hat{u}$ to $\mathbb{C}^n$.

Definition 11.16. *The **Fourier-Laplace transform** of $u \in C_0^\infty(\mathbb{R}^n)$ is defined on $\mathbb{C}^n$ by*

$$z \mapsto (2\pi)^{-\frac{n}{2}} \int_{\mathbb{R}^n} e^{-ix\cdot z} u(x) \, dx \tag{11.27}$$

with $x \cdot z = x_1 z_1 + \cdots + x_n z_n$.

We note that in (11.27) the integral is to be taken over supp u.

Let us discuss briefly the case $n = 1$ first. For $x \in \mathbb{R}$ the function $z \mapsto e^{-ixz}$ is a holomorphic function. Let γ be a simply closed piecewise C^1-curve in $\mathbb{C}$ and suppose that $u \in C_0^\infty(\mathbb{R})$ has a support contained in the interval $[-R, R]$. Applying Fubini's theorem first and then Cauchy's integral theorem we find

$$\int_\gamma \left(\int_{-R}^R e^{-ixz} u(x) \, \mathrm{d}x \right) \mathrm{d}z = \int_{-R}^R \left(\int_\gamma e^{-ixz} \, \mathrm{d}z \, u(x) \right) \mathrm{d}x = 0$$

and by Morera's theorem, see Theorem III.22.1, we conclude that the Fourier-Laplace transform of $u \in C_0^\infty(\mathbb{R})$ is an entire function. This result also holds in n-dimensions, however we first have to introduce holomorphic functions of several variables.

Let $G \subset \mathbb{C}^n$ be an open set and $f : G \to \mathbb{C}$ be a continuous function. Let $a = (a_1, \ldots, a_n) \in G$. For $1 \leq j \leq n$ we can consider the functions

$$g_j(z_j) := f(a_1, \ldots, a_{j-1}, a_j + z_j, a_{j+1}, \ldots, a_n). \tag{11.28}$$

We call $f : G \to \mathbb{C}$ **holomorphic in G** if for every $a \in \mathbb{C}^n$ and all $j = 1, \ldots, n$ the functions g_j are holomorphic in a neighbourhood of $0 \in \mathbb{C}$. If f is holomorphic in $\mathbb{C}^n$ we call f an **entire function** of n complex variables. In many respects the theory of holomorphic functions of several variables differs much from the theory of holomorphic functions of one variable, we just refer to [41], [63], [81] or [102]. However, the definition allows us to transfer certain properties known for holomorphic functions of one complex variable to the general case. For our purposes we need an extension of Morera's theorem and we only give a formulation of the case $\mathbb{C}^n$. For $a \in \mathbb{C}^n$ fixed we set

$$\mathbb{C}_{j,a} := \left\{ z \in \mathbb{C}^n \,\middle|\, z = (a_1, \ldots, a_{j-1}, a_j + z_j, a_{j+1}, \ldots, a_n) \right\} \subset \mathbb{C}^n$$

which is a complex plane lying in $\mathbb{C}^n$. Let γ_j be a simply closed piecewise C^1-curve with trace in $\mathbb{C}_{j,a} \subset \mathbb{C}^n$. If f is an entire function in $\mathbb{C}^n$ then for all these curves and all $j = 1, \ldots, n$ we have

$$\int_{\gamma_j} f(z) \, \mathrm{d}z_j = 0. \tag{11.29}$$

Moreover the converse holds: the vanishing of all integrals implies that f is an entire function. These statements are not deep since we can reduce them to the complex one-dimensional case using (11.28) and the definition of holomorphy in $\mathbb{C}^n$.

Every $z \in \mathbb{C}^n$ we can split according to $z = \operatorname{Re} z + i \operatorname{Im} z = x + iy$, $x, y \in \mathbb{R}^n$, or

$\mathbb{C}^n = \mathbb{R}^n + i\mathbb{R}^n$. In this sense we may consider $\mathbb{R}^n$ as a subset of $\mathbb{C}^n$. Now we can formulate an analogous result to the uniqueness theorem for entire functions on $\mathbb{C}$, see also Problem 10.

Lemma 11.17. *Suppose that the entire function $f : \mathbb{C}^n \to \mathbb{C}$ vanishes on $\mathbb{R}^n$, i.e. $f|_{\mathbb{R}^n} = 0$. Then f is identically 0 in $\mathbb{C}^n$.*

With these preparations we can prove the **Theorem of Paley-Wiener**

Theorem 11.18. *A function $f : \mathbb{C}^n \to \mathbb{C}$ is the Fourier-Laplace transform of $u \in C_0^\infty(\mathbb{R}^n)$ with $\operatorname{supp} u \subset \overline{B_R(0)}$ if and only if f is an entire function and for every $N \in \mathbb{N}$ there exists a constant C_N such that*

$$|f(z)| \le C_N \frac{e^{R|\operatorname{Im} z|}}{(1 + \|z\|)^N} \tag{11.30}$$

holds for all $z \in \mathbb{C}^n$.

Proof. Let $u \in C_0^\infty(\mathbb{R}^n)$ and $\operatorname{supp} u \subset \overline{B_R(0)}$. For $z \in \mathbb{C}^n$ we find that

$$f(z) := (2\pi)^{-\frac{n}{2}} \int_{\mathbb{R}^n} e^{-ix\cdot z} u(x)\,\mathrm{d}x = (2\pi)^{-\frac{n}{2}} \int_{B_R(0)} e^{-ix\cdot z} u(x)\,\mathrm{d}x$$

is a continuous function on $\mathbb{C}^n$ ($\cong \mathbb{R}^{2n}$) and for any simply closed piecewise C^1-curve γ_j in $\mathbb{C}_j = \mathbb{C}_{j,0}$, $1 \le j \le n$, we have

$$\int_{\gamma_j} f(z)\,\mathrm{d}z_j = (2\pi)^{-\frac{n}{2}} \int_{\gamma_j} \left(\int_{B_R(0)} e^{-ix\cdot z} u(x)\,\mathrm{d}x \right) \mathrm{d}z_j$$

$$= (2\pi)^{-\frac{n}{2}} \int_{B_R(0)} \left(\int_{\gamma_j} e^{-ix\cdot z}\,\mathrm{d}z_j \right) u(x)\,\mathrm{d}x = 0,$$

hence f is an entire function. Now let $\alpha \in \mathbb{N}_0^n$ and note

$$(2\pi)^{\frac{n}{2}} z^\alpha f(z) = \int_{B_R(0)} u(x) z^\alpha e^{-ix\cdot z}\,\mathrm{d}x$$

$$= (-1)^{|\alpha|} \int_{B_R(0)} u(x) D_x^\alpha \left(e^{-ix\cdot z} \right)\,\mathrm{d}x$$

$$= \int_{B_R(0)} (D_x^\alpha u(x)) e^{-ix\cdot z}\,\mathrm{d}x$$

where we used again that when integrating by parts the boundary terms vanish. Thus we find

$$|z^\alpha||f(z)| \le (2\pi)^{-\frac{n}{2}} \|D^\alpha u\|_{L^1} e^{R|\operatorname{Im} z|} \tag{11.31}$$

since $\left|e^{-ix\cdot z}\right| = \left|e^{-ix\cdot\operatorname{Re}z}e^{x\cdot\operatorname{Im}z}\right| = e^{x\cdot\operatorname{Im}z}$, with $z = \operatorname{Re}z + i\operatorname{Im}z$, and $e^{x\cdot\operatorname{Im}z} \le e^{R|\operatorname{Im}z|}$ for $x \in \overline{B_R(0)}$. We observe further that

$$1 + \sum_{|\alpha|=N} |z|^\alpha \ge \tilde{C}_N \left(1 + \|z\|\right)^N$$

for some $\tilde{C}_N > 0$ and now (11.31) implies (11.30).

Next we prove the converse statement, i.e. assuming that for an entire function $f : \mathbb{C}^n \to \mathbb{C}$ estimate (11.30) holds, we want to find $u \in C_0^\infty(\mathbb{R}^n)$ such that $f(z) = (2\pi)^{-\frac{n}{2}} \int_{\mathbb{R}^n} e^{-ix\cdot z} u(x)\,\mathrm{d}x$. The natural choice for u is of course

$$u(x) := (2\pi)^{-\frac{n}{2}} \int_{\mathbb{R}^n} e^{ix\cdot\xi} f(\xi)\,\mathrm{d}\xi, \qquad x \in \mathbb{R}^n, \tag{11.32}$$

where $\xi = \operatorname{Re}z$. From (11.30) we deduce that for every $\alpha \in \mathbb{N}_0^n$ the functions $\xi \mapsto \xi^\alpha f(\xi)$ belong to $L^1(\mathbb{R}^n)$ and therefore we find that

$$\partial_x^\alpha u(x) = (2\pi)^{-\frac{n}{2}} \int_{\mathbb{R}^n} \left(\partial_x^\alpha e^{ix\cdot\xi}\right) f(\xi)\,\mathrm{d}\xi$$

$$= (2\pi)^{-\frac{n}{2}} i^{|\alpha|} \int_{\mathbb{R}^n} e^{ix\cdot\xi} \xi^\alpha f(\xi)\,\mathrm{d}\xi < \infty,$$

i.e. $u \in C^\infty(\mathbb{R}^n)$. Next we claim for $1 \le j \le n$

$$\int_{\mathbb{R}} f(z_1, \dots, \xi_j + i\eta_j, \dots, z_n) e^{i(x_1 z_1 + \cdots + x_j(\xi_j + i\eta_j) + \cdots + x_n z_n)}\,\mathrm{d}\xi_j$$

is independent of η_j. To see this we apply the Cauchy integral theorem to the curve $\gamma_j : [0,1] \to \mathbb{C}\ (= \mathbb{C}_j)$ defined by

$$\gamma_j(s) := \begin{cases} 8\xi_j^0 s, & 0 \le s \le \frac{1}{4} \\ 4i\eta_j^0 s + \xi_j^0 - i\eta_j^0, & \frac{1}{4} \le s \le \frac{1}{2} \\ -8\xi_j^0 s + 5\xi_j^0 + i\eta_j^0, & \frac{1}{2} \le s \le \frac{3}{4} \\ -4i\eta_j^0 s - \xi_j^0 + 4i\eta_j^0, & \frac{3}{4} \le s \le 1, \end{cases} \tag{11.33}$$

$\xi_j^0, \eta_j^0 \in \mathbb{R}$, see Figure 11.1:

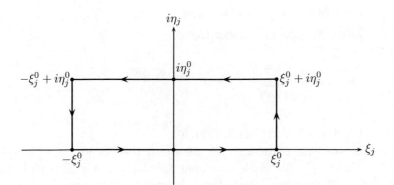

Figure 11.1

Since f is an entire function we find for

$$h(z_j) = f(z_1, \ldots, z_j, \ldots, z_n) \, e^{i(x_1 z_1 + \cdots + x_j z_j + \cdots + x_n z_n)}$$

that

$$\int_{\gamma_j} h(z_j) \, \mathrm{d}z_j = 0$$

or

$$\int_{-\xi_j^0}^{\xi_j^0} h(\xi_j) \, \mathrm{d}\xi_j = \int_{-\xi_j^0}^{\xi_j^0} h(\xi_j + i\eta_j^0) \, \mathrm{d}\xi_j - \int_0^{\eta_j^0} h(\xi_j^0 + i\eta_j) \, \mathrm{d}\eta_j + \int_0^{\eta_j^0} h(-\xi_j^0 + i\eta_j) \, \mathrm{d}\eta_j.$$

From (11.30) we deduce for $z = t + iy$, $t, y \in \mathbb{R}^n$, that

$$\left| h\left(\pm \xi_j^0 + i\eta_j \right) \right| \le C_N \frac{e^{R(y_1^2 + \cdots + \eta_j^2 + \cdots + y_n^2)^{\frac{1}{2}}} e^{-(x_1 y_1 + \cdots + x_j \eta_j + \cdots + x_n y_n)}}{\left(1 + \left(|z_1|^2 + \cdots + (\xi_j^0)^2 + \eta_j^2 + \cdots + |z_n|^2 \right)^{\frac{1}{2}} \right)^N},$$

which implies for x fixed that

$$\left| \int_0^{\eta_j} h\left(\pm \xi_j^0 + i\eta_j \right) \mathrm{d}\eta_j \right| \le C_N \eta_j^0 \frac{e^{R(y_1^2 + \cdots (\eta_j^0)^2 + \cdots + y^2)} e^{-(x_1 y_1 + \cdots + |x_j| \eta_j^0 + \cdots + x_n y_n)}}{\left(1 + \left(|z_1|^2 \cdots + (\xi_j^0)^2 + \cdots + |z_n|^2 \right)^{\frac{1}{2}} \right)^N}.$$

For $\xi_j^0 \to \infty$ we now get

$$\int_{\mathbb{R}} h(\xi_j) \, \mathrm{d}\xi_j = \int_{\mathbb{R}} h(\xi_j + i\eta_j^0) \, \mathrm{d}\xi_j,$$

and therefore we arrive at

$$u(x) = (2\pi)^{-\frac{n}{2}} \int_{\mathbb{R}} f(t+iy)e^{ix\cdot(t+iy)}\, \mathrm{d}t. \tag{11.34}$$

Let $x \in \mathbb{R}^n$, $x \neq 0$, and set $y = \lambda\frac{x}{\|x\|}$, $\lambda > 0$, i.e. $x \cdot y = \lambda\|x\|$ and $\|y\| = \lambda$. It follows from (11.30) that

$$\left| f(t+iy)e^{ix\cdot(t+iy)} \right| \leq C_N e^{(R-\|x\|)\lambda}\,(1+\|t\|)^{-N}$$

and for N sufficiently large, i.e. $N > n$, we deduce further

$$|u(x)| = (2\pi)^{-\frac{n}{2}} \left| \int_{\mathbb{R}^n} f(t+iy)e^{ix\cdot(t+iy)}\,\mathrm{d}t \right|$$

$$\leq C_N' e^{(R-\|x\|)\lambda} \int_{\mathbb{R}^n} (1+\|t\|)^{-N}\,\mathrm{d}t$$

$$= C_N'' e^{(R-\|x\|)\lambda} < \infty.$$

Now if $\|x\| > R$ we find for $\lambda \to \infty$ that $u(x) = 0$, i.e. $\mathrm{supp}\, u \subset \overline{B_R(0)}$, implying that $u \in C_0^\infty(\mathbb{R}^n)$, $\mathrm{supp}\, u \subset \overline{B_R(0)}$. From the first part we can now deduce that

$$z \mapsto (2\pi)^{-\frac{n}{2}} \int_{\mathbb{R}^n} e^{ix\cdot z}u(x)\,\mathrm{d}x$$

is an entire function which restricted to $\mathbb{R}^n$ is the inverser Fourier transform of $u \in C_0^\infty(\mathbb{R}^n) \subset \mathcal{S}(\mathbb{R}^n)$. Thus we have for $t \in \mathbb{R}^n$

$$f(t) = (2\pi)^{-\frac{n}{2}} \int_{\mathbb{R}^n} e^{-it\cdot x}u(x)\,\mathrm{d}x. \tag{11.35}$$

The left hand side of (11.35) is an entire function and the right hand side has by the first part an extension to an entire function. Since they coincide on $\mathbb{R}^n$ they must be equal and the theorem is proved. $\qquad\square$

Remark 11.19. In our presentation of Theorem 11.18 and its proof we follow closely our monograph [69] which in turn picked up ideas from W. Rudin [102].

Problems

1.   a) Prove that $w(x) := \frac{1}{\cosh x}$ defines an element in $\mathcal{S}(\mathbb{R}^n)$.

b) Use the Fourier transform to show $\int_{\mathbb{R}^n} \frac{1}{\cosh\sqrt{\frac{\pi}{2}}x}\,\mathrm{d}x = \sqrt{2\pi}$.

2.   a) Find $F(xe^{-\frac{x^2}{2}})(\xi)$.

b) Use Plancherel's formula to show for $v \in \mathcal{S}(\mathbb{R})$

$$\int_{\mathbb{R}} \xi^2 |\hat{u}(\xi)|^2 \, d\xi = \int_{\mathbb{R}} |v'(x)|^2 \, dx.$$

3.   a) Use the recurrence and differential equations stated in Appendix III for the Hermite polynomials H_n, $n \in \mathbb{N}_0$, to derive for the Hermite functions $h_n(x) := e^{-\frac{x^2}{2}} H_n(x)$ the relations

$$xh_n(x) + h'_n(x) = 2nh_{n-1}(x),$$
$$xh_n(x) - h'_n(x) = h_{n+1}(x),$$
$$h''_n(x) - x^2 h_n(x) + (2n+1)h_n(x) = 0.$$

b) Prove that the Hermite functions $(h_n)_{n \in \mathbb{N}_0}$ form an orthogonal system in $L^2(\mathbb{R})$.

c) Use the relations of part a) to prove that $h_n \in \mathcal{S}(\mathbb{R})$ satisfies $\hat{h}_n = (-i)^n h_n$.

4. For $s \in \mathbb{R}$ and $\Lambda^s : \mathbb{R}^n \to \mathbb{R}$, $\Lambda^s(\xi) = (1 + \|\xi\|^2)^{\frac{s}{2}}$ define on $\mathcal{S}(\mathbb{R}^n)$ the linear operator $\Lambda^s(D)$ by

$$\Lambda^s(D)u(x) := (2\pi)^{-\frac{n}{2}} \int_{\mathbb{R}^n} e^{ix \cdot \xi} \Lambda^s(\xi) \hat{u}(\xi) \, d\xi.$$

Prove that $\Lambda^s(D)$ maps $\mathcal{S}(\mathbb{R}^n)$ continuously and bijectively onto $\mathcal{S}(\mathbb{R}^n)$ with continuous inverse $\Lambda^{-s}(D)$. Furthermore show that $(\Lambda^s(D))_{s \in \mathbb{R}}$ forms an Abelian group of operators on $\mathcal{S}(\mathbb{R}^n)$ and we have $\Lambda^s(D) \circ \Lambda^t(D) = \Lambda^{s+t}(D)$.

5. Let $f \in \mathcal{S}(\mathbb{R}^n)$. Show that $u := F^{-1}\left(\frac{\hat{f}}{(1+\|\cdot\|^2)^k}\right)$, $k \in \mathbb{N}$, solves the partial differential equation $(1 - \Delta_n)^k u = f$.

6. Let $P(D) = \sum_{|\alpha| \leq m} a_\alpha D^\alpha$ be a linear differential operator with constant coefficients $a_\alpha \in \mathbb{C}$. For $u, v \in \mathcal{S}(\mathbb{R}^n)$ prove

$$P(D)(u * v) = (P(D)u) * v = u * (P(D)v).$$

7. For $u \in \mathcal{S}(\mathbb{R}^n)$ and $p \geq 2$ prove the estimate

$$\|\hat{u}\|_{L^p} \leq (2\pi)^{\frac{(2-p)n}{2p}} \|u\|_{L^1}^{1-\frac{2}{p}} \|u\|_{L^2}^{\frac{2}{p}}.$$

8. For which $c > 0$ can we find a solution $u \in \mathcal{S}(\mathbb{R})$ of the integral equation

$$\int_{\mathbb{R}} e^{-\frac{|x-y|^2}{2}} u(y) \, dy = e^{-c|x|^2}?$$

9. Let $f \in \mathcal{S}(\mathbb{R})$ be real-valued. It follows that

$$\hat{f}(\xi) = \frac{1}{\sqrt{2\pi}} \int_{\mathbb{R}} e^{-ix\xi} f(x) \, dx$$
$$= \frac{1}{\sqrt{2\pi}} \int_{\mathbb{R}} f(x) \cos x\xi \, dx - i \frac{1}{\sqrt{2\pi}} \int_{\mathbb{R}} f(x) \sin x\xi \, dx.$$

We call

$$(F_{\cos} f)(\xi) := \frac{1}{\sqrt{2\pi}} \int_{\mathbb{R}} f(x) \cos x\xi \, dx = \sqrt{\frac{2}{\pi}} \int_0^\infty f(x) \cos x\xi \, dx$$

the **cosine-transform** of f and

$$(F_{\sin} f)(\xi) := \frac{1}{\sqrt{2\pi}} \int_{\mathbb{R}} f(x) \sin x\xi \, dx = \sqrt{\frac{2}{\pi}} \int_0^\infty f(x) \sin x\xi \, dx$$

the **sine-transform** of f.

a) Prove that for an even function we have $Ff = F_{\cos} f$ and for an odd function we have $Fg = -iF_{\sin} g$.

b) For $f \in \mathcal{S}(\mathbb{R})$ show that:

$$(F_{\cos} f')(\xi) = \xi F_{\sin}(f)(\xi) - \sqrt{\frac{2}{\pi}} f(0);$$
$$(F_{\sin} f')(\xi) = -\xi F_{\cos}(f)(\xi).$$

10. Let $f : \mathbb{C}^n \to \mathbb{C}$ be an entire function vanishing on $\mathbb{R}^n$. Prove that f is identically equal to 0.

12 The Fourier Transform in L^p-Spaces

For $u \in L^1(\mathbb{R}^n)$ and $\xi \in \mathbb{R}^n$ the function $x \mapsto e^{-ix\cdot\xi}u(x)$ is an element in $L^1(\mathbb{R}^n)$ and therefore we can give

Definition 12.1. *For $u \in L^1(\mathbb{R}^n)$ its **Fourier transform** $\hat{u}$,*

$$\hat{u}(\xi) := (Fu)(\xi) := (F_{x\mapsto\xi}u)(\xi) := (2\pi)^{-\frac{n}{2}} \int_{\mathbb{R}^n} e^{-ix\cdot\xi}u(x)\,\mathrm{d}x, \tag{12.1}$$

defines a function $\hat{u} : \mathbb{R}^n \to \mathbb{C}$.

Since $\mathcal{S}(\mathbb{R}^n) \subset L^1(\mathbb{R}^n)$ this definition extends Definition 11.1. Moreover we find for $u \in L^1(\mathbb{R}^n)$ that

$$\|\hat{u}\|_\infty \le (2\pi)^{-\frac{n}{2}} \|u\|_{L^1}, \tag{12.2}$$

compare with Theorem 11.12, which allows us to prove a version of the **Riemann-Lebesgue lemma**:

Theorem 12.2. *The Fourier transform $\hat{u}$ of $u \in L^1(\mathbb{R}^n)$ is a continuous function vanishing at infinity, i.e. $\hat{u} \in C_\infty(\mathbb{R}^n)$.*

Proof. We know that $\mathcal{S}(\mathbb{R}^n)$ is dense in $L^1(\mathbb{R}^n)$ as well as in $C_\infty(\mathbb{R}^n)$. Let $(u_k)_{k\in\mathbb{N}}$ be a sequence in $\mathcal{S}(\mathbb{R}^n)$ converging in $L^1(\mathbb{R}^n)$ to u. For the Fourier transforms we find

$$\|\hat{u}_k - \hat{u}\|_\infty \le (2\pi)^{-\frac{n}{2}} \|u_k - u\|_{L^1},$$

implying that $\hat{u}$ is the uniform limit of Schwartz space functions $\hat{u}$. Hence $\hat{u}$ must be an element in $C_\infty(\mathbb{R}^n)$, see Lemma 10.12. $\square$

We can introduce formally $F : L^1(\mathbb{R}^n) \to C_\infty(\mathbb{R}^n)$, $Fu = \hat{u}$, which is clearly a linear and continuous operator: the linearity is obvious and if $u_k \to u$ in $L^1(\mathbb{R}^n)$ then by (12.2) we have $Fu_k \to Fu$ in $C_\infty(\mathbb{R}^n)$. The following example shows that for $u \in L^1(\mathbb{R}^n)$ the Fourier transform $\hat{u}$ is not necessarily an element of $L^1(\mathbb{R}^n)$.

Example 12.3. By [46], 17.34.7, we find for $\psi_\lambda : \mathbb{R} \to \mathbb{R}$, $\psi_\lambda(x) = |x|^\lambda e^{-|x|}$, $\lambda > -1$, that

$$\hat{\psi}_\lambda(\xi) = \frac{2\Gamma(\lambda+1)}{\sqrt{2\pi}} \left(\frac{1}{1+\xi^2}\right)^{\frac{\lambda+1}{2}} \cos((\lambda+1)\arctan\xi). \tag{12.3}$$

For $\lambda = -\frac{3}{4}$, i.e. $\psi(x) := \psi_{-\frac{3}{4}}(x) = |x|^{-\frac{3}{4}}e^{-|x|}$, $x \ne 0$, and $\psi(0) = 0$, we have

$$\int_\mathbb{R} |x|^{-\frac{3}{4}}e^{-|x|}\,\mathrm{d}x \le 2\int_0^1 |x|^{-\frac{3}{4}}\,\mathrm{d}x + 2\int_1^\infty e^{-x}\,\mathrm{d}x = 4 + \frac{2}{e},$$

i.e. $\psi \in L^1(\mathbb{R})$. We observe that for $\frac{\pi}{3} \leq \xi < \infty$ it follows that

$$\cos \frac{\sqrt{1}}{4} \leq \cos \left(\frac{1}{4} \arctan \xi \right) \leq \cos \frac{\pi}{8} \tag{12.4}$$

and

$$\left(\frac{1}{2\xi^2} \right)^{\frac{1}{8}} \leq \left(\frac{1}{1+\xi^2} \right)^{\frac{1}{8}} \leq \left(\frac{1}{\xi^2} \right)^{\frac{1}{8}}. \tag{12.5}$$

The lower bounds in (12.4) and (12.5) yield for the Fourier transform $\hat{\psi}$, i.e.

$$\hat{\psi}(\xi) = \frac{2\Gamma \left(\frac{1}{4} \right)}{\sqrt{2\pi}} \left(\frac{1}{1+\xi^2} \right)^{\frac{1}{8}} \cos \left(\frac{1}{4} \arctan \xi \right), \tag{12.6}$$

that

$$\int_{\frac{\pi}{3}}^{\infty} \left| \hat{\psi}(\xi) \right| \mathrm{d}\xi \geq \int_{\frac{\pi}{3}}^{\infty} \frac{1}{(2\xi^2)^{\frac{1}{8}}} \cos \frac{\sqrt{3}}{4} \, \mathrm{d}\xi = \infty,$$

thus $\hat{\psi}$ is not an element in $L^1(\mathbb{R}^n)$. Clearly $\hat{\psi}$ is a continuous function and it vanishes at infinity.

Having in mind the formula for the inverse Fourier transform in $\mathcal{S}(\mathbb{R}^n)$, in light of Example 12.3, we are led to conjecture that $F : L^1(\mathbb{R}) \to C_\infty(\mathbb{R}^n)$ is not surjective. We will deal with this problem as well as with the injectivity of $F : L^1(\mathbb{R}^n) \to C_\infty(\mathbb{R}^n)$ shortly. Without any change of argument we can prove, as for functions in $\mathcal{S}(\mathbb{R}^n)$, the following:

Lemma 12.4. *For $u \in L^1(\mathbb{R}^n)$ we have*

$$(u \circ T_a)\hat{\ }(\xi) = e^{ia\cdot\xi} \hat{u}(\xi), \quad T_a x = a + x; \tag{12.7}$$

$$(u \circ T)\hat{\ }(\xi) = \frac{1}{|\det T|} \hat{u} \circ \left(T^{-1} \right)^t (\xi), \quad T \in GL(n;\mathbb{R}); \tag{12.8}$$

$$(u \circ S)\hat{\ }(\xi) = \hat{u}(-\xi), \quad Sx = -x; \tag{12.9}$$

$$(u \circ H_\lambda)\hat{\ }(\xi) = \lambda^{-n} \hat{u} \left(\frac{\xi}{\lambda} \right), \quad H_\lambda(x) = \lambda x, \, \lambda > 0; \tag{12.10}$$

$$\overline{\hat{u}(\xi)} = F^{-1}(\overline{u})(\xi). \tag{12.11}$$

Note that for $u \in L^1(\mathbb{R}^n)$ the definition

$$\left(F^{-1} u \right)(\xi) = (2\pi)^{-\frac{n}{2}} \int_{\mathbb{R}^n} e^{ix\cdot\xi} u(x) \, \mathrm{d}x$$

makes sense.

Theorem 12.5. *For $u, v \in L^1(\mathbb{R}^n)$ it follows that*

$$(u * v)\hat{\ }(\xi) = (2\pi)^{\frac{n}{2}} \hat{u}(\xi)\hat{v}(\xi) \tag{12.12}$$

holds.

Proof. By Young's inequality we know that $u * v \in L^1(\mathbb{R}^n)$ and Fubini's theorem now yields

$$\begin{aligned}
(u * v)\hat{\ }(\xi) &= (2\pi)^{-\frac{n}{2}} \int_{\mathbb{R}^n} \left(\int_{\mathbb{R}^n} e^{-ix\cdot\xi} u(x-y)v(y)\,\mathrm{d}y \right) \mathrm{d}x \\
&= (2\pi)^{-\frac{n}{2}} \int_{\mathbb{R}^n} \left(\int_{\mathbb{R}^n} e^{-i(x-y)\cdot\xi} u(x-y) e^{-iy\cdot\xi} v(y)\,\mathrm{d}y \right) \mathrm{d}x \\
&= \hat{u}(\xi) \int_{\mathbb{R}^n} e^{-iy\cdot\xi} v(y)\,\mathrm{d}y = (2\pi)^{\frac{n}{2}} \hat{u}(\xi)\hat{v}(\xi).
\end{aligned}$$

$\square$

Recall that for $u, v \in L^1(\mathbb{R}^n)$ the function $u \cdot v$ does not necessarily belong to $L^1(\mathbb{R}^n)$ and therefore $(u \cdot v)\hat{\ }$ is not necessarily defined.

Lemma 12.6. *For $u, v \in L^1(\mathbb{R}^n)$ the equality*

$$\int_{\mathbb{R}^n} \hat{u}(\xi)v(\xi)\,\mathrm{d}\xi = \int_{\mathbb{R}^n} u(x)\hat{v}(x)\,\mathrm{d}x \tag{12.13}$$

holds.

Proof. Again it is Fubini's theorem which gives the result:

$$\begin{aligned}
\int_{\mathbb{R}^n} \hat{u}(\xi)v(\xi)\,\mathrm{d}\xi &= (2\pi)^{-\frac{n}{2}} \int_{\mathbb{R}^n} \left(\int_{\mathbb{R}^n} e^{-ix\cdot\xi} u(x)\,\mathrm{d}x \right) v(\xi)\,\mathrm{d}\xi \\
&= (2\pi)^{-\frac{n}{2}} \int_{\mathbb{R}^n} \left(\int_{\mathbb{R}^n} v(\xi)e^{-ix\cdot\xi}\,\mathrm{d}\xi\, u(x) \right) \mathrm{d}x \\
&= \int_{\mathbb{R}^n} \hat{v}(x)u(x)\,\mathrm{d}x.
\end{aligned}$$

$\square$

We have seen that the Friedrichs mollifier is a rather powerful tool for approximating less smooth functions by rather smooth functions. A similar procedure we encounter in Chapter 4 when discussing "good kernels" and the approximation of identity. In fact, in Example 11.15 we have seen that the heat kernel can also be interpreted as an approximation of identity. The following result is of analogous nature. Let $\Phi \in C_\infty(\mathbb{R}^n) \cap L^1(\mathbb{R}^n)$ such that $\Phi(0) = 1$ and set $\varphi := \hat{\Phi}$. It follows from (12.13) and the calculation made in the proof of Theorem 11.8 that $\int_{\mathbb{R}^n} \varphi(x) \, dx = (2\pi)^{-\frac{n}{2}}$ and with $\varphi_\epsilon(x) = \epsilon^{-n}\varphi\left(\frac{x}{\epsilon}\right)$, $\epsilon > 0$ we find by (12.10) that

$$(\Phi \circ H_\epsilon)\hat{}\,(\xi) = \epsilon^{-n}\hat{\Phi}\left(\frac{\xi}{\epsilon}\right) = \varphi_\epsilon(\xi). \tag{12.14}$$

If we now choose in (12.13) for v the function $\xi \mapsto e^{iy \cdot \xi}(\Phi \circ H_\epsilon)(\xi)$ we find

$$\int_{\mathbb{R}^n} \hat{u}(\xi)e^{iy \cdot \xi}\Phi(\epsilon\xi) \, d\xi = \int_{\mathbb{R}^n} u(y)\varphi_\epsilon(x - y) \, dy. \tag{12.15}$$

With these preparations we can prove

Theorem 12.7. *Let Φ, φ and φ_ϵ be as above and define*

$$M_\epsilon(u)(x) := (2\pi)^{-\frac{n}{2}}\int_{\mathbb{R}^n} \hat{u}(\xi)e^{ix \cdot \xi}\Phi(\epsilon\xi) \, d\xi. \tag{12.16}$$

It follows that

$$\lim_{\epsilon \to 0} \|M_\epsilon(u) - u\|_{L^1} = 0. \tag{12.17}$$

Proof. From (12.15) we derive that $M_\epsilon(u) = (2\pi)^{-\frac{n}{2}} u * \varphi_\epsilon$ and it remains to show that

$$\lim_{\epsilon \to 0} \left\|(2\pi)^{-\frac{n}{2}} u * \varphi_\epsilon - u\right\|_{L^1} = 0.$$

By Minkowski's integral inequality, Theorem III.9.25, we find

$$\left\|(2\pi)^{-\frac{n}{2}} u * \varphi_\epsilon - u\right\|_{L^1} = \int_{\mathbb{R}^n} \left|(2\pi)^{-\frac{n}{2}}\int_{\mathbb{R}^n}(u(x - y) - u(x))\varphi_\epsilon(y) \, dy\right| dx$$

$$\leq (2\pi)^{-\frac{n}{2}}\int_{\mathbb{R}^n}\int_{\mathbb{R}^n} |u(x - y) - u(x)| \, dx \, \epsilon^{-n}\left|\varphi\left(\frac{y}{\epsilon}\right)\right| dy$$

$$= (2\pi)^{-\frac{n}{2}}\int_{\mathbb{R}^n}\int_{\mathbb{R}^n} |u(x - \epsilon y) - u(x)| \, dx \, |\varphi(y)| \, dy.$$

We note that $\int_{\mathbb{R}^n} |u(x - \epsilon y) - u(x)| \, dx \leq 2\|u\|_{L^1}$. For $u \in \mathcal{S}(\mathbb{R}^n) \subset L^1(\mathbb{R}^n)$ we have in addition that

$$\lim_{\epsilon \to 0} \int_{\mathbb{R}^n} |u(x - \epsilon y) - u(x)| \, dx = 0,$$

and the dominated convergence theorem yields (12.17) for $u \in \mathcal{S}(\mathbb{R}^n)$. For a general $u \in L^1(\mathbb{R}^n)$ we use the fact that we can approximate u by functions from $\mathcal{S}(\mathbb{R}^n)$ and (12.17) follows for all $u \in L^1(\mathbb{R}^n)$, see also Problem 6. $\qquad\square$

The Wiener algebra which we are introducing now turns out to be a rather useful tool for handling the Fourier transform in $L^1(\mathbb{R}^n)$, but it is also interesting by its own right.

Definition 12.8. *The* ***Wiener algebra*** $A(\mathbb{R}^n)$ *is defined by*

$$A(\mathbb{R}^n) := \left\{ u \in L^1(\mathbb{R}^n) \,\middle|\, \hat{u} \in L^1(\mathbb{R}^n) \right\}. \tag{12.18}$$

We can use Theorem 12.7 to invert F on $A(\mathbb{R}^n)$.

Proposition 12.9. *For $u \in A(\mathbb{R}^n)$ we have $\lambda^{(n)}$-almost everywhere*

$$u(x) = (2\pi)^{-\frac{n}{2}} \int_{\mathbb{R}^n} e^{ix\cdot\xi} \hat{u}(\xi) \, d\xi. \tag{12.19}$$

Proof. With $M_\epsilon(u)$ as in Theorem 12.7 we know that $M_\epsilon(u)$ converges in $L^1(\mathbb{R}^n)$ to u which implies that for some sequence $(\epsilon_k)_{k\in\mathbb{N}}$, $\lim_{k\to 0} \epsilon_k = 0$, the sequence $(M_{\epsilon_k}(u))_{k\in\mathbb{N}}$ converges $\lambda^{(n)}$-almost everywhere to u, compare with Corollary III.7.37 and use a diagonal argument for the sets $(B_N(0))_{N\in\mathbb{N}}$. However, if $u \in A(\mathbb{R}^n)$, then the dominated convergence theorem implies that $M_{\epsilon_k}(u)$ converges to $(2\pi)^{-\frac{n}{2}} \int_{\mathbb{R}^n} e^{ix\cdot\xi} \hat{u}(\xi) \, d\xi$. $\square$

An immediate consequence of Proposition 12.9 is

Corollary 12.10. *The Fourier transform $F : L^1(\mathbb{R}^n) \to C_\infty(\mathbb{R}^n)$ is injective.*

Proof. Since F is linear we need to prove that $u \in L^1(\mathbb{R}^n)$ and $\hat{u} = 0 \in C_\infty(\mathbb{R}^n) \cap L^1(\mathbb{R}^n)$ then $u = 0$ in $L^1(\mathbb{R}^n)$. But this follows immediately from (12.19). $\square$

From (12.19) we conclude that $u \in A(\mathbb{R}^n)$ is almost everywhere equal to a unique element in $C_\infty(\mathbb{R}^n)$ and in the following we will always work with this function and hence identify $A(\mathbb{R}^n)$ with a subspace of $C_\infty(\mathbb{R}^n)$. The fact that $A(\mathbb{R}^n)$ is a linear space is trivial. As a norm on $A(\mathbb{R}^n)$ we introduce

$$\|u\|_{A(\mathbb{R}^n)} := \|u\|_{L^1} + \|\hat{u}\|_{L^1}. \tag{12.20}$$

Lemma 12.11. A. *A function $u \in L^1(\mathbb{R}^n)$ belongs to $A(\mathbb{R}^n)$ if and only if $\hat{u}$ belongs to $A(\mathbb{R}^n)$ and for $u \in A(\mathbb{R}^n) \subset C_\infty(\mathbb{R}^n)$ we have*

$$\|u\|_\infty \leq (2\pi)^{-\frac{n}{2}} \|u\|_{A(\mathbb{R}^n)}. \tag{12.21}$$

B. *For $1 \leq p \leq \infty$ the inclusion $A(\mathbb{R}^n) \subset L^p(\mathbb{R}^n)$ holds.*

Proof. **A.** Using the inversion formula (12.19) we find

$$u(-x) = (2\pi)^{-\frac{n}{2}} \int_{\mathbb{R}^n} e^{-ix\cdot\xi}\hat{u}(\xi)\,\mathrm{d}\xi = \left(F^2 u\right)(x). \tag{12.22}$$

Moreover, $x \mapsto u(-x)$ belongs to $L^1(\mathbb{R}^n)$ if and only if $u \in L^1(\mathbb{R}^n)$, and therefore $u \in A(\mathbb{R}^n)$ implies that $\hat{u}$ and $F^2 u$ belong to $L^1(\mathbb{R}^n)$, i.e. $\hat{u} \in A(\mathbb{R}^n)$. Conversely, $\hat{u} \in A(\mathbb{R}^n)$ gives $\hat{u} \in L^1(\mathbb{R}^n)$ and $F^2 u \in L^1(\mathbb{R}^n)$ implying by (12.22) that $\hat{u}$ and u are elements of $L^1(\mathbb{R}^n)$, i.e. $u \in A(\mathbb{R}^n)$. The estimate (12.21) follows from (12.19) and (12.20):

$$\|u\|_\infty \le (2\pi)^{-\frac{n}{2}} \|\hat{u}\|_{L^1} \le (2\pi)^{-\frac{n}{2}} \|u\|_{A(\mathbb{R}^n)}.$$

B. For $u \in A(\mathbb{R}^n)$ and $1 < p < \infty$ we have

$$\int_{\mathbb{R}^n} |u(x)|^p\,\mathrm{d}x \le \|u\|_\infty^{p-1} \int_{\mathbb{R}^n} |u(x)|\,\mathrm{d}x \le \|u\|_\infty^{p-1} \|u\|_{A(\mathbb{R}^n)},$$

the cases $p = 1$ and $p = \infty$ are trivial. $\qquad\square$

Since $\mathcal{S}(\mathbb{R}^n) \subset A(\mathbb{R}^n)$, a trivial consequence is

Proposition 12.12. *The Wiener algebra is dense in $L^p(\mathbb{R}^n)$, $1 \le p < \infty$, as well as in $C_\infty(\mathbb{R}^n)$.*

The following result extends to the convolution theorem and it also justifies the name "algebra" for $A(\mathbb{R}^n)$.

Theorem 12.13. *With $u, v \in A(\mathbb{R}^n)$ the functions $u*v$ and $u\cdot v$ belong to $A(\mathbb{R}^n)$, in particular $A(\mathbb{R}^n)$ is an algebra over $\mathbb{C}$, and we have*

$$(u * v)^\wedge(\xi) = (2\pi)^{\frac{n}{2}}\hat{u}(\xi) \cdot \hat{v}(\xi) \tag{12.23}$$

as well as

$$(u \cdot v)^\wedge(\xi) = (2\pi)^{-\frac{n}{2}}\left(\hat{u} * \hat{v}\right)(\xi). \tag{12.24}$$

Proof. From Theorem 12.5 we deduce again (12.23) since $u, v \in L^1(\mathbb{R}^n)$. Next we show that $(u * v)^\wedge \in L^1(\mathbb{R}^n)$. With the help of (12.23) we deduce

$$\|(u * v)^\wedge\|_{L^1} = (2\pi)^{-\frac{n}{2}} \|\hat{u} \cdot \hat{v}\|_{L^1} \le (2\pi)^{-\frac{n}{2}} \|\hat{u}\|_\infty \|\hat{v}\|_{L^1}$$
$$\le (2\pi)^{-\frac{n}{2}} \|u\|_{L^1} \|v\|_{A(\mathbb{R}^n)},$$

i.e. $(u * v)^\wedge \in L^1(\mathbb{R}^n)$ and since $u * v \in L^1(\mathbb{R}^n)$ it follows that $u * v \in A(\mathbb{R}^n)$. Furthermore, combining (12.23) with (12.22) we get

$$(u \cdot v)(-x) = (2\pi)^{-\frac{n}{2}}\left(\hat{u} * \hat{v}\right)^\wedge(x)$$

and the inversion formula yields

$$(2\pi)^{-\frac{n}{2}} \int_{\mathbb{R}^n} (u \cdot v)(-x) e^{ix \cdot \xi} \, \mathrm{d}x = (2\pi)^{-\frac{n}{2}} \left(\hat{u} * \hat{v} \right)(\xi),$$

hence

$$(u \cdot v)\hat{}\,(\xi) = (2\pi)^{-\frac{n}{2}} \left(\hat{u} * \hat{v} \right)(\xi).$$

$\square$

Remark 12.14. On $A(\mathbb{R}^n)$ we have two structures of an algebra, namely $A(\mathbb{R}^n)$ with the multiplication $u \cdot v$ and $A(\mathbb{R}^n)$ with the convolution $u * v$. The general properties of convolutions show indeed that $A(\mathbb{R}^n)$ with the convolution is an algebra. We define $\mathcal{F} : A(\mathbb{R}^n) \to A(\mathbb{R}^n)$ by $\mathcal{F}u = (2\pi)^{-\frac{n}{2}} Fu$ and $\mathcal{F}^{-1}u = (2\pi)^{\frac{n}{2}} F^{-1}u$. For $u \in A(\mathbb{R}^n)$ it follows that

$$\mathcal{F} \circ \mathcal{F}^{-1}u = F \circ F^{-1}u = u \quad \text{and} \quad \mathcal{F}^{-1} \circ \mathcal{F}u = F^{-1} \circ Fu = u,$$

and clearly, both $\mathcal{F}$ and $\mathcal{F}^{-1}$ are linear and bijective on $A(\mathbb{R}^n)$. Moreover we find

$$\begin{aligned}
\mathcal{F}(u \cdot v) &= (2\pi)^{-\frac{n}{2}} F(u \cdot v) \\
&= (2\pi)^{-n} (Fu) * (Fv) \\
&= \mathcal{F}u * \mathcal{F}v
\end{aligned}$$

and

$$\begin{aligned}
\mathcal{F}^{-1}(u * v) &= (2\pi)^{\frac{n}{2}} F^{-1}(u * v) \\
&= (2\pi)^n \left(F^{-1}u \right) \cdot \left(F^{-1}v \right) \\
&= \left(\mathcal{F}^{-1}u \right) \cdot \left(\mathcal{F}^{-1}v \right).
\end{aligned}$$

Denoting by $(A(\mathbb{R}^n), \cdot)$ for a moment the algebra $A(\mathbb{R}^n)$ with multiplication of functions and $(A(\mathbb{R}^n), *)$ the algebra with multiplication given by convolution then

$$\mathcal{F} : (A(\mathbb{R}^n), \cdot) \to (A(\mathbb{R}^n), *)$$

is an algebra homomorphism with inverse $\mathcal{F}^{-1} : (A(\mathbb{R}^n), *) \to (A(\mathbb{R}^n), \cdot)$.

Remark 12.15. Our discussions of the Wiener algebra follows closely [69] and was much influenced by [88].

We want to now consider the Fourier transform on some L^p-spaces for $p > 1$. Our first observation is however that for a generic function $u \in L^p(\mathbb{R}^n)$, $p > 1$, the integral

$$\int_{\mathbb{R}^n} e^{-ix \cdot \xi} u(x) \, \mathrm{d}x \tag{12.25}$$

is not defined. Theorem 11.12, more precisely Plancherel's equality (11.16), opens a way to extend F as a linear and continuous operator from $L^1(\mathbb{R}^n) \cap L^2(\mathbb{R}^n)$ to $L^2(\mathbb{R}^n)$.

Lemma 12.16. *For all $u \in L^1(\mathbb{R}^n) \cap L^2(\mathbb{R}^n)$ we have*

$$\|\hat{u}\|_{L^2} = \|u\|_{L^2}. \tag{12.26}$$

Proof. By Lemma 12.6 we know for all $u, v \in L^1(\mathbb{R}^n) \cap L^2(\mathbb{R}^n)$ that

$$\int_{\mathbb{R}^n} \hat{u}(\xi) v(\xi) \, \mathrm{d}\xi = \int_{\mathbb{R}^n} u(x) \hat{v}(\xi) \, \mathrm{d}\xi.$$

For a sequence $(u_k)_{k \in \mathbb{N}}$, $u_k \in \mathcal{S}(\mathbb{R}^n) \subset L^1(\mathbb{R}^n) \cap L^2(\mathbb{R}^n)$, converging in $L^2(\mathbb{R}^n)$ to u we define

$$v_k(\xi) := (2\pi)^{-\frac{n}{2}} \int_{\mathbb{R}^n} e^{ix \cdot \xi} \overline{u_k(x)} \, \mathrm{d}x.$$

It follows that $\hat{v}_k = \overline{u}_k$ implying that $(\hat{v}_k)_{k \in \mathbb{N}}$ converges in $L^2(\mathbb{R}^n)$ to $\overline{u}$. Therefore we may pass in the identity

$$\int_{\mathbb{R}^n} |\hat{u}_k(x)|^2 \, \mathrm{d}x = \int_{\mathbb{R}^n} |u_k(x)|^2 \, \mathrm{d}x$$

to the limit and (12.26) follows. $\qquad\square$

Now we can prove

Theorem 12.17. *The Fourier transform $F : L^1(\mathbb{R}^n) \cap L^2(\mathbb{R}^n) \to C_\infty(\mathbb{R}^n)$ has an extension as a linear and continuous operator $\tilde{F} : L^2(\mathbb{R}^n) \to L^2(\mathbb{R}^n)$. This extension is an L^2-isometry which is bijective and has a continuous inverse $\tilde{F}^{-1}$ which is an extension of $F^{-1} : L^1(\mathbb{R}^n) \cap L^2(\mathbb{R}^n) \to C_\infty(\mathbb{R}^n)$, and $\tilde{F}^{-1}$ is again an isometry of $L^2(\mathbb{R}^n)$.*

Proof. From Lemma 12.16 we deduce that F maps $L^1(\mathbb{R}^n) \cap L^2(\mathbb{R}^n)$ into $L^2(\mathbb{R}^n)$. Further, since $\mathcal{S}(\mathbb{R}^n) \subset L^1(\mathbb{R}^n) \cap L^2(\mathbb{R}^n)$ the space $L^1(\mathbb{R}^n) \cap L^2(\mathbb{R}^n)$ is dense in $L^2(\mathbb{R}^n)$. For $u \in L^2(\mathbb{R}^n)$ we can choose a sequence $(u_k)_{k \in \mathbb{N}}$, $u_k \in L^1(\mathbb{R}^n) \cap L^2(\mathbb{R}^n)$, converging in $L^2(\mathbb{R}^n)$ to u. We define $\tilde{F}u$ as the L^2-limit of Fu_k, i.e.

$$\lim_{k \to \infty} \left\| Fu_k - \tilde{F}u \right\|_{L^2} = 0.$$

This is justified since $(Fu_k)_{k \in \mathbb{N}}$ forms by (12.26) a Cauchy sequence in $L^2(\mathbb{R}^n)$, hence it has a limit which we denote by $\tilde{F}u$. If for two approximating sequences

$(u_k)_{k\in\mathbb{N}}$ and $(v_k)_{k\in\mathbb{N}}$ of u we denote the corresponding limits of $(Fu_k)_{k\in\mathbb{N}}$ and $(Fv_k)_{k\in\mathbb{N}}$ by $\tilde{F}u$ and $\tilde{F}v$ respectively we find

$$\left\|\tilde{F}u - \tilde{F}v\right\|_{L^2} \leq \left\|\tilde{F}u - Fu_k\right\|_{L^2} + \|Fu_k - Fv_k\|_{L^2} + \left\|\tilde{F}v - Fv_k\right\|_{L^2}$$

and since $\left\|\tilde{F}u - Fu_k\right\|_{L^2} \rightarrow 0$ as well as $\left\|\tilde{F}v - Fv_k\right\|_{L^2} \rightarrow 0$ and $\|Fu_k - Fv_k\|_{L^2} = \|u_k - v_k\|_{L^2} \leq \|u_k - u\|_{L^2} + \|v_k - u\|_{L^2} \rightarrow 0$, we deduce $\tilde{F}u = \tilde{F}v$, i.e. $\tilde{F}u$ is independent of the approximating sequence. Thus we may extend F to $L^2(\mathbb{R}^n)$ by $\tilde{F}$ and $\tilde{F}$ maps $L^2(\mathbb{R}^n)$ into $L^2(\mathbb{R}^n)$. By the converse triangle inequality we get

$$\left|\left\|\tilde{F}u\right\|_{L^2} - \|u_k\|_{L^2}\right| = \left|\left\|\tilde{F}u\right\|_{L^2} - \|Fu_k\|_{L^2}\right| \leq \left\|\tilde{F}u - Fu_k\right\|_{L^2}$$

which yields

$$\left\|\tilde{F}u\right\|_{L^2} = \|u\|_{L^2}$$

implying that $\tilde{F} : L^2(\mathbb{R}^n) \rightarrow L^2(\mathbb{R}^n)$ is an isometry, hence it is injective. Since $A(\mathbb{R}^n) \subset L^1(\mathbb{R}^n) \cap L^2(\mathbb{R}^n)$ we further have

$$\tilde{F}\left(L^1(\mathbb{R}^n) \cap L^2(\mathbb{R}^n)\right) \supset \tilde{F}(A(\mathbb{R}^n)) = F(A(\mathbb{R}^n)) = A(\mathbb{R}^n),$$

thus $\tilde{F}\left(L^1(\mathbb{R}^n) \cap L^2(\mathbb{R}^n)\right)$ is dense in $L^2(\mathbb{R}^n)$ and since $\tilde{F}$ is an isometry $\tilde{F}(L^2(\mathbb{R}^n))$ is a closed subspace of $L^2(\mathbb{R}^n)$, hence $\tilde{F}(L^2(\mathbb{R}^n)) = L^2(\mathbb{R}^n)$ which implies the surjectivity of $\tilde{F}$, i.e. we have proved that $\tilde{F} : L^2(\mathbb{R}^n) \rightarrow L^2(\mathbb{R}^n)$ is a bijective isometry. For $u \in A(\mathbb{R}^n)$ we observe

$$u(x) = (2\pi)^{-\frac{n}{2}} \int_{\mathbb{R}^n} e^{ix\cdot\xi} \hat{u}(\xi)\,d\xi = (2\pi)^{-\frac{n}{2}} \overline{\int_{\mathbb{R}^n} e^{-ix\cdot\xi} \overline{\hat{u}(x)}\,d\xi}$$

and therefore we find on $L^2(\mathbb{R}^n)$

$$\tilde{F}^{-1}(u) = \overline{\tilde{F}(\bar{u})}$$

which proves that $\tilde{F}^{-1}$ is an extension of $\tilde{F}^{-1} : L^1(\mathbb{R}^n) \cap L^2(\mathbb{R}^n) \rightarrow C_\infty(\mathbb{R}^n)$ and it proves that $\tilde{F}^{-1}$ is also an L^2-isometry, in particular it maps $L^2(\mathbb{R}^n)$ continuously onto itself. $\qquad\square$

It is convenient to denote, as we will do in the following, the extensions $\tilde{F}$ and $\tilde{F}^{-1}$ of F and F^{-1} to $L^2(\mathbb{R}^n)$ again by F and F^{-1}, as we will write $\hat{u}$ for Fu, $u \in L^2(\mathbb{R}^n)$.

Remark 12.18. Let $u \in L^2(\mathbb{R}^n)$. Our construction of $\hat{u}$ yields that for every sequence $(v_k)_{k \in \mathbb{N}}$, $v_k \in L^2(\mathbb{R}^n) \cap L^1(\mathbb{R}^n)$, converging in $L^2(\mathbb{R}^n)$ to u we have

$$\lim_{k \to \infty} \|\hat{u} - \hat{v}_k\|_{L^2} = 0.$$

In particular we may take $v_k := u \chi_{B_k(0)}$ which implies that

$$\hat{v}_k(\xi) := (2\pi)^{-\frac{n}{2}} \int_{\|x\| \le k} e^{-ix \cdot \xi} u(x) \, dx \tag{12.27}$$

converges in the quadratic mean, i.e. $L^2(\mathbb{R}^n)$, to $\hat{u}$.

Remark 12.19. In Volume V we will learn about interpolation theory which in particular will allow us to prove using the continuity of $F : L^1(\mathbb{R}^n) \to L^1(\mathbb{R}^n)$ and $F : L^2(\mathbb{R}^n) \to L^2(\mathbb{R}^n)$ that we can extend F as a linear and continuous mapping $F : L^{p_\theta}(\mathbb{R}^n) \to L^{p'_\theta}(\mathbb{R}^n)$ for $p_\theta = \frac{2}{1+\theta}$, $p'_\theta = \frac{2}{1-\theta'}$, i.e. $\frac{1}{p_\theta} + \frac{1}{p'_\theta} = 1$, where $\theta \in [0,1]$, i.e. $1 \le p_\theta \le 2$.

We want to pick up Example 11.15 and discuss it now in the L^2-context.

Example 12.20. As before let

$$h_t(x) := (4\pi t)^{-\frac{n}{2}} e^{-\frac{\|x\|^2}{4t}} \tag{12.28}$$

be the heat kernel in $\mathbb{R}^n$ and recall, see (11.21), that

$$\hat{h}_t(\xi) = (2\pi)^{-\frac{n}{2}} e^{-t\|\xi\|^2}. \tag{12.29}$$

For $t > 0$ we can define again

$$T_t u := h_t * u, \tag{12.30}$$

but we now allow $u \in L^2(\mathbb{R}^n)$ which implies by Young's inequality, Theorem III.10.13, that

$$\|T_t u\|_{L^2} \le \|h_t\|_{L^1} \|u\|_{L^2} = \|u\|_{L^2}, \tag{12.31}$$

since $\|h_t\|_{L^1} = \int_{\mathbb{R}^n} h_t(x) \, dx = 1$. The semigroup property (11.24), i.e.

$$T_{t_2} \circ T_{t_1} = T_{t_2 + t_1}, \quad t_1, t_2 > 0, \tag{12.32}$$

carries over to the new case since $T_t : L^2(\mathbb{R}^n) \to L^2(\mathbb{R}^n)$ is continuous and $\mathcal{S}(\mathbb{R}^n) \subset L^2(\mathbb{R}^n)$ is dense, see also Problem 7. With the help of the convolution theorem we find for $u \in L^1(\mathbb{R}^n) \cap L^2(\mathbb{R}^n)$ that

$$(T_t u)^{\hat{}}(\xi) = (h_t * u)^{\hat{}}(\xi) = (2\pi)^{-\frac{n}{2}} \left(\hat{h}_t * \hat{u}\right)(\xi) = e^{-t\|\xi\|^2} \hat{u}(\xi)$$

which allows the representation

$$(T_t u)(x) = F^{-1}\left(e^{-t\|\cdot\|^2}\hat{u}(\cdot)\right)(x), \quad u \in L^1(\mathbb{R}^n) \cap L^2(\mathbb{R}^n). \tag{12.33}$$

Now we find for a sequence $(u_k)_{k\in\mathbb{N}}$, $u_k \in L^1(\mathbb{R}^n) \cap L^2(\mathbb{R}^n)$, converging in $L^2(\mathbb{R}^n)$ to u, that (12.33) holds for the limit u, but as an equality of elements in $L^2(\mathbb{R}^n)$:

$$T_t u = F^{-1}\left(e^{-t\|\cdot\|^2}\hat{u}\right).$$

Note that for $(T_t u)\hat{}$ we find almost everywhere

$$\frac{\partial}{\partial t}(T_t u)\hat{}(\xi) = -\|\xi\|^2(T_t u)\hat{}(\xi).$$

For those functions for which we can prove $\frac{\partial}{\partial t}(T_t u)\hat{} = \left(\frac{\partial}{\partial t}T_t u\right)\hat{}$ and $-\|\cdot\|^2(T_t u)\hat{} = (\Delta T_t u)\hat{}$, we again obtain a solution of the heat equation $\frac{\partial w}{\partial t} = \Delta w$.

Let $s \geq 0$ and define

$$H^s(\mathbb{R}^n) := \left\{u \in L^2(\mathbb{R}^n)\,\big|\,\|u\|_s < \infty\right\} \tag{12.34}$$

where

$$\|u\|_s^2 = \int_{\mathbb{R}^n}(1+\|\xi\|^2)^s\,|\hat{u}(\xi)|^2\,\mathrm{d}\xi. \tag{12.35}$$

First we note that $H^0(\mathbb{R}^n) = L^2(\mathbb{R}^n)$ and therefore we sometimes will write

$$\|u\|_0 := \|u\|_{L^2}. \tag{12.36}$$

Moreover, since for $u \in \mathcal{S}(\mathbb{R}^n)$ we always have $\|u\|_s < \infty$ it follows that $H^s(\mathbb{R}^n) \neq \emptyset$. It is easy to check that $\|\cdot\|_s$ is a norm and $H^s(\mathbb{R}^n)$ is a normed space. In fact $H^s(\mathbb{R}^n)$ is a Banach space with respect to the norm $\|\cdot\|_s$ but before proving this we want to explore some of the properties of $H^s(\mathbb{R}^n)$ and $\|\cdot\|_s$. From (12.35) we deduce that

$$H^s(\mathbb{R}^n) \subset H^t(\mathbb{R}^n) \quad \text{for } t \leq s, \tag{12.37}$$

and

$$\|u\|_t \leq \|u\|_s. \tag{12.38}$$

Let us define the operator $\Lambda^s(D) : H^s(\mathbb{R}^n) \to L^2(\mathbb{R}^n)$ by

$$\Lambda^s(D)u = F^{-1}\left((1+\|\cdot\|^2)^{\frac{s}{2}}Fu\right). \tag{12.39}$$

Since for $u \in H^s(\mathbb{R}^n)$ the function $(1 + \| \cdot \|^2)^{\frac{s}{2}} Fu$ is an element in $L^2(\mathbb{R}^n)$ the operator $\Lambda^s(D)$ is well defined and we find

$$\|\Lambda^s(D)u\|_0 = \|u\|_s. \tag{12.40}$$

In particular it follows that $\Lambda^s(D) : H^s(\mathbb{R}^n) \to L^2(\mathbb{R}^n)$ is a linear isometry, hence continuous. For $u \in L^2(\mathbb{R}^n)$ we find further that $(1 + \| \cdot \|^2)^{-\frac{s}{2}} \hat{u} \in L^2(\mathbb{R}^n)$ since

$$\left\| (1 + \|x\|^2)^{\frac{-s}{2}} \hat{u} \right\|_{L^2} \leq \|\hat{u}\|_{L^2},$$

and therefore we have that $\Lambda^{-s}(D)u := F^{-1}\left((1 + \| \cdot \|^2)^{-\frac{s}{2}} \hat{u} \right) \in L^2(\mathbb{R}^n)$. We claim that $\Lambda^{-s}(D)u \in H^s(\mathbb{R}^n)$. Indeed we have

$$\begin{aligned}
\left\| \Lambda^{-s}(D)u \right\|_s^2 &= \int_{\mathbb{R}^n} (1 + \|\xi\|^2)^s \left| F\left(\Lambda^{-s}(D)u \right)(\xi) \right|^2 d\xi \\
&= \int_{\mathbb{R}^n} (1 + \|\xi\|^2)^s \left| F\left(F^{-1}\left((1 + \| \cdot \|^2)^{-\frac{s}{2}} \hat{u} \right) \right)(\xi) \right|^2 d\xi \\
&= \int_{\mathbb{R}^n} |\hat{u}(\xi)|^2 d\xi = \|u\|_0^2.
\end{aligned}$$

Hence $\Lambda^{-s}(D) : L^2(\mathbb{R}^n) \to H^s(\mathbb{R}^n)$, $u \mapsto F^{-1}\left((1 + \| \cdot \|^2)^{-\frac{s}{2}} \hat{u} \right)$ is a linear isometry, in particular it is continuous, and on $L^2(\mathbb{R}^n)$ it follows

$$\Lambda^s(D) \circ \Lambda^{-s}(D)u = F^{-1}\left((1 + \| \cdot \|^2)^{\frac{s}{2}} F\left(F^{-1}(1 + \| \cdot \|^2)^{\frac{s}{2}} \hat{u} \right) \right) = u$$

whereas on $H^s(\mathbb{R}^n)$ we have

$$\Lambda^{-s}(D) \circ \Lambda^s(D)u = F^{-1}\left((1 + \| \cdot \|^2)^{-\frac{s}{2}} F\left(F^{-1}(1 + \| \cdot \|^2)^{-\frac{s}{2}} \hat{u} \right) \right) = u.$$

Thus we have a linear continuous and bijective isometry between $H^s(\mathbb{R}^n)$ and $L^2(\mathbb{R}^n)$ with continuous inverse which is again a linear isometry. This however implies the completeness of $H^s(\mathbb{R}^n)$. Indeed, if $(u_k)_{k \in \mathbb{N}}$ is a Cauchy sequence in $H^s(\mathbb{R}^n)$ then $(\Lambda^s(D)u_k)_{k \in \mathbb{N}}$ is a Cauchy sequence in $L^2(\mathbb{R}^n)$ hence it has a limit $w \in L^2(\mathbb{R}^n)$ and $u := \Lambda^{-s}(D)w \in H^s(\mathbb{R}^n)$. It follows that

$$\begin{aligned}
\|u - u_k\|_s &= \|\Lambda^s(D)u - \Lambda^s(D)u_k\|_0 \\
&= \|w - \Lambda^s(D)u_k\|_0 \to 0,
\end{aligned}$$

i.e. $(u_k)_{k \in \mathbb{N}}$ converges in $H^s(\mathbb{R}^n)$ to $u = \Lambda^{-s}(D)w$ and therefore $H^s(\mathbb{R}^n)$ is complete. Moreover we find that $\mathcal{S}(\mathbb{R}^n)$ is dense in $H^s(\mathbb{R}^n)$. For this we note that for every $t \in \mathbb{R}$ and $v \in \mathcal{S}(\mathbb{R}^n)$ the function $(1 + \|\cdot\|^2)^{\frac{t}{2}} v$ belongs again to $\mathcal{S}(\mathbb{R}^n)$. Now let $u \in H^s(\mathbb{R}^n)$. Since $\mathcal{S}(\mathbb{R}^n)$ is dense in $L^2(\mathbb{R}^n)$ we can find a sequence $(w_k)_{k \in \mathbb{N}}$, $w_k \in \mathcal{S}(\mathbb{R}^n)$, converging in $L^2(\mathbb{R}^n)$ to $\Lambda^s(D)u$. For the sequence $(\Lambda^{-s}(D)w_k)_{k \in \mathbb{N}}$ it follows that $\Lambda^{-s}(D)w_k \in \mathcal{S}(\mathbb{R}^n)$ and

$$\left\| u - \Lambda^{-s}(D)w_k \right\|_s = \left\| \Lambda^s(D)u - \Lambda^s(D)\left(\Lambda^{-s}(D)w_k\right) \right\|_0$$
$$= \left\| \Lambda^s(D)u - w_k \right\|_0 \to 0.$$

Eventually we have proved

Theorem 12.21. *The space* $(H^s(\mathbb{R}^n), \|\cdot\|_s)$, $s \geq 0$, *is a Hilbert space and* $\mathcal{S}(\mathbb{R}^n)$ *is a dense subspace.*

Proof. It remains to prove that $\|\cdot\|_s$ is a norm associated with a scalar product. For this we define

$$\langle u, v \rangle_s := \int_{\mathbb{R}^n} (1 + \|\xi\|^2)^s \hat{u}(\xi) \overline{\hat{v}(\xi)} \, d\xi \tag{12.41}$$

which is a scalar product and we just have to note that $\|u\|_s^2 = \langle u, u \rangle_s$. $\square$

Definition 12.22. *For* $s \geq 0$ *we call* $H^s(\mathbb{R}^n)$ *the* **Bessel potential space** *of order* s.

Theorem 12.23 (Sobolev's embedding theorem). *Let* $s > \frac{n}{2}$. *Then the space* $H^s(\mathbb{R}^n)$ *is continuously embedded into the space* $C_\infty(\mathbb{R}^n)$ *and the estimate*

$$\|u\|_\infty \leq c_{s,n} \|u\|_s \tag{12.42}$$

holds for all $u \in H^s(\mathbb{R}^n)$. *In particular we can identify each* $u \in H^s(\mathbb{R}^n)$ *with a unique element of* $C_\infty(\mathbb{R}^n)$ *and in this sense we may write* $H^s(\mathbb{R}^n) \subset C_\infty(\mathbb{R}^n)$.

Proof. For $u \in \mathcal{S}(\mathbb{R}^n)$ we find

$$|u(x)| = \left|\left(F^{-1}Fu\right)(x)\right|$$
$$= (2\pi)^{-\frac{n}{2}} \left| \int_{\mathbb{R}^n} e^{ix \cdot \xi} \hat{u}(\xi) \, d\xi \right|$$
$$= (2\pi)^{-\frac{n}{2}} \left| \int_{\mathbb{R}^n} \frac{1}{(1 + \|\xi\|^2)^{\frac{s}{2}}} (1 + \|\xi\|^2)^{\frac{s}{2}} e^{ix \cdot \xi} \hat{u}(\xi) \, d\xi \right|$$
$$\leq (2\pi)^{-\frac{n}{2}} \left(\int_{\mathbb{R}^n} \frac{1}{(1 + \|\xi\|^2)^s} \, d\xi \right)^{\frac{1}{2}} \|u\|_s,$$

hence

$$\|u\|_\infty \leq (2\pi)^{-\frac{n}{2}} \left(\int_{\mathbb{R}^n} \frac{1}{(1 + \|\xi\|^2)^s} \, d\xi \right)^{\frac{1}{2}} \|u\|_s.$$

It remains to show that the integral on the right hand side is for $s > \frac{n}{2}$ finite. This follows however from

$$\int_{\mathbb{R}^n} \frac{1}{(1+\|\xi\|^2)^s} \, d\xi \leq \int_{B_1(0)} \frac{1}{(1+\|\xi\|^2)^s} \, d\xi + n\omega_n \int_1^\infty r^{-2s+n-1} \, dr.$$

Since the closure of $\mathcal{S}(\mathbb{R}^n)$ in $\| \cdot \|_\infty$ is $C_\infty(\mathbb{R}^n)$ the theorem is proved. $\quad\square$

Next let $s = m \in \mathbb{N}$ and note that in this case we have with suitable constants $c_{n,m}$, $\tilde{c}_{n,m}$, $c_{n,m} > 0$,

$$c_{n,m}(1+\|\xi\|^2)^m \leq \sum_{|\alpha| \leq m} \xi^{2\alpha} \leq \tilde{c}_{n,m}(1+\|\xi\|^2)^m. \tag{12.43}$$

For $u \in \mathcal{S}(\mathbb{R}^n) \subset H^m(\mathbb{R}^n)$ it follows that

$$\sum_{|\alpha| \leq m} \|D^\alpha u\|_{L^2}^2 = \sum_{|\alpha| \leq m} \|(D^\alpha u)\hat{\,}\|_{L^2}^2$$

$$= \sum_{|\alpha| \leq m} \int_{\mathbb{R}^n} \xi^{2\alpha} |\hat{u}(\xi)|^2 \, d\xi$$

$$\leq \tilde{c}_{n,m} \int_{\mathbb{R}^n} (1+\|\xi\|^2)^m |\hat{u}(\xi)|^2 \, d\xi$$

$$\leq \frac{\tilde{c}_{n,m}}{c_{n,m}} \int_{\mathbb{R}^n} \left(\sum_{|\alpha| \leq m} \xi^{2\alpha} \right) |\hat{u}(\xi)|^2 \, d\xi$$

$$= \frac{\tilde{c}_{m,n}}{c_{n,m}} \sum_{|\alpha| \leq m} \|D^\alpha u\|_{L^2}^2 \, ,$$

and we have proved

Proposition 12.24. On $\mathcal{S}(\mathbb{R}^n)$ an equivalent norm to $\| \cdot \|_m$, $m \in \mathbb{N}$, is given by

$$\|u\|_{W^{m,2}} := \left(\sum_{|\alpha| \leq m} \|\partial^\alpha u\|_{L^2}^2 \right)^{\frac{1}{2}}. \tag{12.44}$$

In Volume V we will discuss the completion of $\mathcal{S}(\mathbb{R}^n)$ with respect to the norm (12.44) which will lead to the **Sobolev space** $W^{m,2}(\mathbb{R}^n)$, a space we may identify with a subspace of $L^2(\mathbb{R}^n)$, namely $H^m(\mathbb{R}^n)$. Of course, for a generic $u \in W^{m,2}(\mathbb{R}^n)$ (or $u \in H^m(\mathbb{R}^n)$) the term $\partial^\alpha u$, $|\alpha| \leq m$, has no "classical" meaning. A way to understand $\partial^\alpha u$ for classically non-differentiable functions with the help of the Fourier transform is indicated below.

Definition 12.25. *Let $u \in L^p(\mathbb{R}^n)$, $1 \leq p < \infty$. We say that u has a first order partial derivative with respect to x_k, $1 \leq k \leq n$, in the sense of the space $L^p(\mathbb{R}^n)$ if there exists a function $g_{(k)} \in L^p(\mathbb{R}^n)$ such that*

$$\lim_{h \to 0} \left(\int_{\mathbb{R}^n} \left| \frac{u(x + he_k) - u(x)}{h} - g_{(k)}(x) \right|^p \, dx \right)^{\frac{1}{p}} = 0 \qquad (12.45)$$

holds.

Note that for $u \in L^2(\mathbb{R}^n)$ Plancherel's theorem applied to (12.45) yields

$$\lim_{h \to 0} \left(\int_{\mathbb{R}^n} \left| \left(\frac{e^{ih \cdot \xi_k} - 1}{h} \right) \hat{u}(\xi) - \hat{g}_{(k)}(\xi) \right|^2 \, d\xi \right)^{\frac{1}{2}} = 0. \qquad (12.46)$$

We claim

Theorem 12.26. *Every $u \in H^1(\mathbb{R}^n)$ has for $1 \leq k \leq n$ the first order partial derivative with respect to x_k and it is given by $g_k(x) = F^{-1}_{\xi \to x} (i\xi_k \hat{u}(\xi)) (x)$.*

The proof of this theorem needs some preparation.

Lemma 12.27. *Let $(x_k)_{k \in \mathbb{N}}$ be a sequence in a Hilbert space $(H, \langle \cdot, \cdot \rangle)$ with the property that for some $x \in H$*

$$\lim_{k \to \infty} \langle x_k, y \rangle = \langle x, y \rangle \qquad (12.47)$$

holds for all $y \in H$. Then we have

$$\|x\| \leq \liminf_{k \to \infty} \|x_k\|. \qquad (12.48)$$

Proof. We observe that

$$0 \leq \langle x_k - x, x_k - x \rangle = \langle x_k, x_k \rangle - \langle x, x_k \rangle - \langle x_k, x \rangle + \langle x, x \rangle,$$

or

$$\langle x, x_k \rangle + \langle x_k, x \rangle \leq \langle x_k, x_k \rangle + \langle x, x \rangle$$

implying

$$\|x\|^2 + \|x\|^2 \leq \liminf_{k \to \infty} \|x_k\|^2 + \|x\|^2$$

or (12.48). $\qquad \square$

Lemma 12.28. *Let $(x_k)_{k \in \mathbb{N}}$ be a sequence in a Hilbert space such that $\|x_k\| = 1$. If with some $x \in H$, $\|x\| = 1$, we have $\lim_{k \to \infty} \langle x_k - x, y \rangle = 0$ for all $y \in H$ and $\lim_{k \to \infty} \|x_k + x\| = 2$ then it follows that*

$$\lim_{k \to \infty} \|x_k - x\| = 0. \tag{12.49}$$

Proof. We note that

$$0 \le \langle x_k - x, x_k - x \rangle = \|x_k\|^2 - \langle x, x_k \rangle - \langle x_k, x \rangle + \|x\|^2$$
$$= 2 - \overline{\langle x_k, x \rangle} - \langle x_k, x \rangle$$

implying

$$0 \le \limsup_{k \to \infty} \|x_k - x\|^2 \le 2 - \lim_{k \to \infty} \overline{\langle x_k, x \rangle} - \lim_{k \to \infty} \langle x_k, x \rangle = 0,$$

hence we get (12.49). $\qquad\square$

Proof of Theorem 12.26. For $u \in H^1(\mathbb{R}^n)$ it follows that the function $g_k(x) := F_{\xi \mapsto x}^{-1}(i\xi_k \hat{u}(\xi))(x)$ belongs to $L^2(\mathbb{R}^n)$. We will first prove

$$\lim_{h \to 0} \int_{\mathbb{R}^n} \left(\frac{e^{ih\xi_k} - 1}{h} \right) \hat{u}(\xi) \overline{\hat{v}(\xi)} \, d\xi = \int_{\mathbb{R}^n} g_k(x) \overline{v}(x) \, dx \tag{12.50}$$

for all $v \in L^2(\mathbb{R}^n)$ and

$$\lim_{h \to 0} \int_{\mathbb{R}^n} \left| \left(\frac{e^{ih\xi_k} - 1}{h} \right) \hat{u}(\xi) \right|^2 \, d\xi = \int_{\mathbb{R}^n} |g_k(x)|^2 \, dx. \tag{12.51}$$

The remainder term for the Taylor expansion of the exponential function yields

$$\left| \frac{e^{ih\xi_k} - 1}{h} \right| \le |\xi_k|$$

and furthermore we have

$$\lim_{h \to 0} \frac{e^{ih\xi_k} - 1}{h} = i\xi_k.$$

Since for $u, v \in L^2(\mathbb{R}^n)$ the functions $\hat{u}\overline{\hat{v}}$ and $|\hat{u}|^2$ belong to $L^1(\mathbb{R}^n)$ the dominated convergence theorem applied to the left hand side of (12.50) yields together with Plancherel's theorem

$$\lim_{h \to 0} \int_{\mathbb{R}^n} \left(\frac{e^{ih\xi_k} - 1}{h} \right) \hat{u}(\xi) \overline{\hat{v}(\xi)} \, d\xi = \int_{\mathbb{R}^n} i\xi_k \hat{u}(\xi) \overline{\hat{v}(\xi)} \, d\xi \tag{12.52}$$

$$= \int_{\mathbb{R}^n} g_k(x) \overline{v(x)} \, dx,$$

and with the same type arguments applied to the left hand side of (12.51) we arrive at

$$\lim_{h \to 0} \int_{\mathbb{R}^n} \left| \left(\frac{e^{ih\xi_k} - 1}{h} \right) \hat{u}(\xi) \right|^2 \, d\xi = \int_{\mathbb{R}^n} |g_k(x)|^2 \, dx. \tag{12.53}$$

Our aim is to show that

$$\lim_{h \to 0} \int_{\mathbb{R}^n} \left| \left(\frac{e^{ih\xi_k} - 1}{h} \right) \hat{u}(\xi) - i\xi_k \hat{u}(\xi) \right|^2 \, d\xi = 0, \tag{12.54}$$

or equivalently

$$\lim_{h \to 0} \left\| \left(\frac{e^{ih\xi_k} - 1}{h} \right) \hat{u}(\cdot) - \hat{g}_k(\cdot) \right\|_{L^2} = 0. \tag{12.55}$$

If $g_k = 0$, i.e. $\hat{g}_k = 0$, nothing is left to prove by (12.53). In the case that $g_k \neq 0$ for h sufficiently small we have $w_h := \left(\frac{e^{ih\xi_k} - 1}{h} \right) \hat{u} \neq 0$ in $L^2(\mathbb{R}^n)$ and for $z_h := \frac{w_h}{\|w_h\|_{L^2}}$ we find $\|z_h\|_{L^2} = 1$ as we have for $f_k := \frac{g_k}{\|g_k\|_{L^2}}$ that $\|f_k\|_{L^2} = 1$. From (12.52) we deduce now

$$\lim_{h \to 0} \langle z_h, v \rangle_{L^2} = \langle f_k, v \rangle_{L^2} \tag{12.56}$$

for all $v \in L^2(\mathbb{R}^n)$, and from (12.53) we get

$$\lim_{h \to 0} \|z_h\|_{L^2} = 1 = \|f_k\|_{L^2}.$$

It follows that for all $v \in L^2(\mathbb{R}^n)$ we have $\lim_{h \to 0} \langle z_h + f_k, v \rangle_{L^2} = 2 \langle f_k, v \rangle_{L^2}$ and therefore by Lemma 12.27

$$2 = \|2f_k\| \leq \liminf_{h \to 0} \|z_h + f_k\|_{L^2}$$
$$\leq \limsup_{h \to 0} \|z_h + f_k\|_{L^2}$$
$$\leq \lim_{h \to 0} \|z_h\|_{L^2} + \|f_k\|_{L^2} = 2,$$

and we have proved that $\lim_{h \to 0} \|z_h + f_k\|_{L^2} = 2$. Now we deduce from Lemma 12.28 that

$$\lim_{h \to 0} \|z_h - f_k\|_{L^2} = 0$$

and therefore (12.54) follows. □

Remark 12.29. The reader who is already familiar to the concept of weak convergence in (separable) Hilbert space which we will discuss in Volume V will recognise that our proof of Theorem 12.26 is based on the fact that weak convergence and convergence of the norms imply strong convergence.

Remark 12.30. Of course we can iterate Definition 12.25 and we can introduce for $\alpha \in \mathbb{N}_0^n$ the partial derivative $\partial^\alpha u$ in the sense of the space $L^p(\mathbb{R}^n)$. For $p = 2$ we can prove that if $u \in H^m(\mathbb{R}^n)$ then u has all partial derivatives $\partial^\alpha u$, $|\alpha| \leq m$, in the sense of the space $L^2(\mathbb{R}^n)$.

Example 12.31. We return to Example 12.20 where we have proved for $T_t u = h_t * u$, $u \in L^2(\mathbb{R}^n)$, that almost everywhere

$$\frac{\partial}{\partial t}(T_t u)\hat{}(\xi) = -\|\xi\|^2 (T_t u)\hat{}(\xi).$$

Moreover, for $s \geq 0$ and $u \in \mathcal{S}(\mathbb{R}^n)$ we find

$$\|T_t u\|_s^2 = \int_{\mathbb{R}^n} (1 + \|\xi\|^2)^s |(h_t * u)\hat{}(\xi)|^2 \, d\xi$$

$$= \int_{\mathbb{R}^n} (1 + \|\xi\|^2)^s e^{-2t\|\xi\|^2} |\hat{u}(\xi)|^2 \, d\xi$$

$$\leq c_{s,n} \|u\|_{L^2}^2.$$

Since $\mathcal{S}(\mathbb{R}^n)$ is dense in $L^2(\mathbb{R}^n)$ this estimate holds for all $u \in L^2(\mathbb{R}^n)$ and we find that for $u \in L^2(\mathbb{R}^n)$ we have

$$T_t u \in \bigcap_{s \geq 0} H^s(\mathbb{R}^n) = \bigcap_{m=1}^{\infty} W^{m,2}(\mathbb{R}^n).$$

We claim that $\bigcap_{m=1}^{\infty} W^{m,2}(\mathbb{R}^n) = C_\infty^\infty(\mathbb{R}^n)$ where

$$C_\infty^\infty(\mathbb{R}^n) := \left\{ u \in C^\infty(\mathbb{R}^n) \, \big| \, \partial^\alpha u \in C_\infty(\mathbb{R}^n) \text{ for all } \alpha \in \mathbb{N}_0^n \right\}.$$

Take $\alpha \in \mathbb{N}_0^n$ and $u \in \bigcap_{m=1}^{\infty} W^{m,2}(\mathbb{R}^n)$. It follows that $\partial^\alpha u \in \bigcap_{m=1}^{\infty} W^{m,2}(\mathbb{R}^n)$. Indeed, for any $r \geq 0$ we find

$$\|\partial^\alpha u\|_r^2 = \int_{\mathbb{R}^n} (1 + \|\xi\|^2)^r |\xi^\alpha \hat{u}(\xi)|^2 \, d\xi$$

$$\leq c \int_{\mathbb{R}^n} (1 + \|\xi\|^2)^{r+|\alpha|} |\hat{u}(\xi)|^2 \, d\xi = c\|u\|_{r+|\alpha|}.$$

The Sobolev embedding theorem now yields that $\partial^\alpha u \in C_\infty(\mathbb{R}^n)$. Thus we find $-\|\xi\|^2 (T_t u)\hat{}(\xi) = (\Delta T_t u)\hat{}(\xi)$ and therefore

$$F_{\xi \mapsto x}^{-1} \left(\frac{\partial}{\partial t}(T_t u)\hat{} \right)(x) = (\Delta T_t u)(x).$$

Finally we claim that

$$F_{\xi \mapsto x}^{-1}\left(\frac{\partial}{\partial t}(T_t u)^\wedge\right)(x) = \frac{\partial}{\partial t} T_t u(x), \quad t > 0.$$

This will follow if we can apply Theorem III.8.4 to the function $w(t, \xi) = e^{ix \cdot \xi} e^{-t\|\xi\|^2} \hat{u}(\xi)$, $u \in L^2(\mathbb{R}^n)$, for $x \in \mathbb{R}^n$ fixed. We fix $t_0 > 0$ and note for $t \geq t_0$ that the function $w(t, \cdot)$ is integrable over $\mathbb{R}^n$ since

$$\int_{\mathbb{R}^n} |w(t, \xi)| \, d\xi \leq \int_{\mathbb{R}^n} e^{-t_0 \|\xi\|^2} |\hat{u}(\xi)| \, d\xi$$

$$\leq \left(\int_{\mathbb{R}^n} e^{-2t_0 \|\xi\|^2} \, d\xi\right)^{\frac{1}{2}} \|u\|_{L^2}.$$

Clearly, $t \mapsto w(t, \xi)$ is differentiable with $\frac{\partial w(t,\xi)}{\partial t} = -\|\xi\|^2 e^{ix \cdot \xi} e^{-t\|\xi\|^2} \hat{u}(\xi)$. For $t \geq t_0$ we have further

$$\int_{\mathbb{R}^n} \left|\frac{\partial w(t, \xi)}{\partial t}\right| \, d\xi \leq \int_{\mathbb{R}^n} \|\xi\|^2 e^{-t_0 \|\xi\|^2} |\hat{u}(\xi)| \, d\xi$$

$$\leq \left(\int_{\mathbb{R}^n} \|\xi\|^4 e^{-2t_0 \|\xi\|^2} \, d\xi\right)^{\frac{1}{2}} \|u\|_{L^2}.$$

Thus we find for $u \in L^2(\mathbb{R}^n)$ and $t > 0$ that $\frac{\partial}{\partial t} T_t u = \Delta T_t u$. Finally we observe that for $u \in L^2(\mathbb{R}^n)$ it follows that

$$\lim_{t \to 0} \|T_t u - u\|_{L^2}^2 = \lim_{t \to 0} \int_{\mathbb{R}^n} \left|e^{-t\|\xi\|^2} - 1\right|^2 |\hat{u}(\xi)|^2 \, d\xi = 0,$$

hence

$$\lim_{t \to 0} \|T_t u - u\|_{L^2} = 0.$$

Problems

1. Find the Fourier transform of the following functions defined on $\mathbb{R}$:

a) $f_M(x) = \chi_{[-M,M]}(x)$, $M > 0$;

b) $g_R(x) = \chi_{[-R,R]}(x)\left(1 - \frac{|x|}{R}\right)$, $R > 0$;

c) $h_a(x) = e^{a|x|}$, $a > 0$;

d) $k_t(x) = \frac{1}{x^2+t^2}$, $t > 0$.

2. Use Plancherel's theorem to find the integral

$$\int_{\mathbb{R}} \frac{1}{(a^2 + x^2)(b^2 + x^2)} \, dx, \quad a, b > 0.$$

3. With k_t from Problem 1 d) find for $K_t(x) := \sqrt{\frac{2}{\pi}} t k_t$ the convolution $K_t * K_s$, $s, t > 0$.

4. a) For $\alpha > 0$ verify that $e^{-\alpha} = \frac{1}{\sqrt{\pi}} \int_0^\infty e^{-s} \frac{1}{\sqrt{s}} e^{-\frac{\alpha^2}{4s}} \, ds$.

    b) Find the Fourier transform of $g : \mathbb{R}^n \to \mathbb{R}$, $g(x) = e^{-\|x\|}$.

    c) Use part b) to find the Fourier transform of $g_t(x) := e^{-t\|x\|}$.

5. a) Let $h \in L^1(\mathbb{R})$, $\int_{\mathbb{R}} h(x) \, dx = 1$, and $h_\lambda(t) = \lambda h(\lambda t)$. Further let $f \in L^1(\mathbb{R}) \cap L^\infty(\mathbb{R})$ be continuous at t_0. Prove that $\lim_{\lambda \to \infty} (f * h_\lambda)(t_0) = f(t_0)$. Moreover show that if f is uniformly continuous then $f * h_\lambda$ converges uniformly on $\mathbb{R}$ to f. (This result has certain similarities to our considerations on "good kernels".)

    b) Let $f \in L^1(\mathbb{R}) \cap L^\infty(\mathbb{R})$ be continuous at t_0. Prove

$$\lim_{R \to \infty} \frac{1}{\sqrt{2\pi}} \int_{-R}^{R} e^{ixt_0} \left(1 - \frac{|x|}{R} \right) \hat{f}(x) \, dx = f(t_0).$$

What can we prove for f uniformly continuous on $\mathbb{R}$?

    c) Use Part b) in order to formulate and prove a statement in the Wiener algebra $A(\mathbb{R})$.

6. Give the missing details in the proof of Theorem 12.7.

7. In the situation of Example 12.20 prove that $(T_t)_{t>0}$ satisfies the semigroup property on $L^2(\mathbb{R}^n)$.

8. For $0 < s < 1$ prove the existence of some constant $c = c_{n,s}$ such that for all $u \in \mathcal{S}(\mathbb{R})$ we have

$$\|u\|_{H^s}^2 \sim \|u\|_{L^2}^2 + c_{n,s} \int_{\mathbb{R}^n} \int_{\mathbb{R}^n} \frac{|u(x) - u(y)|^2}{\|x - y\|^{n+s}} \, dx \, dy$$

where for $0 < a, b$ the symbol $a \sim b$ means that $0 < \kappa_1 \leq \frac{b}{a} \leq \kappa_2$ holds.

9. Let $\nu : \mathbb{R}^n \to \mathbb{R}$ be a function such that by

$$\psi(\xi) := \int_{\mathbb{R}^n} (1 - \cos \xi \cdot y) \nu(y) \, dy$$

a continuous and bounded function $\psi : \mathbb{R}^n \to \mathbb{R}$ is defined. For $u \in \mathcal{S}(\mathbb{R}^n)$ prove with a suitable constant c that

$$\int_{\mathbb{R}^n} |\hat{u}(\xi)| \psi(\xi)\, d\xi = \frac{1}{2} \int_{\mathbb{R}^n} \left(\int_{\mathbb{R}^n} (u(x) - u(y))^2 \nu(x - y)\, dy \right) dx.$$

10.   a) For $s \geq 1$ show the estimate

$$|\Lambda^s(\xi) - \Lambda^s(\eta)| \leq c_s (1 + \|\xi - \eta\|^2)^{\frac{s+1}{2}} \Lambda^{s-1}(\eta)$$

(which is of course equivalent to $|\Lambda^s(\xi) - \Lambda^s(\eta)| \leq c_s \Lambda^{\frac{s+1}{2}}(\xi - \eta)\Lambda^{s-1}(\eta)$.)
Hint: Note that

$$|\Lambda^s(\xi) - \Lambda^s(\eta)| = \left| s \int_{\Lambda^1(\eta)}^{\Lambda^1(\xi)} t^{s-1}\, dt \right|$$

and later on use Peetre's inequality.

b)* For $\varphi \in \mathcal{S}(\mathbb{R}^n)$ and $s > 0$ we define the commutator $[\Lambda^s(D), \varphi]$: $\mathcal{S}(\mathbb{R}^n) \to \mathcal{S}(\mathbb{R}^n)$ by $[\Lambda^s(D), \varphi]u := \Lambda^s(D)(\varphi u) - \varphi \Lambda^s(D)u$. Prove that for $t \geq \max(0, 1 - s)$ the estimate

$$\|[\Lambda^s(D), \varphi]u\|_{H^t} \leq c_{s,t,\varphi} \|u\|_{H^{s+t-1}}$$

holds.

11. For $(T_t)_{t>0}$ $T_t u = h_t * u$, where h_t is the heat kernel in $\mathbb{R}^n$, justify for $u \in H^4(\mathbb{R}^n)$ that

$$\lim_{t \to 0} \left\| \frac{T_t u - u}{t} - \Delta u \right\|_{L^2} = 0.$$

13   The Fourier Transform of Bounded Measures

For every bounded Borel measure $\mu \in \mathcal{M}_b^+(\mathbb{R}^n)$ the integral

$$\mu(f) := \int_{\mathbb{R}^n} f(x)\mu(\mathrm{d}x) \tag{13.1}$$

exists for every bounded continuous function $f : \mathbb{R}^n \to \mathbb{C}$. Therefore the following definition makes sense.

Definition 13.1. *The **Fourier transform of a bounded measure** $\mu \in \mathcal{M}_b^+(\mathbb{R}^n)$ is the function $\hat{\mu} : \mathbb{R}^n \to \mathbb{C}$ defined by*

$$\hat{\mu}(\xi) := (2\pi)^{-\frac{n}{2}} \int_{\mathbb{R}^n} e^{-ix\cdot\xi}\mu(\mathrm{d}x). \tag{13.2}$$

Suppose that $\mu = g\lambda^{(n)}$ where $\lambda^{(n)}$ denotes the Lebesgue measure in $\mathbb{R}^n$ and $g \in L^1(\mathbb{R}^n)$, $g \geq 0$ almost everywhere. Since now

$$\|\mu\| = \mu(\mathbb{R}^n) = \int_{\mathbb{R}^n} g(x)\lambda^{(n)}(\mathrm{d}x) = \|g\|_{L^1} \tag{13.3}$$

the measure μ is bounded and we find that

$$\hat{\mu}(\xi) = (2\pi)^{-\frac{n}{2}} \int_{\mathbb{R}^n} e^{-ix\cdot\xi}g(x)\,\mathrm{d}x = \hat{g}(\xi). \tag{13.4}$$

Theorem 13.2. *The Fourier transform of $\mu \in \mathcal{M}_b^+(\mathbb{R}^n)$ is a uniformly continuous function which is bounded by $(2\pi)^{-\frac{n}{2}}\|\mu\|$, i.e.*

$$|\hat{\mu}(\xi)| \leq (2\pi)^{-\frac{n}{2}}\|\mu\| = \hat{\mu}(0). \tag{13.5}$$

Proof. First we note that

$$\left| e^{-ix\cdot\xi_1} - e^{-ix\cdot\xi_2} \right| = \left| i\int_{x\cdot\xi_1}^{x\cdot\xi_2} e^{-it}\,\mathrm{d}t \right| \leq \|x\|\|\xi_1 - \xi_2\|. \tag{13.6}$$

The boundedness of μ implies that for every $\epsilon > 0$ there exists a ball $B_R(0) \subset \mathbb{R}^n$ with radius $R = R(\epsilon) > 0$ such that $\mu\left(B_R^{\complement}(0)\right) < \epsilon$. Now it follows that

$$\left| \int_{\mathbb{R}^n} e^{-ix\cdot\xi_1} \mu(dx) - \int_{\mathbb{R}^n} e^{-ix\cdot\xi_2} \mu(dx) \right| \leq \int_{\mathbb{R}^n} \left| e^{-ix\cdot\xi_1} - e^{-ix\cdot\xi_2} \right| \mu(dx)$$

$$\leq \int_{B_R(0)} \left| e^{-ix\cdot\xi_1} - e^{-x\cdot\xi_2} \right| \mu(dx) + \int_{B_R^C(0)} \left| e^{-ix\cdot\xi_1} - e^{-ix\cdot\xi_2} \right| \mu(dx)$$

$$\leq R\|\xi_1 - \xi_2\| \mu(B_R(0)) + 2\mu\left(B_R^C(0) \right)$$

$$\leq R\|\xi_1 - \xi_2\| \|\mu\| + 2\epsilon.$$

Given $\epsilon > 0$ we choose $\delta = \frac{\epsilon}{R(\epsilon)\|\mu\|}$ to find for all $\xi_1, \xi_2 \in \mathbb{R}^n$ with $\|\xi_1 - \xi_2\| < \delta$ that

$$|\hat{\mu}_1(\xi) - \hat{\mu}_2(\xi)| \leq (2\pi)^{-\frac{n}{2}} R\|\xi_1 - \xi_2\| \|\mu\| + 2(2\pi)^{-\frac{n}{2}} \epsilon$$

$$\leq 3(2\pi)^{-\frac{n}{2}} \epsilon,$$

i.e. we have proved that $\hat{\mu}$ is uniformly continuous. Moreover we get

$$|\hat{\mu}(\xi)| = \left| (2\pi)^{-\frac{n}{2}} \int_{\mathbb{R}^n} e^{-ix\cdot\xi} \mu(dx) \right| \leq (2\pi)^{-\frac{n}{2}} \mu(\mathbb{R}^n)$$

but $\mu(\mathbb{R}^n) = \int_{\mathbb{R}^n} 1 \, d\mu = \int e^{-ix\cdot 0} \mu(dx) = (2\pi)^{\frac{n}{2}} \hat{\mu}(0)$. $\qquad\square$

Example 13.3. Let $\mu = \sum_{k=1}^{\infty} a_k \epsilon_{x_k}$, where $a_k \geq 0$ with $\sum_{k=1}^{\infty} a_k < \infty$ and $(x_k)_{k\in\mathbb{N}}$ is a sequence of points in $\mathbb{R}^n$. It follows that

$$\hat{\mu}(\xi) = (2\pi)^{-\frac{n}{2}} \int_{\mathbb{R}^n} e^{-ix\cdot\xi} \sum_{k=1}^{\infty} a_k \epsilon_{x_k}(dx)$$

$$= (2\pi)^{-\frac{n}{2}} \sum_{j=1}^{\infty} a_k \int_{\mathbb{R}^n} e^{-ix\cdot\xi} \epsilon_{x_k}(dx)$$

$$= (2\pi)^{-\frac{n}{2}} \sum_{k=1}^{\infty} a_k e^{-ix_k\cdot\xi}.$$

For $n = 1$ we may consider as special cases some well-known probability measures:

$\mu = \epsilon_a, \quad a \in \mathbb{R},$ $\qquad\qquad\qquad \hat{\mu}(\xi) = \dfrac{1}{\sqrt{2\pi}} e^{-ia\cdot\xi},$

(degenerate distribution)

$\mu = \dfrac{1}{2}(\epsilon_a + \epsilon_{-a}), \quad a \in \mathbb{R},$ $\qquad \hat{\mu}(\xi) = \dfrac{1}{\sqrt{2\pi}} \cos a\xi,$

(symmetric distribution)

$\mu = \displaystyle\sum_{l=0}^{m} \binom{m}{l} p^l (1-p)^{m-l} \epsilon_l, \quad p \in (0,1),$ $\qquad \hat{\mu}(\xi) = \dfrac{1}{\sqrt{2\pi}} \left((1-p) + p e^{-i\xi} \right)^m,$

(binomial distribution with parameter p)

$\mu = \displaystyle\sum_{l=0}^{\infty} e^{-\sigma} \dfrac{\sigma^l}{l!} \epsilon_l, \quad \sigma > 0,$ $\qquad \hat{\mu}(\xi) = \dfrac{1}{\sqrt{2\pi}} e^{\sigma(e^{-i\xi}-1)}.$

(Poisson distribution with parameter σ)

In Problem 1 we will provide some of the calculations leading to some of these $\hat{\mu}$'s.

The next theorem summarises some basic properties of the Fourier transform of measures.

Theorem 13.4. A. *For the image measure $T(\mu)$ of $\mu \in \mathcal{M}_b^+(\mathbb{R}^n)$ under a linear mapping $T : \mathbb{R}^n \to \mathbb{R}^n$ we have*

$$(T(\mu))\widehat{} = \hat{\mu} \circ T^t. \tag{13.7}$$

In particular for the reflection $S : \mathbb{R}^n \to \mathbb{R}^n$, $Sx = -x$, we find

$$S(\mu)\widehat{} = \overline{\hat{\mu}} = \hat{\mu} \circ S. \tag{13.8}$$

B. *The Fourier transform of the image of $\mu \in \mathcal{M}_b^+(\mathbb{R}^n)$ under the translation T_a, $T_a x = a + x$, is given by*

$$(T_a(\mu))\widehat{}\,(\xi) = e^{-i\xi\cdot a} \hat{\mu}(\xi). \tag{13.9}$$

C. *For $\mu, \nu \in \mathcal{M}_b^+(\mathbb{R}^n)$ the convolution theorem holds, i.e.*

$$(\mu * \nu)\widehat{} = (2\pi)^{\frac{n}{2}} \hat{\mu} \cdot \hat{\nu}. \tag{13.10}$$

D. *Let $\mu \in \mathcal{M}_b^+(\mathbb{R}^n)$ and $\nu \in \mathcal{M}_b^+(\mathbb{R}^m)$. The Fourier transform of their product $\mu \otimes \nu \in \mathcal{M}_b^+(\mathbb{R}^{n+m})$ is*

$$(\mu \otimes \nu)\widehat{}\,(\xi, \eta) = \hat{\mu}(\xi)\hat{\nu}(\eta), \quad \xi \in \mathbb{R}^n, \eta \in \mathbb{R}^m. \tag{13.11}$$

Proof. **A.** Since linear mappings are measurable $T(\mu)$ is a bounded Borel measure on $\mathbb{R}^n$ and we have

$$(T(\mu))\widehat{\ }(\xi) = (2\pi)^{-\frac{n}{2}} \int_{\mathbb{R}^n} e^{-i\xi \cdot Tx} \mu(dx) = (2\pi)^{-\frac{n}{2}} \int_{\mathbb{R}^n} e^{-i(T^t\xi)\cdot x} \mu(dx)$$
$$= \hat{\mu}\left(T^t\xi\right) = \left(\hat{\mu} \circ T^t\right)(\xi).$$

Since $\overline{\hat{\mu}}(\xi) = \hat{\mu}(-\xi)$ we also obtain (13.8).
B. For $a \in \mathbb{R}^n$ it follows that

$$(T_a(\mu))\widehat{\ }(\xi) = (2\pi)^{-\frac{n}{2}} \int_{\mathbb{R}^n} e^{-i\xi \cdot (a+x)} \mu(dx) = e^{-i\xi \cdot a} \hat{\mu}(x).$$

C. Using (III.10.26) and its consequence we find for $f \in C_b(\mathbb{R}^n)$ that

$$\int_{\mathbb{R}^n} f \, d(\mu * \nu) = \int_{\mathbb{R}^n} \int_{\mathbb{R}^n} f(x+y) \mu(dy) \nu(dx)$$

and with $f(x) = e^{-ix \cdot \xi}$ it follows that

$$(\mu * \nu)\widehat{\ }(\xi) = (2\pi)^{-\frac{n}{2}} \int_{\mathbb{R}^n} \int_{\mathbb{R}^n} e^{-i(x+y)\cdot\xi} \mu(dy)\nu(dx)$$
$$= (2\pi)^{\frac{n}{2}} \hat{\mu}(\xi)\hat{\nu}(\xi).$$

D. We just need to observe that

$$(\mu \otimes \nu)(\xi, \eta) = (2\pi)^{\frac{-n-m}{2}} \int_{\mathbb{R}^{n+m}} e^{-i(x\cdot\xi + y\cdot\eta)} \mu(dx)\nu(dy)$$
$$= (2\pi)^{-\frac{n}{2}} \int_{\mathbb{R}^n} e^{-ix\cdot\xi} \mu(dx) \cdot (2\pi)^{-\frac{m}{2}} \int_{\mathbb{R}^m} e^{-iy\cdot\eta} \nu(dy)$$
$$= \hat{\mu}(\xi)\hat{\nu}(\eta).$$

$\square$

Remark 13.5. The reader should note that there is a certain duality. For functions we may consider $(u \circ T)\widehat{\ }$ for certain mappings T. This however does not make sense for measures, but for measures we can consider the Fourier transform $(T(\mu))\widehat{\ }$ of the image measure $(T(\mu))$ for certain mappings T.

In Chapter 8 we have discussed the Fourier series of bounded measures on the torus $\mathbb{T}^1$ and we were led to consider positive definite sequences $(c_k)_{k \in \mathbb{Z}}$, or re-interpreting a sequence as a mapping from $\mathbb{Z}$ to $\mathbb{C}$, we may speak of positive definite functions $\varphi : \mathbb{Z} \to \mathbb{C}$, $\varphi(k) = c_k$. In order to understand the Fourier transform of bounded measures $\mu \in \mathcal{M}_b^+(\mathbb{R}^n)$ better we need

Definition 13.6. *A function* $u : \mathbb{R}^n \to \mathbb{C}$ *is called* **positive definite** *if for all* $k \in \mathbb{N}$ *and any choice of vectors* $\xi^1, \ldots, \xi^k \in \mathbb{R}^n$ *the matrix* $\left(u \left(\xi^j - \xi^l \right) \right)_{j,l=1,\ldots,k}$ *is positive Hermitian, i.e. we have for all* $\lambda_1, \ldots, \lambda_k \in \mathbb{C}$

$$\sum_{j,l=1}^{k} u \left(\xi^j - \xi^l \right) \lambda_j \overline{\lambda}_l \geq 0. \tag{13.12}$$

In full analogy to Lemma 8.5 we now have

Lemma 13.7. *The Fourier transform* $\hat{\mu}$ *of* $\mu \in M_b^+(\mathbb{R}^n)$ *is a positive definite function.*

Proof. For $k \in \mathbb{N}$ and $\xi^1, \ldots, \xi^k \in \mathbb{R}^n$ we have with $\lambda_1, \ldots, \lambda_k \in \mathbb{C}$ that

$$\sum_{j,k=1}^{k} \lambda_j \overline{\lambda}_l \hat{\mu} \left(\xi^j - \xi^l \right) = (2\pi)^{-\frac{n}{2}} \int_{\mathbb{R}^n} \sum_{j,l=1}^{k} \lambda_j \overline{\lambda}_l e^{-i\left(\xi^j - \xi^l \right) \cdot x} \mu(\mathrm{d}x)$$

$$= (2\pi)^{-\frac{n}{2}} \int_{\mathbb{R}^n} \left(\sum_{j=1}^{k} \lambda_j e^{-i\xi^j \cdot x} \right) \overline{\left(\sum_{l=1}^{k} \lambda_l e^{-i\xi^l \cdot x} \right)} \mu(\mathrm{d}x)$$

$$= (2\pi)^{-\frac{n}{2}} \int_{\mathbb{R}^n} \left| \sum_{j=1}^{k} \lambda_j e^{-i\xi^j \cdot x} \right|^2 \mu(\mathrm{d}x) \geq 0.$$

$\square$

Thus we have re-established the role of positive definite functions for the Fourier transform of measures and before proceeding further we want to collect some of their properties.

Lemma 13.8. *Let* $u, v : \mathbb{R}^n \to \mathbb{C}$ *be positive definite functions and* $\lambda \geq 0$. *Then we have*

$$\overline{u}(\xi) = u(-\xi) \quad \text{and} \quad u(0) \geq 0, \tag{13.13}$$

as well as

$$|u(x)| \leq u(0). \tag{13.14}$$

Furthermore the functions $u + v$ *and* λv *are positive definite as is* $u \cdot v$. *If* $(u_k)_{k\in\mathbb{N}}$ *is a sequence of positive definite functions converging pointwisely to* u, *then* u *is positive definite too.*

Proof. When we choose in (13.12) for k the value 2 and $\xi^1 = 0$, $\xi^2 = \xi$ then the fact that $\begin{pmatrix} u(0) & u(\xi) \\ u(-\xi) & u(0) \end{pmatrix}$ is positive Hermitian implies (13.13). Since

$\det \begin{pmatrix} u(0) & u(\xi) \\ u(-\xi) & u(0) \end{pmatrix} \geq 0$ we also obtain (13.14). That $u + v$ and λu are positive definite follows immediately from (13.12) as does the fact that the pointwise limit of positive definite functions is positive definite, we just have to pass in (13.12) to the pointwise limit. Finally with the help of Lemma 8.3 we deduce that $u \cdot v$ is positive definite. □

Example 13.9. Let $f(z) = \sum_{j=0}^{\infty} a_j z^j$ be a power series with non-negative coefficients $a_j \geq 0$ and radius of convergence $R > 0$. If $u : \mathbb{R}^n \to \mathbb{C}$ is a (continuous) positive definite function with $u(0) < R$ then the function $f \circ u$ is a (continuous) positive definite function too. Indeed, first we note that by (13.14) we have $|u(x)| < R$ and therefore $(f \circ u)(x) = \sum_{j=0}^{\infty} a_j(u(x))^j$ is for every $x \in \mathbb{R}^n$ well defined, i.e. convergent. If u is continuous then $f \circ u$ is continuous too. It remains to prove that $f \circ u$ is positive definite. By Lemma 13.8 it follows that for every $j \in \mathbb{N}_0$ the function $(u(\cdot))^j$ is positive definite, hence for $N \in \mathbb{N}$ the function $\sum_{j=1}^{N} a_j(u(\cdot))^j$ is positive definite. Thus $f \circ u$ is a pointwise limit of positive definite functions.

We denote by $P(\mathbb{R}^n)$ the set of all positive definite functions on $\mathbb{R}^n$ and $CP(\mathbb{R}^n)$ denotes the set of all continuous positive definite functions on $\mathbb{R}^n$. Note that if $(u_k)_{k\in\mathbb{N}}$, $u_k \in CP(\mathbb{R}^n)$, converges uniformly on compact sets to u then $u \in CP(\mathbb{R}^n)$ since u must be continuous and by Lemma 13.8 positive definite.

The Theorem of Herglotz-Bochner, Theorem 8.9, states that a sequence in $\mathbb{Z}$ is positive definite if and only if it is the sequence of the Fourier-Stieltjes coefficients of a measure. The proof of Theorem 8.9 was rather lengthy, it needed the Riesz representation theorem, the proof of which we have postponed, and in addition it relied on many non-trivial results we had derived before for Fourier series, e.g. summability results, the Weierstrass approximation theorem, etc. A natural question is whether we can prove the converse to Lemma 13.7. The good news is that we can, the bad news is that we have to postpone the entire proof. First let us state the result:

Theorem 13.10 (Bochner). *A function $u : \mathbb{R}^n \to \mathbb{C}$ is the Fourier transform of a measure $\mu \in \mathcal{M}_b^+(\mathbb{R}^n)$ if and only if u is continuous, $u(0) = \hat{\mu}(0) = (2\pi)^{-\frac{n}{2}}\|\mu\|$ and u is positive definite.*

There are several proofs of Bochner's theorem, each requires more advanced tools. The reason we can understand. A positive definite function is in general not in

$L^1(\mathbb{R}^n)$ (or $L^2(\mathbb{R}^n)$), however we expect that $\mu = F^{-1}(\hat{\mu})$. But F^{-1} should be

$$(2\pi)^{-\frac{n}{2}} \int_{\mathbb{R}^n} e^{ix\cdot\xi} \hat{\mu}(\xi) \,\mathrm{d}\xi. \tag{13.15}$$

Neither needs this integral to exist and if it exists it will give a function, not a measure. We cannot expect, for example, that

$$\frac{1}{2} (\epsilon_a + \epsilon_{-a}) = \frac{1}{\sqrt{2\pi}} \int_{\mathbb{R}} e^{ix\xi} \left(\frac{1}{\sqrt{2\pi}} \cos a\xi \right) \,\mathrm{d}\xi$$

holds with a classical meaning of the right hand side. The natural idea is to try some approximation of the "integral" (13.15). But this requires an approximation in $\mathcal{M}_b^+(\mathbb{R}^n)$ where we need to introduce a topology. Such a strategy works and the idea to introduce a topology in $\mathcal{M}_b^+(\mathbb{R}^n)$ will be picked up soon.

The proof of Bochner's theorem we will provide in Volume V is of a different na-ture. We first will extend the Fourier transform from $\mathcal{S}(\mathbb{R}^n)$ to the space $\mathcal{S}'(\mathbb{R}^n)$ of tempered distributions, the topological dual space of $\mathcal{S}(\mathbb{R}^n)$, by duality. We also will prove that bounded measures as well as bounded continuous functions are tempered distributions, hence in $\mathcal{S}'(\mathbb{R}^n)$ we can find $F^{-1}(\hat{\mu})$. Then we will show that if $u \in \mathcal{S}'(\mathbb{R}^n)$ is a continuous bounded positive definite function $F^{-1}(u)$ is a bounded Borel measure which will imply $\mu = F^{-1}(\hat{\mu})$ for $\mu \in \mathcal{M}_b^+(\mathbb{R}^n)$ when the equality is read in $\mathcal{S}'(\mathbb{R}^n)$. We believe that it is justifiable to state Bochner's theorem already here, but to prove it in a different volume.

Let us return to the problem of a topology on $\mathcal{M}_b^+(\mathbb{R}^n)$, more precisely, we are interested in the concept of convergence of measures. In Definition III.12.9 we in-troduced the notion of weakly convergent sequences in $L^p(G)$. A sequence $(g_k)_{k\in\mathbb{N}}$, $g_k \in L^p(G)$ is said to converge weakly to $g \in L^p(G)$ if for every $\varphi \in L^{p'}(G)$, $\frac{1}{p} + \frac{1}{p'} = 1$, it follows that

$$\lim_{k\to\infty} \int_G g_k \varphi \,\mathrm{d}\mu = \int_G g\varphi \,\mathrm{d}\mu.$$

The idea may be extended to measures.

Definition 13.11. *We call a sequence $(\mu_k)_{k\in\mathbb{N}}$, $\mu_k \in \mathcal{M}_b^+(\mathbb{R}^n)$, **weakly conver-gent** to $\mu \in \mathcal{M}_b^+(\mathbb{R}^n)$ if for all $\varphi \in C_b(\mathbb{R}^n)$*

$$\lim_{k\to\infty} \int \varphi \,\mathrm{d}\mu_k = \int \varphi \,\mathrm{d}\mu \tag{13.16}$$

*holds. If (13.16) holds only for all $\varphi \in C_0(\mathbb{R}^n)$ we call the sequence $(\mu_k)_{k\in\mathbb{N}}$ **vaguely convergent** to μ.*

In Appendix II we give some characterisation of these notions of convergence. An easy consequence of Definition 13.11 is

Corollary 13.12. *If a sequence of measures $(\mu_k)_{k\in\mathbb{N}}$, $\mu_k \in \mathcal{M}_b^+(\mathbb{R}^n)$, converges weakly to $\mu \in \mathcal{M}_b^+(\mathbb{R}^n)$ then the sequence $(\hat{\mu}_k)_{k\in\mathbb{N}}$ of its Fourier transforms converges pointwise to $\hat{\mu}$, i.e. $\lim_{k\to\infty} \hat{\mu}_k(\xi) = \hat{\mu}(\xi)$ for all $\xi \in \mathbb{R}^n$.*

Proof. We just have to choose in (13.16) the function $\varphi_\xi(x) := (2\pi)^{-\frac{n}{2}} e^{-ix\cdot\xi}$. $\square$

Note that by Bochner's theorem $\hat{\mu}$ must be continuous. The following result, **Lévy's continuity theorem**, is of great importance in probability theory since it can be proved without the use of Bochner's theorem and admits the existence of a limit measure.

Theorem 13.13. *Let $(\mu_k)_{k\in\mathbb{N}}$, $\mu_k \in \mathcal{M}_b^+(\mathbb{R}^n)$, be a sequence of bounded measures such that for some function $u : \mathbb{R}^n \to \mathbb{C}$ continuous at 0 it follows that*

$$\lim_{k\to\infty} \hat{\mu}_k(\xi) = u(\xi) \tag{13.17}$$

for all $\xi \in \mathbb{R}^n$. Then there exists a measure $\mu \in \mathcal{M}_b^+(\mathbb{R}^n)$ such that $(\mu_k)_{k\in\mathbb{N}}$ converges weakly to μ.

Let us try to understand this statement. Suppose for a moment that u is continuous on $\mathbb{R}^n$. Since we know that u is a positive definite function (as a limit of positive definite functions) this will imply that u is a continuous positive definite function, hence by Bochner's theorem the Fourier transform of a bounded measure μ, i.e. we have the existence of a limit measure with the property

$$\lim_{k\to\infty} \hat{\mu}_k(\xi) = \hat{\mu}(\xi), \tag{13.18}$$

but it still remains to prove that $(\mu_k)_{k\in\mathbb{N}}$ converges weakly to μ. This we will do in Appendix II as we will prove that under the assumptions of Theorem 13.13 the limit function u in (13.17) must be already continuous.

The Fourier transform of bounded measures becomes a central tool when dealing with convolution semigroups.

Definition 13.14. *Suppose that for every $t \geq 0$ a bounded Borel measure $\mu_t \in \mathcal{M}_b^+(\mathbb{R}^n)$ is given. We call $(\mu_t)_{t\geq 0}$ a **convolution semi-group** on $\mathbb{R}^n$ if*

$$\mu_t(\mathbb{R}^n) \leq 1 \quad \text{for all } t \geq 0; \tag{13.19}$$

$$\mu_t * \mu_s = \mu_{t+s}, \quad s, t \geq 0, \quad \text{and } \mu_0 = \epsilon_0; \tag{13.20}$$

$$\mu_t \to \epsilon_0 \quad \text{vaguely as } t \to 0. \tag{13.21}$$

Note that the last statement means that

$$\lim_{t \to 0} \int \varphi \, d\mu_t = \varphi(0) \tag{13.22}$$

for all $\varphi \in C_0(\mathbb{R}^n)$.

Convolution semi-groups will provide us in Volume V with interesting families of linear operators. Their analysis needs a bit more understanding of topologies on $\mathcal{M}_b^+(\mathbb{R}^n)$ and we will discuss this in Appendix II. The key result will be that for every convolution semigroup there exists a unique continuous function $\psi : \mathbb{R}^n \to \mathbb{C}$ such that

$$\hat{\mu}_t(\xi) = (2\pi)^{-\frac{n}{2}} e^{-t\psi(\xi)}, \quad t \geq 0, \tag{13.23}$$

where $\psi(0) \geq 0$ and clearly $\xi \mapsto e^{-t\psi(\xi)}$ must be positive definite for all $t \geq 0$.

Fourier transforms of probability measures are essentially the characteristic functions of random variables and they play an important role in probability theory. Some of the problems in this chapter are related to probabilistic applications.

Problems

1.   a) For the binomial distribution $\mu = \sum_{l=0}^m \binom{m}{l} p^l (1-p)^{m-l} \epsilon_l$. Show that $\hat{\mu}(\xi) = \frac{1}{\sqrt{2\pi}} ((1-p) + pe^{-i\xi})^m$.

 b) Find the Fourier transform of the Poisson distribution with parameter σ.

2.   Suppose that the probability measure μ on $\mathbb{R}$ has the kth **absolute moment** $M_k(\mu) := \int_{\mathbb{R}} |x|^k \mu(dx) < \infty$. Prove that $\hat{\mu} \in C^k(\mathbb{R}^n)$.

3.   Show that if $\varphi : \mathbb{R} \to \mathbb{C}$ is a continuous positive definite function then $|\varphi|^2 : \mathbb{R} \to \mathbb{R}$ is also a continuous positive definite function.

4.* Prove **Polya's theorem**: suppose that the continuous function $f : \mathbb{R} \to \mathbb{R}$ is even, $f|_{(0,\infty)}$ is decreasing and convex, $f(0) > 0$ and $\lim_{t \to \infty} f(t) = 0$. Then f is a positive definite function.

5.   a) Show that if a continuous positive definite function $u : \mathbb{R} \to \mathbb{R}$ satisfies $\lim_{t \to 0} \frac{u(0)-u(t)}{t^2} = 0$ then it must be constant.
Hint: use Problem 4 of Chapter 8.

 b) Prove that there is no bounded Borel measure μ on $\mathbb{R}$ with Fourier transform $\hat{\mu}(\xi) = ce^{-|\xi|^\alpha}$ for some $\alpha > 2$.

6. Let $u \in C_b(\mathbb{R}^n)$ be a positive definite function. Show that for all $\varphi \in \mathcal{S}(\mathbb{R}^n)$ we have

$$\int_{\mathbb{R}^n} \int_{\mathbb{R}^n} u(\xi - \eta)\overline{\varphi(\eta)}\varphi(\xi)\, d\xi\, d\eta \geq 0.$$

Hint: approximate the integral by Riemann sums.

7. For $\mu \in \mathcal{M}_b^+(\mathbb{R}^n)$ and $u \in L^1(\mathbb{R}^n)$ verify

$$\int_{\mathbb{R}^n} \hat{u}(\xi)\mu(d\xi) = \int_{\mathbb{R}^n} u(x)\hat{\mu}(x)\, dx.$$

8. We call a probability measure μ on $\mathbb{R}^n$ **infinitely divisible** if for every $k \in \mathbb{N}$ there exists a probability measure μ_k on $\mathbb{R}^n$ such that $\mu = \mu_k * \cdots * \mu_k$ (k terms).

a) Prove that if μ is infinitely divisible then for $k \in \mathbb{N}$ and μ_k from the definition we must have

$$(2\pi)^{-\frac{n}{2}}\hat{\mu}(\xi) = \left((2\pi)^{-\frac{n}{2}}\hat{\mu}_k(\xi)\right)^k.$$

b) Suppose that $\psi : \mathbb{R}^n \to \mathbb{C}$ is a continuous function with $\psi(0) = 0$ and for all $t > 0$ the function $\xi \mapsto e^{-t\psi(\xi)}$ is positive definite. Prove that the probability measure μ defined by $\hat{\mu}(\xi) = (2\pi)^{-\frac{n}{2}}e^{-\psi(\xi)}$ is infinite divisible.

c) Construct an infinitely divisible measure μ on $\mathbb{R}$ corresponding to the function $\psi(\xi) := |\xi|$.

14 Selected Topics on the Fourier Transform

In this chapter we will treat a few topics that are independent of one another, but still of interest and with many applications.

First we want to study the Fourier transform of radial symmetric functions and for this we need more results on Bessel functions. The **Bessel function** J_l of order l, $l \in \mathbb{N}_0$, was introduced in Problem 7 to Chapter I.29 as the power series

$$J_l(x) = \sum_{k=0}^{\infty} \frac{(-1)^k x^{2k+l}}{2^{2k+l} k! (k+l)!} = \left(\frac{x}{2}\right)^l \sum_{k=0}^{\infty} \frac{(-1)^k}{k!(k+l)!} \left(\frac{x}{2}\right)^{2k}. \tag{14.1}$$

We have proved that the radius of convergence is ∞, thus we can extend J_l to an entire function on $\mathbb{C}$ and (14.1) holds for all $z \in \mathbb{C}$. Furthermore we have proved that J_l satisfies the differential equation

$$z^2 J_l''(z) + z J_l'(z) + (z^2 - l^2) J_l(z) = 0. \tag{14.2}$$

In addition, in Problem 10 of Chapter III.22, it was shown that

$$J_l(z) = \frac{1}{\Gamma\left(l + \frac{1}{2}\right) \Gamma\left(\frac{1}{2}\right)} \left(\frac{z}{2}\right)^l \int_{-1}^{1} (1 - t^2)^{l - \frac{1}{2}} \cos zt \, dt. \tag{14.3}$$

First we want to allow l in (14.1) and (14.2) to be a negative integer by defining

$$J_{-l}(x) = (-1)^l J_l(x), \quad l \in \mathbb{N}, \tag{14.4}$$

which yields

$$J_{-l}(x) = \sum_{k=0}^{\infty} \frac{(-1)^{l+k}}{2^{2k+l} k! (k+l)!} \left(\frac{x}{2}\right)^{2k}. \tag{14.5}$$

Proposition 14.1. *For all $x \in \mathbb{R}$ and $z \in \mathbb{C} \setminus \{0\}$ the equality*

$$\sum_{k \in \mathbb{Z}} J_k(x) z^k = e^{\frac{x}{2}\left(z - \frac{1}{z}\right)} \tag{14.6}$$

holds.

Proof. (Following [34]) The power series expansion for the exponential function gives

$$e^{\frac{xz}{2}} = \sum_{k=0}^{\infty} \frac{z^k}{k!} \left(\frac{x}{2}\right)^k \quad \text{and} \quad e^{-\frac{x}{2z}} = \sum_{m=0}^{\infty} \frac{(-1)^m}{z^m m!} \left(\frac{x}{2}\right)^m.$$

Since $e^{\frac{xz}{2}-\frac{x}{2z}} = e^{\frac{xz}{2}} e^{-\frac{x}{2z}}$ we find

$$e^{\frac{xz}{2}-\frac{x}{2z}} = \sum_{k,m\in\mathbb{N}_0} \frac{(-1)^m z^{k-m}}{k!m!} \left(\frac{x}{2}\right)^{k+m}.$$

We rearrange this series according to the powers of z to obtain with $n = k - m$, i.e. $k = m + n$,

$$e^{\frac{xz}{2}-\frac{x}{2z}} = \sum_{n\in\mathbb{Z}} \left(\sum_{k=0}^{\infty} \frac{(-1)^k}{k!(k+n)!} \left(\frac{x}{2}\right)^{2k+n}\right) z^n$$

$$= \sum_{n\in\mathbb{Z}} J_n(x) z^n,$$

where we have used the convention $\frac{1}{(k+n)!} = \frac{1}{\Gamma(k+n+1)} = 0$ for $k + n < 0$. $\qquad\square$

Proposition 14.1 states that the **generating function of the Bessel functions** J_l is the function $e^{\frac{xz}{2}-\frac{x}{2z}}$.

Since $e^w \neq 0$ for all $w \in \mathbb{C}$ we can choose $z = e^{i\varphi}$ in (14.5) which gives $\frac{1}{2}\left(z - z^{-1}\right) = \frac{1}{2}\left(e^{i\varphi} - e^{-i\varphi}\right) = i\sin\varphi$, $\varphi \in \mathbb{R}$, and hence we arrive at the Fourier series

$$e^{ix\sin\varphi} = \sum_{k\in\mathbb{Z}} J_k(x) e^{ik\varphi}. \tag{14.7}$$

Therefore we may identify $J_k(x)$ with the corresponding Fourier coefficients, i.e.

$$J_k(x) = \frac{1}{2\pi} \int_{-\pi}^{\pi} e^{ix\sin\varphi - ik\varphi} \, d\varphi. \tag{14.8}$$

Some authors take (14.8) as the definition of J_k. For $x \in \mathbb{R}$ we find

$$|J_k(x)| \leq \frac{1}{2\pi} \int_{-\pi}^{\pi} \left|e^{ix\sin\varphi - ik\varphi}\right| d\varphi = 1$$

and since

$$J_k^{(\nu)}(x) = \frac{d^\nu}{dx^\nu} J_k(x) = \frac{1}{2\pi} \int_{-\pi}^{\pi} (i\sin\varphi)^\nu e^{ix\sin\varphi - ik\varphi} \, d\varphi$$

we obtain also

$$\left|J_k^{(\nu)}(x)\right| \leq 1.$$

Proposition 14.2. *For $z \in \mathbb{C}$ and $k \in \mathbb{Z}$ we have **Bessel's integral formula***

$$J_k(z) = \frac{1}{2\pi} \int_{-\pi}^{\pi} e^{iz\sin\varphi - ik\varphi} \, d\varphi = \frac{1}{\pi} \int_0^{\pi} \cos(z\sin\varphi - k\varphi) \, d\varphi. \tag{14.9}$$

224

Proof. (Following [34]) Changing in (14.8) the variable φ to $-\varphi$ (and replacing x by z) we obtain

$$J_k(z) = \frac{1}{2\pi} \int_{-\pi}^{\pi} e^{-iz \sin \varphi + ik\varphi} \, d\varphi$$

as well as by (14.8)

$$J_k(z) = \frac{1}{2\pi} \int_{-\pi}^{\pi} e^{iz \sin \varphi - ik\varphi} \, d\varphi.$$

Adding these two terms and dividing by 2 we arrive at

$$J_k(z) = \frac{1}{2\pi} \int_{-\pi}^{\pi} \frac{1}{2} \left(e^{iz \sin \varphi - ik\varphi} + e^{-iz \sin \varphi + ik\varphi} \right) d\varphi$$

$$= \frac{1}{2\pi} \int_{-\pi}^{\pi} \cos(z \sin \varphi - k\varphi) \, d\varphi = \frac{1}{\pi} \int_0^{\pi} \cos(z \sin \varphi - k\varphi) \, d\varphi.$$

$\square$

With these preparations we can turn to the Fourier transform of radial symmetric or rotational invariant functions. We recall

Definition 14.3. *A function $f : \mathbb{R}^n \to \mathbb{C}$ is **rotational invariant** or **radial symmetric** if for every $T \in O(n)$ we have $f(Tx) = f(x)$.*

Proposition 14.4. *Let $u \in L^1(\mathbb{R}^n)$ be rotational invariant. Then $\hat{u}$ is rotational invariant too.*

Proof. This follows from Lemma 12.4, in particular (12.8), since for $T \in O(n)$ we have $|\det T| = 1$ and $T^{-1} = T^t$ or $\left(T^{-1}\right)^t = \mathrm{id}$. It follows that

$$(u \circ T)\widehat{\,}(\xi) = \frac{1}{|\det T|} \hat{u} \circ \left(T^{-1}\right)^t (\xi) = \hat{u}\left(\left(T^{-1}\right)^t \xi\right) = \hat{u}(\xi).$$

$\square$

Remark 14.5. If $u \in L^2(\mathbb{R}^n)$ is a rotational invariant function then its Fourier transform is rotational invariant too since $\hat{u}$ can be approximated in $L^2(\mathbb{R}^n)$ by rotational invariant functions.

Since for a rotational invariant function $u(x) = u_0(|x|)$ the Fourier transform $\hat{u}$ is rotational invariant too, i.e. $\hat{u}(\xi) = [\hat{u}]_0 (|\xi|)$, we may try to find a one-dimensional integral representation of $[\hat{u}]_0$ with the help of the function u_0. For $n = 2$ and $u \in L^1(\mathbb{R}^n)$ with $u(x, y) = u_0(r)$, $x = r \cos \varphi$, $y = r \sin \varphi$ and $r^2 = x^2 + y^2$ we have

$$\hat{u}(\xi, \eta) = \frac{1}{2\pi} \int_{\mathbb{R}} \int_{\mathbb{R}} u(x, y) e^{-ix\xi - iy\eta} \, dx \, dy.$$

With $\xi = \rho \cos \vartheta$ and $\eta = \rho \sin \vartheta$, $\rho^2 = \xi^2 + \eta^2$, we now find $x\xi + y\eta = r\rho \cos(\varphi - \vartheta)$ and with $\hat{u}(\xi, \eta) = [\hat{u}]_0 (\rho)$ we have

$$[\hat{u}]_0 (\rho) = \hat{u}(\xi, \eta) = \int_0^\infty \int_0^{2\pi} u_0(r) e^{-ir\rho \cos(\varphi - \vartheta)} r \, d\varphi \, dr.$$

Taking 2π-periodicity into account, the change of variable $\varphi \mapsto \vartheta + \varphi + \frac{\pi}{2}$ yields

$$\int_0^{2\pi} e^{-ir\rho \cos(\varphi - \vartheta)} \, d\varphi = \int_0^{2\pi} e^{ir\rho \sin \varphi} \, d\varphi.$$

Now, with $k = 0$ in Bessel's integral formula (14.9) we obtain

$$\int_0^{2\pi} e^{-ir\rho \cos(\varphi - \vartheta)} \, d\varphi = 2\pi J_0(r\rho)$$

implying that

$$[\hat{u}]_0 (\rho) = \int_0^\infty u_0(r) J_0(r\rho) r \, dr. \tag{14.10}$$

For $n = 3$ we find with $u(x, y, z) = u_0(r)$, $r^2 = x^2 + y^2 + z^2$ and $(x, y, z) = r \cdot w$, $w = (w_1, w_2, w_3) \in S^2$, that for some $[\hat{u}]_0$ we have $\hat{u}(\xi, \eta, \zeta) = [\hat{u}]_0 (\rho)$ where $\rho^2 = \xi^2 + \eta^2 + \zeta^2$. Denoting by $\sigma(dw)$ the surface measure on S^2, compare with Chapter II.24, we find by using spherical coordinates in $\mathbb{R}^3$

$$[\hat{u}]_0 (\rho) = \hat{u}(\xi, \eta, \zeta) = (2\pi)^{-\frac{3}{2}} \int_{\mathbb{R}} \int_{\mathbb{R}} \int_{\mathbb{R}} e^{-i(x\xi + y\eta + z\zeta)} u(x, y, z) \, dx \, dy \, dz$$

$$= (2\pi)^{-\frac{3}{2}} \int_0^\infty u_0(r) \int_{S^2} e^{-ir(w_1\xi + w_2\eta + w_3\zeta)} \sigma(dw) \, r^2 \, dr. \tag{14.11}$$

We want to evaluate the inner integral and start with following observation. If we write $\gamma := (\xi, \eta, \zeta)$ we find $w_1\xi + w_2\eta + w_3\zeta = w\gamma$. For a rotation T it follows that

$$\int_{S^2} e^{-irw \cdot T\gamma} \sigma(dw) = \int_{S^2} e^{-ir\gamma \cdot T^{-1}w} \sigma(dw)$$

$$= \int_{S^2} e^{-ir\gamma \cdot w} \sigma(dw)$$

where in the last step we used the fact that σ is invariant under rotations. In order to evaluate the inner integral we may choose $\gamma = (0, 0, \rho)$, $\rho \neq 0$, by applying an appropriate rotation to (ξ, η, ζ). For $\rho \neq 0$ we find

$$\int_{S^2} e^{-ir\gamma \cdot w} \sigma(dw) = \int_0^{2\pi} \int_0^\pi e^{-ir\rho \cos \vartheta} \sin \vartheta \, d\vartheta \, d\varphi = 2\pi \int_0^\pi e^{-ir\rho \cos \vartheta} \sin \vartheta \, d\vartheta. \tag{14.12}$$

The substitution $t = -\cos\vartheta$ gives

$$\int_0^{2\pi}\int_0^{\pi} e^{-ir\rho\cos\vartheta}\sin\vartheta\,\mathrm{d}\vartheta\,\mathrm{d}\varphi = 2\pi\int_{-1}^{1} e^{ir\rho t}\,\mathrm{d}t$$

$$= \frac{2\pi e^{ir\rho t}}{ir\rho}\bigg|_{t=-1}^{t=1} = 4\pi\frac{\sin r\rho}{r\rho}.$$

Returning to (14.11) we conclude for $\rho \neq 0$ that

$$[\hat{u}]_0(\rho) = (2\pi)^{-\frac{3}{2}}\int_0^{\infty} u_0(r)4\pi\frac{\sin r\rho}{r\rho}r^2\,\mathrm{d}r$$

$$= \frac{2}{\sqrt{2\pi}\rho}\int_0^{\infty} u_0(r)\sin(r\rho)r\,\mathrm{d}r, \tag{14.13}$$

and the case $\rho = 0$ is trivial.

Now we turn to the general case for n being even. In this case we have

Theorem 14.6. *Let $u : \mathbb{R}^n \to \mathbb{C}$, n even, be a radial symmetric function in $L^1(\mathbb{R}^n)$, $u(x) = u_0(r)$, $r = \|x\|$. For its Fourier transform $\hat{u}(\xi) = [\hat{u}]_0(\rho)$, $\rho = \|\xi\|$, we have*

$$[\hat{u}]_0(\rho) = \frac{1}{\rho^{\frac{n}{2}-1}}\int_0^{\infty} u_0(\gamma)\gamma^{\frac{n}{2}} J_{\frac{n}{2}-1}(\rho\gamma)\,\mathrm{d}\gamma. \tag{14.14}$$

Proof. (Following [107] which is based on [14]) Using spherical coordinates in $\mathbb{R}^n$ we find

$$[\hat{u}]_0(\rho) = (2\pi)^{-\frac{n}{2}}\int_{\mathbb{R}^n} e^{-ix\cdot\xi}u(x)\,\mathrm{d}x$$

$$= (2\pi)^{-\frac{n}{2}}\int_0^{\infty} u_0(r)\int_{S^{n-1}} e^{-ir\omega\cdot\xi}\sigma(\mathrm{d}w)\,r^{n-1}\,\mathrm{d}r,$$

where σ is now the surface measure on S^{n-1}. Again we can use the rotational invariance of σ to find a rotation $T \in O(n)$ such that $T\xi = \rho e_n$ and we have

$$\int_{S^{n-1}} e^{-irw\cdot T\xi}\sigma(\mathrm{d}w) = \int_{S^{n-1}} e^{-ir\xi\cdot T^{-1}w}\sigma(\mathrm{d}w) = \int_{S^{n-1}} e^{-ir\xi\cdot w}\sigma(\mathrm{d}w).$$

Hence it follows that

$$[\hat{u}]_0(\rho) = (2\pi)^{-\frac{n}{2}}\int_0^{\infty} u_0(r)\int_{S^{n-1}} e^{-ir\rho w_n}\sigma(\mathrm{d}w)\,r^{n-1}\,\mathrm{d}r$$

$$= (2\pi)^{-\frac{n}{2}}\int_{\mathbb{R}^n} u_0(\|x\|)e^{-ix_n\rho}\,\mathrm{d}x.$$

227

We now write $x = (x_1, \ldots, x_n) = (y, x_n)$ with $y = (x_1, \ldots, x_{n-1})$ to find that $\|x\|^2 = \|y\|^2 + x_n^2$ and therefore

$$
\begin{aligned}
[\hat{u}]_0(\rho) &= (2\pi)^{-\frac{n}{2}} \int_{\mathbb{R}^{n-1}} \left(\int_{\mathbb{R}} u_0 \left(\sqrt{\|y\|^2 + x_n^2} \right) e^{-ix_n\rho} \, \mathrm{d}x_n \right) \mathrm{d}y \\
&= (2\pi)^{-\frac{n}{2}} \int_{\mathbb{R}} e^{-ix_n\rho} \int_0^\infty \int_{S^{n-2}} u_0 \left(\sqrt{\eta^2 + x_n^2} \right) \sigma(\mathrm{d}\tilde{w}) \eta^{n-2} \, \mathrm{d}\eta \, \mathrm{d}x_n \\
&= (2\pi)^{-\frac{n}{2}} (n-1)\omega_{n-1} \int_0^\infty \int_{\mathbb{R}} u_0(\sqrt{\eta^2 + x_n^2}) e^{-ix_n\rho} \eta^{n-2} \, \mathrm{d}x_n \, \mathrm{d}\eta.
\end{aligned}
$$

Next we use for the variables $(\eta, x_n) \in (0, \infty) \times \mathbb{R}$ polar coordinates $(\gamma, \alpha) \in (0, \infty) \times (0, \pi)$ to find

$$
\begin{aligned}
[\hat{u}]_0(\rho) &= (2\pi)^{-\frac{n}{2}} (n-1)\omega_{n-1} \int_0^\infty \int_0^\pi u_0(\gamma) e^{-i\gamma\rho\cos\alpha} \gamma^{n-1} \sin^{n-2}\alpha \, \mathrm{d}\alpha \, \mathrm{d}\gamma \\
&= (2\pi)^{-\frac{n}{2}} \int_0^\infty u_0(\gamma) \gamma^{\frac{n}{2}} \left((n-1)\omega_{n-1} \int_0^\pi e^{-i\gamma\rho\cos\alpha} \gamma^{\frac{n}{2}-1} \sin^{n-2}\alpha \, \mathrm{d}\alpha \right) \mathrm{d}\gamma.
\end{aligned}
$$

The inner integral is similar to the integral (14.12) and we use again the substitution $t = -\cos\alpha$ and the observation that $\sin^2\alpha = 1 - t^2$ to get

$$
\begin{aligned}
\int_0^\pi e^{-i\gamma\rho\cos\alpha} \gamma^{\frac{n}{2}-1} \sin^{n-2}\alpha \, \mathrm{d}\alpha &= \int_{-1}^1 e^{i\gamma\rho t} \gamma^{\frac{n}{2}-1} (1-t^2)^{\frac{n}{2}-1-\frac{1}{2}} \, \mathrm{d}t \\
&= \gamma^{\frac{n}{2}-1} \int_{-1}^1 \cos\gamma\rho t (1-t^2)^{\frac{n}{2}-1-\frac{1}{2}} \, \mathrm{d}t
\end{aligned}
$$

where we used that $t \to \sin\gamma\rho t$ is an odd function and $t \mapsto (1-t^2)^{\frac{n}{2}-1-\frac{1}{2}}$ is an even function. By (14.3) it follows that

$$
\begin{aligned}
\gamma^{\frac{n}{2}-1} \int_{-1}^1 \cos\gamma\rho t (1-t^2)^{\frac{n}{2}-1-\frac{1}{2}} \, \mathrm{d}t &= \frac{2^{\frac{n}{2}-1}\gamma^{\frac{n}{2}-1}}{\rho^{\frac{n}{2}-1}\gamma^{\frac{n}{2}-1}} \Gamma\left(\frac{n}{2} - \frac{1}{2}\right) \Gamma\left(\frac{1}{2}\right) J_{\frac{n}{2}-1}(\rho\gamma) \\
&= \frac{2^{\frac{n}{2}-1}}{\rho^{\frac{n}{2}-1}} \Gamma\left(\frac{n}{2} - \frac{1}{2}\right) \Gamma\left(\frac{1}{2}\right) J_{\frac{n}{2}-1}(\rho\gamma)
\end{aligned}
$$

which yields

$$
[\hat{u}]_0(\rho) = \frac{2^{-\frac{n}{2}}\pi^{-\frac{n}{2}} 2^{\frac{n}{2}-1}(n-1)\omega_{n-1}\Gamma\left(\frac{n}{2} - \frac{1}{2}\right)\Gamma\left(\frac{1}{2}\right)}{\rho^{\frac{n}{2}-1}} \int_0^\infty u_0(\gamma)\gamma^{\frac{n}{2}} J_{\frac{n}{2}-1}(\rho\gamma) \, \mathrm{d}\gamma.
$$

228

Now we observe that

$$2^{-\frac{n}{2}}\pi^{-\frac{n}{2}}2^{\frac{n}{2}-1}(n-1)\omega_{n-1}\Gamma\left(\frac{n}{2}-\frac{1}{2}\right)\Gamma\left(\frac{1}{2}\right)$$

$$= \pi^{-\frac{n}{2}}2^{-1}(n-1)\frac{\pi^{\frac{n-1}{2}}}{\Gamma\left(\frac{n-1}{2}+1\right)}\Gamma\left(\frac{n}{2}-\frac{1}{2}\right)\Gamma\left(\frac{1}{2}\right)$$

$$= \pi^{-\frac{1}{2}}\Gamma\left(\frac{1}{2}\right)\frac{\left(\frac{n-1}{2}\right)\Gamma\left(\frac{n}{2}-\frac{1}{2}\right)}{\Gamma\left(\frac{n}{2}-\frac{1}{2}+1\right)}=1$$

and we arrive at

$$[\hat{u}]_0(\rho) = \frac{1}{\rho^{\frac{n}{2}-1}}\int_0^\infty u_0(\gamma)\gamma^{\frac{n}{2}}J_{\frac{n}{2}-1}(\rho\gamma)\,\mathrm{d}\gamma,$$

proving the theorem. □

In order to get a formula for the Fourier transform of rotational invariant functions in $\mathbb{R}^n$ with n being an odd number we need to introduce **Bessel functions of non-integer order**. We briefly sketch the definition and the main result we need, but for details we refer to [11]. For ν not an integer, say $\nu > 0$, we define

$$J_\nu(x) = \left(\frac{x}{2}\right)^\nu\sum_{k=0}^\infty\frac{(-1)^k}{\Gamma(\nu+k+1)k!}\left(\frac{x}{2}\right)^{2k} = \sum_{k=0}^\infty\frac{(-1)^k}{\Gamma(\nu+k+1)k!}\left(\frac{x}{2}\right)^{2k+\nu}. \tag{14.15}$$

This definition looks natural in light of (14.1) but note that now we cannot extend J_ν to an entire function. One can prove again the integral representation (14.3), i.e.

$$J_\nu(x) = \frac{1}{\Gamma\left(\nu+\frac{1}{2}\right)\Gamma\left(\frac{1}{2}\right)}\left(\frac{x}{2}\right)^\nu\int_{-1}^1(1-t^2)^{\nu-\frac{1}{2}}\cos xt\,\mathrm{d}t \tag{14.16}$$

which holds in $\mathbb{C}\backslash(-\infty,0)$. From here it is now possible to prove (14.14) for n being odd along the same lines as for n being even, but in (14.14) we have now a Bessel function of fractional order.

We have seen with the help of the Gaussian semi-groups $(\mu_t)_{t\geq 0}$,

$$\hat{\mu}_t(\xi) = (2\pi)^{-\frac{n}{2}}e^{-t\|\xi\|^2} \tag{14.17}$$

that we can construct a solution of the heat equation

$$\frac{\partial}{\partial t}w = \Delta w,\quad w(x,0) = u_0(x) \tag{14.18}$$

229

by defining $w(x,t) := (T_t u_0)(x) := (\mu_t * u_0)(x)$, see Chapter 11 and Chapter 12. The effort to prove that $w = \mu_t * u_0$ is indeed a solution depends on the smoothness of u_0. In our previous considerations we started with $T_t u_0 = \mu_t * u_0$ and then we have verified (14.18). However we may derive our solution formally when starting with (14.18) and then take the Fourier transform with respect to x to get

$$\frac{\mathrm{d}}{\mathrm{d}t}\hat{w}(\xi,t) = -\|\xi\|^2 \hat{w}(\xi,t), \quad \hat{w}(\xi,0) = \hat{u}_0(\xi). \tag{14.19}$$

This is a simple ordinary differential equation with respect to t with solution

$$\hat{w}(\xi,t) = e^{-\|\xi\|^2 t}\hat{u}_0(\xi). \tag{14.20}$$

Applying now the inverse Fourier transform to (14.20) (and adjusting the constants) gives our solution to (14.18). Since from (14.18) to (14.20) is a "formal" calculation we have to explicitly verify that we have obtained a solution which we did.

We want to use the idea to reduce, with the help of the Fourier transform, a partial differential equation to an ordinary differential equation for a first study of the **wave equation** in $\mathbb{R}^n$. The problem we want to solve is

$$\frac{\partial^2}{\partial t^2}w(x,t) = \Delta w(x,t), \quad x \in \mathbb{R}^n, \ t > 0, \quad w(x,0) = u_0(x), \quad \frac{\partial w}{\partial t}(x,0) = u_1(x). \tag{14.21}$$

As a first step we apply the Fourier transform to (14.21) and we find

$$\frac{\mathrm{d}^2}{\mathrm{d}t^2}\hat{w}(\xi,t) = -\|\xi\|^2 \hat{w}(\xi,t), \quad \hat{w}(\xi,0) = \hat{u}_0(\xi), \quad \frac{\mathrm{d}\hat{w}}{\mathrm{d}t}(\xi,0) = \hat{u}_1(\xi). \tag{14.22}$$

Of course for this we have to assume that $\hat{u}_0$ and $\hat{u}_1$ exist. For our considerations here we assume that $u_0, u_1 \in \mathcal{S}(\mathbb{R}^n)$. The solution of $v''(r) + c^2 v(r) = 0$ is given by $v(r) = a_1 \cos cr + a_2 \sin cr$ and this yields in our case

$$\hat{w}(\xi,t) = a_1(\xi)\cos(\|\xi\|t) + a_2(\xi)\sin(\|\xi\|t). \tag{14.23}$$

Next we determine a_1 and a_2 by using the intial condition from (14.22) which gives

$$\hat{w}(\xi,0) = a_1(\xi) \quad \text{and} \quad \frac{\mathrm{d}\hat{w}}{\mathrm{d}t}(\xi,0) = \|\xi\|a_2(\xi).$$

Hence we arrive as solution for (14.22) at

$$\hat{w}(\xi,t) = \hat{u}_0(\xi)\cos(\|\xi\|t) + \hat{u}_1(\xi)\frac{\sin(\|\xi\|t)}{\|\xi\|}. \tag{14.24}$$

230

It is easy to see by a direct verification that $\hat{w}$ as given in (14.24) indeed solves (14.22). To obtain a solution to (14.21) we have to take the inverse Fourier transform in (14.24). By assumption $\hat{u}_0, \hat{u}_1 \in \mathcal{S}(\mathbb{R}^n)$ and since $\lim_{r \to 0} \frac{\sin r}{r} = 1$ the right hand side of (14.24) is at least in $L^1(\mathbb{R}^n)$, hence

$$w(x,t) = (2\pi)^{-\frac{n}{2}} \int_{\mathbb{R}^n} e^{ix \cdot \xi} \left(\hat{u}_0(\xi) \cos(\|\xi\|t) + \hat{u}_1(\xi) \frac{\sin(\|\xi\|t)}{\|\xi\|} \right) d\xi \qquad (14.25)$$

is a candidate for a solution to (14.21).

Theorem 14.7. *The function (14.25) is twice continuously differentiable in (x,t) and satisfies (14.21).*

Proof. First we note that

$$\frac{\partial}{\partial t} \left(e^{ix \cdot \xi} \left(\hat{u}_0(\xi) \cos(\|\xi\|t) + \hat{u}_1(\xi) \frac{\sin(\|\xi\|t)}{\|\xi\|} \right) \right) = e^{ix \cdot \xi} \left(-\hat{u}_0(\xi) \|\xi\| \sin(\|\xi\|t) + \hat{u}_1(\xi) \cos(\|\xi\|t) \right) ;$$

$$\frac{\partial^2}{\partial t^2} \left(e^{ix \cdot \xi} \left(\hat{u}_0(\xi) \cos(\|\xi\|t) + \hat{u}_1(\xi) \frac{\sin(\|\xi\|t)}{\|\xi\|} \right) \right) = e^{ix \cdot \xi} \left(-\hat{u}_0(\xi) \|\xi\|^2 \cos(\|\xi\|t) - \hat{u}_1(\xi) \|\xi\| \sin(\|\xi\|t) \right) ;$$

$$\frac{\partial}{\partial x_j} \left(e^{ix \cdot \xi} \left(\hat{u}_0(\xi) \cos(\|\xi\|t) + \hat{u}_1(\xi) \frac{\sin(\|\xi\|t)}{\|\xi\|} \right) \right) = i\xi_j e^{ix \cdot \xi} \left(\hat{u}_0(\xi) \cos(\|\xi\|t) + \hat{u}_1(\xi) \frac{\sin(\|\xi\|t)}{\|\xi\|} \right) ;$$

$$\frac{\partial^2}{\partial x_l \partial x_j} \left(e^{ix \cdot \xi} \left(\hat{u}_0(\xi) \cos(\|\xi\|t) + \hat{u}_1(\xi) \frac{\sin(\|\xi\|t)}{\|\xi\|} \right) \right) = -\xi_j \xi_l e^{ix \cdot \xi} \left(\hat{u}_0(\xi) \cos(\|\xi\|t) + \hat{u}_1(\xi) \frac{\sin(\|\xi\|t)}{\|\xi\|} \right) .$$

Since $\hat{u}_0$ and $\hat{u}_1$ belongs to $\mathcal{S}(\mathbb{R}^n)$ the differentiation under the integral sign is allowed and it follows that

$$\frac{\partial^2}{\partial t^2} w(x,t) = -(2\pi)^{-\frac{n}{2}} \int_{\mathbb{R}^n} e^{ix \cdot \xi} \left(\hat{u}_0(\xi) \|\xi\|^2 \cos(\|\xi\|t) + \hat{u}_1(\xi) \|\xi\| \sin(\|\xi\|t) \right) d\xi$$

and

$$\Delta w(x,t) = -(2\pi)^{-\frac{n}{2}} \int_{\mathbb{R}^n} e^{ix \cdot \xi} \left(\hat{u}_0(\xi) \|\xi\|^2 \cos(\|\xi\|t) + \hat{u}_1(\xi) \|\xi\| \sin(\|\xi\|t) \right) d\xi,$$

implying that w is indeed a solution to the wave equation. Next we have to verify that this solution attains the correct initial data. However

$$w(x,0) = (2\pi)^{-\frac{n}{2}} \int_{\mathbb{R}^n} e^{ix \cdot \xi} \hat{u}_0(\xi) \, d\xi = u_0(x)$$

and

$$\frac{\partial w}{\partial t}(x,0) = (2\pi)^{-\frac{n}{2}} \int_{\mathbb{R}^n} e^{ix \cdot \xi} \hat{u}_1(\xi) \, d\xi = u_1(x).$$

$\square$

We want to show that w given by (14.25) is in fact (in a large class of functions) the unique solution to (14.21). For this we will use the energy associated with a solution of the wave equation. We define $E_w(t)$ as

$$E_w(t) := \frac{1}{2} \int_{\mathbb{R}^n} \left(\left| \frac{\partial w}{\partial t}(x,t) \right|^2 + \| \operatorname{grad} w(x,t) \|^2 \right) dx, \qquad (14.26)$$

and we claim that $t \mapsto E_w(t)$ is a conserved quantity, more precisely we have

Theorem 14.8. *Let w given by (14.25) be a solution to the wave equation. Then we have*

$$E_w(t) = E_w(0). \qquad (14.27)$$

Proof. We may use Plancherel's theorem to find

$$\int_{\mathbb{R}^n} \left| \frac{\partial w(x,t)}{\partial t} \right|^2 dx = \int_{\mathbb{R}^n} \left| -\|\xi\| \hat{u}_0(\xi) \sin(\|\xi\| t) + \hat{u}_1(\xi) \cos(\|\xi\| t) \right|^2 d\xi$$

and

$$\int_{\mathbb{R}^n} \| \operatorname{grad} w(x,t) \|^2 dx = \int_{\mathbb{R}^n} \left| \|\xi\| \hat{u}_0(\xi) \cos(\|\xi\| t) + \hat{u}_1(\xi) \sin(\|\xi\| t) \right|^2 d\xi.$$

We note that, see Problem 5, that

$$\left| -\|\xi\| \hat{u}_0(\xi) \sin(\|\xi\| t) + \hat{u}_1(\xi) \cos(\|\xi\| t) \right|^2 + \left| \|\xi\| \hat{u}_0(\xi) \cos(\|\xi\| t) + \hat{u}_1(\xi) \sin(\|\xi\| t) \right|^2$$
$$= \left| \|\xi\| \hat{u}_0(\xi) \right|^2 + |\hat{u}_1(\xi)|^2,$$

which implies that

$$E_w(t) = \frac{1}{2} \int_{\mathbb{R}^n} \left(\left| \frac{\partial w(x,t)}{\partial t} \right|^2 + \| \operatorname{grad} w(x,t) \|^2 \right) dx$$
$$= \frac{1}{2} \int_{\mathbb{R}^n} \left(|\hat{u}_1(\xi)|^2 + \|\xi\|^2 |\hat{u}_0(\xi)|^2 \right) d\xi$$
$$= \frac{1}{2} \int_{\mathbb{R}^n} \left(\left| \frac{\partial w}{\partial t}(x,0) \right|^2 + \| \operatorname{grad} w(x,0) \|^2 \right) dx$$
$$= E_w(0). \qquad \square$$

With the help of the energy $E(t)$ we may prove some uniqueness results for the wave equation which will follow from

Theorem 14.9. *Let $u \in C^2(\mathbb{R}^n \times [0, T])$ be a solution of the wave equation and let $x_0 \in \mathbb{R}^n$, $t_0 \in (0, T)$. Suppose that on $\tilde{B}_{t_0}(x) := B_{t_0}(x_0) \times \{0\} \subset \mathbb{R}^{n+1}$ we have that $u|_{\tilde{B}_{t_0}(x_0)} = \frac{\partial u}{\partial t}\Big|_{\tilde{B}_{t_0}(x_0)} = 0$. Then u vanishes identically in the cone*
$$K(x_0, t) := \{(x, t) \in \mathbb{R}^n \times [0, \infty) \,|\, 0 \leq t \leq t_0, \, \|x - x_0\| \leq t_0 - t\}.$$

Proof. We consider the energy of u of the set $G_t := B_{t_0-t}(x_0)$ which is given by
$$E_{u,G_t}(t) := \frac{1}{2} \int_{G_t} \left(\left(\frac{\partial u(x,t)}{\partial t} \right)^2 + \| \operatorname{grad} u(x,t) \|^2 \right) dx$$

and we aim to prove that $E_{u,G_t}(t) = 0$ for all $t \leq t_0$ which will imply that $u(x, t) = 0$ in $K(x_0, t_0)$. We note that
$$\frac{dE_{u,G_t}(t)}{dt} = \int_{G_t} \left(\frac{\partial u}{\partial t} \frac{\partial^2 u}{\partial t^2} + \sum_{j=1}^{n} \frac{\partial u}{\partial x_j} \frac{\partial^2 u}{\partial x_j \partial t} \right) - \frac{1}{2} \int_{\partial G_t} \left(\left| \frac{\partial u}{\partial t} \right|^2 + \| \operatorname{grad} u \|^2 \right) \sigma(dx) \qquad (14.28)$$

and the divergence theorem yields
$$\int_{G_t} \left(\frac{\partial u}{\partial t} \frac{\partial^2 u}{\partial t^2} + \sum_{j=1}^{n} \frac{\partial u}{\partial x_j} \frac{\partial^2 u}{\partial x_j \partial t} \right) dx$$
$$= \int_{G_t} \left(\frac{\partial u}{\partial t} \frac{\partial^2 u}{\partial t^2} + \sum_{j=1}^{n} \left(\frac{\partial}{\partial x_j} \left(\frac{\partial u}{\partial x_j} \frac{\partial u}{\partial t} \right) - \frac{\partial^2 u}{\partial x_j^2} \frac{\partial u}{\partial t} \right) \right) dx$$
$$= \int_{G_t} \frac{\partial u}{\partial t} \left(\frac{\partial^2 u}{\partial t^2} - \sum_{j=1}^{n} \frac{\partial^2 u}{\partial x_j^2} \right) dx + \int_{\partial G_t} \sum_{j=1}^{n} \frac{\partial u}{\partial x_j} \cdot \frac{\partial u}{\partial t} \cdot \nu_j \sigma(dx)$$

where ν_j is the jth component of the outer normal unit vector of ∂G_t. Since u is a solution of the wave equation the first integral vanishes and for the integrand of the surface integral we find
$$\sum_{j=1}^{n} \frac{\partial u}{\partial x_j} \frac{\partial u}{\partial t} \nu_j \leq \left| \frac{\partial u}{\partial t} \right| \left(\sum_{j=1}^{n} \left| \frac{\partial u}{\partial x_j} \right|^2 \right)^{\frac{1}{2}} \leq \frac{1}{2} \left(\left| \frac{\partial u}{\partial t} \right|^2 + \| \operatorname{grad} u \|^2 \right).$$

With the help of (14.28) we deduce
$$\frac{dE_{u,G_t}(t)}{dt} = \int_{\partial G_t} \left(\sum_{j=1}^{n} \frac{\partial u}{\partial x_j} \frac{\partial u}{\partial t} \nu_j - \frac{1}{2} \left(\left| \frac{\partial u}{\partial t} \right|^2 + \| \operatorname{grad} u \|^2 \right) \right) \sigma(dx) \leq 0. \qquad (14.29)$$

Thus $t \mapsto E_{u,G_t}(t)$ is decreasing and non-negative and $E_{u,G_t}(0) = 0$, which implies $E_{u,G_t}(t) = 0$ for all $t \leq t_0$ and therefore u is constant in G_t and by the initial conditions this constant must be equal to 0. $\qquad \square$

Corollary 14.10. *There is at most one solution $w \in C^2(\mathbb{R}^n \times [0, \infty))$ to the wave equation satisfying $w(x, 0) = u_0(x)$ and $\frac{\partial w}{\partial t}(x, 0) = u_1(x)$.*

Proof. The difference between two solutions of the wave equation with the same initial data satisfies the wave equation with zero initial data. By Theorem 14.9 it must therefore vanish in every cone $K(x_0, t_0)$, hence in $\mathbb{R}^n \times (0, \infty)$. This proves the uniqueness of the solution. $\square$

Our formula (14.25) for the solution of the wave equation is rather indirect since it involves the Fourier transform of the initial data and not the initial data directly. In Volume VI we will derive and discuss formulae which involve u_0 and u_1 directly.

The next topic we want to discuss is the Poisson summation formula and a few of its applications. Suppose that $u : \mathbb{R} \to \mathbb{C}$ is a continuous function satisfying the estimate $|u(x)| \leq \frac{c}{1+|x|^{1+\epsilon}}$ for some $\epsilon > 0$. Then u is integrable and moreover for every $x \in \mathbb{R}$

$$\sum_{k \in \mathbb{Z}} |u(x + 2\pi k)| \leq \sum_{k \in \mathbb{Z}} \frac{c}{1 + 2\pi |k|^{1+\epsilon}} < \infty,$$

i.e. the series $\sum_{k \in \mathbb{Z}} u(x + 2\pi k)$ converges absolutely and uniformly to a continuous 2π-periodic function. The periodicity follows from

$$\sum_{k \in \mathbb{Z}} u(x + 2\pi k + 2\pi) = \sum_{k \in \mathbb{Z}} u(x + 2\pi(k+1)) = \sum_{k \in \mathbb{Z}} u(x + 2\pi k).$$

Theorem 14.11 (Poisson summation formula). *Suppose that $u : \mathbb{R} \to \mathbb{C}$ is a continuous function such that $|u(x)| \leq \frac{c_1}{1+|x|^{1+\epsilon}}$ for some $\epsilon > 0$ and assume that for $\hat{u}$ we have $|\hat{u}(\xi)| \leq \frac{c_2}{1+|\xi|^{1+\epsilon}}$. Then the equality*

$$\sqrt{2\pi} \sum_{k \in \mathbb{Z}} u(x + 2\pi k) = \sum_{k \in \mathbb{Z}} \hat{u}(k) e^{ixk} \tag{14.30}$$

holds for all $x \in \mathbb{R}$.

Proof. First we note that the condition $|u(x)| \leq \frac{c_1}{1+|x|^{1+\epsilon}}$, $\epsilon > 0$, yields the existence of $\hat{u}$ and the condition $|\hat{u}| \leq \frac{c_2}{1+|\xi|^{1+\epsilon}}$ implies that the right hand side of (14.30) represents a continuous function. In order to prove (14.30) it is therefore sufficient to prove that the nth Fourier coefficient of the 2π-periodic function on the left hand side of (14.30) is equal to $\hat{u}(n)$. For this we note

$$\frac{1}{2\pi}\int_{-\pi}^{\pi}\sqrt{2\pi}\sum_{k\in\mathbb{Z}}u(x+2\pi k)e^{-inx}\,\mathrm{d}x$$

$$=\frac{1}{\sqrt{2\pi}}\sum_{k\in\mathbb{Z}}\int_{-\pi}^{\pi}u(x+2\pi k)e^{-inx}\,\mathrm{d}x$$

$$=\frac{1}{\sqrt{2\pi}}\sum_{k\in\mathbb{Z}}\int_{(2k-1)\pi}^{(2k+1)\pi}u(y)e^{-in(y-2\pi k)}\,\mathrm{d}y$$

$$=\frac{1}{\sqrt{2\pi}}\sum_{k\in\mathbb{Z}}\int_{(2k-1)\pi}^{(2k+1)\pi}u(y)e^{-iny}\,\mathrm{d}y$$

$$=\frac{1}{\sqrt{2\pi}}\int_{\mathbb{R}}u(y)e^{-iny}\,\mathrm{d}y=\hat{u}(n).$$

$\square$

Choosing in (14.30) now $x=0$ we arrive at

Corollary 14.12. *For u satisfying the assumptions of Theorem 14.11 we have*

$$\sqrt{2\pi}\sum_{k\in\mathbb{Z}}u(2\pi k)=\sum_{k\in\mathbb{Z}}\hat{u}(k). \tag{14.31}$$

We note that if $u\in\mathcal{S}(\mathbb{R}^n)$ then it satisfies the assumptions of Theorem 14.11 and in particular we can apply (14.31) to the function $x\mapsto e^{-ax^2}$, $a>0$.

Definition 14.13. *On $(0,\infty)$ the **theta function** $t\mapsto\vartheta(t)$ is defined by*

$$\vartheta(t)=\sum_{k\in\mathbb{Z}}e^{-\pi k^2 t}. \tag{14.32}$$

Theorem 14.14. *The theta function satisfies for $t>0$ the functional equation*

$$t^{-\frac{1}{2}}\vartheta\left(\frac{1}{t}\right)=\vartheta(t). \tag{14.33}$$

Proof. We apply (14.31) to $u(x)=e^{-\frac{x^2}{4\pi t}}$. For the Fourier transform of u we find

$$\hat{u}(\xi)=\frac{1}{\sqrt{2\pi}}\int_{\mathbb{R}}e^{-ix\xi}e^{-\frac{x^2}{4\pi t}}\,\mathrm{d}x=\frac{1}{\sqrt{2\pi}}\int_{\mathbb{R}}e^{-i\sqrt{2\pi}ty\xi}e^{-\frac{y^2}{2}}\sqrt{2\pi t}\,\mathrm{d}y$$

$$=t^{\frac{1}{2}}\int_{\mathbb{R}^n}e^{-iy(\sqrt{2\pi t}\xi)}e^{-\frac{y^2}{2}}\,\mathrm{d}y$$

$$=\sqrt{2\pi}t^{\frac{1}{2}}e^{-\pi\xi^2 t}$$

implying that

$$\sum_{k\in\mathbb{Z}}\hat{u}(k) = \sqrt{2\pi}t^{\frac{1}{2}}\sum_{k\in\mathbb{Z}}e^{-\pi k^2 t} = \sqrt{2\pi}t^{\frac{1}{2}}\vartheta(t). \tag{14.34}$$

On the other hand we have

$$\sqrt{2\pi}\sum_{k\in\mathbb{Z}}u(2\pi k) = \sqrt{2\pi}\sum_{k\in\mathbb{Z}}e^{-\frac{\pi k^2}{t}} = \sqrt{2\pi}\vartheta\left(\frac{1}{t}\right)$$

which implies with (14.34) that $t^{-\frac{1}{2}}\vartheta\left(\frac{1}{t}\right) = \vartheta(t)$. □

Example 14.15. The Fourier transform of the function $u : \mathbb{R} \to \mathbb{R}$, $u(x) = \frac{t}{\pi}\frac{1}{t^2+x^2}$ is calculated as

$$\hat{u}(\xi) = \frac{1}{\sqrt{2\pi}}\frac{t}{\pi}\int_{\mathbb{R}}e^{-ix\xi}\frac{1}{t^2+x^2}\,\mathrm{d}x$$

$$= \frac{2}{\sqrt{2\pi}}\frac{t}{\pi}\int_0^\infty\frac{\cos x\xi}{t^2+x^2}\,\mathrm{d}x = \frac{1}{\sqrt{2\pi}}e^{-|\xi|t}.$$

On the other hand we find

$$\sqrt{2\pi}\sum_{k\in\mathbb{Z}}u(2\pi k) = \sqrt{2\pi}\frac{1}{\pi}\sum_{k\in\mathbb{Z}}\frac{t}{t^2+4\pi^2 k^2}$$

which yields now by the Poisson formula

$$2\sum_{k\in\mathbb{Z}}\frac{t}{t^2+4\pi^2 k^2} = \sum_{k\in\mathbb{Z}}e^{-|k|t} = \frac{2}{1-e^{-t}} - 1$$

where we used that $\sum_{k\in\mathbb{Z}}e^{-|k|t} = 2\sum_{k=0}^\infty\left(e^{-t}\right)^k - 1$. Thus we have for $t > 0$

$$\sum_{k\in\mathbb{Z}}\frac{t}{t^2+4\pi^2 k^2} = \frac{1}{2}\frac{1+e^{-t}}{1-e^{-t}}. \tag{14.35}$$

We want to use (14.35) to calculate once more $\zeta(2)$. We may write the left hand side of (14.35) as

$$\sum_{k\in\mathbb{Z}}\frac{t}{t^2+4\pi^2 k^2} = 2\sum_{k=0}^\infty\frac{t}{t^2+4\pi^2 k^2} - \frac{1}{t}$$

which yields

$$\sum_{k=0}^{\infty} \frac{1}{t^2 + 4\pi^2 k^2} = \frac{1}{2t} \left(\frac{1}{2} \frac{1 + e^{-t}}{1 - e^{-t}} + \frac{1}{t} \right)$$

$$= \frac{1}{4} \left(\frac{t + 2 - 2e^{-t} + te^{-t}}{t^2 - t^2 e^{-t}} \right).$$

For $t \to 0$ the left hand side tends to

$$\frac{1}{4\pi^2} \sum_{k=0}^{\infty} \frac{1}{k^2} = \frac{1}{4\pi^2} \zeta(2)$$

and applying three times the rules of de l'Hospital to the terms in the brackets on the right hand side gives in the limit $t \to 0$ the value $\frac{1}{6}$, i.e. we arrive at

$$\zeta(2) = \frac{4\pi^2}{4 \cdot 6} = \frac{\pi^2}{6}. \tag{14.36}$$

The Poisson summation formula has quite a few applications in the study of (meromorphic) functions related to number theory. In fact the theta-function is also such a function, for example it is related to the ζ-function by the equation

$$\pi^{-\frac{s}{2}} \Gamma\left(\frac{s}{2}\right) \zeta(s) = \frac{1}{2} \int_0^{\infty} t^{\frac{s}{2}-1} (\vartheta(s) - 1) \, dt \tag{14.37}$$

which we will not prove here.

Finally we want to prove a simple inequality which is known as the **uncertainty principle**. We start with $u \in \mathcal{S}(\mathbb{R}^n)$ and assume that $\|u\|_{L^2} = 1$. Since

$$1 = \int_{\mathbb{R}} |u(x)|^2 \, dx = \int_{\mathbb{R}} \frac{dx}{dx} |u(x)|^2 \, dx$$

an integration by parts followed by an application of the Cauchy-Schwarz inequality and Plancherel's theorem yields

$$1 = -\int_{\mathbb{R}} x \frac{\mathrm{d}}{\mathrm{d}x} |u(x)|^2 \, \mathrm{d}x$$

$$= -\int_{\mathbb{R}} \left(x u'(x)\overline{u(x)} + x \overline{u'(x)} u(x) \right) \mathrm{d}x$$

$$\leq 2 \int_{\mathbb{R}} |x||u(x)||u'(x)| \, \mathrm{d}x$$

$$\leq 2 \left(\int_{\mathbb{R}} x^2 |u(x)|^2 \, \mathrm{d}x \right)^{\frac{1}{2}} \left(\int_{\mathbb{R}} |u'(x)|^2 \, \mathrm{d}x \right)^{\frac{1}{2}}$$

$$= 2 \left(\int_{\mathbb{R}} x^2 |u(x)|^2 \, \mathrm{d}x \right)^{\frac{1}{2}} \left(\int_{\mathbb{R}} |F(u')(\xi)|^2 \, \mathrm{d}\xi \right)^{\frac{1}{2}}.$$

However $F(u')(\xi) = i\xi \hat{u}(\xi)$ and we arrive at

$$\frac{1}{4} \leq \left(\int_{\mathbb{R}} x^2 |u(x)|^2 \, \mathrm{d}x \right) \left(\int_{\mathbb{R}} \xi^2 |\hat{u}(\xi)|^2 \, \mathrm{d}\xi \right). \tag{14.38}$$

Proposition 14.16. Let $u \in \mathcal{S}(\mathbb{R}^n)$ such that $\|u\|_{L^2} = 1$ and let $x_0, \xi_0 \in \mathbb{R}$. Then we have

$$\frac{1}{4} \leq \left(\int_{\mathbb{R}} (x - x_0)^2 |u(x)|^2 \, \mathrm{d}x \right) \left(\int_{\mathbb{R}} (\xi - \xi_0)^2 |\hat{u}(\xi)|^2 \, \mathrm{d}\xi \right). \tag{14.39}$$

Proof. We need only apply (14.38) to $v(x) = e^{-ix\xi_0} u(x + x_0)$ to find

$$\frac{1}{4} \left(\int_{\mathbb{R}} x^2 \left| e^{-ix\xi_0} u(x + x_0) \right|^2 \mathrm{d}x \right) \left(\int_{\mathbb{R}} \xi^2 \left| F\left(e^{-i\xi_0 \bullet} u(\bullet + x_0) \right) (\xi) \right|^2 \mathrm{d}\xi \right).$$

The change of variable $x \mapsto x + x_0$ in the first integral gives

$$\int_{\mathbb{R}} x^2 \left| e^{-ix\xi_0} u(x + x_0) \right|^2 \mathrm{d}x = \int_{\mathbb{R}} (x - x_0)^2 |u(x)|^2 \, \mathrm{d}x.$$

To handle the second integral we note first that

$$F\left(e^{-i\xi_0 \bullet} u(\bullet + x_0) \right) (\xi) = e^{ix_0\xi} e^{ix_0\xi_0} \hat{u}(\xi + \xi_0)$$

which yields

$$\int_{\mathbb{R}} \xi^2 \left| F\left(e^{-i\xi_0 \bullet} u(\bullet + x_0) \right) (\xi) \right|^2 \mathrm{d}\xi = \int_{\mathbb{R}} \xi^2 |\hat{u}(\xi + \xi_0)|^2 \, \mathrm{d}\xi$$

and again the change of variable $\xi \mapsto \xi + \xi_0$ leads to

$$\int_{\mathbb{R}^n} \xi^2 |\hat{u}(\xi + \xi_0)|^2 \, \mathrm{d}\xi = \int_{\mathbb{R}} (\xi - \xi_0)^2 |\hat{u}(\xi)|^2 \, \mathrm{d}\xi.$$

$\square$

Remark 14.17. Heisenberg's uncertainty principle states that two conjugate variables in quantum mechanics can only be measured simultaneously with a limited accuracy, a lower bound of which is essentially given by the Planck constant. Once the standard approach to quantum mechanics in the Schrödinger formulation is taken, Heisenberg's uncertainty principle for momentum and position corresponds to Proposition 14.16. However the simple estimate (14.39) does not prove any truth in physics.

Problems

1. Prove that $\frac{d}{dx}(x^l J_l(x)) = x^l J_{l-1}(x), l \in \mathbb{N}$.

2. Use Bessel's differential equation to show that
$$J_l'(x) J_{-l}(x) - J_{-l}'(x) J_l(x) = 0, \ l \in \mathbb{N}, \ x \in \mathbb{R}.$$

3. Use (14.7) to show that
$$J_l(x+y) = \sum_{k \in \mathbb{Z}} J_l(x) J_{k-l}(y).$$

4. Find the Fourier transform of $\chi_{B_1(0)}, B_1(0) \subset \mathbb{R}^2$.

5. Verify for $u_0, u_1 \in \mathcal{S}(\mathbb{R}^n)$ that
$$| - \|\xi\| \hat{u}_0(\xi) \sin(\|\xi\|t) + \hat{u}_1(\xi) \cos(\|\xi\|t)|^2$$
$$+ |\|\xi\| \hat{u}_0(\xi) \cos(\|\xi\|t) + \hat{u}_1(\xi) \sin(\|\xi\|t)|^2 = |\|\xi\| \hat{u}_0(\xi)|^2 + |\hat{u}_1(\xi)|^2.$$

6.   a) For $n = 1$ prove that for $u_0, u_1 \in \mathcal{S}(\mathbb{R})$ the solution w of the initial value problem $\frac{\partial^2 w}{\partial t^2} - \frac{\partial^2 w}{\partial x^2} = 0$, $w(x,0) = u_0(x)$, $\frac{\partial w}{\partial t}(x,0) = u_1(x)$, has the structure

$$(*) \qquad w(x,t) = \frac{g(x+t) + g(x-t)}{2} + \frac{1}{2} \int_{x-t}^{x+t} h(y)\,dy$$

with suitable functions g and h.

b) Show that for any pair of functions $g \in C^2(\mathbb{R})$ and $h \in C^1(\mathbb{R})$ a solution w of the wave equation in $\mathbb{R}$ is given by $(*)$. Find the corresponding initial values $w(x,0)$ and $\frac{\partial w}{\partial t}(x,0)$.

7. Find the solution to the problem
$$\frac{\partial^2 u(x,t)}{\partial t^2} - \frac{\partial^2 u(x,t)}{\partial x^2} = 0, \quad u(x,0) = e^{-\frac{x^2}{2}}, \quad \frac{\partial u}{\partial t}(x,0) = 0.$$

8. Suppose $u : \mathbb{R}^2 \to \mathbb{R}$ is a solution of the problem $\frac{\partial^2 u}{\partial t^2} = \frac{\partial^2 u}{\partial x^2} - u$ with initial data $u(x,0) = u_0(x)$ and $\frac{\partial u}{\partial t}(x,0) = u_1(x), u_0, u_1 \in \mathcal{S}(\mathbb{R})$. Following the method employed for the wave equation find a formula for $\hat{u}(\xi, t)$ and verify that $(F^{-1}\hat{u})(x,t)$ is indeed a solution to the given initial value problem.

9. Use the Poisson summation formula to show for all $t \in \mathbb{R}$ that

$$\sum_{k \in \mathbb{Z}} \frac{1}{(t+k)^2} = \frac{\pi}{(\sin \pi t)^2}, \quad k \notin \mathbb{Z}.$$

Evaluate the formula for $t = \frac{1}{2}$ and $t = \frac{1}{4}$.

10. Use the Poisson summation formula to derive a 2π-periodic solution to the one-dimensional heat equation $\left(\frac{\partial}{\partial t} - \frac{\partial^2}{\partial x^2} \right) u = 0$.

Part 9: Ordinary Differential Equations

15 Some Orientation — First Results

Although we have met ordinary differential equations in several places, only Chapter I.27 was completely devoted to differential equations with the aim to discuss in detail the method of separating the variables, see most of all Theorem I.27.1. In this chapter we want to give some orientation and to study some elementary, not necessarily trivial, methods to solve or integrate some classical types of mainly first order equations. The chapter serves for providing examples, and most of all to give indications where problems or interesting questions arise when trying to establish a type of general theory.

Before starting a further remark is necessary. Advanced topics in analysis, e.g. in Fourier analysis, ordinary or partial differential equations, the calculus of variations etc., were historically developed in order to solve problems in physics, astronomy or mechanics (just to mention the main subject areas of importance in this respect). For good reasons in the past every mathematician, especially when specialising in analysis, was supposed to have not just a basic, but an advanced knowledge in some of these subjects. Unfortunately this tradition has faded and therefore motivating our discussions by first looking at problems in physics etc. does not help most students anymore. Nonetheless, here and there we will relate, in examples, mathematical results to problems in physics or mechanics.

In the following I stands for an interval $I \subset \mathbb{R}$ with non-empty interior, sometimes we are more explicit by writing (a, b) or $[a, b]$, sometimes we write $I(t_0)$ when we want to emphasize that I is an interval containing t_0. Consider the differential equation

$$u'(t) + \alpha u(t) = f(t), \quad t \in I, \, \alpha \in \mathbb{R}, \, f \in C(I). \tag{15.1}$$

First we observe that if v solves this equation and w solves the corresponding homogeneous equation $u'(t) + \alpha u(t) = 0$, then $u := v + w$ is a further solution to (15.1). A solution to the homogeneous equation is given by $w(t) = ce^{-\alpha t}$ for some constant $c \in \mathbb{R}$. In fact, every solution of the homogeneous equation is of this form. Indeed, $w'(t) + \alpha w(t) = 0$ implies under the assumption that $w \neq 0$ that $\int \frac{dw}{w} = -\int \alpha \, dt$ or $\ln |w| = -\alpha t + \gamma$ which yields $w(t) = e^{-\alpha t + \gamma} = ce^{-\alpha t}$, $c = e^{\gamma}$. In Problem 1 we will discuss the case $w(t_1) = 0$ for some $t_1 \in I$. Now we turn to the inhomogeneous equation (15.1) and for $f \in C(I)$ we follow an idea proposed by Lagrange to find a solution to (15.1) by starting with $w(t) = ce^{-\alpha t}$. The Ansatz is to consider $v(t) = c(t)e^{-\alpha t}$, i.e. to allow the constant to vary with t, hence being not anymore a constant. This idea is called the method of **variations of constants** and a simple calculation shows that the assumption $v(t) = c(t)e^{-\alpha t}$ solves (15.1) yields

$$f(t) = v'(t) + \alpha v(t) = c'(t) - c(t)\alpha e^{-\alpha t} + \alpha c(t)e^{-\alpha t} = c'(t)e^{-\alpha t}$$

implying

$$c'(t) = f(t)e^{\alpha t}$$

or

$$c(t) = c(t_0) + \int_{t_0}^{t} f(s)e^{\alpha s}\,ds, \quad t_0 \in I.$$

If we fix an initial value, say $u(t_0) = u_0$ and choose $c(t_0) = u_0 e^{\alpha t_0}$ we may consider

$$u(t) = e^{-\alpha t}\left(u_0 e^{\alpha t_0} + \int_{t_0}^{t} f(s)e^{\alpha s}\,ds\right) \tag{15.2}$$

as a candidate for a solution to (15.1). First we note that

$$u(t_0) = e^{-\alpha t_0}u_0 e^{\alpha t_0} + e^{-\alpha t_0}\int_{t_0}^{t_0} f(s)e^{\alpha s}\,ds = u_0$$

and further it follows that

$$u'(t) = -\alpha e^{-\alpha t}\left(u_0 e^{\alpha t_0} + \int_{t_0}^{t} f(s)e^{\alpha s}\,ds\right) + e^{-\alpha t}\left(f(t)e^{\alpha t}\right)$$
$$= -\alpha u(t) + f(t).$$

Thus, (15.2) is indeed a solution to (15.1) with initial value $u(t_0) = u_0$. It is also unique since for two solutions u and v we find with $w = u - v$ that

$$w'(t) + \alpha w(t) = 0 \quad \text{and} \quad w(t_0) = 0.$$

If $w \neq 0$ then it must be of form $w(t) = ce^{-\alpha t}$, hence $w(t_0) \neq 0$. Thus we must have $w(t) = 0$ for all $t \in I$ implying the uniqueness result, also see Problem 1.

From this example we can pick up a lot of ideas but also warnings:

- When dealing with differential equations it is often useful to exhibit some "formal" calculations in order to derive interesting formulae such as $\ln|w| = -\alpha t + \gamma$ or (15.2). However, since we might have violated certain crucial assumptions for the calculation to hold - note we do calculations with functions we do not know - we have to verify the results of our formal calculations.

- Differential equations in general do not have a unique solution, but certain additional conditions, e.g. an initial value, may, but not necessarily enforce uniqueness.

- The set of all solutions of a given differential equation forms a certain subset in some (vector) space of differentiable functions, e.g. it could be a finite dimensional affine subspace. It is worth investigating the structure of the set of all solutions of a differential equation.

- Solutions of a differential equation may be defined on a larger domain than initially sought, we want to know the maximal domain of solutions, i.e. the maximal domain to which we can extend a solution such that it remains a solution. In our example we may start with any interval, but all solutions are extending to $\mathbb{R}$, in fact even to $\mathbb{C}$.

- Solutions to a differential equation might be much more regular than originally required to solve the equation. To solve $u' + \alpha u = 0$ we need a C^1-function u, but it turns out that all solutions are C^∞-functions, in fact they are holomorphic.

The following discussion shows that the uniqueness problem can be rather complicated. Consider the equation

$$u'(t) = |u(t)|^{\frac{1}{2}}, \quad t \in \mathbb{R}. \tag{15.3}$$

Using separation of variables we arrive at

$$2|u|^{\frac{1}{2}} = \int \frac{du}{\sqrt{u}} = \int 1 \, dt = t + c \tag{15.4}$$

or

$$u(t) = \frac{(t + c)^2}{4}. \tag{15.5}$$

Note that (15.4) does not make sense for $t < -c$, which is however no condition for (15.5) to make sense. Differentiating (15.5) yields

$$u'(t) = \frac{t + c}{2} = \sqrt{\frac{(t + c)^2}{4}} = \sqrt{|u(t)|} \quad \text{for } t + c > 0.$$

We may choose for t_0 a value $u_0 > 0$ to find that

$$u_0 = \frac{(t_0 + c)^2}{4} \quad \text{or} \quad c = 2\sqrt{u_0} - t_0$$

provided $t_0 + c > 0$, i.e. $u_0 > 0$. So far we have that for $u_0 > 0$

$$u(t) = \frac{(t - t_0 + 2\sqrt{u_0})^2}{4} \tag{15.6}$$

is a solution to $u'(t) = |u(t)|^{\frac{1}{2}}$ and $u(t_0) = u_0$. We observe that $v : \mathbb{R} \to \mathbb{R}$, $v(t) = 0$ for all t, is a solution to (15.3) with initial condition $v(t_1) = 0$ for any $t_1 \in \mathbb{R}$. We now look at

$$u'(t) = |u(t)|^{\frac{1}{2}}, \qquad u(0) = 0, \tag{15.7}$$

and define for $t_0 > 0$ and $-b < 0 < a$ the function, see Figure 15.1,

$$u_{a,b}(t) := \begin{cases} \frac{(t-a)^2}{4}, & t > a \\ 0, & t \in [-b, a] \\ -\frac{(t+b)^2}{4}, & t < -b. \end{cases} \tag{15.8}$$

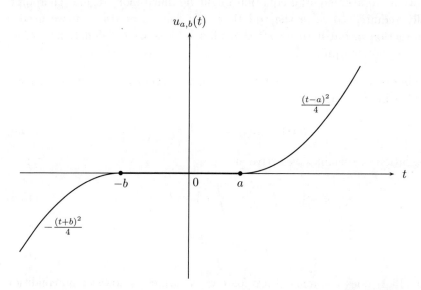

Figure 15.1

This function is differentiable on the whole real line, the only critical points to look at are a and $-b$, but for $t > a$ we have $u'_{a,b}(t) = \frac{t-a}{2}$ hence $u'_{a,b}(a) = 0$, and similarly we find $u'_{a,b}(-b) = 0$. The function also satisfies the differential equation $u'(t) = |u(t)|^{\frac{1}{2}}$, see Problem 2. Thus for any choice of $-b < 0 < a$ the function $u_{a,b}$ defined by (15.8) solves the initial value problem (15.7) and hence we have no uniqueness result. It is the initial value $u_0 = 0$ which causes the problem. If $u_0 = u(t_0) \neq 0$, say $u_0 > 0$, we know that u must be larger than zero in a neighbourhood of t_0, say an open interval $I(t_0)$. Suppose that u and v are solutions to $\frac{d}{dt}w = |w|^{\frac{1}{2}} = w^{\frac{1}{2}}$ in $I(t_0)$ with $w(t_0) > 0$ in $I(t_0)$. It follows that $\frac{u'}{u^{\frac{1}{2}}} = \frac{v'}{v^{\frac{1}{2}}} = 1$ and therefore $2u^{\frac{1}{2}} - 2v^{\frac{1}{2}} = c$, but $u(t_0) = v(t_0) > 0$

246

implying that $c = 0$, and since in $I(t_0)$ both $u(t_0)$ and $v(t_0)$ are strictly positive we conclude that they must coincide in $I(t_0)$. Thus the question to find a maximal interval of uniqueness (and existence) is justified. Before starting to formalize our terminology a further remark is needed. Instead of trying to find solutions to a differential equation in terms of known functions we shall consider a differential equation as a rule or a law which defines in general a family of functions. In this way we have introduced the exponential function, namely as (unique) solution to the equation $u' = u$ with initial condition $u(0) = 1$. Only these two conditions were used as tools to derive all properties of exp. The existence was shown much later and independently by a power series expansion. We have indicated that the same approach is possible to introduce sine and cosine, but one has to use a system of differential equations. Needless to say that Bessel functions were also introduced (in principle) as solutions of second order differential equations. Thus our aim is to assure that a given differential equation has a solution which under additional conditions is unique and then we want to investigate this solution having in mind that our principle tool must be the differential equation. The following example illustrates this point of view by deriving a differential equation for a given family of functions.

Example 15.1. The one parameter family of circles with centre at the origin and radius $r > 0$, r is the parameter, is given by $x^2 + y^2 = r^2$. If we consider $t = x$ as a curve parameter of $\gamma(t) = (t, y(t))$ we find by differentiating $t^2 + y^2(t) = r^2$ that $2t + 2y(t)y'(t) = 0$ or $y'(t) + \frac{t}{y(t)} = 0$ is a differential equation which might characterize all these circles, see Problem 4.

In general, if by $y = h(x, c)$ a family of curves in the plane is given where $c \in \tilde{I}$ and when considering $t = x \in I$ as curve parameter we find

$$y(t) = h(t, c) \quad \text{and} \quad y'(t) = \frac{\partial h}{\partial t}(t, c). \tag{15.9}$$

In some cases we may use these two equations to eliminate the parameter c and we may obtain a differential equation $R(t, y(t), y'(t)) = 0$.

Example 15.2. A. For $y(t, c) = t + ce^t$ we find $y'(t) = 1 + y(t) - t$.
B. For $y(t) = (t + c)^k$, $k > 1$, we find $y'(t) = ky(t)^{\frac{k-1}{k}}$. Note that the function $z(t) = 0$ for all t solves this differential equation but does not belong to the family.
C. In the case where we can solve $y = h(t, c)$ to obtain $c = g(t, y)$ we find as corresponding differential equation

$$\frac{\partial y}{\partial t}(t, y(t)) + \frac{\partial g}{\partial y}(t, y(t))y'(t) = 0.$$

247

We now try to introduce a systematic approach for some of the equations we may encounter. A formal definition of a differential equation covering "all" interesting cases is difficult to give, maybe it is impossible. Informally we may say that a differential equation is a relation

$$R\left(t, u, \ldots, u^{(k)}\right) = 0 \tag{15.10}$$

between a function $u : I \to \mathbb{R}$ and several (but finitely many) of its derivatives and the independent variable $t \in I$. We can look at R as a function $R : I \times \mathbb{R}^{k+1} \to \mathbb{R}$ and a k-times differential function is called a solution to (15.10) if for all $t \in I$ we have $R\left(t, u(t), u'(t), \ldots, u^{(k)}(t)\right) = 0$. A second thought teaches us that we need two modifications. To cover more interesting cases we better assume that $R : I \times G \to \mathbb{R}$, where $G \subset \mathbb{R}^{k+1}$. Further we may allow that (15.10) holds only for all $t \in \tilde{I}$ where $\tilde{I}$ is a sub-interval of I. The order of the highest derivative entering in R is called the **order of the differential equation**. Note that the order may change, i.e. for $t_1 \neq t_2$, $t_1, t_2 \in I$, the differential equation might have a different order.

Example 15.3. Let $\varphi : \mathbb{R} \to \mathbb{R}$ be a continuous function which vanishes for $t < 0$ and is strictly positive for $t > 0$. The differential equation $u(t) + u'(t) + \varphi(t)u''(t) = 0$ has order 1 for $t \leq 0$ and order 2 for $t > 0$. The order of a differential equation may change even at a single point, for example the differential equation $|t|^r u''(t) + u'(t) = 0$, $r > 0$, is for all $t \neq 0$ of order 2, but for $t = 0$ it is of order 1. Such a "singular" behaviour is important to note and will require extra investigations.

Given the differential equation (15.10) on $I \subset \mathbb{R}$. We call this differential equation **explicit** on I if we can rewrite it as

$$u^{(k)}(t) = \tilde{R}\left(t, u(t), \ldots, u^{(k-1)}(t)\right) \tag{15.11}$$

with a suitable function $\tilde{R} : I \times \tilde{G} \to \mathbb{R}$, $\tilde{G} \subset \mathbb{R}^k$. If a differential equation is not explicit we call it **implicit**. The equation $u(t)^2 + (u'(t))^2 - h(t) = 0$ is an implicit equation which we may write as

$$|u'(t)| = \sqrt{|h(t) - u(t)^2|} \tag{15.12}$$

which is still an implicit equation and the explicit equation

$$u'(t) = \sqrt{|h(t) - u(t)^2|} \tag{15.13}$$

is not necessarily equivalent to (15.12). Still we may use (15.13) for some calculations which might lead to solutions to (15.12).

When the function R is, with respect to the variables $u, \ldots, u^{(k)}$, linear we call the corresponding differential equation a **linear differential equation of order k**. A linear differential equation of order k is always explicit and we can write it as

$$a_k(t)u^{(k)}(t) + a_{k-1}u^{(k-1)}(t) + \cdots + a_1(t)u^{(1)}(t) + a_0(t)u(t) = f(t) \qquad (15.14)$$

or

$$\sum_{l=0}^{k} a_l(t)u^{(l)}(t) = f(t) \qquad (15.15)$$

as well as

$$u^{(k)}(t) = \sum_{l=0}^{k-1} \frac{a_l(t)}{a_k(t)}u^{(l)}(t) + f(t), \qquad (15.16)$$

where we used that if the equation is of order k then $a_k(t) \neq 0$ for all t. We call the functions a_l the **coefficients** of the linear equation and f the **right hand side** or **inhomogeneity**. If f is identically 0 we call (15.14) (and (15.15) - (15.16)) a **homogeneous equation** of order k, otherwise it is called an inhomogeneous equation (of order k). The notion of a **non-linear differential equation** is now self explanatory.

Instead of one equation we may consider a **system** of several equations for several unknown functions $u_1, \ldots, u_n : I \to \mathbb{R}$. We can use vector notation, i.e. we write

$$u : I \to \mathbb{R}^n, \; u(t) = \begin{pmatrix} u_1(t) \\ \vdots \\ u_n(t) \end{pmatrix} \quad \text{and instead of } R : I \times G \to \mathbb{R}, \; G \subset \mathbb{R}^{k+1}, \text{ we need}$$

now a function $R : I \times G \to \mathbb{R}^m$, $G \subset \mathbb{R}^{n(k+1)}$, i.e. we get

$$\begin{cases} R_1\left(t, u_1, \ldots, u_n, u_1', \ldots, u_n', \ldots, u_1^{(k)}, \ldots, u_n^{(k)}\right) = 0 \\ \qquad\qquad \vdots \\ R_m\left(t, u_1, \ldots, u_n, u_1', \ldots, u_n', \ldots, u_1^{(k)}, \ldots, u_n^{(k)}\right) = 0. \end{cases} \qquad (15.17)$$

We will only deal with explicit first order systems, i.e. systems of the type

$$\begin{cases} u_1'(t) = R_1(t, u_1, \ldots, u_n) \\ \qquad \vdots \\ u_n'(t) = R_n(t, u_1, \ldots, u_n) \end{cases} \qquad (15.18)$$

with the exception of looking at Euler-Lagrange equations in Part 10. Later, in Chapter 17, we will see that every explicit single equation of order k can be transformed into a system (15.18).

A linear first order system reads as

$$\begin{cases} u_1'(t) = a_{11}(t)u_1(t) + \cdots + a_{1n}(t)u_n(t) + f_1(t) \\ \quad \vdots \\ u_n'(t) = a_{n1}(t)u_1(t) + \cdots + a_{nn}(t)u_n(t) + f_n(t). \end{cases} \tag{15.19}$$

With the matrix $A(t) = (a_{kl}(t))_{k,l=1,\ldots,n}$ we can rewrite (15.19) as

$$u'(t) = A(t)u(t) + f(t), \tag{15.20}$$

where $u(t) = \begin{pmatrix} u_1(t) \\ \vdots \\ u_n(t) \end{pmatrix}$ and $f(t) = \begin{pmatrix} f_1(t) \\ \vdots \\ f_n(t) \end{pmatrix}.$

Example 15.4. Let $H(q,p)$ be a **Hamilton function**, i.e. $H : \mathbb{R}^{2n} \to \mathbb{R}$, $(q,p) \mapsto H(q,p)$ with $q, p \in \mathbb{R}^n$. We associate with H the Hamilton system

$$\frac{dq(t)}{dt} = \frac{\partial H}{\partial p}(q(t), p(t)) \quad \text{and} \quad \frac{dp(t)}{dt} = -\frac{dH}{dq}(q(t), p(t)). \tag{15.21}$$

Since t is here a time parameter we write as the physicists do for $\frac{dq}{dt}(t)$ just $\dot{q}(t)$, hence (15.21) becomes

$$\dot{q}(t) = \frac{\partial H}{\partial p}(q(t), p(t)) \quad \text{and} \quad \dot{p}(t) = -\frac{\partial H}{\partial q}(q(t), p(t)) \tag{15.22}$$

and $\frac{\partial H}{\partial p} = \left(\frac{\partial H}{\partial p_1}, \cdots, \frac{\partial H}{\partial p_n} \right)$, $\frac{\partial H}{\partial q} = \left(\frac{\partial H}{\partial q_1}, \cdots, \frac{\partial H}{\partial q_n} \right)$. In other words, we deal with the system

$$\dot{q}_j(t) = \frac{\partial H}{\partial p_j}(q(t), p(t)) \quad \text{and} \quad \dot{p}_j(t) = -\frac{\partial H}{\partial q_j}(q(t), p(t)), \quad 1 \le j \le n, \tag{15.23}$$

which is a system of order one for $2n$ functions. For the three dimensional harmonic oscillator the Hamilton function is $H(p,q) = \frac{1}{2}\|p\|^2 + \frac{1}{2}\|q\|^2$ which leads to the linear system

$$\dot{q}_j(t) = p_j(t) \quad \text{and} \quad \dot{p}_j(t) = -q_j(t), \quad 1 \le j \le 3. \tag{15.24}$$

It is convenient to adopt the physicists notation. Thus in the following we will write $\dot{u}(t) = \frac{du(t)}{dt}$, $\ddot{u}(t) = \frac{d^2u(t)}{dt^2}$, but for $k \ge 4$ we will continue to write $u^{(k)}(t) = \frac{d^k u(t)}{dt^k}$.

We will now discuss some typical examples of first order explicit equations and the aim is mainly to transform them in such a way that eventually we can find a solution with the method of separating the variables, see Theorem I.27.1. For the reader's convenience we restate this result now

Theorem I.27.1 *Let $h : [t_0, t_1] \to \mathbb{R}$, $t_0 < t_1$ be a continuous function and $u_0 \in \mathbb{R}$. Suppose that $g : \mathbb{R} \to \mathbb{R}$ is a continuous function and $g(y) \neq 0$ for all $y \in \mathbb{R}$. In this case the initial value problem*

$$g(u(t))\dot{u}(t) = h(t), \quad u(t_0) = u_0 \tag{15.25}$$

has the unique solution

$$u(t) = G^{-1}\left(H(t) + G(u_0) - H(t_0)\right) \tag{15.26}$$

where H is a primitive of h and G is a primitive of g.

For variants and first applications of this theorem we refer to Chapter I.27.

Example 15.5. We consider the **homogeneous** differential equation

$$\dot{y} = f\left(\frac{y}{t}\right) \tag{15.27}$$

for a continuous function $f : \mathbb{R} \to \mathbb{R}$. The name homogeneous refers to the fact that the function $g(t, y) = f\left(\frac{y}{t}\right)$ is a homogeneous function of degree 0 since $g(\lambda t, \lambda y) = f\left(\frac{\lambda y}{\lambda t}\right) = f\left(\frac{y}{t}\right) = g(t, y)$, compare with Definition II.6.11. In general we cannot solve (15.27) by separating the variables. However if we introduce the new function $u(t) = \frac{y(t)}{t}$ it follows that $\dot{y} = u + t\dot{u} = f(u)$, i.e.

$$\dot{u} = \frac{f(u) - u}{t} \tag{15.28}$$

or

$$\int \frac{du}{f(u) - u} = \int \frac{1}{t}\,dt. \tag{15.29}$$

Once we can find from (15.29) the function u, a solution to (15.27) is given by the function $y(t) = tu(t)$. For example, the differential equation $\dot{y} = 1 + \frac{y}{t} + \frac{y^2}{t^2}$ is transformed to the differential equation $\dot{u} = \frac{1 + u^2}{t}$ or $\arctan u = \ln t + c$ which yields $u(t) = \tan(\ln t + c)$ or $y(t) = t \tan(\ln t + c)$ as general solution to $\dot{y} = 1 + \frac{y}{t} + \frac{y^2}{t^2}$, but clearly we now need a more detailed discussion of the domain of y.

Example 15.6. We want to study the inhomogeneous linear differential equation

$$\dot{y}(t) = a(t)y(t) + f(t) \tag{15.30}$$

where a and f are continuous functions defined on some interval I and we give the initial value $y(t_0) = y_0$ for some $t_0 \in I$. We first solve the homogeneous equation by separating the variables to find

$$y_h(t) = y_0 e^{\int_{t_0}^{t} a(s)\, ds}.$$

Now we try to find a solution of the inhomogeneous equation by Lagrange's idea of variation of the constant, i.e. we assume that $y(t) = c(t) e^{\int_{t_0}^{t} a(s)\, ds}$ solves the equation and we try to determine c. This leads to

$$\dot{y}(t) = \dot{c}(t) e^{\int_{t_0}^{t} a(s)\, ds} + c(t) a(t) e^{\int_{t_0}^{t} a(s)\, ds} = f(t) + a(t) c(t) e^{\int_{t_0}^{t} a(s)\, ds}$$

or

$$\dot{c}(t) = f(t) e^{-\int_{t_0}^{t} a(s)\, ds},$$

i.e.

$$c(t) - c(t_0) = \int_{t_0}^{t} f(r) e^{-\int_{t_0}^{r} a(s)\, ds}\, dr.$$

We claim that

$$y(t) = \left(y_0 + \int_{t_0}^{t} f(r) e^{-\int_{t_0}^{r} a(s)\, ds}\, dr \right) e^{\int_{t_0}^{t} a(s)\, ds} \tag{15.31}$$

is a solution to (15.30) with initial condition $y(t_0) = y_0$. The latter is easily seen by setting in (15.31) the value of t equal to t_0. Differentiating (15.31) yields

$$\dot{y}(t) = f(t) e^{-\int_{t_0}^{t} a(s)\, ds} e^{\int_{t_0}^{t} a(s)\, ds} + \left(y_0 + \int_{t_0}^{t} f(r) e^{-\int_{t_0}^{r} a(s)\, ds}\, dr \right) a(t) e^{\int_{t_0}^{t} a(s)\, ds}$$

$$= f(t) + a(t) y(t).$$

We now discuss some differential equations which we can transform to the problem treated in Example 15.6.

Example 15.7. The differential equation

$$\dot{y}(t) = a(t) y(t) + b(t) \, (y(r))^r, \quad r \in \mathbb{R}, \tag{15.32}$$

is called the **Bernoulli differential equation**. Assuming $y(t) \geq 0$ and $r \neq 1$, in the latter case the equation would be linear, we set $z(t) := y(t)^{1-r}$ and we find

$$\dot{z}(t) = (1 - r)(y(t))^{-r} \dot{y}(t)$$

leading to

$$\dot{z}(t) = (1 - r) a(t) z(t) + (1 - r) b(t). \tag{15.33}$$

This equation can be solved using Example 15.6, however only solutions $z(t) \geq 0$ are admissible.

252

Example 15.8. The **Riccati differential equation** is given by

$$\dot{y}(t) = a_0(t) + a_1(t)y(t) + a_2(t)y(t)^2. \tag{15.34}$$

We assume that $y_0(t)$ is a known solution to (15.34) in some interval I_0 and we are looking for a further solution $y(t)$. If $y(t)$ is such a solution we find for $z(t) = y(t) - y_0(t)$ the Bernoulli differential equation

$$\dot{z}(t) = (a_1(t) + 2a_2(t)y_0(t))\, z(t) + a_2(t)z(t)^2 \tag{15.35}$$

and we may apply the ideas from Example 15.7. As a concrete example let us study the differential equation

$$\dot{y} = 4 - 2ty - \frac{1}{2}y^2. \tag{15.36}$$

A straightforward calculation shows that $y_0(t) = -\frac{2}{t}$ solves (15.36). Therefore we find with $z(t) = y(t) - y_0(t)$ the equation

$$\dot{z}(t) = \left(-2t + \frac{2}{t}\right)z(t) - \frac{1}{2}z(t)^2,$$

which is of Bernoulli type and leads with $x(t) = (z(t))^{-1}$ to the linear equation

$$\dot{x}(t) = \left(2t - \frac{2}{t}\right)x(t) + \frac{1}{2}.$$

With the help of Example 15.6 we find

$$x(t) = \left(x_0 e^{-t^2 + 2\ln t_0} + \frac{1}{2}\int_{t_0}^{t} e^{-r^2 + 2\ln r}\, dr\right)e^{t^2 + 2\ln t},$$

and here we stop. This shows the limitation of these methods: eventually the problem is whether or not we can evaluate certain integrals, and even when we can, we still have to find $z(t)$ from $x(t)$ and then $y(t)$ from $z(t)$, finally we have to check whether $y(t)$ is indeed a solution.

Consider the differential equation

$$h(t, y) + g(t, y)\dot{y} = 0, \tag{15.37}$$

which is essentially an explicit differential equation. It may happen that we can find a function $F(t, y)$ such that

$$0 = \frac{dF}{dt}(t, y(t)) = \frac{\partial F}{\partial t}(t, y(t)) + \frac{\partial F}{\partial y}(t, y(t))\dot{y}(t) = h(t, y(t)) + g(t, y(t))\dot{y}(t). \tag{15.38}$$

In this case we find

$$F(t, y(t)) = c \tag{15.39}$$

and we obtain $y(t)$ provided we can solve (15.39) for $y(t)$. Of course there might be multiple solutions, restrictions on the range of $y(t)$ or the set I we can choose t from.

Definition 15.9. *We call a differential equation* (15.37) ***exact*** *if there exists a differentiable function $F(t, y(t))$ satisfying* (15.38).

We may consider $(t, y) \mapsto \begin{pmatrix} h(t, y) \\ g(t, y) \end{pmatrix}$ as a vector field defined on some set $G \subset \mathbb{R}^2$.

The differential equation (15.37) is exact if and only if $\begin{pmatrix} h(t, y) \\ g(t, y) \end{pmatrix}$ is a **gradient field** with **potential function** F in the sense of Definition II.15.21, and by Proposition II.15.27 a necessary condition f or $\begin{pmatrix} h(t, y) \\ g(t, y) \end{pmatrix}$ being a gradient field is that g and h are C^1-functions and the **integrability condition**

$$\frac{\partial h}{\partial y}(t, y) = \frac{\partial g}{\partial t}(t, y) \tag{15.40}$$

holds. This condition is not sufficient, but under suitable topological assumptions it is. For example if G is simply connected in the sense of Definition III.19.34 the integrability condition (15.40) is also sufficient for $\begin{pmatrix} h \\ g \end{pmatrix}$ being a gradient field (a result which we will discuss and prove in our final volume of the Course). Further we know by Theorem II.15.24 that if $\begin{pmatrix} h \\ g \end{pmatrix}$ is a gradient field then a potential function F is obtained by

$$F(t, g) = \int_{(a_1, a_2)}^{(t, y)} X(p) \, dp, \quad X(p) = \begin{pmatrix} h(t, y) \\ g(t, y) \end{pmatrix}. \tag{15.41}$$

With (15.40) and (15.41) we have two important tools in our hands. Using (15.40) we may check whether a given differential equation is exact, and when this is true we can use (15.41) to find a potential function F. In practice we do this by finding a primitive of say $h(t, y)$ with respect to t while y is fixed. This gives a function $\tilde{F}(t, y) + c(y)$. Now we adjust c by the condition $\frac{\partial \tilde{F}}{\partial y}(t, y) = g(t, y)$.

Example 15.10. A. We consider the differential equation

$$\left(3t^2 + 6ty^2\right) + \left(6t^2 y + 4y^2\right) \dot{y} = 0, \tag{15.42}$$

254

i.e. $h(t, y) = 3t^2 + 6ty^2$ and $g(t, y) = 6t^2y + 4y^2$. Since

$$\frac{\partial h}{\partial y} = 12ty \quad \text{and} \quad \frac{\partial g}{\partial t} = 12ty$$

the equation is exact. To find a potential function $F(t, y)$ we first integrate $\frac{\partial F}{\partial t}(t, y) = h(t, y) = 3t^2 + 6ty^2$ to obtain $F(t, y) = t^3 + 3t^2y^2 + \gamma(y)$ and now we adjust γ by employing the condition $\frac{\partial F}{\partial y}(t, y) = g(t, y)$ which yields $F(t, y) = t^3 + 3t^2y^2 + \frac{4}{3}y^3$ as a potential function. Indeed it is easy to check that $\frac{\partial F}{\partial t} = h$ and $\frac{\partial F}{\partial y} = g$. A solution to (15.42) we can obtain now by trying to solve

$$F(t, y) = t^3 + 3t^2y^2 + \frac{4}{3}y^3 = c \tag{15.43}$$

for y. This is now a separate problem and we do not discuss it here.

B. The equation

$$2t + 2y\dot{y} = 0 \tag{15.44}$$

is exact since with $h(t) = 2t$ and $g(y) = 2y$ we have $\frac{\partial h}{\partial y} = \frac{\partial g}{\partial t} = 0$. A potential function is obtained from $\frac{\partial F}{\partial t}(t, y) = 2t$, i.e. $F(t, y) = t^2 + \gamma(y)$ with the adjustment $\frac{\partial F}{\partial y} = \gamma'(y) = 2y$, thus $F(t, y) = t^2 + y^2$ is a potential function. The equation $F(t, y) = c$ yields $t^2 + y^2 = c$. For $c > 0$ this relation describes a circle with centre $0 \in \mathbb{R}^2$ and radius $\sqrt{c}$ and we find two solutions for (15.44) by $y(t) = \pm\sqrt{c - t^2}$, $t \in [-\sqrt{c}, \sqrt{c}]$. For $c < 0$ no real-valued solution exists and for $c = 0$ the only real-valued solution of $t^2 + y^2 = c$ is $t = y = 0$.

In general the differential equation

$$h(t, y) + g(t, y)\dot{y} = 0 \tag{15.45}$$

is not exact. However there are cases where we can turn it into an exact equation when multiplying with a function $M = M(t, y)$, i.e. the equation

$$h(t, y)M(t, y) + M(t, y)g(t, y)\dot{y}(t) = 0 \tag{15.46}$$

is exact. If H is a potential function for (15.46) we may obtain a solution to (15.45) by looking at $H(t, y) = 0$. In the case that such a function M exists it is called the **Euler multiplier** or more commonly **integrating factor**.

Example 15.11. The differential equation

$$(ty^2 - y^3) + (1 - ty^2)\dot{y} = 0 \tag{15.47}$$

255

is not exact since with $h(t, y) = ty^2 - y^3$ and $g(t, y) = 1 - ty^2$ we have $\frac{\partial h}{\partial y}(t, y) = 2ty - 3y^2$ but $\frac{\partial g}{\partial t}(t, y) = -y^2$. We choose as integrating factor $M(t, y) = \frac{1}{y^2}$ and we obtain the equation

$$(t - y) + \left(\frac{1}{y^2} - t\right) \dot{y} = 0$$

which is exact since $\frac{\partial(t-y)}{\partial y} = -1 = \frac{\partial\left(\frac{1}{y^2}-t\right)}{\partial t}$. A possible potential function is $F(t, y) = \frac{t^2}{2} - ty - \frac{1}{y}$, indeed $\frac{\partial F}{\partial t}(t, y) = t - y$ and $\frac{\partial F}{\partial y}(t, y) = \frac{1}{y^2} - t$. From $\frac{t^2}{2} - ty - \frac{1}{y} = c$ or

$$t^2 y - 2ty^2 - 2cy - 2 = 0 \tag{15.48}$$

we may find a solution $y = y(t)$ of the original differential equation.

A further tool suitable to solve linear equations with constant coefficients subjected to initial values is the Laplace transform. We start with

Example 15.12. Consider the problem

$$u''(t) + au'(t) + bu(t) = f(t), \quad t > 0, \tag{15.49}$$

and

$$u(0) = u_0, \quad u'(0) = u_1, \tag{15.50}$$

where a, b, u_0, u_1 are real numbers. Multiplying (15.49) by e^{-st}, $s \geq 0$, and integrating from 0 to ∞ we get

$$\int_0^\infty e^{-st} u''(t)\, dt + a \int_0^\infty e^{-st} u'(t)\, dt + b \int_0^\infty e^{-st} u(t)\, dt = \int_0^\infty e^{-st} f(t)\, dt.$$

Of course we have to assume that all integrals exist as improper Riemann integrals. We also assume that we can integrate by parts and then we find

$$\int_0^\infty e^{-st} u'(t)\, dt = e^{-st} u(t)\Big|_0^\infty - \int_0^\infty \frac{d}{dt}\left(e^{-st}\right) u(t)\, dt$$

$$= -u(0) - \int_0^\infty \left(-se^{-st}\right) u(t)\, dt$$

$$= s \int_0^\infty e^{-st} u(t)\, dt - u(0)$$

and

$$\int_0^\infty e^{-st} u''(t)\, dt = s \int_0^\infty e^{-st} u'(t)\, dt - u'(0)$$

$$= s^2 \int_0^\infty e^{-st} u(t)\, dt - u'(0) - su(0)$$

which gives with $\mathcal{L}(u)(s) = \int_0^\infty e^{-st} u(t)\, dt$

$$s^2 \mathcal{L}(u)(s) + as\mathcal{L}(u)(s) + b\mathcal{L}(u)(s) = \mathcal{L}(f)(s) + u(0)(1 + s) + u'(0)$$

or

$$\mathcal{L}(u)(s) = \frac{\mathcal{L}(f)(s) + u(0)(1 + s) + u'(0)}{s^2 + as + b}. \tag{15.51}$$

If we now can find u given $\mathcal{L}(u)$, we may solve (15.49) and (15.50).

Let us try to make some of the calculations in Example 15.12 more rigorous.

Definition 15.13. *We call a continuous function $u : [0, \infty) \to \mathbb{C}$ of **exponential order** A if there exists $M \geq 0$ and $T_0 \geq 0$ such that for all $t \geq T_0$*

$$|u(t)| \leq Me^{At} \tag{15.52}$$

holds.

Lemma 15.14. *For u of exponential order A the improper integral*

$$(\mathcal{L}u)\,(s) := \int_0^\infty e^{-st} u(t)\, dt \tag{15.53}$$

exists for all $s \in \mathbb{C}$, $\operatorname{Re} s > A$.

Proof. We split the integral in (15.53)

$$\int_0^\infty e^{-st} u(t)\, dt = \int_0^{T_0} e^{-st} u(t)\, dt + \int_{T_0}^\infty e^{-st} u(t)\, dt$$

and we note that the first integral is finite as Riemann integral of a continuous function. For the second integral we find

$$\int_{T_0}^\infty e^{-st} u(t)\, dt = \lim_{R \to \infty} \int_{T_0}^R e^{-st} u(t)\, dt$$

and

$$\int_{T_0}^R \left| e^{-st} u(t) \right| dt \leq \int_{T_0}^R e^{(-\operatorname{Re} s)t} |u(t)|\, dt$$

$$\leq M \int_{T_0}^R e^{-(\operatorname{Re} s - A)t}\, dt$$

$$= \left. \frac{Me^{-(\operatorname{Re} s - A)t}}{A - \operatorname{Re} s} \right|_{T_0}^R.$$

257

Since $\operatorname{Re} s > A$ it follows that

$$\int_{T_0}^{R} \left| e^{-st} u(t) \right| dt \leq \frac{M}{\operatorname{Re} s - A} < \infty$$

which implies the existence of (15.53). □

Definition 15.15. *We call $\mathcal{L}(u)(s)$ in (15.53) the **Laplace transforms** of u at s.*

If u is of exponential order A then $\mathcal{L}(u)$ is at least defined in the half plane $\operatorname{Re} s > A$.

Corollary 15.16. *If u is of exponential order A then for $\operatorname{Re} s > A_1 > A$ the integral (15.53) converges uniformly.*

Proof. This follows from the estimate

$$\frac{M e^{-(\operatorname{Re} s - A)t}}{\operatorname{Re} s - A} \leq \frac{M}{A_1 - A} e^{-(A_1 - A)t}, \quad t > T_0$$

and the calculation in the proof of Lemma 15.14. □

Corollary 15.17. *If u is of exponential order A and v is of exponential order B then the Laplace transform of $\alpha u + \beta v$, $\alpha, \beta \in \mathbb{R}$, is defined for all $s \in \mathbb{C}$, $\operatorname{Re} s \geq \max(A, B)$ and we have*

$$\mathcal{L}(\alpha u + \beta v) = \alpha \mathcal{L}(u) + \beta \mathcal{L}(v). \tag{15.54}$$

Moreover for $s \in \mathbb{R}$, $s > A$ we have

$$\lim_{s \to \infty} \mathcal{L}(u)(s) = 0 \tag{15.55}$$

and $\mathcal{L}(u)$ is a C^∞-function.

Proof. It remains to prove that $\mathcal{L}(u)$ is a C^∞-function and that (15.55) holds. The first statement follows from the fact that we may differentiate under the integral together with the observation that if u is of exponential order then for every polynomial p the function pu is also of exponential order. Further, for $s \in \mathbb{R}$, $s > A$ we find

$$\left| \int_0^\infty e^{-st} u(t) \, dt \right| \leq \int_0^{T_0} e^{-st} |u(t)| \, dt + M \int_{T_0}^\infty e^{-st} e^{At} \, dt.$$

Since $\lim_{s \to 0} \int_0^{T_0} e^{-st} |u(t)| \, dt = 0$ and $\int_{T_0}^\infty e^{-st} e^{At} \, dt \leq \frac{1}{s-A}$ it follows that $\lim_{s \to \infty} \mathcal{L}(u)(s) = 0$. □

For $u \in C^1([0, \infty))$ of exponential order A we can now justify the calculations made in Example 15.12.

Corollary 15.18. *If $u \in C^1([0, \infty))$ is of exponential order A we have*

$$\mathcal{L}(u')(s) = s\mathcal{L}(u)(s) - u(0). \tag{15.56}$$

By induction we get further

Corollary 15.19. *For $u \in C^m([0, \infty))$ of exponential order A it follows*

$$\mathcal{L}\left(u^{(m)}\right)(s) = s^m \mathcal{L}(u)(s) - \sum_{j=1}^{m} s^{m-j} u^{(j-1)}(0). \tag{15.57}$$

Thus if $u : [0, \infty) \to \mathbb{R}$ is a C^m-solution of

$$u^{(m)}(t) + \sum_{j=1}^{m} a_{m-j} u^{(m-j)}(t) = f(t) \tag{15.58}$$

which is of exponential type A, then we have

$$s^m \mathcal{L}(u)(s) - \sum_{j=1}^{m} s^{m-1} u^{(j-1)}(0) + \sum_{j=1}^{m} a_{m-j} \left(s^{m-j} \mathcal{L}(u)(s) - \sum_{l=1}^{m-j} s^{m-j-l} u^{(j+l-1)}(0) \right)$$
$$= \mathcal{L}(f)(s)$$

or

$$\left(s^m + \sum_{j=1}^{m} a_{m-j} s^{m-j} \right) \mathcal{L}(u)(s) = \mathcal{L}(f)(s) + \sum_{j=1}^{m} s^{m-1} u^{(j-1)}(0)$$
$$+ \sum_{j=1}^{m} a_{m-j} \sum_{l=1}^{m-j} s^{m-j-l} u^{(j+l-1)}(0),$$

which allows us to isolate $\mathcal{L}(u)(s)$ at least for s large.

A non-trivial question is whether the Laplace transform admits an inverse. For continuous functions the Laplace transform is indeed injective and typical formulae for the inverse Laplace transform are with $h(t) = \mathcal{L}(u)(t)$

$$u(t) = \lim_{k \to \infty} \frac{(-1)^k}{k!} h^{(k)} \left(\frac{k}{x} \right) \left(\frac{k}{x} \right)^{k+1} \tag{15.59}$$

or

$$u(t) = \frac{1}{2\pi i} \int_{\gamma_0 - i\infty}^{\gamma_0 + i\infty} e^{zt} h(z) \, dz, \tag{15.60}$$

where h, i.e. u, of course has to fulfil certain conditions. We refer to [129] or [96].

Here we are (only) interested in sketching a method and therefore we leave out some theoretical details. Now we want to see how the method can work.

First we observe that we can calculate certain Laplace transforms explicitly, see Problem 10. For example we have, compare with [106]

$$\mathcal{L}(1)(s) = \frac{1}{s}, \quad s > 0;$$

$$\mathcal{L}(t^n)(s) = \frac{n!}{s^{n+1}}, \quad s > 0, \quad n \in \mathbb{N};$$

$$\mathcal{L}(t^p)(s) = \frac{\Gamma(p+1)}{s^{p+1}}, \quad s > 0, \quad p > -1;$$

$$\mathcal{L}(e^{at})(s) = \frac{1}{s-a}, \quad s > a;$$

$$\mathcal{L}(\cos \alpha t)(s) = \frac{s}{s^2 + \alpha^2}, \quad s > 0;$$

$$\mathcal{L}(\sin \alpha t)(s) = \frac{\alpha}{s^2 + \alpha^2}, \quad s > 0.$$

Thus, by the injectivity of the Laplace transform for continuous functions we may use tables of Laplace transforms to identify u given $\mathcal{L}(u) = h$. In particular we may use a decomposition into partial fractions to identify Laplace transforms of rational functions.

Example 15.20. A. Since

$$\frac{2s^2 - 4}{(s-2)(s+1)(s-3)} = -\frac{4}{3}\frac{1}{s-2} - \frac{1}{6}\frac{1}{s+1} + \frac{7}{2}\frac{1}{s-3}$$

we have

$$\mathcal{L}^{-1}\left(\frac{2s^2 - 4}{(s-2)(s+1)(s-3)}\right)(t) = -\frac{4}{3}e^{2t} - \frac{1}{6}e^{-t} + \frac{7}{2}e^{3t}.$$

B. We have

$$\frac{3s+1}{(s-1)(s^2+1)} = \frac{2}{s-1} - \frac{2s+1}{s^2+1}$$

and therefore

$$\mathcal{L}\left(\frac{3s+1}{(s-1)(s^2+1)}\right)(t) = 2e^t - 2\cos t + \sin t.$$

(These calculations are taken from [112].)

260

Now we use this idea to solve certain initial value problems. Let us return to (15.49). Taking the Laplace transform in (15.49) yields for a solution u the identity (15.51), i.e.

$$\mathcal{L}(u)(s) = \frac{\mathcal{L}(f)(s) + u(0)(1+s) + u'(0)}{s^2 + as + b}.$$

If we choose for example $f(t) = 0$ for all t, $a = 0$, $b = 4$, $u(0) = 0$ and $u'(0) = 1$ we arrive at

$$\mathcal{L}(u)(s) = \frac{1}{s^2 + 4} = \frac{1}{2} \frac{2}{s^2 + 4}$$

and therefore we find $u(t) = \frac{1}{2} \sin 2t$. However, since there are steps in our derivation of this solution which are not obvious, we have to jutify that $u(t) = \frac{1}{2} \sin 2t$ is indeed a solution to

$$u''(t) + 4u(t) = 0, \quad u(0) = 0, \quad u'(0) = 1.$$

Since $u'(t) = \cos 2t$ and $u''(t) = -2 \sin 2t$ we find first

$$u''(t) + 4u(t) = -2 \sin 2t + 4 \cdot \frac{1}{2} \sin 2t = 0$$

and moreover $u(0) = \frac{1}{2} \sin 0 = 0$ as well as $u'(0) = \cos 0 = 1$. We will discuss more examples in Problem 13.

We close our introductory chapter here with a few remarks. In this chapter we wanted to give ideas related to the problems and difficulties we may encounter and therefore we discussed examples using seemingly ad hoc methods. Some of these methods we will turn into theorems or even theories, but others are just suitable to get special solutions for special problems. We also want to emphasize that given a concrete differential equation, trying to find by some manipulations candidates for solutions is absolutely reasonable and can lead to new insights, however there will always be a need to verify that we have obtained a solution, or even more, that this is the (unique?) solution to our problem.

Problems

1. Prove that a solution $w \in C^1(\mathbb{R})$ of $w'(t) + \alpha w(t) = 0$ satisfying $w(t_0) = 0$ for some $t_0 \in \mathbb{R}$ is identically 0.

2. Prove that the function defined by (15.8) solves problem (15.7).

3. Consider the differential equation $\dot{u}(t) = |u(t)|^{\frac{1}{4}}$. Prove that if u is a solution then z, $z(t) = -u(-t)$, is a further solution. Show that for every pair of real numbers $b < 0 < a$ the function

$$u_{a,b}(t) := \begin{cases} (\frac{3}{4}(t-a))^{\frac{4}{3}}, & t \geq a \\ 0 \\ -(\frac{3}{4}(-t+b))^{\frac{4}{3}}, & t \leq b \end{cases}$$

solves the initial value problem $\dot{u}(t) = |u(t)|^{\frac{1}{4}}$, $u(0) = 0$.

4. Discuss the set of solutions to $y'(t) + \frac{t}{y(t)} = 0$, $y(t) \neq 0$ for all t.

5. Find the general solution for the Hamilton system (15.24) which is a mathematical model for the three dimensional harmonic oscillator.

6. Find the general solution of the homogeneous differential equation

$$y'(t) = \frac{10t + 8y(t)}{2t - y(t)}.$$

7. Find the solution to $\dot{y}(t) = \frac{1}{1+t}y(t) + (1+t)^2$, $t \geq 0$, and $y(0) = 1$.

8. Solve the initial value problem $y(1) = 1$ for the Bernoulli differential equation $\dot{y}(t) = -\frac{1}{t}y(t) + 2y^3(t)$, $t > 0$.

9.   a) Check the following differential equation for exactness and when this is true try to find a potential function:

$$(at^n + y(t)\cos t) + (\sin t - by^m(t))\dot{y}(t), \quad a, b \in \mathbb{R}, \quad n, m \in \mathbb{N}_0.$$

Use your calculation to find a special solution in the case where $m = 0$.

   b) Find an integrating factor for the differential equation

$$(t^3 + ty^2(t) - y(t)) + t\dot{y}(t).$$

After having multiplied the equation by the integrating factor try to find a potential function.

10. Find the Laplace transform of the following functions:

   a) $h(t) = e^{at}$, $a \in \mathbb{R} \setminus \{0\}$;

   b) $h(t) = t^p$, $p > -1$;

   c) $h(t) = \sin \alpha t$, $\alpha \in \mathbb{R} \setminus \{0\}$.

11. Find the Laplace transform of $\cosh(at)$ and $\sinh(at)$, $a \in \mathbb{R} \setminus \{0\}$.

12. Let $u : \mathbb{R} \to \mathbb{R}$ be a continuous function with period $a > 0$, i.e. $u(x + a) = u(x)$ for all $x \in \mathbb{R}$. Prove that the Laplace transform $\mathcal{L}(u)$ exists and that

$$\mathcal{L}(u)(s) = \frac{1}{1 - e^{-sa}} \int_0^a e^{-st} u(t) \, dt.$$

13. Find the inverse Laplace transform of:

   a) $g(s) = \frac{2s-3}{(3s-4)^5}$;

   b) $h(s) = \ln(1 + \frac{1}{s})$.

14. Use the Laplace transform to find a solution to the problem $y'' - 3y' + 2y = e^{-t}$, $y(0) = 2$, $y'(0) = -1$.

16 Basic Existence and Uniqueness Results I

In this and the following chapter we want to discuss for explicit first order differential equations existence and uniqueness results. This chapter is devoted to a single equation while the following chapter will handle first order systems, in general non-linear, and we will see that they also cover the case of higher order equations.

We consider the initial value problem

$$\dot{u}(t) = g(t, u(t)), \ u(t_0) = u_0 \tag{16.1}$$

for a real-valued function $u : I \to \mathbb{R}$ where I is an interval with endpoints $a < b$ and $t_0 \in I$. The function g we assume to be continuous, however the convenient assumption that g is defined for all $(t, y) \in I \times \mathbb{R}$ is not suitable. Just think about the example $\dot{u} = \sqrt{1 - u^2}$ which admits, for example, the solution $u(t) = \sin t$. Thus we must allow $g : I \times J \to \mathbb{R}$ with suitable intervals I and J.

Since we do not assume that we can find u by some special method such as separating variables, we need a more general strategy. A first idea is to replace the derivative by the difference quotient, i.e. to look at

$$\frac{u(t + h) - u(t)}{h} = g(t, u(t)), \ u(t_0) = u_0, \tag{16.2}$$

or

$$u(t + h) = u(t) + hg(t, u(t)). \tag{16.3}$$

Starting with t_0 and fixing h we can calculate

$$u(t_0 + h) = u(t_0) + hg(t_0, u(t_0))$$

which allows us to find $u(t_0 + 2h)$ as

$$u(t_0 + 2h) = u(t_0 + h) + h(g(t_0 + h), u(t_0 + h)),$$

a process we may iterate. We get by this a finite number of approximating values. Suppose we fix a compact interval $[t_0, t_1] \subset I$ which we divide into N intervals of length $\frac{t_1 - t_0}{N}$, i.e. we choose $h = \frac{t_1 - t_0}{N}$ and then we get $N + 1$ values $(t_0 + \frac{k(t_1 - t_0)}{N}, u(t_0 + \frac{k(t_1 - t_0)}{N}))$, $k = 0, \ldots, N$. But more care is needed: we do not know u and we cannot claim that the value $u(t_0 + \frac{k(t_1 - t_0)}{N})$ calculated by our method has some relation to u. Thus we switch to the better notation $\tilde{u}(t_0 + h)$, $\tilde{u}(t_0 + 2h)$, etc. The $N + 1$ values $(t_0 + \frac{(k+1)(t_1 - t_0)}{N}, \tilde{u}(t_0 + \frac{(k+1)(t_1 - t_0)}{N}))$ are now used to construct a piecewise linear, hence continuous function $\tilde{u}_N : [t_0, t_1] \to \mathbb{R}$ by choosing for $\tilde{u}$ on $[t_0 + \frac{k(t_1-t_0)}{N}, t_0 + \frac{(k+1)(t_1-t_0)}{N}]$ the function the graph of which is the line segment connecting $(t_0 + \frac{k(t_1-t_0)}{N}, \tilde{u}(t_0 + \frac{k(t_1-t_0)}{N}))$ and $(t_0 + \frac{(k+1)(t_1-t_0)}{N}, \tilde{u}(t_0 + \frac{(k+1)(t_1-t_0)}{N}))$. Our aim should be to prove that the sequence $(\tilde{u}_N)_{N\in\mathbb{N}}$ converges to a differentiable function which satisfies (16.1).

Such approximation by **difference methods** have some theoretical relevance, most of all they are of utmost importance in the numerical treatment of differential equations.

We have seen in Chapter I.17 a method to find $\sqrt{a}, a > 0$, using only rational numbers for $a \in \mathbb{Q}$. In Example I.17.17 we discussed the iteration $x_0 := x_0$ and $x_{n+1} := \frac{1}{2}(x_n + \frac{a}{x_n})$. For us it is now interesting to note that iterations may lead to solutions of equations, in this case to a solution of $x^2 = a$. We can transform (16.1) into an integral equation by using the fundamental theorem of calculus and we obtain

$$u(t) = u_0 + \int_{t_0}^{t} g(s, u(s)) \, \mathrm{d}s. \tag{16.4}$$

First we note that if g is continuous in both variables and if u is continuous, then (16.4) implies that u must already be a continuously differentiable function solving (16.1). Conversely, a continuously differentiable solution of (16.1) is a solution of (16.4). With the help of (16.4) we can now construct an iteration scheme which might lead to a solution of (16.1). Starting with $u_0(t) := u_0$ we define

$$u_k(t) := u_0 + \int_{t_0}^{t} g(s, u_{k-1}(s)) \, \mathrm{d}s, \ k \in \mathbb{N}. \tag{16.5}$$

In the case that the sequence $(u_k)_{k\in\mathbb{N}_0}$ converges to some continuous function in such a way that we may take the limit under the integral sign we find

$$u(t) = \lim_{k\to\infty} u_k(t) = u_0 + \int_{t_0}^{t} \lim_{k\to\infty} g(s, u_{k-1}(s)) \, \mathrm{d}s$$

$$= u_0 + \int_{t_0}^{t} g(s, \lim_{k\to\infty} u_{k-1}(s)) \, \mathrm{d}s = u_0 + \int_{t_0}^{t} g(s, u(s)) \, \mathrm{d}s,$$

and hence we get a solution to (16.1).

Example 16.1. Consider the initial value problem

$$\dot{u}(t) = u(t), \ u(0) = 1. \tag{16.6}$$

Of course we know that the solution is $u(t) = e^t = \sum_{n=0}^{\infty} \frac{t^n}{n!}$. With the help of (16.5) we find

$$u_0(t) = u_0,$$

$$u_1(t) = u_0 + \int_{t_0}^{t} u_0(s) \, \mathrm{d}s = 1 + \int_{0}^{t} 1 \, \mathrm{d}s = 1 + t,$$

$$u_2(t) = u_0 + \int_{t_0}^{t} u_1(s) \, \mathrm{d}s = 1 + \int_{0}^{t} (1+s) \, \mathrm{d}s = 1 + t + \frac{t^2}{2},$$

and we claim

$$u_N(t) = \sum_{n=0}^{N} \frac{t^n}{n!}.$$

Indeed, for $N = 0, 1, 2$ we have already proved the result and further we have

$$u_{k+1}(t) = u_0 + \int_{t_0}^{t} u_k(s) \, \mathrm{d}s = 1 + \int_{0}^{t} \sum_{n=0}^{k} \frac{s^n}{n!} \, \mathrm{d}s$$

$$= 1 + \sum_{n=0}^{k} \int_{0}^{t} \frac{s^n}{n!} \, \mathrm{d}s = 1 + \sum_{n=0}^{k} \frac{s^{n+1}}{(n+1)!}$$

$$= \sum_{n=0}^{k+1} \frac{s^n}{n!}.$$

Thus in this example the iteration scheme produces the partial sums of the exponential function which converges uniformly with all their derivatives on compact sets to the solution $t \mapsto e^t$.

In order to turn our ideas into a theorem, let us introduce the operator $T : C_b(I) \to C_b(I)$, where $I \subset \mathbb{R}$ is a suitable interval with endpoints a and b, by

$$(Tv)(t) = v(t_0) + \int_{a}^{t} g(s, v(s)) \, \mathrm{d}s. \tag{16.7}$$

A fixed point of this operator, i.e. $u \in C_b(I)$ satisfying $Tu = u$, is a solution to (16.1) where we assume that $g : I \times J \to \mathbb{R}$ is a continuous function and $J \subset \mathbb{R}$ a suitable interval. Thus our problem is reduced to find and investigate fixed points of T, however in some situations we might want to change the space $C_b(I)$ to another subspace of all continuous functions defined on I.

We want to prove up front two fixed point results, namely Banach's fixed point theorem or the **mapping contraction theorem**, and variants of Schauder's fixed point theorem which are directly or indirectly based on Brouwer's fixed point theorem which we discuss in Appendix IV. Both will lead to existence results, the first one also to a uniqueness statement.

Theorem 16.2. *[Banach's fixed point theorem] Let $(X, \| \cdot \|)$ be a Banach space and $D \subset X$ a non-empty closed set. Further let $T : D \to D$ be an operator satisfying in D the Lipschitz condition*

$$\|Tx - Ty\| \leq L\|x - y\| \tag{16.8}$$

for all $x, y \in D$ and some $L \in (0, 1)$. Then T has a unique fixed point $\tilde{x} \in D$. Furthermore, for every $x_0 \in D$ the sequence $(Tx_n)_{n \in \mathbb{N}_0}$, $x_{n+1} := Tx_n$, converges to $\tilde{x}$ and we have

$$\|x_{n+k} - x_n\| \leq \frac{L^n}{1 - L}\|x_1 - x_0\| \tag{16.9}$$

for all $n \in \mathbb{N}$.

Proof. We start with some $x_0 \in D$ and prove that $(x_n)_{n \in \mathbb{N}_0}$, $x_{n+1} := Tx_n$, is a Cauchy sequence in X. Since X is a Banach space this Cauchy sequence must have a limit $\tilde{x}$ and since D is closed in X the limit must belong to D. We will then show that $\tilde{x}$ is a fixed point of T. By assumption $x_{n+1} = Tx_n \in D$ and the Lipschitz condition yields

$$\|x_{n+1} - x_n\| = \|Tx_n - Tx_{n-1}\| \leq L\|x_n - x_{n-1}\|. \tag{16.10}$$

Iterating this observation we arrive at

$$\|x_{n+1} - x_n\| \leq L^n\|x_1 - x_0\|. \tag{16.11}$$

From (16.11) we deduce

$$\|x_{n+k} - x_n\| \leq \sum_{j=1}^{k} \|x_{n+j} - x_{n+j-1}\|$$

$$\leq \left(\sum_{j=1}^{k} L^{n+j} \right) \|x_1 - x_0\|$$

$$= L^n \left(\sum_{j=1}^{k} L^j \right) \|x_1 - x_0\|.$$

Since $L \in (0, 1)$ it follows that

$$\|x_{n+k} - x_n\| \leq L^n \frac{1}{1 - L}\|x_1 - x_0\| \tag{16.12}$$

268

where we used that $0 < \sum_{j=1}^{k} L^j \leq \sum_{j=0}^{\infty} L^j = \frac{1}{1-L}$. Since $0 < L < 1$ we know that $\lim_{n\to\infty} L^n = 0$. Hence, given $\epsilon > 0$ we can find $N \in \mathbb{N}$ such that $n \geq N$ implies

$$\|x_{n+k} - x_n\| < \epsilon, \tag{16.13}$$

i.e. $(x_n)_{n\in\mathbb{N}_0}$ is indeed a Cauchy sequence and we have already proved (16.9). Denote by $\tilde{x} \in D$ the limit of $(x_n)_{n\in\mathbb{N}}$ and passing in (16.12) with k to infinity we get

$$\|\tilde{x} - x_n\| \leq \frac{L^n}{1-L}\|x_1 - x_0\| = L^n\left(\frac{\|x_1 - x_0\|}{1-L}\right). \tag{16.14}$$

Moreover, the Lipschitz condition (16.8) implies that T is continuous, i.e. for every sequence $(y_n)_{n\in\mathbb{N}_0}$, $y_n \in D$, converging in X to $y \in D$ it follows that $\lim_{n\to\infty} Ty_n = Ty$. Thus we may pass in

$$x_{n+1} = Tx_n \tag{16.15}$$

to the limit to obtain eventually that $\tilde{x}$ is indeed a fixed point of T

$$\tilde{x} = \lim_{n\to\infty} x_{n+1} = \lim_{n\to\infty} Tx_n = T(\lim_{n\to\infty} x_n) = T\tilde{x}.$$

It remains to prove the uniqueness of $\tilde{x}$. For this let $\tilde{y}$ be a further fixed point of T and note that

$$\|\tilde{x} - \tilde{y}\| = \|T\tilde{x} - T\tilde{y}\| \leq L\|\tilde{x} - \tilde{y}\| < \|\tilde{x} - \tilde{y}\|$$

which is a contradiction. $\square$

The more general result, Schauder's fixed point theorem is based on a finite dimensional result due to L. Brouwer which we will prove in Appendix IV.

Theorem 16.3 (Brouwer's fixed point theorem). *Every continuous function $f : K \to K$ mapping a compact and convex subset $K \neq \emptyset$ of $\mathbb{R}^n$ into itself has at least one fixed point.*

Note that no uniqueness of the fixed point is claimed. Indeed, the function $g : [-1,1] \to [-1,1]$, $g(t) = t^3$, has the three fixed points $t_1 = -1$, $t_2 = 0$ and $t_3 = 1$.

Theorem 16.4 (Schauder's fixed point theorem). *Let $(X, \|\cdot\|)$ be a Banach space and $K \subset X$ a convex set. Suppose that $C \subset K$ is a non-empty compact set and $T : K \to C$ a continuous mapping. Then T has at least one fixed point.*

Proof. (Following partly [55]) First we note that, as we will see in Problem 6, the convex hull of finitely many elements in a Banach space is always compact. This result is of course not surprising since the convex hull of finitely many points is

269

a subset of a finite dimensional subspace of the given Banach space. Since C is compact, for every $\epsilon > 0$ we can find points $x_1, \ldots, x_m \in C$ such that the open balls $B_\epsilon(x_j)$, $j = 1, \ldots, m$, form an open covering of C. For $1 \le k \le m$ we define the functions $\varphi_k : C \to \mathbb{R}$ by

$$\varphi_k(x) = \begin{cases} 0, & \|x - x_k\| \ge \epsilon \\ \epsilon - \|x - x_k\|, & \|x - x_k\| < \epsilon. \end{cases} \tag{16.16}$$

These functions are non-negative and continuous, moreover

$$\varphi(x) := \sum_{k=1}^{m} \varphi_k(x) > 0 \text{ for all } x \in C. \tag{16.17}$$

We put

$$\psi_k(x) := \frac{\varphi_k(x)}{\varphi(x)}, \quad x \in C, \tag{16.18}$$

and note that $\psi_k(x) \ge 0$ as well as $\sum_{k=1}^{m} \psi_k(x) = 1$. The mapping $g : C \to X$ defined by

$$g(x) := \sum_{k=1}^{m} \psi_k(x) x_k \tag{16.19}$$

is well-defined, continuous and the range of g is conv$\{x_1, \ldots, x_m\}$, where as usual conv$M$ denotes the convex hull of M. For $x \in C$ it follows that

$$g(x) - x = \sum_{k=1}^{m} \psi_k(x)(x_k - x) = \sum_{k \in M(x)} \psi_k(x)(x_k - x),$$

where $M(x) = \{l \in \{1, \ldots, m\} \mid \|x - x_l\| < \epsilon\}$, since for $j \notin M(x)$ it follows that $\psi_j(x) = 0$. This implies

$$\|g(x) - x\| \le \sum_{k \in M(x)} \psi_k(x) \|x_k - x\| < \epsilon \sum_{k \in M(x)} \psi_k(x) = \epsilon. \tag{16.20}$$

The mapping $h := g \circ T$ is a continuous mapping from K to conv$\{x_1, \ldots, x_m\}$ $\subset K$ and $h|_{\text{conv}\{x_1, \ldots, x_m\}}$ is a continuous mapping from the non-empty convex set conv$\{x_1, \ldots, x_m\}$ into itself and by our preceding remark this set is compact. Hence by Brouwer's fixed point theorem $h|_{\text{conv}\{x_1, \ldots, x_m\}}$ has a fixed point $\tilde{x}_\epsilon \in$ conv$\{x_1, \ldots, x_m\}$. For $\tilde{x}_\epsilon$ we find

$$\|T(\tilde{x}_\epsilon) - \tilde{x}_\epsilon\| = \|T(\tilde{x}_\epsilon) - g(T(\tilde{x}_\epsilon))\| < \epsilon,$$

where the last step follows from (16.20). We now choose $\epsilon = \frac{1}{n}$, i.e. for every $n \in \mathbb{N}$ we can find $\tilde{x}_n \in K$ such that

$$\|T(\tilde{x}_n) - \tilde{x}_n\| < \frac{1}{n} \tag{16.21}$$

and $T(\tilde{x}_n) \in C$. Since C is compact there exists a subsequence $(\tilde{x}_{n_l})_{l \in \mathbb{N}}$ converging in X to some $\tilde{y} \in C$, i.e. $\lim_{l \to \infty} T(\tilde{x}_{n_l}) = \tilde{y}$, and (16.21) now implies $\lim_{l \to \infty} \tilde{x}_{n_l} = \tilde{y}$. The continuity of T yields finally that

$$\tilde{y} = \lim_{l \to \infty} \tilde{x}_{n_l} = \lim_{l \to \infty} T(\tilde{x}_n) = T\tilde{y},$$

i.e. $\tilde{y}$ is a fixed point of T. $\qquad\square$

Corollary 16.5. Let $(X, \|\cdot\|)$ be a Banach space and $\emptyset \neq K \subset X$ be a convex set.
A. If K is compact and $T : K \to K$ is continuous then T has at least one fixed point.
B. If K is closed, $T : K \to K$ is continuous and $T(K)$ is relative compact then T has at least one fixed point.

Remark 16.6. The reader of Volume II might have noticed that the proof of the implicit function theorem, Theorem II.10.5, is based on Banach's fixed point theorem. Fixed point theorems are extremely powerful and important tools in non-linear analysis, i.e. problems dealing with non-linear equations. The monograph [130] by E. Zeidler gives a comprehensive treatment of fixed point theorems and their applications. We will discuss further fixed point results and applications of fixed points in Volume V.

Our next goal is to use Theorem 16.3 and Theorem 16.5 to solve the initial value problem (16.1). First we look at applications of Banach's fixed point theorem. The operator we want to study is of the type

$$(Tu)(t) = u_0 + \int_{t_0}^{t} g(s, u(s)) \, ds. \tag{16.22}$$

If the functions are defined on the interval I with endpoints $a < b$ we find

$$\|Tu - Tv\|_\infty = \left\| \int_{t_0}^{\cdot} g(s, u(s)) \, ds - \int_{t_0}^{\cdot} g(s, v(s)) \, ds \right\|_\infty \tag{16.23}$$

$$\leq \int_a^b \|g(\cdot, u(\cdot)) - g(\cdot, v(\cdot))\|_\infty \, ds$$

$$= (b - a)\|g(\cdot, u(\cdot)) - g(\cdot, v(\cdot))\|_\infty.$$

Thus if we can find a constant $L > 0$ such that

$$\|g(\cdot, u(\cdot)) - g(\cdot, v(\cdot))\|_\infty \leq L\|u - v\|_\infty \tag{16.24}$$

we end up with the estimate

$$\|Tu - Tv\|_\infty \leq L(b - a)\|u - v\|_\infty. \tag{16.25}$$

This estimate is helpful since it gives already the Lipschitz continuity of T in the space $(C(I), \|\cdot\|_\infty)$. However the Lipschitz constant $L(b - a)$ will be, in general, not less than 1. Nonetheless (16.25) is sufficient to get a first result, the Theorem of Picard-Lindelöf.

Theorem 16.7 (Picard-Lindelöf). *Let $g : [a, b] \times \mathbb{R} \to \mathbb{R}$ be a continuous function such that with some $L > 0$*

$$|g(t, x) - g(t, y)| \leq L|x - y| \tag{16.26}$$

holds for all $x, y \in \mathbb{R}$ and $t \in [a, b]$ with L independent of x, y and t. Then the initial value problem

$$\dot{u}(t) = g(t, u(t)), \quad u(a) = u_0 \tag{16.27}$$

has a unique solution $u \in C([a, b]) \cap C^1((a, b))$.

Proof. We first note that (16.26) implies for two functions $u, v \in C([a, b])$

$$\|g(\cdot, u(\cdot)) - g(\cdot, v(\cdot))\|_\infty := \sup_{s \in [a,b]} |g(s, u(s)) - g(s, v(s))|$$

$$\leq L \sup_{s \in [a,b]} |u(s) - v(s)| = L\|u - v\|_\infty,$$

i.e. (16.24). Hence it follows that T as defined by (16.22) is Lipschitz continuous with constant $L(b - a)$ which need not be less than 1. However if $L(b - a) < 1$ then Banach's fixed point theorem, Theorem 16.3, already gives the result. Suppose that $L(b - a) \geq 1$. We can find $N \in \mathbb{N}$ such that $\frac{b-a}{N} < \frac{1}{L}$ and we obtain a unique solution u_1 to (16.27) in the interval $[a, \frac{b-a}{N}]$. Now we solve the problem

$$u_2(t) = g(t, u_2(t)) \text{ in } \left[a + \frac{b - a}{N}, a + 2\frac{b - a}{N}\right], \tag{16.28}$$

$$u_2\left(a + \frac{b - a}{N}\right) = u_1\left(a + \frac{b - a}{N}\right).$$

Such a solution exists and is unique since g satisfies on the interval $[a + \frac{b-a}{N}, a + 2\frac{b-a}{N}]$ all assumptions needed to apply Banach's fixed point theorem, in particular the corresponding operator has Lipschitz constant less than 1. We continue this procedure and solve for $0 < l \leq N$

$$\begin{cases} \dot{u}_l(t) = g(t, u_l(t)) \text{ on } [a + (l-1)\frac{b-a}{N}, a + l\frac{b-a}{N}] \\ u_l(a + (l-1)\frac{b-a}{N}) = u_{l-1}(a + (l-1)\frac{b-a}{N}) \end{cases} \tag{16.29}$$

where u_0 is the constant function $t \mapsto u_0$. We now define the function $u : [a, b] \to \mathbb{R}$ by

$$u(t) = u_l(t) \text{ for } t \in \left[a + (l-1)\frac{b-a}{N}, a + l\frac{b-a}{N}\right] \tag{16.30}$$

and we find for $a < t < a + l\frac{b-a}{N}$

$$u(t) - u_0 = u(t) - u(a)$$

$$= u(t) - u\left(a + (l-1)\frac{b-a}{N}\right) + u\left(a + (l-1)\frac{b-a}{N}\right)$$

$$- u\left(a + (l-2)\frac{b-a}{N}\right) + \cdots + u\left(a + \frac{b-a}{N}\right) - u(a)$$

$$= \int_{a+(l-1)\frac{b-a}{N}}^{t} g(s, u_l(s)) \, ds + \int_{a+(l-2)\frac{b-a}{N}}^{a+(l-1)\frac{b-a}{N}} g(s, u_{l-1}(s)) \, ds$$

$$+ \cdots + \int_{a}^{a+\frac{b-a}{N}} g(s, u_1(s)) \, ds$$

$$= \int_{a+(l-1)\frac{b-a}{N}}^{t} g(s, u(s)) \, ds + \int_{a+(l-2)\frac{b-a}{N}}^{a+(l-1)\frac{b-a}{N}} g(s, u(s)) \, ds$$

$$+ \cdots + \int_{a}^{a+\frac{b-a}{N}} g(s, u(s)) \, ds$$

$$= \int_{a}^{t} g(s, u(s)) \, ds$$

which implies that $u \in C([a, b]) \cap C^1((a, b))$ and that u solves (16.27). $\qquad \square$

A closer look at the proof of Theorem 16.7 reveals that we do not need to assume that the right endpoint of the interval I on which u is sought must belong to I. This observation allows us to extend Theorem 16.7 to the interval $[a, \infty)$ provided that g is defined on $[a, \infty) \times \mathbb{R}$ and (16.26) holds with $[a, b]$ replaced by $[a, \infty)$. Indeed, for any $a \leq t$ we can pick $T > t$ and solve (16.12) on $[a, T]$. Since for $T_1 < T_2$ the solution u_{T_1} and u_{T_2} must coincide on $[a, T_1]$ this procedure gives a well defined solution on $[a, \infty)$. However, at least now it becomes evident that condition (16.26) is rather restrictive since it implies

$$|g(t, x)| \leq |g(a, u_0)| + L|a| + L|x|,$$

i.e. the growth of $x \mapsto g(t, x)$ is at most linear. A further problem occurs in the cases where $g(t, \cdot)$ is not defined on $\mathbb{R}$ but just on a proper subinterval of $\mathbb{R}$ which is not necessarily closed or the Lipschitz condition may fail to hold at the endpoints. The following result is a first step to resolve these problems.

Proposition 16.8. *Let $a < b$, $c > 0$ and $u_0 \in \mathbb{R}$. Further let $g : [a, b] \times [u_0 - c, u_0 + c] \to \mathbb{R}$ be a continuous function satisfying the Lipschitz condition*

$$|g(t, x) - g(t, y)| \leq L|x - y|$$

for all $t \in [a, b]$, $x, y \in [u_0 - c, u_0 + c]$ and L is independent of t and x, y. The initial value problem

$$\dot{u}(t) = g(t, u(t)), \ t \in [a, b], \ u(a) = u_0, \tag{16.31}$$

has at least in the interval $[a, A]$, $A := \min\left(b, \frac{c}{\|g\|_\infty}\right)$ a unique solution.

Proof. We consider the following extension G of g

$$G(t, x) := \begin{cases} g(t, u_0 - c), & t \in [a, b], \ x < u_0 - c, \\ g(t, x), & (t, x) \in [a, b] \times [u_0 - c, u_0 + c], \\ g(t, u_0 + c), & t \in [a, b], \ x > u_0 + c. \end{cases} \tag{16.32}$$

The function G is by construction continuous on $[a, b] \times \mathbb{R}$ and satisfies the Lipschitz condition

$$|G(t, x) - G(t, y)| \leq L|x - y|$$

for all $t \in [a, b]$ and $x, y \in \mathbb{R}$. By the theorem of Picard and Lindelöf the initial value problem

$$\dot{v}(t) = G(t, v(t)), \ v(a) = u_0,$$

has a unique solution in $[a, b]$. Moreover we have

$$|\dot{v}(t)| \leq \|G\|_\infty = \|g\|_\infty$$

and this estimate implies that the graph of v must be a subset of the half-cone with vertex u_0 and bounded by the two half-lines passing through (a, u_0) and having slope $\pm\|g\|_\infty$, see Figure 16.1.

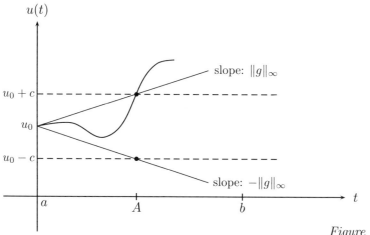

Figure 16.1

For $t \le \min(b, \frac{c}{\|g\|_\infty}) =: A$ the graph of u must stay in this set and the proposition is proved. $\qquad\square$

The reduction of problem (16.31) to the problem (16.27) is achieved by paying a price: the existence of a solution to the original problem (16.31) is only proved for a subinterval $[a, b]$ and we do not yet know whether there exists further solutions defined on a larger interval. We need

Definition 16.9. *Let $D \subset \mathbb{R}^2$ be a set and $g : D \to \mathbb{R}$ be a function. We say that g satisfies a **local Lipschitz condition** or is **locally Lipschitz continuous** in D with respect to the second variable if every point $(t_0, x_0) \in D$ admits a neighbourhood $U = U(t_0, x_0) \subset D$ such that with some $L = L(t_0, x_0) > 0$ we have*

$$|g(t, x) - g(t, y)| \le L|x - y| \tag{16.33}$$

for all $(t, x), (t, y) \in U$.

The mean-value theorem provides us with many examples of locally Lipschitz continuous functions.

Lemma 16.10. *Let $D \subset \mathbb{R}^2$ be open and suppose that $g : D \to \mathbb{R}$ has in D a continuous partial derivative $\frac{\partial g}{\partial x}$ with respect to the second variable. Then g fulfills a local Lipschitz condition in D with respect to the second variable.*

Proof. For $(t_0, x_0) \in D$ there exists $r > 0$ such that $\overline{B_r((t_0, x_0))} \subset D$. The continuous function $(t, x) \mapsto \frac{\partial g}{\partial x}(t, x)$ is bounded on $\overline{B_r((t_0, x_0))}$ and the mean-value theorem yields with ξ belonging to the line segment connecting (t, x) with

(t, y) that

$$|g(t, x) - g(t, y)| \le \left|\frac{\partial g}{\partial x}(t, \xi)\right| |x - y| \le \left\|\frac{\partial g}{\partial x}\right\|_{\infty, B_r((t_0, x_0))} |x - y|.$$

$\square$

For locally Lipschitz continuous functions with respect to the second variable we can now derive a local existence result. In the following $pr_1 : \mathbb{R}^2 \to \mathbb{R}$ is the projection onto the 1st coordinate, i.e. $pr_1(t, x) = t$.

Definition 16.11. *Let $D \subset \mathbb{R}^2$ be an open set and $g : D \to \mathbb{R}$ a continuous function. The initial value problem*

$$\dot{u}(t) = g(t, u(t)) \text{ and } u(t_0) = u_0, \ (t_0, u_0) \in D, \tag{16.34}$$

*is said to be in D **locally solvable** if for every point $(t_0, u_0) \in D$ there exists an interval $[t_0, t_0 + t_1] \subset pr_1(D)$, $t_1 > 0$, such that for all $t \in [t_0, t_0 + t_1]$ problem (16.34) has a solution.*

An immediate consequence of Theorem 16.7 and Definition 16.11 is

Corollary 16.12. *For a continuous, locally Lipschitz continuous function with respect to the second variable $g : D \to \mathbb{R}$, $D \subset \mathbb{R}^2$ open, the initial value problem (16.34) is locally solvable and the local solution is unique.*

Exercise 16.13. *Provide the details of the proof of Corollary 16.12.*

The Picard-Lindelöf theorem and its variants give solution to the initial value problems, i.e. functions solving a differential equation (of first order) defined on an interval with prescribed value at the left endpoint of the interval. The discussion of the equation (15.3) has shown that other problems, problems where the prescribed value is not an endpoint of the existence interval are also of interest. Moreover, we may consider the problem

$$\dot{u}(t) = 2tu^2(t), \ u\left(\frac{1}{3}\right) = \frac{9}{8}, \tag{16.35}$$

and find by separating the variables the solution $u(t) = \frac{1}{1-t^2}$. The function $g(t, x) = 2tx^2$ is clearly locally Lipschitz continuous with respect to the second variable and therefore by Corollary 16.12 we have a unique bounded solution in some interval $[\frac{1}{3}, A]$ where $\frac{1}{3} < A < 1$. Thus Corollary 16.12 will give a solution defined at most on $[\frac{1}{3}, 1)$. However $u(t) = \frac{1}{1-t^2}$ solves the problem in $(-1, 1)$.

A natural question is: given a solution u to (16.34) defined on some interval containing t_0, can we find a maximal interval $I_{\max}$ in which u solves $\dot{u}(t) = g(t, u(t))$ and $t_0 \in I_{\max}$? Extending the existence interval $[t_0, b]$ (or $[t_0, b)$) to values $t < t_0$ should be possible along the lines of the Picard-Lindelöf theorem: in certain situations we may solve

$$\dot{u}(t) = g(t, u(t)) \text{ in } [t_0, c] \text{ and in } [d, t_o]$$

with the condition $u(t_0) = u_0$, and by this we can obtain a solution in $[d, c]$. With this observation in mind we may concentrate on the question to find a larger or even maximal interval of existence for an initial value problem, i.e. a problem where the prescribed value is given at the left endpoint of the existence interval.

Lemma 16.14. *Let $D \subset \mathbb{R}^2$ be an open set and $g : D \to \mathbb{R}$ be a continuous function locally Lipschitz continuous with respect to the second variable. Consider the initial value problem*

$$\dot{u}(t) = g(t, u(t)) \text{ and } u(t_0) = u_0 \tag{16.36}$$

for some $(t_0, u_0) \in D$. Let $Z \neq \emptyset$ be an index set and for $k \in Z$ let $J_k \subset pr_1(D)$ be an interval containing t_0. Further let $u_k : J_k \to \mathbb{R}$ be a solution to (16.36). If

$$u_k|_{J_k \cap J_l} = u_l|_{J_k \cap J_l} \tag{16.37}$$

holds for all $k, l \in Z$ then (16.36) has exactly one solution $u : J \to \mathbb{R}$ where $J = \bigcup_{k \in Z} J_k$ and $u|_{J_k} = u_k$.

Proof. For $t \in J$ we set $u(t) := u_k(t)$, $t \in J_k$. If $t \in J_k \cap J_l$ then $u_k(t) = u_l(t)$ and therefore u is well defined and a solution to (16.36). The uniqueness of u is clear. $\square$

Corollary 16.15. *Suppose that (16.36) has a solution and in addition that for any two solutions (16.37) holds. Then (16.36) has a **maximal solution** in the sense that there exists an interval J and a function $u : J \to \mathbb{R}$ solving (16.36) and u has no extension solving (16.36) and every solution to (16.36) is a restriction of u.*

Proof. We may apply Lemma 16.14 to the set of all solutions to (16.36). $\square$

So far we do not know whether our existence interval is closed or just half-open (containing the initial point).

Proposition 16.16. *Let $D \subset \mathbb{R}^2$ be an open set and $g : D \to \mathbb{R}$ be a continuous function.*
A. *If $u : [a, c] \to \mathbb{R}$ solves the differential equation $\dot{v}(t) = g(t, v(t))$ and Γ_u, the graph of u, is contained in a compact set $K \subset D$ then u admits an extension as a solution of $\dot{v}(t) = g(t, v(t))$ to the closed interval.*
B. *Let $u_1 : [a, c] \to \mathbb{R}$ and $u_2 : [c, d] \to \mathbb{R}$ be solutions to $\dot{v}(t) = g(t, v(t))$ such that $u_1(c) = u_2(c)$ then*

$$u(t) := \begin{cases} u_1(t), & a \leq t \leq c \\ u_2(t), & c \leq t \leq d \end{cases} \tag{16.38}$$

solves $\dot{v}(t) = g(t, v(t))$ in $[a, d]$.

Proof. **A.** On K the function g is bounded and since u satisfies $\dot{u}(t) = g(t, u(t))$ it follows that $\|\dot{u}\|_\infty \leq \|g\|_{\infty, K}$ which implies that u is uniformly continuous on $[a, c)$. Indeed, the mean-value theorem yields

$$|u(t) - u(s)| = |\dot{u}(\xi)||t - s| \leq \|g\|_{\infty, K}|t - s|.$$

It follows that the limit $\beta := \lim_{t \nearrow c} u(t)$ exists and with $u(c) := \beta$ we can extend u to $[a, c]$ as a continuous function, compare also with Problem 6 in Chapter I.20. Furthermore, since

$$u(t) = u(a) + \int_a^t g(s, u(s)) \, \mathrm{d}s, \ a \leq t < c \tag{16.39}$$

we may pass to the limit $t \to c$, $t < c$, to find that (16.39) holds also for $t = c$.
B. Only the differentiability of u at c remains to be proved. We know that u is differentiable at c from the left and from the right and that these one-sided derivatives coincide, they are just $g(c, u(c))$. Hence u is differentiable at c. $\qquad\square$

Eventually we can prove a more general version of the Picard-Lindelöf theorem.

Theorem 16.17. *Let $D \subset \mathbb{R}^2$ be an open set and suppose that the continuous function $g : D \to \mathbb{R}$ satisfies a local Lipschitz condition with respect to the second variable. Then for every $(t_0, u_0) \in D$ the initial value problem (16.36) has a unique maximal solution u.*

Proof. By Corollary 16.12 we know that a local solution exists which is unique. Now, from Lemma 16.14 and Corollary 16.15 the existence of a unique maximal solution follows. $\qquad\square$

Remark 16.18. A. The reader should recollect the need of all the efforts leading from Theorem 16.7 to Theorem 16.17: for many interesting cases the function g

in (16.36) is not defined on $[a, b] \times \mathbb{R}$.

B. It is worth mentioning that Banach formulated and proved his fixed point theorem using the Picard-Lindelöf theorem as a model.

C. Our discussion around the Picard-Lindelöf theorem was much influenced by [125].

While Theorem 16.17 covers a lot of situations and the Picard-Lindelöf iteration can even be used to approximate the solution, see Example 16.1, it does not cover all the cases of interest. There are functions which are continuous but not locally Lipschitz continuous, for example $t \mapsto (\text{sgn}t)|t|^{\frac{1}{3}}$. The following existence theorem of Peano guarantees the existence of a solution to the problem

$$\dot{u}(t) = g(t, u(t)), \ u(t_0) = u_0 \tag{16.40}$$

for g merely continuous. More precisely we have

Theorem 16.19 (Peano). *Let $(t_0, u_0) \in \mathbb{R}^2$ and for some $a, c > 0$ let $g : [t_0 - a, t_0 + a] \times [u_0 - c, u_0 + c] \to \mathbb{R}$ be a continuous function with $\|g\|_\infty > 0$, i.e. g is not identically 0. Then the problem (16.40) has at least one solution in $[t_0 - A, t_0 + A]$ where $A := \min(a, \frac{c}{\|g\|_\infty}) > 0$.*

Proof. We denote by I the interval $[t_0 - A, t_0 + A]$ and we set

$$K := \{v \in C(I) \mid |v(t) - u_0| \le c \text{ for all } t \in I\}.$$

Since the continuous function $t \mapsto u_0$ belongs to K, the set K is not empty. We know that a fixed point $u \in C(I)$ of the operator

$$(Tv)(t) = u_0 + \int_{t_0}^{t} g(s, v(s)) \ \mathrm{d}s \tag{16.41}$$

will give a solution to (16.40) and vice versa. For $v \in K$ we find

$$|(Tv)(t) - u_0| \le \|g\|_\infty |t - t_0| \le \|g\|_\infty A \le c,$$

i.e. T maps K into itself, or in other words, K is invariant under T. We want to prove that T has a fixed point in K which will imply the existence of a solution to (16.40) defined on I. The set K is a closed and convex subset of $C(I)$. Indeed, for a sequence $(v_k)_{k \in \mathbb{N}}$, $v_k \in K$, converging in $C(I)$, i.e. uniformly on I, to $v \in C(I)$ we find $|v_k(t) - u_0| \le c$ for all $t \in I$ implying that K is closed. Further, for $v, w \in K$ and $\lambda \in [0, 1]$ it follows that

$$|\lambda v + (1 - \lambda)w - u_0| = |(\lambda v - \lambda u_0) + (1 - \lambda)(w - u_0)|$$
$$\le \lambda |v - u_0| + (1 - \lambda)|w - u_0| \le \lambda c + (1 - \lambda)c = c,$$

and we have proved that K is convex. Next we claim that $T : K \to K$ is continuous. Since g is continuous on the compact set $[t_0 - a, t_0 + a] \times [u_0 - c, u_0 + c]$ the function g is uniformly continuous. Therefore, given $\epsilon > 0$ we can find $\delta > 0$ such that $|x - y| < \delta$, $x, y \in [u_0 - c, u_0 + c]$, implies

$$|g(t, x) - g(t, y)| \leq \frac{\epsilon}{A}. \tag{16.42}$$

But (16.42) yields for $v, w \in K$ such that $\|v - w\|_\infty < \delta$, hence $|v(t) - w(t)| < \delta$ for all $t \in I$, the estimate

$$|g(t, v(t)) - g(t, w(t))| < \frac{\epsilon}{A}.$$

This estimate gives for $t \in I$

$$|(Tv)(t) - (Tw)(t)| = \left| \int_{t_0}^{t} (g(s, v(s)) - g(s, w(s))) \, ds \right|$$

$$\leq |t - t_0| \frac{\epsilon}{A} \leq A \cdot \frac{\epsilon}{A} = \epsilon$$

implying $\|Tv - Tw\|_\infty \leq \epsilon$, i.e. $T : K \to K$ is continuous and maps the closed convex subset of $C(I)$ into itself. By Corollary 16.15.B, once we can prove that $T(K)$ is relative compact in K (hence in $C(I)$), it follows that T has a fixed point. Now we use the theorem of Arzelà and Ascoli, Theorem II.14.25. For $v \in K$ we have with $t \in I$

$$|Tv(t)| \leq |u_0| + A\|g\|_\infty,$$

thus $T(K)$ is a bounded subset of K. Moreover, for $t_1, t_2 \in I$ we find

$$|(Tv)(t_1) - (Tv)(t_2)| \leq \left| \int_{t_0}^{t_1} g(s, v(s)) \, ds - \int_{t_0}^{t_2} g(s, v(s)) \, ds \right|$$

$$= \left| \int_{t_2}^{t_1} g(s, v(s)) \, ds \right| \leq \|g\|_\infty |t_1 - t_2|$$

which proves the equi-continuity of $T(K) \subset K \subset C(I)$. Thus the set $T(K)$ is relative compact and the theorem is proved. $\square$

The reader should note that the hard work in proving the Peano theorem is hidden in the two theorems used in its proof, the theorem of Arzelà and Ascoli and the fixed point theorem of Schauder. Our proof is one of the standard proofs and we followed the presentation of [56]. A different approach, still depending on the Arzelà-Ascoli theorem uses an approximation of the solution to (16.40) with the help of Cauchy polygons, we refer to [3] or [125].

Problems

1. Use the iteration (16.5) in order to construct a solution the initial value problem $\dot{g}(t) = tg(t)$, $y(0) = 1$.

2. Let $\Delta := \{(y,t) \in \mathbb{R}^2 | 0 \le t \le y \le a\}$ be the triangle with the vertices $(0,0), (0,a)$ and (a,a). Further let $k : \Delta \times \mathbb{R} \to \mathbb{R}$, $(x,t,z) \mapsto k(x,t,z)$ be a continuous function satisfying

$$|k(x,t,z) - k(x,t,z')| \le L|z - z'|$$

for all $(x,t) \in \Delta$ and $z, z' \in \mathbb{R}$ with L independent of x, t and z, z'. Let $g \in C([0,a])$ and for $u \in C([0,a])$ consider

$$(Tu)(x) := g(x) + \int_0^x k(x,t,u(t)) \, dt.$$

Prove that T maps $C([0,a])$ into itself and that for $L < \frac{1}{a}$ the operator $T : C([0,a]) \to C([0,a])$ has a unique fixed point $w \in C([0,a])$ which is a solution of the **Volterra integral equation**

$$w(x) = g(x) + \int_0^x k(x,t,w(t)) \, dt.$$

As usual, we consider $C([0,a])$ as a Banach space equipped with the norm $\|\cdot\|_\infty$.

3. Let $(X, \|\cdot\|_X)$ be a real Banach space which is also an algebra with unit element I_X with respect to multiplication. If $\|1_X\| = 1$ and $\|x \cdot y\|_X \le \|x\|_X \|y\|_X$ we call $(X, \|\cdot\|_X)$ a **Banach algebra**.

   a) Prove that $(C([a,b]), \|\cdot\|_\infty)$ is a Banach algebra.

   b) Let $\sum_{k=0}^\infty c_k s^k$, $c_k \in \mathbb{R}$, be a power series in $\mathbb{R}$ converging in $(-R, R)$. Prove that if $x \in X$ and $\|x\|_X < R$ then $\lim_{N \to \infty} \sum_{k=0}^N c_k x^k = \sum_{k=0}^\infty c_k x^k$ exists in X where $x^k := x \cdot \ldots \cdot x$ is the $k-$times product of x with itself in the Banach algebra X.

4. Let $k \in L^2([0,1] \times [0,1])$, $k(x,y) = \left(\frac{x+y}{2}\right)^{\frac{1}{2}}$. Prove that $K : L^2([0,1]) \to L^2([0,1])$ defined by

$$(Ku)(x) := \int_0^1 k(x,y)u(y) \, dy$$

is a contradiction.

Hint: Minkowski's integral inequality or Problem 11 of Chapter III.9 may be used.

5. Let $p \in L^2(\mathbb{R}^n) \cap L^1(\mathbb{R}), p \geq 0$, with $|\hat{p}(\xi)| < (2\pi)^{-\frac{n}{2}}$ for all $\xi \in \mathbb{R}^n$. Prove that the convolution operator

$$(Tu)(x) := (p * u)(x) = \int_{\mathbb{R}} p(x - y)u(y)\,\mathrm{d}y$$

is a contraction on $L^2(\mathbb{R}^n)$.
Hint: use Fourier analysis.

6. Prove that the convex hull of finitely many points in a Banach space is compact.

7. Solve Exercise 16.13.

17 Basic Existence and Uniqueness Results II

In this chapter we want to discuss explicit first order systems and try to obtain solutions for a corresponding initial value problem along the lines of the Picard-Lindelöf theorem. Thus we want to investigate for $R : I \times D \to \mathbb{R}^n$, $D \subset \mathbb{R}^n$, the problem

$$\dot{u}(t) = R(t, u(t)) \quad \text{and} \quad u(t_0) = u_0, \tag{17.1}$$

but now $u : I \to \mathbb{R}^n$ and $u_0 \in \mathbb{R}^n$ are vector-valued, i.e. (17.1) stands for

$$\begin{cases} \dot{u}_1(t) = R_1(t, u_1(t), \dots, u_n(t)) \\ \quad \vdots \\ \dot{u}_n(t) = R_n(t, u_1(t), \dots, u_n(t)) \end{cases} \tag{17.2}$$

and

$$u_j(t_0) = u_{0,j}, \quad j = 1, \dots, n, \quad u_0 = \begin{pmatrix} u_{0,1} \\ \vdots \\ u_{0,n} \end{pmatrix}, \tag{17.3}$$

where $R_j : I \times D \to \mathbb{R}$, $1 \le j \le n$, is the jth component of R. Already in Volume I we encountered such a system, namely in Chapter I.10 where we have mentioned that we may introduce the functions $u(t) = \sin t$ and $v(t) = \cos t$ as unique solutions to the system

$$\dot{u}(t) = v(t), \quad \dot{v}(t) = -u(t) \tag{17.4}$$

and

$$u(0) = 0, \quad v(0) = 1. \tag{17.5}$$

Before starting with some theoretical considerations we want to show that every explicit single differential equation of order m can be reduced to a system of first order. We restrict ourselves to the case where $G : I \times \mathbb{R}^m \to \mathbb{R}$, $m \ge 1$, is a continuous function and $I \subset \mathbb{R}$ is an interval. We are looking for a solution $u : I \to \mathbb{R}$ to the equation

$$u^{(m)}(t) = G(t, u(t), \dot{u}(t), \dots, u^{(m-1)}(t)), \quad t \in I, \tag{17.6}$$

under the conditions

$$u(t_0) = u_{0,1}, \dot{u}(t_0) = u_{0,2}, \dots, u^{(m-1)}(t_0) = u_{0,m}. \tag{17.7}$$

We introduce new functions $v_1, \ldots, v_m : I \to \mathbb{R}$ by

$$
\begin{cases}
\dot{v}_1(t) = v_2(t) \\
\dot{v}_2(t) = v_3(t) \\
\vdots \\
\dot{v}_{m-1}(t) = v_m(t) \\
\dot{v}_m(t) = G(t, v_1, \ldots, v_m(t))
\end{cases}
\tag{17.8}
$$

and we find the initial conditions

$$
v_1(t_0) = u_{0,1}, \; v_2(t_0) = u_{0,2}, \ldots, v_m(t_0) = u_{0,m}.
\tag{17.9}
$$

Now, if $u \in C^m(I)$ solves (17.6) and (17.7) then

$$
v := \begin{pmatrix} v_1 \\ v_2 \\ \vdots \\ v_m \end{pmatrix} = \begin{pmatrix} u \\ \dot{u} \\ \vdots \\ u^{(m-1)} \end{pmatrix}
\tag{17.10}
$$

solves (17.8) and (17.7) corresponds to (17.9). Conversely, if $v = \begin{pmatrix} v_1 \\ \vdots \\ v_m \end{pmatrix}$ is a

solution to (17.8) under the condition (17.9) then $u \in C^m(I)$ defined as $u := v_1$ solves (17.6) and (17.7). With

$$
R(t, v) := \begin{pmatrix} v_2 \\ v_3 \\ \vdots \\ v_m \\ G(t, v_1, \ldots, v_m) \end{pmatrix}
\tag{17.11}
$$

we can rewrite (17.8), (17.9) as

$$
\dot{v}(t) = R(t, v(t)) \quad \text{and} \quad \dot{v}(t_0) = u_0.
\tag{17.12}
$$

Note that this reduction of a single equation of order m to a first order system also works for explicit systems of order m, i.e. they can be reduced to a first order system, only the notation becomes more complicated. Furthermore, while we can apply results we are going to prove for systems to higher order equations, there are many situations where it is an advantage not to transform a higher order equation to a system, see Chapter 20 or Chapter 21.

In order to transfer the Picard-Lindelöf theorem to systems we need to introduce some spaces of vector-valued functions, see also Chapter II.7. By $C^k([a,b];\mathbb{R}^n)$, $k \in \mathbb{N}_0$, we denote the space of all functions $u : [a,b] \to \mathbb{R}^n$ which are k-times continuously differentiable on $[a,b]$, i.e. $u_j \in C^k([a,b])$, $1 \le j \le n$, $u = \begin{pmatrix} u_1 \\ \vdots \\ u_n \end{pmatrix}$. This space is a vector space over $\mathbb{R}$. However it is important to note that $C^k([a,b];G)$, $G \subset \mathbb{R}^n$ open, $G \ne \mathbb{R}^n$, is not a vector space, where $C^k([a,b];G)$ consists of all k-times continuously differentiable functions $u : [a,b] \to G$. For example all parametric curves with trace in the unit ball $B_1(0) \subset \mathbb{R}^n$ do not form a vector space when considered as the set $C^1([a,b];B_1(0))$.

On $C^k([a,b];\mathbb{R}^n)$ we introduce the norm

$$\|u\|_{k,\infty} := \max_{0 \le l \le k} \left\| u^{(l)} \right\|_\infty \tag{17.13}$$

where $u^{(0)} = u$. This norm is equivalent to the norm

$$\sum_{l=0}^{k} \left\| u^{(l)} \right\|_\infty \tag{17.14}$$

and the space $(C^k([a,b];\mathbb{R}^n), \|\cdot\|_{k,\infty})$ is a Banach space, see Problem 4. We can transform (17.2) and (17.3) to the vector-valued integral equation

$$u(t) = u_0 + \int_{t_0}^{t} R(s, u(s)) \, ds \tag{17.15}$$

and we may consider the operator $T : C([a,b];\mathbb{R}^n) \to C([a,b];\mathbb{R}^n)$ defined by

$$(Tu)(t) = u_0 + \int_{t_0}^{t} R(s, u(s)) \, ds. \tag{17.16}$$

Clearly T maps elements from $C([a,b];\mathbb{R}^n)$ onto elements of $C^1([a,b];\mathbb{R}^n)$. Again we note that a fixed point of T is a solution to (17.2) and (17.3).

As before we can introduce a Lipschitz condition for R with respect to the second argument, i.e. the estimate

$$\|R(s,x) - R(s,y)\| \le L\|x - y\| \tag{17.17}$$

where $\|\cdot\|$ is any norm in $\mathbb{R}^n$. We prefer now to choose the maximum norm $\|x\| = \max_{1 \le j \le n} |x_j|$. In this case (17.17) stands for

$$\max_{1 \le j \le n} |R_j(s, x) - R_j(s, y)| \le L \max_{1 \le j \le n} |x_j - y_j|$$

with $L > 0$ independent of $s \in [a, b]$ and $x, y \in \mathbb{R}^n$. Note that if for all $1 \le j \le n$ and all $1 \le l \le n$ we have

$$\left| \frac{\partial R_j(t, x)}{\partial x_l} \right| \le c$$

with some c independent of t and x, then by the mean-value theorem the mapping R satisfies (17.17). With (17.17) we find in $C([a, b]; \mathbb{R}^n)$ that

$$\begin{aligned}
\|Tu - Tv\|_\infty &= \left\| \int_{t_0}^{\cdot} (R(s, u(s)) - R(s, v(s))) \, ds \right\|_\infty \\
&\le \int_a^b \sup_{s \in [a,b]} \|R(s, u(s)) - R(s, v(s))\| \, ds \\
&\le L \sup_{s \in [a,b]} \|u(s) - v(s)\|(b - a) \\
&= L(b - a)\|u - v\|_\infty.
\end{aligned}$$

We can now use the proof of Theorem 16.7 to obtain an existence and uniqueness result for (17.2) and (17.3). However, we want to give a proof with the help of a weighted norm on $C([a, b]; \mathbb{R}^n)$ and by this we will avoid the problems caused by the fact that in general the Lipschitz constant $L(b - a)$ is not less than 1. We set for $\lambda \in \mathbb{R}$

$$\|u\|_{\infty,\lambda} := \sup_{t \in [a,b]} (\|u(t)\| e^{-\lambda t}), \tag{17.18}$$

the meaning of which is

$$\|u\|_{\infty,\lambda} = \sup_{t \in [a,b]} (\max_{1 \le j \le n} |u_j(t)| e^{-\lambda t}).$$

Since $e^{-\lambda t} \ge e^{-\lambda \max(|a|, |b|)} > 0$ for $t \in [a, b]$ it is clear that (17.18) is a norm on $C([a, b]; \mathbb{R}^n)$ equivalent to the original one, in particular $(C([a, b]; \mathbb{R}^n), \|\cdot\|_{\infty,\lambda})$ is a Banach space. Moreover we find

$$
\begin{aligned}
|(Tu - Tv)(t)| &= \left| \int_{t_0}^{t} (R(s, u(s)) - R(s, v(s)))e^{-\lambda s}e^{\lambda s} \, ds \right| \\
&\leq \int_{t_0}^{t} |(R(s, u(s)) - R(s, v(s)))e^{-\lambda s}|e^{\lambda s} \, ds \\
&\leq L \int_{t_0}^{t} e^{\lambda s} \, ds \, \|u - v\|_{\infty,\lambda} \\
&= \frac{L(e^{\lambda t} - e^{\lambda t_0})}{\lambda} \|u - v\|_{\infty,\lambda},
\end{aligned}
$$

or

$$
\|Tu - Tv\|_{\infty,\lambda} \leq \frac{L}{\lambda} \|u - v\|_{\infty,\lambda}. \tag{17.19}
$$

Now we choose $\lambda = 2L$ to obtain

$$
\|Tu - Tv\|_{\infty,2L} \leq \frac{1}{2} \|u - v\|_{\infty,2L}, \tag{17.20}
$$

i.e.

$$
T : (C([a, b]; \mathbb{R}^n), \| \cdot \|_{\infty,2L}) \to (C([a, b]; \mathbb{R}^n), \| \cdot \|_{\infty,2L})
$$

is a contraction and therefore admits a unique fixed point in $(C([a, b]; \mathbb{R}^n), \| \cdot \|_{\infty,2L})$, hence in $C([a, b]; \mathbb{R}^n)$. We now use Banach's fixed point theorem and the Picard-Lindelöf theorem for systems follows:

Theorem 17.1. *Suppose that* $R : [a, b] \times \mathbb{R}^n \to \mathbb{R}^n$ *is continuous and satisfies* (17.17) *with some* $L > 0$. *Then the system* (17.2) *with initial condition* (17.3) *has a unique solution* $u \in C([a, b]; \mathbb{R}^n) \cap C^1([a, b]; \mathbb{R}^n)$.

Again problems arise when $R(s, \cdot)$ is not well defined on all of $\mathbb{R}^n$ but these problems can be resolved eventually along the lines of the previous chapter. It is of some interest to look at the Picard-Lindelöf iteration for the system (17.4) and (17.5). We start with $\begin{pmatrix} u_0 \\ v_0 \end{pmatrix} = \begin{pmatrix} 0 \\ 1 \end{pmatrix}$ and define

$$
\begin{pmatrix} u_n \\ v_n \end{pmatrix}(t) = T \begin{pmatrix} u_{n-1} \\ v_{n-1} \end{pmatrix}(t) = \begin{pmatrix} 0 \\ 1 \end{pmatrix} + \int_0^t \begin{pmatrix} 0 & 1 \\ -1 & 0 \end{pmatrix} \begin{pmatrix} u_{n-1} \\ v_{n-1} \end{pmatrix}(s) \, ds. \tag{17.21}
$$

Here are the first three iterations:

$$\begin{pmatrix} u_1 \\ v_1 \end{pmatrix} (t) = \begin{pmatrix} 0 \\ 1 \end{pmatrix} + \int_0^t \begin{pmatrix} 0 & 1 \\ -1 & 0 \end{pmatrix} \begin{pmatrix} 0 \\ 1 \end{pmatrix} \, ds = \begin{pmatrix} t \\ 1 \end{pmatrix},$$

$$\begin{pmatrix} u_2 \\ v_2 \end{pmatrix} (t) = \begin{pmatrix} 0 \\ 1 \end{pmatrix} + \int_0^t \begin{pmatrix} 0 & 1 \\ -1 & 0 \end{pmatrix} \begin{pmatrix} s \\ 1 \end{pmatrix} \, ds = \begin{pmatrix} t \\ 1 - \frac{t^2}{2!} \end{pmatrix},$$

$$\begin{pmatrix} u_3 \\ v_3 \end{pmatrix} (t) = \begin{pmatrix} 0 \\ 1 \end{pmatrix} + \int_0^t \begin{pmatrix} 0 & 1 \\ -1 & 0 \end{pmatrix} \begin{pmatrix} s \\ 1 - \frac{s^2}{2!} \end{pmatrix} \, ds = \begin{pmatrix} \frac{t}{1!} - \frac{t^3}{3!} \\ 1 - \frac{t^2}{2!} \end{pmatrix}.$$

A straightforward induction shows

$$\begin{pmatrix} u_n \\ v_n \end{pmatrix} (t) = \begin{pmatrix} \sum_{k=1}^{\frac{n}{2}} (-1)^{k+1} \frac{t^{2k-1}}{(2k-1)!} \\ \sum_{k=0}^{\frac{n}{2}} (-1)^k \frac{t^{2k}}{(2k)!} \end{pmatrix} \quad \text{for } n \text{ even} \tag{17.22}$$

and

$$\begin{pmatrix} u_n \\ v_n \end{pmatrix} (t) = \begin{pmatrix} \sum_{k=1}^{\frac{n+1}{2}} (-1)^{k+1} \frac{t^{2k-1}}{(2k-1)!} \\ \sum_{k=0}^{\frac{n+1}{2}} (-1)^k \frac{t^{2k}}{(2k)!} \end{pmatrix} \quad \text{for } n \text{ odd.} \tag{17.23}$$

In the limit we obtain

$$\lim_{n \to \infty} \begin{pmatrix} u_n \\ v_n \end{pmatrix} (t) = \begin{pmatrix} \sin t \\ \cos t \end{pmatrix}$$

as we expect.

The system (17.4), (17.5) is linear and we may rewrite it with $w = \begin{pmatrix} u \\ v \end{pmatrix}$ as

$$\dot{w}(t) = \begin{pmatrix} 0 & 1 \\ -1 & 0 \end{pmatrix} w(t), \quad w(0) = \begin{pmatrix} 0 \\ 1 \end{pmatrix}. \tag{17.24}$$

In analogy to the single (scalar-valued) equation $\dot{g}(t) = \alpha g(t)$, $g(0) = 1$, we may try to find a solution by using the exponential function of the matrix $A := \begin{pmatrix} 0 & 1 \\ -1 & 0 \end{pmatrix}$. For A we define

$$e^{tA} = \sum_{k=0}^{\infty} \frac{t^k A^k}{k!}. \tag{17.25}$$

We leave aside the convergence problem and look at $(tA)^k$ to find

$$(tA)^k = \begin{pmatrix} (-1)^m t^{2m} & 0 \\ 0 & (-1)^m t^{2m} \end{pmatrix} \quad \text{for } k = 2m$$

and

$$(tA)^k = \begin{pmatrix} 0 & (-1)^{m-1}t^{2m-1} \\ -(-1)^{m-1}t^{2m-1} & 0 \end{pmatrix} \quad \text{for } k = 2m-1.$$

By a formal calculation, i.e. a calculation not taking possible convergence problems into account, we find

$$e^{tA} = \sum_{k=0}^{\infty} \frac{(At)^k}{k!}$$

$$= \sum_{m=0}^{\infty} \begin{pmatrix} \frac{(-1)^m t^{2m}}{(2m)!} & 0 \\ 0 & \frac{(-1)^m t^{2m}}{(2m)!} \end{pmatrix} + \sum_{m=1}^{\infty} \begin{pmatrix} 0 & \frac{(-1)^{m-1}t^{2m-1}}{(2m-1)!} \\ -\frac{(-1)^{m-1}t^{2m-1}}{(2m-1)!} & 0 \end{pmatrix}$$

$$= \begin{pmatrix} \cos t & 0 \\ 0 & \cos t \end{pmatrix} + \begin{pmatrix} 0 & \sin t \\ -\sin t & 0 \end{pmatrix} = \begin{pmatrix} \cos t & \sin t \\ -\sin t & \cos t \end{pmatrix}.$$

For (17.24) this means

$$w(t) = \begin{pmatrix} \cos t & \sin t \\ -\sin t & \cos t \end{pmatrix} \begin{pmatrix} 0 \\ 1 \end{pmatrix} = \begin{pmatrix} \sin t \\ \cos t \end{pmatrix}$$

and we end up with the explicit solution to (17.24). It turns out that for linear systems this approach works in general and indeed very well and we will study linear systems in the next chapter.

Problems

1. Transform the differential equation

$$u'''(t) = (\cos t)u''(t) + \frac{e^{-t}}{1+t^2}u'(t) - (1+t^4)u(t) + t, \quad t \in \mathbb{R},$$

   into a first order system.

2. Given the system

$$y''(t) + y'(t) + a(t)w(t) = 0$$
$$w'(t) + b(t)y(t) = 0,$$

   where $a, b : \mathbb{R} \to \mathbb{R}$ are continuous functions. Add the initial conditions $y(0) = 0$, $y'(0) = 1$ and $w(0) = 1$, and try to transform this system into a first order system.

3. Let $f : \mathbb{R}^n \to \mathbb{R}^n$ be a continuously differentiable mapping with components f_j such that

$$\left\| \frac{\partial f_j}{\partial x_k} \right\|_\infty \leq M \quad \text{for} \ 1 \leq j, k \leq n. \quad (*)$$

Prove that f is Lipschitz continuous with respect to the norm $\| \cdot \|_\infty$ in $\mathbb{R}^n$, i.e. we have

$$\| f(x) - f(y) \|_\infty \leq L \| x - y \|_\infty \quad (**)$$

for all $x, y \in \mathbb{R}^n$ and some $L > 0$ independent of x and y. Note that while in $(**)$ on both sides we have the sup-norm in $\mathbb{R}^n$, i.e. $\| z \|_\infty = \max_{1 \leq j \leq n} |z_j|$, we are dealing in $(*)$ with the norm $\left\| \frac{\partial f_j}{\partial x_k} \right\|_\infty = \sup \left| \frac{\partial f_j}{\partial x_k} \right|$.

4. Prove that $(C^k([a, b]; \mathbb{R}^n), \| \cdot \|_{k,\infty})$ is a Banach space.

5. Prove that the norms (17.13), (17.14) and (17.18) are equivalent.

6. Consider the system

$$\dot{u}(t) = v(t)$$
$$\dot{v}(t) = -u(t)$$
$$\dot{w}(t) = w(t)$$

with initial conditions $u(0) = 0$, $v(0) = 1$ and $w(0) = 1$. Use the Picard-Lindelöf iteration to find the solution of this initial value problem.

18 Linear Systems of First Order. Constant Coefficients

Let $I \subset \mathbb{R}$ be an interval with endpoints $a < b$ and $A : I \to M(n;\mathbb{R})$ be a matrix-valued function. Since we can identify $M(n;\mathbb{R})$ with $\mathbb{R}^{n^2}$, basic calculus results do hold for such a mapping, the reader may compare with Chapter II.7. For $f : I \to \mathbb{R}^n$ we want to discuss the system

$$\dot{u}(t) = Au(t) + f(t) \tag{18.1}$$

and

$$u(t_0) = u_0, \; t_0 \in I, \; u_0 \in \mathbb{R}^n, \tag{18.2}$$

where $u : I \to \mathbb{R}^n$ is a sought function. The matrix

$$A = \begin{pmatrix} a_{11}(t) & \cdots & a_{1n}(t) \\ \vdots & & \vdots \\ a_{n1}(t) & \cdots & a_{nn}(t) \end{pmatrix}$$

has elements $a_{kl}(t)$ where for $1 \le k,l \le n$ the functions $a_{kl} : I \to \mathbb{R}$ are assumed to be at least continuous. Written in terms of components, (18.1) and (18.2) read as

$$\begin{cases} \dot{u}_1(t) = a_{11}(t)u_1(t) + \cdots + a_{1n}(t)u_n(t) + f_1(t) = \sum_{j=1}^{n} a_{1j}(t)u_j(t) + f_1(t) \\ \vdots \\ \dot{u}_n(t) = a_{n1}(t)u_1(t) + \cdots + a_{nn}(t)u_n(t) + f_n(t) = \sum_{j=1}^{n} a_{nj}(t)u_j(t) + f_n(t) \end{cases} \tag{18.3}$$

and

$$\begin{pmatrix} u_1(t_0) \\ \vdots \\ u_n(t_0) \end{pmatrix} = \begin{pmatrix} u_{01} \\ \vdots \\ u_{0n} \end{pmatrix}. \tag{18.4}$$

As indicated in the last chapter we want to use matrix-valued exponential functions to handle (18.1)/(18.2) and for this we need some preparations.

Since we can identify $M(n;\mathbb{R})$ with $\mathbb{R}^{n^2}$ we can turn the vector space $M(n;\mathbb{R})$ into a normed space by picking any norm in $\mathbb{R}^n$. However we also want to interpret matrices as linear mappings as we want to deal with their products, for example, interpreted as composition of mappings. For these reasons we need to introduce the notion of a matrix norm and we do it for $M(n;\mathbb{R})$ as well as for $M(n;\mathbb{C})$.

Definition 18.1. *Let* $\mathbb{K} = \mathbb{R}, \mathbb{C}$ *and* $M(n; \mathbb{K})$ *be the* $\mathbb{K}$*-vector space of all* $n \times n$*-matrices with elements from* $\mathbb{K}$*. We call* $\| \cdot \| : M(n; \mathbb{K}) \to \mathbb{R}$ *a **matrix-norm** on* $M(n; \mathbb{K})$ *with respect to the norm* $\| \cdot \|_{\mathbb{K}^n}$ *given on* $\mathbb{K}^n$ *if for all* $A, B \in M(n; \mathbb{K})$ *and* $\lambda \in \mathbb{K}$ *we have*

$$\|A\| \geq 0 \text{ and } \|A\| = 0 \text{ if and only if } A = 0 \in M(n; \mathbb{K}); \tag{18.5}$$

$$\|\lambda A\| \leq |\lambda| \|A\|; \tag{18.6}$$

$$\|A + B\| \leq \|A\| + \|B\|; \tag{18.7}$$

$$\|AB\| \leq \|A\| \|B\| \tag{18.8}$$

and if for all $x \in \mathbb{K}^n$

$$\|Ax\|_{\mathbb{K}^n} \leq \|A\| \|x\|_{\mathbb{K}^n} \tag{18.9}$$

holds.

Remark 18.2. In Problem 2 we will give a generalisation of the notion of a matrix norm by taking into account that in (18.9) we may have on the left and the right hand side a different norm in $\mathbb{K}^n$. For $A \in M(n; \mathbb{K})$ we set

$$\|A\|_1 := \sum_{k,l=1}^{n} |a_{kl}| \tag{18.10}$$

and

$$\|A\|_2 := \left(\sum_{k,l=1}^{n} |a_{kl}|^2 \right)^{\frac{1}{2}}. \tag{18.11}$$

Lemma 18.3. A. *By* $\| \cdot \|_1$ *a matrix norm on* $M(n; \mathbb{K})$ *is given with respect to the norm* $\|x\|_\infty := \max_{1 \leq j \leq n} |x_j|, \ x \in \mathbb{K}^n$.
B. *By* $\| \cdot \|_2$ *a matrix norm on* $M(n; \mathbb{K})$ *is given with respect to the norm* $\|x\|_2 := (\sum_{j=1}^{n} |x_j|^2)^{\frac{1}{2}}, \ x \in \mathbb{K}^n$.

Proof. In both cases (18.5)-(18.7) are clear since $\| \cdot \|_1$ and $\| \cdot \|_2$ are norms on $M(n; \mathbb{K}) \cong \mathbb{K}^{n^2}$. Now we prove that $\| \cdot \|_1$ is a matrix norm, i.e. that (18.8) and (18.9) hold. With $C = AB$ we find

$$\|C\|_1 \leq \sum_{i,j=1}^{n} \sum_{k=1}^{n} |a_{ik} b_{kj}| \leq \sum_{i,j,k,l} |a_{ik} b_{lj}| = \|A\|_1 \|B\|_1$$

and further

$$|(Ax)_i| \leq \sum_{j=1}^{n} |a_{ij}x_j| \leq \|x\|_\infty \sum_{j=1}^{n} |a_{ij}|$$

$$\leq \sum_{i,j=1}^{n} |a_{ij}|\|x\|_\infty = \|A\|_1 \|x\|_\infty$$

implying that

$$\|Ax\|_\infty \leq \|A\|_1 \|x\|_\infty.$$

Finally we show that $\|\cdot\|_2$ is a matrix norm. First we note that by the Cauchy-Schwarz inequality

$$\|AB\|_2 = \left(\sum_{i,j=1}^{n} \left| \sum_{k=1}^{n} a_{ik}b_{kj} \right|^2 \right)^{\frac{1}{2}}$$

$$\leq \left(\sum_{i,j=1}^{n} \left(\sum_{k=1}^{n} |a_{ik}|^2 \right) \left(\sum_{k=1}^{n} |b_{kj}|^2 \right) \right)^{\frac{1}{2}}$$

$$= \left(\sum_{i=1}^{n} \sum_{j=1}^{n} \left(\sum_{k=1}^{n} |a_{ik}|^2 \right) \left(\sum_{l=1}^{n} |b_{lj}|^2 \right) \right)^{\frac{1}{2}}$$

$$= \left(\left(\sum_{i=1}^{n} \sum_{k=1}^{n} |a_{ik}|^2 \right) \left(\sum_{j=1}^{n} \sum_{l=1}^{n} |b_{lj}|^2 \right) \right)^{\frac{1}{2}}$$

$$= \left(\sum_{i,k=1}^{n} |a_{ik}|^2 \right)^{\frac{1}{2}} \left(\sum_{j,l=1}^{n} |b_{lj}|^2 \right)^{\frac{1}{2}} = \|A\|_2 \|B\|_2.$$

Moreover we find for $x \in \mathbb{K}^n$ again by the Cauchy-Schwarz inequality

$$\|Ax\|_2 = \left(\sum_{j=1}^{n} \left| \sum_{k=1}^{n} a_{jk}x_k \right|^2 \right)^{\frac{1}{2}}$$

$$\leq \left(\sum_{j=1}^{n} \left(\sum_{k=1}^{n} |a_{jk}|^2 \right) \left(\sum_{k=1}^{n} |x_k|^2 \right) \right)^{\frac{1}{2}}$$

$$= \left(\sum_{j,k=1}^{n} |a_{jk}|^2 \right)^{\frac{1}{2}} \left(\sum_{k=1}^{n} |x_k|^2 \right)^{\frac{1}{2}} = \|A\|_2 \|x\|_2$$

and the lemma is proved. $\square$

Since for $A \in M(n; \mathbb{K})$ it follows that $A^k \in M(n; \mathbb{K})$, $k \in \mathbb{N}_0$, we can form polynomials of A, i.e. for $p(z) = \sum_{k=0}^{N} a_k z^k$, $a_k \in \mathbb{K}$, $z \in \mathbb{K}$, we can consider the matrix

$$p(A) := \sum_{k=0}^{N} a_k A^k \in M(n; \mathbb{K}). \tag{18.12}$$

Moreover, assume that $\sum_{k=0}^{\infty} a_k z^k$ is a power series with coefficients $a_k \in \mathbb{K}$ and radius of convergence $\rho > 0$. Suppose that $A \in M(n; \mathbb{K})$ and $\|A\| < \rho$ for some matrix norm $\| \cdot \|$ on $M(n; \mathbb{K})$. It follows that

$$\left\| \sum_{k=0}^{\infty} a_k A^k \right\| \leq \sum_{k=0}^{\infty} |a_k| \|A^k\| \leq \sum_{k=0}^{\infty} |a_k| \|A\|^k < \infty,$$

implying that with respect to the matrix norm $\| \cdot \|$ the series $\sum_{k=0}^{\infty} a_k A^k$ converges in $(M(n; \mathbb{K}), \| \cdot \|)$. Thus we may define matrix-valued functions with the help of power series. In particular for every entire function $f(z) = \sum_{k=0}^{\infty} a_k z^k$ and every matrix $A \in M(n; \mathbb{K})$ the matrix

$$f(A) := \sum_{k=0}^{\infty} a_k A^k$$

is well defined as convergent series in $M(n; \mathbb{K})$ with respect to every matrix norm.

Now let $A \in M(n; \mathbb{K})$ and $t \in \mathbb{R}$. We define

$$\exp(tA) := e^{tA} := \sum_{k=0}^{\infty} \frac{(tA)^k}{k!} = \sum_{k=0}^{\infty} \frac{t^k A^k}{k!} \tag{18.13}$$

as well as

$$\sin tA := \sum_{k=1}^{\infty} (-1)^{k+1} \frac{(tA)^{2k-1}}{(2k-1)!} \tag{18.14}$$

and

$$\cos tA := \sum_{k=0}^{\infty} (-1)^k \frac{(tA)^{2k}}{(2k)!}. \tag{18.15}$$

We can consider the functions exp, sin and cos as functions from $\mathbb{R}$ (or a subset of $\mathbb{R}$) to $M(n; \mathbb{K})$ and we will see in Problem 5 (as well as in Lemma 18.4 below) that they are differentiable functions with respect to t.

Lemma 18.4. *For $x \in \mathbb{K}$ and $A \in M(n; \mathbb{K})$ define $T_t^A : \mathbb{K}^n \to \mathbb{K}^n$ by $T_t^A x := \exp(tA)x = e^{tA}x$. Then it follows that*

$$\frac{\mathrm{d}}{\mathrm{d}t} T_t^A x = A T_t^A x. \tag{18.16}$$

294

Proof. We note that due to the convergence properties of $\sum_{k=0}^{\infty} \frac{(tA)^k}{k!}$ we have

$$\left(\sum_{k=0}^{\infty} \frac{\mathrm{d}}{\mathrm{d}t}\left(\frac{(tA)^k}{k!} \right) \right) x = \left(\sum_{k=0}^{\infty} \frac{\mathrm{d}}{\mathrm{d}t}\left(\frac{t^k A^k}{k!} \right) \right) x$$

$$= \left(\sum_{k=0}^{\infty} A^k \frac{kt^{k-1}}{k!} \right) x$$

$$= \left(A \left(\sum_{k=1}^{\infty} \frac{(tA)^{k-1}}{(k-1)!} \right) \right) x$$

$$= A \left(\sum_{k=0}^{\infty} \frac{(tA)^k}{k!} \right) x = AT_t^A x,$$

and therefore we may change the order of summation and differentiation to find

$$\frac{\mathrm{d}}{\mathrm{d}t} T_t^A x = \frac{\mathrm{d}}{\mathrm{d}t} \left(\sum_{k=0}^{\infty} \frac{(tA)^k}{k!} \right) x$$

$$= \left(\sum_{k=0}^{\infty} \frac{\mathrm{d}}{\mathrm{d}t}\left(\frac{(tA)^k}{k!} \right) \right) x = AT_t^A x.$$

$\square$

A consequence of Lemma 18.4 is

Theorem 18.5. *Let $A \in M(n;\mathbb{R})$ and T_t^A, $t \geq 0$, be defined by (18.13). Then $u(t) := T_t^A u_0$, $u_0 \in \mathbb{R}^n$, is the unique solution to the system*

$$\dot{u}(t) = Au(t), \ u(0) = u_0. \tag{18.17}$$

Proof. Lemma 18.4 yields that $t \mapsto T_t^A u_0$ solves the equation $\dot{u}(t) = Au(t)$ and since $T_0^A u_0 = u_0$ we also obtain that $u(0) = T_0^A u_0 = u_0$. The uniqueness statement follows from Theorem 17.1.

$\square$

Thus for all linear systems with real constant coefficients we obtain a unique solution to (18.17) defined for all $t \in \mathbb{R}$.

Example 18.6. Let $A = (\lambda_j \delta_{jl})_{j,l=1,\ldots,n}$ be a diagonal matrix. The system (18.17) becomes now

$$\dot{u}(t) = \begin{pmatrix} \lambda_1 & & \mathbf{0} \\ & \ddots & \\ \mathbf{0} & & \lambda_n \end{pmatrix} u(t), \ u(0) = u_0.$$

Since $A^k = \begin{pmatrix} \lambda_1^k & & 0 \\ & \ddots & \\ 0 & & \lambda_n^k \end{pmatrix}$ we find

$$\exp tA = \sum_{k=0}^{\infty} \frac{1}{k!} \begin{pmatrix} (t\lambda_1)^k & & 0 \\ & \ddots & \\ 0 & & (t\lambda_n)^k \end{pmatrix} = \begin{pmatrix} \exp t\lambda_1 & & 0 \\ & \ddots & \\ 0 & & \exp t\lambda_n \end{pmatrix}$$

which yields for the solution u

$$u(t) = \begin{pmatrix} e^{\lambda_1 t} u_{01} \\ \vdots \\ e^{\lambda_n t} u_{0n} \end{pmatrix}. \tag{18.18}$$

Of course this is not a surprise. The system (18.18) is completely decoupled, we have n completely independent equations $\dot{u}_j(t) = \lambda_j u_j(t)$ with independent initial conditions $u_j(0) = u_{0j}$, each having the solution $u_j(t) = e^{\lambda_j t} u_{0j}$.

Example 18.7. Consider the system

$$\begin{cases} \dot{u}_1(t) = u_2(t) \text{ and } u_1(0) = 0 \\ \dot{u}_2(t) = -u_1(t) \text{ and } u_2(0) = 1 \end{cases} \tag{18.19}$$

This is just the system discussed at the end of the last chapter with $A = \begin{pmatrix} 0 & 1 \\ -1 & 0 \end{pmatrix}$ and $\exp tA = \begin{pmatrix} \cos t & \sin t \\ -\sin t & \cos t \end{pmatrix}$ which yields the known solution $u(t) = \begin{pmatrix} \sin t \\ \cos t \end{pmatrix}$.

Let us return to e^A. In the one-dimensional case, i.e. $A = a$, $a \in \mathbb{K}$, the exponential function satisfies the functional equation $e^{a+b} = e^a e^b$. We want to transfer this relation to matrices $A, B \in M(n; \mathbb{K})$. However there we encounter a problem: we need to take into account that the product of matrices is in general not commutative. We introduce the **commutator** of two matrices $A, B \in M(n; \mathbb{K})$ by

$$[A, B] = AB - BA. \tag{18.20}$$

Obviously $[A, B] = 0$ if and only if $AB = BA$, i.e. the two matrices commute. In Proposition I.29.22 we proved for $a, b \in \mathbb{R}$ the functional equation for the exponential function by using its power series expansion and the Cauchy product of series. The crucial step was an application of the binomial theorem to find

$$\sum_{n=0}^{\infty} \left(\sum_{k=0}^{\infty} \frac{a^{n-k}}{(n-k)!} \frac{b^k}{k!} \right) = \sum_{n=0}^{\infty} \frac{1}{n!} \left(\sum_{k=0}^{n} \binom{n}{k} a^{n-k} b^k \right)$$

$$= \sum_{n=0}^{\infty} \frac{1}{n!} (a+b)^k.$$

If we try to replace in the identity

$$\sum_{k=0}^{n} \binom{n}{k} a^{n-k} b^k = (a+b)^k,$$

i.e. the binomial theorem, a and b by matrices A and B we need that they commute as the case $n=2$ already shows

$$(A+B)^2 = A^2 + AB + BA + B^2 \neq A^2 + 2AB + B^2$$

for $[A, B] \neq 0$. However, if $[A, B] = 0$ then the proof of Proposition I.29.22 works in the new situation and we have

Lemma 18.8. *If $A, B \in M(n; \mathbb{K})$ commute, i.e. $[A, B] = 0$, then we have*

$$e^{A+B} = e^A e^B = e^B e^A. \tag{18.21}$$

Clearly we may replace in (18.21) the term $A + B$ by $(A+B)t$, $t \in \mathbb{R}$, to get

$$e^{(A+B)t} = e^{At} e^{Bt}. \tag{18.22}$$

In Example 18.6 we have discussed a linear system with constant coefficients where the matrix A was a diagonal matrix and in this case solving the corresponding initial value problem was quite easy. Suppose that A is a real symmetric $n \times n$-matrix. It follows that there exists an orthogonal matrix $U \in O(n)$ such that

$$U^{-1} A U = \begin{pmatrix} \lambda_1 & & \mathbf{0} \\ & \ddots & \\ \mathbf{0} & & \lambda_n \end{pmatrix} \tag{18.23}$$

where the eigenvalues λ_j of A are counted according to their multiplicity. Suppose that $v : I \to \mathbb{R}^n$ solves

$$\dot{v}(t) = A v(t) \text{ and } v(t_0) = v_0. \tag{18.24}$$

With $w(t) := U^{-1} v(t)$ we find

$$\dot{w}(t) = U^{-1} \dot{v}(t) = U^{-1} A v(t) = U^{-1} A U w(t)$$

and

$$w(t_0) = U^{-1} v(t_0) = U^{-1} v_0 =: w_0.$$

Thus we are dealing with the diagonal system

$$\dot{w}(t) = \begin{pmatrix} \lambda_1 & & \mathbf{0} \\ & \ddots & \\ \mathbf{0} & & \lambda_n \end{pmatrix} w(t), \ w(t_0) = w_0 \tag{18.25}$$

which has the solution

$$w(t) = \begin{pmatrix} e^{\lambda_1 t} w_{01} \\ \vdots \\ e^{\lambda_n t} w_{0n} \end{pmatrix} \tag{18.26}$$

and therefore the solution to (18.24) is given by

$$v(t) = Uw(t) = U \begin{pmatrix} e^{\lambda_1 t} w_{01} \\ \vdots \\ e^{\lambda_n t} w_{0n} \end{pmatrix} = U \begin{pmatrix} e^{\lambda_1 t} & & \mathbf{0} \\ & \ddots & \\ \mathbf{0} & & e^{\lambda_n t} \end{pmatrix} U^{-1} v_0. \tag{18.27}$$

From (18.27) we can obtain interesting information about the behaviour of $v(t)$ as t tends to infinity. First we stay with a diagonal system. If $\lambda_j > 0$ then for $t \to \infty$ the jth component of the solution tends to $+\infty$. For $\lambda_l < 0$ the lth component of the solution tends to 0 and for $\lambda_k = 0$ the kth component of the solution tends to u_{0k}. In particular we find that if all eigenvalues are less than or equal to 0, then the solution has a limit for $t \to \infty$. Now, in the case that A is symmetric and the solution to $\dot{v}(t) = A(t)v(t)$, $v(0) = v_0$ has the representation (18.27) we find if all eigenvalues are non-positive

$$\lim_{t\to\infty} v(t) = U \left(\lim_{t\to\infty} \begin{pmatrix} e^{\lambda_1 t} & & \mathbf{0} \\ & \ddots & \\ \mathbf{0} & & e^{\lambda_n t} \end{pmatrix} \right) U^{-1} v_0$$

$$= U \begin{pmatrix} \lim_{t\to\infty} e^{\lambda_1 t} & & \mathbf{0} \\ & \ddots & \\ \mathbf{0} & & \lim_{t\to\infty} e^{\lambda_n t} \end{pmatrix} U^{-1} v_0$$

and $\lim_{t\to\infty} e^{\lambda_j t} \begin{cases} 0, & \lambda_j < 0 \\ 1, & \lambda_j = 0. \end{cases}$

Example 18.9. The eigenvalues of the matrix $A = \begin{pmatrix} -\frac{1}{2} & 0 & -\frac{1}{2} \\ 0 & 1 & 0 \\ -\frac{1}{2} & 0 & -\frac{1}{2} \end{pmatrix}$ are $-1, -2$ and

0 and with $U = \begin{pmatrix} \frac{1}{2}\sqrt{2} & 0 & \frac{1}{2}\sqrt{2} \\ 0 & 1 & 0 \\ -\frac{1}{2}\sqrt{2} & 0 & \frac{1}{2}\sqrt{2} \end{pmatrix}$ we find $UAU^{-1} = \begin{pmatrix} -1 & 0 & 0 \\ 0 & -2 & 0 \\ 0 & 0 & 0 \end{pmatrix}$. For a solution of

$\dot{v}(t) = Av(t), \; v(t) = v_0$ it follows that

$$\lim_{t \to \infty} v(t) = \frac{1}{2} \begin{pmatrix} v_{01} + v_{03} \\ 0 \\ v_{01} + v_{03} \end{pmatrix}.$$

This observation raises the question to which extent the general solution of the normal form problem for matrices can be used for studying solutions of linear systems with constant coefficients. Before turning to this question, here is a further observation with far reaching consequences. For $A \in M(n; \mathbb{R})$ the family of matrices $(T_t^A)_{t \geq 0}$, $T_t^A = e^{tA}$, has the **semi-group property**, i.e.

$$T_{t+s}^A = T_t^A \circ T_s^A, \; t, s \geq 0, \; T_0^A = id_n, \tag{18.28}$$

where $id_n \in M(n; \mathbb{R})$ is the unit matrix $(\delta_{kl})_{k,l=1,\dots,n}$. For some $s > 0$ we obtain a solution to

$$\dot{u}(t) = Au(t), \; u(0) = u_0$$

in the interval $[0, s]$ by $T_t^A u_0, \; t \in [0, s]$. Now we may solve for $t > s$

$$\dot{v}(t) = Av(t), \; v(s) = v_s := T_s^A u_0. \tag{18.29}$$

This means that we first solve the equation $\dot{u}(t) = Au(t)$ starting at $t_0 = 0$ with initial condition u_0 and follow the solution until $t_1 = s$. Then we solve the same equation for $t > s$ with initial condition at $t_1 = s$ given by v_s. The solution to the latter problem is for $t > s$ given by $v(t) = T_{t-s}^A v_s$ since

$$\frac{d}{dt} v(t) = \frac{d}{dt}(T_{t-s}^A v_s) = \frac{d}{dt}(e^{(t-s)A} v_s) = A e^{(t-s)A} v_s = A T_{t-s}^A v_s$$

and $v(s) = T_{s-s}^A v_s = v_s$. On the other hand we have for $t > s$ that

$$T_t^A u_0 = T_{t-s}^A \circ T_s^A u_0.$$

Thus, just solving $\dot{u}(t) = Au(t)$ with initial value $u(0) = u_0$ until $s > 0$ and then solving this equation for $t > s$ with the initial condition $v(s) = T_s^A u_0$ is the same as directly solving the equation until $t > s > 0$ under the initial condition $u(0) = u_0$.

We return to the considerations preceding Example 18.7 and we need to recollect basic results on the normal form problem for $(n \times n)$-matrices over $\mathbb{K}$, $\mathbb{K} \in \{\mathbb{R}, \mathbb{C}\}$. We do not give proofs of these results but refer to [31], [52], [61] or [83]. Two matrices $A, B \in M(n; \mathbb{K})$ are called equivalent if

$$B = C^{-1} A C \tag{18.30}$$

for some $C \in GL(n; \mathbb{K})$. The **normal form problem** is to find in the equivalence class of a given matrix a particular "simple" representative. Diagonal matrices are considered as the most simple matrices, thus the first question is whether we can diagonalise a given matrix A, i.e. whether we can find $C \in GL(n; \mathbb{K})$ such that B in (18.30) is a diagonal matrix. For symmetric real matrices and Hermitian matrices over $\mathbb{C}$ such a result holds and we have already often made use of it in our Course. However, in general we cannot diagonalise every matrix. We need to look at the **characteristic polynomial** $\chi(A)$ associated with A defined by

$$\chi(A)(\lambda) = \det(A - \lambda id_n), \ A \in M(n; \mathbb{K}). \tag{18.31}$$

The zeroes of $\chi(A)$ are the **eigenvalues** of A. Of importance is whether or not $\chi(A)$ factorizes into linear factors, i.e. whether we can find eigenvalues $\lambda_1, \dots, \lambda_m$ and integers $\alpha_1, \dots, \alpha_m$ such that

$$\chi(A)(\lambda) = c \prod_{k=1}^{m} (\lambda - \lambda_k)^{\alpha_k}. \tag{18.32}$$

Note that for $\mathbb{K} = \mathbb{C}$ such a factorization is a consequence of the fundamental theorem of algebra, see Theorem III.23.15, but over $\mathbb{R}$ we cannot expect such a result to hold in general. We call α_k the **algebraic multiplicity** of the eigenvalue λ_k. The eigenspace associated with λ_k is a subspace of $\mathbb{K}^n$ and has a dimension $\gamma_k \le \alpha_k$. We call γ_k the **geometric multiplicity** of λ_k. The fundamental result is

Theorem 18.10. *The matrix $A \in M(n; \mathbb{K})$ is diagonalizable if and only if its characteristic polynomial factorizes into linear factors and for every eigenvalue λ_k of A the algebraic multiplicity α_k is equal to the geometric multiplicity γ_k. Furthermore, this is equivalent to the property that $\mathbb{K}^n$ is the direct sum of the eigenspaces corresponding to λ_k, $1 \le k \le m$.*

Example 18.11. Consider the matrix $A = \left(\begin{smallmatrix} 1 & \alpha \\ 0 & 1 \end{smallmatrix}\right)$, $\alpha \neq 0$. The eigenvalue of A is $\lambda = 1$ with algebraic multiplicity 2. The geometric multiplicity of λ is the dimension of the space determined by the linear system $(1 - \lambda)x + \alpha y = 0$ and $(1-\lambda)y = 0$ for $\lambda = 1$, i.e. by the equations $\alpha y = 0$. Hence, for $\alpha \neq 0$ the matrix A is not diagonalizable, while for $\alpha = 0$ the matrix A is of course a diagonal matrix.

This next simplest type of a matrix we may think about is that of an upper (or lower) triangular matrix.

$$B = \begin{pmatrix} \lambda_1 & & & \alpha_{kl} \\ & \lambda_2 & & \\ & & \ddots & \\ \mathbf{0} & & & \lambda_n \end{pmatrix}, \ \alpha_{kl} \in \mathbb{K}.$$

Since $\det(B - \lambda id_n) = \prod_{k=1}^{n}(\lambda - \lambda_k)$ the λ_k's are the eigenvalues of B. It is more convenient to look for block matrices

$$
B = \begin{pmatrix} \boxed{B_1} & & & \mathbf{0} \\ & \boxed{B_2} & & \\ & & \ddots & \\ \mathbf{0} & & & \boxed{B_m} \end{pmatrix}
$$

where $B_k = \begin{pmatrix} \lambda_k & & \alpha_{ij}^{(k)} \\ & \ddots & \\ \mathbf{0} & & \lambda_k \end{pmatrix}$, and it turns out that we can reduce our investigation to the case where each B_k is a Jordan matrix, where we call a matrix F a **Jordan matrix** if it is of form

$$
F = \begin{pmatrix} \lambda & 1 & & & \mathbf{0} \\ & \lambda & 1 & & \\ & & \ddots & \ddots & \\ & & & \lambda & 1 \\ \mathbf{0} & & & & \lambda \end{pmatrix}, \quad \lambda \in \mathbb{K}. \tag{18.33}
$$

Here the central result is

Theorem 18.12. *Let $A \in M(n; \mathbb{K})$. If $\chi(A)$ factorizes into linear factors,*

$$
\chi(A)(\lambda) = c \prod_{k=1}^{m}(\lambda - \lambda_k)^{\alpha_k},
$$

then A is equivalent to a block matrix B. The blocks of B are the $(\alpha_k \times \alpha_k)$-Jordan

301

matrices $\begin{pmatrix} \lambda_k & 1 & & \mathbf{0} \\ & \ddots & \ddots & \\ & & \ddots & 1 \\ \mathbf{0} & & & \lambda_k \end{pmatrix}$. *Up to an enumeration of eigenvalues we have*

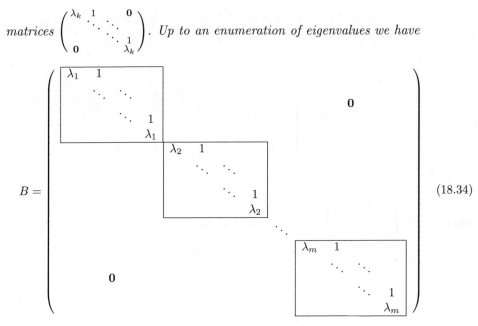

$$(18.34)$$

*which we call the **Jordan normal form** of A.*

In the case where $\chi(A)$ factorizes into linear factors we can find $C \in GL(n; \mathbb{K})$ such that $B = C^{-1}AC$ with B as in (18.34), and now we may return to the problem $\dot{u}(t) = Au(t)$ and consider the simplified equation $\dot{w} = Bw$, $w := C^{-1}u$. However, there are still some algebraic problems we had better resolve first. For $\mathbb{K} = \mathbb{C}$ the characteristic polynomial $\chi(A)$ of every matrix $A \in M(n; \mathbb{C})$ factorizes into linear factors and Theorem 18.12 is applicable. For a matrix $A \in M(n; \mathbb{R})$ the situation is different, but such a matrix corresponds to a linear system of first order differential equations with real constant coefficients, i.e. is of great interest to us. We may consider $A \in M(n; \mathbb{R})$ as an element in $M(n; \mathbb{C})$ and then $\chi(A)$ factorizes into linear factors, but over $\mathbb{C}$, i.e. the zeroes, which means the eigenvalues, might be complex, a fact which is, at the moment, disturbing for our strategy to solve $\dot{u}(t) = Au(t)$.

Still, over $\mathbb{C}$, we can find $B \in M(n; \mathbb{C})$ and $C \in GL(n; \mathbb{C})$ such that for $A \in M(n; \mathbb{R})$ we have

$$C^{-1}AC = B = \begin{pmatrix} \boxed{B_1} & & & \\ & \boxed{B_2} & & \\ & & \ddots & \\ & & & \boxed{B_m} \end{pmatrix}$$

where $B_k = \begin{pmatrix} \lambda_k & 1 & & \mathbf{0} \\ & \ddots & \ddots & \\ & & \ddots & 1 \\ \mathbf{0} & & & \lambda_k \end{pmatrix}$ is a Jordan matrix with $\lambda_k \in \mathbb{C}$. We want to analyse

this situation more carefully. First we note that if $A \in M(n; \mathbb{R})$ then, if $\mu \in \mathbb{C} \setminus \mathbb{R}$ is an eigenvalue of A so is $\bar{\mu}$. Indeed for every polynomial p with real coefficients we can find complex numbers $\lambda_1, \ldots, \lambda_k$ and integers $\alpha_1, \ldots, \alpha_k$ such that

$$p(\lambda) = c \prod_{k=1}^{m} (\lambda - \lambda_k)^{\alpha_k}, \ \lambda \in \mathbb{R}, \ c \in \mathbb{R}.$$

This implies however

$$c \prod_{k=1}^{m} (\lambda - \overline{\lambda_k})^{\alpha_k} = \overline{c \prod_{k=1}^{m} (\lambda - \lambda_k)^{\alpha_k}},$$

and therefore $p(\lambda_k) = 0$ implies $p(\overline{\lambda_k}) = 0$, i.e. if a complex number $\mu \in \mathbb{C} \setminus \mathbb{R}$ is a zero of p then $\bar{\mu}$ is a further zero of p. Thus $A \in M(n; \mathbb{R})$ may have real eigenvalues $\lambda_1, \ldots, \lambda_r$ with some algebraic multiplicities $\alpha_1, \ldots, \alpha_r$, and q pairs of conjugate complex eigenvalues $\mu_1, \overline{\mu_1}, \ldots, \mu_q, \overline{\mu_q}$ with algebraic multiplicities $\beta_1, \ldots, \beta_q$.

For $\lambda \in \mathbb{R}$ and $\mu_l, \overline{\mu_l} \in \mathbb{C} \setminus \mathbb{R}$, $\mu_l = \xi_l + i\eta_l$, $\xi_l, \eta_l \in \mathbb{R}$, we find

$$(\lambda - \mu_l)^{\beta_l} (\lambda - \overline{\mu_l})^{\beta_l} = ((\lambda - \xi_l - i\eta_l)(\lambda - \xi_l + i\eta_l))^{\beta_l}$$
$$= (\lambda^2 - 2\lambda\xi_l + \xi_l^2 + \eta_l^2)^{\beta_l}$$

and therefore, for $A \in M(n; \mathbb{R})$ the characteristic polynomial factorizes over $\mathbb{R}$ according to

$$\chi(A)(\lambda) = c \prod_{k=1}^{r} (\lambda - \lambda_k)^{\alpha_k} \prod_{l=1}^{q} (\lambda^2 - 2\lambda\xi_l + \xi_l^2 + \eta_l^2)^{\beta_l}$$

with $\xi_l, \eta_l \in \mathbb{R}$, $\eta_l \neq 0$. The Jordan blocks corresponding to λ_k, $1 \leq k \leq r$ causes no problem for our considerations, they are just real $(\alpha_k) \times (\alpha_k)$-matrices $\begin{pmatrix} \lambda_k & 1 & & \mathbf{0} \\ & \ddots & \ddots & \\ & & \ddots & 1 \\ \mathbf{0} & & & \lambda_k \end{pmatrix}$. The Jordan blocks corresponding to μ_l or $\overline{\mu_l}$ cause problems since they are of the type

$$\begin{pmatrix} \mu_l & 1 & & \mathbf{0} \\ & \ddots & \ddots & \\ & & \ddots & 1 \\ \mathbf{0} & & & \mu_l \end{pmatrix}$$

303

with $\mu_l \in \mathbb{C} \setminus \mathbb{R}$.

Let us try to understand the role of a non-real eigenvalue for the system $\dot{u}(t) = Au(t)$. We have discussed the system $\dot{u}(t) = \begin{pmatrix} 0 & 1 \\ -1 & 0 \end{pmatrix} u(t)$ in detail at the end of Chapter 17 and its solutions were real linear combinations of the functions $\cos$ and $\sin$. If the matrix $\begin{pmatrix} 0 & 1 \\ -1 & 0 \end{pmatrix}$ is not diagonalizable over $\mathbb{R}$, the roots of the characteristic polynomial $\det \begin{pmatrix} -\lambda & 1 \\ -1 & -\lambda \end{pmatrix} = \lambda^2 + 1$ are $\pm i$.

Example 18.13. We want to discuss the system

$$\dot{w}(t) = \begin{pmatrix} 1 & -5 \\ 2 & 3 \end{pmatrix} w(t), \quad w(0) = \begin{pmatrix} -2 \\ 4 \end{pmatrix}, \tag{18.35}$$

along the lines we have discussed systems with a symmetric real-valued matrix. The eigenvalues of $A = \begin{pmatrix} 1 & -5 \\ 2 & 3 \end{pmatrix}$ are $\lambda_{1,2} = 2 \mp 3i$ and considered as a matrix of $\mathbb{C}$ we can diagonalize A with the help of $C = \begin{pmatrix} -1-3i & -1+3i \\ 2 & 2 \end{pmatrix}$, $C^{-1} = \frac{i}{12} \begin{pmatrix} 2 & 1-3i \\ -2 & -1-3i \end{pmatrix}$ to obtain

$$\begin{pmatrix} 2-3i & 0 \\ 0 & 2+3i \end{pmatrix} = B = C^{-1}AC.$$

Introducing $z(t) = \begin{pmatrix} z_1(t) \\ z_2(t) \end{pmatrix} := C^{-1} \begin{pmatrix} w_1(t) \\ w_2(t) \end{pmatrix}$ we find for a solution w of (18.35)

$$\dot{z}(t) = C^{-1}\dot{w}(t) = C^{-1}Aw(t) = C^{-1}ACz(t)$$

or

$$\begin{pmatrix} \dot{z}_1(t) \\ \dot{z}_2(t) \end{pmatrix} = \begin{pmatrix} 2-3i & 0 \\ 0 & 2+3i \end{pmatrix} \begin{pmatrix} z_1(t) \\ z_2(t) \end{pmatrix}, \quad \begin{pmatrix} z_1(0) \\ z_2(0) \end{pmatrix} = \begin{pmatrix} 1 \\ 1 \end{pmatrix}. \tag{18.36}$$

which gives the solution

$$z_1(t) = e^{2t}(\cos 3t - i \sin 3t) \text{ and } z_2(t) = e^{2t}(\cos 3t + i \sin 3t). \tag{18.37}$$

From (18.36) we obtain now

$$\begin{pmatrix} w_1(t) \\ w_2(t) \end{pmatrix} = C \begin{pmatrix} z_1(t) \\ z_2(t) \end{pmatrix} = e^{2t} \begin{pmatrix} -1-3i & -1+3i \\ 2 & 2 \end{pmatrix} \begin{pmatrix} \cos 3t - i \sin 3t \\ \cos 3t + i \sin 3t \end{pmatrix}$$
$$= \begin{pmatrix} -e^{2t}(2\cos 3t + 6 \sin 3t) \\ 4e^{2t} \cos 3t \end{pmatrix}.$$

The interesting fact is that passing to the complexified systems, working then in $\mathbb{C}$ and transforming after solving the problem in $\mathbb{C}$ to the original problem, we obtain a real-valued solution. This is not accidental as we will see soon.

Finally, as preparation, we have a look at the matrix $A = \begin{pmatrix} a & b \\ -b & a \end{pmatrix}$, $a, b \in \mathbb{R}$. The characteristic polynomial of A is $\chi(A)(\lambda) = \lambda^2 - 2a\lambda + a^2 + b^2$, which gives the eigenvalues $\lambda_{1,2} = a \pm i|b|$. Further we note that

$$\begin{pmatrix} a & b \\ -b & a \end{pmatrix} = \begin{pmatrix} a & 0 \\ 0 & a \end{pmatrix} + \begin{pmatrix} 0 & b \\ -b & 0 \end{pmatrix} \text{ and } \left[\begin{pmatrix} a & 0 \\ 0 & a \end{pmatrix}, \begin{pmatrix} 0 & b \\ -b & 0 \end{pmatrix} \right] = 0.$$

Therefore we find

$$e^{\begin{pmatrix} a & b \\ -b & a \end{pmatrix} t} = e^{\begin{pmatrix} a & 0 \\ 0 & a \end{pmatrix} t} e^{\begin{pmatrix} 0 & b \\ -b & 0 \end{pmatrix} t} = \begin{pmatrix} e^{at} & 0 \\ 0 & e^{at} \end{pmatrix} \begin{pmatrix} \cos|b|t & \sin|b|t \\ -\sin|b|t & \cos|b|t \end{pmatrix}$$

where the calcuation from the end of Chapter 17 is used. Assuming for simplicity $b > 0$ we find

$$e^{\begin{pmatrix} a & b \\ -b & a \end{pmatrix} t} = \begin{pmatrix} e^{at}\cos bt & e^{at}\sin bt \\ -e^{at}\sin bt & e^{at}\cos bt \end{pmatrix}$$

and now we can easily solve $\dot{w} = Aw(t)$, $w(0) = w_0$. We also can understand the role which the real and the imaginary part of a complex eigenvalue have for a solution to the system $\dot{w}(t) = Aw(t)$. The real part determines the growth or decay of the amplitude whereas the imaginary part determines the frequency of oscillations of the solution.

The following result, a proof of which can be found in [61], clarifies the algebraic background.

Theorem 18.14. *Let $A \in M(2; \mathbb{R})$ have the eigenvalues $\mu = a + ib$ and $\bar{\mu} = a - ib$, $b \neq 0$. Then we can find a basis in $\mathbb{R}^2$ such that*

$$A = \begin{pmatrix} a & b \\ -b & a \end{pmatrix}. \tag{18.38}$$

Now let $A \in M(n; \mathbb{R})$ have the eigenvalues $\lambda_1, \ldots, \lambda_r \in \mathbb{R}$, and $\mu_1, \bar{\mu_1}, \ldots, \mu_q$, $\bar{\mu_q} \in \mathbb{C} \setminus \mathbb{R}$ and suppose that A is diagonalizable over $\mathbb{C}$. Using that this leads to a decomposition of $\mathbb{R}^n$ into a direct sum of invariant subspaces, one can prove that now A is equivalent to a matrix of the form

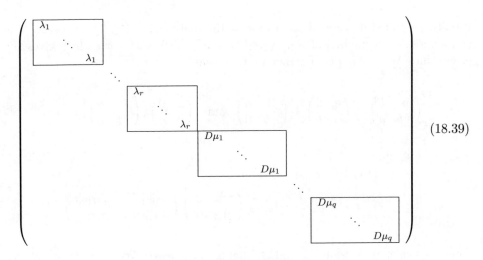

$$(18.39)$$

where for λ_j, $1 \leq j \leq r$, the corresponding block matrix is an $(\alpha_j \times \alpha_j)$-matrix where α_j denotes the multiplicity of λ_j, and $D\mu_k = \begin{pmatrix} \mathrm{Re}(\mu_k) & \mathrm{Im}(\mu_k) \\ -\mathrm{Im}(\mu_k) & \mathrm{Re}(\mu_k) \end{pmatrix}$ and the corresponding block matrix is a $(2\beta_k) \times (2\beta_k)$-matrix with β_k being the multiplicity of μ_k, and hence of $\overline{\mu_k}$.

Example 18.15. Consider the initial value problem

$$\dot{u}(t) = Au(t), \ u(0) = u_0, \ A = \begin{pmatrix} 4 & -9 & 0 & 0 \\ -1 & 4 & 0 & 0 \\ 1 & -2 & 1 & -5 \\ 2 & -4 & 2 & 3 \end{pmatrix}.$$

The characteristic polynomial of A is given by

$$\chi(A)(\lambda) = \det(A - \lambda id_4)$$
$$= (4 - \lambda)^2 \det \begin{pmatrix} 1 - \lambda & -5 \\ 2 & 3 - \lambda \end{pmatrix} - 9 \det \begin{pmatrix} 1 - \lambda & -5 \\ 2 & 3 - \lambda \end{pmatrix}$$
$$= (7 - \lambda)(\lambda - 1)(2 - 3i - \lambda)(2 + 3i - \lambda)$$

which leads to the transformed problem

$$\dot{w}(t) = Bw(t), \ w(0) = C^{-1}u_0$$

where $B = C^{-1}AC = \begin{pmatrix} 7 & & & \mathbf{0} \\ & 1 & & \\ & & 2-3i & \\ \mathbf{0} & & & 2+3i \end{pmatrix}$. This yields

$$w(t) = \begin{pmatrix} e^{7t} & 0 & 0 & 0 \\ 0 & e^t & 0 & 0 \\ 0 & 0 & e^{2t}\cos 3t & e^{2t}\sin 3t \\ 0 & 0 & -e^{2t}\sin 3t & e^{2t}\cos 3t \end{pmatrix} C^{-1} u_0$$

and we leave it to the reader to determine C which is a standard problem in linear algebra.

Eventually we state the result of Theorem 18.12 for a real matrix

Theorem 18.16. *Let $A \in M(n; \mathbb{R})$ have the real eigenvalues $\lambda_1, \ldots, \lambda_r$ with algebraic multiplicities $\alpha_1, \ldots, \alpha_r$ and the complex eigenvalues $\mu_1, \overline{\mu_1}, \ldots, \mu_q, \overline{\mu_q}$ with algebraic multiplicities $\beta_1, \ldots, \beta_q$, i.e. $\sum_{k=1}^r \alpha_k + 2\sum_{j=1}^q \beta_j = n$. Then the matrix is equivalent to a matrix of the structure*

$$B = \begin{pmatrix} \boxed{B_1} & & & & & & & \mathbf{0} \\ & \boxed{B_2} & & & & & & \\ & & \ddots & & & & & \\ & & & \boxed{B_r} & & & & \\ & & & & \boxed{F_1} & & & \\ & & & & & \boxed{F_2} & & \\ & & & & & & \ddots & \\ \mathbf{0} & & & & & & & \boxed{F_q} \end{pmatrix}. \tag{18.40}$$

Here B_k is the $(\alpha_k \times \alpha_k)$-block matrix given by

$$B_k = \begin{pmatrix} \lambda_k & 1 & & \\ & \ddots & \ddots & \\ & & \ddots & 1 \\ & & & \lambda_k \end{pmatrix}$$

and F_l is the $(2\beta_l \times 2\beta_l)$-block matrix given by

$$F_l = \begin{pmatrix} D\mu_l & id_2 & & & \mathbf{0} \\ & D\mu_l & id_2 & & \\ & & \ddots & \ddots & \\ & & & \ddots & id_2 \\ \mathbf{0} & & & & D\mu_l \end{pmatrix}$$

where $id_2 = \begin{pmatrix} 1 & 0 \\ 0 & 1 \end{pmatrix}$ and as before $D\mu_l = \begin{pmatrix} \mathrm{Re}(\mu_k) & \mathrm{Im}(\mu_k) \\ -\mathrm{Im}(\mu_k) & \mathrm{Re}(\mu_k) \end{pmatrix}$.

In principle, Theorem 18.16 enables us to solve any linear system $\dot{w}(t) = Aw(t)$, $w(0) = w_0$, for $A \in M(n; \mathbb{R})$. We first have to determine the Jordan normal form (18.40) of A as well as the matrix C and its inverse transforming A to (18.40). Next we write (B_k) and (F_l) for the $n \times n$-matrices having only the blocks B_k and F_l as non-trivial entries so that

$$B = \sum_{k=1}^{r}(B_k) + \sum_{l=1}^{q}(F_l)$$

and we observe that $[(B_k),(B_j)] = 0$ for $k \neq j$, $[(F_l),(F_m)] = 0$ for $l \neq m$ and $[(B_k),(F_l)] = 0$ for all k and l. This implies that

$$e^{Bt} = e^{(B_1)t}e^{(B_2)t}\ldots e^{(B_r)t}e^{(F_1)t}e^{(F_2)t}\ldots e^{(F_q)t} \tag{18.41}$$

and it remains to understand $e^{B_k t}$ and $e^{F_l t}$. For $\alpha_k = 1$ or $\beta_l = 1$ this is already done.

First we look at a matrix $B = \begin{pmatrix} \lambda & 1 & & \mathbf{0} \\ & \ddots & \ddots & \\ & & & 1 \\ \mathbf{0} & & & \lambda \end{pmatrix} \in M(m; \mathbb{R})$. We split B according to

$B = \lambda id_m + N$ where $N = \begin{pmatrix} 0 & 1 & & \mathbf{0} \\ & \ddots & \ddots & \\ & & & 1 \\ \mathbf{0} & & & 0 \end{pmatrix} \in M(m; \mathbb{R})$. Since $[\lambda id_n, N] = 0$ we find

$$e^{Bt} = e^{\lambda id_m t}e^{Nt}. \tag{18.42}$$

The first factor is just the matrix $\begin{pmatrix} e^{\lambda t} & & \mathbf{0} \\ & \ddots & \\ \mathbf{0} & & e^{\lambda t} \end{pmatrix}$. The matrix N however is **nilpotent of degree** m, i.e. $N^m = 0 \in M(m; \mathbb{R})$ and therefore we know already that

$$e^{Nt} = \sum_{k=0}^{m} \frac{(tN)^k}{k!} \tag{18.43}$$

and as we will see in Problem 12 we have

$$e^{Nt} = \begin{pmatrix} 1 & t & \frac{t^2}{2} & \cdots & \cdots & \frac{t^{m-1}}{(m-1)!} \\ 0 & 1 & t & \frac{t^2}{2} & \cdots & \frac{t^{m-2}}{(m-2)!} \\ \vdots & & & & & \vdots \\ 0 & 0 & \cdots & \cdots & 1 & t \\ 0 & 0 & \cdots & \cdots & 0 & 1 \end{pmatrix}. \tag{18.44}$$

Example 18.17. We consider the initial value problem

$$\dot{u}(t) = Au(t), \ u(0) = u_0, \ A = \begin{pmatrix} 2 & -1 & 1 \\ -1 & 2 & -1 \\ 2 & 2 & 3 \end{pmatrix}.$$

The characteristic polynomial of A is $\chi(A)(\lambda) = (1 - \lambda)(3 - \lambda)^2$ and the rank of $A - 3id_3$ is equal to 2, i.e. the geometric multiplicity of the eigenvalue 3 is 1 but its algebraic multiplicity is 2. Thus we cannot diagonalize A. Its Jordan normal form is given by $B = \begin{pmatrix} 1 & 0 & 0 \\ 0 & 3 & 1 \\ 0 & 0 & 3 \end{pmatrix}$. If $B = C^{-1}AC$ we find for the solution of the transformed problem $\dot{w}(t) = Bw(t), \ w(0) = Cu_0$, the solution

$$w(t) = \begin{pmatrix} e^t & 0 & 0 \\ 0 & e^{3t} & te^{3t} \\ 0 & 0 & e^{3t} \end{pmatrix} Cu_0.$$

The matrix C is given by $C = \begin{pmatrix} -1 & 1 & 0 \\ 0 & -1 & 0 \\ 1 & 0 & 1 \end{pmatrix}$.

It remains to study the case of a block matrix

$$F = \begin{pmatrix} D\mu & id_2 & & & \mathbf{0} \\ & D\mu & \ddots & & \\ & & \ddots & \ddots & \\ & & & D\mu & id_2 \\ \mathbf{0} & & & & D\mu \end{pmatrix}$$

with $\mu = a + ib, \ b \neq 0$, and F is assumed to be a $2m \times 2m$-matrix. We denote by D the matrix

$$D = \begin{pmatrix} D\mu & & \mathbf{0} \\ & \ddots & \\ \mathbf{0} & & D\mu \end{pmatrix}$$

and by $N(2, \mu)$ the matrix

$$N(2, \mu) = \begin{pmatrix} 0 & id_2 & & \mathbf{0} \\ & \ddots & \ddots & \\ & & \ddots & id_2 \\ \mathbf{0} & & & 0 \end{pmatrix}.$$

Both matrices are built up by blocks of 2×2-matrices. Since $[D, N(2, \mu)] = 0$ we find

$$e^{tF} = e^{tD}e^{tN(2,\mu)},$$

where $N(2, \mu)$ is nilpotent of degree m in the sense that $N(2, \mu)^k = 0$.

From our previous considerations we deduce

$$e^{tD} = \begin{pmatrix} \begin{array}{cc} e^{at} \cos |b|t & e^{at} \sin |b|t \\ -e^{at} \sin |b|t & e^{at} \cos |b|t \end{array} & & \\ & \begin{array}{cc} e^{at} \cos |b|t & e^{at} \sin |b|t \\ -e^{at} \sin |b|t & e^{at} \cos |b|t \end{array} & \\ & & \ddots \end{pmatrix}$$

with m blocks $\begin{pmatrix} e^{at} \cos |b|t & e^{at} \sin |b|t \\ -e^{at} \sin |b|t & e^{at} \cos |b|t \end{pmatrix}$ and further

$$e^{tN(2,\mu)} = \sum_{k=0}^{m} \frac{t^k}{k!} N(2, \mu)^k$$

where $N(2, \mu)^k$ is a $2m \times 2m$-matrix.

Now we can give a solution to the general problem

$$\dot{u}(t) = Au(t), \ u(0) = u_0.$$

We denote by $\lambda_1, \ldots, \lambda_r$ the real eigenvalues of A with multiplicity α_k, $1 \le k \le r$, and by $\mu_1, \overline{\mu_1}, \ldots, \mu_q, \overline{\mu_q}$ the eigenvalues in $\mathbb{C} \backslash \mathbb{R}$ with multiplicity β_l, $1 \le l \le q$. The Jordan block corresponding to λ_k we denote by B_k and the block corresponding to μ_l we denote by F_l, i.e.

$$B_k = \begin{pmatrix} \lambda_k & 1 & & \mathbf{0} \\ & \ddots & \ddots & \\ & & \ddots & 1 \\ \mathbf{0} & & & \lambda_k \end{pmatrix} \in M(\alpha_k; \mathbb{R})$$

and

$$F_l = \begin{pmatrix} D\mu_l & id_2 & & \mathbf{0} \\ & \ddots & \ddots & \\ & & \ddots & id_2 \\ \mathbf{0} & & & D\mu_l \end{pmatrix} \in M(2\beta_l; \mathbb{R}).$$

With (B_k) and (F_l) we denote the $n \times n$-matrix having as only non-zero entries the entries of the block B_k and F_l respectively, in their natural position, e.g.

$$(B_k) = \left(\begin{array}{c|c|c} O_1 & 0 & 0 \\ \hline 0 & B_k & 0 \\ \hline 0 & 0 & O_2 \end{array} \right)$$

310

where O_1 is the $(\alpha_1 + \cdots + \alpha_{k-1}) \times (\alpha_1 + \cdots + \alpha_{k-1})$-matrix with all entries equal to zero and O_2 is the $(\alpha_{k+1} + \cdots + \alpha_r + 2\beta_1 + \cdots + 2\beta_q) \times (\alpha_{k+1} + \cdots + \alpha_r + 2\beta_1 + \cdots + 2\beta_q)$-matrix with all entries equal to zero. With these notations, we have

Theorem 18.18. *Let $A \in M(n; \mathbb{R})$ and $C \in GL(n; \mathbb{R})$ be the matrix transforming A onto its Jordan normal form, i.e.*

$$B = C^{-1}AC = \begin{pmatrix} B_1 & & & & & & \mathbf{0} \\ & \ddots & & & & & \\ & & B_r & & & & \\ & & & F_1 & & & \\ & & & & \ddots & & \\ \mathbf{0} & & & & & & F_q \end{pmatrix}.$$

The initial value problem $\dot{u}(t) = Au(t)$, $u(0) = u_0$, can be transformed by $w := C^{-1}u$ to

$$\dot{w}(t) = Bw(t), \quad w(0) = C^{-1}u(0) \tag{18.45}$$

and (18.45) is solved by $w(t) = e^{tB}C^{-1}u(0)$. The matrix e^{tB} can be calculated as

$$e^{tB} = e^{t(B_1)} \dots e^{t(B_r)} e^{t(F_1)} \dots e^{t(F_q)} \tag{18.46}$$

with

$$e^{t(B_k)} = (e^{\lambda_k t} id_{\alpha_k})(L_{\alpha_k})$$

where L_{α_k} is equal to the right hand side of (18.44) with m replaced by α_k, and with

$$e^{t(F_l)} = e^{t(\tilde{D}_{\mu_l})} e^{tN(2,\mu_l)}$$

where

$$\tilde{D}_{\mu_l} = \begin{pmatrix} D\mu_l & & \mathbf{0} \\ & \ddots & \\ \mathbf{0} & & D\mu_l \end{pmatrix} \in M(2\beta_l, \mathbb{R})$$

and

$$N(2, \mu_l) = \begin{pmatrix} 0 & id_2 & & \mathbf{0} \\ & \ddots & \ddots & \\ & & \ddots & id_2 \\ \mathbf{0} & & & 0 \end{pmatrix} \in M(2\beta_l, \mathbb{R}).$$

Remark 18.19. Applications of Theorem 18.18 require to determine the matrix C and C^{-1}, and of course the eigenvalues of A. This is nowadays best done by using computational software. For this reason, we do not include some problems related to Theorem 18.18.

Problems

1. a) Find

$$\frac{d}{dt} \begin{pmatrix} \cos t & t^3 & e^t \\ \frac{1}{1+t^2} & 2t & \sin 5t \\ e^{t^2} & \cosh t & 3 \end{pmatrix}.$$

   b) Evaluate

$$\int_0^t \begin{pmatrix} \cos s & s^2 \\ e^s & 4s \end{pmatrix} ds.$$

   c) For $n \times n$-matrices $A(t)$ and $B(t)$ with differentiable elements find

$$\frac{d}{dt} A(t)B(t).$$

2. a) Let $A \in M(n; \mathbb{R})$ and $\| \cdot \|_\alpha, \| \cdot \|_\beta$ be two norms on $\mathbb{R}^n$. Prove that

$$\inf\{\kappa \geq 0 | \|Ax\|_\beta \leq \kappa \|x\|_\alpha\} = \sup\left\{ \frac{\|Ax\|_\beta}{\|x\|_\alpha} \Big| x \in \mathbb{R}^n \setminus \{0\}\right\}.$$

   b) Use part a) to show that $\|A\|_{\alpha,\beta} := \inf\{\kappa \geq 0 | \|Ax\|_\beta \leq \kappa \|x\|_\alpha\}$ is a norm on $M(n; \mathbb{R})$ satisfying $\|Ax\|_\beta \leq \|A\|_{\alpha,\beta} \|x\|_\alpha$.

3. a) Let $\| \cdot \|$ be a metric norm on $M(n; \mathbb{R})$. Find $A, B \in M(n; \mathbb{R})$ such that $\|A\| > 0$ and $\|B\| > 0$, but $\|AB\| = 0$.

   b) Let $\| \cdot \|$ be a matrix norm on $M(n; \mathbb{R})$ and $A \in M(n, \mathbb{R})$ such that $\|A\| < 1$. Prove that $id_n - A$ is invertible with inverse $\sum_{k=0}^\infty A^k$.

4. Let $\sum_{k=0}^\infty c_k s^k$, $c_k \in \mathbb{R}$, be a power in $\mathbb{R}$ with positive radius of convergence. Let $A \in M(n; \mathbb{R})$ such that $A^N = 0$. Prove that $s \mapsto \sum_{k=0}^\infty c_k (sA)^k$ is a matrix-valued polynomial in s of degree less than N.

5. Prove that for every $A \in M(n; \mathbb{R})$ we have

$$\frac{d}{dt} \cos At = -A \sin At \quad \text{and} \quad \frac{d}{dt} \sin At = A \cos At.$$

6. a) For $A \in M(n; \mathbb{R})$ prove that $e^{iA} = \cos A + i \sin A$.

   b) For $A, B \in M(n; \mathbb{R})$ such that $[A, B] = 0$ show that

$$\cos(A + B) = \cos A \cos B - \sin A \sin B.$$

7. By finding $e^{\begin{pmatrix} 0 & 1 \\ 1 & 0 \end{pmatrix}t}$ find a solution to the initial value problem

$$\begin{pmatrix} u'(t) \\ v'(t) \end{pmatrix} = \begin{pmatrix} 0 & 1 \\ 1 & 0 \end{pmatrix} \begin{pmatrix} u(t) \\ v(t) \end{pmatrix}, \quad \begin{pmatrix} u(0) \\ v(0) \end{pmatrix} = \begin{pmatrix} 1 \\ 0 \end{pmatrix}$$

8. Find a solution to the initial value problem

$$\dot{y}_1 = y_1 - 3y_2 + 3y_3$$
$$\dot{y}_2 = 3y_1 - 5y_2 + 3y_3$$
$$\dot{y}_3 = 6y_1 - 6y_2 + 4y_3$$

with $y_1(0) = 1, y_2(0) = 2$ and $y_3(0) = 3$.

9. Let $A \in M(n; \mathbb{R})$ be a symmetric positive definite matrix with largest eigen-value λ_1. Prove, in this case, that for $\lambda > \lambda_1$

$$\int_0^\infty e^{-\lambda t} e^{tA} \, dt := \lim_{N \to \infty} \int_0^N e^{-\lambda t} e^{tA} \, dt$$

exists and find its value.

10. Find the solution of the initial value problem

$$\dot{y}_1 = 9y_1 - 6y_2 - 2y_3$$
$$\dot{y}_2 = 18y_1 - 12y_2 - 3y_3$$
$$\dot{y}_3 = 18y_1 - 9y_2 - 6y_3$$

with $y_1(0) = 1, y_2(0) = 1$ and $y_3(0) = 0$. You may take for granted that $B = C^{-1}AC$ where A is the matrix of the system under consideration,

$$C = \begin{pmatrix} -2 & 0 & 1 \\ -3 & 0 & 0 \\ -3 & 1 & 6 \end{pmatrix} \text{ and } B = \begin{pmatrix} -3 & 1 & 0 \\ 0 & -3 & 0 \\ 0 & 0 & -3 \end{pmatrix}.$$

11. Prove that the matrix $A = \begin{pmatrix} 0 & 1 & 0 & 1 \\ 0 & 0 & 1 & 1 \\ 0 & 0 & 0 & 0 \\ 0 & 0 & 0 & 0 \end{pmatrix}$ is nilpotent and find $\sin At$ and

$\cosh At, t \in \mathbb{R}$.

12. Verify formula (18.44).

19 Linear Systems of First Order. Variable Coefficients

We now want to study linear systems with variable coefficients

$$\begin{cases} \dot{u}_1(t) = a_{11}(t)u_1(t) + \cdots + a_{1n}(t)u_n(t) + f_1(t) \\ \quad\vdots \\ \dot{u}_n(t) = a_{n1}(t)u_1(t) + \cdots + a_{nn}(t)u_n(t) + f_n(t). \end{cases} \tag{19.1}$$

The functions a_{kl} and f_j need to have a common domain which we assume to be an interval $I \subset \mathbb{R}$ with endpoints $a < b$. We prefer to rewrite (19.1) in matrix form

$$\dot{u}(t) = A(t)u(t) + f(t) \tag{19.2}$$

where $A : I \to M(n; \mathbb{R})$ and $f : I \to \mathbb{R}^n$ are given and $u : I \to \mathbb{R}^n$ is a sought function. As before we add to (19.2) an initial value, i.e. for some $t_0 \in I$ we require

$$u(t_0) = u_0 \in \mathbb{R}^n. \tag{19.3}$$

We will discuss two approaches to (19.2) and (19.3). The first one will use our general existence result from Chapter 17, while the second one will investigate the question to which extent we can transfer the methods of Chapter 18 to systems with variable coefficients, which eventually will bring us back to the first approach.

We want to make use of Theorem 17.1 and for this we need to fix some norms. Let $I' \subset \mathbb{R}$ be a compact interval. As before we use on $C^k(I'; \mathbb{R}^n)$ the norm

$$\|u\|_{k,\infty} := \max_{0 \leq l \leq k} \|u^{(l)}\|_{\infty, I'}$$

where $\|u^{(l)}\|_{\infty, I'} = \max_{1 \leq j \leq n} \sup_{t \in I'} |u_j^{(l)}(t)|$. By $\|\cdot\|$ we denote the matrix norm on $M(n; \mathbb{R})$ with respect to the norm $\|x\|_\infty = \max_{1 \leq j \leq n} |x_j|$ and for $A : I' \to M(n; \mathbb{R})$ we set $\|A\|_\infty = \|A\|_{\infty, I'} = \sup_{t \in I'} \|A(t)\|$. Now we can prove as a consequence of Theorem 17.1

Proposition 19.1. *Let $A : I \to M(n; \mathbb{R})$ and $f : I \to \mathbb{R}^n$ be continuous. Suppose that $I' \subset I$ is compact and $t_0 \in I'$. Then the initial value problem $\dot{u}(t) = A(t)u(t)$, $t \in I'$, and $u(0) = u_0 \in \mathbb{R}^n$ has on I' a unique solution.*

Proof. On the compact interval I' the continuous functions $t \mapsto \|A(t)\|$ and $t \mapsto \max_{1 \leq j \leq n} |f_j(t)|$ are bounded, i.e. $\|A\|_{\infty, I'} \leq L < \infty$ and $\|f\|_{\infty, I'} \leq K < \infty$. This implies for $R(t, u) := A(t)u + f$ and $t \in I'$ that

$$\|R(t, u(t)) - R(t, v(t))\|_\infty = \|A(t)u(t) + f(t) - A(t)v(t) - f(t)\|_\infty$$
$$\leq \|A(t)\|\|u(t) - v(t)\|_\infty$$
$$\leq L\|u - v\|_{\infty, I'},$$

hence R satisfies (17.17) on I' and the result follows. $\qquad\square$

Note that by assuming A and f to be continuous and bounded on I we obtain a unique solution to (19.2) and (19.3) on I, and in the following these will be our assumptions. Our first aim is to study in more detail the homogeneous system

$$\dot{u}(t) = A(t)u(t). \tag{19.4}$$

Theorem 19.2. *Let $A : I \to M(n; \mathbb{R})$ be continuous and bounded. The set of all solutions to (19.4) is an n-dimensional subspace of $C(I; \mathbb{R}^n)$. Furthermore, if we denote the solution to $\dot{u}(t) = A(t)u(t)$, $u(t_0) = u_0$, at $t \in I$ by $u(t; t_0, u)$, then the mapping*

$$u_0 \mapsto u(t; t_0, u_0) \tag{19.5}$$

is a vector space isomorphism from $\mathbb{R}^n$ to the solution space.

Proof. The existence of a solution is already proved. Once we have shown that by (19.5) a vector space isomorphism is given, the theorem is proved. We first show that $u_0 \mapsto u(t; t_0, u_0)$ is linear. By definition $u(t; t_0, \lambda u_0 + \mu v_0)$, $\lambda, \mu \in \mathbb{R}$, is the unique solution to

$$\dot{u}(t) = A(t)u(t), \quad u(t_0) = \lambda u_0 + \mu v_0.$$

Further let $u(t; t_0, u_0)$ and $u(t; t_0, v_0)$ be the unique solution to $\dot{u}(t) = A(t)u(t)$, $u(t_0) = u_0$, and $\dot{u}(t) = A(t)u(t)$, $u(t_0) = v_0$. It follows that

$$\lambda \dot{u}(t; t_0, u_0) + \mu \dot{u}(t; t_0, v_0) = A(t)(\lambda u(t; t_0, u_0) + \mu u(t; t_0, v_0))$$

and

$$\lambda u(t_0; t_0, u_0) + \mu u(t_0; t_0, v_0) = \lambda u_0 + \mu v_0$$

which yields by the uniqueness result

$$u(t; t_0, \lambda u_0 + \mu v_0) = \lambda u(t; t_0, u_0) + \mu u(t; t_0, v_0), \tag{19.6}$$

hence $u_0 \mapsto u(t; t_0, u_0)$ is a linear mapping. The injectivity of this mapping is trivial since $u_0 \neq 0$ implies that $t \mapsto u(t; t_0, u_0)$ is not the zero-function. Finally, given $u_0 \in \mathbb{R}^n$ we can find a solution to $\dot{u}(t) = A(t)u(t)$, $u(t_0) = u_0$, implying the surjectivity of the mapping. Thus we have proved that (19.5) defines a bijective mapping and the solution space is an n-dimensional subspace of $C(I; \mathbb{R}^n)$. $\qquad\square$

Corollary 19.3. A. *If $u_{(k)} : I \to \mathbb{R}^n$, $1 \leq k \leq m$, solve (19.4) then $u = \sum_{j=1}^m \lambda_j u_{(j)}$, $\lambda_j \in \mathbb{R}$, is a further solution.*
B. *If $u : I \to \mathbb{R}^n$ solves (19.4) and if for some $t_0 \in I$ we have $u(t_0) = 0$ then $u(t) = 0$ for all $t \in I$.*

We need to clarify the notion of linear dependence for solutions to (19.4).

Definition 19.4. *We call m solutions $u_{(1)}, \dots, u_{(m)}$ to (19.4) **linearly dependent solutions** if there exist constants $\lambda_1, \dots, \lambda_m \in \mathbb{R}$ not all zero such that*

$$\lambda_1 u_{(1)}(t) + \cdots + \lambda_m u_{(m)}(t) = 0 \tag{19.7}$$

*holds for one, hence by Corollary 19.3.B, for all $t \in I$. If $u_{(1)}, \dots, u_{(m)}$ are not linearly dependent, then we call them **linearly independent solutions**.*

Since $C(I; \mathbb{R}^n)$ is a vector space the notion of linear dependence and linear independence are well defined as a relation of functions. Due to Corollary 19.3.B we can reduce this for solutions to (19.4) to a relation of vectors in $\mathbb{R}^n$.

Corollary 19.5. A. *If $A(t) \in M(n; \mathbb{R})$ and $m > n$ then m solutions to (19.4) must be linearly dependent.*
B. *The system (19.4) must have n linearly independent solutions.*

From Corollary 19.5 we deduce that the following definition makes sense:

Definition 19.6. *A set $\{w_{(1)}, \dots, w_{(n)}\}$ of n linearly independent solutions to (19.4) is called a **fundamental system** to (19.4) or for $A(t)$.*

Given n solutions $u_{(1)}, \dots, u_{(n)}$ to (19.4) we can form the matrix

$$U(t) = \begin{pmatrix} u_{(1),1}(t) & \cdots & u_{(n),1}(t) \\ \vdots & & \vdots \\ u_{(1),n}(t) & \cdots & u_{(n),n}(t) \end{pmatrix} \tag{19.8}$$

where $u_{(k),j}$ denotes the jth component of the solution $u_{(k)}$. The n equations in $\mathbb{R}^n$

$$\dot{u}_{(j)}(t) = A(t) u_{(j)}(t), \tag{19.9}$$

i.e.

$$\dot{u}_{(j),l}(t) = \sum_{k=1}^n a_{lk}(t) u_{(j),k}(t), \ 1 \leq j, l \leq n, \tag{19.10}$$

can be combined into one matrix differential equation

$$\dot{U}(t) = A(t) U(t) \tag{19.11}$$

or

$$\begin{pmatrix} \dot{u}_{(1),1}(t) & \cdots & \dot{u}_{(n),1}(t) \\ \vdots & & \vdots \\ \dot{u}_{(1),n}(t) & \cdots & \dot{u}_{(n),n}(t) \end{pmatrix} = \begin{pmatrix} a_{11}(t) & \cdots & a_{1n}(t) \\ \vdots & & \vdots \\ a_{n1}(t) & \cdots & a_{nn}(t) \end{pmatrix} \begin{pmatrix} u_{(1),1}(t) & \cdots & u_{(n),1}(t) \\ \vdots & & \vdots \\ u_{(1),n}(t) & \cdots & u_{(n),n}(t) \end{pmatrix}. \tag{19.12}$$

We can consider (19.11) or (19.12) as an $n^2 \times n^2$-system and we may add an initial condition of type

$$U(t_0) = C \in M(n; \mathbb{R}). \tag{19.13}$$

Equation (19.11) together with (19.13) should be read as a linear system with variable coefficients for the n^2 unknown functions $u_{(j),l} : I \to \mathbb{R}$ under the initial condition $u_{(j),l}(t_0) = c_{jl}$, $C = (c_{ij})$. Therefore the problem (19.11)/(19.13) has a unique solution and in the case where C is regular, i.e. $\det C \neq 0$, the vector valued functions $u_{(1)}, \ldots, u_{(n)}$ are linearly independent, i.e. they form a fundamental system of $A(t)$. Indeed if C is regular then $\{u_{(1)}(t_0), \ldots, u_{(n)}(t_0)\}$ is an independent set, hence $\{u_{(1)}(t), \ldots, u_{(n)}(t)\}$ is an independent set for all $t \in I$. The converse is also true, if $\{u_{(1)}(t), \ldots, u_{(n)}(t)\}$ is for every $t \in I$ an independent set, then it is an independent set for t_0 and therefore $\det C \neq 0$, i.e. C must be regular.

Suppose that $U(t)$ is a matrix corresponding to a fundamental system of $A(t)$, or $\dot{u}(t) = A(t)u(t)$. Then all solutions of $\dot{u}(t) = A(t)u(t)$ must be linear combinations of the columns of $U(t)$, i.e. all solutions are of the form

$$u(t) = U(t)c, \ c \in \mathbb{R}^n,$$

where $U(t)$ is given by (19.8). Of particular interest is the case when we choose for c the unit vectors e_j, $j = 1, \ldots, n$, which means when considering the fundamental system given by

$$\dot{V}(t) = A(t)V(t), \ V(t_0) = id_n. \tag{19.14}$$

Recall that (19.14) corresponds to the n initial value problems

$$\dot{v}_{(j)}(t) = A(t)v_{(j)}(t), \ v_{(j)}(t_0) = e_j, \ j = 1, \ldots, n. \tag{19.15}$$

We claim

$$\dot{u}(t) = A(t)u(t) \text{ and } u(t_0) = u_0 \text{ if and only if } u(t) = V(t)u_0. \tag{19.16}$$

Indeed, since $V(t_0) = id_n$ we find $u(t_0) = u_0$ and furthermore we have

$$\dot{u}(t) = \frac{\mathrm{d}}{\mathrm{d}t}(V(t)u_0) = \dot{V}(t)u_0 = A(t)V(t)u_0 = A(t)u(t).$$

318

Thus solving (19.14) will lead to a solution to every initial value problem $\dot{u}(t) = A(t)u(t)$ and $u(t_0) = u_0$.

Suppose that $U(t)$ solves (19.11) and let $C \in M(n;\mathbb{R})$ be given. It follows for $Z(t) := U(t)C$ that

$$\dot{Z}(t) = \dot{U}(t)C = A(t)U(t)C = A(t)Z(t), \tag{19.17}$$

i.e. $Z(t)$ is a further solution to (19.11). If in addition $U(t)$ corresponds to a fundamental system and $C \in GL(n;\mathbb{R})$, then $Z(t)$ corresponds to a further fundamental system. In particular we have $U(t) = V(t)U(t_0)$, since $V(t_0) = id_n$ and

$$\dot{U}(t) = \dot{V}(t)U(t_0) = A(t)V(t)U(t_0) = A(t)U(t). \tag{19.18}$$

These considerations make it clear that studying a matrix $U(t)$ built up by n solutions to $\dot{u}(t) = A(t)u(t)$ is of great interest.

Definition 19.7. *Let $U(t) = (u_{(1)}(t), \ldots, u_{(n)}(t))$ be a solution to (19.11), i.e. $\dot{U}(t) = A(t)U(t)$. We call*

$$W(t) := \det U(t) \tag{19.19}$$

*the **Wronski determinant** of $\{u_{(1)}, \ldots, u_{(n)}\}$.*

It turns out that the Wronski determinant is a very powerful tool to investigate (19.11), or the initial value problem (19.11)/(19.13), and hence the original problem for the linear system $\dot{u}(t) = A(t)u(t)$, $u(t_0) = u_0$. The next theorem is due to Liouville and needs the notion of the trace of a matrix. Recall that for $D = (d_{kl}) \in M(n;\mathbb{R})$ the real number $\operatorname{tr}(D) := \sum_{k=1}^{n} d_{kk}$ is called the **trace** of D.

Theorem 19.8. *Let $A : I \to M(n;\mathbb{R})$ be a continuous and bounded matrix-valued function and let $\{u_{(1)}, \ldots, u_{(n)}\}$ be n solutions to $\dot{u}(t) = A(t)u(t)$ with corresponding matrix $U(t)$. The associated Wronski determinant $W(t) = \det U(t)$ solves on I the differential equation*

$$\dot{W}(t) = (\operatorname{tr}A(t))W(t), \tag{19.20}$$

i.e.

$$\dot{W}(t) = \left(\sum_{j=1}^{n} a_{jj}(t) \right) W(t). \tag{19.21}$$

Remark 19.9. Note that $W(\cdot)$ depends by Theorem 19.8 only on $A(\cdot)$ and the initial conditions, however it is independent of the concrete solutions $\{u_{(1)}, \ldots, u_{(n)}\}$.

Corollary 19.10. *In the situation of Theorem 19.8 we have*

$$W(t) = W(t_0)e^{\int_{t_0}^{t} \operatorname{tr}(A(s))\ ds} \tag{19.22}$$

and for $V(t)$ defined by (19.14) we find

$$\det V(t) = e^{\int_{t_0}^{t} \operatorname{tr}(A(s))\ ds}. \tag{19.23}$$

Proof of Theorem 19.8. First we note that if $B : I \to M(n; \mathbb{R})$ is differentiable and has the column vectors $b_j(t)$, $B(t) = (b_1(t), \ldots, b_k(t))$, we find for the derivative of the determinant of $B(t)$, see Problem 4,

$$(\det B)^{\cdot}(t) = \sum_{j=1}^{n} \det(b_1(t), \ldots, b_{j-1}(t), \dot{b}_j(t), b_{j+1}(t), \ldots, b_n(t))$$

which follows by expanding $\det B(t)$ and applying Leibniz's rule. It follows for $V(t)$ as defined in (19.14) that

$$(\det V)^{\cdot}(t) = \sum_{j=1}^{n} \det(V_1(t), \ldots, V_{j-1}(t), \dot{V}_j(t), V_{j+1}(t), \ldots, V_n(t))$$

and since $V_j(t_0) = e_j$ and $\dot{V}_j(t_0) = A(t_0)e_j$ we get

$$(\det V)^{\cdot}(t_0) = \sum_{j=1}^{n} \det(e_1, \ldots, e_{j-1}, A(t_0)e_j, e_{j+1}, \ldots, e_n)$$

$$= \sum_{j=1}^{n} a_{jj}(t_0) = \operatorname{tr} A(t_0).$$

By (19.18) we find for $W(t) := \det U(t)$ that

$$\dot{W}(t) = \det(V(\cdot)W(t_0))^{\cdot}(t) = (\det V)^{\cdot}(t)\det U(t_0)$$
$$= (\det V)^{\cdot}(t)W(t_0).$$

Therefore we have for every $t_0 \in I$

$$\dot{W}(t_0) = (\det V)^{\cdot}(t_0)W(t_0) = \operatorname{tr} A(t_0)W(t_0),$$

and since $t_0 \in I$ was arbitrary we arrive at

$$\dot{W}(t) = \operatorname{tr} A(t)W(t).$$

$\square$

Corollary 19.11. *In the situation of Theorem 19.8 the Wronski determinant is either identically equal to 0 on I or it has no zero on I. The solutions $\{u_{(1)}, \dots, u_{(n)}\}$ form a fundamental system for $\dot{u}(t) = A(t)u(t)$ if and only if $\det U(t_0) = W(t_0) \neq 0$.*

Now we turn to the inhomogeneous system

$$\dot{u}(t) = A(t)u(t) + f(t). \tag{19.24}$$

Our first result is the standard one for inhomogeneous linear systems and the proof is left to the reader.

Theorem 19.12. *Let $A : I \to M(n; \mathbb{R})$ and $f : I \to \mathbb{R}$ be continuous and bounded. If $\tilde{u} : I \to \mathbb{R}^n$ is a special solution to (19.24), then all solutions to (19.24) are given by*

$$\{u = \tilde{u} + v \mid \dot{v}(t) = A(t)v(t)\}. \tag{19.25}$$

Thus we obtain all solutions to (19.24) by adding to one special solution of the inhomogeneous system all solutions of the corresponding homogeneous system and therefore the set of all solutions of the inhomogeneous system is an affine subspace of $C(I; \mathbb{R}^n)$.

The question is how do we get a special solution to (19.24), and an adaptation of Lagrange's method of the variation of constants is successful. For this let $U(t) = (u_{(1)}(t), \dots, u_{(n)}(t))$ be a fundamental system of $\dot{u}(t) = A(t)u(t)$. We know that we can obtain all solutions of the homogeneous system by looking at the set $U(t)v$, $v \in \mathbb{R}^n$. Now we try to find a special solution to (19.24) in the form

$$z(t) = U(t)v(t) \tag{19.26}$$

where $v : I \to \mathbb{R}^n$ is a function we want to determine. Differentiating in (19.26) yields

$$\begin{aligned} \dot{z}(t) &= \dot{U}(t)v(t) + U(t)\dot{v}(t) \\ &= A(t)U(t)v(t) + U(t)\dot{v}(t), \end{aligned}$$

and we have solved our problem if we can find a solution to

$$A(t)U(t)v(t) + U(t)\dot{v}(t) = A(t)U(t)v(t) + f(t),$$

or

$$U(t)\dot{v}(t) = f(t). \tag{19.27}$$

By assumption $U(t)$ corresponds to a fundamental system and hence its Wronski determinant is never zero, i.e. $\det U(t) \neq 0$ for all $t \in I$, implying that $U^{-1}(t)$ exists for all $t \in I$ and (19.27) becomes

$$\dot{v}(t) = U^{-1}(t)f(t)$$

with solution

$$v(t) = v(t_0) + \int_{t_0}^{t} U^{-1}(s)f(s) \, ds. \tag{19.28}$$

Choosing $v(t_0) = 0$, this gives for $z(t)$ the function

$$z(t) = U(t) \int_{t_0}^{t} U^{-1}(s)f(s) \, ds. \tag{19.29}$$

Thus we have proved as a general result

Theorem 19.13. *Let $A : I \to M(n; \mathbb{R})$ and $f : I \to \mathbb{R}^n$ be continuous and bounded. Denote by $V(t)$ the matrix corresponding to the fundamental system $\dot{V}(t) = A(t)V(t)$, $V(t_0) = id_n$. Then the unique solution to*

$$\dot{u}(t) = A(t)u(t) + f(t), \ u(t_0) = u_0, \tag{19.30}$$

is given by

$$u(t) = V(t)u_0 + \int_{t_0}^{t} V(t)V^{-1}(s)f(s) \, ds. \tag{19.31}$$

Remark 19.14. Let $U(t)$ correspond to any fundamental system of $\dot{u}(t) = A(t)u(t)$. In this case we have $U(t) = V(t)U(t_0)$ and therefore $V(t)V^{-1}(s) = U(t)U^{-1}(s)$ which yields a further representation of u in (19.31) as

$$u(t) = U(t)U^{-1}(t_0)u_0 + \int_{t_0}^{t} U(t)U^{-1}(s)f(s) \, ds. \tag{19.32}$$

Remark 19.15. Note that in the proof of Theorem 19.13 we have used that if $\dot{w}(t) = A(t)w(t)$ and $w(t_0) = u_0$ and if $\dot{v}(t) = A(t)v(t) + f(t)$ with $v(t_0) = 0$, then $u := w + v$ solves $\dot{u}(t) = A(t)u(t) + f(t)$ and $u(t_0) = u_0$. For this reason it makes sense to seek for a special solution of the inhomogeneous system under the initial condition 0.

Example 19.16. Let $A : I \to M(n; \mathbb{R})$ be a diagonal matrix, i.e.

$$A(t) = \begin{pmatrix} a_{11}(t) & & \mathbf{0} \\ & \ddots & \\ \mathbf{0} & & a_{nn}(t) \end{pmatrix} \tag{19.33}$$

with continuous and bounded functions $a_{kk} : I \to \mathbb{R}$. In this case the system

$$\dot{u}(t) = A(t)u(t)$$

splits into n uncoupled equations

$$\dot{u}_k(t) = a_{kk}(t)u_k(t)$$

with general solution

$$u_k(t) = u_{0,k}e^{\int_{t_0}^t a_{kk}(s)\,ds}$$

and $u_{0,k} = u_k(t_0)$. Moreover, the inhomogeneous problem

$$\dot{u}(t) = A(t)u(t) + f(t), \; u(t_0) = u_0$$

has for a continuous and bounded function $f : I \to \mathbb{R}^n$ the solution $u(t) = \begin{pmatrix} u_1(t) \\ \vdots \\ u_n(t) \end{pmatrix}$
where

$$u_k(t) = u_{0,k}e^{\int_{t_0}^t a_{kk}(s)\,ds} + e^{\int_{t_0}^t a_{kk}(s)\,ds}\int_{t_0}^t f_k(s)e^{-\int_{t_0}^s a_{kk}(r)\,dr}\,ds. \tag{19.34}$$

Example 19.17. Let $A \in M(n; \mathbb{R})$ be a matrix with n distinct real eigenvalues $\lambda_1, \dots, \lambda_n$ and let $\{b_1, \dots, b_n\} \subset \mathbb{R}^n$ be a basis of eigenvectors, i.e. $Ab_k = \lambda_k b_k$. For $u_{(k)}(t) = b_k e^{\lambda_k t}$ it follows that

$$\dot{u}_k(t) = b_k \lambda_k e^{\lambda_k t} = Ab_k e^{\lambda_k t} = Au_k(t)$$

and further we have $u_k(0) = b_k$. Hence $\{b_1 e^{\lambda_1 t}, \dots, b_n e^{\lambda_n t}\}$ is a fundamental system for A. For the corresponding Wronski determinant we find

$$W(t) = \det(b_1 e^{\lambda_1 t}, \dots, b_n e^{\lambda_n t})$$
$$= e^{(\lambda_1 + \dots + \lambda_n)t}\det(b_1, \dots, b_n) \neq 0.$$

Note that $\operatorname{tr}(A) = \lambda_1 + \dots + \lambda_n$ and $\det(b_1, \dots, b_n) = W(0)$. Using formula (19.32) we find for the unique solution to the inhomogeneous system

$$\dot{u}(t) = Au(t) + f(t), \; u(0) = u_0, \tag{19.35}$$

the representation

$$u(t) = U(t)U^{-1}(0)u_0 + \int_0^t U(t)U^{-1}(s)f(s)\,ds \tag{19.36}$$

where

$$U(t) = (b_1 e^{\lambda_1 t}, \dots, b_n e^{\lambda_n t}). \tag{19.37}$$

Example 19.18. Let $A \in M(n;\mathbb{R})$ and $\mu, \bar{\mu} \in \mathbb{C}$ be two complex eigenvalues of A, i.e. $\mu = \alpha + i\beta$, $\bar{\mu} = \alpha - i\beta$, $\beta \neq 0$. The corresponding eigenvectors z_1 and z_2 are also complex conjugated, i.e. $z_1 = a + ib$ and $z_2 = a - ib$ with $a, b \in \mathbb{R}^n$. Since z_1 and z_2 are linearly independent, a and b are linearly independent too. We claim that

$$u_1(t) = e^{\alpha t}(a \cos \beta t - b \sin \beta t) = \frac{1}{2}(z_1 e^{\mu t} + z_2 e^{\bar{\mu} t}) \tag{19.38}$$

and

$$u_2(t) = e^{\alpha t}(a \sin \beta t + b \cos \beta t) = \frac{1}{2i}(z_1 e^{\mu t} + z_2 e^{\bar{\mu} t}) \tag{19.39}$$

are two independent solutions to $\dot{u}(t) = Au(t)$. For u_1 we find

$$\dot{u}_1(t) = \frac{1}{2}(\mu z_1 e^{\mu t} + \bar{\mu} z_2 e^{\bar{\mu} t})$$
$$= \frac{1}{2}(Az_1 e^{\mu t} + Az_2 e^{\bar{\mu} t})$$
$$= A\left(\frac{1}{2}(z_1 e^{\mu t} + z_2 e^{\bar{\mu} t})\right) = Au_1(t)$$

and analogously we see that $\dot{u}_2(t) = Au_2(t)$. Next we suppose that for real numbers λ_1, λ_2 not both equal to 0 we have for all t that $\lambda_1 u_1(t) + \lambda_2 u_2(t) = 0$ or

$$e^{\alpha t}(\lambda_1 a \cos \beta t - \lambda_1 b \sin \beta t + \lambda_2 a \sin \beta t + \lambda_2 b \cos \beta t) = 0,$$

i.e.

$$(\lambda_1 \cos \beta t + \lambda_2 \sin \beta t)a + (\lambda_2 \cos \beta t - \lambda_1 \sin \beta t)b = 0$$

for all t. Since a and b are independent it follows that

$$\lambda_1 \cos \beta t + \lambda_2 \sin \beta t = 0 \text{ and } \lambda_2 \cos \beta t - \lambda_1 \sin \beta t = 0$$

for all t. Suppose that $\lambda_2 \neq 0$, i.e. $\cos \beta t = \frac{\lambda_1}{\lambda_2} \sin \beta t$ for all t, which is a contradiction. But $\lambda_1 \neq 0$ implies $\sin \beta t = -\frac{\lambda_2}{\lambda_1} \cos \beta t$ for all t which is again a contradiction. Thus u_1 and u_2 are independent in $C(I;\mathbb{R}^n)$. We can now combine Example 19.17 with these considerations and have a tool to find a fundamental system for A if A has r distinct real eigenvalues and q distinct pairs of complex and conjugate complex eigenvalues, $r + 2q = n$.

Example 19.19. We want to find the solution to the system

$$\dot{u}(t) = \begin{pmatrix} 0 & 1 \\ -1 & 0 \end{pmatrix} u(t) + f(t), \ u(0) = \begin{pmatrix} 1 \\ 0 \end{pmatrix}. \tag{19.40}$$

For this we solve the homogeneous system

$$\dot{u}(t) = \begin{pmatrix} 0 & 1 \\ -1 & 0 \end{pmatrix} u(t) \tag{19.41}$$

for the two initial conditions $u_{(1)}(0) = \begin{pmatrix} 1 \\ 0 \end{pmatrix}$ and $u_{(2)}(0) = \begin{pmatrix} 0 \\ 1 \end{pmatrix}$, which gives

$$u_{(1)}(t) = \begin{pmatrix} \cos t \\ -\sin t \end{pmatrix} \quad \text{and} \quad u_{(2)}(t) = \begin{pmatrix} -\sin t \\ \cos t \end{pmatrix}$$

and since $\det\begin{pmatrix} \cos t & \sin t \\ -\sin t & \cos t \end{pmatrix} = 1$ these two solutions form a fundamental system for $\dot{u}(t) = \begin{pmatrix} 0 & 1 \\ -1 & 0 \end{pmatrix} u(t)$. The inverse matrix to $V(t) = \begin{pmatrix} \cos t & \sin t \\ -\sin t & \cos t \end{pmatrix}$ is $V^{-1}(t) = \begin{pmatrix} \cos t & -\sin t \\ \sin t & \cos t \end{pmatrix}$ and therefore the solution to (19.40) is

$$u(t) = V(t)u_0 + \int_0^t V(t)V^{-1}(s)f(s)\,ds$$

$$= \begin{pmatrix} \cos t & \sin t \\ -\sin t & \cos t \end{pmatrix}\begin{pmatrix} 1 \\ 0 \end{pmatrix} + \begin{pmatrix} \cos t & \sin t \\ -\sin t & \cos t \end{pmatrix}\int_0^t \begin{pmatrix} f_1(s)\cos s - f_2(s)\sin s \\ f_1(s)\sin s + f_2(s)\cos s \end{pmatrix} ds.$$

If we choose for $f(t) = \begin{pmatrix} \sin t \\ 0 \end{pmatrix}$ we find

$$u(t) = \begin{pmatrix} \cos t \\ -\sin t \end{pmatrix} + \begin{pmatrix} \cos t & \sin t \\ -\sin t & \cos t \end{pmatrix}\begin{pmatrix} \frac{1}{4} - \frac{1}{4}\cos 2t \\ \frac{1}{2}t - \frac{1}{4}\sin 2t \end{pmatrix}$$

$$= \begin{pmatrix} \frac{5}{4}\cos t - \frac{1}{4}\cos t \cos 2t + \frac{1}{2}t\sin t - \frac{1}{4}\sin t \sin 2t \\ -\frac{5}{4}\sin t + \frac{1}{4}\sin t \cos 2t - \frac{1}{4}\sin 2t \cos t + \frac{1}{2}t\cos t \end{pmatrix}.$$

So far, all of our examples dealt with a matrix A having constant elements. In the case of a system with 2 or more equations with variables coefficients (and not in diagonal form) we can rarely find explicit solutions. However, once we know one solution, for example by guessing, we can reduce a system for n unknown functions to a system for $n-1$ unknown functions. This procedure is often called **d'Alembert's method of reduction**. We follow in our presentation essentially [125].

Given the $n \times n$-system

$$\dot{u}(t) = A(t)u(t) \tag{19.42}$$

and suppose that $v : I \to \mathbb{R}^n$ is a solution not identically zero. We try to find a further solution in the form

$$w(t) = \varphi(t)v(t) + z(t) \tag{19.43}$$

325

where $z_1(t) = 0$ for all $t \in I$ and $\varphi : I \to \mathbb{R}$ is an appropriate function. Inserting (19.43) into (19.42) we find

$$\dot{w}(t) = \dot{\varphi}(t)v(t) + \varphi(t)\dot{v}(t) + \dot{z}(t)$$
$$= A\varphi(t)v(t) + Az(t) = \varphi(t)Av(t) + Az(t)$$

or

$$\dot{z}(t) = Az(t) - \dot{\varphi}(t)v(t). \tag{19.44}$$

The first component of (19.44) reads as

$$\sum_{j=2}^{n} a_{1j}(t)z_j(t) = \dot{\varphi}(t)v_1(t), \tag{19.45}$$

and for the other components we find

$$\dot{z}_k(t) = \sum_{l=2}^{n} a_{kl}(t)z_l(t) - \dot{\varphi}(t)v_k(t), \ \ k = 2, \ldots, n. \tag{19.46}$$

Solving (19.45) for $\dot{\varphi}$ yields for $k = 2, \ldots, n$

$$\dot{z}_k(t) = \sum_{l=2}^{n} \left(a_{kl}(t) - \frac{v_k(t)}{v_1(t)} a_{1l}(t) \right) z_l(t) \tag{19.47}$$

which is a linear homogeneous $(n-1) \times (n-1)$ system. Of course we need to assume $v_1(t) \neq 0$ for $t \in I$, but instead of singling out the first component we may use any other component. Now suppose that $z_2, \ldots, z_n$ are solutions to (19.47). We take φ as a primitive of $\frac{1}{v_1(\cdot)} \sum_{l=2}^{n} a_{1l}(\cdot)z_l(\cdot)$ and by (19.43) we can find w. Suppose that $\{z^{(2)}, \ldots, z^{(n)}\}$ is a fundamental system for (19.47) and let us define $w^{(2)}, \ldots, w^{(n)}$ by (19.43). We claim that in this case $\{v(\cdot), w^{(2)}(\cdot), \ldots, w^{(n)}(\cdot)\}$ is a fundamental system for (19.42). In order to verify this claim we only need to check the linear independence of $\{v, w^{(2)}, \ldots, w^{(n)}\}$. With

$$w^{(j)} = \varphi_j v + z^{(j)}, \ j = 2, \ldots, n$$

where φ_j is the primitive as constructed above, but with z_l replaced by $z_l^{(j)}$, we consider

$$\lambda_1 v + \lambda_2 w^{(2)} + \cdots + \lambda_n w^{(n)} = 0, \ \lambda_j \in \mathbb{R}. \tag{19.48}$$

For the first component in (19.48) we obtain

$$\lambda_1 + \lambda_2 \varphi_2 + \cdots + \lambda_n \varphi_n = 0,$$

hence

$$\lambda_1 v + \lambda_2 \varphi_2 v + \cdots + \lambda_n \varphi_n v = 0. \tag{19.49}$$

Subtracting (19.49) from (19.48) we obtain

$$\lambda_2 z^{(2)} + \cdots + \lambda_n z^{(n)} = 0$$

and since $\{z^{(2)}, \ldots, z^{(n)}\}$ is linearly independent we deduce that $\lambda_2 = \cdots = \lambda_n = 0$, which now leads to $\lambda_1 = 0$, and therefore the set $\{v, z^{(2)}, \ldots, z^{(n)}\}$ is linearly independent, i.e. a fundamental system to (19.42).

Example 19.20. We consider the homogeneous system

$$\begin{pmatrix} \dot{u}_1(t) \\ \dot{u}_2(t) \end{pmatrix} = \begin{pmatrix} \frac{3}{t} & -1 \\ \frac{1}{2t^2} & \frac{1}{2t} \end{pmatrix} \begin{pmatrix} u_1(t) \\ u_2(t) \end{pmatrix}, \quad t > 0. \tag{19.50}$$

One may guess that $\begin{pmatrix} v_1(t) \\ v_2(t) \end{pmatrix} = \begin{pmatrix} t^2 \\ t \end{pmatrix}$ is a special solution. In fact we have

$$\begin{pmatrix} \dot{v}_1(t) \\ \dot{v}_2(t) \end{pmatrix} = \begin{pmatrix} 2t \\ 1 \end{pmatrix} = \begin{pmatrix} \frac{3}{t} & -1 \\ \frac{1}{2t^2} & \frac{1}{2t} \end{pmatrix} \begin{pmatrix} t^2 \\ t \end{pmatrix} = \begin{pmatrix} \frac{3}{t} & -1 \\ \frac{1}{2t^2} & \frac{1}{2t} \end{pmatrix} \begin{pmatrix} v_1(t) \\ v_2(t) \end{pmatrix}.$$

We are now searching for a second solution

$$\begin{pmatrix} w_1(t) \\ w_2(t) \end{pmatrix} = \varphi(t) \begin{pmatrix} v_1(t) \\ v_2(t) \end{pmatrix} + \begin{pmatrix} 0 \\ z_2(t) \end{pmatrix}$$

which leads to

$$\dot{z}_2(t) = \frac{3}{2t} z_2(t)$$

or $z_2(t) = t^{\frac{3}{2}}$. Since now $\frac{1}{v_1(t)} \sum_{l=2}^{2} a_{12}(t) z_2(t) = -t^{-\frac{1}{2}}$ we find $\varphi(t) = -2t^{\frac{1}{2}}$ and therefore

$$\begin{pmatrix} w_1(t) \\ w_2(t) \end{pmatrix} = -2t^{\frac{1}{2}} \begin{pmatrix} t^2 \\ t \end{pmatrix} + \begin{pmatrix} 0 \\ t^{\frac{3}{2}} \end{pmatrix} = \begin{pmatrix} -2t^{\frac{5}{2}} \\ -t^{\frac{3}{2}} \end{pmatrix}.$$

The reader can easily check that this is indeed a second solution to (19.50). Hence $\left\{ \begin{pmatrix} t^2 \\ t \end{pmatrix}, \begin{pmatrix} -2t^{\frac{5}{2}} \\ -t^{\frac{3}{2}} \end{pmatrix} \right\}$ is a fundamental system for $A(t) = \begin{pmatrix} \frac{3}{t} & -1 \\ \frac{1}{2t^2} & \frac{1}{2t} \end{pmatrix}$. In Problem 6 we will use this fundamental system to solve an inhomogeneous problem corresponding to (19.50).

In Chapter 17 we have seen that we can always transform a differential equation of order m into an $m \times m$ first order system. Obviously, we may apply this to a linear (homogeneous) equation of order m, say

$$u^{(m)}(t) = a_{m-1}(t) u^{(m-1)}(t) + a_{m-2}(t) u^{(m-2)}(t) + \cdots + a_0(t) u(t) \tag{19.51}$$

to obtain the system

$$
\begin{cases}
\dot{v}_1(t) = v_2(t) \\
\quad\vdots \qquad \vdots \\
\dot{v}_{m-1}(t) = v_m(t) \\
\dot{v}_m(t) = a_{m-1}(t)v_m(t) + a_{m-2}(t)v_{m-1}(t) + \cdots + a_0(t)v_1(t).
\end{cases}
\tag{19.52}
$$

Further we have seen that a solution $v = \begin{pmatrix} v_1 \\ \vdots \\ v_m \end{pmatrix}$ of (19.52) gives a solution u to (19.51) when setting $u := v_1$. Then it follows, see (17.10), that

$$
\begin{pmatrix} v_1 \\ v_2 \\ \vdots \\ v_m \end{pmatrix} = \begin{pmatrix} u \\ \dot{u} \\ \vdots \\ u^{(m-1)} \end{pmatrix}.
$$

This enables us to transfer some of our results for linear systems of first order equations to a single linear equation of higher order. In particular, for a fundamental system $\{v_{(1)}, \ldots, v_{(m)}\}$ of (19.52) there corresponds m solutions to (19.51) and we want to consider these solutions as independent solutions. This leads also to

Definition 19.21. A. *Let $y_j : I \to \mathbb{R}$, $1 \le j \le n$, be $(n-1)$-times continuously differentiable functions. We call*

$$
W(y_1, \ldots, y_n)(t) := \det \begin{pmatrix} y_1(t) & \cdots & y_n(t) \\ \dot{y}_1(t) & & \dot{y}_n(t) \\ \vdots & & \vdots \\ y_1^{(n-1)}(t) & \cdots & y_n^{(n-1)}(t) \end{pmatrix}
\tag{19.53}
$$

*the **Wronski determinant** of $\{y_1, \ldots, y_n\}$.*
***B.** The system $\{y_1, \ldots, y_n\}$ is said to be **linearly independent** in I if $W(y_1, \ldots, y_n)(t) \neq 0$ for all $t \in I$.*

Remark 19.22. If $\{y_1, \ldots, y_n\} \subset C^{n-1}(I; \mathbb{R})$ is linearly independent in the sense of Definition 19.21 then the set $\{y_1, \ldots, y_n\}$ is linearly independent in the ordinary sense and, see Problem 8.

We will pick up these considerations in Chapter 21.

Now we want to investigate to which extent we can use matrix-exponential functions to discuss linear systems with variable coefficients.

The linear system with constant coefficients $\dot{u}(t) = Au(t)$ and with initial condition $u(0) = u_0$ we can solve with the help of a family of linear operators, i.e. matrices on $\mathbb{R}^n$, namely with the help of $(T_t^A)_{t\geq 0}$, $T_t^A = e^{tA}$. This family has the semigroup property, i.e. $T_t^A \circ T_s^A = T_{t+s}^A$ and $T_0^A = id_n$, and in addition we have $\lim_{t\to 0} T_t^A = id_n$ for every matrix norm on $\mathbb{R}^n$. Moreover we can recover A from $(T_t^A)_{t\geq 0}$ by

$$A = \frac{\mathrm{d}}{\mathrm{d}t} T_t^A \bigg|_{t=0}. \tag{19.54}$$

The solution to the initial value problem is given by $u(t) = T_t^A u_0$. In addition, if the initial value is given at $t_0 \neq 0$ we can find the solution to $\dot{u}(t) = Au(t)$, $u(t_0) = u_0$ as $u(t) = T_{t-t_0}^A u_0$ for $t \geq t_0$.

For a single linear equation with a variable coefficient, i.e. for $\dot{v}(t) = a(t)v(t)$, $v(0) = v_0$, we have the solution

$$v(t) = v_0 e^{\int_0^t a(s)\,\mathrm{d}s},$$

a result which extends to diagonal systems. Therefore we might be tempted to introduce the operator

$$U(t) = e^{\int_{t_0}^t A(s)\,\mathrm{d}s} \tag{19.55}$$

and try to solve $\dot{u}(t) = A(t)u(t)$, $u(0) = u_0$, by $U(t)u_0$. With $\tilde{A}$ being a primitive of A we derive from (19.55)

$$U(t) = e^{(\tilde{A}(t) - \tilde{A}(0))}, \tag{19.56}$$

but now we encounter a serious problem: although we can define the right hand side of (19.56) by

$$\sum_{k=0}^{\infty} \frac{(\tilde{A}(t) - \tilde{A}(0))^k}{k!}$$

provided $\tilde{A}(t)$ is bounded, we fail to obtain "nice" properties such as the semigroup property since the matrices $(\tilde{A}(t_1) - \tilde{A}(0))$ and $(\tilde{A}(t_2) - \tilde{A}(0))$ will in general not commute. Moreover, differentiating this series with respect to t yields formally

$$\frac{\mathrm{d}}{\mathrm{d}t} \sum_{k=0}^{\infty} \frac{(\tilde{A}(t) - \tilde{A}(0))^k}{k!} = \sum_{k=0}^{\infty} \frac{1}{k!} \frac{\mathrm{d}}{\mathrm{d}t} (\tilde{A}(t) - \tilde{A}(0))^k$$

and already for $k = 2$ we have to be rather careful:

$$\frac{d}{dt}(\tilde{A}(t) - \tilde{A}(0))^2 = \frac{d}{dt}((\tilde{A}(t) - \tilde{A}(0))(\tilde{A}(t) - \tilde{A}(0)))$$
$$= A(t)(\tilde{A}(t) - \tilde{A}(0)) + (\tilde{A}(t) - \tilde{A}(0))A(t)$$

where we used that $(\tilde{A})^{\cdot} = A$. Since we cannot expect $[\tilde{A}(t) - \tilde{A}(0), A(t)] = 0$ for all t, we cannot expect $\frac{d}{dt}(\tilde{A}(t) - \tilde{A}(0))^2 = 2A(t)(\tilde{A}(t) - \tilde{A}(0))$, or more generally $\frac{d}{dt}(\tilde{A}(t) - \tilde{A}(0))^k = kA(t)(\tilde{A}(t) - \tilde{A}(0))^{k-1}$ to hold. Thus in general (19.55) cannot provide us with solutions to $\dot{u}(t) = A(t)u(t)$. A by-product of our considerations so far is

Proposition 19.23. *Let $A : I \to M(n; \mathbb{R})$ be a continuous and bounded matrix-function with primitive $\tilde{A} : I \to M(n; \mathbb{R})$. If for all $t \in I$ we have $[A(t), \tilde{A}(t)] = 0$ then by*

$$u(t) = e^{\int_{t_0}^t A(s)\, ds} u_0 \tag{19.57}$$

a solution to $\dot{u}(t) = A(t)u(t)$ and $u(t_0) = u_0$ is given.

Proof. For $t = t_0$ we have $u(t_0) = e^{\int_{t_0}^{t_0} A(s)\, ds} u_0 = u_0$. In order to show that $u(\cdot)$ solves the differential equation we use the power series expansion of the exponential and we write $\tilde{A}(t) = \int_{t_0}^t A(s)\, ds$ and note that $(\tilde{A})^{\cdot}(t) = A(t)$ to find

$$\frac{d}{dt}u(t) = \frac{d}{dt}\sum_{k=0}^{\infty} \frac{\tilde{A}^k(t)}{k!} u_0 = \sum_{k=0}^{\infty} \frac{1}{k!}\frac{d}{dt}\tilde{A}^k(t)u_0$$
$$= \sum_{k=0}^{\infty} \frac{1}{k!}kA(t)\tilde{A}^{k-1}(t)u_0$$

where in the last step we have used that $[A(t), \tilde{A}(t)] = 0$. Now it is clear that

$$\frac{d}{dt}u(t) = A(t)\sum_{k=0}^{\infty} \frac{\tilde{A}^k(t)}{k!}u_0 = A(t)u(t).$$

$\square$

The question remains whether we can find in the general case a type of "solution matrix" which plays in the situation of variable coefficients the role of the exponential function. A key observation is formula (19.31) or even better for our purposes (19.32). Let $U(t)$ be a matrix corresponding to a fundamental system of $A(t)$. To solve $\dot{u}(t) = A(t)u(t) + f(t)$ under the initial condition $u(t_0) = u_0$ operators of type $U(t)U^{-1}(s)$ turn out to be of central importance. We set

$$\begin{cases} \Phi : I \times I \to GL(n; \mathbb{R}) \\ \quad \Phi(t, s) := U(t)U^{-1}(s). \end{cases} \tag{19.58}$$

Lemma 19.24. *For Φ as defined by (19.58) we have for $r, s, t \in I$*

$$\frac{\partial}{\partial t}\Phi(t, s) = A(t)\Phi(t, s); \tag{19.59}$$

$$\frac{\partial}{\partial s}\Phi(t, s) = -\Phi(t, s)A(s); \tag{19.60}$$

$$\Phi(t, t) = id_n; \tag{19.61}$$

$$\Phi(t, s)\Phi(s, r) = \Phi(t, r). \tag{19.62}$$

In particular we have

$$\Phi(t, s)\Phi(s, t) = id_n \text{ or } \Phi(s, t) = \Phi^{-1}(t, s). \tag{19.63}$$

Proof. To see (19.59) we just have to use the fact that $U(t)$ is a matrix corresponding to a fundamental system for $A(t)$ and therefore

$$\frac{\partial}{\partial t}\Phi(t, s) = \frac{\partial}{\partial t}U(t)U^{-1}(s) = \frac{\partial U(t)}{\partial t}U^{-1}(s)$$
$$= A(t)U(t)U^{-1}(s) = A(t)\Phi(t, s).$$

In order to prove (19.60) we note that $\Phi(t, s)U(s) = U(t)$ and therefore $\frac{\partial}{\partial s}(\Phi(t, s)U(s)) = 0$. However

$$\frac{\partial}{\partial s}\Phi(t, s)U(s) = \frac{\partial \Phi(t, s)}{\partial s}U(s) + \Phi(t, s)\frac{\partial U(s)}{\partial s}$$
$$= \frac{\partial \Phi(t, s)}{\partial s}U(s) + \Phi(t, s)A(s)U(s) = 0$$

and (19.60) follows. The relation (19.61) is trivial since $U(t, t) = U(t)U^{-1}(t)$. Furthermore we find

$$\Phi(t, s)\Phi(s, r) = U(t)U^{-1}(s)U(s)U^{-1}(r) = U(t)U^{-1}(r) = \Phi(t, r),$$

i.e. (19.62) is proved and (19.61) now implies (19.63). $\qquad \square$

Definition 19.25. *Let $A : I \to M(n; \mathbb{R})$ be a continuous and bounded matrix-function. A function $\Phi : I \times I \to GL(n; \mathbb{R})$ satisfying the conditions (19.59)-(19.63) of Lemma 19.24 is called a **fundamental solution** to $\dot{u}(t) = A(t)u(t)$.*

Thus in order to deal with linear systems with variable coefficients in an analogous way as we dealt with linear systems having constant coefficients, namely by using solutions operators corresponding to the semi-group $(T_t^A)_{t \geq 0}$ we have to switch to fundamental solutions. In fact Theorem 19.13 or Remark 19.14 give solutions to

331

the initial value problem $\dot{u}(t) = A(t)u(t) + f(t)$, $u(0) = u_0$ in terms of fundamental solutions. Note that in the case where $A(t)$ is a constant matrix A we may use as a fundamental solution for $t > s$ the operators $\Phi^A(t, s) = T^A_{t-s}$. Now it is obvious that (19.62) generalizes the semi-group property. We will encounter one-parameter operator semi-groups and fundamental solutions in later volumes when dealing with certain partial differential equations.

Problems

1. a) For $0 < a_1 < a_2$ prove that $\{f, g\} \subset C(\mathbb{R})$ is an independent set of functions where $f(t) = a_1^t$ and $g(t) = a_2^t$.

    b) Prove that $\{f, g, h\}$ is an independent set in $C(\mathbb{R})$ where $f(x) = \cos 4x$, $g(x) = \sin^2 2x$ and $h(x) = \cos^2 2x$.

    c) Prove that the mappings $t \mapsto \begin{pmatrix} \cosh t \\ \sinh t \end{pmatrix}$ and $t \mapsto \begin{pmatrix} \sinh t \\ \cosh t \end{pmatrix}$ are in $C(\mathbb{R}, \mathbb{R}^2)$ independent solutions to the system $\dot{y}_1 = y_2$ and $\dot{y}_2 = y_1$.

2. For the constant coefficient system $\dot{y}_1 = -y_2$ and $\dot{y}_2 = y_1$ verify (19.12) for the independent solutions $t \mapsto \begin{pmatrix} \cos t \\ \sin t \end{pmatrix}$ and $t \mapsto \begin{pmatrix} -\sin t \\ \cos t \end{pmatrix}$.

3. Consider for $t \in (0, \infty)$ the system

$$\begin{pmatrix} u \\ v \end{pmatrix}^{\cdot}(t) = \begin{pmatrix} 0 & 1 \\ -4t^2 & \frac{1}{t} \end{pmatrix} \begin{pmatrix} u \\ v \end{pmatrix}(t) = A(t) \begin{pmatrix} u \\ v \end{pmatrix}(t).$$

Prove that $\begin{pmatrix} u_1 \\ v_1 \end{pmatrix}(t) = \begin{pmatrix} \cos t^2 \\ -2t \sin t^2 \end{pmatrix}$ and $\begin{pmatrix} u_2 \\ v_2 \end{pmatrix}(t) = \begin{pmatrix} \sin t^2 \\ 2t \cos t^2 \end{pmatrix}$ are two solutions and calculate the corresponding Wronski determinant to deduce that they are independent. Verify (19.12) and (19.20).

4. Let $B : (a, b) \to M(n; \mathbb{R})$ be a differentiable matrix-valued function with column vectors $b_j(t)$, $B(t) = (b_1(t), \ldots, b_n(t))$. Prove that

$$(\det B)^{\cdot}(t) = \sum_{j=1}^{n} \det(b_1(t), \ldots, b_{j-1}(t), \dot{b}_j(t), b_{j+1}(t), \ldots, b_n(t)).$$

5. Sketch the proof of Theorem 19.12.

6. From Example 19.20 we know that $\{w_1, w_2\}$ with $w_1(t) = \begin{pmatrix} t^2 \\ t \end{pmatrix}$, $w_2(t) = \begin{pmatrix} -2t^{\frac{5}{2}} \\ -t^{\frac{3}{2}} \end{pmatrix}$ is a fundamental system to

$$\dot{u}(t) = \frac{3}{t}u(t) - v(t)$$

$$\dot{v}(t) = \frac{1}{2t^2}u(t) + \frac{1}{2t}v(t)$$

for $t > 0$. Use this fundamental system to find a solution to the problem

$$\dot{u}(t) = \frac{3}{t}u(t) - v(t) + t^4$$

$$\dot{v}(t) = \frac{1}{2t^2}u(t) + \frac{1}{2t}v(t) + t^{\frac{5}{2}}$$

for $t > 0$ and $u(1) = v(1) = 1$.

7.   a) Let $u_1, \ldots, u_N \in C([0, 1])$ such that $\int_0^1 |u_j(t)|^2 \, dt = 1$ and $\int_0^1 u_k(t)u_l(t) \, dt = 0$ for $k \neq l$. Prove that the set $\{u_1, \ldots, u_N\}$ is linearly independent in $C([0, 1])$.

   b) Let $a \in C^1([0, 1])$, $a(t) \geq \kappa > 0$ for all $t \in [0, 1]$. Suppose that $u, v \in C^2([0, 1])$, $\int_0^1 |u(t)|^2 \, dt = \int_0^1 |v(t)|^2 \, dt = 1$, and we have for some $\lambda, \mu \in \mathbb{R}, \lambda \neq \mu$, that

$$\frac{d}{dx}\left(a(x)\frac{du(x)}{dx}\right) = \lambda u(x), \quad u(0) = u(1) = 0,$$

$$\frac{d}{dx}\left(a(x)\frac{dv(x)}{dx}\right) = \mu v(x), \quad v(0) = v(1) = 0.$$

Prove that the set $\{u, v\}$ is linearly independent in $C([0, 1])$.

8. Prove that N functions $u_1, \ldots, u_N \in C^{N-1}([a, b])$ are linearly independent in the sense of Definition 19.21 then they are also linearly independent in the sense of linear algebra. Conversely, if $\{u_1, \ldots, u_N\} \subset C^{N-1}([a, b])$ is an independent set in the sense of linear algebra and each function u_j solves the linear equation $P(D)u = a_1 u^{(N-1)} + \cdots + a_N u = 0$ in $[a, b]$, $a_j \in \mathbb{R}$, then $u_1, \ldots, u_N$ are linearly independent in the sense of Definition 19.21.

9. Given the differential equation $a_1 u^{(N-1)} + \cdots + a_N u = 0$ in $[a, b]$, $a_j \in C([a, b])$. Suppose that $u_1, \ldots, u_N$ are N solutions of this equation and for some $t_0 \in [a, b]$ we have $u_j(t_0) = 0$ for $1 \leq j \leq N$ or that $u_j^{(k)}(t_0) = 0$ for some $1 \leq k \leq N - 1$. Prove that the function $u_1, \ldots, u_N$ are not linearly independent in the sense of Definition 19.21.

10. Use Proposition 19.23 to find a solution to the system

$$\dot{u}(t) = tu(t) + tv(t)$$
$$\dot{v}(t) = tv(t)$$

with $u(0) = v(0) = 1$.

11. Let $y_1, y_2 : I \to \mathbb{R}$ be solutions to

$$(*) \quad y''(x) + p(x)y'(x) + q(x)y(x) = 0$$

with suitable functions $p, q : I \to \mathbb{R}$. For the Wronski determinant

$$W(x) = y_1(x)y_2'(x) - y_2(x)y_1'(x)$$

show that

$$W(x) = ce^{P(x)},$$

where P is a primitive of p and $c \in \mathbb{R}$.
Hint: consider $(*)$ for y_1 and y_2, then multiply the equation for y_1 by y_2 and the equation for y_2 by y_1. Next form the difference to find

$$y_1 y_2'' - y_2 y_1'' + p(y_1 y_2' - y_2 y_1') = 0$$

and deduce

$$\frac{dW}{dx} + pW = 0.$$

20 Second Order Linear Differential Equations with Real Analytic Coefficients

In this chapter we discuss a class of differential equations which looks rather special, namely equations of the type

$$a(t)\ddot{u}(t) + b(t)\dot{u}(t) + c(t)u(t) - \lambda u(t) = 0, \tag{20.1}$$

where $\lambda \in \mathbb{R}$ is a parameter and $a, b, c : I \to \mathbb{R}$, $I \subset \mathbb{R}$, are functions admitting a convergent power series expansion about $\tau_0 \in I$ with a radius of convergence $r > 0$. By shifting the origin we may achieve that $\tau_0 = 0$ which we will do later on, but in general we have

$$a(t) = \sum_{k=0}^{\infty} a_k(t - \tau_0)^k, \quad b(t) = \sum_{k=0}^{\infty} b_k(t - \tau_0)^k \text{ and } c(t) = \sum_{k=0}^{\infty} c_k(t - \tau_0)^k,$$

with $a_k, b_k, c_k \in \mathbb{R}$ and these series converge at least for $|t - \tau_0| < r$. We call a function $a : I \to \mathbb{R}$ **real analytic** in $(\tau_0 - r, \tau_0 + r)$ if it admits a convergent power series representation in $(\tau_0 - r, \tau_0 + r)$. In this case a has a holomorphic extension at least to the open disc $D_r(\tau_0) = \{z \in \mathbb{C} \mid |z - \tau_0| < r\}$ and we can apply the results developed in Part 7 Complex Analysis to a when considering it as a holomorphic function in $D_r(\tau_0)$.

Although the class of differential equations (20.1) is very special, many important differential equations are covered by (20.1), for example Bessel's differential equation, the Legendre and the Hermite differential equation. Also included is the hypergeometric differential equation, and of course the differential equations with constant coefficients such as $\dot{u}(t) - u(t) = 0$ or $\ddot{u}(t) + u(t) = 0$ leading to exponential and trigonometric functions, respectively.

First we want to study (20.1) in a neighbourhood of a point t_0 which belongs to the interval of convergence of the power series representing a, b and c. For a moment we may assume $\lambda = 0$ (or replacing c by $c - \lambda$). It may happen that $a(t_0) = 0$ and then (20.1) changes its order at t_0. Otherwise we may divide in a neighbourhood of t_0 in (20.1) by $a(t)$ to obtain

$$\ddot{u}(t) + \frac{b(t)}{a(t)}\dot{u}(t) + \frac{c(t)}{a(t)}u(t) = 0. \tag{20.2}$$

It turns out that a certain behaviour of b and c at the zeroes of a still allows a good theory. Since a, b, c have holomorphic extensions we may consider $\frac{b}{a}$ and $\frac{c}{a}$ as complex-valued functions on $D_\rho(\tau_0) \setminus \{z \mid a(z) = 0\}$. We need

Definition 20.1. *A point $t_0 \in (\tau_0 - r, \tau_0 + r)$ is called a **regular singular point** or a **singularity of first kind** for (20.1) if $\frac{b(\cdot)}{a(\cdot)}$ has a pole of order 1 and $\frac{c(\cdot)}{a(\cdot)}$ has a pole of order 2 at t_0.*

For a regular singular point t_0 of (20.1) there exists $r_0 > 0$ such that

$$\frac{b(t)}{a(t)} = (t - t_0)^{-1} \sum_{k=0}^{\infty} \alpha_k (t - t_0)^k \tag{20.3}$$

and

$$\frac{c(t)}{a(t)} = (t - t_0)^{-2} \sum_{k=0}^{\infty} \beta_k (t - t_0)^k \tag{20.4}$$

where both series converge in $|t - t_0| < r_0$ (or $|z - t_0| < r_0$). In the case where $\alpha_0 = 0$ and $\beta_0 = \beta_1 = 0$ the functions $\frac{b(\cdot)}{a(\cdot)}$ and $\frac{c(\cdot)}{a(\cdot)}$ have a removable singularity at t_0. Such a point we call a **regular point** for (20.1). For simplicity we now set

$$\alpha(t) := \frac{b(t)}{a(t)} \text{ and } \beta(t) := \frac{c(t)}{a(t)}, \tag{20.5}$$

i.e. we consider the equation

$$\ddot{u}(t) + \alpha(t)\dot{u}(t) + \beta(t)u(t) = 0. \tag{20.6}$$

It turns out that the roots ρ_1 and ρ_2 of the polynomial

$$\chi(\rho) := \rho^2 + (\alpha_0 - 1)\rho + \beta_0 \tag{20.7}$$

are of fundamental importance in the theory. They are called the **characteristic exponents** of (20.6) at t_0 while χ is called the **characteristic polynomial** of (20.6). (Some authors call (20.7) the **indicial equation** and ρ_1 and ρ_2 the **indices** or **characteristic indices** of the equation (20.6).) We order the characteristic exponents ρ_1 and ρ_2 such that $\mathrm{Re}(\rho_1 - \rho_2) \geq 0$ and if $\mathrm{Re}(\rho_1 - \rho_2) = 0$ we assume that $\mathrm{Im}(\rho_1 - \rho_2) \leq 0$. Now we can discuss the basic

Theorem 20.2 (Frobenius). *Let t_0 be a regular singular point for $\ddot{u} + \alpha\dot{u} + \beta u = 0$, i.e. (20.6). Then (20.6) admits a unique fundamental system u_1 and u_2 in $|t - t_0| < r_0$ where*

$$u_1(t) = (t - t_0)^{\rho_1} \sum_{k=0}^{\infty} \gamma_k (t - t_0)^k \tag{20.8}$$

and

$$u_2(t) = (t - t_0)^{\rho_2} \sum_{k=0}^{\infty} \delta_k (t - t_0)^k + c u_1(t) \ln |t - t_0|. \tag{20.9}$$

Both series converge in $|t - t_0| < r_0$ *and we have*

$$\gamma_0 = \delta_0 = 1 \text{ and } c = 0 \text{ if } \rho_1 - \rho_2 \notin \mathbb{N}_0; \tag{20.10}$$

$$\gamma_0 = 1, \ \delta_0 = 0, \ c = -1 \text{ if } \rho_1 - \rho_2 = 0; \tag{20.11}$$

$$\gamma_0 = 1, \ \delta_0 = 1 \text{ and } \delta_{m_0} = 0 \text{ if } \rho_1 - \rho_2 = m_0 \in \mathbb{N}. \tag{20.12}$$

Proof. We do not give the complete proof. We derive the existence of the solutions (20.8) and (20.9), in particular we prove the convergence of the series, but we do not prove the independence of u_1 and u_2. This is usually done by transforming the problem to a problem in complex analysis by showing that non-trivial winding numbers associated with u_1 and u_2 will lead in a linear combination $\lambda u_1 + \mu u_2 = 0$ to trivial coefficients. The reader may find this argument in detail in [64]. In [110], a proof by studying the Wronski determinant is given relying on investigations of the asymptotic behaviours of solutions at a regular singular point.

We may assume that $t_0 = 0$ and we seek first solutions to (20.6) having the representation

$$v(t) = t^\rho \sum_{k=0}^{\infty} \gamma_k t^k. \tag{20.13}$$

It follows that

$$\dot{v}(t) = \rho t^{\rho-1} \sum_{k=0}^{\infty} \gamma_k t^k + t^\rho \sum_{k=0}^{\infty} k \gamma_k t^{k-1}$$

and

$$\ddot{v}(t) = \rho(\rho-1) t^{\rho-2} \sum_{k=0}^{\infty} \gamma_k t^k + \rho t^{\rho-1} \sum_{k=0}^{\infty} k \gamma_k t^{k-1}$$

$$+ \rho t^{\rho-1} \sum_{k=0}^{\infty} k \gamma_k t^{k-1} + t^\rho \sum_{k=0}^{\infty} k(k-1) \gamma_k t^{k-2}$$

which implies

$$\ddot{v}(t) + \alpha(t)\dot{v}(t) + \beta(t)v(t) = t^{\rho-2} \sum_{k=0}^{\infty} ((\rho+k)^2 + (\alpha_0 - 1)(\rho+k) + \beta_0)\gamma_k$$

$$+ \sum_{l=1}^{k} (((\rho+k-l)\alpha_l + \beta_l)\gamma_{k-l}) t^k.$$

337

Note that $(\rho + k)^2 + (\alpha_0 - 1)(\rho + k) + \beta_0 = \chi(\rho + k)$. Thus v given by (20.13) satisfies (20.6) if and only if for $k \in \mathbb{N}_0$

$$\chi(\rho + k)\gamma_k + \sum_{l=1}^{k}((\rho + k - l)\alpha_l + \beta_l)\gamma_{k-l} = 0. \tag{20.14}$$

We now choose in (20.13) the value ρ_1 for ρ and $\gamma_0 = 1$. In this case the equation (20.14) is satisfied for $k = 0$ and moreover we find since $\chi(\rho) = (\rho - \rho_1)(\rho - \rho_2)$ that

$$\chi(\rho_1 + k) = ((\rho_1 + k) - \rho_1)((\rho_1 + k) - \rho_2) \tag{20.15}$$
$$= k(k + \rho_1 - \rho_2) \neq 0,$$

and therefore (20.14) yields for $k \in \mathbb{N}$

$$\gamma_k = -\frac{1}{\chi(\rho_1 + k)} \sum_{l=1}^{k}((\rho_1 + k - l)\alpha_l + \beta_l)\gamma_{k-l}, \tag{20.16}$$

i.e. all coefficients $\gamma_0, \gamma_1, \ldots$ are already uniquely determined. It remains to prove the convergence of the series in

$$u_1(t) := t^{\rho_1} \sum_{k=0}^{\infty} \gamma_k t^k \tag{20.17}$$

with $\gamma_0 = 1$ and γ_k as in (20.16). For this we have of course to use the convergence of the series in (20.3) and (20.4) in $|t - t_0| < r_0$. Let $r_1 < r_2 < r_0$. It follows the existence of a constant $\kappa > 0$ such that for $k \in \mathbb{N}_0$

$$|\alpha_k| \leq \kappa r_2^{-k} \text{ and } |\beta_k| \leq \kappa r_2^{-k}.$$

Given r_1 and r_2 we can find $k_1 \in \mathbb{N}$ such that $k \geq k_1$ implies

$$\frac{\kappa \left(1 + \frac{|\rho_1|}{k}\right) r_2}{k(r_2 - r_1)} \leq 1, \tag{20.18}$$

and we determine c_1 such that for $0 \leq k \leq k_1 - 1$ we have

$$|\gamma_k| r_1^k \leq c_1. \tag{20.19}$$

From (20.15) we deduce that

$$|\chi(\rho_1 + k)| \geq \operatorname{Re}(k(k + \rho_1 - \rho_2)) \geq k^2 \tag{20.20}$$

338

since by our assumptions $\text{Re}(\rho_1 - \rho_2) \geq 0$. We claim that

$$|\gamma_l| \leq c_1 r_1^{-l} \text{ for all } l \in \mathbb{N}_0. \tag{20.21}$$

By construction, this estimate holds for $l < k_1$. Now assuming (20.21) for $l < k$ with some $k \geq k_1$ we may use (20.16) to find

$$|\gamma_k| \leq \frac{1}{|\chi(\rho_1 + k)|} \sum_{l=1}^{k} ((|\rho_1| + k - l)|\alpha_l| + |\beta_l|)|\gamma_{k-l}|$$

$$\leq \frac{1}{k^2} \sum_{l=1}^{k} (|\rho_1| + k)\kappa r_2^{-l} c_1 r_1^{l-k}$$

$$= \frac{1}{k}\kappa\left(1 + \frac{|\rho_1|}{k}\right) \sum_{l=1}^{k} (r_1 r_2^{-1})^l c_1 r_1^{-k}$$

$$= \frac{\kappa}{k} \frac{\left(1 + \frac{|\rho_1|}{k}\right)}{r_2 - r_1} c_1 r_1^{-k} \leq c_1 r_1^{-k},$$

where in the last step we used (20.18). Since $r_1 < r_0$ was arbitrary it follows that $\sum_{k=0}^{\infty} \gamma_k t^k$ converges for all $|t| < r_0$. The convergence properties of power series allow us to differentiate term by term implying eventually that

$$u_1(t) = t^{\rho_1} \sum_{k=0}^{\infty} \gamma_k t^k$$

is a solution to (20.6).

Now we search for a second (independent) solution to (20.6). If $\rho_1 - \rho_2 \notin \mathbb{N}_0$ we can use again the factorization of χ, i.e. $\chi(\rho) = (\rho - \rho_1)(\rho - \rho_2)$, to derive as in (20.15) $\chi(\rho_2 + k) = k(k + \rho_2 - \rho_1)$ implying

$$|\chi(\rho_2 + k)| \geq k^2 - k\text{Re}(\rho_2 - \rho_1) \geq \frac{1}{2}k^2$$

for k sufficiently large. Now we can employ the arguments made above to prove that with $\delta_0 = 1$ and $c = 0$ the function

$$u_2(t) = t^{\rho_2} \sum_{k=0}^{\infty} \delta_k t^k$$

is a solution to (20.6). To determine and estimate the coefficients δ_k we have of course to now replace ρ_1 by ρ_2 in (20.16).

Next we turn to the case $\rho_1 = \rho_2$. Then the two solutions obtained by the method from above coincide. Therefore we make the Ansatz

$$u_2(t) = v(t) + cu_1(t)\ln t \tag{20.22}$$

to obtain a second (independent) solution to (20.6). Substituting (20.22) into (20.6) yields

$$\ddot{v}(t) + \alpha(t)\dot{v}(t) + \beta(t)v(t) + c\left(\frac{2}{t}\dot{u}_1(t) - \frac{1}{t^2}u_1(t) + \frac{1}{t}\alpha(t)u_1(t)\right) = 0, \tag{20.23}$$

see Problem 3. In (20.23) we can expand the function in the brackets into a series of the form

$$t^{\rho_1-2}\sum_{k=0}^{\infty}\eta_k t^k \tag{20.24}$$

which must converge for $|t| < r_0$ and for $\eta_0 = 2\rho_1 - 1 + \alpha_0 = 0$, recall that we are dealing with the case where $\rho_1 = \rho_2$. With

$$v(t) = t^{\rho_2}\sum_{k=0}^{\infty}\delta_k t^k \tag{20.25}$$

we now find from (20.23)

$$\chi(\rho_2 + k)\delta_k + \sum_{l=1}^{k}((\rho_2 + k - l)\alpha_l + \beta_l)\delta_{k-l} = -c\eta_k. \tag{20.26}$$

For $k = 0$ the equation (20.26) is always satisfied and therefore we can choose $\delta_0 = 0$ and $c = -1$ to obtain from (20.26) all other coefficients δ_k, $k \geq 1$. As in the first case we can estimate the coefficients δ_k to get $|\delta_k| \leq c_1 r_1^{-k}$ with r_1 as above. Thus we have now with

$$u_2(t) = t^{\rho_2}\sum_{k=0}^{\infty}\delta_k t^k - u_1(t)\ln t$$

a second solution to (20.6).

Finally we look at the case $\rho_1 - \rho_2 = k_0 \in \mathbb{N}$. In this case we have $\chi(\rho_0 + k_0) = 0$. Again we try to find u_2 with the help of the Ansatz

$$u_2(t) = v(t) + cu_1(t)\ln t \tag{20.27}$$

and consequently we arrive once again at (20.23) and (20.24). Now the recurrence formula becomes

$$\chi(\rho_2 + k)\delta_k + \sum_{l=1}^{k}((\rho_2 + k - l)\alpha_l + \beta_l)\delta_{k-l} = \begin{cases} 0, & k < k_0 \\ -c_1\eta_{k-k_0}, & k \geq k_0. \end{cases} \qquad (20.28)$$

This equation holds for $k = 0$ when we choose $\delta_0 = 1$. For $k = 1, \ldots, k_0 - 1$ the equation (20.28) has unique solutions $\delta_1, \ldots, \delta_{k_0-1}$. We further put $\delta_{k_0} = 0$ and determine c_1 from (20.28) for $k = k_0$. Now we can determine all other coefficients δ_k, $k > k_0$ and with r_1 as above we obtain the estimates $|\delta_k| \leq c_1 r_1^{-k}$ which entails the convergence of v and u_2 becomes a second solution. $\qquad\square$

Let us have a closer look at the content of Frobenius' theorem. At a regular singular point t_0 at least one of the solutions u_1 and u_2 is singular and we should consider t_0 as endpoint of the interval I where we seek solutions. In many cases we will have $t_0 = 0$ and choose t_0 as the left endpoint of I. In the case that $a(t)$ in (20.1) has no further zeroes we will consider (20.1) or (20.6) on (t_0, ∞), e.g. $(0, \infty)$. However if $a(t)$ has a further zero $t_1 > t_0$ which corresponds to a regular singular point of (20.1) or (20.6) we will try to solve these equations on $I = (t_0, t_1)$. Note that we may allow the point ∞ as a regular singular point when considering $\alpha(t)$ and $\beta(t)$ as meromorphic functions on the Riemann sphere S^2.

In the following examples we are sometimes brief when determining the coefficients of the power series entering in the solutions u_1 and u_2. Often we will refer to the literature, sometimes we will leave this combinatorial exercise to the reader. Our standard references are [34], [56], [85] and [110]. Once we have arrived at a class of "special functions" we also will make much use of [11].

Our first examples are differential equations with regular points, thus all their coefficients are holomorphic functions in some open interval and in principle we may apply the theory of Chapter 19. Thus they serve now mainly to see how a power series approach will work and to introduce further interesting examples of special functions.

Example 20.3. We consider the **Airy differential equation**

$$\ddot{u}(t) - tu(t) = 0. \qquad (20.29)$$

The coefficients are entire functions, namely $a(t) = 1$, $b(t) = -t$ and $c(t) = 0$ which yields $\alpha(t) = -t$ and $\beta(t) = 0$. A power series Ansatz

$$u(t) = \sum_{k=0}^{\infty} \gamma_k t^k$$

341

first yields

$$\dot{u}(t) = \sum_{k=0}^{\infty} k\gamma_k t^{k-1} = \sum_{k=0}^{\infty} (k+1)\gamma_{k+1} t^k$$

and

$$\ddot{u}(t) = \sum_{k=0}^{\infty} k(k-1)\gamma_k t^{k-2} = \sum_{k=0}^{\infty} (k+1)(k+2)\gamma_{k+2} t^k$$

which gives the equation

$$2\gamma_2 + \sum_{k=1}^{\infty} ((k+1)(k+2)\gamma_{k+2} - \gamma_{k-1})t^k = 0 \qquad (20.30)$$

and the recurrence formulae

$$2\gamma_2 = 0 \text{ and } (k+1)(k+2)\gamma_{k+2} - \gamma_{k-1} = 0, \ k \in \mathbb{N}.$$

From $\gamma_2 = 0$ we deduce that $\gamma_{3l-1} = 0$, $l \in \mathbb{N}_0$, and further

$$\gamma_{3l} = \frac{\gamma_0}{2 \cdot 5 \cdot 8 \cdots (3l-1)3^l l!}, \ l \in \mathbb{N},$$

and

$$\gamma_{3l+1} = \frac{\gamma_1}{4 \cdot 7 \cdot 10 \cdots (3l+1)3^l l!}, \ l \in \mathbb{N},$$

which gives as general solution to (20.29)

$$u(t) = \gamma_0 \left(1 + \sum_{l=1}^{\infty} \frac{t^{3l}}{2 \cdot 5 \cdot 8 \cdots (3l-1)3^l l!} \right) \qquad (20.31)$$

$$+ \gamma_1 \left(t + \sum_{l=1}^{\infty} \frac{t^{3l+1}}{4 \cdot 7 \cdot 10 \cdots (3l+1)3^l l!} \right).$$

It is easy to see that u in (20.31) is an entire function, see Problem 4. Airy's differential equation can be related to a (modified) Bessel differential equation and consequently u in (20.31) admits a representation with the help of certain Bessel functions. Note that the two free coefficients γ_0 and γ_1 allow us to determine two independent solutions to (20.29).

Example 20.4. The differential equation

$$\ddot{u}(t) - 2t\dot{u}(t) + \lambda u(t) = 0 \qquad (20.32)$$

342

is called **Hermite differential equation**. Again we note that all points are regular. Inserting the power series Ansatz $u(t) = \sum_{k=0}^{\infty} \gamma_k t^k$ into (20.32) yields

$$2\gamma_2 + \lambda\gamma_0 + \sum_{k=1}^{\infty}((k+1)(k+2)\gamma_{k+2} - 2k\gamma_k + \lambda\gamma_k)t^k = 0. \tag{20.33}$$

Again, as expected we have two "free" coefficients γ_0 and γ_1 and the recurrence relation

$$\gamma_{k+2} = \frac{2k - \lambda}{(k+1)(k+2)}\gamma_k, \quad k \in \mathbb{N}_0 \tag{20.34}$$

which gives the general solution

$$u(t) = \gamma_0\left(1 - \frac{\lambda}{2!}t^2 - \sum_{k=2}^{\infty}\frac{(4(k-1)-\lambda)(4(k-2)-\lambda)\dots(4-\lambda)}{(2k)!}t^{2k}\right) \tag{20.35}$$

$$+ \gamma_1\left(t + \sum_{k=1}^{\infty}\frac{(4k-2-\lambda)(4(k-1)-2-\lambda)\dots(2-\lambda)}{(2k+1)!}t^{2k+1}\right).$$

Again one can prove that u converges on $\mathbb{R}$ and solves (20.32). Further we note that u is a linear combination of an odd and an even function. So far we kept $\lambda \in \mathbb{R}$ general. However, if we choose $\lambda = 2l$, $l \in \mathbb{N}_0$, then the recurrence relation (20.34) implies $\gamma_{l+2} = \gamma_{l+4} = \gamma_{l+6} = \dots = 0$. This has a consequence that one of the series in (20.35) becomes a finite sum, i.e. a polynomial, in fact up to a normalization factor we obtain the Hermite polynomials H_l already studied in Chapter 9. Now we can also deduce that they satisfy the differential equation

$$H_l''(x) - 2xH_l'(x) + 2lH_l(x) = 0, \tag{20.36}$$

also see Appendix III.

Example 20.5. Consider the **Legendre differential equation**

$$(1 - t^2)\ddot{u}(t) - 2t\dot{u}(t) + \lambda(\lambda+1)u(t) = 0. \tag{20.37}$$

First we note that we can rewrite (20.37) as

$$((1 - t^2)\dot{u}(t))^{\cdot} + \lambda(\lambda+1)u(t) = 0 \tag{20.38}$$

and in this form we encountered Legendre's differential equation in Problem 6 c) in Chapter 9. Next we observe that (20.37) is of form (20.1) with $a(t) = (1 - t^2)$, $b(t) = -2t$ and $c(t) - \lambda = \lambda(\lambda+1)$. But now a has a zero at $t_0 = -1$ and $t = 1$ and therefore we consider (20.37) on the interval $(-1, 1)$. We note that for

$$\alpha(t) = \frac{-2t}{1 - t^2} \text{ and } \beta(t) = \frac{\lambda(\lambda+1)}{1 - t^2}$$

but we still get a representation by a power series about 0 converging in $(-1,1)$ since $\frac{1}{1-t^2} = \sum_{k=0}^{\infty} t^{2k}$. The Ansatz $u(t) = \sum_{k=0}^{\infty} \gamma_k t^k$ yields after some long calculation (which we omit) the formulae

$$\gamma_{2k} = -\frac{\lambda(\lambda+1)}{2k}\left(1 - \frac{\lambda(\lambda+1)}{2 \cdot 3}\right)\left(1 - \frac{\lambda(\lambda+1)}{4 \cdot 5}\right)\cdots\left(1 - \frac{\lambda(\lambda+1)}{(2k-2)(2k-1)}\right)\gamma_0, \quad (20.39)$$

for $k = 2, 3, \ldots$, and for $k \in \mathbb{N}$

$$\gamma_{2k+1} = -\frac{1}{2k+1}\left(1 - \frac{\lambda(\lambda+1)}{1 \cdot 2}\right)\left(1 - \frac{\lambda(\lambda+1)}{3 \cdot 4}\right)\cdots\left(1 - \frac{\lambda(\lambda+1)}{(2k-1)(2k)}\right)\gamma_1, \quad (20.40)$$

where we have again γ_0 and γ_1 as "free" coefficients. We do not want to look at the corresponding power series, but we note that for $\lambda = 2l$ the lth factor in (20.39) is zero and for $\lambda = 2l+1$ the $(l+1)^{\text{th}}$ factor in (20.40) is 0. Thus we obtain polynomial solutions (as in the case of the Hermite differential equation) and these polynomials are of course upto a normalization the Legendre polynomial.

Remark 20.6. Since $a(t)\ddot{u}(t) = (a(t)\dot{u}(t))^{\cdot} - \dot{a}(t)u(t)$ we can always rewrite (20.1) as

$$(a(t)\dot{u}(t))^{\cdot} + (b(t) - \dot{a}(t))\dot{u}(t) + c(t)u(t) - \lambda u(t) = 0. \quad (20.41)$$

This observation becomes of importance in the case where $b(t) = \dot{a}(t)$. In this case (20.41) becomes

$$(a(t)\dot{u}(t))^{\cdot} + c(t)u(t) = \lambda u(t) \quad (20.42)$$

and this is the standard for Sturm-Liouville operators which we will investigate in Chapter 21 as well as in Volume V. They are of utmost importance in physics and other areas of applications and provide important examples in spectral theory.

Eventually we want to study **Bessel's differential equation(s)** in a context more general than before. The equation under discussion is

$$t^2\ddot{u}(t) + t\dot{u}(t) + (t^2 - \nu^2)u(t) = 0. \quad (20.43)$$

We have already discussed the case $\nu \in \mathbb{N}$ in Problem 7 of Chapter I.29, in Chapter III.22, as well as for $\nu = l + \frac{1}{2}$, $l \in \mathbb{N}_0$, in Chapter 14. With $a(t) = t^2$, $b(t) = t$ and $c(t) = t^2 - \nu^2$ this equation is of the type (20.1) and with

$$\alpha(t) = \frac{1}{t} \text{ and } \beta(t) = \frac{t^2 - \nu^2}{t^2}$$

we can transform it to (20.6) noting that at $t_0 = 0$ we have a regular singular point. Therefore we study (20.43) or

$$\ddot{u}(t) + \frac{1}{t}\dot{u}(t) + \left(1 - \frac{\nu^2}{t^2}\right)u(t) = 0 \quad (20.44)$$

on the interval $(0, \infty)$. Since in the corresponding expansion (20.3) and (20.4) we have $\alpha_0 = 1$ and $\beta_0 = -\nu^2$ the equation

$$\chi(\rho) = \rho^2 + (\alpha - 1)\rho + \beta_0 = \rho^2 - \nu^2$$

has the roots $\rho_1 = \nu$ and $\rho_2 = -\nu$, $\nu \geq 0$. As we have seen before (in special cases with rather explicit calculations), one solution to (20.43) is given by

$$u_{1,\nu}(t) = t^\nu \sum_{k=0}^{\infty} \frac{(-1)^k}{2^{2k} k! (\nu + 1) \ldots (\nu + n)} t^{2k}, \quad t > 0. \tag{20.45}$$

Changing the normalization we arrived at the Bessel function of first kind and order ν, J_ν, defined by

$$J_\nu(t) = \frac{1}{2^\nu \Gamma(\nu + 1)} u_{1,\nu}(t) \tag{20.46}$$

or

$$J_\nu(t) = \sum_{k=0}^{\infty} \frac{(-1)^k}{2^{2k+\nu} k! \Gamma(k + \nu + 1)} t^{2k+\nu} \tag{20.47}$$

$$= \sum_{k=0}^{\infty} \frac{(-1)^k}{k! \Gamma(k + \nu + 1)} \left(\frac{t}{2}\right)^{2k+\nu}.$$

The functions J_l, $l \in \mathbb{N}_0$, and $J_{l+\frac{1}{2}}$ we have already seen. In the case that $\nu \geq 0$ is not an integer, i.e. $\nu \notin \mathbb{N}_0$, a second and independent solution to (20.43) is given by

$$J_{-\nu}(t) = \sum_{k=0}^{\infty} \frac{(-1)^k}{2^{2k-\nu} k! \Gamma(k - \nu + 1)} t^{2k-\nu} \tag{20.48}$$

$$= \sum_{k=0}^{\infty} \frac{(-1)^k}{k! \Gamma(k - \nu + 1)} \left(\frac{t}{2}\right)^{2k-\nu}.$$

Note that the differential equation does not change when ν is replaced by $-\nu$ and the convergence of the power series part in (20.47) and (20.48) is easily established as is done for the case $\nu = l \in \mathbb{N}_0$. Using (20.43) (or (20.44)) we find

$$J_\nu(t) \dot{J}_{-\nu}(t) - \dot{J}_\nu(t) J_{-\nu}(t) = -\frac{2\nu}{t} \left(\frac{1}{t^\nu} J_\nu(t) \cdot \frac{1}{t^{-\nu}} J_{-\nu}(t)\right)$$

$$+ \frac{1}{t^\nu} J_\nu(t) \left(\frac{1}{t^{-\nu}} J_{-\nu}(t)\right)^{\cdot} - \left(\frac{1}{t^\nu} J_\nu(t)\right)^{\cdot} \left(\frac{1}{t^{-\nu}} J_{-\nu}(t)\right)$$

and the right hand side has a pole of finite order at $t = 0$. Thus

$$J_\nu(t)\dot{J}_{-\nu}(t) - \dot{J}_\nu(t)J_{-\nu}(t) \neq 0,$$

but this expression is just the Wronski determinant of J_ν and $J_{-\nu}$, thus they are independent functions for $\nu \notin \mathbb{N}_0$. In the case where $\nu \in \mathbb{N}$ we discuss a second independent solution in the problems.

Our next aim is to study some properties of solutions to (20.1). Typical questions we are interested in are related to the zeroes of solutions, oscillatory behaviour and maybe the asymptotic behaviour of solutions to equations of type (20.1) or (20.6). Often we are not interested in single solutions but either in two solutions forming a fundamental system or in Chapter 21 in the behaviour of so called eigenfunctions. We start with

Definition 20.7. *Let u_1 and u_2 be two solutions to (20.1) on some interval $I \subset \mathbb{R}$ with end points $a < b$. Suppose that both u_1 and u_2 have at most countably many zeroes ξ_j and η_l, i.e. $u_1(\xi_j) = 0$ and $u_2(\eta_l) = 0$ where j and l are running through an at most countable set. We call the sets of zeroes of u_1 and u_2 **separating** if*

$$\cdots < \xi_j < \eta_j < \xi_{j+1} < \eta_{j+1} < \xi_{j+2} < \ldots . \tag{20.49}$$

The following theorem is due to Charles Sturm

Theorem 20.8. *Suppose that the coefficients α and β in*

$$\ddot{u}(t) + \alpha(t)\dot{u}(t) + \beta(t)u(t) = 0 \tag{20.50}$$

are continuous on I. Then every solution to (20.50) which is not identically equal to zero has at most countably many zeroes which are simple and which do not have an accumulation point in I. If u_1 and u_2 are independent, i.e. they form a fundamental system, then the zeroes of u_1 and u_2 are separating.

Proof. If $u(\xi) = 0$ and $u'(\xi) = 0$ we conclude from the results of Chapter 19 that u must be identically zero. Suppose that $(\xi_j)_{j\in\mathbb{N}}$ is a sequence of zeroes of u in I with limit $\xi_0 \in I$, but $\xi_j \neq \xi_0$ for all j. First we note that this implies

$$0 = \lim_{j\to\infty} u(\xi_j) = u(\xi_0),$$

and furthermore we find

$$0 = \lim_{j\to\infty} \frac{u(\xi_j) - u(\xi_0)}{\xi_j - \xi_0} = u'(\xi_0),$$

i.e. ξ_0 is a zero of u but it cannot be a simple zero, thus $(\xi_j)_{j\in\mathbb{N}}$ cannot have an accumulation point in I. Next let $\eta_1 < \eta_2$ be two zeroes of u_2 and suppose that u_1 and u_2 form a fundamental system for (20.50). In this case their Wronski determinant is never zero and without loss of generality we may assume that

$$W(t) = u_1(t)\dot{u}_2(t) - \dot{u}_1(t)u_2(t) > 0$$

for all $t \in I$. It follows for $k = 1, 2$ that

$$W(\eta_k) = u_1(\eta_k)\dot{u}_2(\eta_k) > 0, \qquad (20.51)$$

implying $u_1(\eta_1)$, $u_1(\eta_2)$, $\dot{u}_2(\eta_1)$, $\dot{u}_2(\eta_2) \neq 0$. We assume $\dot{u}_2(\eta_1) > 0$. Since $\dot{u}_2$ is continuous $\dot{u}_2 > 0$ in a neighbourhood of η_1 and therefore u_2 passes strictly increasing through η_1, i.e. $u_2(t) > 0$ for $\eta_1 < t < \tilde{t}_1$. Now suppose $\dot{u}_2(\eta_2) > 0$ too. Then $u_2(t) < 0$ for $\tilde{t}_2 < t < \eta_2$ and by the intermediate value theorem u_2 must have a further zero between $\eta_1 < \eta_2$ which is a contradiction. Hence $\dot{u}_2(\eta_1) > 0$ implies $\dot{u}_2(\eta_2) < 0$. From (20.51) we conclude that $u_1(\eta_1) > 0$ and $u_2(\eta_2) < 0$ and again the intermediate value theorem implies the existence of a zero $\tilde{\xi}$ of u_1 with $\eta_1 < \tilde{\xi} < \eta_2$. The same result follows in the case where $\dot{u}_2(\eta_1) < 0$. We claim that $\tilde{\xi}$ is the only zero ξ of u_1 with $\eta_1 < \xi < \eta_2$. Suppose that ξ^* is a second zero of u_1 with $\eta_1 < \xi^* < \eta_2$. With the arguments given above we can conclude that then u_2 must have a further zero $\tilde{\eta}$ with $\eta_1 < \min(\tilde{\xi}, \xi^*) < \tilde{\eta} < \max(\tilde{\xi}, \xi^*) < \eta_2$ which is again a contradiction. $\square$

In the following we will discuss differential equations in the **Sturm-Liouville form**

$$Lu = (p\dot{u})^{\cdot} + qu = 0 \qquad (20.52)$$

or

$$Lu = p\ddot{u} + \dot{p}\dot{u} + qu = 0.$$

We have already seen that a differential equation of the type (20.1) can be transformed into such equation if $b(t) = \dot{a}(t)$. Moreover, if (20.50) is given with α and β continuous we may introduce $p(t) = e^{A(t)}$ where A is a primitive of α and we find with (20.50)

$$p(t)\ddot{u}(t) + \alpha(t)p(t)\dot{u}(t) + \beta(t)p(t)u(t) = 0,$$

or

$$(p(t)\dot{u}(t))^{\cdot} - \dot{p}(t)\dot{u}(t) + \alpha(t)p(t)\dot{u}(t) + \beta(t)p(t)u(t) = 0,$$

but $\dot{p}(t)\dot{u}(t) + \dot{A}(t)p(t)\dot{u}(t) = \alpha(t)p(t)\dot{u}(t)$ and (20.52) follows with $q(t) = p(t)\beta(t)$. We always assume that p is a C^1 function on I which is in addition positive, i.e. $p > 0$, and q is assumed to be continuous on I.

Proposition 20.9 (Lagrange). *Let p and q be as above and let $u, v \in C^2(I)$. For $Lu = (p\dot{u})^{\cdot} + qu$ the relation*

$$u(t)Lv(t) - v(t)Lu(t) = \frac{d}{dt}(p(t)u(t)\dot{v}(t) - v(t)\dot{u}(t)) \qquad (20.53)$$

$$= \frac{d}{dt}(p(t)W(u,v)(t))$$

holds where $W(u,v)$ denotes the Wronski determinant of u and v.

Proof. We have

$$uLv - vLu = u((p\dot{v})^{\cdot} + qv) - v((p\dot{u})^{\cdot} + qu)$$

$$= u(p\dot{v})^{\cdot} - v(p\dot{u})^{\cdot} = \frac{d}{dt}(p(u\dot{v} - v\dot{u}))$$

$$= \frac{d}{dt}(pW(u,v)).$$

$\square$

Now we can prove

Theorem 20.10 (Sturm). *Let u and v be non-trivial solutions on the interval I to the equations*

$$L_1 u = (p\dot{u})^{\cdot} + q_1 u = 0 \qquad (20.54)$$

and

$$L_2 v = (p\dot{v})^{\cdot} + q_2 v = 0, \qquad (20.55)$$

respectively. If $q_2 < q_1$ on I then between two consecutive zeroes of v there is at least one zero of u.

Proof. From $vL_1 u = 0$ and $uL_2 v = 0$ we obtain

$$uL_2 v - vL_1 u = u(p\dot{v})^{\cdot} - v(p\dot{u})^{\cdot} + (q_2 - q_1)uv = 0. \qquad (20.56)$$

Now let $\xi_1 < \xi_2$ be two consecutive zeroes of v. Integrating (20.56) yields

$$\int_{\xi_1}^{\xi_2} ((u(p\dot{v})^{\cdot} - v(p\dot{u})^{\cdot}) + (q_2 - q_1)uv)\, dt = 0$$

and in light of Proposition 20.9 we arrive at

$$p(\xi_2)u(\xi_2)\dot{v}(\xi_2) - p(\xi_1)u(\xi_1)\dot{v}(\xi_1) = \int_{\xi_1}^{\xi_2} (q_1(s) - q_2(s))u(s)v(s)\, ds, \qquad (20.57)$$

where we have used that $v(\xi_1) = v(\xi_2) = 0$. Since ξ_1 and ξ_2 are by assumption two consecutive zeroes of v, v must have in (ξ_1, ξ_2) a constant sign which we may assume to be positive, otherwise we replace v by $-v$. Since $q_1 > q_2$ we deduce from (20.57) that

$$p(\xi_2)u(\xi_2)\dot{v}(\xi_2) - p(\xi_1)u(\xi_1)\dot{v}(\xi_1) > 0. \tag{20.58}$$

Furthermore, we must have $\dot{v}(\xi_1) > 0$ and $\dot{v}(\xi_2) < 0$, note that ξ_j are simple zeroes of v. By assumption we have $p > 0$. Now we assume that u has no zero in (ξ_1, ξ_2). Then u must have the same sign in (ξ_1, ξ_2) which we may assume to be positive, otherwise we replace u by $-u$. Thus $u(\xi_1) \geq 0$ and $u(\xi_2) \geq 0$. This however implies that the left hand side of (20.58) is less than or equal to 0 which is a contradiction, so u must have a zero in (ξ_1, ξ_2) □

Zeroes of solutions to $Lu = 0$ are often important in eigenvalue problems, see Chapter 21. We want to see how Theorem 20.10 can be used to find information about the zeroes of Bessel functions.

Example 20.11. We consider Bessel's differential equation

$$t^2\ddot{u}(t) + t\dot{u}(t) + (t^2 - \nu^2)u(t) = 0$$

for $\nu > 0$ and $t > 0$. Introducing the function $u(t) = \frac{w(t)}{\sqrt{t}}$ we find

$$\dot{u}(t) = -\frac{1}{2}t^{-\frac{3}{2}}w(t) + t^{-\frac{1}{2}}\dot{w}(t)$$

and

$$\ddot{u}(t) = \frac{3}{4}t^{-\frac{5}{2}}w(t) - t^{-\frac{3}{2}}\dot{w}(t) + t^{-\frac{1}{2}}\ddot{w}(t),$$

which gives

$$t^2\ddot{u}(t) + t\dot{u}(t) + (t^2 - \nu^2)u(t) = t^{\frac{3}{2}}\left(\ddot{w}(t) + \left(1 + \frac{1-\nu^2}{4t^2}\right)w(t)\right) = 0.$$

Hence, for $t > 0$ the function w satisfies

$$\ddot{w}(t) + \left(1 + \frac{1-\nu^2}{4t^2}\right)w(t) = 0 \tag{20.59}$$

and on $(0, \infty)$ it follows that $u(\xi) = 0$ if and only if $w(\xi) = 0$. We restrict ourselves to the case $0 \leq \nu < \frac{1}{2}$ and we compare (20.59) with the equation $\ddot{v}(t) + v(t) = 0$, i.e. we choose $q_1(t) = 1 + \frac{1-\nu^2}{4t^2}$ and $q_2(t) = 1$. Note that for $0 \leq \nu < \frac{1}{2}$ we have $q_2(t) < q_1(t)$ for all $t > 0$. We know that the general solution to $\ddot{v}(t) + v(t) = 0$ is a linear combination of $\sin t$ and $\cos t$, or $v(t) = \gamma\sin(t - \varphi)$ for some φ. The zeroes of v are given by $\xi_k = \varphi + k\pi$, $k \in \mathbb{Z}$, and therefore the Bessel function J_ν, $0 \leq \nu < \frac{1}{2}$, has in every open interval $(a, a + \pi)$, $a > 0$, a zero. For $\nu = \frac{1}{2}$ the equation (20.59) is just $\ddot{v} + v = 0$.

So far we do not know whether in general a solution to (20.1) or (20.6) or (20.52) has a zero. This problem is treated in the following **oscillation theorem** which we state without proof.

Theorem 20.12. *Let $p, q : (0, \infty) \to \mathbb{R}$ be as above, i.e. $p \in C^1((0, \infty))$, $p > 0$, and $q \in C((0, \infty))$. In addition assume $q > 0$ as well as*

$$\int_0^\infty \frac{dt}{p(t)} = \infty \text{ and } \int_0^\infty q(t) \, dt = 0.$$

Under these assumptions every solution to (20.52) has countably many zeroes in $(0, \infty)$.

A proof is for example given in [56] or [60]. For many considerations in this chapter so far, the monograph [56] was influential.

We want to close this chapter by giving some indications of the origin of many of the equations covered by (20.1), (20.6) or (20.52). This leads to consider certain partial differential equations of mathematical physics in curvilinear coordinates, compare with Chapter II.12. We start with the Laplace operator Δ_3 in $\mathbb{R}^3$ in cylinder coordinates $(x_1, x_2, x_3) = (r \cos \varphi, r \sin \varphi, x_3)$. This leads for $g(r, \varphi, x_3)$ to

$$\Delta_3 g(r, \varphi, x_3) = \frac{\partial^2 g(r, \varphi, x_3)}{\partial r^2} + \frac{1}{r} \frac{\partial g(r, \varphi, x_3)}{\partial r} \tag{20.60}$$
$$+ \frac{1}{r^2} \frac{\partial^2 g(r, \varphi, x_3)}{\partial \varphi^2} + \frac{\partial^2 g(r, \varphi, x_3)}{\partial x_3^2} = 0,$$

see (II.12.16). We now make the Ansatz $g(r, \varphi, x_3) = R(r)Q(\varphi)Z(x_3)$ and we find

$$\frac{1}{g(r, \varphi, x_3)} \Delta_3 g(r, \varphi, x_3) = \frac{1}{R(r)} \frac{d^2 R(r)}{dr^2} + \frac{1}{rR(r)} \frac{dR(r)}{dr}$$
$$+ \frac{1}{r^2 Q(\varphi)} \frac{d^2 Q(\varphi)}{d\varphi^2} + \frac{1}{Z(x_3)} \frac{d^2 Z(x_3)}{dx_3^2} = 0,$$

or

$$\frac{1}{R(r)} \frac{d^2 R(r)}{dr^2} + \frac{1}{rR(r)} \frac{dR(r)}{dr} + \frac{1}{r^2 Q(\varphi)} \frac{d^2 Q(\varphi)}{d\varphi^2} = -\frac{1}{Z(x_3)} \frac{d^2 Z(x_3)}{dx_3^2}. \tag{20.61}$$

Since the left hand side of (20.61) depends only on r and φ while the right hand side depends only on x_3 we must have with some constant (which we may assume to be non-negative)

$$\frac{1}{Z(x_3)} \frac{d^2 Z(x_3)}{dx_3^2} = \kappa^2 \text{ or } \frac{d^2 Z(x_3)}{dx_3^2} - \kappa^2 Z(x_3) = 0, \tag{20.62}$$

350

and

$$\frac{1}{R(r)}\frac{d^2R(r)}{dr^2} + \frac{1}{rR(r)}\frac{dR(r)}{dr} + \frac{1}{r^2Q(\varphi)}\frac{d^2Q(\varphi)}{d\varphi^2} = -\kappa^2. \tag{20.63}$$

Multiplying in (20.63) with r^2 and rearranging the equation afterwards we get

$$\frac{r^2}{R(r)}\frac{d^2R(r)}{dr^2} + \frac{r}{R(r)}\frac{dR(r)}{dr} + r^2\kappa^2 = -\frac{1}{Q(\varphi)}\frac{d^2Q(\varphi)}{d\varphi^2}, \tag{20.64}$$

and now we conclude that for some ν^2 we must have

$$-\frac{1}{Q(\varphi)}\frac{d^2Q(\varphi)}{d\varphi^2} = \nu^2 \text{ or } \frac{d^2Q(\varphi)}{d\varphi^2} + \nu^2 Q(\varphi) = 0, \tag{20.65}$$

and

$$\frac{1}{R(r)}\frac{d^2R(r)}{dr^2} + \frac{1}{rR(r)}\frac{dR(r)}{dr} + \kappa^2 = \frac{\nu^2}{r^2}, \tag{20.66}$$

or

$$\frac{d^2R(r)}{dr^2} + \frac{1}{r}\frac{dR(r)}{dr} + \left(\kappa^2 - \frac{\nu^2}{r^2}\right)R(r) = 0. \tag{20.67}$$

Thus we arrive at three differential equations each of the type we have studied in this chapter, in particular (20.67) is Bessel's differential equation.

As a further example we look at the Laplace operator Δ_3 in spherical coordinates. According to (II.12.20) we look at the equation

$$\Delta_3 g(r, \vartheta, \varphi) = \frac{1}{r^2}\frac{\partial}{\partial r}\left(r^2\frac{\partial g(r, \vartheta, \varphi)}{\partial r}\right)$$
$$+ \frac{1}{r^2\sin\vartheta}\frac{\partial}{\partial\vartheta}\left(\sin\vartheta\frac{\partial g(r, \vartheta, \varphi)}{\partial\varphi^2}\right) + \frac{1}{r^2\sin^2\vartheta}\frac{\partial^2 g(r, \vartheta, \varphi)}{\partial\varphi^2}\right) = 0,$$

where $(x_1, x_2, x_3) = (r\sin\vartheta\cos\varphi, r\sin\vartheta\sin\varphi, r\cos\vartheta)$. The Ansatz

$$g(r, \vartheta, \varphi) = \frac{U(r)}{r}P(\vartheta)Q(\varphi)$$

now leads to

$$\Delta_3 g(r, \vartheta, \varphi) = P(\vartheta)Q(\varphi)\frac{d^2U(r)}{dr^2} + \frac{U(r)Q(\varphi)}{r^2\sin\vartheta}\frac{d}{d\vartheta}\left(\sin\vartheta\frac{dP(\vartheta)}{d\vartheta}\right)$$
$$+ \frac{U(r)P(\vartheta)}{r^2\sin^2\vartheta}\frac{d^2Q(\varphi)}{d\varphi^2} = 0.$$

351

From here we first conclude that with some $m \in \mathbb{Z}$ we must have

$$\frac{1}{Q(\varphi)} \frac{\mathrm{d}^2 Q(\varphi)}{\mathrm{d}\varphi^2} = -m^2 \quad \text{or} \quad \frac{\mathrm{d}^2 Q(\varphi)}{\mathrm{d}\varphi^2} + m^2 Q(\varphi) = 0 \qquad (20.68)$$

and

$$r^2 \sin^2 \vartheta \left(\frac{1}{U(r)} \frac{\mathrm{d}^2 U(r)}{\mathrm{d}r^2} + \frac{1}{r^2 \sin \vartheta P(\vartheta)} \frac{\mathrm{d}}{\mathrm{d}\vartheta} \left(\sin \vartheta \frac{\mathrm{d}P(\vartheta)}{\mathrm{d}\vartheta} \right) \right) = m^2, \qquad (20.69)$$

see also Problem 9. From the last equation we derive

$$\frac{1}{\sin \vartheta} \frac{\mathrm{d}}{\mathrm{d}\vartheta} \left(\sin \vartheta \frac{\mathrm{d}P(\vartheta)}{\mathrm{d}\vartheta} \right) + \left(l(l+1) - \frac{m^2}{\sin^2 \vartheta} \right) P(\vartheta) = 0 \qquad (20.70)$$

as well as

$$\frac{\mathrm{d}^2 U(r)}{\mathrm{d}r^2} - \frac{l(l+1)}{r^2} U(r) = 0 \qquad (20.71)$$

with $l \in \mathbb{N}_0$. Again we arrive at differential equations belonging to the class considered in this chapter. Moreover, if we set $x = \cos \vartheta$ then equation (20.70) becomes

$$\frac{\mathrm{d}}{\mathrm{d}x} \left((1 - x^2) \frac{\mathrm{d}P(x)}{\mathrm{d}x} \right) + \left(l(l+1) - \frac{m^2}{1 - x^2} \right) P(x) = 0 \qquad (20.72)$$

which is a generalized Legendre differential equation.

As a general rule we may state that separating the Laplace equation in orthogonal curvilinear coordinates into ordinary differential equations for functions depending only on one of these coordinates will lead to equations such as (20.1), (20.6) or (20.52).

Problems

1. By using a power series Ansatz, find a solution to the equation

$$(1 - x^2) y''(x) - xy'(x) + \lambda y(x) = 0.$$

Deduce that for $\lambda = n \in \mathbb{N}_0$ we have polynomial solutions.

2. Find two independent solutions of the equation

$$(1 + t^2) \ddot{u}(t) + 2t \dot{u}(t) - u(t) = 0$$

which are defined on $\mathbb{R}$.

Hint: one solution is easy to guess. A second solution can be obtained on $(-1, 1)$ with the help of a power series which, in fact, extends to a function defined on $\mathbb{R}$.

3. Substitute (20.22) into (20.6) in order to verify (20.23).

4. Prove that (20.31) defines an entire function.

5. The Bessel differential equation of order 0, i.e. the equation $t^2\ddot{u}(t) + t\dot{u}(t) + t^2 u(t) = 0$ has the Bessel function J_0 as a solution. Find a second solution independent of J_0 in the interval $(0, \infty)$ by using the Ansatz

$$y_0(t) = J_0(t) \ln t + \sum_{k=1}^{\infty} c_k t^k.$$

6. Find two independent solutions of the Bessel differential equation of order $\frac{1}{2}$, i.e. the equation $x^2 y''(x) + x y'(x) + \left(x^2 - \frac{1}{4}\right) y(x) = 0$, $x > 0$.

7.   a) For $\nu \in \mathbb{R}$ we consider the function

$$J_\nu(t) = t^\nu \sum_{k=0}^{\infty} \frac{(-1)^k}{2^{2k+\nu} k! \Gamma(k+\nu+1)} t^{2k}, \quad t > 0,$$

which is indeed well defined for $t \in \mathbb{C} \setminus (-\infty, 0)$ and $\nu \in \mathbb{R}$. Note that for $t \in \overline{D_R(0)} \subset \mathbb{C}$ and $|\nu| \leq N$ the estimate

$$\frac{|t|^2}{4(k+1)|k+1+\nu|} \leq \frac{R^2}{4(k+1)(k+1-N)}$$

holds for sufficiently large k and deduce that $\nu \mapsto J_\nu(t)$ is an entire function for every $t \in \overline{D_R(0)} \setminus (-\infty, 0)$.

b) For $\nu \notin \mathbb{Z}$, define the **Bessel function of second kind Y_ν** by

$$Y_\nu(z) := \frac{J_\nu(z) \cos \nu\pi - J_{-\nu}(z)}{\sin \nu\pi}, \quad z \in \mathbb{C} \setminus (-\infty, 0)$$

and prove with the help of the rules of l'Hospital that for $n \in \mathbb{Z}$

$$Y_n(z) := \lim_{\nu \to n} \frac{J_\nu(z) \cos \nu\pi - J_{-\nu}(z)}{\sin \nu\pi}$$

is well defined and independent of J_n.

c) Prove that Y_ν solves the Bessel differential equation of order ν, i.e. the equation $t^2\ddot{u}(t) + t\dot{u}(t) + (t^2 - \nu^2)u(t) = 0$, $t > 0$.

8. Consider the generalized Bessel differential equation

$$(*) \qquad t^2 u''(t) + tu'(t) + (\mu t^2 - \nu^2)u(t) = 0, \quad t > 0.$$

For $\nu \notin \mathbb{Z}$ and $\mu = i$, prove that $I_\nu(t) := i^{-\nu} J_\nu(it)$ is a solution to $(*)$ as is

$$K_\nu(t) := \frac{\pi}{2} \frac{I_{-\nu}(t) - I_\nu(t)}{\sin \nu \pi}.$$

The function I_ν and K_ν are called **modified Bessel functions** of first and second kind, respectively.

9. Prove that the general solution to the equation $xy''(x) + y'(x) + ay(x) = 0$, $a > 0$, is given by

$$y(x) = \lambda_1 J_0(\sqrt{4ax}) + \lambda_2 Y_0(\sqrt{4ax}), \quad x > 0.$$

Hint: consider the substitution $x = \frac{s^2}{4a}$ and $y(x) = g(s) = g(s(x))$.

10. Give a proof for (20.68) and (20.69), as well as for (20.70) and (20.71).

21 Boundary Value and Eigenvalue Problems. First Observations

This chapter is devoted to some first observations on boundary value and eigenvalue problems. These are problems which, however, are better dealt with when a certain functional analytic framework is at our disposal and we will do so in Volume V. Here we want to introduce some first problems and ideas of how to solve them.

Let us consider the equation $\ddot{u} + w^2 u = 0$ in I where $w > 0$ is a real number and $I \subset \mathbb{R}$ an interval. We know that $\cos wt$ and $\sin wt$ form a fundamental system for this equation, hence every solution to $\ddot{u} + w^2 u = 0$ is of the form $\mu \cos wt + \nu \sin wt$. In order to determine a unique solution we need two specific conditions which allow us to determine μ and ν in a unique way. One possibility is to prescribe u and $\dot{u}$ at a fixed point, say $t_0 \in I$, in our case we may choose $t_0 = 0$ if $0 \in I$. We may then add the initial conditions $u(0) = A$ and $\dot{u}(0) = B$. This yields the two equations

$$A = \mu \cos w0 + \nu \sin w0 = \mu$$

and

$$B = -w\mu \sin w0 + w\nu \cos w0 = w\nu.$$

It follows that $u(t) = A \cos wt + \frac{B}{w} \sin wt$ is the unique solution to

$$\ddot{u} + w^2 u = 0 \text{ in } I, \ u(0) = A \text{ and } \dot{u}(0) = B \tag{21.1}$$

(provided $t_0 = 0 \in I$). Since $\cos wt$ and $\sin wt$ form a fundamental system for any I, $\mathring{I} \neq \emptyset$, the condition $0 \in I$ is no restriction.

However, there are other ways to prescribe values for a solution to $\ddot{u} + w^2 u = 0$. Let $I = [a, b]$, $a < b$, $a, b \in \mathbb{R}$. We may ask for solutions to $\ddot{u} + w^2 u = 0$ satisfying, for example, one of the following sets of conditions

$$u(a) = \alpha \text{ and } u(b) = \beta, \tag{21.2}$$
$$u(a) = \alpha \text{ and } \dot{u}(b) = \beta, \tag{21.3}$$
$$\dot{u}(a) = \alpha \text{ and } u(b) = \beta, \tag{21.4}$$
$$\dot{u}(a) = \alpha \text{ and } \dot{u}(b) = \beta. \tag{21.5}$$

Since now values of the sought solution are prescribed at the boundary of the existence interval $I = [a, b]$ we call such a problem a **boundary value problem**. We want to discuss in more detail

$$\ddot{u}(t) + u(t) = 0, \ u(a) = \alpha \text{ and } u(b) = \beta. \tag{21.6}$$

Since the general solution to the equation is given by $\mu \cos t + \nu \sin t$ we have to solve the boundary conditions

$$\mu \cos a + \nu \sin a = \alpha$$
$$\mu \cos b + \nu \sin b = \beta$$

or

$$\begin{pmatrix} \cos a & \sin a \\ \cos b & \sin b \end{pmatrix} \begin{pmatrix} \mu \\ \nu \end{pmatrix} = \begin{pmatrix} \alpha \\ \beta \end{pmatrix}. \tag{21.7}$$

In the case where $\det\begin{pmatrix} \cos a & \sin a \\ \cos b & \sin b \end{pmatrix} = \cos a \sin b - \sin a \cos b \neq 0$ we obtain by Cramer's rule a unique solution by

$$\mu = \frac{\det\begin{pmatrix} \alpha & \sin a \\ \beta & \sin b \end{pmatrix}}{\det\begin{pmatrix} \cos a & \sin a \\ \cos b & \sin b \end{pmatrix}} \quad \text{and} \quad \nu = \frac{\det\begin{pmatrix} \cos a & \alpha \\ \cos b & \beta \end{pmatrix}}{\det\begin{pmatrix} \cos a & \sin a \\ \cos b & \sin b \end{pmatrix}}.$$

In Problem 1 we will discuss some of the other cases. But here comes an interesting and far reaching observation: since

$$\cos a \sin b - \sin a \cos b = \sin(b - a)$$

it follows that whenever $b - a = \pi k$, $k \in \mathbb{N}$, the determinant $\det\begin{pmatrix} \cos a & \sin a \\ \cos b & \sin b \end{pmatrix}$ is zero and either we cannot solve the system (21.7) or the solution will not be unique.

Let us return to the **regular Sturm-Liouville operator**

$$Lu(t) = (p\dot{u})^{\cdot}(t) + q(t)u(t) \tag{21.8}$$

which we consider for $p, q : [a, b] \to \mathbb{R}$, $p \in C^1([a, b])$, $p > 0$, and $q \in C([a, b])$. In some cases we also allow a right hand side $f : [a, b] \to \mathbb{R}$, $f \in C([a, b])$. In addition we introduce the two boundary conditions

$$B^a(u) := \alpha_1 u(a) + \alpha_2 p(a)\dot{u}(a) = \eta_1 \tag{21.9}$$

and

$$B^b(u) := \beta_1 u(b) + \beta_2 p(b)\dot{u}(b) = \eta_2. \tag{21.10}$$

We also pose the non-triviality conditions $\alpha_1^2 + \alpha_2^2 > 0$ and $\beta_1^2 + \beta_2^2 > 0$. The homogeneous boundary value problem

$$Lu(t) = 0, \ B^a(u) = 0 \text{ and } B^b(u) = 0 \tag{21.11}$$

356

may have non-trivial solutions. For example the equation $\ddot{u} + u = 0$ in $[0, \pi]$ and $B^0(u) = 0$, $B^\pi(u) = 0$ with $\alpha_2 = \beta_2 = 0$ and $\alpha_1 = \beta_1 = 1$, has the solutions $A \sin t$, $A \in \mathbb{R}$, and therefore we cannot expect a general uniqueness result for (21.11). However, we can derive from our previous considerations that L always admits a fundamental system. We claim

Theorem 21.1. *Let L and B^a, B^b be as above. Then the problem $Lu = f$ and $B^a(u) = \eta_1$, $B^b(u) = \eta_2$ has for every $f \in C([a, b])$ a unique solution $u \in C^2([a, b])$ if and only if for a fundamental system $\{u_1, u_2\}$ of L we have*

$$\det \begin{pmatrix} B^a(u_1) & B^a(u_2) \\ B^b(u_1) & B^b(u_2) \end{pmatrix} \neq 0. \tag{21.12}$$

In this case the homogeneous problem $Lu = 0$, $B^a(u) = B^b(u) = 0$ has only the trivial solution.

Proof. If v is a special solution to $Lu = f$ then all solutions to $Lu = f$ are of the form

$$w = v + \mu u_1 + \nu u_2, \ \mu, \nu \in \mathbb{R}.$$

The boundary conditions lead to

$$B^a(w) = B^a(v) + \mu B^a(u_1) + \nu B^a(u_2) = \eta_1$$

and

$$B^b(w) = B^b(v) + \mu B^b(u_1) + \nu B^b(u_2) = \eta_2.$$

Thus we have to investigate the linear system

$$\mu B^a(u_1) + \nu B^a(u_2) = \eta_1 - B^a(v)$$
$$\mu B^b(u_1) + \nu B^b(u_2) = \eta_2 - B^b(v)$$

which has under condition (21.12) a unique solution μ and ν. $\qquad\square$

Example 21.2. Consider the problem $Lu = \ddot{u} + u = f$ in $[0, \pi]$. A fundamental system for L is of course given by $u_1 = \cos$ and $u_2 = \sin$. We add the boundary value conditions

$$B^0(u) = u(0) + \dot{u}(0) = \eta_1 \text{ and } B^\pi(u) = u(\pi) = \eta_2,$$

and we note that

$$\det \begin{pmatrix} B^0(u_1) & B^0(u_2) \\ B^\pi(u_1) & B^\pi(u_2) \end{pmatrix} = \det \begin{pmatrix} 1 & 1 \\ -1 & 0 \end{pmatrix} = 1.$$

If we take as inhomogeneity $f(t) = t$ we find for $v(t) = t$ that $\ddot{v}(t) = 0$ and therefore $\ddot{v}(t) + v(t) = t = f(t)$. Thus the general solution to $Lu = f$ is given by $w(t) = t + \mu \cos t + \nu \sin t$. This implies

$$B^0(w) = w(0) + \dot{w}(0) = \mu + 1 + \nu = \eta_1$$

and

$$B^\pi(w) = w(\pi) = \pi - \mu = \eta_2$$

or

$$\begin{pmatrix} 1 & 1 \\ -1 & 0 \end{pmatrix} \begin{pmatrix} \mu \\ \nu \end{pmatrix} = \begin{pmatrix} \eta_1 - 1 \\ \eta_2 - \pi \end{pmatrix}$$

which yields $\mu = \pi - \eta_2$ and $\nu = \eta_1 + \eta_2 - 1 - \pi$. If we specialize to $\eta_1 = 1$ and $\eta_2 = \pi$ we find as the solution to the boundary value problem

$$\ddot{u}(t) + u(t) = t \text{ in } [0, \pi]$$
$$u(0) + \dot{u}(0) = 1 \text{ and } u(\pi) = \pi$$

the function $u(t) = t$. Indeed, a direct calculation gives $\ddot{u}(t) + u(t) = t$ and $u(0) + \dot{u}(0) = 1$ as well as $u(\pi) = \pi$.

We now turn to cases where the uniqueness condition fails to hold and we start our considerations by discussing the partial differential equation modelling a vibrating string of length 2π. If the displacement of the string from the equilibrium position is $u(x, t)$ and the string is fixed for all times at the endpoints $x = 0$ and $x = 2\pi$, and if in addition the initial displacement is $u_0(x)$ while the initial velocity is 0, the problem we have to solve is

$$\frac{\partial^2 u(x, t)}{\partial t^2} - \frac{\partial^2 u(x, t)}{\partial x^2} = 0 \text{ for } x \in [0, 2\pi], \ t > 0, \tag{21.13}$$

$$u(0, t) = 0 \text{ and } u(2\pi, t) = 0 \text{ for all } t > 0, \tag{21.14}$$

$$u(x, 0) = u_0(x) \text{ and } \dot{u}(x, 0) = 0 \text{ for all } x \in [0, 2\pi], \tag{21.15}$$

with a continuous function $u_0 \in C([0, 2\pi])$. We start with a separation Ansatz amd seeking $u(x, t)$ as a product $u(x, t) = w(x)v(t)$. This gives for the equation (21.13)

$$\frac{1}{w(x)v(t)} \frac{\partial^2(w(x)v(t))}{\partial t^2} - \frac{1}{w(x)v(t)} \frac{\partial^2(w(x)v(t))}{\partial x^2} = 0$$

or

$$\frac{1}{v(t)} \frac{d^2 v(t)}{dt^2} = \frac{1}{w(x)} \frac{d^2 w(x)}{dx^2} \text{ for all } x \text{ and } t.$$

Hence we must have with some constant λ

$$\frac{d^2v(t)}{dt^2} = \lambda v(t) \text{ and } \frac{d^2w(x)}{dx^2} = \lambda w(x). \tag{21.16}$$

The boundary conditions (21.14) lead to

$$w(0)v(t) = 0 \text{ and } w(2\pi)v(t) = 0 \text{ for all } t > 0. \tag{21.17}$$

In order to avoid a trivial solution, which in general will not satisfy (21.15), we have to assume that

$$w(0) = 0 \text{ and } w(2\pi) = 0.$$

Thus we are seeking $w : [0, 2\pi] \to \mathbb{R}$ solving

$$\frac{d^2w(x)}{dx^2} = \lambda w(x), \ w(0) = w(2\pi) = 0. \tag{21.18}$$

The value of λ is also to be determined and only when $\lambda = -k^2$, $k \in \mathbb{N}$, can we find a non-trivial solution to (21.18), namely for $A_k \in \mathbb{R}$ the function $A_k \sin kx$. Indeed:

$$\frac{d^2}{dx^2} A_k \sin kx = -k^2 A_k \sin kx \text{ and } A_k \sin k0 = 0 = A \sin 2\pi k.$$

Now we may insert $\lambda = -k^2$ in the equation to find v. Thus we have to look at $\frac{d^2v(t)}{dt^2} = -k^2 v(t)$ which is solved by any linear combination of $\cos kt$ and $\sin kt$. So far we have obtained for every $k \in \mathbb{N}$ functions $u_k(x, t) = w_k(x)v_k(t)$, namely

$$u_k(x, t) = \gamma_k \sin kx \sin kt + \delta_k \sin kx \cos kt$$

solving (21.13) with $u_k(0, t) = u_k(2\pi, t) = 0$ and $\frac{\partial^2 u_k(x,t)}{\partial t^2} - \frac{\partial^2 u_k(x,t)}{\partial x^2} = 0$. The wave equation is a linear equation and the boundary conditions are linear and homogeneous. Thus every finite linear combination of solutions gives a further solution to $\frac{\partial^2 u}{\partial t^2} - \frac{\partial^2 u}{\partial x^2} = 0$ and $u(0, t) = u(2\pi, t) = 0$. Therefore we may look at

$$u(x, t) = \sum_{k=1}^{N} (\gamma_k \sin kx \sin kt + \delta_k \sin kx \cos kt).$$

Differentiation yields

$$\frac{\partial^2 u(x, t)}{\partial t^2} = -\left(\sum_{k=1}^{N} (k^2 \gamma_k \sin kx \sin kt + k^2 \delta_k \sin kx \cos kt) \right) = \frac{\partial^2 u(x, t)}{\partial x^2},$$

i.e. $\frac{\partial^2 u(x,t)}{\partial t^2} - \frac{\partial^2 u(x,t)}{\partial x^2} = 0$. Further we have $u(0,t) = u(2\pi, t) = 0$ for all $t > 0$. So far we have not yet looked at the initial conditions. They become

$$u_0(x) = \sum_{k=1}^{N} (\gamma_k \sin kx \sin k0 + \delta_k \sin kx \cos k0) = \sum_{k=1}^{N} \delta_k \sin kx \qquad (21.19)$$

and

$$0 = \sum_{k=1}^{N} k\gamma_k \sin kx \cos k0 - \sum_{k=1}^{N} k\delta_k \sin kx \sin k0 = \sum_{k=1}^{N} k\gamma_k \sin kx. \qquad (21.20)$$

From (21.20) we can deduce that $\gamma_k = 0$ for all $k = 1, \ldots, N$. However, a given continuous function u_0 defined on $[0, 2\pi]$ will in general not admit a representation (21.19) with finitely many values $\delta_1, \ldots, \delta_N$.

At this point several aspects are coming together: we have a sequence $(\lambda_k)_{k \in \mathbb{N}}$ of values, namely $\lambda_k = -k^2$, for which

$$\frac{d^2 w(x)}{dx^2} = \lambda_k w(x) = -k^2 w(x), \quad x \in [0, 2\pi] \qquad (21.21)$$

has non-trivial solutions which satisfy

$$w(0) = w(2\pi) = 0. \qquad (21.22)$$

Furthermore, if $u_0 \in C([0, 2\pi])$ admits a pointwise representation as a Fourier series $u_0(x) = \sum_{k=1}^{\infty} \delta_k \sin kx$, then we may expect that in some situations

$$u(x, t) = \sum_{k=1}^{\infty} \delta_k \sin kx \cos kt$$

solves (21.13)-(21.15), where δ_k are the Fourier coefficients of u_0 when expanding u_0 into a sine-Fourier series. The problem is that we need to have a certain decay of the coefficients δ_k so that we can differentiate term by term. Thus solving problem (21.13)-(21.15) leads to (21.21)-(21.22) and a question in Fourier analysis.

We now return to general Sturm-Liouville operators (21.8) and the conditions $B^a(u) = 0$, $B^b(u) = 0$ with $B^a(u)$ and $B^b(u)$ as in (21.9) and (21.10) respectively.

Definition 21.3. *Let* $Lu = (p\dot{u})^{\cdot} + qu$ *be a Sturm-Liouville operator with* $p \in C^1([a,b])$, $p > 0$, *and* $q \in C([a,b])$. *The **eigenvalue problem** for* L *under the* *homogeneous boundary conditions (21.9) and (21.10) is the problem to find all values* λ *for which*

$$Lu = \lambda u \text{ in } [a, b], \tag{21.23}$$

$$B^a(u) = B^b(u) = 0 \tag{21.24}$$

has a non-trivial solution, i.e. a solution not identically zero. We call such a λ *an* **eigenvalue** *of L and a non-trivial solution to (21.23) an* **eigenfunction** *of L to the eigenvalue* λ.

Remark 21.4. A. From linear algebra we are used to calling, for a linear mapping $A : \mathbb{R}^n \to \mathbb{R}^n$, the problem $Ax = \lambda x$, $x \neq 0$, the eigenvalue problem for A, λ an eigenvalue and x an eigenvector. In this sense the notions defined in Definition 21.3 are an extension of those known from linear algebra.
B. Problem $(21.23)/(21.24)$ always has the trivial solution and linear combinations of eigenfunctions to one and the same eigenvalue are again eigenfunctions. Thus the eigenfunctions for a fixed eigenvalue form a linear subspace in the space $C([a, b])$ and we call this space the **eigenspace** corresponding to λ. (We will soon however switch from $C([a, b])$ to $L^2([a, b])$.)

In order to get some first ideas of how to handle eigenvalue problems for Sturm-Liouville operators we restrict ourselves for the moment to the boundary conditions $B^a(u) = u(a) = 0$ and $B^b(u) = u(b) = 0$, which we may refer to as **homogeneous Dirichlet conditions**. In linear algebra we have seen that eigenvalue problems are in particular nice to handle when a scalar product is at our disposal. On $C([a, b])$ we may always use the Riemann integral as a scalar product. For $Lu = \lambda u$ we now find under homogeneous Dirichlet conditions for an eigenfunction $u_\lambda \in C^2([a, b])$ and $v \in C^2([a, b])$ that

$$\int_a^b (Lu_\lambda)v \, \mathrm{d}x = \int_a^b \left(\frac{\mathrm{d}}{\mathrm{d}x}\left(p(x)\frac{\mathrm{d}u_\lambda(x)}{\mathrm{d}x} \right)v(x) + q(x)u_\lambda(x)v(x) \right) \, \mathrm{d}x$$

$$= -\int_a^b \left(p(x)\frac{\mathrm{d}u_\lambda(x)}{\mathrm{d}x}\frac{\mathrm{d}v(x)}{\mathrm{d}x} - q(x)u_\lambda(x)v(x) \right) \, \mathrm{d}x$$

$$= \lambda \int_a^b u_\lambda(x)v(x) \, \mathrm{d}x,$$

where the homogeneous Dirichlet conditions assure that no boundary term enters when integrating by parts. We are allowed to take as v the function u_λ and then we find

$$-\int_a^b \left(p(x)\left(\frac{\mathrm{d}u_\lambda(x)}{\mathrm{d}x}\right)^2 - q(x)u_\lambda^2(x) \right) \, \mathrm{d}x = \lambda \int_a^b u_\lambda^2(x) \, \mathrm{d}x. \tag{21.25}$$

This identity has an interesting consequence: if $q < 0$ then every eigenvalue of L must be a negative number, recall that $p > 0$.

Before we continue to investigate Sturm-Liouville operators we need to add a remark to the framework in which we shall operate. We have seen on several occasions that complete normed spaces, i.e. Banach spaces, are desirable for dealing with problems in analysis, especially for solving equations. The space $C([a, b])$ is a Banach space with respect to the sup-norm but not with respect to the norm induced by the scalar product

$$< u, v >_{L^2} = \int_a^b u(t)v(t) \, dt.$$

Therefore it makes sense to switch to the Hilbert space $L^2([a, b])$ or even to a weighted L^2-space over $[a, b]$ where the underlying measure becomes $p\lambda^{(1)}$. Of course for $u \in L^2([a, b])$ a pointwise statement such as $u(a) = 0$ does not make sense. However, as in the case of $C([a, b])$ we can anyway not define the Liouville operator L on all of $L^2([a, b])$ (or $C([a, b])$) due to the lack of differentiability of general elements of $L^2([a, b])$ (or $C([a, b])$). In the following we consider $Lu = (p\dot{u})^{\cdot} + qu$ with $p \in C^1([a, b])$, $p > 0$, and $q \in C([a, b])$, but we consider L only for certain functions belonging to $C^2([a, b]) \subset L^2([a, b])$. This allows us to already obtain first interesting results for eigenvalue problems; it enables us to link our results to our consideration in Part 8 Fourier Analysis, and at the same time we can prepare the functional analytic treatment of eigenvalue problems for differential operators which we will discuss in Volume V.

Later on, in Volume V, we will also understand why for eigenvalue problems we should prefer to work in the complex L^2-space, i.e. $\{u : [a, b] \to \mathbb{C} \mid u$ is measurable and $\int_a^b |u(x)|^2 \, dx < \infty\}$, even when dealing with Sturm-Liouville operators L with real-valued coefficients. For such operators L we find for $w = u + iv$, u, v real-valued L^2-functions $Lw = Lu + iLv$. The scalar product in the complex L^2-space we denote for the moment by $< \cdot, \cdot >_{\mathbb{C}, L^2}$.

An important observation is that the operator $Lu = (p\dot{u})^{\cdot} + qu$ with domain $\{u \in C^2([a, b]) \mid u(a) = u(b) = 0\}$ is **symmetric** in the sense that

$$< Lu, v >_{L^2} = < u, Lv >_{L^2} \tag{21.26}$$

holds for all real-valued $u, v \in \{w \in C^2([a, b]) \mid w(a) = w(b) = 0\}$. Indeed, integrating by parts twice yields, due to the vanishing boundary conditions,

$$
\begin{aligned}
< Lu, v >_{L^2} &= \int_a^b \left(\frac{\mathrm{d}}{\mathrm{d}x}\left(p(x)\frac{\mathrm{d}u(x)}{\mathrm{d}x} \right) v(x) + q(x)u(x)v(x) \right) \, \mathrm{d}x \\
&= \int_a^b \left(-p(x)\frac{\mathrm{d}u(x)}{\mathrm{d}x}\frac{\mathrm{d}v(x)}{\mathrm{d}x} + q(x)u(x)v(x) \right) \, \mathrm{d}x \\
&= \int_a^b u(x)\frac{\mathrm{d}}{\mathrm{d}x}\left(p(x)\frac{\mathrm{d}v(x)}{\mathrm{d}x} \right) + q(x)u(x)v(x) \, \mathrm{d}x \\
&= < u, Lv >_{L^2} .
\end{aligned}
$$

This observation allows us to prove certain results for the eigenvalue problem $Lu = \lambda u$, $u(a) = u(b) = 0$, which are analogous to results known for symmetric matrices.

Lemma 21.5. *Let $\lambda \neq \mu$ be two eigenvalues for the Sturm-Liouville operator $Lu = (p\dot{u})^{\cdot} + qu$ where p and q satisfy our standard conditions. If u_λ is an eigenfunction corresponding to λ and u_μ is an eigenfunction corresponding to μ then they are orthogonal in $L^2([a,b])$, i.e. $< u_\lambda, u_\mu >_{L^2} = 0$.*

Proof. We note that

$$
\begin{aligned}
(\lambda - \mu) < u_\lambda, u_\mu >_{L^2} &= < \lambda u_\lambda, u_\mu > - < u_\lambda, \mu u_\mu > \\
&= < Lu_\lambda, u_\mu > - < u_\lambda, Lu_\mu > \\
&= \int_a^b \left(-p(x)\frac{\mathrm{d}u_\lambda(x)}{\mathrm{d}x}\frac{\mathrm{d}u_\mu(x)}{\mathrm{d}x} + q(x)u_\lambda(x)u_\mu(x) \right) \, \mathrm{d}x \\
&\quad + \int_a^b p(x)\frac{\mathrm{d}u_\lambda(x)}{\mathrm{d}x}\frac{\mathrm{d}u_\mu(x)}{\mathrm{d}x} - q(x)u_\lambda(x)u_\mu(x) \, \mathrm{d}x \\
&= 0.
\end{aligned}
$$

Since $\lambda \neq \mu$ it follows that $< u_\lambda, u_\mu >_{L^2} = 0$. $\qquad\square$

Lemma 21.6. *The Sturm-Liouville operator $Lu = (p\dot{u})^{\cdot} + qu$ has, under our standard assumptions considered as operator acting on complex-valued functions, only real-valued eigenvalues.*

Proof. First we note that $\lambda u_\lambda = Lu_\lambda$ implies

$$
\begin{aligned}
\overline{\lambda}\overline{u}_\lambda = \overline{Lu_\lambda} &= \overline{(p\dot{u}_\lambda)^{\cdot}} + \overline{qu_\lambda} \\
&= (p\overline{\dot{u}}_\lambda)^{\cdot} + q\overline{u}_\lambda = (p\dot{\overline{u}}_\lambda)^{\cdot} + q\overline{u}_\lambda \\
&= L\overline{u}_\lambda.
\end{aligned}
$$

Next, assuming that $\|u_\lambda\|_{C,L^2} = 1$ we find

$$\lambda - \overline{\lambda} = < \lambda u_\lambda, u_\lambda >_{C,L^2} - < \lambda \overline{u}_\lambda, \overline{u}_\lambda >_{C,L^2}$$
$$= < Lu_\lambda, u_\lambda >_{C,L^2} - < L\overline{u}_\lambda, \overline{u}_\lambda >_{C,L^2}$$
$$= \int_a^b (-p(x)|\dot{u}_\lambda(x)|^2 + |u_\lambda(x)|^2) \, dx$$
$$- \int_a^b (-p(x)|\dot{\overline{u}}_\lambda|^2 + q|\overline{u}_\lambda(x)|^2) \, dx$$
$$= 0,$$

and it follows that $\lambda = \overline{\lambda}$, i.e. $\lambda \in \mathbb{R}$. □

It is clear by now that the symmetry of L is an important property, a more careful analysis shows that it does not only depend on the structure of L but also on the boundary conditions. For $p, q : [a, b] \to \mathbb{R}$, $a < b$, $p > 0$, and $p \in C^1([a, b])$, $q \in C([a, b])$ we find for $u, v \in C^2([a, b])$ that

$$\int_a^b (Lu)(x)v(x) \, dx = \int_a^b \left(\frac{d}{dx}\left(p(x)\frac{du(x)}{dx}\right)v(x) + q(x)u(x)v(x) \right) dx$$
$$= \int_a^b \left(-p(x)\frac{du(x)}{dx}\frac{dv(x)}{dx} + q(x)u(x)v(x) \right) dx$$
$$+ p(x)\frac{du(x)}{dx}v(x)\Big|_a^b$$
$$= \int_a^b \left(u(x)\frac{d}{dx}\left(p(x)\frac{dv(x)}{dx}\right) + q(x)u(x)v(x) \right) dx$$
$$+ p(x)\frac{du(x)}{dx}v(x)\Big|_a^b - p(x)u(x)\frac{dv(x)}{dx}\Big|_a^b$$
$$= \int_a^b u(x)Lv(x) \, dx + p(x)\left(\frac{du(x)}{dx}v(x) - u(x)\frac{dv(x)}{dx} \right)\Big|_a^b.$$

Thus, a domain of symmetry for L would be any subset of $C^2([a, b]) \subset L^2([a, b])$ for which the boundary term vanishes, i.e.

$$p(x)\left(\frac{du(x)}{dx}v(x) - u(x)\frac{dv(x)}{dx} \right)\Big|_a^b = 0 \qquad (21.27)$$

holds. We may investigate (21.27) further. The idea is that u is an element in the sought domain of L while $v \in C^2([a, b])$ is arbitrary. Therefore we can interpret in (21.27) the values $v(a)$, $v(b)$, $\frac{dv}{dx}(a)$ and $\frac{dv}{dx}(b)$ as free constants. Now (21.27) becomes

$$\alpha_1 p(a)u(a) + \alpha_2 p(a)\dot{u}(a) - (\beta_1 p(b)u(b) + \beta_2 p(b)\dot{u}(b)) = 0, \qquad (21.28)$$

however, we still have freedom in the constants α_1, α_2, β_1, β_2 and we may pose several of these conditions, for example in the form (21.9) and (21.10) separating the behaviour at a from that at b. Besides the conditions $u(a) = u(b) = 0$, we can find the more general conditions

$$\alpha_1 u(a) + \alpha_2 \dot{u}(a) = 0 \text{ and } \beta_1 u(b) + \beta_2 \dot{u}(b) = 0 \qquad (21.29)$$

which covers essentially most of the interesting cases. Summarizing these considerations we can give

Definition 21.7. *Given a Sturm-Liouville operator $Lu = (p\dot{u})^{\cdot} + qu$ with $p, q :$ $[a, b] \to \mathbb{R}$ satisfying our standard conditions. We call homogeneous boundary conditions (21.9)/(21.10) **symmetric boundary conditions for L** if L with the domain $\{u \in C^2([a, b]) \mid B^{(a)}u = B^{(b)}u = 0\}$ is a symmetric operator on $L^2([a, b])$, i.e. for all $u, v \in \{u \in C^2([a, b]) \mid B^{(a)}u = B^{(b)}u = 0\}$ we have*

$$< Lu, v >_{L^2} = < u, Lv >_{L^2} . \qquad (21.30)$$

Corollary 21.8. *For a regular Sturm-Liouville operator subjected to symmetric boundary conditions the assertions of Lemma 21.5 and Lemma 21.6 hold.*

Proof. By inspection we see that we only need (21.30) on the domain $\{u \in C^2([a, b]) \mid B^{(a)}u = B^{(b)}u = 0\}$ to hold in order to carry the proofs over to the general case. $\qquad \square$

Lemma 21.9. *Let L be a regular Sturm-Liouville operator with $p, q : [a, b] \to \mathbb{R}$ satisfying the standard conditions. If $Lu = \lambda u$ is subjected to the conditions (21.29) each eigenspace is one-dimensional.*

Proof. We know that every solution to $Lu - \lambda u = 0$ must be a linear combination of two solutions v_1 and v_2 forming a fundamental system. Thus every eigenfunction u_λ of the eigenvalue λ is of the type $u_\lambda = \gamma_1 v_1 + \gamma_2 v_2$, which implies that the eigenspace is at most two-dimensional. Given $\tilde{\gamma}_1$ and $\tilde{\gamma}_2$ such that $u_\lambda = \tilde{\gamma}_1 v_1 + \tilde{\gamma}_2 v_2$ with $u_\lambda(a) = \tilde{\gamma}_1$ and $\dot{u}_\lambda(a) = \tilde{\gamma}_2$. Now we obtain a linear relation between α_1, α_2 in (21.19) and $\tilde{\gamma}_1$, $\tilde{\gamma}_2$ by

$$\alpha_1 \tilde{\gamma}_1 + \alpha_2 \tilde{\gamma}_2 = 0$$

and therefore the dimension of the eigenspace corresponding to λ is reduced by 1. Since we have assumed to have a non-trivial solution u_λ, i.e. an eigenfunction, the second condition cannot reduce the dimension of the solution space further. $\qquad \square$

Before continuing our theoretical considerations we want to discuss an example

Example 21.10. The problem $\ddot{u} + \mu u = 0$, $\dot{u}(0) = 0$ and $u(1) = 0$ for a sought function $u : [0,1] \to \mathbb{R}$ is of some importance in elasticity theory. Considerations from physics imply that we need to assume $\mu > 0$. This follows also from the following pure mathematical consideration which is valid for any solution:

$$\mu \int_0^1 |u(t)|^2 \, dt = - \int_0^1 u(t) \ddot{u}(t) \, dt$$

$$= - \left\{ u\dot{u}\big|_0^1 - \int_0^1 |\dot{u}(t)|^2 \, dt \right\} = \int_0^1 |\dot{u}(t)|^2 \, dt.$$

Note that in our convention the eigenvalue is $-\mu < 0$. The solution to the problem must be of the form

$$u(t) = \gamma_1 \cos \sqrt{\mu}t + \gamma_2 \sin \sqrt{\mu}t$$

since $\cos \sqrt{\mu}t$ and $\sin \sqrt{\mu}t$ form a fundamental system for $\ddot{u} + \mu u = 0$. It follows that

$$\dot{u}(t) = -\gamma_1 \sqrt{\mu} \sin \sqrt{\mu}t + \gamma_2 \sqrt{\mu} \cos \sqrt{\mu}t$$

and since $0 = \dot{u}(0) = \gamma_2 \sqrt{\mu}$ we deduce that $\gamma_2 = 0$. It remains to determine μ from

$$u(1) = \gamma_1 \cos \sqrt{\mu} = 0$$

which gives $\mu = \mu_k = \frac{(2k+1)^2 \pi^2}{4}$, $k \in \mathbb{N}_0$. Thus for $\gamma_1 \in \mathbb{R}$ the function

$$u_k(t) := \gamma_1 \cos \frac{(2k+1)\pi}{2}$$

is an eigenfunction to $\ddot{u} = -\mu_k u$ in $[0,1]$, $\dot{u}(0) = 0$, $u(1) = 0$. (For an interpretation of μ_k we refer to [56].)

When discussing problem (21.13) - (21.15) for the one-dimensional wave equation, eigenfunctions to the operators $\frac{d^2}{dx^2}$ or $\frac{d^2}{dt^2}$, i.e. the operators which we obtain from a separation of variables Ansatz, turned out to be of importance. In particular the question whether we can expand some data or maybe the solution with respect to these eigenfunctions may open the way to solve equations. However, so far we do not even know whether a regular Sturm-Liouville operator has eigenvalues, hence eigenfunctions, nor do we know whether in the affirmative case these eigenfunctions allow us to expand a given function into a convergent series of eigenfunctions. With the considerations of Chapter 9 in mind we may think to obtain, for a given regular Sturm-Liouville operator, a set of eigenfunctions which are complete in $L^2([a,b])$.

Before we continue our studies of Sturm-Liouville operators we want to derive a representation formula for solutions of the Dirichlet boundary value problem with non-trivial right-hand side. Then we will return to Sturm-Liouville operators and more general boundary conditions.

Let $[a, b] \subset \mathbb{R}$ be a bounded interval and $p_0, p_1, p_2, f : [a, b] \to \mathbb{R}$ be continuous functions. We assume in addition that $p_0(x) \neq 0$, i.e. either $p_0(x) > 0$ or $p_0(x) < 0$ for all $x \in [a, b]$, and typically we will assume that $p_0(x) > 0$. In some considerations we may add the assumption that p_0 is a C^1-function. We are interested in solving the boundary value problem which is often called the **Dirichlet problem**, i.e.

$$Lu(x) = p_0(x)u''(x) + p_1'(x)u'(x) + p_2(x)u(x) = f(x) \quad \text{in } [a, b] \qquad (21.31)$$

and

$$u(a) = \alpha \quad \text{and} \quad u(b) = \beta. \qquad (21.32)$$

The reader might have spotted a small change in our notation, namely we write u' instead of $\dot{u}$ and we take x as a symbol for the independent variable and not t. Since some of our examples, especially in Volume V and Volume VI, will be partial differential equations from mathematical physics (such as (21.13)), and in physics t is reserved for the time-variable and $\dot{u}$ as derivative with respect to time, we prefer now to use when dealing with boundary value problems the other notation.

By our assumptions there exists a fundamental system $\{u_1, u_2\}$ for

$$p_0 u''(x) + p_1 u'(x) + p_2(x)u(x) = 0 \qquad (21.33)$$

and the general solution to (21.31) is given by

$$u(x) = \gamma_1 u_1(x) + \gamma_2 u_2(x) + u_0(x) \qquad (21.34)$$

where u_0 is a special solution to (21.31). The Wronski determinant of u_1 and u_2 is denoted by

$$W(x) := W(u_1, u_2)(x) := u_1(x)u_2'(x) - u_1'(x)u_2(x) \qquad (21.35)$$

and from our considerations in Chapter 19, see in particular Remark 19.22 and Problem 8, we know that $W(x) \neq 0$ for all $x \in [a, b]$ and

$$u_0(x) = \int_a^x \frac{u_1(y)u_2(x) - u_1(x)u_2(y)}{p_0(y)W(y)} f(y) \, dy \qquad (21.36)$$

is a special solution to (21.31). Combining (21.34) with (21.31) / (21.32) we arrive at the following linear system to determine γ_1 and γ_2

$$\begin{cases} u_1(a)\gamma_1 + u_2(a)\gamma_2 = \alpha - u_0(a) \\ u_1(b)\gamma_1 + u_2(b)\gamma_2 = \beta - u_0(b). \end{cases} \tag{21.37}$$

Assuming that

$$u_1(a)u_2(b) - u_1(b)u_2(a) \neq 0 \tag{21.38}$$

we can find γ_1 and γ_2 by Cramer's rule as

$$\gamma_1 = \frac{u_2(b)\alpha - u_2(b)u_0(a) - u_2(a)\beta + u_2(a)u_0(b)}{u_1(a)u_2(b) - u_1(b)u_2(a)} \tag{21.39}$$

and

$$\gamma_2 = \frac{u_1(a)\beta - u_1(a)u_0(b) - u_1(b)\alpha + u_1(b)u_0(a)}{u_1(a)u_2(b) - u_1(b)u_2(a)}. \tag{21.40}$$

With the helpful notation

$$\omega(x,y) = u_1(x)u_2(y) - u_1(y)u_2(x), \quad x, y \in I, \tag{21.41}$$

we find now from (21.34)

$$u(x) = \frac{1}{\omega(a,b)}\Big\{(u_2(b)\alpha - u_2(b)u_0(a) - u_2(a)\beta + u_2(a)u_0(b))u_1(x)+$$

$$(u_1(a)\beta - u_1(a)u_0(b) - u_1(b)\alpha + u_1(b)u_0(a))u_2(x)\Big\} + u_0(x)$$

$$= \frac{\omega(x,b)}{\omega(a,b)}\alpha - \frac{\omega(x,a)}{\omega(a,b)}\beta + \frac{\omega(x,a)}{\omega(a,b)}\int_a^b \frac{\omega(y,b)}{p_0(y)W(y)}f(y)\,dy + \int_a^x \frac{\omega(x,y)}{p_0(y)W(y)}f(y)\,dy,$$

where we have used that

$$u_0(a) = \int_a^a \frac{\omega(y,a)}{p_0(y)W(y)}f(y)\,dy = 0.$$

Since

$$\int_a^b \frac{\omega(y,b)}{p_0(y)W(y)}f(y)\,dy = \int_a^x \frac{\omega(y,b)}{p_0(y)W(y)}f(y)\,dy + \int_x^b \frac{\omega(y,b)}{p_0(y)W(y)}f(y)\,dy$$

we find

$$\frac{\omega(x,a)}{\omega(a,b)}\int_a^b \frac{\omega(y,b)}{p_0(y)W(y)}f(y)\,dy + \int_a^x \frac{\omega(y,x)}{p_0(y)W(y)}f(y)\,dy$$

$$= \int_a^x \left(\frac{\omega(x,a)\omega(y,b)}{\omega(a,b)} + \omega(y,x)\right)\frac{f(y)}{p_0(y)W(y)}\,dy + \int_x^b \frac{\omega(x,a)\omega(y,b)}{\omega(a,b)}\frac{f(y)}{p_0(y)W(y)}\,dy.$$

368

A straightforward calculation, see Problem 3, yields

$$\frac{\omega(x,a)\omega(y,b)}{\omega(a,b)} + \omega(y,x) = \frac{\omega(x,b)\omega(y,a)}{\omega(a,b)}. \tag{21.42}$$

Definition 21.11. *The function* $G : [a,b] \times [a,b] \to \mathbb{R}$ *defined by*

$$G(x,y) := \begin{cases} \frac{\omega(x,b)\omega(y,a)}{\omega(a,b)} \frac{1}{p_0(y)W(y)}, & a \le y \le x \le b \\ \frac{\omega(x,a)\omega(y,b)}{\omega(a,b)} \frac{1}{p_0(y)W(y)}, & a \le x \le y \le b \end{cases} \tag{21.43}$$

is called the **Green function** *for the Dirichlet problem for L in the interval $[a,b]$.*

Thus we have proved

Theorem 21.12. *Let* $u : [a,b] \to \mathbb{R}$ *be a solution to* (21.31)/(21.32) *and let* $\{u_1, u_2\}$ *be a fundamental system for* (21.33) *which satisfies* $u_1(a)u_2(b) - u_1(b)u_2(a) \ne 0$. *Then u has the representation*

$$u(x) = \frac{\alpha u_2(b) - \beta u_2(a)}{\omega(a,b)} u_1(x) + \frac{\beta u_1(a) - \alpha u_1(b)}{\omega(a,b)} u_2(x)$$

$$+ \int_a^b G(x,y)f(y)\,dy. \tag{21.44}$$

We want to study Green's function G for the Dirichlet problem in more detail. The following figure is helpful to understand G:

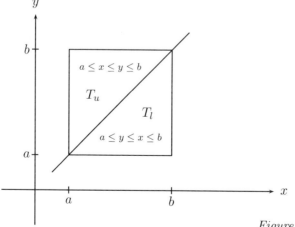

Figure 21.1

369

In Figure 21.1 we have $T_l = \{(x,y) \in [a,b] \times [a,b] \,|\, a \le y \le x \le b\}$ and $T_u = \{(x,y) \in [a,b] \times [a,b] \,|\, a \le x \le y \le b\}$.

First we note that G is continuous on $[a,b] \times [a,b]$ which follows immediately from (21.43). Moreover, for all points (a,y), $y \in [a,b]$, it follows that $G(a,y) = 0$ and for (b,y), $y \in [a,b]$, we also have $G(b,y) = 0$ since $\omega(a,a) = \omega(b,b) = 0$. In $\mathring{T}_u$ and $\mathring{T}_l$ the function G has a first and second order continuous partial derivative with respect to x and it satisfies as a function of x the differential equation (21.33). We remark that the existence of partial derivatives of G with respect to y depends on differentiability properties of p_0. We want to study the first order partial derivative of G with respect to x at the diagonal of $[a,b] \times [a,b]$, i.e. at points $x = y$. For $x > y$ we find

$$\frac{\partial}{\partial x} G(x,y) = \frac{\partial}{\partial x} \left(\frac{\omega(x,b)\omega(y,a)}{\omega(a,b)} \frac{1}{p_0(y)W(y)} \right)$$
$$= \frac{(u_1'(x)u_2(b) - u_1(b)u_2'(x))\,\omega(y,a)}{\omega(a,b)} \frac{1}{p_0(y)W(y)}$$

and for $x \to y$, $x > y$, we find

$$\lim_{\substack{x \to y \\ x>y}} \frac{\partial G}{\partial x}(x,y) = \frac{(u_1'(y)u_2(b) - u_1(b)u_2'(y))\,\omega(y,a)}{\omega(a,b)} \frac{1}{p_0(y)W(y)}, \qquad (21.45)$$

whereas for $x < y$ it follows that

$$\frac{\partial}{\partial x} G(x,y) = \frac{\partial}{\partial x} \left(\frac{\omega(x,a)\omega(y,b)}{\omega(a,b)} \frac{1}{p_0(y)W(y)} \right)$$
$$= \frac{(u_1'(x)u_2(a) - u_1(a)u_2'(x))\omega(y,b)}{\omega(a,b)} \frac{1}{p_0(y)W(y)}$$

and therefore

$$\lim_{\substack{x \to y \\ x<y}} \frac{\partial G}{\partial x}(x,y) = \frac{(u_1'(y)u_2(a) - u_1(a)u_2'(y))\omega(y,b)}{\omega(a,b)} \frac{1}{p_0(y)W(y)}. \qquad (21.46)$$

From (21.45) and (21.46) we conclude that

$$\lim_{\substack{x \to y \\ x>y}} \frac{\partial G}{\partial x}(x,y) - \lim_{\substack{x \to y \\ x<y}} \frac{\partial G}{\partial y}(x,y)$$
$$= \frac{(u_1'(y)u_2(b) - u_1(b)u_2'(y))\,\omega(y,a) - (u_1'(y)u_2(a) - u_1(a)u_2'(y))\,\omega(y,b)}{\omega(a,b)p_0(y)W(y)}.$$

However

$$\big(u_1'(y)u_2(b) - u_1(b)u_2'(y)\big)\,\omega(y,a) - \big(u_1'(y)u_2(a) - u_1(a)u_2'(y)\big)\,\omega(y,b) = \omega(a,b)W(y)$$

which implies

$$\lim_{\substack{x \to y \\ x > y}} \frac{\partial G}{\partial x}(x,y) - \lim_{\substack{x \to y \\ x < y}} \frac{\partial G}{\partial x}(x,y) = \frac{1}{p_0(y)} \neq 0, \tag{21.47}$$

i.e. $\frac{\partial G}{\partial x}$ has a jump discontinuity on the diagonal of $[a,b] \times [a,b]$. We summarise our results in

Lemma 21.13. *The Green function $(x,y) \mapsto G(x,y)$ for the Dirichlet problem for L on $[a,b]$ is a continuous function on $[a,b] \times [a,b]$ and it has continuous first and second order partial derivatives with respect to x in $\mathring{T}_u$ and $\mathring{T}_l$ where it also satisfies the homogeneous equation (21.33). Further, G fulfills the boundary conditions $G(a,y) = G(b,y) = 0$ for $y \in [a,b]$. Moreover $\frac{\partial G}{\partial x}$ has a jump discontinuity on the diagonal of $[a,b] \times [a,b]$ given by (21.47).*

Theorem 21.14. *Let G be the Green function for the Dirichlet problem for L in $[a,b]$ and let $f \in C([a,b])$. Then*

$$u(x) := \int_a^b G(x,y)f(y)\,\mathrm{d}y \tag{21.48}$$

belongs to $C^2([a,b])$ and is a solution to the equation

$$Lu(x) = p_0(x)u''(x) + p_1(x)u'(x) + p_2u(x) = f(x)$$

under homogeneous Dirichlet conditions $u(a) = 0$ and $u(b) = 0$.

Proof. Since G is continuous on $[a,b] \times [a,b]$ and f on $[a,b]$ the integral in (21.48) is for all $x \in [a,b]$ well defined and further we have

$$u(a) = \int_a^b G(a,y)f(y)\,\mathrm{d}y = 0$$

and

$$u(b) = \int_a^b G(b,y)f(y)\,\mathrm{d}y = 0.$$

Now we split the integral and we write

$$u(x) = \int_a^x G(x,y)f(y)\,\mathrm{d}y + \int_x^b G(x,y)f(y)\,\mathrm{d}y. \tag{21.49}$$

Differentiation in (21.49) yields

$$u'(x) = G(x, x)f(x) + \int_a^x \frac{\partial G}{\partial x}(x, y)f(y)\,dy - G(x, x)f(x) + \int_x^b \frac{\partial G}{\partial x}(x, y)f(y)\,dy$$

$$= \int_a^b \frac{\partial G}{\partial x}(x, y)f(y)\,dy,$$

and further

$$u''(x) = \frac{d}{dx}\left(\int_a^x \frac{\partial G}{\partial x}(x, y)f(y) + \int_x^b \frac{\partial G}{\partial x}(x, y)f(y)\,dy\right)$$

$$= \left(\lim_{\substack{\xi \to x \\ x > \xi}} \frac{\partial G}{\partial x}(\xi, x)\right)f(x) + \int_a^x \frac{\partial^2 G}{\partial x^2}(x, y)f(y)\,dy$$

$$- \left(\lim_{\substack{\xi \to x \\ x < \xi}} \frac{\partial G}{\partial x}(\xi, x)\right)f(x) + \int_x^b \frac{\partial^2 G}{\partial x^2}(x, y)f(y)\,dy$$

$$= \frac{f(x)}{p_0(x)} + \int_a^b \frac{\partial^2 G}{\partial x^2}(x, y)f(y)\,dy,$$

which leads to

$$p_0(x)u''(x) + p_1(x)u'(x) + p_2(x)u(x)$$

$$= f(x) + \int_a^b \left(p_0(x)\frac{\partial^2 G}{\partial x^2}(x, y) + p_1(x)\frac{\partial G}{\partial x}(x, y) + p_2(x)G(x, y)\right)f(y)\,dy$$

$$= f(x)$$

where we have used the fact that G satisfies the equation $Lu = 0$ in $\mathring{T}_u \cup \mathring{T}_l$. $\quad\square$

Formally G depends on the choice of the fundamental system $\{u_1, u_2\}$. However with the help of Theorem 21.14 we obtain the following uniqueness result for G:

Corollary 21.15. *Let G_1 and G_2 be two functions satisfying the conclusions of Lemma 21.13. Then we have $G_1 = G_2$.*

Proof. Let $f \in C([a, b])$ be an arbitrary function. It follows that

$$h(x) := \int_a^b (G_1(x, y) - G_2(x, y))\,f(y)\,dy$$

solves the equation $Lh(x) = p_0(x)h''(x) + p_1 h'(x) + p_2(x)h(x) = 0$ and $h(a) = 0$ as well as $h(b) = 0$. Hence $h(x) = 0$ for all $x \in [a, b]$, or

$$0 = \int_a^b (G_1(x, y) - G_2(x, y)) \, f(y) \, dy$$

for all $f \in C([a, b])$. For $x \in [a, b]$ fixed this implies that $G_1(x, y) = G_2(x, y)$ for all $y \in [a, b]$. Since x is arbitrary we conclude that $G_1(x, y) = G_2(x, y)$ for all $x, y \in [a, b]$. □

We now consider the eigenvalue problem

$$\begin{cases} Lu(x) = p_0(x)u''(x) + p_1(x)u'(x) + p_2(x)u(x) = \lambda u(x), & x \in [a, b] \\ u(a) = u(b) = 0. \end{cases} \tag{21.50}$$

Using Green's function for the Dirichlet problem as well as Theorem 21.14 we find for a solution u of (21.50) the identity

$$u(x) = \int_a^b G(x, y)\lambda u(y) \, dy, \tag{21.51}$$

or under the assumption $\lambda \neq 0$ with $\mu = \frac{1}{\lambda}$

$$\int_a^b G(x, y)u(y) \, dy = \mu u(x). \tag{21.52}$$

Denote for a moment by $C_0([a, b])$ the space of all elements $u \in C([a, b])$ with the property $u(a) = u(b) = 0$, and introduce the linear mapping (operator)

$$G_{\mathrm{op}} : C_0([a, b]) \to C_0([a, b]) \tag{21.53}$$
$$v \mapsto G_{\mathrm{op}} v$$

where

$$(G_{\mathrm{op}} v)(x) = \int_a^b G(x, y)v(y) \, dy. \tag{21.54}$$

In light of Theorem 21.14 this is a well-defined linear operator and now (21.52) reads as

$$G_{\mathrm{op}} u = \mu u \tag{21.55}$$

as an equation in $C_0([a, b])$. Thus, if $\lambda \neq 0$ is an eigenvalue for L under Dirichlet conditions then $\mu = \frac{1}{\lambda}$ is an eigenvalue for G_{op}. It turns out that the eigenvalue problem for G_{op} is much easier to handle than our original problem. Note that we have once again transformed a problem for a differential equation into a problem for an integral operator. In Volume V we will develop the tools to solve the eigenvalue problem (21.55) even in a much wider context which shall allow us to include also singular Sturm-Liouville operators, i.e. Sturm-Liouville operators $Lu = \frac{d}{dx}\left(p\frac{du}{dx}\right) + qu$ where p might have a zero at the boundary points a or b, or where $[a, b]$ is replaced by an unbounded interval, say $[0, \infty)$. The Bessel differential equation is of this type as is the Legendre differential equation. We will establish the existence of eigenvalues and prove that the corresponding eigenfunctions form a complete orthonormal system in an appropriate Hilbert space which in turn allows us to give series solutions to boundary value problems. This shows how powerful the idea of introducing Green's function is and we want to extend this idea to regular Sturm-Liouville operators under more general boundary conditions. For this let

$$Lu = \frac{d}{dx}\left(p(\cdot)\frac{du}{dx}\right) + q(\cdot)u \tag{21.56}$$

be a regular Sturm-Liouville operator with $p, q : [a, b] \to \mathbb{R}$ satisfying our standard conditions. We start with

Definition 21.16. *A function* $E : [a, b] \times [a, b] \to \mathbb{R}$ *is called a* **fundamental solution** *to the regular Sturm-Liouville operator* (21.56) *if*

i) *E is continuous;*

ii) *$E|_{T_u \cup T_l}$ has continuous partial derivatives of order one and two with respect to x where T_u and T_l are as in Figure 21.1;*

iii) *for every $y \in (a, b)$ fixed $x \mapsto E(x, y)$ solves the equation $LE(x, y) = 0$ for $x \neq y$;*

iv) *on the diagonal of $[a, b] \times [a, b]$ the partial derivative $\frac{\partial E}{\partial x}$ has the jump $\frac{1}{p(\cdot)}$, i.e.*

$$\lim_{\substack{\xi \to x \\ \xi > x}} \frac{\partial E}{\partial \xi}(\xi, x) - \lim_{\substack{\xi \to x \\ \xi < x}} \frac{\partial E}{\partial \xi}(\xi, x) = \frac{1}{p(x)}, \quad x \in (a, b). \tag{21.57}$$

First we note that the Green function for the Dirichlet problem is a fundamental solution to L given by (21.56). Further, an inspection of the proof of Theorem 21.14 yields that for every fundamental solution E and $f \in C([a, b])$ a solution to the equation

$$Lu = f \tag{21.58}$$

is given by

$$u(x) = \int_a^b E(x, y)f(y)\,dy. \tag{21.59}$$

374

In addition we note that if E is a fundamental solution to L and u_0 is any solution to $Lu = 0$ then for every continuous function $h \in C([a, b])$ the function $E_1 : [a, b] \times [a, b] \to \mathbb{R}$ defined by

$$E_1(x, y) = E(x, y) + h(y)u_0(x) \tag{21.60}$$

is a further fundamental solution to L.

Example 21.17. A. The function $E(x, y) = -\frac{1}{2}|x - y|$ is for every interval $[a, b]$ a fundamental solution to $\frac{d^2}{dx^2}$. Indeed $i)$ and $ii)$ from Definition 21.16 are trivial as is $iii)$. To see $iv)$ we note that $E(x, y) = -\frac{1}{2}|x - y| = -\frac{1}{2}(y - x)$ for $y > x$ and therefore $\lim_{\xi \to x \atop x > 0} \frac{\partial E}{\partial \xi}(\xi, x) = \frac{1}{2}$, and for $x > y$ we have $E(x, y) = -\frac{1}{2}(x - y)$ and $\lim_{\xi \to x \atop x < \xi} \frac{\partial E}{\partial \xi}(\xi, x) = -\frac{1}{2}$ implying that the jump of $\frac{\partial E}{\partial x}$ on the diagonal is 1 - as we expect.

B. We claim that $E_\lambda(x, y) = -\frac{1}{2\lambda} \sin \lambda |x - y|$, $\lambda \neq 0$, is a fundamental solution to $\frac{d^2}{dx^2} + \lambda^2 \mathrm{id}$. Again $i)$- $iii)$ are trivial. For $y > x$ we have $E_\lambda(x, y) = -\frac{1}{2\lambda} \sin(-\lambda(y - x))$ which yields $\lim_{\xi \to x \atop \xi > x} \frac{\partial E}{\partial \xi}(\xi, x) = \frac{1}{2}$ and for $x > y$ it follows that $E_\lambda(x, y) = -\frac{1}{2\lambda} \sin \lambda(x - y)$ and therefore $\lim_{\xi \to x \atop \xi < x} \frac{\partial E}{\partial \xi} \lambda(\xi, x) = -\frac{1}{2}$ implying again that the jump of $\frac{\partial E_\lambda}{\partial x}$ on the diagonal has the value 1.

Definition 21.18. *Let G_{B^a, B^b} be a fundamental solution to the regular Sturm-Liouville operators (21.56). We call G_{B^a, B^b} **Green's function** for L under the boundary conditions B^a and B^b if $B^a G_{B^a, B^b}(x, y) = B^b G_{B^a, B^b}(x, y) = 0$ for all $y \in (a, b)$ where the boundary operators act on the variable x.*

Theorem 21.19. *Under condition (21.12) the Green function G_{B^a, B^b} exists and is unique.*

Proof. We fix a fundamental system $\{u_1, u_2\}$ for L and we search for functions $g_j, h_j, j = 1, 2$, such that

$$\begin{cases} G_{B^a, B^b}(x, y) = (g_1(y) + h_1(y))u_1(x) + (g_2(y) + h_2(y))u_2(x), & (x, y) \in T_l \\ G_{B^a, B^b}(x, y) = (g_1(y) + h_1(y))u_1(x) + (g_2(y) - h_2(y))u_2(x), & (x, y) \in T_u. \end{cases} \tag{21.61}$$

The continuity of $G := G_{B^a, B^b}$ in $[a, b] \times [a, b]$ yields

$$h_1(y)u_1(y) + h_2(y)u_2(y) = 0 \tag{21.62}$$

and the jump discontinuity of $\frac{\partial G}{\partial x}$ on the diagonal implies

$$h_1(y)u_1'(y) + h_2(y)u_2'(y) = \frac{1}{2p(y)}. \tag{21.63}$$

The system $(21.62)/(21.63)$ has a unique solution $h_1(y)$ and $h_2(y)$ since the determinant of the system is just the Wronski determinant of the fundamental system $\{u_1, u_2\}$. Once h_1 and h_2 are determined we can find g_1 and g_2 by looking at

$$(g_1(y) - h_2(y))B^a(u_1) + (g_2(y) - h_1(y))B^a(u_2) = 0$$

and

$$(g_1(y) + h_2(y))B^a(u_2) + (g_2(y) + h_1(y))B^b(u_2) = 0.$$

From these two equations we can uniquely determine g_1 and g_2 provided that $\det \begin{pmatrix} B^a(u_1) & B^a(u_2) \\ B^b(u_1) & B^b(u_2) \end{pmatrix} \neq 0$ which is however condition (21.12). Then $G = G_{B^a, B^b}$ as defined by (21.61) is the Green function to L under the boundary conditions B^a and B^b. $\square$

From the proof of Theorem 21.14 we can now deduce

Corollary 21.20. *The problem $Lu = f$ in $[a, b]$ and $B^a u = B^b u = 0$ has for every $f \in C([a, b])$ a unique solution given by*

$$u(x) = \int_a^b G_{B^a, B^b}(x, y)f(y)\,\mathrm{d}y. \tag{21.64}$$

Finally we prove a non-trivial property of the Green function when the boundary conditions are symmetric in the sense of Definition 21.7.

Theorem 21.21. *The Green function $G = G_{B^a, B^b}$ to the regular Sturm-Liouville operator L given by (21.56) is under symmetric boundary conditions symmetric, i.e.*

$$G(x, y) = G(y, x). \tag{21.65}$$

Proof. For $f, g \in C([a, b])$ we consider

$$u(x) = \int_a^b G(x, y)f(y)\,\mathrm{d}y$$

and

$$v(x) = \int_a^b G(x, y)g(y)\,\mathrm{d}y.$$

The symmetry of the boundary conditions implies

$$\int_a^b (uLv - vLu)\,\mathrm{d}x = 0,$$

or, with $Lu = f$ and $Lv = g$,

$$\int_a^b \int_a^b f(x)G(x,y)g(y)\,\mathrm{d}y\,\mathrm{d}x = \int_a^b \int_a^b g(x)G(x,y)f(y)\,\mathrm{d}y\,\mathrm{d}x,$$

i.e.

$$\int_a^b \int_a^b (G(x,y) - G(y,x))f(y)g(x)\,\mathrm{d}y\,\mathrm{d}x = 0$$

for all $f, g \in C([a,b])$ which implies $G(x,y) = G(y,x)$. □

In writing this chapter we have benefitted much from the monographs [32], [34], [50], [56] and [125].

Problems

1. Discuss the unique solvability of the equation $\ddot{u} + w^2 u = 0$ in $[a,b]$ under the boundary conditions

   a) $u(a) = \alpha, \dot{u}(b) = \beta$

   and

   b) $\dot{u}(a) = \alpha, \dot{u}(b) = \beta$.

2. Discuss the wave equation $\frac{\partial^2 u}{\partial t^2} - \frac{\partial^2 u}{\partial x^2} = 0$ for $x \in [0, 2\pi]$ and $t > 0$ under the conditions

   $$\frac{\partial u}{\partial x}(0,t) = 0, \quad \frac{\partial u}{\partial x}(2\pi, t) = 0,$$

   $$u(x,0) = u_0(x), \quad u_0 \in C^4(\mathbb{R}), \ 2\pi\text{-periodic and even},$$

   $$\frac{\partial u}{\partial t}(x,0) = 0.$$

3. Prove identity (21.42).

4. Consider the equation $\ddot{u}(t) + b\dot{u}(t) = \lambda u(t)$, $b \neq 0$, in $[0,1]$ subjected to the boundary condition $u(0) = u(1) = 0$. For which values of λ does a non-trivial solution exist?

5. Find the Green function to the boundary value problem $(xy'(x))' = f(x)$, $x \in [1, e]$, $y(1) = y(e) = 0$. For $f(x) = x^m$, $m \in \mathbb{N}$, give an explicit solution.

6. Find the Green function for the Dirichlet problem $y'' - y = 0$ in $[0,1]$ and $u(0) = u(1) = 0$.

7. Find differential operators $L_1(x, D), L_2(x, D)$ and $L_3(x, D)$ such that for some λ_1, λ_2 and λ_3 the identity

$$L_j(x, D)u_j = \lambda_j u_j$$

holds, where $u_1 = P_n$ is the nth Legendre polynomial, $u_2 = T_n$ is the nth Chebyshev polynomial and $u_3 = J_n, n \in \mathbb{N}$ is the nth Bessel function.

22 The Hypergeometric and the Confluent Hypergeometric Differential Equation

We have encountered the **hypergeometric differential equation**

$$z(1-z)u''(z) + (\gamma - (\alpha + \beta + 1)z)u'(z) - \alpha\beta u(z) = 0, \quad z \in \mathbb{C}, \qquad (22.1)$$

on several occasions, for example in Chapter III.22, and one of its solutions the (**Gaussian**) **hypergeometric series** $F(\alpha, \beta; \gamma; z) = {}_2F_1(\alpha, \beta; \gamma; z)$,

$$_2F_1(\alpha, \beta; \gamma; z) = \sum_{k=0}^{\infty} \frac{(\alpha)_k (\beta)_k}{k!(\gamma_k)} z^k, \quad \gamma \notin -\mathbb{N}_0, \qquad (22.2)$$

see Chapter III.16. We also started to discuss the **generalised hypergeometric series**

$$_pF_q(\alpha_1, \ldots, \alpha_p; \gamma_1, \ldots, \gamma_q; z) = \sum_{k=0}^{\infty} \frac{(\alpha_1)_k \cdots (\alpha_p)_k}{k!(\gamma_1)_k \cdots (\gamma_q)_k} z^k. \qquad (22.3)$$

Here we used the **Pochhammer symbols**

$$(\alpha)_0 := 1 \quad \text{and} \quad (\alpha)_k := \alpha(\alpha + 1) \cdots (\alpha + k - 1). \qquad (22.4)$$

In Problem 9 of Chapter III.16 we have also proved the formulae

$$(\alpha)_k = \frac{\Gamma(\alpha + k)}{\Gamma(\alpha)} \qquad (22.5)$$

and

$$\frac{(\alpha)_k}{(\beta)_k} = \frac{B(\alpha + k, \beta - \alpha)}{B(\alpha, \beta - \alpha)} \qquad (22.6)$$

where B is the beta-function.

In this chapter we want to study the hypergeometric differential equation and related equations as well as their solutions and associated special functions in more details. There are at least two reasons for this. The first reason is that solutions to the hypergeometric differential equation and related equations cover a lot of important functions needed in pure mathematics, mechanics or physics and other subjects. Secondly, any study of (22.1) will necessarily use methods from complex analysis and therefore we use (22.1) also to provide some ideas of how complex analysis can and should be used to handle certain ordinary differential equations. In this chapter we assume the reader to have mastered a first course in complex analysis, for example as covered by Volume III Part 7 of our Course.

The results we have so far proved for hypergeometric series include Theorem III.22.17 which we recall as

Theorem 22.1. *Let $p, q \in \mathbb{N}_0$, $\alpha_1, \ldots, \alpha_p \in \mathbb{R}$, $\gamma_1, \ldots, \gamma_q \in \mathbb{R}\setminus\{-n \mid n \in \mathbb{N}_0\}$.*

   i) For $p > q+1$ the series $_pF_q$ converges only at $z = 0$.

   ii) For $p = q+1$ the series $_pF_q$ has radius of convergence 1.

   iii) For $p \le q$ the series $_pF_q$ represents an entire function.

In Theorem III.22.16 we have proved that $_2F_1$ satisfies in $B_1(0)$ the differential equation (22.1). Examples relating hypergeometric series to special functions are given by (III.16.31)-(III.16.36), (III.22.30)-(III.22.31) and (III.22.33)-(III.22.37). These examples include certain power functions, logarithmic functions and exponential functions, trigonometric and hyperbolic functions, and Bessel functions.

In the first part of this chapter we stay with (22.1) and $_2F_1$, and therefore we adopt the convention and use $F := {_2F_1}$ as notation for the Gauss hypergeometric series. Our first observation is that (22.1) is (when z is restricted to real values) of the type (20.1) and we can apply Frobenius' theorem, Theorem 20.2. Before doing this a short remark is helpful. We may consider the differential equation (22.1) (and later on related equations) not only on $\mathbb{C}$ but also on the Riemann sphere $S^2 = \hat{\mathbb{C}}$, see Chapter III.15. In this case the singular points of (22.1) are at $z = 0$, $z = 1$ and in addition at the point $\infty \in \hat{\mathbb{C}}$, i.e. the north pole of the Riemann sphere.

We are looking at the singular point $z = 0$ of (22.1) and we find

$$\frac{(\gamma - (\alpha + \beta + 1)z)}{z(1-z)} = \frac{\gamma}{z(1-z)} - \frac{(\alpha + \beta + 1)}{(1-z)}$$

$$= \frac{1}{z}\sum_{k=0}^{\infty}\gamma z^k - \sum_{k=0}^{\infty}(\alpha + \beta + 1)z^k$$

$$= \frac{1}{z}\sum_{k=0}^{\infty}\alpha_k z^k \qquad (22.7)$$

where $\alpha_0 = \gamma$, $\alpha_k = \gamma - (\alpha + \beta + 1)$ for $\alpha \ge 1$, and

$$\frac{-\alpha\beta}{z(z-1)} = \frac{1}{z}\sum_{k=0}^{\infty}(-\alpha\beta)z^k = \frac{1}{z^2}\sum_{k=0}^{\infty}\beta_k z^k$$

where $\beta_0 = 0$ and $\beta_k = -\alpha\beta$ for $k \ge 1$. This gives for the indicial equation

$$\chi(\rho) = \rho^2 + (\alpha_0 - 1)\rho + \beta_0 = \rho^2 + (\gamma - 1)\rho = \rho(\rho + \gamma - 1) = 0$$

with the zeroes $\rho_1 = 0$ and $\rho_2 = 1 - \gamma$. Thus we are seeking solutions $u_1(z) = \sum_{k=0}^{\infty} a_k z^k$ and $u_2(z) = z^{1-\gamma} \sum_{k=0}^{\infty} b_k z^k$. From Theorem III.22.16 we deduce that

$$u_1(z) = F(\alpha, \beta; \gamma; z) = {}_2F_1(\alpha, \beta; \gamma; z). \tag{22.8}$$

Since for $\gamma \notin -\mathbb{N}_0$ we have $\rho_1 - \rho_2 = \gamma - 1 \notin \mathbb{N}$ the proof of Frobenius' theorem, Theorem 20.2, allows to calculate the coefficients b_k as

$$b_k = \frac{(k - \gamma + \alpha)(k - \gamma + \beta)}{k(k + 1 - \gamma)} b_{k-1}$$

and with $b_0 = 1$ we obtain

$$b_k = \frac{(1 - \gamma + \alpha)_k (1 - \gamma + \beta)_k}{k!(2 - \gamma)_k}, \qquad k \in \mathbb{N}_0,$$

which yields

$$u_2(z) = z^{1-\gamma} F(1 - \gamma + \alpha, 1 - \gamma + \beta; 2 - \gamma; z). \tag{22.9}$$

The series in (22.9) converges, provided $\gamma \in \mathbb{N}\backslash\{1\}$, in $B_1(0)$. However we must take care of the term $z^{1-\gamma}$. We agree to take the principal branch of $z^{1-\gamma}$, i.e. we are cutting out from the complex plane the imaginary axis, thus we exclude $|\arg z| = \pi$. Assuming γ is such that u_1 and u_2 are both defined, i.e. $\gamma \notin \mathbb{Z}$, then u_1 and u_2 restricted to $(0, 1)$ are independent since $u_1 = {}_2F_1$ is holomorphic in $B_1(0)$ and u_2 not. We summarise our considerations in

Theorem 22.2. *For $\gamma \notin \mathbb{Z}$ the hypergeometric differential equation (22.1) has on $(0, 1)$ the solutions $u_1(t) = {}_2F_1$ and $u_2(t) = t^{1-\gamma} F(1 - \gamma + \alpha, 1 - \gamma + \beta; 2 - \gamma; t)$. These two functions are independent, hence they form (on $(0, 1)$) a fundamental system for (22.1).*

Our next aim is to extend $F(\alpha, \beta; \gamma; z)$, $\gamma \notin -\mathbb{N}_0$, to the cut plane $\mathbb{C}\backslash[1, \infty)$ and to derive some identities for $F(\alpha, \beta; \gamma; z)$. From Chapter III.26 we know that the beta-function has for $\operatorname{Re} a > 0$ and $\operatorname{Re} b > 0$ a separately holomorphic extension to $\{\operatorname{Re} a > 0\} \times \{\operatorname{Re} b > 0\}$ with integral representation

$$B(a, b) = \int_0^1 t^{a-1}(1 - t)^{b-1} \, dt \tag{22.10}$$

where the integral converges uniformly and absolutely on compact sets. In addition we have the formula

$$B(a, b) = \frac{\Gamma(a)\Gamma(b)}{\Gamma(a + b)}. \tag{22.11}$$

We now replace in

$$F(\alpha, \beta; \gamma; z) = \sum_{k=0}^{\infty} \frac{(\alpha)_k (\beta)_k}{k!(\gamma)_k} z^k \tag{22.12}$$

the term $\frac{(\beta)_k}{(\gamma)_k}$ with the help of (22.6) and (22.11). More precisely we write

$$\frac{(\beta)_k}{(\gamma)_k} = \frac{\Gamma(\gamma)}{\Gamma(\beta)\Gamma(\gamma - \beta)} \int_0^1 t^{\beta - 1 + k}(1 - t)^{\gamma - \beta - 1} \, dt$$

which yields

$$F(\alpha, \beta; \gamma; z) = \frac{\Gamma(\gamma)}{\Gamma(\beta)\Gamma(\gamma - \beta)} \sum_{k=0}^{\infty} \frac{(\alpha)_k}{k!} z^k \int_0^1 t^{\beta - 1 + k}(1 - t)^{\gamma - \beta - 1} \, dt$$

$$- \frac{\Gamma(\gamma)}{\Gamma(\beta)\Gamma(\gamma - \beta)} \int_0^1 t^{\beta - 1}(1 - t)^{\gamma - \beta - 1} \, dt \sum \frac{(\alpha)_k}{k!}(tz)^k.$$

From (III.16.32) we deduce that

$$\sum_{k=0}^{\infty} \frac{(\alpha)_k}{k!}(tz)^k = F(\alpha, 1; 1; z) = (1 - tz)^{-\alpha}, \quad 0 \le t \le 1,$$

where the binomial series has radius of convergence equal to 1. Thus for $\operatorname{Re}\gamma > \operatorname{Re}\beta > 0$ and $|z| < 1$ we find

$$F(\alpha, \beta; \gamma; z) = \frac{\Gamma(\gamma)}{\Gamma(\beta)\Gamma(\gamma - \beta)} \int_0^1 t^{\beta - 1}(1 - t)^{\gamma - \beta - 1}(1 - tz)^{-\alpha} \, dt. \tag{22.13}$$

Now we investigate the integral in (22.13). For $0 < r_0 < R$ and $\delta > 0$ the integrand is with respect to z a holomorphic function in $r_0 \le |z - 1| \le R$ and $|\arg(1 - z)| \le \pi - \delta$ depending continuously on $0 < t < 1$. For these values of t and z we have the estimate

$$\left| t^{\beta - 1}(1 - t)^{\gamma - \beta - 1}(1 - tz)^{-\alpha} \right| \le \kappa t^{\operatorname{Re}\beta - 1}(1 - t)^{\operatorname{Re}\gamma - \operatorname{Re}\beta - 1} \tag{22.14}$$

where

$$\kappa = \sup \left| (1 - tz)^{-\alpha} \right| \tag{22.15}$$

and the sup is taken over $r_0 \le |z - 1| \le R$, $|\arg(1 - z)| \le \pi - \delta$ and $0 \le t \le 1$. Since $\operatorname{Re}\gamma > \operatorname{Re}\beta > 0$ the integral (22.13) converges absolutely because the function (22.14) is integrable over $(0, 1)$. So far we have proved

Proposition 22.3. *For $\operatorname{Re}\gamma > \operatorname{Re}\beta > 0$ the Gauss hypergeometric series has a holomorphic extension to the cut plane $|\arg(1 - z)| < \pi$ given by (22.13).*

In order to weaken the restrictions on $\operatorname{Re}\gamma$ and $\operatorname{Re}\beta$, one first has to represent the series (22.12) with the help of the residue theorem by a certain line integral, we refer to [104] or [85].

From the definition of $F(\alpha, \beta; \gamma; z)$ it is clear that

$$F(\alpha, \beta; \gamma; z) = F(\beta, \alpha; \gamma; z) \tag{22.16}$$

holds. Differentiating $F(\alpha, \beta; \gamma; z)$ with respect to z we find

$$\frac{\mathrm{d}}{\mathrm{d}z}F(\alpha, \beta; \gamma; z) = \sum_{k=0}^{\infty} \frac{(\alpha)_k(\beta)_k}{k!(\gamma)_k} k z^{k-1}$$

$$= \sum_{k=0}^{\infty} \frac{(\alpha)_{k+1}(\beta)_{k+1}}{k!(\gamma)_{k+1}} z^k$$

$$= \frac{\alpha\beta}{\gamma} \sum_{k=0}^{\infty} \frac{(\alpha+1)_k(\beta+1)_k}{k!(\gamma+1)_k} z^k,$$

or

$$\frac{\mathrm{d}}{\mathrm{d}z}F(\alpha, \beta; \gamma; z) = \frac{\alpha\beta}{\gamma}F(\alpha+1, \beta+1; \gamma+1; z) \tag{22.17}$$

and a straightforward induction yields for $n \in \mathbb{N}$

$$\frac{\mathrm{d}^n}{\mathrm{d}z^n}F(\alpha, \beta; \gamma; z) = \frac{(\alpha)_n(\beta)_n}{(\gamma)_n}F(\alpha+n, \beta+n; \gamma+n; z). \tag{22.18}$$

As in our derivation of (22.17) we can use the power series representing $F(\alpha, \beta; \gamma; z)$ to obtain many relations between hypergeometric series with different parameters, e.g.

$$(\gamma - \alpha - \beta)F(\alpha, \beta; \gamma; z) + \alpha(1-z)F(\alpha+1, \beta; \gamma; z) - (\gamma-\beta)F(\alpha, \beta-1; \gamma; z) = 0, \tag{22.19}$$

$$(\gamma - \alpha - 1)F(\alpha, \beta; \gamma; z) + \alpha F(\alpha+1, \beta; \gamma; z) - (\gamma-1)F(\alpha, \beta; \gamma-1; z) = 0, \tag{22.20}$$

or

$$\gamma(1-z)F(\alpha, \beta; \gamma; z) - \gamma F(\alpha-1, \beta; \gamma; z) + (\alpha-\beta)zF(\alpha, \beta; \gamma+1; z) = 0. \tag{22.21}$$

We refer to Problem 2 where we will deal with some of these identities. Note that the appearance of z as a further factor in some of these identities opens a different way for holomorphic extensions.

More interesting in the context of differential equations is the following: since

$F(\alpha, \beta; \gamma; z)$ is at least in $B_1(0)$ holomorphic we can compose F with a Möbius transformation and we will obtain a further holomorphic function. When we choose the Möbius transformation $w(z) = \frac{az+b}{cz+d}$, $ab - bc \neq 0$, in such a way that the set of singular points of the differential equation is invariant under w, i.e. $w(\{0, 1, \infty\}) = \{0, 1, \infty\}$, we shall expect interesting new solutions of the hypergeometric differential equation. We note that the Möbius transformations

$$w(z) = 1 - z, \quad w(z) = \frac{1}{z}, \quad w(z) = \frac{z}{z-1}, \quad w(z) = \frac{1}{1-z}, \quad w(z) = \frac{z-1}{z}$$

leave the set of singular points of (22.1) invariant. We refer the reader to Chapter III.15, Chapter III.16 and Appendix III.III where we have discussed Möbius transformations.

Our starting point is the integral representation (22.13) and we assume again $\operatorname{Re}\gamma > \operatorname{Re}\beta > 0$ and $|z| < 1$. The substitution $t = 1 - s$ yields with $\beta' = \gamma - \beta$ that

$$F(\alpha, \beta; \gamma; z) = \frac{\Gamma(\gamma)}{\Gamma(\beta)\Gamma(\gamma - \beta)} \int_0^\infty s^{\gamma-\beta-1}(1-s)^{\beta-1}(1-z+sz)^{-\alpha}\, ds$$

$$= (1-z)^{-\alpha} \frac{\Gamma(\gamma)}{\Gamma(\beta')\Gamma(\gamma - \beta')} \int_0^1 s^{\beta'-1}(1-s)^{\gamma-\beta'-1}\left(1 - s\frac{z}{z-1}\right) ds,$$

or

$$F(\alpha, \beta; \gamma; z) = (1-z)^{-\alpha} F\left(\alpha, \gamma - \beta; \gamma; \frac{z}{z-1}\right), \qquad |\arg(1-z)| < \pi. \qquad (22.22)$$

Since $z \mapsto F(\alpha, \beta; \gamma; z)$ is a solution to (22.1) it follows that $z \mapsto (1-z)^{-\alpha} F\left(\alpha, \gamma - \beta; \gamma; \frac{z}{z-1}\right)$ is a further solution to (22.1). The symmetry $F(\alpha, \beta; \gamma; z) = F(\beta, \alpha; \gamma; z)$ gives a further functional equation, namely

$$F(\alpha, \beta; \gamma; z) = (1-z)^{-\beta} F\left(\gamma - \alpha, \beta; \gamma; \frac{z}{z-1}\right), \qquad |\arg(1-z)| < \pi, \qquad (22.23)$$

and a further solution to (22.1). Note that we may invert now the functions on the right hand side of (22.22) or (22.23), in particular we may search for its maximal holomorphy domain connected with $(0, 1)$ in which the extension then must satisfy (22.1). However we do not go into details. We may of course iterate the transformations leading to (22.22) and (22.23), respectively, and we obtain, see Problem 5, a further solution to (22.1) by

$$F(\alpha, \beta; \gamma; z) = (1-z)^{\gamma-\alpha-\beta} F(\gamma - \alpha; \gamma - \beta; z), \qquad |\arg(1-z)| < \pi. \qquad (22.24)$$

The formulae (22.22) and (22.23) are functional equations linking the arguments z and $\frac{z}{z-1}$. We may also seek functional equations involving the argument $1 - z$.

To obtain such a functional equation one may start with the fact that $u_1(z) = F(\alpha, \beta; \gamma; z)$ and $u_2(z) = z^{1-\gamma} F(1 - \gamma + \alpha, 1 - \gamma + \beta; 2 - \gamma; z)$ form a fundamental system for (22.1), hence every solution is a linear combination $u(z) = \mu u_1(z) + \lambda u_2(z)$.

We may substitute in (22.1) the variable z by $\tilde{z} = z - 1$ and we obtain as a new differential equation

$$\tilde{z}(1 - \tilde{z}) v''(\tilde{z}) + \left(\tilde{\gamma} - \left(\tilde{\alpha} + \tilde{\beta} + 1\right)\tilde{z}\right) v'(\tilde{z}) - \tilde{\alpha}\tilde{\beta} v(\tilde{z}) = 0$$

where $\tilde{\alpha} = \alpha$, $\tilde{\beta} = \beta$ and $\tilde{\gamma} = 1 + \alpha + \beta - \gamma$, which yields

$$u(z) = \tilde{v}(\tilde{z}) = \mu F(\alpha, \beta; 1+\alpha+\beta-\gamma, 1-z) + \lambda(1-z)^{\gamma-\alpha-\beta} F(\gamma-\alpha, \gamma-\beta; 1-\alpha-\beta+\gamma; 1-z)$$

for $|\arg(1 - z)| < \pi$, $|\arg z| < 1$, $\alpha + \beta - \gamma \notin \mathbb{Z}$. We can choose as a solution $u(z)$ of course $F(\alpha, \beta; \gamma; z)$ and it follows that

$$\begin{aligned}
F(\alpha, \beta; \gamma; z) &= \eta_1 F(\alpha, \beta; 1 + \alpha + \beta - \gamma; 1 - z) \\
&\quad + \eta_2 (1 - z)^{\gamma-\alpha-\beta} F(\gamma - \alpha, \gamma - \beta; 1 - \alpha - \beta + \gamma; 1 - z),
\end{aligned} \tag{22.25}$$

provided $\alpha + \beta - \gamma \notin \mathbb{Z}$. The constants η_1 and η_2 we can determine when looking at the limits $z \to 1$ and $z \to 0$, we refer to [85], where the calculations are done, leading to

$$\eta_1 = \frac{\Gamma(\gamma)\Gamma(\gamma - \alpha - \beta)}{\Gamma(\gamma - \alpha)\Gamma(\gamma - \beta)} \quad \text{and} \quad \eta_2 = \frac{\Gamma(\gamma)\Gamma(\gamma - \alpha - \beta)}{\Gamma(\alpha)\Gamma(\beta)}.$$

In a landmark result E. E. Kummer [82] had proved that for every regular singular point $z_0 \in \{0, 1, \infty\}$ of (22.1) there exists in each case eight solutions to (22.1) which can be expressed by hypergeometric series which are well behaved at this point. We refer to [104].

On the Riemann sphere $\hat{\mathbb{C}}$ the points $0, 1$ and ∞ are not anymore "special" points. For example in light of Theorem III.A.1 we can map them by a Möbius transformation onto any three distinct points ζ_0, ζ_1 and ζ_∞. For this reason it makes sense to consider a modification of (22.1) with three regular singular points $\zeta_1, \zeta_2, \zeta_3 \in \hat{\mathbb{C}}$. In his treatise [74], F. Klein follows this approach consequently. One advantage of this approach is that it allows us to study the case where one singularity, say ζ_1, moves to another, say ζ_2. This process was called the "confluence of singularity" and hence the resulting equation is the **confluent hypergeometric differential equation** with **confluent hypergeometric series or functions** forming the solutions. We will derive and study the confluent hypergeometric differential equation by introducing a change of variable, i.e. by using a pre-Klein approach.

Suppose that in (22.1) the parameter β is not zero, $\beta \neq 0$, and that u is a solution to (22.1). We suggest the change of variable

$$x := \beta z \quad w(x) := u(z). \tag{22.26}$$

In the new variable the equation (22.1) reads as,

$$x \left(1 - \frac{1}{\beta}x\right) w''(x) + \left(\gamma - \left(1 + \frac{\alpha + 1}{\beta}x\right)\right) w'(x) - \alpha w(x) = 0. \tag{22.27}$$

This equation has now the regular singular points $0, \beta$ and ∞, hence we may apply Frobenius' theorem, Theorem 20.2. It is easy to check that one solution to (22.27) is given by

$$u_{1,\beta}(x) := {}_2F_1\left(\alpha, \beta; \gamma; \frac{x}{\beta}\right) \tag{22.28}$$

and a further independent solution is

$$u_{2,\beta}(x) = x^{1-\gamma} {}_2F_1\left(1 - \gamma + \alpha, 1 - \gamma + \beta; 2 - \gamma; \frac{x}{\beta}\right). \tag{22.29}$$

If we pass formally in (22.27) to the limit as β tends to ∞ we obtain the equation

$$xw''(x) + (\gamma - x)w'(x) - \alpha w(x) = 0 \tag{22.30}$$

which is called the **confluent hypergeometric differential equation** or **Kummer differential equation**. But a word of caution is needed: a solution of (22.27) will in general depend on β, compare with (22.28) and (22.29), and therefore much more care is needed to justify (22.30) when looking at it as a limit of (22.27). Nonetheless, we may study (22.30) as a new differential equation in its own right. We may use a power series Ansatz to solve (22.30) and since at $x = 0$ the equation (22.30) has a regular singular point we may use Frobenius' theorem to obtain a second independent solution. The results are

$$\Phi(\alpha; \gamma; x) = \sum_{k=0}^{\infty} \frac{(\alpha)_k}{k!(\gamma)_k} x^k, \quad \gamma \notin -\mathbb{N}_0, \tag{22.31}$$

and

$$\tilde{\Psi}(\alpha; \gamma; x) = x^{1-\gamma} \Phi(\alpha - \gamma + 1; 2 - \gamma; x). \tag{22.32}$$

(The reader is invited to verify by a direct calculation that (22.31) and (22.32) are solutions to (22.30).) The function Φ is often called **Kummer function** or

Kummer series and it is a **confluent hypergeometric function** in the sense that it solves (20.30). Meanwhile the standard notation for $\Phi(a;\gamma;x)$ is $M(\alpha;\gamma;x)$, see [2] or [11]. However, we also can identify $\Phi(\alpha;\gamma;z)$ with the generalised hypergeometric series $_1F_1(\alpha;\gamma;z)$, see (22.3). By Theorem 22.1 we conclude that Φ and $\tilde{\Psi}$ are with respect to x entire functions.

Remark 22.4. As second independent solution to (22.27) often instead of $\tilde{\Psi}$ the function

$$U(\alpha;\gamma;z) := \frac{\pi}{\sin \pi\gamma} \left(\frac{M(\alpha;\gamma;z)}{\Gamma(1+\alpha-\gamma)\Gamma(\gamma)} - \frac{z^{1-\gamma}M(1+\alpha-\gamma;2-\gamma;z)}{\Gamma(\alpha)\Gamma(2-\gamma)} \right) \quad (22.33)$$

is chosen which is called **Kummer function of second kind**. Unfortunately, the notation is here not unique. Some authors use $\Psi(\alpha;\gamma;z)$ as a further notation for $U(\alpha;\gamma;z)$, in some books $\Psi(\alpha;\gamma;z)$ is the symbol used for the function we have denoted by $\tilde{\Psi}(\alpha;\gamma;z)$. Moreover, with some effort one can prove

$$U(\alpha;\gamma;z) = z^{-\alpha}\,_2F_0\left(\alpha, 1+\alpha-\gamma, _-, \frac{1}{z}\right), \quad (22.34)$$

where $_-$ reminds us that for functions $_pF_0$ no coefficient γ enters as $(\gamma)_k$, the denominator of the terms in the series (22.3).

We now want to study the behaviour of $_2F_1\left(\alpha,\beta;\gamma;\frac{x}{\beta}\right)$ for $\beta \to \infty$ and we expect to prove that in the limit we obtain $\Phi(\alpha;\gamma;x)$, or

$$\lim_{\beta\to\infty} \,_2F_1\left(\alpha,\beta;\gamma;\frac{x}{\beta}\right) = \,_1F_1(\alpha;\gamma;x). \quad (22.35)$$

We split the series $_2F_1\left(\alpha,\beta;\gamma;\frac{x}{\beta}\right)$ according to

$$_2F_1\left(\alpha,\beta;\gamma;\frac{x}{\beta}\right) = \sum_{k=0}^{N-1} \frac{(\alpha)_k(\beta)_k}{k!(\gamma)_k}\left(\frac{x}{\beta}\right)^k + \sum_{k=N}^{\infty} \frac{(\alpha)_k(\beta)_k}{k!(\gamma)_k}\left(\frac{x}{\beta}\right)^k, \quad (22.36)$$

and we study the second term in (22.36). Let $|x| \leq A$ and $\left|\frac{1}{\beta}\right| \leq B$ where $AB < 1$, i.e. $\left|\frac{x}{\beta}\right| \leq AB < 1$ which is needed for the convergence of $_2F_1\left(\alpha,\beta;\gamma;\frac{x}{\beta}\right)$. We denote by M the smallest integer such that

$$\sup_{l\geq M}\left|\frac{\alpha+l}{\gamma+l}\right| = a_M \quad \text{and} \quad a_M\left(\frac{1}{M}+B\right)A < 1. \quad (22.37)$$

We write

$$\sum_{k=N}^{\infty} \frac{(\alpha)_k (\beta)_k}{k!(\gamma)_k} \left(\frac{x}{\beta}\right)^k = \sum_{k=N}^{\infty} \frac{(\alpha)_k}{(\gamma)_k} \frac{x^k}{k} \left(1 + \frac{1}{\beta}\right) \left(\frac{1}{2} + \frac{1}{\beta}\right) \cdots \left(\frac{1}{k-1} + \frac{1}{\beta}\right)$$

and observe that under the conditions on $|x|$ and β as well as (22.37) we have the estimate

$$\sum_{k=N}^{\infty} \frac{(\alpha)_k (\beta)_k}{k!(\gamma)_k} \left|\frac{x}{\beta}\right|^k \leq \frac{1}{N} \left|\frac{(\alpha)_M}{(\gamma)_M}\right| (1+B) \cdots \left(\frac{1}{M-1} + B\right) A^M \cdot$$

$$\cdot \sum_{k=N}^{\infty} \left(a_M \left(\frac{1}{M} + B\right) A\right)^{k-M}. \tag{22.38}$$

By (22.38) we have proved

Lemma 22.5. *For every bounded region in $\mathbb{C}$ we have*

$$\lim_{\beta \to \infty} {}_2F_1\left(\alpha, \beta; \gamma; \frac{x}{\beta}\right) = {}_1F_1(\alpha; \gamma; x), \tag{22.39}$$

where the convergence is uniform on bounded sets of $\mathbb{C}$.

Remark 22.6. A. In order that the term ${}_2F_1\left(\alpha; \beta; \gamma; \frac{x}{\beta}\right)$ makes sense we need of course $\left|\frac{x}{\beta}\right| < 1$. Thus for $|x| < R$ we need $R < \beta$ in order that ${}_2F_1\left(\alpha, \beta; \gamma; \frac{x}{\beta}\right)$ converges. Since we are interested in the limit $\beta \to \infty$, the assumption that x belongs to a bounded set implies that the statement (22.39) makes sense.
B. In deriving Lemma 22.5 we followed closely [104].

Note that with Lemma 22.5 at hand we can pass from (22.27) to (22.30) with w being either $u_{1.\beta}$ or $u_{2,\beta}$. Since $\{u_{1,\beta}, u_{2,\beta}\}$ is a fundamental system for (22.30) we can eventually allow any solution w when passing in (22.27) to the limit $\beta \to \infty$.

As in the case of the hypergeometric function ${}_2F_1(\alpha, \beta; \gamma; z)$ we can derive an integral representation for the confluent hypergeometric function $\Phi(\alpha; \gamma; z)$. Using (22.6) we start with

$$\frac{(\alpha)_k}{(\gamma)_k} = \frac{\Gamma(\gamma)}{\Gamma(\alpha)\Gamma(\gamma - \alpha)} \int_0^1 t^{\alpha-1+k}(1-t)^{\gamma-\alpha-1}\, dt$$

which holds for $\operatorname{Re}\gamma > \operatorname{Re}\alpha > 0$ and $k \in \mathbb{N}_0$. It follows that

$$\Phi(\alpha; \gamma; z) = \frac{\Gamma(\gamma)}{\Gamma(\alpha)\Gamma(\gamma - \alpha)} \int_0^1 t^{\alpha-1}(1-t)^{\gamma-\alpha-1}\, dt \sum_{k=0}^{\infty} \frac{(tz)^k}{k!}$$

and since $\sum_{k=0}^{\infty} \frac{(tz)^k}{k!} = e^{tz}$ we arrive at

$$\Phi(\alpha; \gamma; z) = \frac{\Gamma(\gamma)}{\Gamma(\alpha)\Gamma(\gamma - \alpha)} \int_0^1 e^{zt} t^{\alpha-1} (1 - t)^{\gamma-\alpha-1} \, dt \qquad (22.40)$$

which holds for $\operatorname{Re} \gamma > \operatorname{Re} \alpha > 0$. As in the case of the hypergeometric function we use the substitution $s = 1 - t$ to get

$$\Phi(\alpha; \gamma; z) = \frac{\Gamma(\gamma)}{\Gamma(\alpha)\Gamma(\gamma - \alpha)} e^z \int_0^1 e^{-zs} s^{\gamma-\alpha-1} (1 - s)^{\alpha-1} \, ds \qquad (22.41)$$

and comparing with (22.40) yields

$$\Phi(\alpha; \gamma; z) = e^z \Phi(\gamma - \alpha; \gamma; -z). \qquad (22.42)$$

Of course we may now try to find further transformations and to derive more functional equations for $\Phi(\alpha; \gamma; z)$, but for this we refer to [85] or [11].

We close this chapter by identifying some important functions as confluent hypergeometric series (or functions). The most obvious one we obtain for $\alpha = \gamma$, namely

$$\Phi(\alpha; \alpha; z) = \sum_{k=0}^{\infty} \frac{z^k}{k!} = e^z, \qquad (22.43)$$

a further one is

$$\Phi(1; 2; z) = \sum_{k=0}^{\infty} \frac{z^k}{(k + 1)!} = \frac{e^z - 1}{z}. \qquad (22.44)$$

More interesting is the error function. The error integral or the Gauss integral we studied in detail in Volume I, i.e. the integral

$$\int_{-\infty}^{\infty} e^{-x^2} \, dx = 2 \int_0^{\infty} e^{-x^2} \, dx = \sqrt{\pi}. \qquad (22.45)$$

The function

$$x \mapsto \operatorname{erf}(x) := \frac{2}{\sqrt{\pi}} \int_0^x e^{-t^2} \, dt \qquad (22.46)$$

is called the **error function**. If we expand in (22.46) the exponential function into a Taylor series and integrate term by term (which is allowed due to the convergence properties of the integral) we obtain

$$\operatorname{erf}(x) = \frac{2}{\sqrt{\pi}} \int_0^x \sum_{k=0}^{\infty} \frac{(-1)^k t^{2k}}{k!} \, dt = \frac{2}{\sqrt{\pi}} \sum_{k=1}^{\infty} \frac{(-1)^k x^{2k+1}}{k!(2k + 1)} \qquad (22.47)$$

which converges for $|x| < \infty$, hence $z \mapsto \operatorname{erf}(z)$ as defined by (22.47) is an entire function. We may identify

$$\frac{1}{2k+1} = \frac{\left(\frac{1}{2}\right)_k}{\left(\frac{3}{2}\right)_k}$$

and then we find

$$\operatorname{erf}(x) = \frac{2}{\sqrt{\pi}} x \sum_{k=0}^{\infty} \frac{\left(\frac{1}{2}\right)_k}{k! \left(\frac{3}{2}\right)_k} \left(-x^2\right)^k,$$

or even for $z \in \mathbb{C}$

$$\operatorname{erf}(z) = \frac{2}{\sqrt{\pi}} z \Phi\left(\frac{1}{2}; \frac{3}{2}; -z^2\right). \qquad (22.48)$$

It is possible to show that Hermite polynomials or Bessel functions also admit representations involving confluent hypergeometric functions. For example we have

$$J_\nu(z) = \frac{\left(\frac{z}{2}\right)^\nu}{\Gamma(\nu+1)} e^{-iz} \Phi\left(\nu + \frac{1}{2}; 2\nu + 1; 2iz\right), \qquad |\arg z| < \pi. \qquad (22.49)$$

For more details we refer to books on special functions such as [4], [11], [85] or [104]. We also refer to [60] and [64] as good sources to learn more about ordinary differential equations in complex domains.

Problems

1.   a) Prove the following identities for the Pochhammer symbol $(\alpha)_k$:

$$(\alpha + 1)_k = \frac{\alpha + k}{\alpha} (\alpha)_k$$

$$(\alpha - 1)_k = \frac{\alpha - 1}{\alpha - 1 + k} (\alpha)_k.$$

b) Show for $\beta > 0$ and $\gamma - \alpha - \beta > 0$ that

$$\sum_{k=0}^{\infty} \frac{(\alpha)_k (\beta)_k}{(\gamma)_k k!} = \frac{\Gamma(\gamma)\Gamma(\gamma - \alpha - \beta)}{\Gamma(\gamma - \alpha)\Gamma(\gamma - \beta)},$$

where α, β and γ satisfy all assumptions of (22.13).

2.   a) Show (22.19).
     b) Prove (22.20).

3. By comparing power series, prove that

$$\ln(1+z) = z {}_2F_1(1,1;2;-z)$$

and

$$\arctan z = z {}_2F_1(\frac{1}{2},1;\frac{3}{2},-z^2).$$

4. Prove that ${}_2F_1(-n,\beta;\gamma;z)$ is a polynomial or order n.

5. With the help of (22.13), prove the identity

$$ {}_2F_1(\alpha,\beta;\gamma;z) = (1-z)^{-\alpha} {}_2F_1(\alpha,\gamma-\beta;\gamma;\frac{z}{z-1}).$$

6. Use (22.23) and then (22.22) to prove (22.24).

23 Continuous Dependence on Data and Stability

We want to return to the general theory and to continue our studies of systems

$$\dot{u}(t) = R(t, u(t)) \qquad u(t_0) = u_0, \qquad (23.1)$$

where $R : I \times D \to \mathbb{R}^n, D \subset \mathbb{R}^n$, is at least a continuous function and $I \subset \mathbb{R}$ is an interval with $t_0 \in I, \overset{\circ}{I} \neq \phi$. So far we have with Theorem 17.1 a basic existence and uniqueness theorem of Picard-Lindelöf type which we proved for $D = \mathbb{R}^n$. However, we have indicated that we can argue as we did when deriving Theorem 16.17 that $\mathbb{R}^n$ can be replaced by D. For convenience and reference purpose we state this result here in detail.

Theorem 23.1. *Let $D \subset \mathbb{R}^n$ be a connected open set and $I \subset \mathbb{R}$ an interval with non-empty interior. Suppose that $R : I \times D \to \mathbb{R}^n$ is a continuous mapping which satisfies with respect to the second variable $x \in D$ a local Lipschitz condition. Then for every $(t_0, u_0) \in I \times D$ the initial value problem (23.1) has a unique maximal solution u.*

The problems we want to investigate in this chapter are of the following type. Suppose that we change in (23.1) the value u_0 to v_0 but $||u_0 - v_0||$ is small, say for some $\delta > 0$ we have $||u_0 - v_0|| < \delta$. Can we prove that in this case the corresponding solutions u and v to (23.1) are also "close" to each other? Of course, this question needs some precision: in which norm or metric do we want to compare u and v? Further, do we want to compare u and v pointwisely or uniformly, i.e. are we seeking for each t in the joint existence interval I_1 a control on $|u(t) - v(t)|$ or do we want to control $||u - v||_{\infty, I_1}$? Similarly, consider for simplicity the linear system

$$\dot{u}(t) = A(t)u(t) + f(t). \qquad (23.2)$$

We now may replace the function f by g where $||f - g||$ is small in some norm and we may ask whether corresponding solutions u and v to (23.1) (with either the same initial value u_0 or with two different initial values u_0 and v_0 which are however close to each other) are close when measured by some appropriate norm. In fact we may even replace $A(t)$ by a new matrix $B(t)$ such that in some norm $||A(\cdot) - B(\cdot)||$ is small and we may ask the same question for corresponding solutions.

These are not artificial problems. Firstly, when applying differential equations to problems in mechanics or physics, initial conditions, source terms (the function f in (23.2)) or even the mapping A or the mapping R in (23.1) are in general subjected to inaccuracies in measurements. Secondly, in most cases we can solve differential equations "only" by numerical methods. Then we have to discretise f, A or R, i.e. we approximate these functions and we need to know to which extent the solution of the approximate problem is in fact an approximation of the solution of the original problem. Problems of this type are referred to as **continuous dependence** (of solutions) **on data**. A generalisation of these problems will lead us to stability problems, mainly in the sense of Lyapunov, which we will discuss in the second part of this chapter.

Before going into details a word of caution is necessary. The "theory" of ordinary differential equations as the "theory" of partial differential equations is not such a polished theory as complex analysis, groups theory or point set theory, etc., are. Different structures of equations, different regularity assumptions on the data, different constraints such as initial value or boundary value conditions will allow and often even require a variety of techniques and methods as well as assumptions. There is no optimal existence or uniqueness theorem and there are many versions of results addressing the continuous dependence on data or stability problems. Our aim here is to present some quite useful results which are applicable and by this give the flavour of this topic. We do not long for a comprehensive treatment of these problems.

As a preparation of our investigations on continuous dependence on data and stability we collect some auxiliary results.

Lemma 23.2. *Let $\alpha, \beta : (a, b] \to \mathbb{R}, a < b$, be two differentiable functions and suppose that there exists $\eta > 0$ but $a + \eta < b$ such that $\alpha(t) < \beta(t)$ for all $t \in (a, a + \eta)$. In this case we have either $\alpha(t) < \beta(t)$ for all $t \in (a, b]$ or there exists $t_0 \in (a, b]$ such that $\alpha(t) < \beta(t)$ for $t \in (a, t_0)$, $\alpha(t_0) = \beta(t_0)$ and $\alpha'(t_o) \geq \beta'(t_0)$.*

Proof. Suppose that $\alpha(t) < \beta(t)$ does not hold for all $t \in (a, b]$. Then there exists a smallest $t_0 \in (a, b]$ such that $\alpha(t_0) = \beta(t_0)$ and for $h > 0$ sufficiently small we must have

$$\frac{\alpha(t_0) - \alpha(t_0 - h)}{h} > \frac{\beta(t_0) - \beta(t_0 - h)}{h}$$

implying $\alpha'(t_0) \geq \beta'(t_0)$. $\qquad\square$

Lemma 23.3. *Let $I \subset \mathbb{R}$ be an interval with non-empty interior and $f : I \to \mathbb{R}^n$ be a function differentiable at $t_0 \in I$. It follows that the left derivative g'_- of $g := \|f\|$ exists at t_0 and*

$$g'_-(t_0) \leq \|f'(t_0)\|. \qquad (23.3)$$

Proof. For $h > 0$ we find

$$\frac{g(t_0) - g(t_0 - h)}{h} = \frac{\|f(t_0)\| - \|f(t_0 - h)\|}{h}$$

$$\leq \left\| \frac{f(t_0) - f(t_0 - h)}{h} \right\|$$

and the result follows for $h \to 0$. $\qquad\qquad\qquad\qquad\qquad\qquad\qquad\square$

Our final preparation is a variant of the **Gronwall Lemma** which has widespread applications in the theory of differential equations and beyond.

Proposition 23.4. *Let $I \subset \mathbb{R}, \overset{\circ}{I} \neq \phi$, be an interval and $a, b, f : I \to \mathbb{R}_+$ continuous functions and let $t_0 \in I$. The estimate*

$$f(t) \leq a(t) + |\int_{t_0}^t b(s)f(s)\,\mathrm{d}s|, t \in I, \qquad\qquad (23.4)$$

entails

$$f(t) \leq a(t) + |\int_{t_0}^t a(s)b(s)e^{|\int_s^t b(r)\mathrm{d}r|}\,\mathrm{d}s| \qquad\qquad (23.5)$$

for all $t \in I$.

Proof. We set $g(t) := \int_{t_0}^t b(s)f(s)\,\mathrm{d}s$ and we observe that by (23.4)

$$\dot{g}(t) = b(t)f(t) \leq b(t)a(t) + sgn(t - t_0)b(t)g(t).$$

with

$$c(t) = e^{-|\int_{t_0}^t b(s)\,\mathrm{d}s|} = e^{-\int_{t_0}^t sgn(s-t_0)b(s)\,\mathrm{d}s}.$$

We find

$$\dot{c}(t) = -sgn(t - t_0)b(t)c(t)$$

which yields

$$c(t)\dot{g}(t) = c(t)b(t)a(t) - \dot{c}(t)g(t),$$

or

$$(c(t)g(t))^{\cdot} - c(t)b(t)a(t) \leq 0. \qquad\qquad (23.6)$$

Noting that $g(t_0) = 0$ and integrating (23.6) gives

$$sgn(t - t_0)c(t)g(t) = sgn(t - t_0)e^{-\int_{t_0}^t sgn(s-t_0)b(s)\,\mathrm{d}s} \int_{t_0}^t b(s)f(s)\,\mathrm{d}s$$

$$\leq sgn(t - t_0)\int_{t_0}^t c(s)b(s)a(s)\,\mathrm{d}s,$$

or

$$sgn(t - t_0)g(t) \leq \frac{sgn(t - t_0)}{c(t)} \int_{t_0}^t c(s)b(s)a(s)\,\mathrm{d}s$$

$$\leq |\int_{t_0}^t \frac{c(s)b(s)a(s)}{c(t)}\,\mathrm{d}s|.$$

Using the definitions of c we now obtain

$$f(t) \le a(t) + sgn(t - t_0)g(t)$$

$$\le a(t) + |\int_{t_0}^t a(s)b(s)e^{|\int_s^t b(r)\,dr|}\,ds|$$

which is however (23.5). $\qquad\square$

As a corollary suitable for our purposes, we find immediately

Corollary 23.5. *In the situation of Proposition 23.4 let the function a be given by $a(t) = a_0(|t - t_0|)$ where we assume that a_0 is a continuous and monotone increasing function. Then the estimate*

$$f(t) \le a(t) + |\int_{t_0}^t b(s)f(s)\,ds| \tag{23.7}$$

implies

$$f(t) \le a(t)e^{|\int_{t_0}^t b(r)\,dr|}. \tag{23.8}$$

Proof. We note that by our assumption $|s - t_0| \le |t - t_0|$ implies $a(s) \le a(t)$ and therefore (23.5) yields

$$f(t) \le a(t)(1 + |\int_{t_0}^t b(s)e^{|\int_s^t b(r)\,dr|}\,ds|)$$

$$= a(t)(1 + sgn(t - t_0)\int_{t_0}^t b(s)e^{sgn(t-t_0)\int_s^t b(r)\,dr}\,ds).$$

Since for $t > s$

$$b(s)e^{\int_s^t b(r)\,dr} = -\frac{d}{ds}e^{\int_s^t b(r)\,dr}$$

and therefore

$$\int_{t_0}^t b(s)e^{\int_s^t b(r)\,dr}\,ds = -\int_{t_0}^t \frac{d}{ds}e^{\int_s^t b(r)\,dr}\,ds$$

$$= e^{\int_s^t b(r)\,dr} - 1$$

we eventually arrive at

$$f(t) \le a(t)e^{|\int_{t_0}^t b(s)\,ds|}.$$

$\qquad\square$

Remark 23.6. As we have mentioned before, there are many variants of Gronwall's lemma available. We found the one in [3] most suitable for our purposes.

We consider the initial value problem (23.1) and assume that for some interval $I_0 \subset I, t_0 \in I_0$, the function $u : I_0 \to \mathbb{R}^n$ is a solution to (23.1). Let $v : I_0 \to \mathbb{R}^n$ be a further differentiable function. We call this pair

$$v(t_0) - u_0 \quad \text{and} \quad \dot{v}(\cdot) - R(\cdot, v(\cdot))$$

the **defect** of v with respect to (23.1). Obviously, for a solution to (23.1) the defect is zero and vice versa. Our aim is to find a bound for $\|u(t) - v(t)\|$.

Theorem 23.7. *Let $R : I_0 \times D \to \mathbb{R}^n, D \subset \mathbb{R}^n$, satisfy the Lipschitz condition*

$$\|R(t, x) - R(t, y)\| \leq L\|x - y\|. \tag{23.9}$$

Suppose further that $u : I_0 \to \mathbb{R}^n$ is a solution to (23.1) in I_0, $t_0 \in I_0$, and let $v : I_0 \to \mathbb{R}^n$ be a differentiable function such that $\Gamma(v) \subset I_0 \times D$ such that

$$\|v(t_0) - u_0\| < \delta_1 \quad \text{and} \quad \|\dot{v}(t) - R(t, v(t))\| < \delta_2 \tag{23.10}$$

holds where $\delta_1, \delta_2 > 0$ are constants. Then we have for all $t \in [t_0, t_1] \subset I_0$, $t_0 < t_1$, the estimate

$$\|u(t) - v(t)\| \leq \delta_1 \, e^{L|t-t_0|} + \frac{\delta_2}{L}(e^{L|t-t_0|} - 1). \tag{23.11}$$

Remark 23.8. In Theorem 23.7 we denote by $\|.\|$ a suitable norm in $\mathbb{R}^n$, for example the Euclidean norm or the maximum norm. Further, as usual $\Gamma(v)$ denotes the graph of v.

We will derive Theorem 23.7 from the following more general result:

Theorem 23.9. *Let $u, v : I_0 \to \mathbb{R}^n$ and $r : I_0 \to \mathbb{R}$ be differentiable functions and further let $d : I_0 \to \mathbb{R}$ and $Q : I_0 \times \mathbb{R} \to \mathbb{R}$ be continuous functions. Suppose that u solves (23.1) and*

$$\|u_0 - v(t_0)\| < r(t_0); \tag{23.12}$$
$$\|\dot{v}(t) - R(t, v(t))\| \leq d(t) \text{ in } [t_0, t_1]; \tag{23.13}$$
$$\|R(t, x) - R(t, y)\| \leq Q(t, \|x - y\|), t \in I_0, x, y \in D, \tag{23.14}$$

and

$$d(t) + Q(t, r(t)) < \dot{r}(t). \tag{23.15}$$

Then we have

$$\|u(t) - v(t)\| \leq r(t) \text{ in } [t_0, t_1]. \tag{23.16}$$

Proof. (Following [125]) We now choose α to be the function $\alpha(t) := ||u(t) - v(t)||$ and β to be the function $\beta(t) = r(t)$ in Lemma 23.2. By (23.12) we have $\alpha(t) < \beta(t)$ in some interval $[t_0, t_0 + \eta] \subset [t_0, t_1]$. If we can prove that the second case of Lemma 23.2 is excluded, we have established (23.16). Suppose that there exists $\tau \in [t_0, t_1]$ such that $\alpha(\tau) = ||u(\tau) - v(\tau)|| = r(\tau) = \beta(\tau)$. With the help of Lemma 23.3 we conclude using (23.13)-(23.15) that

$$\begin{aligned}
\alpha_-'(\tau) &\leq ||\dot{u}(\tau) - \dot{v}(\tau)|| = ||\dot{v}(\tau) - \dot{u}(\tau)|| \\
&= ||\dot{v}(\tau) - R(\tau, v(\tau)) + R(\tau, v(\tau)) - R(\tau, u(\tau))|| \\
&\leq d(\tau) + Q(\tau, ||v(\tau) - u(\tau)||) \\
&< \dot{r}(\tau) = \beta'(\tau)
\end{aligned}$$

which is however a contradiction to Lemma 23.2 and the estimate (23.16) is proved.

$\square$

Proof of Theorem 23.7. We want to apply Theorem 23.9 and need to establish (23.12) - (23.15) for appropriate functions r, d and Q. For the function Q we may choose $Q(t, ||y||) = L||y||$ and for d we take $d(t) = ||\dot{v}(t) - R(t, v(t))||$. Given δ_1 and δ_2 as in (23.10) we are searching for a suitable function r. With $\delta_1 < \eta_1$ and $\delta_2 < \eta_2$ we solve

$$\dot{r}_1(t) = \eta_2 + Lr(t) \text{ in } I_0 \text{ and } r_1(t_0) = \eta_1. \tag{23.17}$$

For $t \geq t_0$ a solution to (23.17) is given by

$$r_1(t) = \eta_1 \, e^{L(t-t_0)} + \frac{\eta_2}{L}(e^{L(t-t_0)} - 1)$$

and for $\eta_1 \to \delta_1$ and $\eta_2 \to \delta_2$ we obtain for $t \geq t_0$

$$r_2(t) = \delta_1 \, e^{L(t-t_0)} + \frac{\delta_2}{L}(e^{L(t-t_0)} - 1).$$

For $t < t_0$ we can use the same argument and reflect at t_0. Thus we arrive as a candidate for r at

$$r(t) = \delta_1 \, e^{L|t-t_0|} + \frac{\delta_2}{L}(e^{L|t-t_0|} - 1) \tag{23.18}$$

and we claim that (23.12)-(23.15) hold. The estimates (23.12) and (23.14) follow from (23.10) and (23.9) as does (23.13), recall $d(t) = ||\dot{v}(t) - R(t, v(t))||$. Furthermore we have for $t > t_0$

$$\dot{r}(t) = \delta_1 \, Le^{L(t-t_0)} + \delta_2 \, e^{L(t-t_0)}$$

and therefore

$$d(t) + Q(t, r(t)) = ||\dot{v}(t) - R(t, v(t))|| + Lr(t)$$
$$< \delta_2 + Lr(t) = \delta_2 + L(\delta_1 \, e^{L(t-t_0)} + \frac{\delta_2}{L}(e^{L(t-t_0)} - 1)$$
$$= \dot{r}(t),$$

which is however (23.15) for the case $t > t_0$. The calculations for $t < t_0$ goes analogously. Hence by Theorem 23.9 we conclude that (23.11) holds.   □

We now want to apply Theorem 23.7 to the question whether solutions of initial value problems do depend continuously on data. For this let $R : I \times D \to \mathbb{R}^n, D \subset \mathbb{R}^n$, be a continuous function satisfying with respect to the second variable a Lipschitz condition. Consider the initial value problem (23.1), i.e.

$$\dot{u}(t) = R(t, u(t)) \quad \text{and} \quad u(t_0) = u_0$$

where $(t_0, u_0) \in I \times D$. For some compact interval $I_0 \subset I$, $t_0 \in I_0$, $\mathring{I}_0 \neq \emptyset$, let u be a solution such that $(t, u(t)) \in I_0 \times D$ for all $t \in I_0$. We call

$$W_\alpha(u) := \{(t, x) \in I_0 \times D \, | \, t \in I_0, ||x - u(t)|| < \alpha\} \qquad (23.19)$$

an α-**admissible neighbourhood** of $\Gamma_{I_0}(u)$, where $\Gamma_{I_0}(u)$ denotes the graph of $u : I_0 \to \mathbb{R}^n$. We call $\tilde{R} : W_\alpha(u) \to \mathbb{R}^n$ α-**admissible** with respect to u if $\tilde{R}$ is continuous and Lipschitz continuous with respect to the second variable. If $\tilde{R}$ is α-admissible with respect to u for every $(t_0, v_0) \in W_\alpha(u)$ a unique maximal solution to the problem

$$\dot{v}(t) = \tilde{R}(t, v(t)) \quad \text{and} \quad v(t_0) = v_0 \qquad (23.20)$$

exits.

Definition 23.10. *Let R be as in Theorem 23.7 and $u : I_0 \to \mathbb{R}^n$ be a solution to (23.1) where we assume now that I_0 is compact. We say that the solution u to (23.1) **depends continuously on** u_0 **and** R if for every $\epsilon > 0$ there exists $\delta > 0$ and $\alpha > 0$ such that for all α-admissible mappings $\tilde{R}$ and all solutions to (23.20) defined on I_0.*

$$||R(t, y) - \tilde{R}(t, y)|| \leq \delta \text{ in } W_\alpha(n) \text{ and } ||u_0 - v_0|| < \delta$$

implies for all $t \in I_0$ the estimate

$$||v(t) - u(t)|| < \epsilon. \qquad (23.21)$$

Note that (23.21) can be replaced by $||u - v||_{\infty, I_0} < \epsilon$.

Theorem 23.11. *In the situation of Theorem 23.7 and a compact interval I_0 the solution u to (23.1) depends continuously on u_0 and R.*

Proof. The assumption that I_0 is compact allows us to apply Theorem 23.9 and hence Theorem 23.7 for I_0. Let $R, \tilde{R}, u$ and v as in Definition 23.10, in particular we assume

$$\dot{v}(t) = \tilde{R}(t, v(t)) \quad \text{and} \quad v(t_0) = v_0$$

and

$$||R(t, v(t)) - \tilde{R}(t, v(t))|| < \delta \text{ in } W_\alpha(u) \text{ and } ||u_0 - v_0|| < \delta.$$

For $(t, v(t)) \in W_\alpha(u)$ this implies (23.10) for $\delta_1 = \delta_2 = \delta$, and hence (23.11) holds with $\delta_1 = \delta_2 = \delta$. Since $t \in I_0$ and I_0 is compact we can choose in (23.11) $\delta(= \delta_1 = \delta_2)$ such that

$$||u(t) - v(t)|| < \frac{\alpha}{2},$$

i.e. $||u(t) - v(t)|| < \alpha$ implies $||u(t) - v(t)|| < \frac{\alpha}{2}$ and again the compactness of I_0 implies that $(t, v(t)) \in W_\alpha(u)$ for all $t \in I_0$. This yields that (23.11) holds for $\delta_1 = \delta_2 = \delta$ in I_0. Now, given $\epsilon > 0$ choose $\alpha > 0$ such that $W_\alpha(u)$ admits α-admissible functions and choose $\delta > 0$ such that the right hand side of (23.11) becomes less than ϵ, i.e.

$$\delta < \frac{\epsilon}{\max\limits_{t \in I_0} H(t)}$$

with $H(t) = e^{L|t-t_0|} + \frac{1}{L}(e^{L|t-t_0|} - 1)$. This implies the theorem. $\qquad\square$

Corollary 23.12. *Let R be as in Theorem 23.7 and let u and v be the two solutions to (23.1) defined on the same compact interval I_0 corresponding to the initial values u_0 and v_0, respectively. Then we can find for every $\epsilon > 0$ some $\delta > 0$ such that $||u_0 - v_0|| < \delta$ implies $||u(t) - v(t)|| < \epsilon$ for all $t \in I_0$.*

Corollary 23.13. *Let $A : I_0 \to M(n; \mathbb{R})$ be continuous and $I_0 \subset \mathbb{R}$ compact. Further let $f, g : I_0 \to \mathbb{R}$ be two continuous functions and $u_0, v_0 \in \mathbb{R}^n$. For every $\epsilon > 0$ there exists $\delta > 0$ such that $||f - g||_{\infty, I_0} < \delta$ and $||u_0 - v_0|| < \delta$ implies that $||u(t) - v(t)|| < \epsilon$ for all $t \in I_0$, where u solves $\dot{u}(t) = A(t)u(t) + f(t), u(t_0) = u_0$, and v solves $\dot{v}(t) = A(t)v(t) + g(t), v(t_0) = v_0$.*

In many situations data in a given differential equation depend on parameters. A natural question is whether this dependence is smooth, say as a first attempt continuous. We know that for certain (eigenvalue) problems we can not expect such a result. Consider the problem

$$\dot{u}(t) + \lambda^2 u(t) = 0 \text{ in } [0, 2\pi], u(0) = u(2\pi) = 0.$$

For all $\lambda \in \mathbb{R}$ we have always the trivial solution $u(t) = 0$ for all $t \in [0, 2\pi]$. However for $\lambda^2 \in \mathbb{N}$, and only for $\lambda^2 \in \mathbb{N}$, we have the additional solution $t \mapsto \sin |\lambda| t$. Thus the solutions u_λ do not depend continuous on λ. However for initial value problems we can prove under mild conditions continuous dependence of the solution on parameters.

We consider the parameter dependent initial value problem

$$\dot{u}(t, \lambda) = R(t, \lambda, u(t)) \text{ in } [t_0, T] \text{ and } u(t_0) = u_0(\lambda) \tag{23.22}$$

where $\dot{u}(t, \lambda)$ denotes the derivative of $u(t, \lambda)$ with respect to t.

Here $R : [t_0, T] \times K \times \mathbb{R}^n \to \mathbb{R}^n$ is a continuous function where $K \subset \mathbb{R}^m$ is a compact set. We may rewrite (23.22) as an equivalent integral equation (depending on λ), namely as

$$u(t, \lambda) = u_0(\lambda) + \int_{t_0}^t R(s, \lambda, u(s, \lambda)) \, ds \tag{23.23}$$

In light of the proof of Theorem 17.1, the following theorem is no surprise.

Theorem 23.14. *Suppose that $R : [t_0, T] \times K \times \mathbb{R}^n \to \mathbb{R}^n$ is continuous and satisfies*

$$||R(t, \lambda, x) - R(t, \lambda, y)|| \leq L ||x - y|| \tag{23.24}$$

for all $(t, \lambda) \in [t_0, T] \times K$ and all $x, y \in \mathbb{R}^n$. Then the equation (23.23) has a unique continuous solution $u : [t_0, T] \times K \to \mathbb{R}^n$.

Remark 23.15. The important point in Theorem 23.14 is the continuity of u on $[t_0, T] \times K$, especially the continuity of $\lambda \mapsto u(t, \lambda)$.

Sketch of the proof of Theorem 23.14. We introduce that weighted norm

$$||u||_{\infty, L, K} := \sup_{(t, \lambda) \in [t_0, T] \times K} ||u(t, \lambda) e^{-2L|t - t_0|}|| \tag{23.25}$$

and consider the operator

$$Tu(t, \lambda) = u_0(\lambda) + \int_{t_0}^t R(s, \lambda, u(s, \lambda)) \, ds$$

and as before we can prove that T is a contraction with respect to the norm (23.25). Now Banach's fixed point theorem will give the result. (The reader is invited to fill in the details, also see Problem 3.) $\square$

Remark 23.16. A further natural question is whether solutions dependent in a differentiable way on parameter or initial data. For results of this type we refer to [3], [56] or [125].

Our results on continuous dependence on data are proved for compact intervals. We now want to go further and allow unbounded existence intervals and to develop first ideas of stability theory. We start with a lengthy example.

Example 23.17. Let $A = \begin{pmatrix} a_{11} & a_{12} \\ a_{21} & a_{22} \end{pmatrix} \in M(2; \mathbb{R})$ and consider the system

$$\dot{u}(t) = Au(t), \quad u(0) = u_0 \tag{23.26}$$

where we look at u as a function defined on $[0, \infty)$. By Chapter 18 we know that the unique solution to (23.26) is given by

$$u(t) = e^{At}u_0. \tag{23.27}$$

For $u_0 = 0$ the solution is trivial, namely $u(t) = 0$ for all $t \in [0, \infty]$. Now we suppose that $u_0 \neq 0$ but $\|u_0\|$ is small. We are interested in the behaviour of $u(t)$ as t tends to ∞. Suppose that A has two distinct real eigenvalues $\lambda_1 \neq \lambda_2$ and let C be a matrix such that

$$\begin{pmatrix} \lambda_1 & 0 \\ 0 & \lambda_2 \end{pmatrix} = C^{-1}AC.$$

For $w(t) = C^{-1}u(t)$ we find

$$\dot{w}(t) = \begin{pmatrix} \lambda_1 & 0 \\ 0 & \lambda_2 \end{pmatrix} w(t) \text{ and } w(0) = C^{-1}u_0,$$

hence,

$$w(t) = \begin{pmatrix} w_1(0)\, e^{\lambda_1 t} \\ w_2(0)\, e^{\lambda_2 t} \end{pmatrix}$$

If $\lambda_1 < 0$ and $\lambda_2 < 0$ then we find as already observed in Chapter 18 that $\lim\limits_{t \to \infty} w(t) = 0$ and therefore $\lim\limits_{t \to \infty} u(t) = \lim\limits_{t \to \infty} C\, w(t) = 0$. But now we want to interpret 0 as the solution to $\dot{u}(t) = A\, u(t)$, $u(0) = 0$. Then the result says that for every initial data u_0 the solution to (23.26) tends as $t \to \infty$ to the solution of $\dot{u}(t) = A\, u(t)$, $u(0) = 0$. If $\lambda_j = 0$ then $w_j(t) = w_j(0)$ or $u_j(t) = (Cw(0))_j = u_{0j}$. Still we have that $|u_j(t)| < \epsilon$ for all $t \in [0, \infty]$ provided $|u_{0,j}| < \delta = \epsilon$. Thus for $\lambda_1 \leq 0$ and $\lambda_2 \leq 0$ we find that given $\epsilon > 0$ there exists $\delta > 0$ such that $|u_0 - 0| < \delta$ implies that $\lim\limits_{t \to \infty} |u(t) - 0| < \epsilon$. However for $\lambda_j > 0$ we find that

$$\lim\limits_{t \to \infty} u(t) = \lim\limits_{t \to \infty} C \begin{pmatrix} w_1(0)\, e^{\lambda_1 t} \\ w_2(0)\, e^{\lambda_2 t} \end{pmatrix}$$

will not exist since $\lim\limits_{t \to \infty} \begin{pmatrix} w_1(0) & e^{\lambda_1 t} \\ w_2(0) & e^{\lambda_2 t} \end{pmatrix} = \begin{pmatrix} \zeta_1 \\ \zeta_2 \end{pmatrix}$ where $\zeta_j = +\infty$ if $w_j(0) > 0$ and $\zeta_j = -\infty$ if $w_j(0) < \infty$. For $\lambda_1 \in \mathbb{C} \backslash \mathbb{R}$ we know that $\lambda_2 = \overline{\lambda_1}$ and $\mathrm{Im}\, \lambda_1$ gives oscillatory terms (*sine* and *cosine* functions) whereas $\operatorname{Re} \lambda_1$ determines the growth or the amplitude of the solution. For $\operatorname{Re} \lambda_1 < 0$ we will have again decay to 0, but for $\operatorname{Re} \lambda_1 \geq 0$ we shall not expect that $\lim\limits_{t \to \infty} u(t)$ exists.

For our general consideration we restrict ourselves to a system

$$\dot{u} = R(u) \tag{23.28}$$

with initial condition $u(0) = u_0$, i.e. $R : D \to \mathbb{R}^n, D \subset \mathbb{R}^n$, does not depend on t. Such a system is called an **autonomous system**. We always assume that the initial value problem $\dot{u} = R(u), u(0) = u_0$, has a unique solution $u : [0, T_{\max}) \to \mathbb{R}^n$, i.e. the solution exists for all "times" $0 \leq t \leq T_{\max}$. We call $u_0 \in \mathbb{R}^n$ an **equilibrium point** or a **critical point** for $\dot{u} = R(u)$ if $R(u_0) = 0$. An equilibrium point $u_0 \in \mathbb{R}^n$ is always a solution to $\dot{u} = R(u), u(0) = u_0$. In Example 23.17 the point $0 \in \mathbb{R}^2$ was an equilibrium point. The aim is to study the stability of (23.28) at an equilibrium point. In the following it is convenient to denote the solution to $\dot{u} = R(u)$ and $u(0) = u_0$ by $\phi(t, u_0)$, i.e. $\phi : [0, \infty) \times D \to \mathbb{R}^n, (t, u_0) \mapsto \phi(t, u_0)$ and $\frac{\partial \phi}{\partial t} = R(\phi), \phi(0, u_0) = u_0$. Instead of $\frac{\partial \phi}{\partial t}$ we will also write $\dot{\phi}$.

Definition 23.18. *Let $u_0 \in \mathbb{R}^n$ be an equilibrium point of the autonomous system $\dot{u} = R(u)$.*

A. *We call u_0 **stable** or a **stable equilibrium point** or we call $\dot{u} = R(u)$ a **stable system** at u_0 if for every $\epsilon > 0$ there exists $\delta > 0$ such that $\|v_0 - u_0\| < \delta$ implies for all $t \in [0, T_{\max})$ that $\|\phi(t, v_0) - u_0\| < \epsilon$. If $\dot{u} = R(u)$ is not stable at u_0 we call it **unstable** at the equilibrium point u_0.*

B. *An equilibrium point u_0 of $\dot{u} = R(u)$ is called **attractive** or an **attractor** of the system if $T_{\max} = \infty$ and there exists $\rho > 0$ such that $\|v_0 - u_0\| < \rho$ implies $\lim\limits_{t \to \infty} \phi(t, v_0) = u_0$.*

C. *If u_0 is stable and attractive we call u_0 or the system $\dot{u} = R(u)$ **asymptotically stable** (at u_0).*

For linear system with constant coefficients we can prove with the help of the results of Chapter 18

Theorem 23.19. *For every linear system with constant coefficients, i.e. $\dot{u} = Au, A \in M(n; \mathbb{R})$, the point $0 \in \mathbb{R}^n$ is an equilibrium point.*

A. *The system $\dot{u} = Au$ is asymptotically stable at 0 if all eigenvalues of A have negative real part.*

B. *If all eigenvalues λ of A have non-positive real part and in the case where $\operatorname{Re}\lambda = 0$ the geometric multiplicity of λ is equal to the algebraic multiplicity of λ, then the system $\dot{u} = \lambda u$ is stable at 0.*

C. *If one eigenvalue of A has positive real part, then the system is unstable at 0.*

Proof. We apply Theorem 18.18. Let $C \in M(n; \mathbb{R})$ such that $B = C^{-1}AC$ is the Jordan form of A. For e^{tB} we have the representation (18.46). Let λ_j be the eigenvalue corresponding to B_j and $\operatorname{Re}\lambda_j < 0$. In this case we know that $\lim\limits_{t\to\infty} e^{t(B_j)}y_0 = 0$ for all $y_0 \in \mathbb{R}^n$. If λ_j is the eigenvalue corresponding to $\tilde{D}_{\lambda j}$ then we know that $e^{t(N(2,\lambda_j))}$ grows in each component as a polynomial, recall that $N(2, \lambda_j)$ is nilpotent. Hence we have for all $y_0 \in \mathbb{R}^n$ that

$$\lim_{t\to\infty} e^{t(F_j)}y_0 = \lim_{t\to\infty} e^{t(\tilde{D}_{\lambda j})}e^{t(N(2,\lambda_j))}y_0 = 0.$$

This implies part A. Part B follows in a similar way to part A by going back to Theorem 18.18. We have to note that the condition "if $\operatorname{Re}\lambda = 0$ then the geometric multiplicity of λ is equal to the algebraic multiplicity of λ" implies that the corresponding block matrix is a diagonal matrix with pure imaginary diagonal elements all equal to $i\operatorname{Im}\lambda$. While the components corresponding to an eigenvalue with negative real part behave as asymptotically stable components, those corresponding to the eigenvalues with $\operatorname{Re}\lambda = 0$ are stable since *sine* and *cosine* are globally bounded. Finally, let $\operatorname{Re}\lambda_j > 0$. If λ_j is real then $\lim\limits_{t\to\infty} e^{t(B_j)}y_0$ does not exist and if λ_j is complex then $\lim\limits_{t\to\infty} e^{t(F_j)}y_0$ does not exist. So far we have proved the theorem for the case of a matrix in Jordan form. If $\dot{w} = Bw$ then $u = Cw$ solves the original system $\dot{u} = Au$ and the statements about the limits are still valid for u. $\qquad\square$

A further simple consequence of Theorem 18.18 is

Lemma 23.20. *Let A be as in Theorem 23.19 and suppose that $\operatorname{Re}\lambda \le \rho$ for all eigenvalues λ of A. Then there exists a constant $\kappa > 0$ such that*

$$\|e^{tA}y\| \le \kappa\, e^{\rho t}\|y\| \tag{23.29}$$

holds for all $y \in \mathbb{R}^n$ and $t \ge 0$.

Proof. We just look at (18.46) and it is clear that the estimate (23.29) holds for each term e^{tB_j} or e^{tF_j}. Now we can merge the corresponding estimates to the block matrix e^{tB}. Finally we just note that $A = CBC^{-1}$. $\qquad\square$

404

We now turn to general autonomous systems, i.e. not necessarily linear systems $\dot{u} = R(u)$ with an equilibrium point u_0. We assume that $R : D \to \mathbb{R}^n$ is a C^1-mapping and therefore we may approximate $R(\cdot)$ at the point u_0 by its differential

$$R(u) - R(u_0) = (\mathrm{d}R(u_0))(u - u_0) + \text{error term},$$

or since $R(u_0) = 0$

$$R(u) = (\mathrm{d}R(u_0))(u - u_0) + \text{error term}.$$

The idea is that the stability properties of the equilibrium point $0 \in \mathbb{R}^n$ of $\mathrm{d}R(u_0)$ will be the same as those of the equilibrium point u_0 of R. The following results are essentially due to A.M.Ljapunov and H.Poincaré.

Theorem 23.21. *Let $R : D \to \mathbb{R}^n$ be a C^1-mapping and $u_0 \in \mathbb{R}^n$ be an equilibrium point of $\dot{u} = R(u)$. If all eigenvalues of $\mathrm{d}R(u_0)$ have a negative real part then u_0 is asymptotically stable.*

Proof. (Following [32]) Shifting from R to $R - R(u_0)$ if necessary we may assume that $u_0 = 0 \in \mathbb{R}^n$. Since all eigenvalues of $\mathrm{d}R(u_0)$ have negative real part we get

$$2\rho := \max\{\mathrm{Re}\,\lambda \mid \lambda \text{ is eigenvalue of } \mathrm{d}R(u_o)\} < 0,$$

hence $\mathrm{Re}\,\lambda < \rho < 0$ for all eigenvalues of $\mathrm{d}R(u_0)(= \mathrm{d}R(0))$. By Corollary 23.20 there exists $\kappa > 1$ such that

$$||e^{t\,\mathrm{d}R(0)}y|| \leq \kappa e^{\rho^t}||y|| \text{ for all } y \in \mathbb{R}^n \text{ and } t \geq 0. \tag{23.30}$$

The differentiability of R at $u_0 = 0$ implies the existence of $h : \mathbb{R}^n \to \mathbb{R}^n$ such that

$$R(y) = \mathrm{d}\,R(0)y + h(y) \text{ and } \lim_{y \to 0} \frac{h(y)}{||y||} = 0,$$

which means that for $\epsilon > 0$ we can find $0 < \delta < (\kappa + 1)\delta < \epsilon$ such that $B_\delta(0) \subset D$ and $||y|| \leq \delta$ implies $||h(y)|| \leq \epsilon ||y||$. Let now $0 < \epsilon < \frac{-\rho}{2\kappa}$ and δ be chosen as above. Take $y_0 \in B_{\frac{\delta}{\kappa}}(0) \subset B_\delta(0)$ and choose t_1 such that $y(t) = \phi(t, y_0) \in B_\delta(0)$ for all $0 \leq t \leq t_1$. The function y satisfies

$$\dot{y}(t) = (\mathrm{d}R(0))y(t) + h(g(t))$$

and by Theorem 19.13 we find

$$y(t) = e^{t\,\mathrm{d}R(0)}\,y(t) + h(y(t)), \quad (y(0) = y_0). \tag{23.31}$$

Using (23.30) we have

$$||y(t)|| \leq ||e^{t\, dR(0)} y_0|| \leq \int_0^t ||e^{(t-s)dR(0)}\, h(y(s))\, ds$$

$$\leq \kappa e^{\rho t} + \kappa \epsilon \int_0^t e^{(t-s)\rho}\, ||y(s)||\, ds$$

or

$$e^{-\rho t}||y(t)|| \leq \kappa ||y_0|| + \kappa \epsilon \int_0^t e^{-\rho \epsilon}||y(s)||\, ds. \tag{23.32}$$

Thus we may apply Gronwall's lemma, Proposition 22.4, to $f(t) = e^{-\rho t}||y(t)||$ with $a(t) = \kappa||y_0||$ and $b(t) = \kappa \epsilon$ to obtain

$$e^{-\rho t}||y(t)|| \leq \kappa ||y_0|| + \int_0^t \kappa||y_0||\, \epsilon \kappa e^{\int_s^t \epsilon \kappa\, dr}\, ds$$

$$= \kappa ||y_0||\, e^{\epsilon \kappa t}.$$

By our assumptions we have $\kappa \epsilon \leq -\frac{\rho}{2}$ which yields

$$||y(t)|| \leq \kappa e^{\frac{\rho}{2}t}||y_0|| < \delta < \epsilon \text{ for } t \in [0, t_1], \tag{23.33}$$

where we used $\rho < 0$ and $||y_0|| \leq \frac{\delta}{\kappa} < \delta$. Hence for all $0 \leq t \leq t_1$ we have $y(t) \notin \partial B_\delta(0)$ implying that the maximal existence interval must contain $[0, \infty)$. Moreover, since $\rho < 0$ we deduce from (23.33) that 0 is attractive and stable. $\square$

Without proof, compare [125], we state

Theorem 23.22. *In the situation of Theorem 23.21 the equilibrium point u_0 is unstable if one of the eigenvalues of $dR(0)$ has a positive real part.*

While Theorem 23.21 and Theorem 23.22 are in some sense natural results, they are not always applicable. The following method going back to A.M.Ljapunov is in many situations quite powerful. We start with

Definition 23.23. *Let $R : D \to \mathbb{R}^n, D \subset \mathbb{R}^n$, be a C^1- mapping and $u_0 \in D$ be an equilibrium point of $\dot{u} = R(u)$. We call a C^1-function $V : D_0 \to \mathbb{R}$ where D_0 is a neighbourhood of u_0 a **Ljapunov function** for the system $\dot{u} = R(u)$ at u_0 if*

$$V(u_0) = 0; \tag{23.34}$$

$$V(y) > 0 \text{ for all } y \in D_0 \backslash \{u_0\}; \tag{23.35}$$

$$\langle grad\, V(y), R(y) \rangle \leq 0 \text{ in } D_0. \tag{23.36}$$

If instead of (23.36) we have

$$\langle grad\, V(y), R(y) \rangle < 0 \text{ in } D_0 \backslash \{u_0\} \tag{23.37}$$

*then we call V a **strict Ljapunov function**.*

Theorem 23.24 (Ljapunov). *Let R and u_0 be as in Definition 23.23.*

A. *If u_0 admits a Ljapunov function V then u_0 is a stable point for $\dot{u} = R(u)$.*

B. *If we can find a strict Ljapunov function for u_0 then u_0 is asymptotic stable.*

Proof. (Following [32]) As in the proof of Theorem 23.21 we may assume that $u_0 = 0$.

A. Let $V : D_0 \to \mathbb{R}$ be a Ljapunov function at $u_0 = 0$ and choose $r > 0$ such that $\overline{B_r(0)} \subset D_0$. For every $0 < \epsilon \leq r$ the set $\partial B_\epsilon(0) \subset B_r(0)$ is compact and $V|_{\partial B_\epsilon(0)} > 0$ by (23.35). This implies that

$$m(\epsilon) := \min_{y \in \partial B_\epsilon(0)} V(y) > 0.$$

By the continuity of V we can find $\delta > 0$, $0 < \delta < \epsilon$, such that $||y|| < \delta$ implies $V(y) < m(\epsilon)$, note that $V(0) = 0$. For $u(t) = \phi(t, y)$ we find now

$$\frac{d}{dt} V(u(t)) = \langle (grad\, V)(y(t)), \dot{y}(t) \rangle$$
$$= \langle (grad\, V)(y(t)), R(y(t)) \rangle \leq 0,$$

i.e. $t \mapsto V(u(t))$ is monotone decreasing which implies in particular that $V(u(t)) = V(\phi(t, y)) < m(\epsilon)$ for all $t \geq 0$ provided $||y|| < \delta$. Thus the existence interval of $u(t) = \phi(t, y)$ contains $[0, \infty)$ and for $t \geq 0$ we have $||u(t)|| \leq \epsilon$ implying the stability of $u_0 = 0$.

B. We may use the result of part A and for $\epsilon = r$ there exists ρ, $0 < \rho < r$, such that

$$||y|| < \rho \text{ implies } ||u(t)|| < r \text{ for all } t \geq 0, \tag{23.38}$$

where again $u(t) = \phi(t, y)$ is the solution to $\dot{u} = R(u)$ and $u(0) = y$. We claim now that $||y|| < \rho$ implies $\lim_{t \to \infty} \phi(t, y) = \lim_{t \to \infty} u(t) = 0$. Given $\epsilon \in (0, \rho)$ we choose $\delta > 0$ as in part A such that $0 < \delta < \epsilon < \rho < r$. Suppose that $||y|| < \rho$. If we have already $||y|| < \epsilon$ then $||u(t)|| = ||\phi(t, y)|| < \epsilon$ for all $t \geq 0$. Suppose now that $\delta \leq ||y|| < \rho < r$. We consider

$$M := \max\{ \langle (grad\, V)(x), R(x) \rangle | \delta \leq ||x|| \leq r \}$$

and by (23.37) we have $M < 0$. This implies for $\phi(t, y)$

$$\frac{d}{dt} V(\phi(t, y)) = \langle (grad\, V)(\phi(t, y)), \dot{\phi}(t, y) \rangle$$
$$= \langle (grad\, V)\phi(t, y), R(\phi(t, y)) \rangle \leq M.$$

Integrating this inequality implies for $||\phi(t, y)|| > \delta$ that

$$V(\phi(t, y)) \le V(y) + t M.$$

Since $V(x) > 0$ for $x \ne 0$ and since $M < 0$ it follows that the left hand side tends to $-\infty$ as $t \to \infty$. Thus, since $V(\phi(t, y)) \ge 0$ we can find $T > 0$ such that $||\phi(T, y)|| < \delta$. But $\phi(t + T, y) = \phi(t, \phi(T, x))$ by the uniqueness result for $\dot{u} = R(u)$ and we are back in the case with initial data having norm less than ρ which yields $||\phi(t + T, y)|| < \epsilon$ for $t > 0$. Hence in both cases we arrive at $||\phi(t, x)|| < \epsilon$ for t sufficiently large and the asymptotic stability of $u_0 = 0$ is proved. □

A proof of the following instability result due to N.G.Chetaev is given in [32].

Theorem 23.25. *Let $R : D \to \mathbb{R}^n, D \subset \mathbb{R}^n$, be a C^1 - mapping and $u_0 \in D$ an equilibrium point of $\dot{u} = R(u)$. Suppose that there exists a neighbourbood $D_0 \subset D$ of u_0 and C^1 - mapping $V : D_0 \to \mathbb{R}$ such that for some domain $D_1 \subset D_0$ we have*

i) $u_0 \in \partial D$;

ii) $V > 0$ and $\langle grad\, V(y), R(y) \rangle > 0$ in D_1;

iii) $V = 0$ on $D_0 \cap \partial D_1$.

Then u_0 is an unstable equilibrium point.

The crucial question is of course: how can we find for $\dot{u} = R(u)$ with equilibrium point u_0 a (strict) Ljapunov function? Instead giving some ad hoc examples we will discuss an example which has some relevance in mechanics and physics.

Example 23.26. Let $D \subset \mathbb{R}^n$ and $h : D \to \mathbb{R}$ be a C^2-function. We consider the **gradient system**

$$\dot{u} = -(grad\, h)(u). \tag{23.39}$$

For a solution u of (23.39) we find

$$\dot{h}(u(t)) = \langle grad\, h(u(t)), \dot{u}(t) \rangle$$
$$= -\langle grad\, h(u(t)), grad\, h(u(t)) \rangle$$
$$= -||grad\, h(u(t))||^2.$$

Thus we may try as a Lyapunov function at u_0 the function $h : D \to \mathbb{R}$. We have to add the assumptions $h(u_0) = 0$ and $h(y) > 0$ for all $y \in D_0 \backslash \{u_0\}$ where $D_0 \subset D$ is a neighbourhood of u_0. For example if u_0 is an isolate minimum of h with $h(u_0) = 0$ these condition are satisfied. The function $h(y) = \frac{1}{2}||y||^2$ is a concrete example, and in some sense the typical example.

Problems

1. Let $f, a, b, c : [t_0, T] \to \mathbb{R}$ be non-negative continuous functions satisfying the estimate

$$f(t) \leq a(t) + c(t) \int_{t_0}^t b(s) f(s)\, ds, \quad t \in [t_0, T].$$

Show that for f we have the bound

$$f(t) \leq a(t) + c(t) \int_{t_0}^t b(s) f(s) e^{\int_s^t b(r) c(r)\, dr}\, ds.$$

Hint: start with Corollary 23.5.

2. Let $u : [-T, T] \to \mathbb{R}$ be a solution to the problem

$$\ddot{u}(t) + \omega^2 u(t) = 0, \quad u(0) = 1, u'(0) = 0$$

and let $v_{\epsilon,\eta} : [-T, T] \to \mathbb{R}$ be a solution to the problem

$$\ddot{v}(t) + (\omega + \epsilon)^2 v(t) = 0, \quad v(0) = 1, v'(0) = \eta,$$

where $\omega, \epsilon, \eta > 0$. Prove the estimate

$$|u(t) - v_{\epsilon,\eta}(t)| \leq (\epsilon + \eta) T.$$

3. Prove that the operator

$$(Tu)(t, \lambda) = u_0(\lambda) + \int_{t_0}^t R(s, \lambda, u(s, \lambda))\, ds$$

with R as in Theorem 23.14 is a contraction with respect to the norm (23.25).

4. Let $A_1, A_2 \in M(n; \mathbb{R})$ be positive definite as well as symmetric and assume that $[A_1, A_2] = 0$. Prove that $\dot{u} = -A_1 A_2 u$ is asymptotically stable at 0.

5. Consider the autonomous system

$$\dot{u} = v, \quad \dot{v} = -\sin u.$$

Find its equilibrium points and prove that the equilibrium point $(\pi, 0)$ is unstable.

6. Prove that the system

$$\dot{x} = -2x^3 + y, \quad \dot{y} = -x - 2y^5$$

has only one equilibrium point, namely $(0,0)$, and that this equilibrium point admits a Ljapunov function.

7. Let $h : \mathbb{R}^2 \to \mathbb{R}$, $h(x,y) = \sin^2(x^2 + y^2)$, and consider the gradient system $\dot{u} = -(\mathrm{grad}h)u$, $u = \binom{u_1}{u_2}$. Find a Ljapunov function for the equilibrium point $(0,0)$.

24 Tangent Spaces, Tangent Bundles, and Vector Fields

So far we have been searching for methods from analysis to solve and to investigate differential equations. However, every differential equation has in some sense a very natural geometric meaning, and indeed geometry is crucial to understand some of the qualitative properties of solutions or the set of all solutions of a given differential equation. We just need to think about stability problems which we will continue to discuss further below. This chapter and the next one are devoted to some geometric aspects of the theory of ordinary differential equation, but at the moment we are limited in our investigation since we do not yet have the theory of differentiable manifolds at our disposal.

Let $G \subset \mathbb{R}^n$ be an open set and $R : G \to \mathbb{R}^n$ be a C^1-mapping. We consider the first order system

$$\dot{u}(t) = R(u(t)). \tag{24.1}$$

A solution u of (24.1) can be interpreted as a curve $\gamma : I \to G$ with the property that for every $t \in I$ the tangent direction $\dot{\gamma}(t)$ to γ at $t \in I$ is given by $R(\gamma(t))$. Before we can continue we need to clarify some notions. In Chapter II.7 we have already discussed differentiable curves $\gamma : I \to \mathbb{R}^n$ and some notion of tangent vectors. While for our considerations there, in Chapter II.7, it was sufficient and helpful to reduce all of our discussions to curves being parametrised with respect to arc length, and hence every tangent vector was a unit vector, see Definition II.7.29, we now need to move closer to the general theory of differentiable manifolds and we need to have a different view of tangent vectors. What is going to follow in this chapter is a short discussion of tangent vectors, tangent spaces and the tangent bundle of a very special class of differentiable manifolds, namely open sets $G \subset \mathbb{R}^n$. (A reader being familiar with the theory of differentiable manifolds will have no problems to identify G as a basic example of differentiable manifold where the differentiable structure, i.e. the maximal atlas, is just the one induced from $\mathbb{R}^n$ carrying its natural differentiable structure.)

We start with an algebraic preparation.

Definition 24.1. A. A **derivation** X on an $\mathbb{R}$-algebra $\mathcal{A}$ is an $\mathbb{R}$-linear mapping $X : \mathcal{A} \to \mathcal{A}$ which satisfies the Leibniz rule, i.e. for $u, v \in \mathcal{A}$ and $\lambda, \mu \in \mathbb{R}$ we have

$$X(\lambda u + \mu v) = \lambda\, Xu + \mu\, Xv \tag{24.2}$$

and

$$X(uv) = (Xu)\, v + u\, (Xv). \tag{24.3}$$

B. The commutator $[X, Y] = XY - YX$ of two derivations X and Y on $\mathcal{A}$ is called the **Lie product** of X and Y.

Lemma 24.2. *For two derivations X and Y on $\mathcal{A}$ their Lie product $[X, Y]$ is a further derivation.*

Proof. The linearity of $[X, Y] : \mathcal{A} \to \mathcal{A}$ is obvious since the composition of linear mappings as well as the difference of linear mappings are linear. For $u, v \in \mathcal{A}$ we have further

$$
\begin{aligned}
[X, Y](uv) &= X(Y(uv)) - Y(X(uv)) \\
&= X(uYv + vYu) - Y(uXv + vXu) \\
&= uXYv + (Xu)(Yv) + vXYu + (Xv)Yu) \\
&\quad - uYXv - (Yu)(Xv) - vYXu - (Yv)(Xu) \\
&= u[X, Y]v + v[X, Y]u.
\end{aligned}
$$

$\square$

Let $G \subset \mathbb{R}^n$ be an open set and denote by $C^\infty(G; \mathbb{R}^m)$ the set of all arbitrarily often diffentiable mappings $f : G \to \mathbb{R}^m$. (Sometimes we have to consider C^∞- mappings from G to $H \subset \mathbb{R}^m$ and then we will write $C^\infty(G; H)$.) The set $C^\infty_{q,loc}(G; \mathbb{R}^m)$ consists of all C^∞- mappings $g : \mathcal{U} \to \mathbb{R}^m$ which are defined in a neighbourhood $\mathcal{U} \subset G$ of $g \in G$. Clearly, $C^\infty_{q,loc}(G; \mathbb{R}^m)$ is a vector space and $C^\infty_{q,loc}(G) := C^\infty_{q,loc}(G; \mathbb{R})$ is an algebra. We call $f, g \in C^\infty_{q,loc}(G; \mathbb{R}^m)$ equivalent and we write $f \sim_q g$ if there exists a neighbourhood $\mathcal{U}$ of q such that $f|_\mathcal{U} = g|_\mathcal{U}$. It is easy to check that $f \sim_q g$ is an equivalence relation.

Definition 24.3. *An equivalence class $[f] \in C^\infty_{q,loc}(G; \mathbb{R}^m)/ \sim_q$ is called a **germ of a mapping** from G to $\mathbb{R}^m$. The family of all such germs, i.e. $C^\infty_{q,loc}(G, \mathbb{R}^m)/ \sim_q$, is denoted by $C^\infty_q(G; \mathbb{R}^m)$ at q. If $m = 1$ we call $[f] \in C^\infty_q(G; \mathbb{R})$ a **germ of a function** and we write $C^\infty_q(G) := G^\infty_q(G; \mathbb{R})$.*

On $C^\infty_q(G; \mathbb{R}^m)$ we have the natural vector space operations

$$[f] + [g] := [f + g] \tag{24.4}$$

and

$$\lambda[f] := [\lambda f]. \tag{24.5}$$

In Problem 1 we will see that (24.4) and (24.5) are well defined, i.e. independent of representatives. In addition, on $C^\infty_q(G)$ we can define

$$[f] \cdot [g] := [fg] \tag{24.6}$$

412

which is again independent of representatives and (24.4)-(24.6) turn $C_q^\infty(G)$ into an algebra. On $C_q^\infty(G)$ we consider $\mathbb{R}$ - **derivations** which are by definition $\mathbb{R}$-linear mappings $X_q : C_q^\infty(G) \to \mathbb{R}$ satisfying the following form of Leibniz rule

$$X_q([f][g]) = f(q)X_q([g]) + g(q)X_q([f]). \tag{24.7}$$

A typical $\mathbb{R}$-derivation on $C_q^\infty(G)$ we will denote by X_q or Y_q etc. We denote by f_c the function $f_c : \mathbb{R}^n \to \mathbb{R}, f(x) = c$ for all $x \in \mathbb{R}^n$, and we also use f_c as a symbol for $f_c|_G$. For $[f_1](= [1])$ we find

$$X_q[f_1] = X_q[f_1^2] = X_q([f_1][f_1])$$
$$= 2X_q[f_1]$$

implying $X_q[f_1] = 0$. The linearity of X_q now yields

$$X_q([f_c]) = c\, X_q[f_1] = 0. \tag{24.8}$$

By $\Delta_q G$ we denote the set of all derivations on $C_q^\infty(G)$. For $[f] \in C_q^\infty(G; \mathbb{R}^m)$ we define

$$f^* : C_{f(q)}^\infty(\mathbb{R}^m) \to C_q^\infty(G) \tag{24.9}$$

by

$$f^*[\phi] := [\phi \circ f]. \tag{24.10}$$

We claim that f^* is well defined, i.e. independent of representatives of $[f]$ and $[\phi]$. Suppose $f_1, f_2 \in [f]$ and $\phi_1, \phi_2 \in [\phi]$. Then there exists a neighbourhood W of $f(q)$ such that $\phi_1|_W = \phi_2|_W$. Further, we can find a neighbourhood V of q such that $f_1|_V = f_2|_V$ and $f_1(V) \subset W$ as well as $f_2(V) \subset W$. Now it follows

$$\phi_1 \circ f_1 = \phi_1 \circ f_2 = \phi_2 \circ f_2 = \phi_2 \circ f_1 \text{ in } V$$

which proves the claim.

Proposition 24.4. *The mapping f^* is an homomorphism between the algebras $C_{f(q)}^\infty(\mathbb{R}^m)$ and $C_q^\infty(G)$.*

Proof. The linearity follows from

$$f^*(\lambda[\phi] + \mu[\psi]) = f^*([\lambda\phi + \mu\psi])$$
$$= [(\lambda\phi + \mu\psi) \circ f] = \lambda[\phi \circ f] + \mu[\psi \circ f]$$
$$= \lambda f^*([\phi]) + \mu f^*([\psi]).$$

Furthermore we have

$$f^*([\phi] \cdot [\psi]) = f^*([\phi \cdot \psi]) = [(\phi \cdot \psi) \circ f]$$
$$= [(\phi \circ f)(\psi \circ f)] = f^*([\phi])f^*([\psi]).$$

$\square$

For $\mathrm{id}_G : G \to G, \mathrm{id}_G(x) = x$ and $q \in G$ we find that

$$(\mathrm{id}_G)^* = \mathrm{id}_{C_q^\infty(G)} \tag{24.11}$$

and with $G \subset \mathbb{R}^n$ and $H \subset \mathbb{R}^m$ we find for $[f] \in C_q^\infty(G; H)$ and $[g] \in C_{f(q)}^\infty(h; \mathbb{R}^k)$ that

$$(g \circ f)^* = f^* \circ g^*, \tag{24.12}$$

see Problem 4. In particular we have for $[g] \in C_q^\infty(G; H)$ induced by a bijective mapping $g : G \to H$ that

$$(g^{-1})^* = (g^*)^{-1}. \tag{24.13}$$

Definition 24.5. *Let $G \subset \mathbb{R}^n$ and $f : G \to \mathbb{R}^m$ be a C^∞- mapping. We call the mapping*

$$d_q f : \Delta_q G \to \Delta_{f(q)} \mathbb{R}^m \tag{24.14}$$

$$X_q \mapsto Y_{f(q)} : X_q \circ f^*,$$

where

$$(X_q \circ f^*)[\phi] = X_q(f^*[\phi]) = X_q[\phi \circ f], \tag{24.15}$$

*the **differential** of $[f]$ (or f) and q.*

Note that (24.25) also implies that $d_p f$ is independent of the representative. Indeed, for $f_1, f_2 \in [f]$ with $f_1|_V = f_2|_V$ it follows that

$$X_q([\phi \circ (f_1 - f_2)]) = X_q[\phi \circ f_1] - X_q[\phi \circ f_2]$$

but $\phi \circ f_1|_V = \phi \circ f_2|_V$ implying $X_q[\phi \circ (f_1 - f_2)] = 0$.

Proposition 24.6. *Let $G \subset \mathbb{R}^n, H \subset \mathbb{R}^m$ be open and $f : G \to H$ and $g : H \to \mathbb{R}^k$ be C^∞ - mappings. For the differential $d_q(g \circ f) : \Delta_q G \to \Delta_{g(f(q))} \mathbb{R}^k$ the **chain rule**

$$d_q(g \circ f) = (d_{f(q)} g) \circ d_q f \tag{24.16}$$

holds.*

Proof. For $X_q \in \Delta_q G$ it follows that

$$\begin{aligned} d_q(g \circ f)(X_q) &= X_q((g \circ f)^*) = X_q \circ (f^* \circ g^*) \\ &= (X_q \circ f^*) \circ g^* = (d_{f(g)} g)(X_q \circ f^*) \\ &= ((d_{f(q)} g) \circ d_q f)(X_q). \end{aligned}$$

$\square$

Theorem 24.7. *Let* $f : G \to H$, $G, H \subset \mathbb{R}^n$ *open, be a bijective* C^∞ *- mapping and suppose that* $f(q) = 0 \in H \subset \mathbb{R}^n$. *Then the differential* $d_q f : \Delta_q G \to \Delta_0 \mathbb{R}^n$ *is a vector space isomorphism.*

Proof. First we note that since $f : G \to H$ implies $f : G \to \mathbb{R}^n$ we have for $q \in G$ that $\Delta_{f(q)} H = \Delta_{f(q)} \mathbb{R}^n$. We know that $d_q f$ is linear, which follows from Definition 24.5. First we prove that $d_q f$ is injective. Let $X_q, Y_q \in \Delta_q G$, $X_q \neq Y_q$. Hence for some $[\phi] \in C_q^\infty(G)$ we must have $X_q[\phi] \neq Y_q[\phi]$. We set $[\psi] := (f^*)^{-1}[\phi]$ and it follows that

$$(d_q f)(X_q - Y_q)[\psi] = (X_q - Y_q) \circ f^*((f^*)^{-1}[\phi])$$
$$= (X_q - Y_q)[\phi] \neq 0,$$

i.e. $(d_q f)(X_q) \neq (d_q f)(Y_q)$. Next we prove that $d_q f$ is surjective. Let $Z_0 \in \Delta_0 \mathbb{R}^n$ and consider $Z_0 \circ (f^*)^{-1}$. For $\lambda, \mu \in \mathbb{R}$ and $[\phi], [\psi] \in C_q^\infty(G)$ we have

$$Z_0(f^*)^{-1}(\lambda[\phi] + \mu[\psi])$$
$$= Z_0(\lambda(f^*)^{-1}[\phi] + \mu(f^*)^{-1}[\psi])$$
$$= \lambda Z_0 \circ (f^*)^{-1}[\phi] + \mu Z_0 \circ (f^*)^{-1}[\psi],$$

and further

$$Z_0 \circ (f^*)^{-1}([\phi][\psi]) = Z_0((f^*)^{-1}([\phi][\psi]))$$
$$= Z_0((f^*)^{-1}([\phi])(f^*)^{-1}([\psi]))$$
$$= ((f^*)^{-1}[\phi])(0) Z_0(f^*)^{-1}[\psi] + ((f^*)^{-1}([\psi])(0)) Z_0(f^*)^{-1}[\phi].$$

Thus $Z_0 \circ (f^*)^{-1} \in \Delta_q G$ and

$$(d_q f)(Z_0 \circ (f^*)^{-1}) = Z_0,$$

implying that $d_q f$ is bijective and therefore a vector space isomorphism. □

We want to determine the dimension of $\Delta_0 \mathbb{R}^n$. As preparation we give

Lemma 24.8. *For an arbitrarily often differentiable function* $\phi : \mathbb{R}^n \to \mathbb{R}$ *there exists arbitrarily often differentiable functions* $\phi_1, ..., \phi_n : \mathbb{R}^n \to \mathbb{R}$ *such that*

$$\phi(x) = \phi(0) + \sum_{j=1}^{n} x_j \phi_j(x) \tag{24.17}$$

hold.

Proof. This follows from the fundamental theorem of calculus by looking at

$$\phi(x) - \phi(0) = \int_0^1 \frac{d}{dt} \phi(tx_1, ..., tx_n) \, dt$$

and defining

$$\phi_j(x) := \int_0^1 \frac{\partial}{\partial x_j} \phi(tx_1, ..., tx_n) \, dt.$$

$\square$

Remark 24.9. In Lemma 24.8 we may replace ϕ by a C^k - function and then the functions ϕ_j are C^{k-1} -functions, $k \geq 1$.

To proceed further we note that on $C_{\{0\}}^\infty(\mathbb{R}^n)$ a derivation is given by $\frac{\partial}{\partial x_j}\big|_0, 1 \leq j \leq n$. Here we have written as we will do in the following $C_{\{0\}}^\infty(\mathbb{R}^n)$ for $C_0^\infty(\mathbb{R}^n)$ in order to have a distinction from the set of all arbitrarily often differentiable functions on $\mathbb{R}^n$ with compact support. For $[\phi] \in C_{\{0\}}^\infty(\mathbb{R}^n)$ we have

$$\frac{\partial}{\partial x_j}\bigg|_0 [\phi] := \left(\frac{\partial \phi}{\partial x_j}\right)(0). \tag{24.18}$$

Since two representatives of $[\phi]$ coincide in a neighbourhood of $0 \in \mathbb{R}^n$, (24.17) is independent of the representative, i.e. $\frac{\partial}{\partial x_j}\big|_0 [\phi]$ is well defined. The linearity of $\frac{\partial}{\partial x_j}\big|_0$ is clear and the Leibniz rule holds for $\frac{\partial}{\partial x_j}\big|_0$ since it holds for every first order partial derivative.

Theorem 24.10. *The vector space $\Delta_0 \mathbb{R}^n$ has dimension n.*

Proof. We prove that $\{\frac{\partial}{\partial x_j}\big|_0 | j = 1, ..., n\}$ is a basis of $\Delta_0 \mathbb{R}^n$. First we claim that $\{\frac{\partial}{\partial x_j}\big|_0 | j = 1, ..., n\}$ is linearly independent. Suppose that

$$\sum_{j=1}^n a_j \frac{\partial}{\partial x_j}\bigg|_0 = 0, \qquad a_j \in \mathbb{R}.$$

For the projection $pr^k : \mathbb{R}^n \to \mathbb{R}, x \mapsto x_k$, we have

$$\left(\frac{\partial}{\partial x_j}\bigg|_0\right)[pr^k] = \frac{\partial x_k}{\partial x_j}\bigg|_{j=0} = \delta_{jk}$$

and therefore we find for $1 \leq k \leq n$

$$0 = \left(\sum_{j=1}^n a_j \frac{\partial}{\partial x_j}\bigg|_0\right)[pr^k] = a_k,$$

416

implying the independence of $\{\frac{\partial}{\partial x_j}|_0 | j = 1, ..., n\}$. Now let $X_0 \in \Delta_0 \mathbb{R}^n$, define $a_k := X_0 [pr^k]$, and consider

$$Y_0 := X_0 - \sum_{j=1}^{n} a_j \left. \frac{\partial}{\partial x_j} \right|_0.$$

If follows that

$$Y_0 [pr^k] = X_0 [pr^k] - \sum_{j=1}^{n} a_j \left. \frac{\partial}{\partial x_j} \right|_0 [pr^k] = 0.$$

Now we consider for $\phi \in C^\infty(\mathbb{R}^n)$ the germ $[\phi] \in C^\infty_{\{0\}}(\mathbb{R}^n)$. By the previous lemma we have

$$\phi(x) = \phi(0) + \sum_{j=1}^{n} x_j \phi_j(x) = \phi(0) + \sum_{j=0}^{h} pr^j(x)\, \phi_j(x),$$

and therefore

$$[\phi] = [\phi(0)] + \sum_{j=1}^{n} [pr^j]\, [\phi_j]$$

which implies

$$Y_0[\phi] = Y_0[\phi(0)] + \sum_{j=1}^{n} Y_0([pr^j][\phi_j])$$

$$= Y_0[\phi(0)] + \sum_{j=1}^{n} [pr^j](0)\, Y_0([\phi_j]) + \sum_{j=1}^{n} [\phi_j] Y_0[pr^j]$$

$$= 0,$$

where we have used that $Y_0[c] = 0, [pr^j](0) = 0$ and $Y_0[pr^j] = 0$. Thus we have proved that $Y_0 = \sum_{j=1}^{n} a_j \frac{\partial}{\partial x_j}|_0$ proving the theorem. $\square$

Corollary 24.11. Let $G \subset \mathbb{R}^n$ be open. Then all the space $\Delta_q G = \Delta_q \mathbb{R}^n, q \in G$, are n-dimensional vector spaces.

Proof. We only need to combine the chain rule and the fact that translations are bijective C^∞-mappings with C^∞-inverse with Theorem 24.10. $\square$

We continue our investigations by giving a geometric definition of the tangent space at a point q of an open set $G \subset \mathbb{R}^n$. This definition is such that we later in Volume VII can extend it to parametric surfaces or more generally to differentiable manifolds. We begin with

Definition 24.12. *We call two differentiable mappings $f_1, f_2 : G \to \mathbb{R}^m, G \subset \mathbb{R}^n$ open, **tangent** at $q \in G$ if $f_1(q) = f_2(q)$ and $Df_1(q) = Df_2(q)$ where Df denotes the differential of the mapping f.*

For an open set $G \subset \mathbb{R}^n$ we denote by $C_{0,q}^\infty(\mathbb{R}; G)$ the set of all germs $[\gamma]$ of arbitrarily often differentiable mappings $\gamma : I \to G$ where $I \subset \mathbb{R}$ is an open interval with $0 \in I$ and $\gamma(0) = q$. Thus γ is a C^∞-parametric curve in the sense of Chapter II.7 (not necessarily regular) with trace $\mathrm{tr}(\gamma) = \gamma(I) \subset G$. On $C_{0,q}^\infty(\mathbb{R}; G)$ we introduce the equivalence relation $[\gamma_1] \approx [\gamma_2]$ if and only if $\phi_1 \in [\gamma_1]$ and $\phi_2 \in [\gamma_2]$ are tangent at $0 \in \mathbb{R}^n$, note that $\phi_1(0) = \phi_2(0) = q$. It is easy to see that " $\approx$ " is an equivalence relation on $C_{0,q}^\infty(\mathbb{R}; G)$.

Definition 24.13. *The **tangent space** $T_q G$ to G at $q \in G$ is the set consisting of all equivalence classes of $C_{0,q}^\infty(\mathbb{R}; G)$ with respect to " $\approx$ ", i.e. $T_q G = C_{0,q}^\infty(\mathbb{R}; G)/\approx$.*

For some time we will denote elements of $C_{0,q}^\infty(\mathbb{R}; G)/\approx$, i.e. elements of $T_q G$, by $\overline{[\gamma]}$ or $\overline{[\gamma]}_q$ if we want to emphasise the point q. Thus for $\overline{[\gamma]}_q$ there exists a mapping $\phi : I \to G$ where I is an open interval, $0 \in I$ and $\phi(0) = q$, and $[\phi]$ generates $\overline{[\gamma]}$, i.e. $[\phi] \in \overline{[\gamma]}$. While for clarifying the situation it was helpful to denote the mapping by ϕ and not by γ, in the following we will often use γ as a symbol for a mapping generating first $[\gamma]$ and then $\overline{[\gamma]}$.

Our aim is to prove that $T_q G$ carries the structure of an n - dimensional real vector space. This would immediately establish a one-to-one correspondence between $\Delta_q G$ and $T_q G$. For this we define the following mapping. Let $U \subset G$ be an open neighbourhood of q and $[\gamma] \in C_0^\infty(\mathbb{R}; G), \gamma : I \to U \subset G$, where $0 \in I$ and $\gamma(0) = q$. Then $[\gamma]$ induces an equivalence class $\overline{[\gamma]}$, i.e. $\overline{[\gamma]} \in T_q G$, and we define $\theta_{U,q} : T_q G \to \mathbb{R}^n$ by

$$\theta_{U,q}(\overline{[\gamma]}) := (D\gamma)(0). \tag{24.19}$$

First we note that this definition is independent of the representative since for $[\gamma_1], [\gamma_2] \in \overline{[\gamma]}$ with corresponding mapping γ_1 and γ_2 we have $\gamma_1(0) = \gamma_2(0) = q$ and $(D_{\gamma_1})(0) = (D_{\gamma_2})(0)$. We claim that $\theta_{U,q}$ is bijective. For $h \in \mathbb{R}^n$ we consider $\gamma_h : I \to G, \gamma_h(t) = q + th$ where $I \subset \mathbb{R}$ is an open interval such that $0 \in I$ and $\gamma_h(I) \subset U$. The curve γ_h generates a germ $[\gamma_h] \in C_{0,q}^\infty(\mathbb{R}; G)$ and $\theta_{U,q}(\overline{[\gamma_h]}) = h$. Hence $\theta_{U,q}$ is surjective. The injectivity of $\theta_{U,q}$ is trivial since $\theta_{U,q}(\overline{[\gamma_1]}) = \theta_{U,q}(\overline{[\gamma_2]})$ implies that γ_1 and γ_2 are tangent at $0 \in I$, recall we assume that $\theta_{U,q}(\overline{[\gamma_1]}) = \theta_{U,q}(\overline{[\gamma_2]})$ which means $(D_{\gamma_1})(0) = (D_{\gamma_2})(0)$. Therefore γ_1 and γ_2 induce the same equivalence class $\overline{[\gamma_1]} = \overline{[\gamma_2]}$, and we have eventually proved that $\theta_{U,q}$ is bijective. Next we remark that if V is a further open neighbourhood of q and $\theta_{V,q}$ the corresponding mapping (24.19), then we have

$$\theta_{V,q}(\overline{[\gamma]}) = \theta_{U,q}(\overline{[\gamma]})$$

which follows since $U \cap V$ is an open neighbourhood of q. Furthermore

$$\theta_{V,q} \circ \theta_{U,q}^{-1} : T_q G \to T_q G$$

and for $h = \theta_{U,q}(\gamma_h)$ we have

$$\theta_{V,q} \circ \theta_{U,q}^{-1}(h) = D(\theta_{U,q}^{-1}(h))(0) = D(q + th)|_{t=0} = h. \qquad (24.20)$$

We now introduce on $T_q G$ a vector space structure. Let $\overline{[\gamma_1]}, \overline{[\gamma_2]} \in T_q G$. Then there exists uniquely determined vectors $h_1, h_2 \in \mathbb{R}^n$ with $\theta_{U,q}(\overline{[\gamma_j]}) = h_j$. We now define

$$\overline{[\gamma_1]} + \overline{[\gamma_2]} = \theta_{U,q}^{-1}(h_1) + \theta_{U,q}^{-1}(h_2) := \theta_{U,q}^{-1}(h_1 + h_2) \qquad (24.21)$$

and

$$\lambda \overline{[\gamma_1]} = \lambda \theta_{U,q}^{-1}(h_1) := \theta_{U,q}^{-1}(\lambda h_1). \qquad (24.22)$$

These definitions are clearly independent of the representatives and therefore we have

Theorem 24.14. *On $T_q G$ a vector space structure is induced by (24.20), (24.21) and with these operations $T_q G$ is an n-dimensional real vector space.*

Corollary 24.15. *The spaces $T_q G$ and $\Delta_q G$ are isomorphic vector spaces.*

Let $\{e_1, ..., e_n\}$ be the canonical basis in $\mathbb{R}^n$. The set $\{\theta_{U,q}^{-1}(e_1), ..., \theta_{U,q}^{-1}(e_n)\}$ is a basis in $T_q G$ and from (24.20) we deduce that this set is unchanged when we replace $\mathcal{U}$ by another open set $W, q \in W$. With the help of $\theta_{U,q}$ we obtain for $\overline{[\gamma]} \in T_q G$ the vector $V_q \in \mathbb{R}^n$ as

$$V_q(\gamma) := (V_q^1(\gamma), ..., V_q^n(\gamma)) := \theta_{U,q}(\overline{[\gamma]}). \qquad (24.23)$$

For $[\phi] \in C_q^\infty(G)$ we consider

$$V_q(\gamma)[\phi] := \sum_{j=1}^n V_q^j(\gamma) \left(\frac{\partial}{\partial x_j} \phi \right)(q)$$

$$= \sum_{j=1}^n V_q^j(\gamma) \left(\frac{\partial}{\partial x_j} \right)_q \phi$$

where $\left(\frac{\partial}{\partial x_j} \right)_q \phi := \left(\frac{\partial \phi}{\partial x_j} \right)(q)$ and $\left(\frac{\partial}{\partial x_j} \right)_q$ is an $\mathbb{R}$-derivation on $C_q^\infty(G)$. With the projection $pr^k : \mathbb{R}^n \to \mathbb{R}, x \mapsto x_k$ it follows further that

$$V_q^j(\gamma) = V_q(\gamma)([pr^j])$$

419

and therefore

$$V_q(\gamma)[\phi] = \sum_{j=1}^{n} V_q(\gamma)[pr^j]\left(\frac{\partial}{\partial x_j}\right)_q \phi \qquad (24.24)$$

i.e.

$$V_q(\gamma) = \sum_{j=1}^{n} V_q(\gamma)[pr^j]\left(\frac{\partial}{\partial x_j}\right)_q.$$

Thus with the help of (24.23) and (24.24) we have established a linear mapping from $T_q G$ into $\Delta_q G$. Since $\overline{[\gamma_1]} \neq \overline{[\gamma_2]}$ implies $\theta_{U,q}(\overline{[\gamma_1]}) \neq \theta_{U,q}(\overline{[\gamma_2]})$ and therefore $V_q(\gamma_1)([pr^{j_0}]) \neq V_q(\gamma_2)([pr^{j_0}])$ for at least one $j_0, 1 \leq j_0 < n$, it follows that $V_q(\gamma_1) \neq V_q(\gamma_2)$. Hence this correspondence is linear and injective. Since $T_q G$ and $\Delta_q G$ are both n-dimensional the mapping $\overline{[\gamma]}_q \to V_q(\gamma)$ must be bijective and we have established

Theorem 24.16. *The spaces $T_q G$ and $\Delta_q G$ are isomorphic vector spaces and a bijective linear mapping from $T_q G$ to $\Delta_q G$ is given by $\overline{[\gamma]} \to V_q(\gamma)$.*

In light of Theorem 24.16 we will identify $T_q G$ and $\Delta_q G$ and in particular we will consider $\left\{ \left(\frac{\partial}{\partial x_1}\right)_q, \ldots, \left(\frac{\partial}{\partial x_n}\right)_q \right\}$ as basis in $T_q G$ and a term such as $\sum_{j=1}^{n} c_j \left(\frac{\partial}{\partial x_j}\right)_q$ we will consider as a tangent vector to G at q.

In Definition 24.5 we introduced for a C^∞ - mapping $f : G \to \mathbb{R}^m$ the differential $d_q f : \Delta_q G \to \Delta_{f(q)} \mathbb{R}^m$ by $X_q \mapsto Y_{f(q)} := X_q \circ f^*$.

For $\gamma : I \to G$, $\gamma(0) = q$, we define $\gamma_f : I \to \mathbb{R}^m$, $\gamma_f = f \circ \gamma$, and it follows $\gamma_f(0) = f(q)$. Therefore $\overline{[\gamma_f]}_{f(q)} \in I_{f(q)} \mathbb{R}^m$ and we find

$$V_{f(q)}(\gamma_f) = \sum_{k=1}^{m} V_{f(q)}(\gamma_f)[pr^k_{\mathbb{R}^m}]\left(\frac{\partial}{\partial y_k}\right)_{f(q)}$$

whereas we have originally

$$V_q(\gamma) = \sum_{j=1}^{n} V_q(\gamma)[pr^j_G]\left(\frac{\partial}{\partial x_j}\right)_q$$

or

$$V_{f(q)}(\gamma_f) = \sum_{k=1}^{m} \theta_{W,f(q)}\overline{[\gamma_f]}_k \left(\frac{\partial}{\partial y_k}\right)_{f(q)}$$

and

$$V_q(\gamma) = \sum_{j=1}^{n} \theta_{U,q}\overline{[\gamma]}_j \left(\frac{\partial}{\partial x_j}\right)_q$$

where $U \subset G$ is an open neighbourhood of $q \in G$ and $W \subset \mathbb{R}^m$ is an open neighbourhood of $f(q)$ and we may assume $f(U) \subset W$. For $\phi : \mathbb{R}^m \to \mathbb{R}$ we set $\phi_f := \phi \circ f : G \to \mathbb{R}$ and $[\phi] \in C_{f(q)}(\mathbb{R}^n)$ induces an element $\phi_f \in C_q^\infty(G)$. If follows that

$$
\begin{aligned}
V_q(\gamma)[\phi_f] &= \sum_{j=1}^n \theta_{U,q}(\overline{[\gamma]})_j \left(\frac{\partial}{\partial x_j} \right)_q \phi_f \\
&= \sum_{j=1}^n \sum_{l=1}^m \left(\theta_{U,q}(\overline{[\gamma]})_j \right) \left(\frac{\partial f^l}{\partial x_j}(q) \right) \left(\frac{\partial}{\partial y_l} \right)_{f(q)} \phi \\
&= \sum_{l=1}^m \left(\sum_{j=1}^n \left(\frac{\partial f^l}{\partial x_j} \right)(q) \, \theta_{U,q}(\overline{[\gamma]})_j \right) \left(\frac{\partial}{\partial y_l} \right)_{f(q)} \phi.
\end{aligned}
$$

However $\left(\frac{\partial f^l}{\partial x_j} \right)_{\substack{l=1,\dots,m \\ j=1,\dots,n}}$ is the Jacobi matrix of f at q. This now relates the differential $d_p f : \Delta_q G \to \Delta_{f(q)} \mathbb{R}^n$ of germs to the differential $(df)_{(p)}$ of $f : G \to \mathbb{R}^m$ which induces a mapping $\overline{(df)}_p : T_q G \to T_{f(q)} \mathbb{R}^n$ which we will just denote by $d_p f$. We end this remark here, but we will pick up the discussion in Volume VII when introducing differentiable manifolds.

Let $G \subset \mathbb{R}^n$ be open. For every $q \in G$ we have defined its tangent space $T_q G$. For $v \in T_q G$ we will usually write v_q in order to indicate to which point $q \in G$ the element v_q is "attached" to. Now we can consider the set of all tangent vectors to G and we introduce

$$
TG := \{(q, v_q) | q \in G, v_q \in T_q\}. \tag{24.25}
$$

It is convenient and rather common to write for $v = (q, v_q) \in TM$ just v_q and to identify $\{q\} \times T_q G$ with $T_q G$. In this sense we may write for $TM = \bigcup_{q \in G} \{q\} \times T_q G$

just $\bigcup_{q \in G} T_q G$ which is of course some abuse of notation.

Since $T_q G$ can be identified with $\mathbb{R}^n$ we can identify TG with $G \times \mathbb{R}^n$ which allows us to speak of continuous or even differentiable mappings $X : G \to TG$ or $h : TG \to G$. In particular we can introduce the **projection** from TG to G by

$$
\pi : TG \to G \tag{24.26}
$$

$$
v = (q, v_q) \mapsto \pi(v) = q
$$

(or in short $\pi(v_q) = q$) which is a C^∞ - mapping.

Definition 24.17. *We call (TG, π, G) the **tangent bundle** of G. The set TG is called the **total space** of the tangent bundle and G is called its **base** (or the* *base space*). *For an element* $v = (q, v_q) \in TG$ *we call* q *the* **base point** *and* $\pi^{-1}(\{q\}) = T_q G$ *the* **fibre** *over* q.

Remark 24.18. A. Again we follow the custom to write TG for the tangent bundle of G.

B. The tangent bundle is our first example of a vector bundle (or better a vector space bundle) and hence of a fibred space. Vector bundles are of utmost importance in analysis, geometry and topology and we will discuss them in detail in Volume VII.

Definition 24.19. *We call a differentiable mapping* $X : G \to TG$ *a* **differen-tiable section** *of* TG *or a* **vector field** *on* G *if for all* $g \in G$ *it follows that* $X(q) \in T_q G$, *or more precisely* $X(q) \in \{q\} \times T_q G$.

Remark 24.20. A. The so called **zero section** $Z : G \to TG, Z(q) = (q, 0) \in TG$, allows us to embed G into TG, or to identify G with a subset of TG.

B. Typically we write X_q for $X(q)$.

C. In the theory of differentiable manifolds vector fields are usually required to be C^∞-mappings. For us it will be more convenient to allow C^1-mappings, see below.

We want to study vector fields in more details and for this we assume now that they are C^∞-mappings. Later on we will make a few adjustments to allow C^k-mappings. The set of all C^∞-vector fields on G are denoted by $\Gamma(TG)$. Vector fields allow quite different interpretations and we want to study some of them. In order to reduce the introduction of special notations we always will use the same symbol, say X, for a given vector field. Thus some care is needed at the beginning in interpreting formulae - again we just follow the common practice.

Every $X \in \Gamma(TG)$ induces a mapping (denoted again by X)

$$X : C^\infty(G) \to C^\infty(G) \tag{24.27}$$

$$f \mapsto Xf : G \to \mathbb{R}$$
$$q \mapsto X_q[f]_q,$$

where as before $[f]_q$ denotes the germ generated by f at q, i.e. $[f]_q \in C_q^\infty(G)$. Using the representation of X_q with respect to the basis $\left\{ \left(\frac{\partial}{\partial x_1} \right)_q, \ldots, \left(\frac{\partial}{\partial x_n} \right)_q \right\}$ we find by (24.24) that

$$(Xf)(q) = \sum_{j=1}^{n} X_q([pr^j]) \left(\frac{\partial}{\partial x_j} \right)_q f \tag{24.28}$$

$$= \sum_{j=1}^{n} a_j(q) \left(\frac{\partial}{\partial x_j} \right)_q f$$

where we set $a_j(q) = X_q([pr^j])$ and these coefficients are for C^∞-vector fields C^∞-functions. From (24.28) it is clear that $X : C^\infty(G) \to C^\infty(G)$ is a linear mapping and furthermore we have

$$X(fg) = f\,Xg + g\,Xf \qquad (24.29)$$

since X_q is an $\mathbb{R}$-derivation. Hence vector fields induce derivations on the algebra $C^\infty(G)$.

In general for $X, Y \in \Gamma(TG)$ we can not expect that the composition of the induced mappings $X : C^\infty(G) \to C^\infty(G)$ and $Y : C^\infty(G) \to C^\infty(G)$ is a mapping induced by a vector field since in general

$$\begin{aligned}
(X \circ Y)(fg) &= X(Yfg) = X(fYg) + X(gYf) \\
&= XfYg + fX(Yg) + XgYf + gX(Yf) \\
&\neq f(Y \circ X)g + g(Y \circ X)f.
\end{aligned}$$

However for $[X, Y] = X \circ Y - Y \circ X : C^\infty(G) \to C^\infty(G)$ we can easily show linearity as well as Leibniz' rule. For simplicity we write in the following XY for $X \circ Y$. For $\alpha, \beta \in \mathbb{R}$ and $f, g \in C^\infty(G)$ we have

$$\begin{aligned}
[X, Y](\alpha f + \beta g) &= (XY - YX)(\alpha f + \beta g) \\
&= \alpha XYf - \alpha YXf + \beta XYg - \beta YXg \\
&= \alpha(XY - YX)f + \beta(XY - YX)g \\
&= \alpha[X, Y]f + \beta[X, Y]g,
\end{aligned}$$

and

$$\begin{aligned}
[X, Y](fg) &= (XY - YX)(fg) \\
&= fXYg + gXYf + XfYg + XgYf \\
&\quad - fYXg - gYXf - YfXg - YgXf \\
&= f[X, Y]g + g[X, y]f.
\end{aligned}$$

Proposition 24.21. *For $X, Y, Z \in \Gamma(TG)$ and $f, g \in C^\infty(G)$ we have*

$$[X, Y] = -[Y, X] \qquad (24.30)$$

and in particular

$[X, X] = 0;$ \hfill (24.31)

$[X + Y, Z] = [X, Z] + [Y, Z]$ *and* $[X, Y + Z] = [X, Y] + [X, Z];$ \hfill (24.32)

$[X, [Y, Z]] + [Y, [Z, X]] + [Z, [X, Y]] = 0$ *(**Jacobi identity**);* \hfill (24.33)

$[fX, Y] = f[X, Y] - (Yf)X;$ \hfill (24.34)

and

$$[fX, gY] = fg[X, Y] + f(Xg)Y - g(Yf)X. \tag{24.35}$$

Proof. The statements (24.30) - (24.32) are trivial. For (24.33) we note

$$[X, [Y, Z]] = XYZ - XZY - YZX + ZYX,$$
$$[Y, [Z, X]] = YZX - YXZ - ZXY + XZY,$$
$$[Z, [X, Y]] = ZXY - ZYX - XYZ + YXZ,$$

and adding these three terms up gives the Jacobi identity. For $f, g \in C^\infty(G)$ we have

$$\begin{aligned}[fX, Y]g &= fXYg - Y(fX)g \\ &= f(XY)g - (Yf)(Xg) - f(YX)g \\ &= f[X, Y]g - (Yf)(Xg)\end{aligned}$$

or

$$[fX, Y] = f[X, Y] - (Yf)X,$$

i.e. (24.34) is proved. Finally we note using (24.34) that

$$\begin{aligned}[fX, gY] &= f[X, gY] - (gY)(f)X \\ &= -f[gY, X] - (gY)(f)X \\ &= -fg[Y, X] + fX(g)Y - (gY)(f)X \\ &= fg[X, Y] + f(Xg)Y - g(Yf)X.\end{aligned}$$

$\square$

Using for $X, Y \in \Gamma(TG)$ the representations

$$X_q = \sum_{j=1}^n a_j(q) \left(\frac{\partial}{\partial x_j} \right)_q \text{ and } Y_q = \sum_{j=1}^n b_j(q) \left(\frac{\partial}{\partial x_j} \right)_q,$$

with C^∞ - functions $a_j, b_j : G \to \mathbb{R}$ we find

$$[X, Y]_q = \sum_{j=1}^n \left(\sum_{l=1}^n a_l(q) \frac{\partial b_j}{\partial x_l}(q) - b_l(q) \frac{\partial a_j(q)}{\partial x_l} \right) \left(\frac{\partial}{\partial x_j} \right)_q, \tag{24.36}$$

See Problem 8. This implies in particular that $[X, Y]$, which so far is defined for the mappings $X, Y : C^\infty(G) \to C^\infty(G)$ associated with the vector fields X and Y, is indeed a mapping associated with a vector field which we denote of course by $[X, Y]$. Therefore the following definition makes sense

Definition 24.22. *The vector field* $[X, Y]$ *is called the* **Lie product** *of the vector fields* X *and* Y.

Before we will return to differential equations we want to transport vector fields on $G \subset \mathbb{R}^n$ to vector fields on $H \subset \mathbb{R}^m$. Let $G \subset \mathbb{R}^n$ and $H \subset \mathbb{R}^m$ be open sets and $F : G \to H$ a C^∞ - mapping. Further let $X \in \Gamma(TG)$ and $Y \in \Gamma(TH)$. We say that Y is associated to X via F if for all $q \in G$ we have

$$(dF)_q(X_q) = Y_{F(q)} \tag{24.37}$$

which means that for all $g \in C^\infty(H)$

$$X_q(g \circ F) = Y_{F(q)} g \tag{24.38}$$

holds. For $F : G \to H$ and $X \in \Gamma(TG)$ we can of course use (24.38) to define the vector field Y.

So far we have assumed that all vector fields are of class C^∞. If we require less regularity of a vector field, say we assume that it is of class C^k, we must note that the coefficients a_j in (24.28) are only of class C^k and hence Xf, even for $f \in C^\infty(G)$, will in general be only of class C^k. Moreover, having (24.36) in mind, when considering the Lie product of two C^k-vector fields, in general we will end up with a C^{k-1}-vector field. Thus more care is needed, for example, when looking at the Jacobi identity and other expressions involving Lie products. Not only do we need to check the "algebraic" part of the calculation, but we have to assume that all differentiations are permitted.

Let $\gamma : I \to G \subset \mathbb{R}^n, G$ open, be a C^1 -curve where $I \subset \mathbb{R}$ is an open interval. At $t \in I$ there exists the tangent space $T_t I$ and we know that every $v_t \in T_t I$ has a representation $a(t) \left(\dfrac{\mathrm{d}}{\mathrm{d}s} \right)_t$ with $a(t) \in \mathbb{R}$. We can use the differentiable mapping $\gamma : I \to G$ to transport $a(t) \left(\dfrac{\mathrm{d}}{\mathrm{d}s} \right)_t$ using (24.37) to $T_{\gamma(t)}G$, i.e. we can consider

$$(\mathrm{d}\gamma)_t \left(a(t) \left(\frac{\mathrm{d}}{\mathrm{d}s} \right)_t \right) \in T_{\gamma(t)} G.$$

In particular we have

$$\dot{\gamma}(t) = (\mathrm{d}\gamma)_t \left(\frac{\mathrm{d}}{\mathrm{d}s} \right)_t \in T_\gamma(t)G. \tag{24.39}$$

Thus the following problem is natural: given a vector field $X : G \to TG$. Does there exists a C^1- curve $\gamma : I \to G, I \subset \mathbb{R}$ open, such that

$$\dot{\gamma}(t) = X_{\gamma(t)} \text{ for all } t \in I \tag{24.40}$$

holds?

Definition 24.23. *A C^1-curve $\gamma : I \to G$ is called an **integral curve** of the vector field $X : G \to TG$ if for all $t \in I$ equality (24.40) holds.*

We know that with some function β_j we have

$$\dot{\gamma}(t) = \sum_{j=1}^{n} \beta_j(t) \left(\frac{\partial}{\partial x_j} \right)_{\gamma(t)}$$

and interpreting $\dot{\gamma}(t)$ as derivation we find for the projection pr^k

$$\dot{\gamma}(t)(pr^k)(\gamma(t)) = \sum_{j=1}^{n} \beta_j(t) \left(\frac{\partial\, pr^k}{\partial x_j} \right) \gamma(t) = \beta_k(t)$$

and on the other hand we have

$$\dot{\gamma}(t)(pr^k)(\gamma(t)) = \frac{\mathrm{d}}{\mathrm{d}t}(pr^k \circ \gamma)(t)$$

and therefore

$$\dot{\gamma}(t) = \sum_{j=1}^{n} \frac{\mathrm{d}}{\mathrm{d}t}(pr^j \circ \gamma)(t) \left(\frac{\partial}{\partial x_j} \right)_{\gamma(t)}. \tag{24.41}$$

For the vector field X we know the representation

$$X_{\gamma(t)} = \sum_{j=1}^{n} R_j(\gamma(t)) \left(\frac{\partial}{\partial x_j} \right)_{\gamma(t)} \tag{24.42}$$

with some functions R_j, and since $\left\{ \left(\frac{\partial}{\partial x_1} \right)_{\gamma(t)}, \ldots, \left(\frac{\partial}{\partial x_n} \right)_{\gamma(t)} \right\}$ is a basis of $T_{\gamma(t)}G$ it follows from (24.41) and (24.42) that

$$\frac{\mathrm{d}}{\mathrm{d}t}(pr^j \circ \gamma) = R_j(\gamma(t))$$

which we can write as

$$\dot{\gamma}_j(t) = R_j(\gamma(t)), \;\; 1 \leq j \leq n. \tag{24.43}$$

Reading now $\dot{\gamma}(t) = \begin{pmatrix} \dot{\gamma}_1(t) \\ \vdots \\ \dot{\gamma}_n(t) \end{pmatrix}$ and $R = \begin{pmatrix} R_1 \\ \vdots \\ R_n \end{pmatrix}$ we arrive at

$$\dot{\gamma}(t) = R(\gamma(t)) \tag{24.44}$$

426

which is a first order system of ordinary differential equations. Thus necessary for γ to be an integral curve for $X = \sum_{j=1}^{n} R_j(\frac{\partial}{\partial x_j})$ is that γ solves (24.44). Conversely, every C^1-solution of (24.44) is an integral curve of X. Our basic existence theorem now yields that for every C^1-vector field X on G and every point $q \in G$ we can find a maximal integral curve for X passing through q. We will use this geometric interpretation of a differential equation to study the differential equation further with the help of geometry.

Problems

1. Prove that for $[f], [g] \in C_q^{\infty}(G; \mathbb{R}^n)$ the vector space operations (24.4) and (24.5) are well defined.

2. Let $[f] \in C_0^{\infty}((-1, 1))$ be a germ of a function at $0 \in \mathbb{R}$. Suppose that there exists $g \in [f]$ such that for some $0 < r < 1$ we have the power series representation $g(x) = \sum_{k=0}^{\infty} a_k x^k$ for $|x| < r$. Prove that for every $h \in [f]$ there exists $0 < \rho < 1$ such that $h(x) = \sum_{k=0}^{\infty} a_k x^k$, $|x| < \rho$.

3. Let $\mathcal{P} := \{p : \mathbb{R}^n \to \mathbb{R} | p$ is a polynomial$\}$ i.e. $\mathcal{P}$ is the algebra of all polynomials on $\mathbb{R}^n$. For $p \in \mathcal{P}$, $p = \sum_{|\alpha| \leq k} a_\alpha x^\alpha$, $a_\alpha \in \mathbb{R}$, $\alpha \in \mathbb{N}_0^n$, we define $D_j : \mathcal{P} \to \mathcal{P}$ by $D_j p = \sum_{|\alpha| \leq k} a_\alpha \alpha_j x^{\alpha - \epsilon_j}$, $1 \leq j \leq n$, and $\alpha - \epsilon_j = 0$ for $\alpha_j = 0$. Here ϵ_j is the jth unit multi-index. Prove that D_j is a derivation.

4. Verify (24.12).

5. A $\mathbb{K}$-vector space V is called a **Lie algebra** if on V a bilinear mapping $[\cdot, \cdot] : V \times V \to V$, $(X, Y) \mapsto [X, Y]$, is defined with the properties that $[X, Y] = -[Y, X]$ and $[X, [Y, Z]] = [[X, Y], Z] + [Y, [X, Z]]$ (**Jacobi identity**). Prove that $M(n; \mathbb{K})$ with $[A, B] := AB - BA$, $A, B \in M(n; \mathbb{R})$ is a Lie algebra.

6. In the Lie algebra $M(3; \mathbb{R})$, consider the elements

$$X = \begin{pmatrix} 0 & 1 & 0 \\ -1 & 0 & 0 \\ 0 & 0 & 0 \end{pmatrix}, \quad Y = \begin{pmatrix} 0 & 0 & 1 \\ 0 & 0 & 0 \\ -1 & 0 & 0 \end{pmatrix} \text{ and } Z = \begin{pmatrix} 0 & 0 & 0 \\ 0 & 0 & -1 \\ 0 & 1 & 0 \end{pmatrix}.$$

Prove that $\{X, Y, Z\} \subset M(3; \mathbb{R})$ is an independent set in $M(3; \mathbb{R})$ and that $\text{span}\{X, Y, Z\}$ is a Lie algebra.
Hint: show that $[X, Y] = Z, [Z, X] = Y$ and $[Y, Z] = X$.

7.   a) Let H be a Lie algebra such that for $X, Y, Z \in H$ it always follows that $[X, [Y, Z]] = 0$. Define $\circ : H \times H \to H$ by

$$X \circ Y := X + Y + \frac{1}{2}[X, Y]$$

and prove that $(H, \circ)$ is a (non-commutative) group.

b) Consider $\tilde{H} := \left\{ \begin{pmatrix} 0 & x & y \\ 0 & 0 & z \\ 0 & 0 & 0 \end{pmatrix} \middle| x, y, z \in \mathbb{R} \right\} \subset M(3; \mathbb{R})$.

Prove that $\tilde{H}$ satisfies the assumption of part a).
The algebra $\tilde{H}$ is called the three dimensional **Heisenberg algebra** and the corresponding group is the **Heisenberg group** of dimension three.

8. Prove (24.36).

9. On $\mathbb{R}^2$ we consider the vector fields $X = \frac{\partial}{\partial x}$ and $Y = x\frac{\partial}{\partial y}$, $(x, y) \in \mathbb{R}^2$. Show that there exists some vector field Z on $\mathbb{R}^2$ such that for all C^∞-functions $\alpha, \beta \in C^\infty(\mathbb{R}^2)$ we have $Z \neq \alpha X + \beta Y$.
Now prove that there exists $a, b, c \in C^\infty(\mathbb{R}^2)$ such that $Z = aX + bY + c[X, Y]$.

10. For $p = (x, y) \in \mathbb{R}^2$ find an integral curve for the vector field $p \mapsto y\left(\frac{\partial}{\partial x}\right)_p - x\left(\frac{\partial}{\partial y}\right)_p$.

25 Phase Diagrams and Flows

We now want to use our geometric tools, e.g. tangent spaces or vector fields, to study a first order system of ordinary differential equations from a more geometric point of view. We restrict ourselves to autonomous systems, i.e. systems of the type

$$\dot{q}(t) = R(q(t)) \tag{25.1}$$

which is often considered together with an initial condition

$$q(t_0) = q_0, \quad t_0 \in I, \tag{25.2}$$

where $I \subset \mathbb{R}$ is an interval, $\mathring{I} \neq \emptyset$, $q : I \to G$ is the sought solution and $R : G \to \mathbb{R}^n$ a given mapping. We know that we can interpret R_j as the coefficients of a vector field on G, i.e. R corresponds to X where

$$X_q = \sum_{j=1}^{n} R_j(q) \left(\frac{\partial}{\partial x_j} \right)_q, \tag{25.3}$$

and a solution q to (25.1) can be interpreted as an integral curve of X.

The reader will have noticed the change of notation, we now use $q : I \to \mathbb{R}^n$ to denote a solution to (25.1) which fits better to our understanding of $q = (q_1, \ldots, q_n) \in G$ as a point in G and that of $t \mapsto q(t)$ as the motion of a point.

We can read (25.1) in the following way: even when we do not know whether a point q lies on a solution curve of (25.1) (maybe the one determined by the initial condition (25.2)) we know at least that if it were a point on a solution curve the tangent vector to the curve at this point would be $R(q)$. Thus we may start our investigations of (25.1) by looking at the vector field defined by R or associated with (25.1). We also call this field the **direction field** associated with (25.1). For $n = 2$ we can provide nice graphic interpretations of this idea which are best explained by diagrams.

Example 25.1. Let $R : \mathbb{R}^2 \to \mathbb{R}^2$ be the constant mapping $R(q) = a = \begin{pmatrix} a_1 \\ a_2 \end{pmatrix} \in \mathbb{R}^2$. The corresponding direction field is of course constant, i.e. at every point q we have the same vector a, see Figure 25.1

Example 25.2. We consider the mapping $R : \mathbb{R}^2 \to \mathbb{R}^2$, $R(q) = \begin{pmatrix} q_2 \\ -q_1 \end{pmatrix}$ which leads to the direction field shown in the following figure

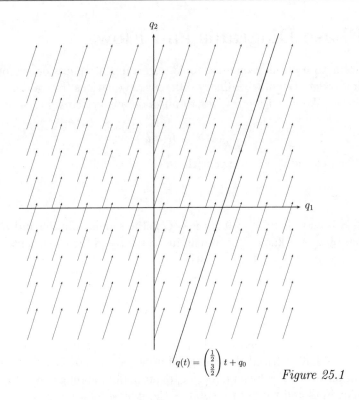

$q(t) = \begin{pmatrix} \frac{1}{2} \\ \frac{3}{2} \end{pmatrix} t + q_0$

Figure 25.1

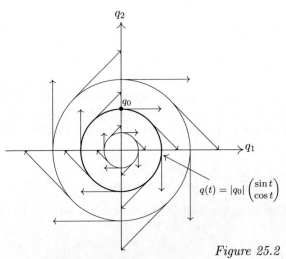

$q(t) = |q_0| \begin{pmatrix} \sin t \\ \cos t \end{pmatrix}$

Figure 25.2

In both examples we can easily solve the corresponding differential equation. In Example 25.1 the solution to $\dot{q}(t) = a$ is $q(t) = at + b$ where $b \in \mathbb{R}^2$ is a fixed

430

vector, in Example 25.2 the general solution to $\dot{q}(t) = \begin{pmatrix} q_2(t) \\ -q_1(t) \end{pmatrix}$ is given by
$q(t) = \begin{pmatrix} r\sin(t + \varphi) \\ r\cos(t + \varphi) \end{pmatrix}$, $r \geq 0$ and $\varphi \in \mathbb{R}$. These two examples suggest the following strategy to get an idea of what a solution might be and what might be its
properties. Given $\dot{q} = R(q)$. Then we can draft the direction field and the trace
of a solution curve $t \mapsto q(t)$ should have at every point a tangent vector belonging
to the direction field. We call the figures in which we draw the curves fitting onto
the direction field the **phase diagram** corresponding to the differential equation
(25.1) and we often will indicate the direction in which we are "moving", i.e. the
orientation of the curve, see Figure 25.3. Before we continue with our investigations, we want to introduce the notion of an **orbit** to be the oriented trace of a
curve $\gamma : I \to \mathbb{R}^n$ and in corresponding figures we will indicate the orientation as
in Figure 25.3. Thus the trace of γ is just the point set $\gamma(I) \subset \mathbb{R}^n$, while the orbit
contains also information on the orientation, thus the phase diagram shows the
orbits. The name orbit has of course its origin in mechanics.

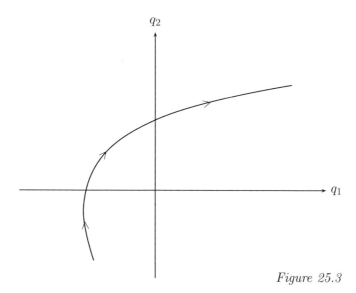

q_2

q_1

Figure 25.3

We now look at some phase diagrams and try to understand what type of information they may provide. Since we can draw in the plane the trace of a curve we first
discuss some orbits of curves together with their tangent vectors. The most simple
one is the curve, the trace of which is a point, i.e. $q(t) = q_0$ for all $t \in I$. This
implies $\dot{q}(t) = 0$. Now, if we discuss $\dot{u} = R(u)$ and u is an equilibrium point of this
equation then by definition $\dot{u} = R(u) = 0$ implying $u(t) = $ constant. Moreover,

no tangent vector (if we exclude the vector 0 for geometric reasons) is attached to an equilibrium point. In a phase diagram an equilibrium point shows up as an isolated point. However it may happen that some or even all tangent vectors attached to non-equilibrium solutions to $\dot{q} = R(q)$ are directed to an equilibrium point.

Example 25.3. Let $h : [0, \infty) \to [0, \infty)$ be a continuous function such that $h(0) = 0$ but $h(\rho) > 0$ for $\rho \neq 0$. It follows that $\lim_{|q| \to 0} h(|q|) \frac{q}{|q|} = 0$ and therefore $R_1(q) = -h(|q|) \frac{q}{|q|}$ and $R_2(|q|) = h(|q|) \frac{q}{|q|}$ are continuous functions in the plane. Since $h(|q|) > 0$ for $|q| \neq 0$ the two differential equations $\dot{q} = R_j(q)$, $j = 1, 2$, have each only the point $q_0 = 0 \in \mathbb{R}^2$ as an equilibrium point. For $q \in \mathbb{R}^2 \backslash \{0\}$ each vector of the direction field of $\dot{q} = R_1(q)$ points towards the equilibrium point, see Figure 25.4, whereas in the case $\dot{q} = R_2(q)$ each vector of the direction field points radially away from the equilibrium point, see Figure 25.5

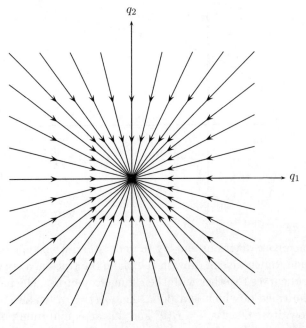

Figure 25.4

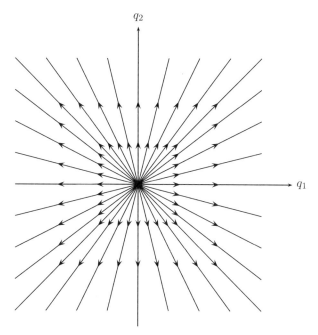

Figure 25.5

Next we consider the case where $q : \mathbb{R} \to \mathbb{R}^2$ is periodic, i.e. with some minimal $T > 0$ we have $q(t + T) = q(t)$ for all $t \in \mathbb{R}$. All further periods of q are given by kT, $k \in \mathbb{N}$. The trace of q must be a closed curve, compare with Definition II.7.12. If we restrict q to a half open interval of length $k_0 T$, $k_0 \in \mathbb{N}$, every point of the trace of q is a k_0-fold point. Thus when a phase diagram suggests that closed curves may occur as solutions, we might try to establish that these are periodic solutions. However, it is not correct to assume that every closed curve in the plane corresponds to a period curve. Consider the curve $q : [0, \infty) \to \mathbb{R}^2$, $q(t) = \begin{pmatrix} \cos t^2 \\ \sin t^2 \end{pmatrix}$. In order that q has a period $T > 0$ we need

$$\cos\left((t + T)^2\right) = \cos\left(t^2\right) \quad \text{and} \quad \sin\left((t + T)^2\right) = \sin\left(t^2\right) \quad \text{for all } t \geq 0,$$

implying that $2tT + T^2 = 2m\pi$, $m \in \mathbb{N}$, for all $t \geq 0$ which is not possible for T independent of t. We note that

$$\begin{pmatrix} \cos t^2 \\ \sin t^2 \end{pmatrix}^{\bullet} = 2t \begin{pmatrix} -\sin t^2 \\ \cos t^2 \end{pmatrix}$$

or with $q(t) = \begin{pmatrix} \cos t^2 \\ \sin t^2 \end{pmatrix}$ we find

$$\begin{pmatrix} \dot{q}_1(t) \\ \dot{q}_2(t) \end{pmatrix} = \begin{pmatrix} -2tq_2(t) \\ 2tq_1(t) \end{pmatrix} \qquad (25.4)$$

and (25.4) is **not** an autonomous system since $R = R(t, q) = \begin{pmatrix} -2tq_2 \\ 2tq_1 \end{pmatrix}$. However, for autonomous systems we have

Theorem 25.4. *Let $\dot{q} = R(q)$ be an autonomous system and R satisfies the assumptions of our basic existence and uniqueness theorem. Denote by $t \mapsto \varphi(t, q_0)$ the solution to $\dot{q} = R(q)$ with initial condition $\varphi(t_0, q_0) = q_0$. Then φ must satisfy one of the following alternatives*

  i) $t \mapsto \varphi(t, q_0)$ is injective;

  ii) $t \mapsto \varphi(t, q_0)$ is periodic;

  iii) $t \mapsto \varphi(t, q_0)$ is constant.

Furthermore, in the first two cases $t \mapsto \varphi(t, q_0)$ is a regular curve, i.e. $\dot{\varphi}(t, q_0) \neq 0$ for all t, and the solution of cases ii) and iii) is defined for all $t \in \mathbb{R}$.

Proof. We already know that if $\dot{\varphi}(t_1, q_0) = 0$ for some t_1 then $\varphi(t_1, q_0)$ must be an equilibrium point of the system and therefore $\varphi(t, q_0) = \varphi(t_1, q_0) = q_0$ for all t. It remains to prove that if $t \mapsto \varphi(t, q_0)$ is neither constant nor injective then it must be periodic. So suppose that $t \mapsto \varphi(t, q_0)$ is neither constant nor injective and denote by I_0 its maximal open existence interval, $t_0 \in I_0$. Then there must exist $t_1 < t_2$, $t_1, t_2 \in I_0$, such that $\varphi(t_1, q_0) = \varphi(t_2, q_0)$. For $\psi(t, q_0) := \varphi(t_1 + t, q_0)$, $t \in I_0 - t_1$ we obtain a further solution to $\dot{q} = R(q)$ with $\psi(0, q_0) = \psi(\tau, q_0)$ where $\tau := t_2 - t_1$. We set

$$T := \inf \{t \,|\, t > 0, \, \psi(t, q_0) = \psi(0, q_0)\}$$

and we choose a sequence $(t_k)_{k \in \mathbb{N}}$, $t_k > 0$, such that $\lim_{k \to 0} t_k = T$ and $\psi(t_k, q_0) = \psi(0, q_0)$. If $\lim_{k \to \infty} t_k = 0$ then

$$\dot{\psi}(0, q_0) = \lim_{k \to \infty} \frac{1}{t_k} (\psi(t_k, q_0) - \psi(0, q_0)) = 0$$

and $\psi(\cdot, q_0)$ must be a constant or stationary solution of the system $\dot{q} = R(q)$. Therefore we must have $T > 0$ and

$$\psi(T, q_0) = \lim_{k \to \infty} \psi(t_k, q_0) = \psi(0, q_0).$$

434

Thus, in an open interval containing 0 we have a solution y of

$$\dot{y}(t) = R(y(t)) \quad \text{and} \quad y(0) = y(T) = q_0.$$

Now the existence and uniqueness theorem implies in an open neighbourhood of $t \in I_0$ that $y(t) = \psi(t, q_0) = \psi(t+T, q_0)$ which allows us to extend y as a T-periodic function to $\mathbb{R}$ satisfying $\dot{q} = R(q)$. $\square$

Remark 25.5. The proof of Theorem 25.4 shows that its assertion holds whenever for $\dot{q} = R(q)$, $q(t_0) = q_0$, an existence and uniqueness result holds, but it does not need to be "our" standard result.

Let $\dot{q} = R(q)$ be an autonomous system satisfying the conditions of Theorem 25.4. Suppose that R maps $G \subset \mathbb{R}^2$ onto $G' \subset \mathbb{R}^2$ and pick $q_0 \in G$. In the phase diagram corresponding to $\dot{q} = R(q)$ the point q_0 might be an isolated point in the sense that no orbit with points other than q_0 passes through q_0. In this case q_0 is an equilibrium point. By the existence and uniqueness result it cannot happen that two distinct orbits intersect at q_0, otherwise the initial value problem $\dot{q} = R(q)$, $q(t_0) = q_0$, would have two distinct solutions. Thus the phase diagram consists of isolated (= equilibrium) points and non-intersecting orbits. If an orbit is closed the corresponding solution is periodic.

Example 25.6. The **Volterra-Lotka system**

$$\dot{q}(t) = \begin{pmatrix} \dot{q}_1(t) \\ \dot{q}_2(t) \end{pmatrix} = \begin{pmatrix} (\alpha - \beta q_2(t))q_1(t) \\ (-\gamma + \delta q_1(t))q_2(t) \end{pmatrix}, \qquad q(0) = \begin{pmatrix} q_{01} \\ q_{02} \end{pmatrix}, \qquad (25.5)$$

serves as a mathematical framework for a basic model in population dynamics. It is an autonomous system with $R(q) = \begin{pmatrix} (\alpha - \beta q_2)q_1 \\ (-\gamma + \delta q_1)q_2 \end{pmatrix}$ where $q_1 = q_1(t)$ is the size of a prey population and $q_2 = q_2(t)$ is the size of a predator population at time t. We may choose $t_0 = 0$ and q_{01}, q_{02} are the initial populations. It is clear that the model requires $q_{01} > 0$ and $q_{02} > 0$ as well as $q_1(t), q_2(t) > 0$ with the limit case of extinction when $q_j(t_1) = 0$ for some $t_1 > 0$. The parameters $\alpha, \beta, \gamma, \delta$ are "rates" of "growth" (reproduction, death) and they are entering the model as positive numbers. The system has two equilibrium points, namely $\begin{pmatrix} 0 \\ 0 \end{pmatrix}$ and $\begin{pmatrix} \frac{\gamma}{\delta} \\ \frac{\alpha}{\beta} \end{pmatrix}$.

The first equilibrium point refers to no prey and predator. The second point is of more interest and is located in the first quadrant. Figure 25.6 indicates in principle the phase diagram in $\mathbb{R}^2$, but as we know, for the biological model only the first quadrant is of interest.

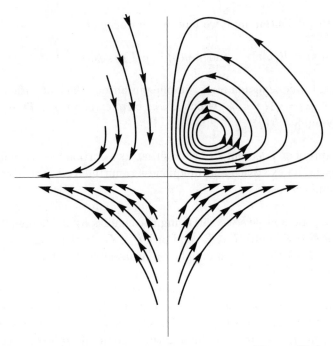

Figure 25.6

It turns out that (25.5) is not a very realistic model and the system

$$\dot{q} = \begin{pmatrix} \dot{q}_1 \\ \dot{q}_2 \end{pmatrix} = \begin{pmatrix} (\alpha - \beta q_2 - \lambda q_1)q_1 \\ (-\gamma + \delta q_1 - \mu q_2)q_2 \end{pmatrix}, \quad q(0) = \begin{pmatrix} q_{01} \\ q_{02} \end{pmatrix}, \tag{25.6}$$

with $\alpha, \beta, \gamma, \delta, \lambda, \mu > 0$ is suggested as an improvement. We refer to H. Amann [3] for a detailed discussions of this system.

Looking at the phase diagrams we understand the potential to transfer this qualitative information into quantitative and qualitative statements for solutions which is an important topic in the theory of dynamical systems. In Chapter 23 we have discussed how a linearization of R allows us to handle stability problems for $\dot{q} = R(q)$ by discussing the system

$$\dot{q} = dR(0)q$$

where we assume that the linearization is at $u_0 = 0$. For this reason it is interesting to study the phase diagram of the system $\dot{q} = Aq$, $A \in M(n; \mathbb{R})$ in a neighbourhood of $q_0 = 0$. We know that spectral properties of A will determine this behaviour and given A we may transform this matrix with the help of $C \in GL(n; \mathbb{R})$ onto some normal form B,

$$B = C^{-1}AC.$$

436

The polar decomposition of C, compare with Theorem II.A.I.28, allows us to write $C = WS$ where $W \in O(n)$ and S is a positive definite matrix, hence $S = V \begin{pmatrix} \mu_1 & & 0 \\ & \ddots & \\ 0 & & \mu_n \end{pmatrix} V^{-1}$, $\mu_j > 0$ are the eigenvalues of S and $V \in O(n)$. Thus we find

$$A = CBC^{-1} = WSBS^{-1}W^{-1}$$

$$= WV \begin{pmatrix} \mu_1 & & 0 \\ & \ddots & \\ 0 & & \mu_n \end{pmatrix} V^{-1}BV \begin{pmatrix} \frac{1}{\mu_1} & & 0 \\ & \ddots & \\ 0 & & \frac{1}{\mu_n} \end{pmatrix} V^{-1}W^{-1}$$

$$= U \begin{pmatrix} \mu_1 & & 0 \\ & \ddots & \\ 0 & & \mu_n \end{pmatrix} V^{-1}BV \begin{pmatrix} \frac{1}{\mu_1} & & 0 \\ & \ddots & \\ 0 & & \frac{1}{\mu_n} \end{pmatrix} U^{-1}$$

with $U, V \in O(n)$. This does mean that we obtain the phase diagram for $\dot{q} = Aq$ from that of $\dot{q} = Bq$ by rotations and distortions along the axes and therefore we need only discuss the case where A has already normal form. In our discussion we follow [32] and [61].

Example 25.7. Suppose that $A = \begin{pmatrix} \lambda_1 & 0 \\ 0 & \lambda_2 \end{pmatrix}$, $\lambda_1, \lambda_2 \in \mathbb{R}$, is a diagonal matrix with $\lambda_1 \cdot \lambda_2 \neq 0$. The solutions to $\dot{q} = Aq$ are $q_1(t) = q_{01}e^{\lambda_1 t}$ and $q_2(t) = q_{02}e^{\lambda_2 t}$, $q_{01}, q_{02} \in \mathbb{R}$. In order to find orbits in the $q_1 - q_2$-plane we try to eliminate t. For $q_{01}, q_{02} > 0$ we have

$$q_1^{\lambda_2} = q_{01}^{\lambda_2}e^{\lambda_1\lambda_2 t} \quad \text{and} \quad q_2^{\lambda_1} = q_{02}^{\lambda_1}e^{\lambda_1\lambda_2 t}$$

or, since $q_{01} \neq 0$,

$$q_{02}^{\lambda_1}q_1^{\lambda_2} = q_{01}^{\lambda_2}q_2^{\lambda_1}$$

and in the first quadrant we find q_2 as function of q_1 as

$$q_2 = \frac{q_{02}}{q_{01}^{\lambda_2/\lambda_1}}q_1^{\frac{\lambda_2}{\lambda_1}}. \tag{25.7}$$

Discussing the three other cases for the signs of q_{01} and q_{02} we eventually obtain the phase diagrams below which of course must take the signs of λ_1 and λ_2 into account.

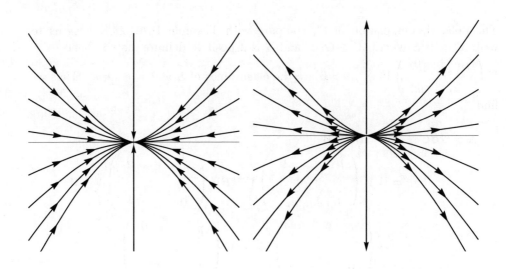

Figure 25.7   Figure 25.8

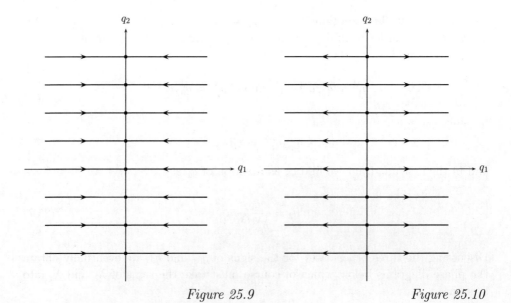

Figure 25.9   Figure 25.10

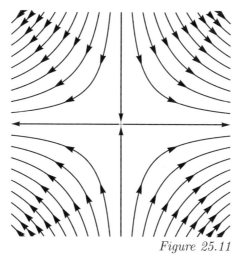

Figure 25.11

Example 25.8. We now suppose that $A = \begin{pmatrix} \lambda & 1 \\ 0 & \lambda \end{pmatrix}$ which corresponds to

$$\begin{cases} \dot{q}_1 = \lambda q_1 + q_2 \\ \dot{q}_2 = \lambda q_2. \end{cases} \tag{25.8}$$

The second equation has the solution $q_2(t) = q_{02}e^{\lambda t}$ and the first equation becomes now

$$\dot{q}_1(t) = \lambda q_1(t) + q_{02}e^{\lambda t}$$

which has the general solution $q_1(t) = (q_{01} + q_{02}t)\, e^{\lambda t}$. In order to find the phase diagram we may try to eliminate t from

$$\begin{cases} q_1(t) = (q_{01} + q_{02}t)e^{\lambda t} \\ q_2(t) = q_{02}e^{\lambda t} \end{cases} \tag{25.9}$$

which however is not possible in general. Still we can derive the typical behaviour of the phase diagram from (25.8) and (25.9). We can use (25.8) to find the direction field and we can use (25.9) to add some asymptotic behaviour of orbits for $t \to 0$ and $t \to \infty$. For this we assume $q_{02} \neq 0$ to find

$$\frac{\dot{q}_1(t)}{\dot{q}_2(t)} = \frac{q_1(t)}{q_2(t)} + \frac{1}{\lambda} = \frac{q_{01} + q_{02}t}{q_{02}} + \frac{1}{\lambda}.$$

It follows that for $t \to \infty$ we have $\frac{\dot{q}_1(t)}{\dot{q}_2(t)} = \frac{q_1(t)}{q_2(t)} + \frac{1}{\lambda} \to \infty$ and therefore $q_1(t)$ and $q_2(t)$ must have the same sign, namely that of q_{02}, as $t \to \infty$. Moreover, for $t \to \infty$ we find $\frac{\dot{q}_2(t)}{\dot{q}_1(t)} > 0$ and $\frac{\dot{q}_2(t)}{\dot{q}_1(t)} \to 0$. Now, if $\lambda < 0$ and $q_2 > 0$, then the tangent vector

439

at $q = (q_1, q_2)$ must have a negative second component while the sign of its first component is that of $\lambda q_1 + q_2$, i.e. depends also on the size of $|\lambda|$. In a similar way we can discuss the other case, $q_2 < 0$, and eventually we arrive at Figure 25.17 as the principle picture of the phase diagram of (25.8) provided $\lambda < 0$. As we know, for $\lambda > 0$ we obtain the phase diagram for (25.8) by reversing the orientation of the orbits.

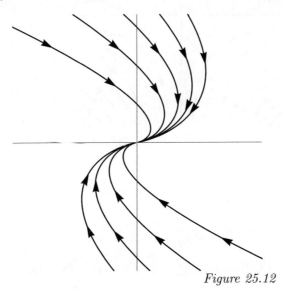

Figure 25.12

Example 25.9. Finally we assume that A has a complex eigenvalue $\lambda = \mu + i\nu$, $\nu > 0$. In this case a second eigenvalue is $\bar\lambda$ and over $\mathbb{C}$ the matrix is diagonalizable. Its normal form over $\mathbb{R}$ is $\begin{pmatrix} \mu & -\nu \\ \nu & \mu \end{pmatrix}$ and we now consider

$$\dot{q}(t) = \begin{pmatrix} \mu & -\nu \\ \nu & \mu \end{pmatrix} q(t). \tag{25.10}$$

The solutions are

$$q_1(t) = re^{\mu t} \cos(\nu t + \varphi) \quad \text{and} \quad q_2(t) = re^{\mu t} \sin(\nu t + \varphi)$$

with $r \geq 0$ and $\varphi \in [0, 2\pi)$. For $r = 0$ we have the equilibrium point $0 \in \mathbb{R}^2$. For $\mu = 0$ we obtain

$$q_1(t) = r \cos(\nu + \varphi) \quad \text{and} \quad q_2(t) = r \sin(\nu t + \varphi)$$

and the orbits are circles. Assuming $\varphi = 0$ the phase diagram is given in Figure 25.18. In the case where $\mu < 0$, a comparison with Chapter II.7, in particular Example II.7.35, shows that each orbit must be a spiral running into the equilibrium point 0, and for $\mu > 0$ we have spirals running to infinity, see Figure 25.19 and Figure 25.20.

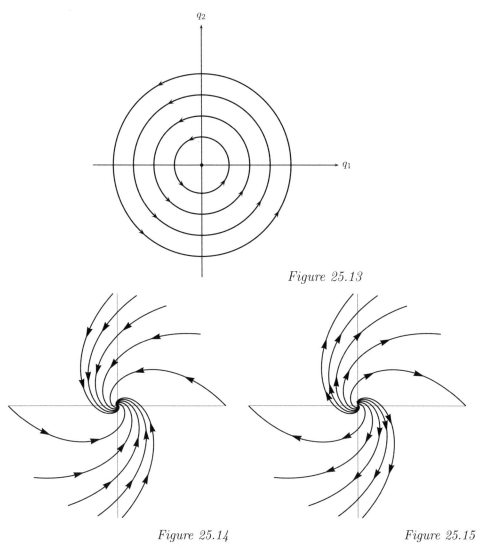

Figure 25.13

Figure 25.14   Figure 25.15

So far we have considered the "evolution" $q_0 \mapsto \varphi(t, q_0)$ of a point under the system $\dot{q} = R(q)$, or equivalently the single integral curve of the vector field $X = \sum_{j=1}^{n} R_j \left(\frac{\partial}{\partial x_j} \right)$ passing through q_0. Here $t \mapsto \varphi(t, q_0)$ denotes the solution to $\dot{q} = R(q)$, $q(t_0) = q_0$, and $R : G \to \mathbb{R}^n$, $G \subset \mathbb{R}^n$ open, is for example a locally Lipschitz continuous mapping. In this case there exists a maximal open interval

$I(q_0) \subset \mathbb{R}$, $t_0 \in I(q_0)$, for which a unique solution exists. It is natural to ask whether we can describe the evolution of an entire subset $H \subset G$ under the system $\dot{q} = R(q)$. For getting some first ideas we look at the initial value problem

$$\dot{q} = R(q), \quad q(t_0) = q_0 \quad \text{for all} \quad q_0 \in H. \tag{25.11}$$

It is convenient for a moment to choose $t_0 = 0$. For each $q_0 \in H$ there exists a maximal open existence interval $I(q_0) \subset \mathbb{R}$, $0 \in I(q_0)$. As an additional condition we require now that for some $T > 0$

$$(-T, T) \subset \bigcap_{q_0 \in H} I(q_0). \tag{25.12}$$

For $t \in (-T, T)$ we may consider the set

$$H_t := \{\varphi(t, q) \mid q \in H\} \subset G. \tag{25.13}$$

The set H_t can be viewed as the set into which H evolved under $\dot{q} = R(q)$. A priori we know of course that $H_t \subset G$. Suppose that $s \in (-T, T)$ such that $s + t \in (-T, T)$. In this case we can form H_{t+s} and $(H_t)_s$ where

$$(H_t)_s = \{\tilde{\varphi}(s, \varphi(t, q)) \mid q \in H\} \subset G$$

and $\tilde{\varphi}(s, \varphi(t, q_0))$ is the unique solution at s to the initial value problem $\dot{q} = R(q)$, $q(t) = \varphi(t, q_0)$, $q_0 \in H$. Of course we expect that $(H_t)_s = H_{t+s}$. In order to prove this statement we want to introduce some new notation. We denote by $\psi(t, t_0; q_0)$ the unique maximal solution at t to the problem

$$\dot{q} = R(q), \quad q(t_0) = q_0.$$

Thus we have $\varphi(t, q_0) = \psi(t, 0; q_0)$ or $\tilde{\varphi}(s, \varphi(t, q_0)) = \psi(s, t; \varphi(t, q_0))$. Now let $R : G \to \mathbb{R}^n$ and H be as above and $(-T, T) \subset \bigcap_{q \in H} I(q)$, $T > 0$. Further let $t, s \in (-T, T)$ such that $t + s \in (-T, T)$. We define

$$\Phi_t : H \to H_t \quad \text{by} \quad \Phi_t(y) = \varphi(t, y) = \psi(t, 0; y) \tag{25.14}$$

and we find

$$(\Phi_t)_s : H_t \to H_{t+s}, \quad \Phi_t(\varphi(t, y)) \mapsto \psi(s, t; y). \tag{25.15}$$

On the other hand we have

$$\Phi_{t+s} : H \to H_{t+s}, \quad \Phi_{t+s}(y) = \psi(t + s, 0; y). \tag{25.16}$$

The existence and uniqueness theorem however implies

$$\psi(s, t; y) = \psi(t + s, 0; y) \tag{25.17}$$

and therefore we have

$$(\Phi_t)_s = \Phi_{t+s}. \tag{25.18}$$

Definition 25.10. *The mapping* $\Phi : \{(t, y) \in \mathbb{R} \times G \,|\, t \in I(y)\} \to G$ *defined by*

$$\Phi(t, y) := \Phi_t(y) := \psi(t, 0, y) = \psi(t, y) \tag{25.19}$$

*is called the **flow** associated with the system* $\dot{q} = R(q)$ *or equivalently with the vector field* $X = \sum_{j=1}^{n} R_j \left(\frac{\partial}{\partial x_j} \right)$.

The full power of the concept "flow of a vector field" we can only unfold when proving that $\Phi_t : H \to H_t$ is for every t an orientation preserving diffeomorphism and that for the Jacobi matrix $\mathcal{J}_{\Phi_t}(y)$ we have

$$\frac{\mathrm{d}}{\mathrm{d}t} \mathcal{J}_{\Phi_t}(y) = \mathcal{J}_R \left(\varphi(t, y) \right) \mathcal{J}_{\Phi_t}(y). \tag{25.20}$$

From (25.20) we can derive **Liouville's theorem** stating that if $\operatorname{div} R = 0$ then

$$\lambda^{(n)}(H_t) = \lambda^{(n)}(H). \tag{25.21}$$

Indeed, since

$$\lambda^{(n)}(H_t) = \int_{H_t} 1 \, \lambda^{(n)}(\mathrm{d}x) = \int_H |\det \mathcal{J}_{\Phi_t}(x)| \, \lambda^{(n)}(\mathrm{d}x) = \int_H \det \mathcal{J}_{\Phi_t}(x) \, \mathrm{d}x \tag{25.22}$$

and since $\det \mathcal{J}_{\Phi_t}(x)$ is the Wronski determinant $W(t)$ associated with $\mathcal{J}_{\Phi_t}(x)$ we have by Theorem 19.8

$$\dot{W}(t) = \operatorname{tr}\left(D\mathcal{J}_{\Phi_t}(y) \right) = (\operatorname{div} R)\left(\varphi(t, y) \right) = 0, \tag{25.23}$$

which implies that

$$\frac{\mathrm{d}}{\mathrm{d}t} \lambda^{(n)}(H_t) = 0$$

or (25.21).

Another result worth mentioning is the Poincaré-Bendixson theorem which gives sufficient conditions for the existence of periodic orbits for two dimensional systems.

A proper treatment of these and related results needs more knowledge in differential geometry and the theory of differentiable manifolds and we do not want to go into details here. We refer to [1], [3], [6] or [61].

Problems

1. Draft the direction field for the following equation

$$\dot{q}(t) = \begin{pmatrix} q_1(t)q_2(t) \\ q_2(t) - q_1(t) \end{pmatrix}.$$

2. Find the phase diagram corresponding to the system

$$\dot{u} = v, \quad \dot{v} = -4u.$$

Hint: first find the general solution to this system.

3. Consider the system

$$\dot{x} = -y + \frac{x}{\sqrt{x^2 + y^2}}(1 - \sqrt{x^2 + y^2})$$

$$\dot{y} = x + \frac{y}{\sqrt{x^2 + y^2}}(1 - \sqrt{x^2 + y^2}).$$

a) Prove that $r(t) = \sqrt{x^2(t) + y^2(t)}$ satisfies the equation $\dot{r} = 1 - r^2$ and deduce that the original system has the solution $u(t) = \begin{pmatrix} x(t) \\ y(t) \end{pmatrix} = \begin{pmatrix} \cos t \\ \sin t \end{pmatrix}.$

b) Suppose that v is a further solution to the system with $\|v(t_0)\| < 1$. Prove that for all t_1, t_2 we must have $u(t_1) \neq v(t_2)$ and interpret the result.

4. Let $\varphi : \mathbb{R}^n \to \mathbb{R}$ be a harmonic function and consider the system $\dot{q}(t) = \mathrm{grad}\varphi(q(t))$. Assume that for some $p_0 \in \mathbb{R}^n$ and $\rho > 0$ condition (25.12) holds for some $T > 0$ and all $q_0 \in H := B_\rho(p_0)$. Why is H_t, $|t| < T$, a compact and connected set of volume $w_n \rho^n$ where w_n is the volume of the unit ball in $\mathbb{R}^n$.

5. Let $A \in M(n; \mathbb{R})$. Show that the flow corresponding to the system $\dot{u} = Au$ corresponds to $\phi(t, q_0) = e^{At}q_0$. Deduce now that $\lambda^{(n)}(H_t) = |\det e^{At}|\lambda^{(n)}(H)$ for $H \subset \mathbb{R}^n$. Suppose that A is symmetric and that $\mathrm{tr}A = 0$. What can we say about $\lambda^{(n)}(H_t)$?

6. Suppose that Liouville's theorem holds for all compact sets $K \subset \mathbb{R}^n$ for the system $\dot{u} = R(u)$, $R : \mathbb{R}^n \to \mathbb{R}^n$. Consider the two sets $\overline{B_1(y)}$ and $\overline{B_1(y)}$, $\|x - y\| > 4$. Let $H_t^x := \phi_t(\overline{B_1(x)})$ and $H_t^y := \phi_t(\overline{B_1(y)})$ where ϕ_t is the flow mapping corresponding to $\dot{u} = R(u)$. Prove that $\lambda^{(n)}(H_t^x \cap H_t^y) = 0$.

Part 10: Introduction to the Calculus of Variations

26 The Calculus of Variations — Setting the Scene

Many scientists will endorse the statement "Nature is governed by extremal principles". Taking more modern epistemology into account and adding a bit of Popper's ideas, the statement "It seems that extremal principles are convenient to model many natural phenomena" is the more agreeable hypothesis. Classical examples are the

Isoperimetric principle: The circle in the plane and the sphere in space are the figures with the smallest perimeter enclosing a given area and a given volume, respectively.

or

Fermat's principle: Light from a point A to a point B always propagates in the quickest way.

These and some other classical principles can be embedded (when adding some additional constraints) into the calculus of one or several (but finitely many) variables. The problem proposed in 1696 by Johann I Bernoulli [13], the so called **brachistochrone problem** or the "problem of quickest descent" is of different nature: it requires some calculus in function spaces, i.e. infinite dimensional spaces. The problem is the following one:
"For two given points A and B in a vertical plane, find a line connecting them on which a movable point M descends under the influence of gravitation from A to B in the quickest way."
We have quoted here the translation from the Latin as provided in [59]. This problem is for good reasons considered as the starting point of the calculus of variations; one of the most beautiful subjects in Mathematics linking analysis, geometry and physics. Stefan Hildebrandt and Mariano Giaquinta, two contemporary masters of the field, have provided us with a very scholarly and actual account of the history of the subject, see [44] and [40] as well as [59] which gives a wonderful readable more popular treatment of the subject. Here we will turn immediately to the mathematical problem which we may in a first attempt describe as finding the minimum (the extremals) of

$$\int_a^b F(t, u(t), \dot{u}(t)) \, dt \tag{26.1}$$

where $F : [a, b] \times \mathbb{R}^n \times \mathbb{R}^n \to \mathbb{R}$ is a given function and $u : [a, b] \to \mathbb{R}^n$ is a curve. However we know by now that we cannot expect to make much progress with this rather general and imprecise formulation. We have to decide what regularity do we require from u and F, and closely related, which notion of integral (Riemann or Lebesgue) do we want or do we have to consider?

Example 26.1. We may ask to find the monomial $p_k(t) = t^k, k \in \mathbb{N}_0$, for which the integral $\int_0^1 p_k(t)\, dt$ becomes minimal. Note that we always have $\int_0^1 p_k(t)\, dt \geq 0$, thus the question is reasonable. Since

$$\int_0^1 p_k(t) dt = \int_0^1 t^k\, dt = \frac{1}{k+1}$$

it follows that $\left(\int_0^1 p_k(t)\, dt \right)_{k \in \mathbb{N}_0}$ is monotone decreasing to 0. However the polynomial $t \mapsto p(t) = 0$ does not belong to our class. Since

$$\lim_{k \to \infty} p_k(t) = \chi_{\{1\}}(t) = \begin{cases} 1, & t = 1 \\ 0, & t \in [0, 1) \end{cases}$$

we find that

$$\lim_{k \to \infty} \int_0^1 p_k(t)\, dt = 0 = \int_0^1 \chi_{\{1\}}(t)\, dt$$

but $\chi_{\{1\}}$ does not belong to our class either. We may extend the class of functions for which we want to minimize $\int_0^1 u(t)\, dt$ to all continuous functions $u : [0, 1] \to \mathbb{R}, u(t) \geq 0, u(0) = 0$ and $u(1) = 1$. Again we find one minimizing sequence, namely $(p_k)_{k \in \mathbb{N}_0}$ which has a limit not belonging to the class. Indeed no function of this class can minimize the integral. Since $u(1) = 1$ there exists $0 < r < 1$ such that $u(t) > \frac{1}{2}$ for all $x \in [r, 1]$ and since $u(t) \geq 0$ for all $t \in [0, 1]$ we have

$$\int_0^1 u(t)\, dt \geq \int_r^1 u(t)\, dt > 0.$$

Thus we cannot expect a solution of the original problem

$$\min \left\{ \int_0^1 u(t)\, dt \,\middle|\, u \in C([0, 1]), u(0) = 0, u(1) = 1 \right\},$$

however if we allow a discontinuous solution it is trivial, but we will lose uniqueness.

We want to restrict our discussion of the calculus of variations to a context which is suitable for many problems in classical mechanics. This means that we will work in classes of continuous functions being (piecewise) differentiable up to a certain order and the notion of integral used is that of the Riemann integral. However, here and there, we will leave the framework and indicate results for more general functions and/or the Lebesgue integral.

Let $u : (a, b) \to \mathbb{R}^n$ be a $C^k - mapping, t_0 \in (a, b)$ and $0 \leq l \leq k$. We call $(u(t_0), u'(t_0), ..., u^{(l)}(t_0))$ the **l-jet** of u at t_0 and $(u(\cdot), u'(\cdot), ..., u^{(l)}(\cdot))$ the l-jet of u. Moreover we call $(t_0, u(t_0), u'(t_0), ..., u^{(l)}(t_0))$ the **extended l-jet** of u at t_0.

Let $F : [a, b] \times \mathbb{R}^n \times \mathbb{R}^n \to \mathbb{R}$ be a C^1- function. We consider now the functional or **variational integral** $\mathcal{F} : C^1([a, b]; \mathbb{R}^n) \to \mathbb{R}$ defined by

$$\mathcal{F}(u) := \int_a^b F(t, u(t), u'(t)) \, dt. \tag{26.2}$$

For $A, B \in \mathbb{R}^n, A \neq B$, we set

$$D_{A,B}(\mathcal{F}) : \{u \in C^1([a, b]; \mathbb{R}^n) \,|\, u(a) = A, u(b) = B\}. \tag{26.3}$$

Elements of $D_{A,B}(\mathcal{F})$ are C^1-curves $u : [a, b] \to \mathbb{R}^n$ with the property that $u(a) = A$ and $u(b) = B$. The central problem is now

Problem 26.2. Find

$$\inf\{\mathcal{F}(u) \,|\, u \in D_{A,B}(\mathcal{F})\}. \tag{26.4}$$

We call any $u_0 \in D_{A,B}(\mathcal{F})$ a **minimiser** of $\mathcal{F}$ on $D_{A,B}(\mathcal{F})$ if

$$\mathcal{F}(u_0) \leq \mathcal{F}(u) \text{ for all } u \in D_{A,B}(\mathcal{F}). \tag{26.5}$$

Since $D_{A,B}(\mathcal{F})$ is not contained in any finite dimensional vector space we cannot use the methods of Chapter II.9 to deal with Problem 26.2, but some related ideas might lead to new and appropriate tools. Suppose that $u_0 \in D_{A,B}(\mathcal{F})$ is a minimiser of $\mathcal{F}$ and $v : [a, b] \to \mathbb{R}^n$ is a C^1-curve with $v(a) = v(b) = 0$. Then for every $\eta \in (-\eta_0, \eta_0), \eta_0 > 0$, the curve $u_0 + \eta v$ belongs to $D_{A,B}(\mathcal{F})$ and we must have

$$\mathcal{F}(u_0) \leq \mathcal{F}(u_0 + \eta v). \tag{26.6}$$

We can now turn our attention to the function of one variable defined on $(-\eta_0, \eta_0)$ by

$$f_{u_0,v}(\eta) := \mathcal{F}(u_0 + \eta v) \tag{26.7}$$

which must have a minimum for $\eta = 0$. By our assumptions, $\eta \mapsto f_{u_0,v}(\eta)$ is differentiable and therefore we must have

$$\frac{d}{d\eta} f_{u_0,v}(\eta)|_{\eta=0} = 0 \tag{26.8}$$

449

for every $v \in C^1([a,b]; \mathbb{R}^n)$, $v(a) = v(b) = 0$. We try to calculate the left hand side in (26.8) explicitly:

$$\frac{\mathrm{d}}{\mathrm{d}\eta} f_{u_0,v}(\eta) = \frac{\mathrm{d}}{\mathrm{d}\eta} \int_a^b F(t, u_0(t) + \eta v(t), u_0'(t) + \eta v'(t)) \, \mathrm{d}t$$

and since we are allowed to differentiate under the integral sign we find

$$\frac{\mathrm{d}}{\mathrm{d}\eta} \int_a^b F(t, u_0(t) + \eta v(t), u_0'(t) + \eta v'(t)) \, \mathrm{d}t = \int_a^b \frac{\mathrm{d}}{\mathrm{d}\eta} F(t, u_0(t) + \eta v(t), u_0'(t) + \eta v'(t)) \, \mathrm{d}t.$$

With $F = F(t, x, y)$, $x, y \in \mathbb{R}^n$, we find

$$\frac{\mathrm{d}}{\mathrm{d}\eta} F(t, u_0(t) + \eta v(t), u_0'(t) + \eta v'(t)) \, \mathrm{d}t$$

$$= \sum_{j=1}^n \frac{\partial F}{\partial x_j}(t, u_0(t) + \eta v(t), u_0'(t) + \eta v'(t)) v_j(t)$$

$$+ \sum_{j=1}^n \frac{\partial F}{\partial y_j}(t, u_0(t) + \eta v(t), u_0'(t) + \eta v'(t)) v_j'(t).$$

This yields

$$\frac{\mathrm{d}}{\mathrm{d}\eta} f_{u_0,v}(\eta)|_{\eta=0} = \int_a^b \left(\sum_{j=1}^n \frac{\partial F}{\partial x_j}(t, u_0(t), u_0'(t)) v_j(t) \right. \tag{26.9}$$

$$\left. + \sum_{j=1}^n \frac{\partial F}{\partial y_j}(t, u_0(t), u_0'(t)) v_j'(t) \right) \mathrm{d}t$$

and by (26.8) we arrive at

$$\int_a^b \left(\sum_{j=1}^n \left(\frac{\partial F}{\partial x_j}(t, u_0(t), u_0'(t)) v_j(t) + \frac{\partial F}{\partial y_j}(t, u_0(t), u_0'(t)) v_j'(t) \right) \right) \mathrm{d}t = 0 \tag{26.10}$$

which must hold for all $v \in C^1([a,b]; \mathbb{R}^n)$ with $v(a) = v(b) = 0$. The question is whether (26.10) allows us to deduce a necessary condition for u_0 to be a minimiser of $\mathcal{F}$. We have seen on several occasions the argument that if for $w \in C([a,b])$

$$\int_a^b w(x) \varphi(x) \, \mathrm{d}x = 0 \tag{26.11}$$

for all $\varphi \in C([a,b])$ then $w = 0$. We want to extend this argument to be applicable to (26.10), but first we need to transform (26.10) to an integral of the type (26.11).

For this we now assume that $\frac{\partial F}{\partial y_j}$, $1 \leq j \leq n$, is a C^1 - function and u_0 is a C^2-function so that we can integrate by parts to obtain

$$\int_a^b \sum_{j=1}^n \frac{\partial F}{\partial y_j}(t, u_0(t), u_0'(t)) v_j'(t)\, \mathrm{d}t$$

$$= -\int_a^b \left(\frac{\mathrm{d}}{\mathrm{d}t} \sum_{j=1}^n \frac{\partial F}{\partial y_j}(t, u_0(t), u_0'(t)) \right) v_j(t)\, \mathrm{d}t$$

$$+ \sum_{j=1}^n \left(\frac{\partial F}{\partial y_j}(t, u_0(t), u_0'(t)) \right) v_j(t) \Big|_a^b$$

$$= -\int_a^b \left(\frac{\mathrm{d}}{\mathrm{d}t} \sum_{j=1}^n \frac{\partial F}{\partial y_j}(t, u_0(t), u_0'(t)) \right) v_j(t)\, \mathrm{d}t,$$

where we used that $v_j(a) = v_j(b) = 0$. Thus for $F \in C^1([a,b]; \mathbb{R})$ such that $\frac{\partial F}{\partial y_j} \in C^1([a,b], \mathbb{R})$ for $1 \leq j \leq n$ we arrive at

$$\int_a^b \left(\sum_{j=1}^n \left(\frac{\partial F}{\partial x_j}(t, u_0(t), u_0'(t)) - \frac{\mathrm{d}}{\mathrm{d}t}\frac{\partial F}{\partial y_j}(t, u_0(t), u_0'(t)) \right) \right) v_j(t)\, \mathrm{d}t = 0 \quad (26.12)$$

for all $v \in C^1([a,b]; \mathbb{R}^n)$ such that $v(a) = v(b) = 0$. To continue our investigations we need the fundamental lemma of the calculus of variations. Let $\varphi : [a,b] \to \mathbb{R}^n$ be continuous such that $\operatorname{supp} \varphi_j \subset (a,b)$, $j = 1, ..., n$. If $\varphi|_{(a,b)}$ is a C^1 - function then φ is also a C^1-function and for the one-sided derivatives we have $\varphi'(a) = \varphi'(b) = 0$ since $\varphi|_{[a,a+\epsilon]} = 0$ and $\varphi|_{[b-\epsilon,b]} = 0$ for some $\epsilon > 0$. In this sense we may identify $C_0^1((a,b); \mathbb{R}^n)$ as a subspace of $C^1([a,b]; \mathbb{R}^n)$.

Lemma 26.3 (Fundamental lemma of the calculus of variations). *Let $f \in C([a,b]; \mathbb{R}^n)$ be such that*

$$\int_a^b \sum_{j=1}^n f_j(t)\varphi_j(t)\, \mathrm{d}t = 0 \quad (26.13)$$

for all $\varphi \in C_0^\infty((a,b); \mathbb{R}^n)$. Then $f(t) = 0$ for all $t \in [a,b]$.

Proof. First we note that it is sufficient to prove the result for $n = 1$. Indeed, with $\varphi(t) = \psi(t)\, e_j$, $\psi \in C_0^1((a,b); \mathbb{R})$ we can reduce the statement to an equivalent statement for all components of f. Suppose now that (26.13) holds for $n = 1$ and that $f(t_0) \neq 0$ for some $t_0 \neq 0$. We may assume that $f(t_0) > 0$ and by continuity we find some $\epsilon > 0$ such that with suitable points $a < \alpha < \beta < b$ we have $f|_{[\alpha,\beta]} > \epsilon > 0$. The function

$$\varphi(t) := \begin{cases} (t-\alpha)^2(t-\beta)^2, & \alpha \le t \le \beta \\ 0, & t \in [a,b]\setminus[\alpha,\beta] \end{cases}$$

has compact support in (a,b), $\operatorname{supp}\varphi = [\alpha,\beta]$, and it is a C^1-function since $\varphi'(t) = 2(t-\alpha)(t-\beta)^2 + 2(t-\alpha)^2(t-\beta)$, $t \in (\alpha,\beta)$, and therefore $\varphi'(\alpha) = \varphi(\beta) = 0$. Moreover we have

$$\int_a^b f(t)\,\varphi(t)\,\mathrm{d}t \ge \epsilon \int_a^b \varphi(t)\,\mathrm{d}t > 0$$

which is a contradiction. $\qquad\square$

Before applying the fundamental lemma of the calculus of variations to (26.12) we want to introduce some notions and notations.

Definition 26.4. A. We call $u_0 \in D_{A,B}(\mathcal{F})$ a **local minimiser** of $\mathcal{F}$ if for some $\rho_0 > 0$ we have $\mathcal{F}(u_0) \le \mathcal{F}(u)$ for all $u \in D_{A,B}(\mathcal{F})$ such that $\|u_0 - u\|_{\infty,[a,b]} < \rho_0$ where

$$\|u_0 - u\|_{\infty,[a,b]} = \max_{1\le j\le n} \sup_{t\in[a,b]} |u_{0_j}(t) - u_j(t)|.$$

B. The **first variation** of $\mathcal{F}$ at $u \in C^1([a,b];\mathbb{R}^n)$ in the direction $v \in C^1([a,b];\mathbb{R}^n)$ is defined as

$$(\delta\mathcal{F})(u;v) := \frac{\mathrm{d}}{\mathrm{d}\eta}\mathcal{F}(u+\eta v)|_{\eta=0}. \tag{26.14}$$

Theorem 26.5. Let $F \in C^1([a,b]\times\mathbb{R}^n\times\mathbb{R}^n)$ such that $\frac{\partial F}{\partial y_j}(t,x,\cdot) = F_{y_j}(t,x,\cdot) \in C^1(\mathbb{R}^n)$ for $1 \le j \le n$ and let $u \in C^2([a,b];\mathbb{R}^n)$.

A. If $(\delta\mathcal{F})(u;v) = 0$ for all $v \in C_0^1((a,b);\mathbb{R}^n)$ then u satisfies the system of ordinary differential equations

$$\frac{\mathrm{d}}{\mathrm{d}t}F_{y_j}(t,u(t),u'(t)) = F_{x_j}(t,u(t),u'(t)), \ t \in (a,b), \ j = 1,...,n. \tag{26.15}$$

B. If $(\delta\mathcal{F})(u;v) = 0$ for all $v \in C^1([a,b];\mathbb{R}^n)$ then u satisfies the system (26.15) and in addition

$$F_{y_j}(a,u(a),u'(a)) = F_{y_j}(b,u(b),u'(b)) = 0 \tag{26.16}$$

holds for $1 \le j \le n$.

Proof. **A.** We can use our derivation of (26.12) and an application of the fundamental lemma, Lemma 26.3, yields (26.15).
B. Since $C_0^1((a, b); \mathbb{R}^n) \subset C^1([a, b]; \mathbb{R}^n)$ we obtain again (26.15). On the other hand, integration by parts now yields

$$\int_a^b \sum_{j=1}^n \frac{\partial}{\partial y_j} F(t, u(t), u'(t)) \, v_j'(t) \, dt$$

$$= -\int_a^b \left(\frac{d}{dt} \sum_{j=1}^n \frac{\partial}{\partial y_j} F(t, u(t), u'(t)) \right) v_j(t) \, dt$$

$$+ \sum_{j=1}^n \frac{\partial}{\partial y_j} F(t, u(t), u'(t)) \, v_j(t) \Big|_a^b$$

and therefore we have

$$(\delta F)(u; v) = \int_a^b \left(\sum_{j=1}^n \frac{\partial F}{\partial x_j}(t, u(t), u'(t)) \, v_j(t) + \sum_{j=1}^n \frac{\partial F}{\partial y_j}(t, u(t), u'(t)) \, v_j'(t) \right) dt$$

$$\tag{26.17}$$

$$= \int_a^b \left(\sum_{j=1}^n \frac{\partial F}{\partial x_j}(t, u(t), u'(t)) - \frac{d}{dt} \frac{\partial F}{\partial y_j}(t, u(t), u'(t)) \, v_j(t) \right) dt$$

$$+ \sum_{j=1}^n \frac{\partial F}{\partial y_j}(t, u(t), u'(t)) v_j(t) \Big|_a^b = 0.$$

Since (26.15) holds we arrive at

$$0 = \sum_{j=1}^n \frac{\partial F}{\partial y_j}(t, u(t), u'(t)) v_j(t) \Big|_a^b = \sum_{j=1}^n \frac{\partial F}{\partial y_j}(b, u(b), u'(b)) v_j(b) \tag{26.18}$$

$$- \sum_{j=1}^n \frac{\partial F}{\partial y_j}(a, u(a), u'(a)) v_j(a)$$

for all $v \in C^1([a, b]; \mathbb{R}^n)$. For arbitrary $z \in \mathbb{R}^n$ we choose $v_z \in C^1([a, b]; \mathbb{R}^n)$ such that $v_z(b) = z$ and $v_z(a) = 0$. Then (26.18) implies

$$\sum_{j=1}^n \frac{\partial F}{\partial y_j}(b, u(b), u'(b)) \, z_j = 0$$

for all $z = (z_1, ..., z_n) \in \mathbb{R}^n$ which means

$$\left(\frac{\partial F}{\partial y_1}(b, u(b), u'(b)), ..., \frac{\partial F}{\partial y_n}(b, u(b), u'(b)) \right) \perp z$$

for all $z \in \mathbb{R}^n$ implying that $\frac{\partial F}{\partial y_j}(b, u(b), u'(b)) = 0$ for $1 \leq j \leq n$. Next we choose $w_z \in C^1([a, b]; \mathbb{R}^n)$ such that $w_z(a) = z$ and $w_z(b) = 0$ and we deduce that $\frac{\partial F}{\partial y_j}(a, u(a), u'(a)) = 0$ for $1 \leq j \leq n$. $\qquad\square$

Definition 26.6. A. *The system (26.15) is called the **Euler-Lagrange differential equations** associated with $\mathcal{F}$ and each of their C^2-solutions is called an $\mathcal{F}$-extremal.*
B. *The conditions (26.16) are called the **natural boundary conditions** for $\mathcal{F}$.*

As a first central result we can now prove

Theorem 26.7. *If $u_0 \in D_{A,B}(\mathcal{F})$ is a local minimiser of $\mathcal{F}$ then*

$$(\delta\mathcal{F})(u_0; v) = 0 \text{ for all } v \in C^1([a, b]; \mathbb{R}^n) \text{ with } v(a) = v(b) = 0. \qquad (26.19)$$

If in addition $\frac{\partial F}{\partial y_j} \in C^1([a, b] \times \mathbb{R}^n \times \mathbb{R}^n)$, $1 \leq j \leq n$, and $u_0|_{(a,b)} \in C^2((a, b); \mathbb{R}^n)$ then u_0 satisfies the Euler-Lagrange equations.

Proof. Since $u_0 \in D_{A,B}(\mathcal{F})$ is a local minimiser of $\mathcal{F}$ there exists $\rho_0 > 0$ such that for all $w \in D_{A,B}(\mathcal{F})$ with $||u_0 - w||_{\infty, [a,b]} < \rho_0$. We have that $\mathcal{F}(u_0) \leq \mathcal{F}(w)$. Now let $v \in C^1([a, b]; \mathbb{R}^n)$ be such that $v(a) = v(b) = 0$. It follows that for all $\eta \in \mathbb{R}$ we have $w = u_0 + \eta v \in D_{A,B}(\mathcal{F})$ and

$$|u_0(x) - w(x)| = |u_0(x) - (u_0(x) + \eta v(x))| = |\eta| \, |v(x)|$$

and therefore

$$||u_0 - (u_0 + \eta v)||_{\infty, [a,b]} \leq |\eta| \, ||v||_{\infty, [a,b]}.$$

For $|\eta| \leq \eta_0 := \frac{\rho_0}{||v||_{\infty, [a,b]}}$ we find $||u_0 - (\eta_0 + \eta v)||_{\infty, [a,b]} < \rho_0$ and therefore

$$\mathcal{F}(u_0) \leq \mathcal{F}(u_0 + \eta v), \qquad |\eta| < \eta_0,$$

i.e. we return to (26.8) as in our new notation to find

$$(\delta\mathcal{F})(u_0; v) = 0 \text{ for all } v \in C^1([a, b]; \mathbb{R}^n), \, v(a) = v(b) = 0.$$

Now Theorem 26.5 implies the second statement of the theorem. $\qquad\square$

Corollary 26.8. *Let $u_0 \in D_{A,B}(\mathcal{F})$ be a local minimiser of $\mathcal{F}$ such that $u_0 \in C^2((a, b); \mathbb{R}^n)$ and assume that $\frac{\partial F}{\partial y_j} \in C^1([a, b] \times \mathbb{R}^n \times \mathbb{R}^n)$ for $1 \leq j \leq n$. Then u_0 is a solution of the boundary value problem.*

$$\begin{cases} \frac{d}{dt} F_{y_j}(t, u(t), u'(t)) - F_{x_j}(t, u(t), u'(t)) = 0, \, 1 \leq j \leq n, \, t \in (a, b) \\ u(a) = A \text{ and } u(b) = B. \end{cases} \qquad (26.20)$$

454

Remark 26.9. Differentiating in (26.15) or (26.20) yields

$$\frac{\mathrm{d}}{\mathrm{d}t}F_{y_j}(t, u(t), u'(t)) = \frac{\partial F_{y_j}}{\partial t}(t, u(t), u'(t)) + \sum_{l=1}^{n} F_{y_j\,x_l}(t, u(t), u'(t))u'_l(t)$$

$$+ \sum_{l=1}^{n} F_{y_j\,y_l}(t, u(t), u'(t))\, u''_l(t)$$

and therefore the Euler-Lagrange equations have the form

$$\frac{\partial F_{y_j}}{\partial t}(t, u(t), u'(t)) + \sum_{l=1}^{n} F_{y_j\,y_l}(t, u(t), u'(t))\, u''_l(t) +$$

$$+ \sum_{l=1}^{n} F_{y_j\,x_l}(t, u(t), u'(t))\, u'_l(t) - F_{x_j}(t, u(t), u'(t)) = 0, \;\; 1 \le j \le n.$$

Thus they are second order differential equations in which the highest order derivatives, i.e. u''_l, $1 \le l \le n$, enters in a linear manner. Such equations are called second order **quasi-linear equations**.

We want to discuss a few examples and it is helpful to recollect some ideas from mechanics first. If a point of mass $m > 0$ is moving on a trajectory $q, t \mapsto q(t) \in \mathbb{R}^n$, its kinetic energy is given by $T = \frac{m}{2}||\dot{q}(t)||^2$. Suppose that the point's movement is subjected to a potential with resulting potential energy $V = V(t, q)$. The function

$$L(t, q(t), \dot{q}(t)) = \frac{m}{2}||\dot{q}(t)||^2 - V(t, q(t)) \tag{26.21}$$

is called the **Lagrange function** or the **Lagrangian** of this mechanical system and a basic principle of mechanics states that the trajectory of the particle connecting $q_1 = q(t_1)$ and $q_2 = q(t_2)$, $t_1 < t_2$, is a minimiser of

$$L(q) := \int_{t_1}^{t_2} L(t, q(t), \dot{q}(t))\, \mathrm{d}t.$$

Due to the close connection of the development of the calculus of variations with mechanics the function F in (26.1) is also often called the Lagrange function even in the case where no interpretation from mechanics is suitable.
In the following we will always assume that V is independent of t, i.e. $V = V(q(t))$.

Example 26.10. Suppose that $V(q) = 0$, thus the particle is moving without any influence of a potential, i.e. we are dealing with a free particle. The trajectory of such a particle from q_1 to q_2 shall be a minimiser of

$$\int_{t_1}^{t_2} \frac{m}{2} ||\dot{q}(t)||^2 \, dt = \frac{m}{2} \int_{t_1}^{t_2} \sum_{j=1}^{n} \dot{q}_j^2(t) \, dt, \quad q_1 = q(t_1), \, q_2 = q(t_2). \quad (26.22)$$

Now we have $L(t, q, \dot{q}) = \frac{m}{2} \sum_{j=1}^{n} \dot{q}_j^2$ which is of course a C^2 - function and the Euler-Lagrange equations become

$$\frac{m}{2} \frac{d}{dt} \left(\frac{\partial}{\partial \dot{q}_j} \sum_{j=1}^{u} \dot{q}_j^2(t) \right) = m \frac{d}{dt} \dot{q}_j(t) = m \ddot{q}_j(t) = 0 \quad (26.23)$$

with the boundary conditions $q(t_1) = q_1$ and $q(t_2) = q_2$, or

$$\ddot{q}_j(t) = 0, \, 1 \le j \le n, \text{ and } q_j(t_1) = q_{1j}, \, q_j(t_2) = q_{2j}. \quad (26.24)$$

Since the general solution to $\ddot{q}(t) = 0$ is a straight line $q(t) = \alpha t + \beta$ the boundary conditions imply

$$q_{ex}(t) := q(t) = \frac{q_1(t - t_2) + q_2(t_1 - t)}{t_1 - t_2}. \quad (26.25)$$

We have proved that the only function satisfying (26.24), i.e. the Euler-Lagrange equation and the corresponding boundary condition, is given by (26.25). However we have **not proved** that among all C^1 - functions $q \in D_{q_1, q_2}(L)$, $L(q, \dot{q}) = \frac{m}{2} ||\dot{q}||^2$, the minimiser of (26.22) is the function (26.25). Note that the value of $L(q_{ex})$ is

$$L(q_{ex}) = \int_{t_1}^{t_2} \frac{m}{2} ||\dot{q}_{ex}||^2 \, dt = \frac{m}{2} \frac{||q_2 - q_1||^2}{t_2 - t_1}. \quad (26.26)$$

Of course we expect the trajectory to be the straight line connecting q_1 and q_2, i.e. q_{ex}. We may try to solve the problem by minimizing the length l_q

$$\inf \int_{t_1}^{t_2} ||\dot{q}(t)|| \, dt, \, q(t_1) = q_1, \, q(t_2) = q_2, \quad (26.27)$$

but the function $\dot{q} \mapsto ||\dot{q}|| = \left(\sum_{j=1}^{n} \dot{q}^2 \right)^{\frac{1}{2}}$ is not a C^2-function and we can not derive the Euler-Lagrange equations for (26.27). We refer to Problem 8 where we sketch that a minimiser of (26.22) must also be a minimiser of (26.27).

Example 26.11. Let $[a, b] \subset \mathbb{R}$ and let $h : [a, b] \to \mathbb{R}$ be a C^1-function. Then $\gamma : [a, b] \to \mathbb{R}^2$, $\gamma(t) = \begin{pmatrix} t \\ h(t) \end{pmatrix}$ is a regular curve the length of which is given by

$$\int_a^b \|\dot{\gamma}(t)\| \, dt = \int_a^b \sqrt{1 + \dot{h}^2(t)} \, dt. \tag{26.28}$$

We want to minimise the functional (26.28) in the class of all C^1- functions $u :$ $[a, b] \to \mathbb{R}$, $u(a) = A, u(b) = B$. In other words we are looking for the function connecting (a, A) and (b, B), the graph of which considered as the trace of a curve has the shortest length, see Figure 26.1. Of course we expect the line segment connecting (a, A) and (b, B) to be the solution.

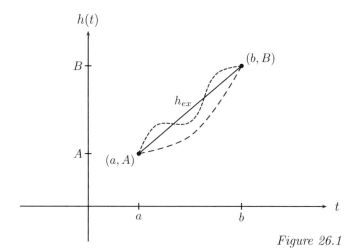

Figure 26.1

Since $F(t, x, y) = \sqrt{1 + y^2}$ and $\frac{d^2}{dy^2}\sqrt{1 + y^2} = \frac{1}{(1+y^2)^{\frac{3}{2}}}$ we find as the Euler-Lagrange equation

$$\frac{\ddot{h}(t)}{(1 + \dot{h}^2(t))^{\frac{3}{2}}} = 0, \, h(a) = A \text{ and } h(b) = B, \tag{26.29}$$

which is equivalent to

$$\ddot{h}(t) = 0, h(a) = A \text{ and } h(b) = B. \tag{26.30}$$

The solution is given by $h_{ex}(t) = h(t) = \frac{A(t-b)+B(a-t)}{a-b}$. Again, we can find the solution of the Euler-Lagrange equation in $C^2([a, b])$ under the corresponding boundary values, but we cannot, at the moment, claim that this solution is also the minimiser in the class of all C^1- functions subjected to the boundary conditions. For h_{ex} we find

$$\int_a^b \sqrt{1 + \dot{h}_{ex}^2(t)}\, dt = \int_a^b \sqrt{1 + \left(\frac{A - B}{a - b}\right)^2}\, dt$$

$$= \sqrt{1 + \left(\frac{A - B}{a - b}\right)^2}\,(b - a) = \sqrt{(b - a)^2 + (B - A)^2}$$

which is of course the value we have expected.

Note that we have not discussed the problem of finding the shortest curve connecting the two points (a, A) and (b, B) in the plane. We have looked for the shortest curve of the type $\gamma(t) = \begin{pmatrix} t \\ h(t) \end{pmatrix}$ with this property. Equation (26.29) has a nice geometric interpretation: the curvature of the minimiser, if it is a C^2-function, must be zero, compare with Example II.7.38.

Example 26.12. It can be shown, see [80] or [71], that solving the **brachistochrone problem** is the problem to minimise

$$\mathcal{F}(u) := \int_{t_1}^{t_2} \sqrt{\frac{1 + \dot{u}^2(t)}{2g\, u(t)}}\, dt,\, u(t_1) = h_1 \text{ and } u(t_2) = h_2 \tag{26.31}$$

where g is the gravitational acceleration. We must assume $h_1 > h_2$, and if we assume $h_2 > 0$, which we may achieve by a proper choice of the origin of the coordinate system, it is plausible - but not proved - to assume that $u(t) \geq h_2 > 0$. In Problem 3, we will establish the Euler-Lagrange equation for (26.31) under the assumption that F, F_x, F_{xy} and F_{yy} are defined for $x > 0$ and $y \in \mathbb{R}$.

Example 26.13. Let $F(t, q, \dot{q}) = q^2(4t - \dot{q})^2$, $t \in [-1, 1]$, and consider

$$\mathcal{F}(q) = \int_{-1}^1 F(t, q(t), \dot{q}(t))\, dt = \int_{-1}^1 q^2(t)(4t - \dot{q}(t))^2\, dt \tag{26.32}$$

for $q \in C^1([-1, 1]; \mathbb{R})$ satisfying the boundary conditions $q(-1) = 0$ and $q(1) = 1$. Since $F(t, q, \dot{q}) \geq 0$, the functional $\mathcal{F}(q)$ is bounded from below by 0. The function

$$u(t) = \begin{cases} 0, & -1 \leq t \leq 0 \\ 2t^2, & 0 \leq t \leq 1 \end{cases} \tag{26.33}$$

is a C^1-function since $u(0-) = u(0+) = u(0)$ and $\dot{u}(0-) = \dot{u}(0+) = 0$. Moreover we have

$$\dot{u}(t) = \begin{cases} 0, & -1 \le t \le 0 \\ 4t, & 0 \le t \le 1 \end{cases}$$

and therefore

$$F(t, u(t), \dot{u}(t)) = \begin{cases} 0, & -1 \le t \le 0 \\ 4t^4(4t - 4t)^2 = 0, & 0 \le t \le 1, \end{cases}$$

i.e. $F(t, u(t), \dot{u}(t)) = 0$ for all $t \in [-1, 1]$ implying that u is a minimiser of $\mathcal{F}(u)$ on $D_{0,2}(\mathcal{F})$. However, u is not a C^2-function since $\dot{u}(0-) = \dot{u}(0+) = 0$, but $\ddot{u}(0-) = 0$ and $\ddot{u}(0+) = 4$. Thus $\mathcal{F}$ has a minimiser which is not a C^2-function and therefore this minimiser cannot solve the Euler-Lagrange equation.

Example 26.14. The one-dimensional harmonic oscillator for a point with $m = 1$ has the Lagrange function $L(q, \dot{q}) = \frac{\dot{q}^2}{2} - \frac{q^2}{2}$ and we want to minimise

$$\mathcal{L}(q) = \int_{t_1}^{t_2} L(q(t), \dot{q}(t)) \, dt = \int_{t_1}^{t_2} \left(\frac{\dot{q}^2(t)}{2} - \frac{q^2(t)}{2} \right) dt \tag{26.34}$$

under the condition that $q(t_1) = q_1$ and $q(t_2) = q_2$. Now it is not clear at all whether a minimum of $\mathcal{L}(q)$ exists since $\mathcal{L}(q)$ need not be bounded from below. We leave this problem aside for the moment and since $L(q, \dot{q}) = \frac{\dot{q}^2}{2} - \frac{q^2}{2}$ is a C^2-function we can formally write down the Euler-Lagrange equation as

$$\ddot{q}(t) + q(t) = 0 \tag{26.35}$$

to which we may add the boundary conditions $q(t_1) = q_1$ and $q(t_2) = q_2$. The general solution to (26.35) is $q(t) = \alpha \cos t + \beta \sin t$ and the boundary conditions yield

$$\alpha(\sin t_2 \cos t_1 - \sin t_1 \cos t_2) = q_1 \sin t_2 - q_2 \sin t_1$$
$$\beta(\sin t_2 \cos t_1 - \sin t_1 \cos t_2) = q_2 \cos t_1 - q_1 \cos t_2.$$

The condition $\sin t_2 \cos t_1 - \sin t_1 \cos t_2 \neq 0$ which is necessary to obtain a unique solution q is not always fulfilled. Thus we may run into some problems, and indeed we have to handle a non-trivial eigenvalue problem of the type we have encountered in Chapter 21. Let us return to (26.34), but let us take into account the information that the Euler-Lagrange equation may have (periodic) eigenfunctions. Take as boundary conditions $q(0) = q(T) = 0$ for some $T > 0$. In this case, we have proved in Proposition I.26.16 a first version of **Poincaré's inequality**

$$\int_0^T q^2(t) \, dt \le 4T^2 \int_0^T \dot{q}^2(t) \, dt \tag{26.36}$$

459

from which we derive

$$\int_0^T \left(\frac{\dot{q}^2(t)}{2} - \frac{q^2(t)}{2} \right) dt \geq \int_0^T \frac{\dot{q}^2(t)}{2} dt - 4T^2 \int_0^T \frac{\dot{q}^2(t)}{2} dt$$

and for $\frac{1}{2} - 2T^2 > 0$, i.e. $T < \frac{1}{2}$ we find that $\mathcal{L}(q)$ is bounded from below, namely

$$\int_0^T \left(\frac{\dot{q}^2(t)}{2} - \frac{q^2(t)}{2} \right) dt \geq 0. \tag{26.37}$$

The corresponding boundary value problem for the Euler-Lagrange equation is

$$\ddot{q}(t) + q(t) = 0, \quad q(0) = q(T) = 0 \tag{26.38}$$

which leads for $q(t) = \alpha \cos t + \beta \sin t$ to the conditions $\alpha = 0$ and either $\beta = 0$ or $\sin T = 0$. Since $T < \frac{1}{2}$, we have to choose $\beta = 0$ and we arrive at $\alpha = \beta = 0$, i.e. $q(t) = 0$ for all $t \in [0, T]$. Of course, we are not surprised that the trivial solution minimises (26.37). A non-trivial question is whether there are further solutions for certain $T > 0$.

These examples show that, on the one hand, the Euler-Lagrange equations are a very useful tool, however it seems that, in general, variational problems cannot be reduced to a study of the Euler-Lagrange equations.

With the help of $L(t, q, \dot{q}) = \frac{1}{2} \sum_{j=1}^n \dot{q}_j^2$, as introduced in Example 26.10, we want to follow quite a different approach. We are searching for

$$\inf \left\{ \frac{1}{2} \int_a^b \dot{q}^2(t) \, dt + \int_a^b f(t) q(t) \, dt \,\middle|\, q \in C^1([a, b]; \mathbb{R}), \; q(a) = q(b) = 0 \right\} \tag{26.39}$$

where we assume that $f \in C([a, b]; \mathbb{R})$. By inspection, the reader may easily verify that a local C^2-solution of the minimising problem (26.39) will satisfy

$$\ddot{q}(t) = f(t) \text{ in } (a, b) \text{ and } q(a) = q(b) = 0, \tag{26.40}$$

see Problem 5. Using the Cauchy-Schwarz inequality, we find for every $\epsilon > 0$ that

$$\mathcal{L}_f(q) := \frac{1}{2} \int_a^b \dot{q}^2(t) \, dt + \int_a^b f(t) q(t) \, dt \tag{26.41}$$

$$\geq \frac{1}{2} \int_a^b \dot{q}^2(t) \, dt - \left(\int_a^b q^2(t) \, dt \right)^{\frac{1}{2}} \left(\int_a^b f^2(t) \, dt \right)^{\frac{1}{2}}$$

$$\geq \frac{1}{2} \int_a^b \dot{q}^2(t) \, dt - \epsilon \int_a^b q^2(t) \, dt - \frac{1}{4\epsilon} \int_a^b f^2(t) \, dt$$

where the last step uses the estimate $|\alpha|\,|\beta| \leq \epsilon\,|\alpha|^2 + \frac{1}{4\epsilon}\,|\beta|^2$, $\epsilon > 0$.
By Poincaré's inequality, Proposition I.26.16, we know further that

$$\int_a^b q^2(t)\,\mathrm{d}t \leq (2\max(|a|,|b|))^2 \int_a^b \dot{q}^2(t)\,\mathrm{d}t \tag{26.42}$$

which yields

$$\frac{1}{2}\int_a^b \dot{q}^2(t)\,\mathrm{d}t + \int_a^b f(t)q(t)\,\mathrm{d}t \geq \left(\frac{1}{2} - 4\epsilon\max(|a|^2,|b|^2)\right)\int_a^b \dot{q}^2(t)\,\mathrm{d}t - \frac{1}{4\epsilon}\int_a^b f^2(t)\,\mathrm{d}t. \tag{26.43}$$

If we choose $\epsilon = \frac{1}{8\max(|a|^2,|b|^2)}$, we arrive at

$$\mathcal{L}_f(q) = \frac{1}{2}\int_a^b \dot{q}^2(t)\,\mathrm{d}t + \int_a^b f(t)q(t)\,\mathrm{d}t \geq -2\max(|a|^2,|b|^2)\int_a^b f^2(t)\,\mathrm{d}t, \tag{26.44}$$

hence $\mathcal{L}_f(q)$ is bounded from below.

A further consequence of the Poincaré inequality (26.42) is that on
$\{\varphi \in C^1([a,b];\mathbb{R}) \mid \varphi(a) = \varphi(b) = 0\}$ a norm is given by $\|\dot{q}\|_{L^2} = \left(\int_a^b \dot{q}^2(t)\,\mathrm{d}t\right)^{\frac{1}{2}}$.
Homogeneity, as well as the triangle inequality, is trivial as is the fact that $\|\dot{q}\|_{L^2} \geq 0$ for all $q \in C^1([a,b];\mathbb{R})$. Moreover, if $\|\dot{q}\|_{L^2} = 0$ then by (26.42) it follows that $\|q\|_{L^2} = 0$ and since q is continuous this implies that $q = 0$ in $[a,b]$. Hence $\|\dot{q}\|_{L^2}$ is indeed a norm on $C^1([a,b];\mathbb{R})$.

Definition 26.15. A. We call $\|\dot{q}\|_{L^2}$ the **energetic norms** of $q \in C^1([a,b];\mathbb{R})$.
B. A sequence $(u_k)_{k\in\mathbb{N}}$, $u_k \in C^1([a,b];\mathbb{R})$ and $u_k(a) = u_k(b) = 0$, is said to be a **minimising sequence** for $\mathcal{L}_f(q)$ if

$$\lim_{k\to\infty} \mathcal{L}_f(u_k) = \mu := \inf\{\mathcal{L}_f(u) \mid u \in C^1([a,b];\mathbb{R}), u(a) = u(b) = 0\}.$$

We claim

Theorem 26.16. *Every minimising sequence for $\mathcal{L}_f(q)$ is a Cauchy sequence with respect to the energetic norm on $\{u \in C^1([a,b];\mathbb{R}) \mid u(a) = u(b) = 0\}$.*

Proof. Let $(u_k)_{k\in\mathbb{N}}$ be a minimising sequence for $\mathcal{L}_f(q)$ and we may assume that

$\mu \leq \mathcal{L}_f(u_k) \leq \mu + \frac{1}{k}$. It follows that

$$||\dot{u}_k - \dot{u}_l||_{L^2}^2 = 2||\dot{u}_k||_{L^2}^2 + 2||\dot{u}_l||_{L^2}^2 - 4\int_a^b \left(\frac{1}{2}(\dot{u}_k(t) + \dot{u}_l(t))\right)^2 dt$$

$$= 2||\dot{u}_k||_{L^2}^2 + 2||\dot{u}_l||_{L^2}^2 - 4\int_a^b \left(\frac{1}{2}(\dot{u}_k(t) + \dot{u}_l(t))\right)^2 dt$$

$$+ 4\int_a^b f(t)\, u_k(t)\, dt + 4\int_a^b f(t)\, u_l(t)\, dt$$

$$- 8\int_a^b f(t)\left(\frac{1}{2}(u_k(t) + u_l(t))\right) dt$$

$$= 4\mathcal{L}_f(u_k) + 4\mathcal{L}_f(u_l) - 8\mathcal{L}_f\left(\frac{1}{2}(u_k + u_l)\right) \leq 4\left(\frac{1}{k} + \frac{1}{l}\right),$$

where we used the definition of u_k and the fact that $\mathcal{L}_f(\frac{1}{2}(u_k + u_l)) \geq \mu$. Thus we have proved that $(u_k)_{k\in\mathbb{N}}$ is indeed a Cauchy sequence with respect to the energetic norm. □

Suppose that a minimising sequence $(u_k)_{k\in\mathbb{N}}$ for $\mathcal{L}_f$ has, with respect to the energetic norm, a limit $u \in C^1([a, b]; \mathbb{R})$. For this limit we find first of all

$$\lim_{k\to\infty} ||\dot{u}_k||_{L^2}^2 = ||\dot{u}||_{L^2}^2$$

and since

$$\left|\int_a^b f(t)\, u_k(t)\, dt - \int_a^b f(t)\, u(t)\, dt\right| \leq ||f||_{L^2}\, ||u_k - u||_{L^2}$$

$$\leq 2\max(|a|, |b|)\, ||f||_{L^2}\, ||\dot{u}_k - \dot{u}||_{L^2}$$

it follows that

$$\mathcal{L}_f(u) = \lim_{k\to\infty} \mathcal{L}_f(u_k).$$

Note that, a priori, we do not know whether $u(a) = u(b) = 0$, i.e. we cannot decide whether u is indeed a minimiser. However, much more important is the fact that, in general, we cannot expect a minimising sequence to have a limit in $C^1([a, b]; \mathbb{R})$ with respect to the energetic norm: the space $C^1([a, b]; \mathbb{R})$ equipped with the energetic norm is not complete, i.e. not a Banach space and hence Cauchy sequences in $C^1([a, b]; \mathbb{R})$ with respect to the energetic norm need not, and will in general not, converge to a limit in $C^1([a, b]; \mathbb{R})$.

Let us summarise our investigation so far. Under additional regularity assumptions on $F \in C^1([a, b] \times \mathbb{R}^n \times \mathbb{R}^n)$, we can prove as a necessary condition for

462

$u \in C^2([a,b]; \mathbb{R}^n) \cap D_{A,B}(\mathcal{F})$ to be a minimiser of Problem 26.2 that the Euler-Lagrange equations hold. They are not sufficient conditions. In fact, according to Example 26.13, minimisers may exist which do not satisfy the Euler-Lagrange equations due to a lack of smoothness. Note that the formulation of the Euler-Lagrange equations requires from F as well as from a minimiser more smoothness or regularity than the original problem. On the other hand, we can try to minimise $\mathcal{F}(u)$ directly by studying minimising sequences. In some cases they even form a Cauchy sequence in a suitable norm, however the corresponding normed space is (in general) not complete, i.e. a minimising sequence although being a Cauchy sequence need not converge. In the first approach, we try to find a solution to the minimising problem indirectly by solving certain differential equations while, in the second approach, we try to minimise the functional $\mathcal{F}$ directly. For this reason the second approach is called a **direct method** in the calculus of variations.

Problems

1. Let $(a,0), (-a,0) \in \mathbb{R}^2$, $a > 0$, be two points in the set $\mathbb{R}^2 \setminus (c,d)$, $c < 0 < d$, see Figure 26.2 below.

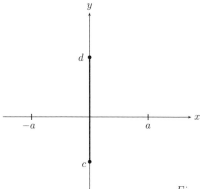

Figure 26.2

   Assuming that the shortest curve connecting two points in the plane is a line segment, find, by elementary geometric reasoning, the shortest curve connecting $(-a,0)$ and $(a,0)$. Does this problem always have a unique solution?

2. Interpret the fundamental lemma of the calculus of variations as an orthogonality result in $L^2([a,b])$.

3. Find the Euler-Lagrange equation corresponding to the **brachistochrone problem**, see Example 26.12.

4. For a C^2-function $\rho : [a, b] \times \mathbb{R}^2 \to \mathbb{R}$, $\rho > 0$, find the Euler-Lagrange equations corresponding to the functional

$$\int_a^b \frac{\sqrt{1 + \dot{u}^2(t) + \dot{v}^2(t)}}{\rho(t, u(t), v(t))}\, dt.$$

5. Prove (26.40).

6. Let $\mathcal{F} : [t_1, t_2] \times \mathbb{R}^n \times \mathbb{R}^n \to \mathbb{R}$ and $G : \mathbb{R}^n \to \mathbb{R}$ be two C^2-functions. Prove that a necessary condition for a local extremal of $\mathcal{F} = \int_{t_1}^{t_2} F(t, u(t), \cdot u(t))\, dt$, $u(t_1) = A$, $u(t_2) = B$, under the constraint $G(u(t)) = 0$ for all $t \in [t_1, t_2]$ is

$$\frac{d}{dt} \frac{\partial F}{\partial y}(t, u(t), \dot{u}(t)) - \frac{\partial F}{\partial x}(t, u(t), \dot{u}(t)) - \lambda \frac{\partial G}{\partial x}(u(t)) = 0,$$

where $\lambda : [t_1, t_2] \to \mathbb{R}$ is a continuous function.

7. Assume that for a simply closed parametric curve $\gamma : [a, b] \to \mathbb{R}^2$, $\gamma(t) = \begin{pmatrix} t \\ y(t) \end{pmatrix}$, the area of the region enclosed by the curve is given by

$$A(\gamma) := \frac{1}{2} \int_a^b (t\dot{y}(t) - y(t))\, dt,$$

compare with (II.26.22). For the length of γ we have

$$l(\gamma) := \int_a^b \sqrt{1 + \dot{y}^2(t)}\, dt.$$

Prove that a necessary condition for a C^2-curve as given above, for which $A(\gamma)$ is maximal while $l(\gamma)$ is fixed, is given by the Euler-Lagrange equation corresponding to the functional

$$\int_a^b \left\{ \left(\frac{1}{2} t\dot{u}(t) - \frac{1}{2} u(t) \right) + \lambda \sqrt{1 + \dot{u}^2(t)} \right\}\, dt,$$

where $\lambda > 0$ is a parameter. Find the equation and try to solve it.

8. The length of a parametric curve $\gamma : [0, 1] \to \mathbb{R}^n$ is defined by $l(\gamma) = \int_0^1 \|\dot{\gamma}(t)\}\, dt$ and it is invariant under a change of parameter, see Chapter II.7. We define the **energy of the curve** γ as $E(\gamma) := \int_0^1 \|\dot{\gamma}(t)\|^2\, dt$. Prove:

   a) $l^2(\gamma) \le E(\gamma)$;

b) in general, $E(\gamma)$ is not invariant under a change of parameter;

c) if γ_0 is a minimiser of the energy in the class of all C^1-curves $\gamma :$ $[0,1] \to \mathbb{R}^n$, $\gamma(0) = A$ and $\gamma(1) = B$, and if γ_0 is parametrized proportional to the arc length, then γ_0 is also a minimiser of $l(\gamma)$. Note that γ_0 is parametrized proportional to the arc length if for some $c_0 > 0$ we have $\|\dot{\gamma}_0(t)\| = c_0$ for all t.

9. Let $G \subset [a,b] \times \mathbb{R}^{n-1} \subset \mathbb{R}^n$ and for $u \in C_0^1(G)$ we define

$$|||u|||^2 := \int_G \|\mathrm{grad}\, u(x)\|^2 \mathrm{d}x.$$

Prove that $||| \cdot |||$ is a norm on $C_0^1(G)$ which we shall consider as an energetic norm on $C_0^1(G)$.

Hint: prove that

$$\langle u, v \rangle_E := \int_G \langle \mathrm{grad}\, u(x), \mathrm{grad}\, v(x) \rangle \, \mathrm{d}x$$

is a scalar product on $C_0^1(G)$.

27 Classical Solutions of the Euler-Lagrange Equations

Let $F : [a, b] \times \mathbb{R}^n \times \mathbb{R}^n \to \mathbb{R}$, $(t, x, y) \mapsto F(t, x, y)$, be a C^1-Lagrange function such that F_{y_j}, $j = 1, \ldots, n$, is a C^1-function too. If $u \in C^2([a, b]; \mathbb{R}^n)$ is a C^2-minimiser of $\mathcal{F}(u) = \int_a^b F(t, u(t), \dot{u}(t)) \, dt$ under the boundary conditions $u(a) = A$ and $u(b) = B$ then u is a solution of the Euler-Lagrange equations

$$\frac{d}{dt} F_{y_j}(t, u(t), u'(t)) - F_{x_j}(t, u(t), u'(t)) = 0, \; j = 1, \ldots, n \qquad (27.1)$$

or more explicitly

$$\frac{\partial F_{y_j}}{\partial t}(t, u(t), u'(t)) + \sum_{l=1}^n F_{y_j y_l}(t, u(t), u'(t)) u_l''(t) + \sum_{l=1}^n F_{y_j x_l}(t, u(t), u'(t)) \qquad (27.2)$$

$$- F_{x_j}(t, u(t), u'(t)) = 0,$$

and of course it satisfies the boundary conditions $u(a) = A$ and $u(b) = B$.

It is convenient to introduce the following notations:

$$F_x(t, x, y) := \text{grad}_x \, F(t, x, y) := \left(\frac{\partial F}{\partial x_1}(t, x, y), \ldots, \frac{\partial F}{\partial x_n}(t, x, y) \right); \qquad (27.3)$$

and

$$F_y(t, x, y) := \text{grad}_y \, F(t, x, y) := \left(\frac{\partial F}{\partial y_1}(t, x, y), \ldots, \frac{\partial F}{\partial y_n}(t, x, y) \right). \qquad (27.4)$$

Using the scalar product $\langle \cdot, \cdot \rangle$ in $\mathbb{R}^n$ we may write for example

$$\langle u'(t), F_y(t, u(t), u'(t)) \rangle = \sum_{j=1}^n F_{y_j}(t, u(t), u'(t)) \, u_j'(t) \qquad (27.5)$$

or instead of (27.1)

$$\frac{d}{dt} F_y(t, u(t), u'(t)) - F_x(t, u(t), u'(t)) = 0. \qquad (27.6)$$

Recall that every C^2-solution of (27.1) is called an $\mathcal{F}$-**extremal**, thus $\mathcal{F}$-extremals are **stationary points** of $\mathcal{F}(u)$. In this chapter we want to study (27.1) as a classical system of ordinary differential equations of second order. In particular, we are seeking classical, i.e. C^2-solutions of (27.1). Often we will add boundary conditions, i.e. we will prescribe the values for a solution $u : [a,b] \to \mathbb{R}^n$ at a and b, $u(a) = A$ and $u(b) = B$. However, we shall not expect to find explicit formulae for solutions often and therefore it is worth deriving general properties of $\mathcal{F}$-extremals.

We start our investigations by looking at some special cases. First let $F : [a,b] \times \mathbb{R} \to \mathbb{R}, (t,y) \mapsto F(t,y)$, be independent of x, i.e. $F_x = 0$. Then (27.1) becomes

$$\frac{\mathrm{d}}{\mathrm{d}t} F_y(t, u'(t)) = 0 \tag{27.7}$$

which implies for some constant $c \in \mathbb{R}$ that

$$F_y(t, u'(t)) = c. \tag{27.8}$$

This is an implicit first order differential equation which does not contain u itself. Whenever we can solve (27.8) for $u'(t)$, i.e. whenever we can write

$$u'(t) = G(t, c) \tag{27.9}$$

then we immediately obtain u by integrating (27.9). The implicit function theorem, Theorem II.10.5, gives sufficient conditions for solving (27.8) locally for $u'(t)$. As a reminder, we state a one-dimensional version of Theorem II.10.5 here.

Theorem 27.1. *Let I and J be two open intervals and $H : I \times J \to \mathbb{R}$ be a differentiable function. Let $(t_0, y_0) \in I \times J$ such that $H(t_0, y_0) = 0$ and $\frac{\partial H}{\partial y}(t_0, y_0) \neq 0$. Then there exists an open interval $I_1 \subset I$, $t_0 \in I_1$, an open interval $J_1 \subset J$, $y_0 \in J_1$, and a continuous function $g : I_1 \to J_2$ such that $H(t, g(t)) = 0$ for $t \in I_1$, and if $(t, y) \in I_1 \times J_1$ such that $H(t, y) = 0$, then $y = g(t)$.*

We may apply Theorem 27.1 to $H(t, y) = F_y(t, y) - c$ and the essential condition becomes $F_{yy}(t, y) \neq 0$. Thus for $F_{yy}(t_0, y_0) \neq 0$ we obtain, at least locally, solutions to (27.8) by a simple integration.

Example 27.2. We consider $F : [a,b] \times \mathbb{R} \to \mathbb{R}$, $F(t, y) = \gamma(t)\sqrt{1 + y^2}$, and the functional $\mathcal{F}(u) = \int_a^b \gamma(t)\sqrt{1 + u'^2(t)}\, \mathrm{d}t$. We now obtain for (27.8)

$$\frac{\gamma(t)\, u'(t)}{\sqrt{1 + u'^2(t)}} = c$$

and for $\gamma(t) \neq 0$

$$\gamma^2(t)\, u'^2(t) = c\,(1 + u'^2(t)) \tag{27.10}$$

which implies $\gamma^2(t) < c$ and

$$u'(t) = \pm \frac{c}{\sqrt{\gamma^2(t) - c^2}}, \tag{27.11}$$

where the sign in (27.11) depends on the value of $u(a)$ and $u(b)$. We return to this example after Example 27.4.

It seems that in the case of Example 27.2 we are not dealing with a second order differential equation, but with a first order differential equation. However, this is not quite correct. The original equation is (27.7), i.e.

$$\frac{\mathrm{d}}{\mathrm{d}t} F_y(t, u'(t)) = \frac{\partial F_y}{\partial t}(t, u'(t)) + F_{yy}(t, u'(t)) \, u''(t) = 0$$

and only the fact that along every solution of (27.7) we have (27.8) allows us to simplify the original problem. This leads to the concept of a first integral of (27.1).

Definition 27.3. *A function $H : [a,b] \times \mathbb{R}^n \times \mathbb{R}^n \to \mathbb{R}$ is called a **first integral** of the Euler-Lagrange equations (27.1) if for every C^2-solution u of (27.1) we have*

$$H(t, u(t), u'(t)) = c, \qquad c \in \mathbb{R}. \tag{27.12}$$

Example 27.4. A. For $F = F(t, y)$ the function F_y is a first integral of (27.1).
B. Suppose that $F : \mathbb{R}^n \times \mathbb{R}^n \to \mathbb{R}$, i.e. F is independent of t. We claim that

$$E(u(t), u'(t)) := \langle u'(t), F_y(u(t), u'(t)) \rangle - F(u(t), u'(t)) \tag{27.13}$$

is a first integral of the Euler-Lagrange equations corresponding to F. Indeed, for $u \in C^2([a,b]; \mathbb{R}^n)$ solving (27.1) we find

$$\frac{\mathrm{d}}{\mathrm{d}t} E(u(t), u'(t)) = \frac{\mathrm{d}}{\mathrm{d}t} \langle u'(t), F_y(u(t), u'(t)) \rangle - \frac{\mathrm{d}}{\mathrm{d}t} F(u(t), u'(t))$$

$$= \langle u''(t), F_y(u(t), u'(t)) \rangle + \langle u'(t), \frac{\mathrm{d}}{\mathrm{d}t} F_y(u(t), u'(t)) \rangle$$

$$- \langle F_x(u(t), u'(t)), u'(t) \rangle - \langle F_y(u(t), u'(t)), u''(t) \rangle$$

$$= \left\langle \frac{\mathrm{d}}{\mathrm{d}t} F_y(u(t), u'(t)) - F_x(u(t), u'(t)), u'(t) \right\rangle = 0$$

since $\left(\frac{\mathrm{d}}{\mathrm{d}t} F_y - F_x \right)(u(t), u'(t)) = 0$. The function E in (27.13) is often called the **energy** or the **energy integral** corresponding to F. The case where F is independent of t is the most important one and we will discuss it in more detail below.

469

The case where $F : [a, b] \times \mathbb{R} \to \mathbb{R}, (t, x) \mapsto F(t, x)$ leads to the Euler-Lagrange equation

$$F_x(t, u(t)) = 0 \tag{27.14}$$

which is not a differential equation and again the implicit function theorem is a tool to decide whether we can solve (locally) (27.14) for $u(t)$.

Many geometric problems lead to variational integrals of the type

$$\int_a^b h(t, u(t)) \sqrt{1 + u'^2(t)} \, dt. \tag{27.15}$$

Such a type of integral arises when we consider the graph of a function $u : [a, b] \to \mathbb{R}$ as a curve γ in the plane $\mathbb{R}^2$ parametrized by $\gamma(t) = \begin{pmatrix} t \\ u(t) \end{pmatrix}$ and h is a function on γ, see (II.15.43), or when we consider a vector field $X(\gamma(t))$ on γ which is parallel to γ. In this case we find

$$\langle X(\gamma(t)), \dot{\gamma}(t) \rangle = \|X(\gamma(t))\| \, \|\dot{\gamma}(t)\| \cos \sphericalangle (X(\gamma(t)), \dot{\gamma}(t)) = \|X(\gamma(t))\| \, \|\dot{\gamma}(t)\|$$

and $\|\dot{\gamma}(t)\| = \sqrt{1 + u'^2(t)}$. The case $h(t, x) = 1$ was already discussed in Example 26.11.

The Euler-Lagrange equation corresponding to (27.15) is

$$
\begin{aligned}
0 &= \frac{d}{dt} \left(h(t, u(t)) \frac{u'(t)}{\sqrt{1 + u'^2(t)}} \right) - \frac{\partial h}{\partial x}(t, u(t)) \sqrt{1 + u'^2(t)} \\
&= \frac{\partial h}{\partial t}(t, u(t)) \frac{u'(t)}{\sqrt{1 + u'^2(t)}} + \frac{\partial h}{\partial x}(t, u(t)) \frac{u'^2(t)}{\sqrt{1 + u'^2(t)}} \\
&\quad + (h(t, u(t))) \frac{u''(t)}{(1 + u'^2(t))^{\frac{3}{2}}} - \frac{\partial h}{\partial x}(t, u(t)) \sqrt{1 + u'^2(t)} \\
&= \frac{1}{\sqrt{(1 + u'^2(t))}} \left(\frac{\partial h}{\partial t}(t, u(t)) u'(t) + h(t, u(t)) \frac{u''(t)}{1 + u'^2(t)} - \frac{\partial h}{\partial x}(t, u(t)) \right)
\end{aligned}
$$

or

$$\frac{h(t, u(t))}{1 + u'^2(t)} u''(t) + \frac{\partial h}{\partial t}(t, u(t)) u'(t) - \frac{\partial h}{\partial x}(t, u(t)) = 0. \tag{27.16}$$

Example 27.5. A. Let $F : [1, 3] \times \mathbb{R} \to \mathbb{R}, F(t, y) = \frac{1}{t} \sqrt{1 + \dot{y}^2}$, i.e. $h(t, u(t)) = \frac{1}{t}$. The Euler-Lagrange equation corresponding to the minimising problem

$$\inf \{ \mathcal{F}(u) \mid u \in D_{0,1}(\mathcal{F}) \}$$

470

reads for a C^2-extremal $u \in C^2([1,3])$ as

$$\frac{1}{t}\frac{u''(t)}{1+u'^2(t)} - \frac{1}{t^2}u'(t) = \frac{\mathrm{d}}{\mathrm{d}t}\left(\frac{u'(t)}{t\sqrt{1+u'^2(t)}}\right) = 0. \qquad (27.17)$$

In fact, we could have derived (27.17) from (27.7). From (27.17) we deduce that

$$\frac{u'(t)}{t\sqrt{1+u'^2(t)}} = c$$

which leads to

$$u'(t) = \pm\frac{ct}{\sqrt{1-c^2t^2}}$$

or

$$u(t) = \pm\frac{1}{c}\sqrt{1-c^2\,t^2} + c_1. \qquad (27.18)$$

We may now use the boundary conditions $u(1) = 0$ and $u(3) = 2$ to find $c_1 = \frac{3}{2}$ and $c = \frac{2}{\sqrt{13}}$, as these conditions enforce the plus sign in (27.18).

B. We consider in $\mathbb{R}^3$ a surface of revolution, compare with Chapter II.8, given by

$$S(t,\varphi) = \begin{pmatrix} u(t)\,\cos\varphi \\ u(t)\,\sin\varphi \\ t \end{pmatrix}, \ u(t) \geq 0, t \in [a,b], \varphi \in [0, 2\pi]. \text{ The surface area of } S \text{ is}$$

given by

$$A(S) = \mathcal{F}(u) = 2\pi \int_a^b u(t)\sqrt{1+u'^2(t)}\,\mathrm{d}t, \qquad (27.19)$$

see Problem 3. Suppose that $u(a) = A$ and $u(b) = B$ are fixed.

We may ask for the surface of revolution given by $S(t,\varphi)$ with the boundary conditions stated above which has minimal area. The Lagrange function in (27.19) is $F(t,x,y) = 2\pi x\sqrt{1+y^2}$ and it is independent of t. We find that the energy $u'(t)F_y(u(t), u'(t)) - F(u(t), u'(t))$ is a first integral of the Euler-Lagrange equation which yields

$$u(t)\sqrt{(1+u'^2(t))} - u(t)\frac{u'^2(t)}{\sqrt{1+u'^2(t)}} = c$$

or

$$u(t) = c\sqrt{1+u'^2(t)}$$

which gives

$$u'(t) = \sqrt{\frac{u^2(t)-c^2}{c^2}}. \qquad (27.20)$$

This differential equation can be solved by separating variables and we obtain the general solution for an F-extremal

$$u(t) = C \cosh\left(\frac{t + c_1}{c}\right).$$ (27.21)

Thus F-extremals are catenoids, compare with Problem 5 of Chapter II.8. The constants c and c_1 can be determined with the help of the boundary condition $u(a) = A$ and $u(b) = B$, but same care is needed.

Satisfying the Euler-Lagrange equations is a necessary condition for a local C^2-minimiser u of $\mathcal{F}(u)$ if the Lagrange function F and the mapping F_y are C^1-functions. The regularity required for u is often fulfilled.

Theorem 27.6. *Let $F : [a, b] \times \mathbb{R}^n \times \mathbb{R}^n \to \mathbb{R}$ be a C^1-function with the property that F_y is also a C^1-mapping. Suppose that in addition*

$$\det\left(\frac{\partial^2 F}{\partial y_k \, \partial y_l}(t, u(t), u'(t))\right)_{k,l=1,\dots,n} \neq 0$$ (27.22)

for all $t \in [a, b]$ where for $u \in C^1([a, b]; \mathbb{R}^n)$

$$(\delta\mathcal{F})(u, \varphi) = 0$$

holds for all $\varphi \in C_0^1((a, b); \mathbb{R}^n)$. Then $u \in C^2([a, b]; \mathbb{R}^n)$.

Proof. (Following [71]). The implicit function theorem, now in its full generality, i.e. Theorem II.10.5, is the key ingredient. We consider the function $G : (a, b) \times \mathbb{R}^n \times \mathbb{R}^n \times \mathbb{R}^n \to \mathbb{R}$ defined by $G(t, x, y, z) = F_y(t, x, y) - z$. By (27.22) the implicit function theorem implies that we can solve the equation $G(t, x, y, z) = 0$ for y for every t_0 near a point $(t_0, x_0, y_0, z_0) = (t_0, u(t_0), u'(t_0), F(t_0, u(t_0), u'(t_0)))$, i.e. there exists an open neighbourhood $\mathcal{U}$ of (t_0, x_0, z_0) such that in $\mathcal{U}$ the equation $G(t, u(t_0), u'(t_0), F_p(t_0, u(t_0), u'(t_0))) = 0$ has a unique solution $\varphi : \mathcal{U} \to \mathbb{R}^n$ which is of class C^1. However, by assumption we have $G(t, u(t), u'(t), F_p(t, u(t), u'(t))) = 0$ and therefore the uniqueness part of the implicit function theorem yields

$$u'(t) = \varphi(t, u(t), F_p(t, u(t), u'(t)))$$

in a neighbourhood of t_0. The function φ is of class C^1 and this implies that u is of class C^2. $\qquad\square$

In mechanics, typical Lagrange functions are of the type $L(t, x, y) = T(y) - V(x)$ where T is the kinetic energy and V is the potential energy. For a given system of points, the kinetic energy is often of the form

$$T(y) = \sum_{j=1}^{N} \frac{m_j}{2} y_j^2 \tag{27.23}$$

and since V depends often only on x, or sometimes on t and x, the Lagrange function $L(x, y) = T(y) - V(x)$ with T as in (27.23) satisfies (27.22). Thus in this case a local C^1-minimiser for the corresponding minimising problem is a C^2-L-extremal. Moreover, suppose that the minimising problem has at most one C^1-solution and that (27.22) holds. If, in addition, the Euler-Lagrange equations have a unique C^2-solution under the given boundary conditions, then this solution must be the minimiser. We will return to the uniqueness problem later on.

We have seen that first integrals are quite helpful to find solutions or at least to obtain some information about solutions of the Euler-Lagrange equations. The natural question is, how can we find first integrals? We have encountered a first integral for the first time in Example 27.2 for the Lagrange function $F(t, y) = \gamma(t)\sqrt{1 + y^2}$, and we found that F_y is a first integral whenever F is independent of x, see Example 27.4.A. Moreover, the independence of F with respect to t in Example 27.4.B led to the energy $E = \langle y, F_y \rangle - F$ as a first integral. In both cases, the Lagrange function admits a certain invariance. In the first case, $F(t, y)$ is invariant under the translation $(t, x, y) \mapsto (t, x + z, y)$, in the second case we have invariance under the translation $(t, x, y) \mapsto (t + s, x, y)$. It turns out that certain invariances or symmetries are key to the understanding of first integrals. In full generality this result is the content of Noether's theorem and is due to Emmy Noether. A proof is given in [71]. We follow [58] for a first version of Noether's theorem. In order to have a more readable presentation let us state some of the basic assumptions separately.

Assumptions for Theorem 27.7. Let $F \in C^2(\mathbb{R} \times \mathbb{R}^n \times \mathbb{R}^n)$ be a Lagrange function, $(t, x, y) \mapsto F(t, x, y)$. Let $\varphi \in C^2(\mathbb{R} \times \mathbb{R}^2 \times (-\epsilon_0, \epsilon_0); \mathbb{R}^n)$, $\epsilon_0 > 0$, and assume that for every fixed pair $(t, \epsilon) \in \mathbb{R} \times (-\epsilon_0, \epsilon_0)$ the mapping $\varphi(t, \cdot, \epsilon) : \mathbb{R}^n \to \mathbb{R}^n$ is a diffeomorphism and $\varphi(t, x, 0) = x$ for all $t \in \mathbb{R}^n$. We set

$$\tau(t, x) := \frac{\partial \varphi}{\partial \epsilon}(t, x, \epsilon)\bigg|_{\epsilon=0} = \varphi_\epsilon(t, x, 0). \tag{27.24}$$

By Taylor's formula it follows that

$$\varphi(t, x, \epsilon) = x + \epsilon\tau(t, x) + \sigma(t, x, \epsilon), \tag{27.25}$$

where

$$\lim_{\epsilon \to 0} \frac{\sigma(t, x, \epsilon)}{\epsilon} = 0.$$

In addition, we assume the existence of $W \in C^2(\mathbb{R} \times \mathbb{R}^n \times (-\epsilon_0, \epsilon_0); \mathbb{R})$ such that for every $u \in C^2(I; \mathbb{R}^n)$ we have

$$F(t, \varphi(t, u(t), \epsilon), \dot{\varphi}(t, u(t), \epsilon)) = F(t, u(t), \dot{u}(t)) + \frac{\mathrm{d}}{\mathrm{d}t} W(t, u(t), \epsilon) \qquad (27.26)$$

where $\dot{\varphi}(t, u(t), \epsilon) = \frac{\mathrm{d}}{\mathrm{d}t} \varphi(t, u(t), \epsilon)$ and $\dot{u}(t) = \frac{\mathrm{d}}{\mathrm{d}t} u(t)$. We further set

$$w(t, x) := \frac{\partial}{\partial \epsilon} W(t, x, \epsilon)\Big|_{\epsilon=0} = W_\epsilon(t, x, 0). \qquad (27.27)$$

Theorem 27.7. *(E.Noether) Suppose that the above stated assumptions hold. Then the function*

$$G(t, x, y) := \langle \tau(t, x), F_y(t, x, y) \rangle - w(t, x) \qquad (27.28)$$

is a first integral of the Euler-Lagrange equations associated with F.

Proof. (Following [58]). We have to prove that for every solution $u \in C^2(I; \mathbb{R}^n)$ of the Euler-Lagrange equations it follows that

$$\frac{\mathrm{d}}{\mathrm{d}t} G(t, u(t), \dot{u}(t)) = 0, \quad t \in I.$$

The proof requires a lot of differentiations of functions with rather complicated arguments. Therefore we introduce some short hand notations, namely

$$\alpha := (t, u(t), \epsilon) \text{ and } \beta = (t, \varphi(\alpha), \dot{\varphi}(\alpha)).$$

We differentiate (27.26) with respect to ϵ and we obtain using the right hand side

$$\frac{\partial}{\partial \epsilon} F(\beta) = \frac{\partial}{\partial \epsilon} \frac{\mathrm{d}}{\mathrm{d}t} W(\alpha) = \frac{\mathrm{d}}{\mathrm{d}t} \frac{\partial}{\partial \epsilon} W(t).$$

Differentiating the left hand side yields

$$\frac{\partial}{\partial \epsilon} F(\beta) = \left\langle F_x(\beta), \frac{\partial}{\partial \epsilon} \varphi(\alpha) \right\rangle + \left\langle F_y(\beta), \frac{\partial \dot{\varphi}}{\partial \epsilon}(\alpha) \right\rangle$$

$$= \left\langle F_x(\beta), \frac{\partial}{\partial \epsilon} \varphi(\alpha) \right\rangle + \left\langle F_y(\beta), \frac{\mathrm{d}}{\mathrm{d}t} \frac{\partial}{\partial \epsilon} \varphi(\alpha) \right\rangle$$

and therefore we get

$$\frac{\partial}{\partial \epsilon} F(\beta) = \left\langle \frac{\partial}{\partial \epsilon} \varphi(\beta), F_x(\beta) - \frac{\mathrm{d}}{\mathrm{d}t} F_y(\beta) \right\rangle$$
$$+ \left\langle \frac{\partial}{\partial \epsilon} \varphi(\alpha), \frac{\mathrm{d}}{\mathrm{d}t} F_y(\beta) \right\rangle + \left\langle \frac{\mathrm{d}}{\mathrm{d}t} \frac{\partial}{\partial \epsilon} \varphi(\alpha), F_y(\beta) \right\rangle.$$

By assumption u is an F-extremal and $\varphi(t, u(t), 0) = u(t)$ which gives

$$F_x(\beta) - \frac{\mathrm{d}}{\mathrm{d}t} F_y(\beta) \bigg|_{\epsilon=0} = 0.$$

(Note that $\beta|_{\epsilon=0} = (t, \varphi(t, u(t), 0), \dot\varphi(t, u(t), 0)) = (t, u(t), \dot u(t))$). Moreover, we have

$$\varphi(t, u(t), \epsilon) = u(t) + \epsilon \tau(t, u(t)) + \sigma(t, u(t), \epsilon)$$

and therefore we find $\frac{\partial}{\partial \epsilon} \varphi(t, u(t), \epsilon) \big|_{\epsilon=0} = \tau(t, u(t))$ and it follows that

$$\frac{\mathrm{d}}{\mathrm{d}t} \left\langle \frac{\partial \varphi}{\partial \epsilon}(\alpha), F_y(\beta) \right\rangle \bigg|_{\epsilon=0} = \frac{\mathrm{d}}{\mathrm{d}t} \left(\left\langle \frac{\partial \varphi}{\partial \epsilon}(\alpha), F_y(\beta) \right\rangle \bigg|_{\epsilon=0} \right)$$
$$= \frac{\mathrm{d}}{\mathrm{d}t} \langle \tau(t, u(t)), F_y(t, u(t), \dot u(t)) \rangle.$$

Now we find

$$\frac{\partial F}{\partial \epsilon}(\beta) \bigg|_{\epsilon=0} = \frac{\mathrm{d}}{\mathrm{d}t} \langle \tau(t, u(t)), F_y(t, u(t), \dot u(t)) \rangle \qquad (27.29)$$

and since $\frac{\partial}{\partial \epsilon} W(t, u(t), \epsilon) \big|_{\epsilon=0} = w(t, u(t))$ we arrive at

$$\frac{\partial}{\partial \epsilon} F(\beta) \bigg|_{\epsilon=0} = \left(\frac{\mathrm{d}}{\mathrm{d}t} \frac{\partial}{\partial \epsilon} W(t, u(t), \epsilon) \right) \bigg|_{\epsilon=0} \qquad (27.30)$$
$$= \frac{\mathrm{d}}{\mathrm{d}t} \left(\frac{\partial}{\partial \epsilon} W(t, u(t), \epsilon) \bigg|_{\epsilon=0} \right) = \frac{\mathrm{d}}{\mathrm{d}t} w(t, u(t)).$$

From (27.29) and (27.30) we deduce

$$\frac{\mathrm{d}}{\mathrm{d}t} (\langle \tau(t, u(t)), F_y(t, u(t), \dot u(t)) \rangle - w(t, u(t))) = 0,$$

or

$$\frac{\mathrm{d}}{\mathrm{d}t} G(t, u(t), \dot u(t)) = 0$$

which implies that G is a first integral. $\qquad \square$

Example 27.8. We consider $F : \mathbb{R}^3 \times \mathbb{R}^3 \to \mathbb{R}, F(x,y) = \frac{m}{2}||y||^2 - V(||x||)$ where $m > 0$ and $V : [0, \infty] \to \mathbb{R}$ is a C^1-function. By Example 27.4 we know that the energy

$$E(u(t), u'(t)) = \langle u'(t), F_y(u(t), u'(t)) \rangle - F(u(t), u'(t))$$
$$= m \langle u'(t), u'(t) \rangle - \frac{m}{2}||u'(t)||^2 + V(||u(t)||)$$
$$= \frac{m}{2}||u'(t)||^2 + V(||u(t)||)$$

is a first integral for F. We now consider the mapping

$$\varphi_1 : \mathbb{R} \times \mathbb{R}^3 \times (-\epsilon_0, \epsilon_0) \to \mathbb{R}^3, \ \varphi_1(t, x, \epsilon) = \begin{pmatrix} x_1 \\ x_2 \cos \epsilon - x_3 \sin \epsilon \\ x_2 \sin \epsilon + x_3 \cos \epsilon \end{pmatrix}.$$

Note that

$$\varphi_1(t, x, \epsilon) = \begin{pmatrix} 1 & 0 & 0 \\ 0 & \cos \epsilon & -\sin \epsilon \\ 0 & \sin \epsilon & \cos \epsilon \end{pmatrix} \begin{pmatrix} x_1 \\ x_2 \\ x_3 \end{pmatrix}$$

is a rotation around the x_1-axis by an angle ϵ which implies that $\varphi(t, \cdot, \epsilon) : \mathbb{R}^3 \to \mathbb{R}^3$ is a diffeomorphism and $\varphi(t, x, 0) = x$. Further we find

$$\frac{\partial \varphi_1}{\partial \epsilon}(t, x, \epsilon) = \begin{pmatrix} 0 \\ -x_2 \sin \epsilon - x_3 \cos \epsilon \\ x_2 \cos \epsilon - x_3 \sin \epsilon \end{pmatrix}$$

which yields

$$\tau_1(t, x) = \frac{\partial \varphi_1}{\partial \epsilon}(t, x, \epsilon) \Big|_{\epsilon=0} = \begin{pmatrix} 0 \\ -x_3 \\ x_2 \end{pmatrix}.$$

Moreover, we have

$$\dot{\varphi}_1(t, u(t), \epsilon) = \frac{d}{dt} \begin{pmatrix} u_1(t) \\ u_2(t) \cos \epsilon - u_3(t) \sin \epsilon \\ u_2(t) \sin \epsilon + u_3(t) \cos \epsilon \end{pmatrix} = \begin{pmatrix} \dot{u}_1(t) \\ \dot{u}_2(t) \cos \epsilon - \dot{u}_3(t) \sin \epsilon \\ \dot{u}_2(t) \sin \epsilon + \dot{u}_3(t) \cos \epsilon \end{pmatrix}.$$

This leads to

$$F(\varphi_1(t, u(t), \epsilon), \dot{\varphi}_1(t, u(t), \epsilon)) = \frac{m}{2}||\dot{\varphi}_1(t, u(t), \epsilon)||^2 - V(||\varphi_1(t, u(t), \epsilon)||).$$

However

$$||\dot{\varphi}_1(t, u(t), \epsilon)||^2 = \dot{u}_1^2(t) + \dot{u}_2^2(t) \cos^2 \epsilon - 2\dot{u}_2(t)\dot{u}_3(t) \cos \epsilon \sin \epsilon + \dot{u}_3^2(t) \sin^2 \epsilon$$
$$+ \dot{u}_2^2(t) \sin^2 \epsilon + 2\dot{u}_2(t) \dot{u}_3(t) \cos \epsilon \sin \epsilon + \dot{u}_3^2(t) \cos^2 \epsilon$$
$$= ||\dot{u}(t)||^2$$

and

$$\|\varphi_1(t, u(t), \epsilon)\|^2 = u_1^2(t) + u_2^2(t)\cos^2\epsilon - 2u_2(t)u_3(t)\cos\epsilon\sin\epsilon + u_3^2(t)\sin^2\epsilon$$
$$+ u_2^2(t)\sin^2\epsilon + 2u_2(t)\,u_3(t)\cos\epsilon\sin\epsilon + u_3^2(t)\cos^2\epsilon$$
$$= \|u(t)\|^2.$$

Thus we find

$$F(t, \varphi_1(t, u(t), \epsilon), \dot{\varphi}_1(t, u(t), \epsilon)) = F(t, u(t), \dot{u}(t))$$

and we may choose $W = 0$ in Noether's theorem which implies $w(t, x) = 0$ for all t and x. The corresponding first integral is obtained by (27.28) as

$$\langle \tau_1(t, u(t)), F_y(u(t), \dot{u}(t)) \rangle - w(t, u(t)) = \left\langle \begin{pmatrix} 0 \\ -u_3(t) \\ u_2(t) \end{pmatrix}, \begin{pmatrix} m\,\dot{u}_1(t) \\ m\,\dot{u}_2(t) \\ m\,\dot{u}_3(t) \end{pmatrix} \right\rangle$$
$$= m(u_2(t)\,\dot{u}_3(t) - u_3(t)\dot{u}_2(t)).$$

Analogously, we find for $\varphi_2(t, x, \epsilon) = \begin{pmatrix} \cos\epsilon & 0 & -\sin\epsilon \\ 0 & 1 & 0 \\ \sin\epsilon & 0 & \cos\epsilon \end{pmatrix} \begin{pmatrix} x_1 \\ x_2 \\ x_3 \end{pmatrix}$ the first integral

$$\langle \tau_2(t, u(t)), F_y(u(t), u'(t)) \rangle - w(t, x) = m(u_3(t)\,\dot{u}_1(t) - u_1(t)\dot{u}_3(t)),$$

and for $\varphi_3(t, x, \epsilon) = \begin{pmatrix} \cos\epsilon & -\sin\epsilon & 0 \\ \sin\epsilon & \cos\epsilon & 0 \\ 0 & 0 & 1 \end{pmatrix} \begin{pmatrix} x_1 \\ x_2 \\ x_3 \end{pmatrix}$ we get as a first integral

$$\langle \tau_3(t, u(t)), F_y(u(t), u'(t)) \rangle - w(t, x) = m(u_1(t)\dot{u}_2(t) - u_2(t)\dot{u}_1(t)).$$

With $q(t) = \begin{pmatrix} u_1(t) \\ u_2(t) \\ u_3(t) \end{pmatrix}$ and $p(t) = m \begin{pmatrix} \dot{u}_1(t) \\ \dot{u}_2(t) \\ \dot{u}_3(t) \end{pmatrix}$ we conclude that

$$A(t) := q(t) \times p(t)$$

is a vector-valued function where each of its components is a first integral and A corresponds, of course, to the **angular momentum** in mechanics. Thus we have deduced from Noether's theorem that for a classical particle (point mass) with mass m moving in a rotational invariant potential the corresponding angular momentum provides three first integrals.

477

Our derivation of the Euler-Lagrange equations depends on additional regularity properties of a local minimiser and the Lagrange function. The following approach reduces these assumptions, however as a consequence we will have to be careful when interpreting the equation. We start with the important **Lemma of du Bois-Reymond**, the proof of which is taken from [80].

Lemma 27.9. *Let $u \in C([a, b]; \mathbb{R})$ and assume that*

$$\int_a^b u(t)\, \varphi'(t)\, \mathrm{d}t = 0 \tag{27.31}$$

holds for all $\varphi \in C^1([a, b]; \mathbb{R})$ such that $\varphi(a) = \varphi(b) = 0$. Then the function u is a constant.

Proof. We denote by $m := \frac{1}{b-a} \int_a^b u(t)\, \mathrm{d}t$ the mean-value of u and define

$$\varphi(t) = \int_a^t (u(s) - m)\, \mathrm{d}s.$$

The function φ is a C^1-function with $\varphi'(t) = u(t) - m$ and further we have $\varphi(a) = 0$ and $\varphi(b) = 0$ since

$$\int_a^b (u(s) - m)\, \mathrm{d}s = 0. \tag{27.32}$$

It follows that

$$0 = \int_a^b u(t)\, \varphi'(t)\, \mathrm{d}t = \int_a^b u(t)\, (u(t) - m)\, \mathrm{d}t. \tag{27.33}$$

Combining (27.32) and (27.33) yields

$$\int_a^b u(t)\, (u(t) - m)\, \mathrm{d}t - m \int_a^b (u(t) - m)\, \mathrm{d}t = 0$$

or

$$\int_a^b (u(t) - m)^2\, \mathrm{d}t = 0. \tag{27.34}$$

Since u is continuous (27.34) implies $u(t) = m$ for all $t \in [a, b]$, i.e. u is a constant. $\square$

Corollary 27.10. *Let $u \in C([a,b]; \mathbb{R}^n)$ and assume that*

$$\int_a^b \langle u(t), \varphi'(t) \rangle \, \mathrm{d}t = 0 \qquad (27.35)$$

for all $\varphi \in C^1([a,b]; \mathbb{R}^n)$ such that $\varphi(a) = \varphi(b) = 0 \in \mathbb{R}^n$. Then there exists $m \in \mathbb{R}^n$ such that $u(t) = m$ for all $t \in [a,b]$.

Proof. We choose $\varphi_j(x) = \psi(x) \, e_j$ where $\psi \in C^1([a,b]; \mathbb{R})$ with $\psi(a) = \psi(b) = 0$ and $\{e_1, \ldots, e_n\}$ is the canonical basis in $\mathbb{R}^n$. With this choice (27.35) becomes

$$\int_a^b u_j(t) \, \psi(t) \, \mathrm{d}t = 0$$

for all $\psi \in C^1([a,b]; \mathbb{R})$ with $\psi(a) = \psi(b) = 0$ and Lemma 27.9 gives the result. $\square$

Now we can prove

Theorem 27.11. *Let $F \in C^1([a,b] \times \mathbb{R}^n \times \mathbb{R}^n; \mathbb{R})$, $(t,x,y) \mapsto F(t,x,y)$, and assume that $u \in C^1([a,b]; \mathbb{R}^n)$ satisfies the condition $(\delta F)(u; \varphi) = 0$ for all $\varphi \in C^1([a,b]; \mathbb{R}^n)$ such that $\varphi(a) = \varphi(b) = 0$, i.e. we have*

$$\int_a^b \left(\langle F_x(t, u(t), u'(t)), \varphi(t) \rangle + \langle F_y(t, u(t), u'(t)), \varphi'(t) \rangle \right) \mathrm{d}t = 0 \qquad (27.36)$$

for all $\varphi \in C^1([a,b]; \mathbb{R}^n)$ with $\varphi(a) = \varphi(b) = 0$. Then there exists $m \in \mathbb{R}^n$ such that for all $t \in [a,b]$ we have

$$F_y(t, u(t), u'(t)) = m + \int_a^t F_x(s, u(s), u'(s)) \, \mathrm{d}s. \qquad (27.37)$$

Moreover the equation

$$\frac{\mathrm{d}}{\mathrm{d}t} F_y(t, u(t), u'(t)) - F_x(t, u(t), u'(t)) = 0 \qquad (27.38)$$

holds for all $t \in [a,b]$.

Remark 27.12. We must be careful when reading (27.38). Since u is just a C^1-function and not a C^2-function, and since F_y is just continuous and not differentiable with respect to x or y, in general, we cannot use the chain rule on the left hand side of (27.38) to derive (27.2).

Proof of Theorem 27.11. For $\varphi \in C^1([a,b]; \mathbb{R}^n)$, $\varphi(a) = \varphi(b) = 0$, we have

$$\int_a^b \langle F_x(t, u(t), u'(t)), \varphi(t) \rangle \, dt = \int_a^b \langle \left(\frac{d}{dt} \left(\int_a^t F_x(s, u(s), u'(s)) \, ds \right) \right), \varphi(t) \rangle \, dt$$

$$= -\int_a^b \langle \int_a^t F_x(s, u(s), u'(s)) \, ds, \, \varphi'(t) \rangle \, dt$$

and now (27.36) implies

$$\int_a^b \langle (F_y(t, u(t), u'(t)) - \int_a^t F_x(s, u(s), u'(s)) \, ds), \varphi'(t) \rangle \, dt = 0.$$

From Corollary 27.10 we may deduce with a suitable $m \in \mathbb{R}^n$ that

$$F_y(t, u(t), u'(t)) = m + \int_a^t F_x(s, u(s), u'(s)) \, ds. \tag{27.39}$$

Since the right hand side of (27.39) is a differentiable function in t, the left hand side is also a differentiable function in t and we find

$$\frac{d}{dt} F_y(t, u(t), u'(t)) = F_x(t, u(t), u'(t)),$$

i.e. (27.38) is proved. □

Remark 27.13. It is possible to prove the conclusion of Lemma 27.9 if (27.31) holds only for all $\varphi \in C_0^1((a,b); \mathbb{R})$, and this implies that Corollary 27.10 and Theorem 27.11 still hold when $\varphi \in C^1([a,b]; \mathbb{R}^n)$, $\varphi(a) = \varphi(b) = 0$, is replaced by $\varphi \in C_0^1([a,b]; \mathbb{R}^n)$, see Problem 8.

Problems

1. Prove that in the class $D := \{u \in C^1([0,1]) | u(0) = u(1) = 0\}$ the minimising problem

$$\inf \left\{ \int_0^1 e^{-u'^2(t)} \, dt \, | \, u \in D \right\}$$

has no solution.

2. Prove that $u(t) := \frac{\sinh t}{\sinh 1}$ is the unique solution to the Euler-Lagrange equation corresponding to the problem

$$\inf\left\{\int_0^1 (v^2(t) + \dot{v}(t))\,dt \mid v \in C^1([0,1]), v(0) = 0, v(1) = 1\right\}.$$

3. Prove that the surface area of a surface of revolution S is given by

$$A(S) = \mathcal{F}(u) = 2\pi\int_a^b u(t)\sqrt{1 + \dot{u}^2(t)}\,dt,$$

where S is the trace of $S(t,\varphi) = \begin{pmatrix} u(t)\cos\varphi \\ u(t)\sin\varphi \\ t \end{pmatrix}$, $u(t) \geq 0$, $t \in [a,b]$ and

$\varphi \in [0, 2\pi]$, $u(a) = A$, $u(b) = B$. (Compare with Example II.24.8.)

4. Prove that for a C^2-function $F : \mathbb{R} \to \mathbb{R}$ extremals of $\int_a^b F(\dot{u}(t))\,dt$ must be straight lines.

5. Consider the sphere S^2 as the trace of the parametric surface

$$g : [0,\pi] \times [0, 2\pi] \to \mathbb{R}^3, \ g(\vartheta,\varphi) = \begin{pmatrix} \sin\vartheta\cos\varphi \\ \sin\vartheta\sin\varphi \\ \cos\vartheta \end{pmatrix}$$

Suppose now that a curve γ is given on S^2 with the help of the C^2-function $\varphi : [\vartheta_1, \vartheta_2] \to [0, 2\pi]$ as $\gamma : [\vartheta_1, \vartheta_2] \to S^2$,

$$\gamma(\vartheta) = \begin{pmatrix} \sin\vartheta\cos\varphi(\vartheta) \\ \sin\vartheta\sin\varphi(\vartheta) \\ \cos\vartheta \end{pmatrix}.$$

In Volume VII, we will prove that the length of γ is

$$l(\gamma) = \int_{\vartheta_1}^{\vartheta_2} \sqrt{1 + \dot{\varphi}^2(\vartheta)\sin^2\vartheta}\,d\vartheta.$$

Consider the variational problem: among all C^2-curves on the sphere joining $A = \gamma(\vartheta_1)$ and $B = \gamma(\vartheta_2)$ find the one with minimal length. Use Example 27.4.A to find a first integral for this problem. Deduce that meridians on S^2 are extremals. Recall that a curve on S^2 is called a meridian if it is of the type $\vartheta \mapsto g(\vartheta, \varphi_0)$.

481

6. For $F(x,y) = \frac{1}{p}\|y\|^p - \frac{1}{p}\|x\|^p$, $F : \mathbb{R}^n \times \mathbb{R}^n \to \mathbb{R}$, $p \geq 2$, find the energy. For $n = 2$, $C > \frac{1}{p}$ and $\alpha = \left(\frac{pC-1}{p-1}\right)^{\frac{1}{p}}$ verify that for $u(t) = \begin{pmatrix} \cos \alpha t \\ \sin \alpha t \end{pmatrix}$ we have $E(u(t), u'(t)) = C$.

7. Use Noether's theorem to show that by

$$G(u(t), \dot{u}(t)) = (u_1(t)\dot{u}_3(t) - u_3(t)\dot{u}_1(t))\|\dot{u}(t)\|^2$$

a first integral is given for the Lagrange function $F : \mathbb{R}^3 \times \mathbb{R}^3 \to \mathbb{R}$, $F(x,y) = \frac{1}{4}\|y\|^4 - \frac{1}{4}\|x\|^4$.

Hint: consider the transformation $\varphi(t, x, \epsilon) = \begin{pmatrix} \cos \epsilon & 0 & -\sin \epsilon \\ 0 & 1 & 0 \\ \sin \epsilon & 0 & \cos \epsilon \end{pmatrix} \begin{pmatrix} x_1 \\ x_2 \\ x_3 \end{pmatrix}$.

8. Prove the statement of Lemma 27.9 if φ belongs to the set $C_0^1((a,b); \mathbb{R})$.

9. Let $u \in C^1([a,b])$ and assume that for all $\varphi \in C^2([a,b])$ such that $\varphi(a) = \varphi(b) = 0$ and $\varphi'(a) = \varphi'(b) = 0$ it follows that $\int_a^b u(x)\varphi''(x)\,dx = 0$. Prove that for two constants $c_0, c_1 \in \mathbb{R}$ we have $u(x) = c_0 x + c_1$.

28 More on Local Minimisers

We want to find further necessary, and maybe sufficient, criteria for a minimum of $\mathcal{F}(u)$ to exist. For this we take up the considerations of Chapter 26 and discuss

$$\mathcal{F}(u + \eta v) = \int_a^b F(t, u(t) + \eta \, v(t), u'(t) + \eta \, v'(t)) \, \mathrm{d}t \tag{28.1}$$

in more detail. We assume now that $F : [a, b] \times \mathbb{R}^n \times \mathbb{R}^n \to \mathbb{R}$ is a C^3-function, $\eta \in [-\eta_0, \eta_0]$, $\eta_0 > 0$, $u \in D_{A,B}(\mathcal{F})$ and $v \in C_0^1((a, b); \mathbb{R}^n)$. For a local minimiser u_0 we can consider the function $g : [-\eta_0, \eta_0] \to \mathbb{R}$ defined for $v \in C_0^1((a, b); \mathbb{R}^n)$ by

$$g(\eta) = \mathcal{F}(u_0 + \eta \, v). \tag{28.2}$$

Our regularity assumptions imply that g is a C^3-function and since g has a local minimum at u_0 we find

$$g'(0) = 0 \tag{28.3}$$

and

$$g''(0) \geq 0. \tag{28.4}$$

Of course, (28.3) is nothing but the known statement that the first variation $(\delta \mathcal{F})(u_0; v) = 0$ for all $v \in C_0^2((a, b); \mathbb{R}^n)$. Now we calculate $g''(\eta)$ and as before we write $\frac{\partial}{\partial x_j} F = F_{x_j}$, $\frac{\partial}{\partial y_j} F = F_{y_j}$, etc, where $F = F(t, x, y)$. It follows that

$$\frac{\mathrm{d}^2}{\mathrm{d}\eta^2} g(\eta) = \frac{\mathrm{d}^2}{\mathrm{d}\eta^2} \int_a^b F(t, u_0(t) + \eta \, v(t), u_0'(t) + \eta \, v'(t)) \, \mathrm{d}t$$

$$= \int_a^b \frac{\mathrm{d}^2}{\mathrm{d}\eta^2} F(t, u_0(t) + \eta \, v(t), u_0'(t) + \eta \, v'(t)) \, \mathrm{d}t.$$

From Chapter 26 we know that

$$\frac{\mathrm{d}}{\mathrm{d}\eta} F(t, u_0(t) + \eta \, v(t), u_0'(t) + \eta \, v'(t))$$

$$= \sum_{j=1}^n \frac{\partial F}{\partial x_j}(t, u_0(t) + \eta \, v(t), u_0'(t) + \eta \, v'(t)) \, v_j(t)$$

$$+ \sum_{j=1}^n \frac{\partial F}{\partial y_j}(t, u_0(t) + \eta \, v(t), u_0'(t) + \eta \, v'(t)) \, v_j'(t)$$

483

which yields

$$\frac{\mathrm{d}^2}{\mathrm{d}\eta^2} F(t, u_0(t) + \eta\, v(t), u_0'(t) + \eta\, v'(t))$$

$$= \sum_{k=1}^{n} \sum_{j=1}^{n} \frac{\partial^2}{\partial x_k\, \partial x_j} F(t, u_0(t) + \eta\, v(t), u_0'(t) + \eta\, v'(t))\, v_j(t)\, v_k(t)$$

$$+ \sum_{k=1}^{n} \sum_{j=1}^{n} \frac{\partial^2}{\partial y_k\, \partial x_j} F(t, u_0(t) + \eta\, v(t), u_0'(t) + \eta\, v'(t))\, v_j(t)\, v_k'(t)$$

$$+ \sum_{k=1}^{n} \sum_{j=1}^{n} \frac{\partial^2}{\partial x_k\, \partial y_j} F(t, u_j(t) + \eta\, v(t), u_0'(t) + \eta\, v'(t))\, v_j'(t)\, v_k(t)$$

$$+ \sum_{k-1}^{n} \sum_{j=1}^{n} \frac{\partial^2}{\partial y_k\, \partial y_j} F(t, u_0(t) + \eta\, v(t), u_0'(t) + \eta\, v'(t))\, v_j'(t)\, v_k'(t).$$

Therefore, we find

$$\frac{\mathrm{d}^2}{\mathrm{d}\eta^2} g(0) = \int_a^b \frac{\mathrm{d}^2}{\mathrm{d}\eta^2} F(t, u_0(t) + \eta\, v(t), u_0'(t) + \eta\, v'(t)) \Big|_{\eta=0} \mathrm{d}t$$

$$= \int_a^b \sum_{k,j=1}^{n} \{ F_{x_j\, x_k}(t, u_0(t), u_0'(t))\, v_j(t)\, v_k(t) + 2 F_{x_j\, y_k}(t, u_0(t), u_0'(t))\, v_j(t)\, v_k'(t)$$

$$+ F_{y_j\, y_k}(t, u_0(t), u_0'(t))\, v_j'(t)\, v_k'(t) \} \, \mathrm{d}t.$$

Definition 28.1. *We call*

$$(\delta^2 \mathcal{F})(u_0; v) := \frac{\mathrm{d}^2}{\mathrm{d}\eta^2} \mathcal{F}(u_0 + \eta v) \Big|_{\eta=0} = \int_a^b \sum_{k,j=1}^{n} \{ F_{x_j\, x_k}(t, u_0(t), u_0'(t))\, v_j(t)\, v_k(t) \qquad (28.5)$$

$$+ 2 F_{x_j\, y_k}(t, u_0(t), u_0'(t))\, v_j(t)\, v_k'(t) + F_{y_j\, y_k}(t, u_0(t), u_0'(t))\, v_j'(t)\, v_k'(t) \} \, \mathrm{d}t$$

*the **second variation** of $\mathcal{F}$ at u_0 in direction of v.*

From our discussion we deduce

Theorem 28.2. *Let $F \in C^3([a,b] \times \mathbb{R}^n \times \mathbb{R}^n)$ and $u_0 \in D_{A,B}(\mathcal{F})$ be a local minimiser of $\mathcal{F}$. Then we have for all $v \in C_0^1((a,b))$*

$$(\delta^2 \mathcal{F})(u_0; v) \geq 0. \qquad (28.6)$$

In a next step we want to derive a first sufficient criterion for a local minimum of $\mathcal{F}(u)$ to exist. According to Theorem I.22.16, if the function $g : (-\eta_0, \eta_0) \to \mathbb{R}$ satisfies $g'(0) = 0$ and $g''(0) > 0$ then g has a local minimum at $\eta = 0$. Making this criterion work in our situation, i.e. in the study of $\eta \mapsto g(\eta) := \mathcal{F}(u + \eta v)$, we need some preparation.

Let $G : [a, b] \times \mathbb{R}^n \times \mathbb{R}^n \to \mathbb{R}$ be a C^3-function, $(t, x, y) \mapsto G(t, x, y)$ and we use, as before, $G_{x_j}, G_{y_k x_l}$ etc. to denote the corresponding partial derivatives. The Taylor expansion of G up to order 2 is given by

$$G(t, x_0 + x, y_0 + y) = G(t, x_0, y_0) + <G_x(t, x_0, y_0), x> + <G_y(t, x_0, y_0), y> \quad (28.7)$$

$$+ \frac{1}{2} \sum_{k,l=1}^n \{G_{x_k x_l}(t, x_0, y_0) x_k x_l + 2 G_{x_k y_l}(t, x_0, y_0) x_k y_l$$

$$+ G_{y_k y_l}(t, x_0, y_0) y_k y_l\} + \tau^G(t, x_0, y_0, x, y),$$

where

$$\lim_{(x,y) \to 0} \frac{\tau^G(t, x_0, y_0, x, y)}{||x||^2 + ||y||^2} = 0. \quad (28.8)$$

Suppose that for some $\kappa > 0$ we have

$$\frac{1}{2} \sum_{k,l=1}^n \{G_{x_k x_l}(t, x_0, y_0) x_k x_l + 2 G_{x_k y_l}(t, x_0, y_0) x_k y_l + G_{y_k y_l}(t, x_0, y_0) y_k y_l)\}$$

$$\geq \kappa(||x||^2 + ||y||^2)$$

where $||x||^2 + ||y||^2 = \sum_{j=1}^n (x_j^2 + y_j^2)$. By (28.8) we can find $\delta > 0$ such that $||x||^2 + ||y||^2 > \delta$ implies

$$|\tau^G(t, x_0, y_0, x, y)| \leq \frac{\kappa}{2}(||x||^2 + ||y||^2).$$

If now $G_x(t, x_0, y_0) = G_y(t, x_0, y_0) = 0$ we can deduce that

$$G(t, x_0 + x, y_0 + y) - G(t, x_0, y_0) \geq \frac{\kappa}{2}(||x||^2 + ||y||^2)$$

and hence G has a local minimum at (t, x_0, y_0). Of course, we just have given a version of the proof of Theorem II.9.14.A.

Now let us consider

$$\mathcal{F}(u) = \int_a^b F(t, u(t), u'(t)) \, dt$$

for $F \in C^3([a,b] \times \mathbb{R}^n \times \mathbb{R}^n)$. In order to simplify the notation in the following calculations we will write $F(t, u, u')$ or $F_{x_j}(t, u, u')$, etc., for $F(t, u(t), u'(t))$ or $F_{x_j}(t, u(t), u'(t))$, etc. For $u \in D_{A,B}(\mathcal{F})$ and $v \in C_0^1((a,b); \mathbb{R}^n)$ we find

$$\mathcal{F}(u + \eta\, v) - \mathcal{F}(u) = \int_a^b \left(F(t, u + \eta\, v, u' + \eta\, v') - F(t, u, u') \right) \mathrm{d}t$$

$$= \int_a^b \left(< \mathrm{grad}_x\, F(t, u, u'), \eta\, v > + < \mathrm{grad}_y\, F(t, u, u'), \eta\, v' > \right) \mathrm{d}t$$

$$+ \frac{1}{2} \int_a^b \sum_{k,l=1}^n \{ F_{x_k\, x_l}(t, u, u')\eta^2\, v_k\, v_l + 2\, F_{x_k\, y_l}(t, u, u')\eta^2\, v_k\, v_l'$$

$$+ F_{y_k\, y_l}(t, u, u')\eta^2\, v_k'\, v_l'\} \, \mathrm{d}t + \int_a^b \tau^F(t, u, u', \eta\, v, \eta\, v') \, \mathrm{d}t$$

$$= (\delta\mathcal{F})(u, v)\, \eta + \frac{1}{2}(\delta^2\mathcal{F})(u; v)\, \eta^2 + \int_a^b \tau^F(t, u, u', \eta\, v, \eta\, v') \, \mathrm{d}t.$$

Let us assume that

$$\frac{1}{2}\sum_{k,l=1}^n \{ F_{x_k\, x_l}(t, u, u')w_k\, w_l + 2\, F_{x_k\, y_l}(t, u, u')w_k\, z_l + F_{y_k\, y_l}(t, u, u')z_k\, z_l \} \quad (28.9)$$

$$\geq \kappa(||w||^2 + ||z||^2)$$

holds for all $w, z \in \mathbb{R}^n$; and some $\kappa > 0$. Of course, (28.9) is the statement that for t fixed $(x, y) \mapsto F(t, x, y)$ has a positive definite Hesse-matrix which is a $2n \times 2n$-matrix. Since we require κ to be independent of t we have a uniform (with respect to t) lower bound for the eigenvalues of the Hesse-matrix. For the reminder τ^F we know that for u and v fixed we can find $\tilde{\eta} > 0$, $0 < \tilde{\eta} < \tilde{\eta}_0$, such that for $|\eta| < \tilde{\eta}$ it follows that

$$|\tau^F(t, u, u', \eta\, v, \eta\, v')| \leq \frac{\kappa}{2}\eta^2(||v||^2 + ||v'||^2). \quad (28.10)$$

Finally, we can prove

Theorem 28.3. *Let $F \in C^3([a,b] \times \mathbb{R}^n \times \mathbb{R}^n)$ and $u_0 \in D_{A,B}(\mathcal{F})$. Assume that for $F(t, u_0, u_0')$ the estimate (28.9) holds for all $w, z \in \mathbb{R}^n$. Moreover assume $(\delta\mathcal{F})(u_0; v) = 0$ for all $v \in C_0^1((a,b); \mathbb{R}^n)$. Then u_0 is a local minimiser of $\mathcal{F}$.*

Proof. From the previous calculation we deduce for $\eta_0 > 0$ sufficiently small that

$$\mathcal{F}(u_0 + \eta v) - \mathcal{F}(u_0) = (\delta \mathcal{F})(u_0; v)\eta + \frac{1}{2}(\delta^2 \mathcal{F})(u_0; v)\eta^2 + \int_a^b \tau^F(t, u_0, u_0', \eta v, \eta v')\,dt$$

$$\geq \kappa \int_a^b (||v(t)||^2 + ||v'(t)||^2)\,dt\,\eta^2 - \frac{\kappa}{2}\int_a^b (||v(t)||^2 + ||v'(t)||^2)\,dt\,\eta^2$$

$$= \frac{\kappa}{2}\eta^2 \int_a^b (||v(t)||^2 + ||v'(t)||^2)\,dt > 0.$$

$\square$

In the case where F depends only on y the conditions of Theorem 28.3 become more simple, we only need the condition that $(F_{y_k y_l})_{k,l=1,\ldots n}$ is positive definite.

Example 28.4. For $F : \mathbb{R} \to \mathbb{R}, F(y) = \sqrt{1+y^2}$, we have $F_{yy}(y) = \frac{1}{(1+y^2)^{\frac{3}{2}}} > 0$

for all $y \in \mathbb{R}$. If $u_0 \in D_{A,B}(\mathcal{F})$ is a critical point for $\mathcal{F}(u) = \int_a^b \sqrt{1 + u'^2(t)}\,dt$, i.e. $(\delta \mathcal{F})(u_0; v) = 0$ for all $v \in C_0^1((a,b); \mathbb{R}^n)$, then u_0 is already a local minimiser of $\mathcal{F}(u)$.

We want to study for $F \in C^3([a,b] \times \mathbb{R}^n \times \mathbb{R}^n)$ and $u \in D_{A,B}(\mathcal{F})$ fixed the term

$$H(u; t, w, z) := \sum_{k,l=1}^n \{F_{x_k x_l}(t, u, u')w_k w_l + 2F_{x_k y_l}(t, u, u')w_k z_l + F_{y_k y_l}(t, u, u')z_k z_l\}$$

(28.11)

which is for $u \in D_{A,B}(\mathcal{F})$ and $t \in [a,b]$ fixed a bilinear form in $(w, z) \in \mathbb{R}^n \times \mathbb{R}^n$. With H we can associate the variational problem

$$\inf \int_a^b H(u; t, v, v')\,dt$$

(28.12)

$$= \inf \int_a^b \sum_{k,l=1}^n \{F_{x_k x_l}(t, u(t), u'(t))\,v_k(t)v_l(t) + 2\,F_{x_k y_l}(t, u(t), u'(t))v_k(t)\,v_l'(t)$$

$$+ F_{y_k y_l}(t, u(t), u'(t))v_k'(t)\,v_l'(t)\}\,dt$$

where the infimum is taken over all $v \in C_0^1((a,b); \mathbb{R}^n)$. If $u_0 \in D_{A,B}(\mathcal{F})$ is a local minimiser of $\mathcal{F}(u)$ then we know by Theorem 28.2 that

$$\inf \int_a^b H(u_0; t, v, v') \, dt \geq 0$$

and therefore $v(t) = 0 \in \mathbb{R}^n$ for all $t \in [a,b]$ is a minimiser of (28.12). A natural question is whether (28.12) admits a further minimiser?

We believe that it is more convenient for the reader to first discuss in detail the case $n = 1$ and then we will indicate the results for the general case. While in the general case we will follow [71], the one-dimensional case uses mainly [111] and [44].

Let $F : [a,b] \times \mathbb{R} \times \mathbb{R} \to \mathbb{R}$, $(t, x, y) \mapsto F(t, x, y)$, be a C^3-function and consider on $D_{A,B}(\mathcal{F})$

$$\mathcal{F}(u) := \int_a^b F(t, u(t), u'(t)) \, dt. \tag{28.13}$$

It follows for $v \in C_0^1((a,b); \mathbb{R})$ and $\eta \in (-\eta_0, \eta_0)$ that

$$\mathcal{F}(u + \eta v) - \mathcal{F}(u) = \eta \int_a^b \left(F_x(t, u, u')v + F_y(t, u, u')v' \right) dt$$

$$+ \frac{\eta^2}{2} \int_a^b \left(F_{xx}(t, u, u')v^2 + 2F_{xy}(t, u, u')vv' + F_{yy}(t, u, u')v'^2 \right) dt$$

$$+ \int_a^b \tau^F(t, u, u', \eta v, \eta v') \, dt.$$

We note that

$$2F_{xy}(t, u, u')vv' = F_{xy}(t, u, u') \frac{d}{dt} v^2$$

and therefore an integration by parts yields

$$\int_a^b 2F_{xy}(t, u, u')vv' \, dt = \int_a^b F_{xy}(t, u, u') \left(\frac{d}{dt} v^2 \right) dt$$

$$= - \int_a^b \left(\frac{d}{dt} F_{xy}(t, u, u') \right) v^2 \, dt$$

and we find

$$\frac{\eta^2}{2} \int_a^b \left(F_{xx}(t, u, u')v^2 + 2F_{xy}(t, u, u')vv' + F_{yy}(t, u, u')v'^2 \right) dt$$

$$= \frac{\eta^2}{2} \int_a^b \left(\left(F_{xx}(t, u, u') - \frac{d}{dt} F_{xy}(t, u, u') \right) v^2 + F_{yy}(t, u, u')v'^2 \right) dt.$$

488

With

$$S(t, u, u') = F_{xx}(t, u, u') - \frac{\mathrm{d}}{\mathrm{d}t} F_{xy}(t, u, u') \quad \text{and} \quad R(t, u, u') = F_{yy}(t, u, u') \quad (28.14)$$

we find

$$\frac{\eta^2}{2} \int_a^b \left(F_{xx}(t, u, u') v^2 + 2 F_{xy}(t, u, u') v v' + F_{yy}(t, u, u') v'^2 \right) \mathrm{d}t \qquad (28.15)$$

$$= \frac{\eta^2}{2} \int_a^b \left(S(t, u, u') v^2 + R(t, u, u') v'^2 \right) \mathrm{d}t$$

for all $v \in C_0^1((a, b); \mathbb{R})$. Now we choose u_0 to be a local minimum of $\mathcal{F}(u)$ and by Theorem 28.2 we must have $(\delta \mathcal{F})(u_0; v) = 0$ and $(\delta^2 \mathcal{F})(u_0; v) \geq 0$.
To proceed further we need

Lemma 28.5. *Let $h, g : [a, b] \to \mathbb{R}$ be two continuous functions and assume for all $v \in C_0^1((a, b))$ that*

$$\int_a^b \left(g(t) v^2(t) + h(t) v'^2(t) \right) \mathrm{d}t \geq 0. \qquad (28.16)$$

Then we have $h(t) \geq 0$ for all $t \in [a, b]$.

Proof. Suppose that for some $t_0 \in [a, b]$ we have $h(t_0) < 0$. The continuity of h then implies that $h(t) < 0$ for all t in a neighbourhood of t_0. Hence we may assume $t_0 \in (a, b)$ and that for some $\epsilon > 0$ we have $h(t) < -h_0$, $h_0 > 0$, for $t \in (-\epsilon + t_0, t_0 + \epsilon) \subset (a, b)$. The function

$$w(t) := \begin{cases} \sin^2 \frac{\pi(t - t_0)}{\epsilon}, & t \in (-\epsilon + t_0, t_0 + \epsilon) \\ 0, & t \in [a, b] \setminus (-\epsilon + t_0, t_0 + \epsilon) \end{cases}$$

is continuous on $[a, b]$ and has compact support $[-\epsilon + t_0, t_0 + \epsilon] \subset (a, b)$. Since g is continuous on $[a, b]$ we can find $M > 0$ such that $|g(t)| \leq M$ for all $t \in [a, b]$ and it follows that

$$\int_a^b \left(g(t) w^2(t) + h(t) w'^2(t) \right) \mathrm{d}t$$

$$= \int_{t_0 - \epsilon}^{t_0 + \epsilon} g(t) \sin^4 \left(\frac{\pi(t - t_0)}{\epsilon} \right) \mathrm{d}t + \int_{t_0 - \epsilon}^{t_0 + \epsilon} h(t) \frac{\pi^2}{\epsilon^2} \left(2 \cos \frac{\pi(t - t_0)}{\epsilon} \sin \frac{\pi(t - t_0)}{\epsilon} \right)^2 \mathrm{d}t$$

$$\leq 2 \epsilon M + \int_{t_0 - \epsilon}^{t_0 + \epsilon} h(t) \frac{\pi^2}{\epsilon^2} \sin^2 \frac{2 \pi(t - t_0)}{\epsilon} \mathrm{d}t$$

$$< 2 \epsilon M - \frac{2 h_0 \pi^2}{\epsilon}.$$

For $\epsilon < \sqrt{\frac{h_0}{M}}\pi$ we obtain

$$\int_a^b \left(g(t)w^2(t) + h(t)w'^2(t) \right) dt < 0,$$

which is a contradiction. $\qquad\square$

From Lemma 28.5 we deduce the **necessary condition of Legendre** for a local minimiser of $\mathcal{F}$.

Theorem 28.6. *Let $F : [a,b] \times \mathbb{R} \times \mathbb{R} \to \mathbb{R}$, $(t, x, y) \mapsto F(t, x, y)$, be a C^3-function and $u_0 \in D_{A,B}(\mathcal{F})$ be a local minimiser of $\mathcal{F}(u)$. Then we must have*

$$F_{yy}(t, u_0(t), u_0'(t)) \geq 0. \tag{28.17}$$

Proof. We know that for $v \in C_0^1((a,b); \mathbb{R})$ we must have

$$\int_a^b \left(S(t, u_0(t), u_0'(t))v^2(t) + R(t, u_0(t), u_0'(t))v'^2(t) \right) dt \geq 0$$

and by Lemma 28.5 this implies

$$R(t, u_0(t), u_0'(t)) = F_{yy}(t, u_0(t), u_0'(t)) \geq 0.$$

$\qquad\square$

Our next step is to discuss whether the assumption $F_{yy}(t, u_0, u_0') > 0$ leads to a sufficient condition for a local minimum of $\mathcal{F}$. For $u_0 \in D_{A,B}(\mathcal{F})$ a local minimiser of $\mathcal{F}(u)$, we consider the variational integral with respect to $v \in C^1((a,b); \mathbb{R})$ given by

$$\int_a^b \left(S(t, u_0, u_0')v^2(t) + R(t, u_0, u_0')v'^2(t) \right) dt \tag{28.18}$$

under the boundary conditions $v(a) = v(b) = 0$. With (28.18) we can associate the bilinear form

$$\mathcal{B}(v, w) := \int_a^b \left(S(t, u_0, u_0')v(t)w(t) + R(t, u_0, u_0')v'(t)w'(t) \right) dt \tag{28.19}$$

and we may ask when the corresponding quadratic form

$$\mathcal{B}(v) := \mathcal{B}(v, v) := \int_a^b \left(S(t, u_0, u_0')v^2(t) + R(t, u_0, u_0')v'^2(t) \right) dt \tag{28.20}$$

490

is positive definite (for u_0 fixed)?

By Theorem 28.6 we know that $R(t, u_0, u_0') \geq 0$ for $t \in [a, b]$. We now assume the stronger condition

$$R(t, u_0, u_0') > 0 \quad \text{for all} \quad t \in [a, b]. \tag{28.21}$$

The Euler-Lagrange equation for the minimising problem $\inf \mathcal{B}(v)$ is

$$-\frac{\mathrm{d}}{\mathrm{d}t} \left(R\left(t, u_0(t), u_0'(t)\right) v'(t) \right) + S\left(t, u_0(t), u_0'(t)\right) v(t) = 0 \tag{28.22}$$

which is a linear second order equation which we consider under the boundary conditions $v(a) = 0$ and $v(b) = 0$.

Definition 28.7. *We call* $c \in (a, b)$ *a **conjugate point** to a if there exists a solution v to (28.22) not identically zero which vanishes at a and at c, i.e. $v(a) = v(c) = 0$.*

Remark 28.8. A. For a solution v to (28.22) with $v(a) = 0$ and $v(d) = 0$ for some $d \in (a, b]$ the function $w(t) = \gamma v(t)$, $\gamma \in \mathbb{R}$, is a further solution to (28.22) under the same boundary value conditions at a and d and therefore it is helpful to add the normalisation $v'(a) = 1$.
B. Some authors call c in Definition 28.7 a conjugate parameter value and $(c, v(c))$ the conjugate point.

Theorem 28.9. *Suppose that (28.21) holds and that $[a, b]$ does not contain a conjugate point to a. Then the bilinear form $\mathcal{B}$ is positive definite on $\{v \in C^1([a, b]; \mathbb{R}) \mid v(a) = v(b) = 0\}$.*

Proof. Since $R(t, u_0(t), u_0'(t)) > 0$ for all $t \in [a, b]$ we try to represent $\mathcal{B}(v)$ in the form

$$\mathcal{B}(v) = \int_a^b R(t, u_0(t), u_0'(t)) g^2(v(t), v'(t)) \, \mathrm{d}t \tag{28.23}$$

where $g^2(v(t), v'(t)) = 0$ for all $t \in [a, b]$, $v \in \{C^1([a, b]; \mathbb{R}) \mid v(a) = v(b) = 0\}$ if and only if $v(t) = 0$ for all $t \in [a, b]$.

We observe that with u_0 fixed it follows for every $w \in C^1([a, b]; \mathbb{R})$ that

$$\int_a^b \left(R(t, u_0, u_0') v'^2(t) + S(t, u_0, u_0') v^2(t) \right) \mathrm{d}t$$

$$= \int_a^b \left(R(t, u_0, u_0') v'^2(t) + S(t, u_0, u_0') v^2(t) + \frac{\mathrm{d}}{\mathrm{d}t} \left(w(t) v^2(t) \right) \right) \mathrm{d}t.$$

Suppose that w satisfies

$$R(t, u_0, u_0') \left(S(t, u_0, u_0') + w'(t) \right) = w^2(t) \tag{28.24}$$

491

then we find

$$\int_a^b \left(R(t, u_0, u_0')v'^2(t) + S(t, u_0, u_0')v^2(t) \right) dt = \int_a^b R(t, u_0, u_0') \left(v'(t) + \frac{w(t)}{R(t, u_0, u_0')} v(t) \right)^2 dt.$$

Moreover, if now $\mathcal{B}(v) = 0$ for some $v \in C^1([a, b]; \mathbb{R})$ such that $v(a) = v(b) = 0$ then the fact that $R(t, u_0, u_0') > 0$ implies

$$v'(t) + \frac{w(t)}{R(t, u_0, u_0')} v(t) = 0$$

and since $v(a) = v(b) = 0$ implies $v(t) = 0$ for all $t \in [a, b]$ it follows $\mathcal{B}(v) > 0$, i.e. $\mathcal{B}$ is positive definite.

It remains to prove that if no point conjugate to a exists in $[a, b]$ then we can find a function $w \in C^1([a, b]; \mathbb{R})$ which solves (28.24). This is, however, a Riccati type differential equation, see Example 15.8, and with the change

$$w(t) := \frac{-z'(t)}{z(t)} R(t, u_0(t), u_0'(t)) \tag{28.25}$$

it becomes the linear differential equation for the function $z(t)$:

$$-\frac{d}{dt} \left(R(t, u_0, u_0')z'(t) \right) + S(t, u_0, u_0')z(t) = 0, \tag{28.26}$$

which is the Euler-Lagrange equation associated with $\mathcal{B}$ given in (28.20). Now our assumption that no point conjugate to a exists in $[a, b]$ implies that (28.26) has a non-vanishing solution in $[a, b]$ which implies the existence of a solution to (28.24) defined on $[a, b]$, and the theorem is proved. $\qquad \square$

The following results state a type of converse to Theorem 28.9 and they are proved in detail in [43].

Theorem 28.10. *Suppose that $R(t, u_0, u_0') > 0$ and that $\mathcal{B}(\cdot)$ is positive definite on $\{v \in C^1([a, b]; \mathbb{R}) \mid v(a) = v(b) = 0\}$. Then there is no point in $[a, b]$ conjugate to a.*

Theorem 28.11. *Suppose that $R(t, u_0, u_0') > 0$ and $\mathcal{B}(v) \geq 0$ for all $v \in C^1([a, b]; \mathbb{R})$ such that $v(a) = v(b) = 0$. Then (a, b) contains no point conjugate to a.*

Definition 28.12. *Let u_0 be a local minimiser of $\mathcal{F}$ and R and S be defined as above. The equation*

$$-\frac{d}{dt} \left(R(t, u_0(t), u_0'(t))v'(t) \right) + S(t, u_0(t), u_0'(t))v(t) = 0 \tag{28.27}$$

*is called the **Jacobi equation** corresponding to $\mathcal{F}$.*

Now we can deduce

Theorem 28.13. *Let $F \in C^3([a,b] \times \mathbb{R} \times \mathbb{R}; \mathbb{R})$ and u_0 an extremal to $\mathcal{F}$, i.e. a solution of the corresponding Euler-Lagrange equation. Suppose that we have $R(t, u_0(t), u_0'(t)) > 0$ for all $t \in [a,b]$ and the Jacobi-equation (28.27) has a solution $v : [a,b] \to \mathbb{R}$ satisfying $v(t) \neq 0$ for all $a \leq t \leq b$ then u_0 is a minimiser of $\mathcal{F}$.*

We need some clarification of the notion of a local minimiser in order to be compatible with the existing literature. When speaking about a **local minimiser** we mean a minimiser in the sense of Definition 26.4.A. Some authors call such a minimiser a **weak minimiser**, see, for example, [43] or [80], and this seems to be a traditional name to make a distinction from a **strong minimiser** for which the norm in Definition 26.4.A is replaced by

$$\max_{1 \leq j \leq n} \sup_{t \in [a,b]} |u_{0j}(t) - u_j(t)| + \max_{1 \leq j \leq n} \sup_{t \in [a,b]} |u_{0j}'(t) - u_j'(t)|.$$

Unfortunately, this classical terminology can be confused with notions arising in the direct methods of the calculus of variations. In this context, a function $u \in C^1([a,b]; \mathbb{R}^n)$ (or in some Sobolev space) is called a **weak extremal** of $\mathcal{F}$ if

$$\int_a^b \left(\langle F_x(t, u(t), u'(t)), v(t) \rangle + \langle F_y(t, u(t), u'(t)), v'(t) \rangle \right) \mathrm{d}t = 0$$

holds for all $\varphi \in C_0^\infty((a,b); \mathbb{R}^n)$ (or $\varphi \in C_0^1((a,b); \mathbb{R}^n)$), we refer, for example, to [18]. In order to avoid any possible confusion we strictly stay with our definition of a local minimiser in the sense of Definition 26.4.A and the reader using the cited references (or further literature) is advised to be more careful at this point and to check the exact terminology used.

With this remark to different notions of a minimiser we shall consider Theorem 28.13 as our basic sufficiency condition for a local minimiser to exist, see also [80], p. 149.

In order to apply Theorem 28.13 we need to find certain solutions of the Jacobi equation (28.27). However for this we refer to the existence monographs on the calculus of variations, e.g. [18], [43], [71] or [111].

Before turning to examples in the one-dimensional case, we make a short remark to the case $n > 1$. The principal idea remains unchanged, i.e. we first have to study the second variation $\delta^2 \mathcal{F}$ which, however, now leads to a more complicated form $\mathcal{B}$ which we need to investigate. As in the case $n = 1$, $\mathcal{B}$ is a bilinear form, but now for functions $v, w \in C^1([a,b]; \mathbb{R}^n)$ and again the coefficients of $\mathcal{B}$ depend on a given (local) extremal of $\mathcal{F}$. For a detailed discussion (in a more general regularity setting) we refer to [71].

Example 28.14. In Example 26.10 we have considered the Lagrange function $F(t, x, y) = \frac{m}{2}y^2$, $m > 0$, which leads to the Euler-Lagrange equation $mu''(t) = 0$ with the general solution $u(t) = mt + \beta$. The corresponding function R is given by $R = m > 0$ and S is the function $S = 0$. Thus for every extremal the Jacobi equation becomes

$$-\frac{d}{dt}(Rv'(t) + Sv(t)) = -mv''(t) = 0,$$

with general solution $v(t) = -mt + \gamma$ and the Jacobi equation admits an everywhere non-zero solution. Hence the extremal $u_0(t) = \frac{A-B}{t_1-t_2}t + \frac{Bt_1-At_2}{t_1-t_2}$ of the minimising problem $\inf_{u \in D_{A,B}(\mathcal{F})} \overrightarrow{\mathcal{F}}(u)$ where $D_{A,B}(\mathcal{F}) = \{u \in C^1([a,b]; \mathbb{R}) \,|\, u(t_1) = A, u(t_2) = B\}$ is indeed a minimiser.

Closely related to this example is

Example 28.15. We consider the Lagrange function $F(t, x, y) = \frac{y^2}{2} - \frac{x^2}{2}$ as in Example 26.14 and we find as the Euler-Lagrange equation $u''(t) + u(t) = 0$ as well as $R = 1 > 0$ and $S = -1$, which gives the Jacobi equation

$$-\frac{d}{dt}(Rv'(t)) + (-v(t)) = -(v''(t) + v(t)) = 0,$$

or $v''(t) + v(t) = 0$. This equation has the general solution $v(t) = \alpha \cos t + \beta \sin t$. Now the question of a never vanishing solution becomes more delicate. For simplicity, let us consider the problem on the interval $[0, \eta]$, $\eta > 0$. Choosing, for example, $\beta = 0$ and $\alpha > 0$ we have a strictly positive solution v on $[0, \frac{\pi}{2})$, i.e. for every closed interval $[0, \eta]$, $\eta < \frac{\pi}{2}$. In order to have a never vanishing solution we must have $\tan t \neq -\frac{\alpha}{\beta}$, $\beta \neq 0$, or $\cot t \neq -\frac{\beta}{\alpha}$, $\alpha \neq 0$. However, the range of tan and the range of cot is $\mathbb{R}$ in each case. Thus once α and β are chosen we only obtain in certain intervals solutions to the Jacobi equation which are everywhere on these intervals non-zero. This resonates well with our use of the Poincaré inequality in Examples 26.14.

Example 28.16. For the Lagrange function $F(t, x, y) = \sqrt{1 + y^2}$ from Example 26.11 we derive the Euler-Lagrange equation $\frac{u''(t)}{(1+u'^2(t))^{\frac{3}{2}}} = 0$ which admits the general solution $u(t) = \alpha t + \beta$ and for $u(t_1) = A$ and $u(t_2) = B$ we obtain $\alpha = \frac{A-B}{a-b}$. Further we find $R = \frac{1}{(1+y^2)^{\frac{3}{2}}}$ and $S = 0$ which first yields for $u_0(t) = \alpha t + \beta$ that $R = \frac{1}{(1+\alpha^2)^{\frac{3}{2}}}$ and we find the Jacobi equation

$$-\frac{d}{dt}(Rv') + Sv = -\frac{1}{(1+\alpha^2)^{\frac{3}{2}}}v''(t) = 0$$

which has the general solution $v(t) = \gamma t + \delta$. Hence the Jacobi equation has a solution which is nowhere zero implying that extremals are minimisers.

Problems

1. Given the Lagrange function $F : \mathbb{R} \times \mathbb{R} \to \mathbb{R}$, $F(x, y) = \frac{\alpha}{2}y^2 + \frac{\beta}{2}x^2$, $\alpha, \beta > 0$. Prove that if $\frac{\beta}{\alpha}$ is not an eigenvalue λ for the problem $\ddot{w}(t) = \lambda w(t), t \in [a, b]$, $w(a) = w(b) = 0$, then the minimising problem

$$\inf \left\{ \int_a^b \left(\frac{\alpha}{2}\dot{u}^2(t) + \frac{\beta}{2}u^2(t) \right) dt \mid u \in C^1([a, b]), u(a) = A, u(b) = B \right\}$$

has at most one solution in $C^2([a, b])$.

2. Let $F : \mathbb{R}^n \times \mathbb{R}^n \to \mathbb{R}$ be a C^2-Lagrange function of the type $F(x, y) = h(y) + g(x)$. Assume that the Hesse matrices of h and g are positive define. Prove that (28.9) holds.

3. Use Theorem 28.3 to prove that $u_0 : [0, \frac{1}{2}\ln 2] \to \mathbb{R}$ defined by $u(t) = \frac{1}{3}e^{2t} + \frac{2}{3}e^{-2t}$ is a local minimiser of

$$\inf \left\{ \int_0^{\frac{1}{2}\ln 2} \left(\frac{1}{2}\dot{u}^2(t) + 2u^2(t) \right) dt \mid u \in C^1([0, \frac{1}{2}\ln 2]), u(0) = u(\frac{1}{2}\ln 2) = 1 \right\}.$$

4. For which values of $\lambda \in \mathbb{R} \backslash \{0\}$ can we prove the existence of a local minimiser of

$$\inf \left\{ \int_0^1 \left(\frac{1}{2}\dot{u}^2(t) + \frac{\lambda}{2}u^2(t) \right) dt \mid u \in C^1([0, 1]), u(0) = u(1) = 1 \right\}$$

by using Theorem 28.3.

5. Give a simple geometric interpretation of the necessary condition of Legrendre.

6. Consider the functional

$$\int_0^1 \left(\frac{1}{2}\dot{u}^2(t) - \frac{(2m\pi)^2}{2}u^2(t) \right) dt$$

and the corresponding equation (28.22) which turns out to be independent of u_0. Prove that

$$-\frac{d}{dt}(R(t, u_0(t), \dot{u}_0(t))\dot{v}(t)) + S(t, u_0(t), \dot{u}_0(t))v(t) = 0$$

subjected to the conditions $v(0) = v(1) = 0$ has the solution $v(t) = \sin(2m\pi)t$ which has a conjugate point at $t_0 = \frac{1}{2}$.

7. Find the Jacobi equation for the functional $\mathcal{F}(u) := \int_0^1 \dot{u}^3(t)\,dt$, $u \in C^1([0,1])$, $u(0) = 0$ and $u(1) = 1$, and try to apply Theorem 28.13 to the corresponding minimising problem.

   Hint: the solution to Problem 4 of Chapter 27 may be used.

8. Let $F : \mathbb{R} \times \mathbb{R} \to \mathbb{R}$ be a C^3-function and define

$$\mathcal{F}(u) := \int_0^1 F(u(t), \dot{u}(t))\,dt, \quad u(0) = A, u(1) = B.$$

   Now consider the functional $\mathcal{H}(u) = e^{\mathcal{F}(u)} > 0$ and find its second variation $(\delta^2 \mathcal{H})(u, v)$. Deduce that $(\delta^2 \mathcal{F})(u, v) > 0$ implies $(\delta^2 \mathcal{H})(u, v) > 0$.

29 Partial Differential Equations of 1st Order

The calculus of variations is naturally linked to the theory of first order partial differential equations. Historically, the link was established in analytical mechanics or Lagrange-Hamilton-Jacobi mechanics as well as in the discussion of the relation of geometric optics to wave optics. The latter is related to the propagation of singularities of solutions of higher order partial differential equations and this is a topic we will handle in Volume VI. Thus, this chapter will serve us to prepare some further investigations on the calculus of variations and, at the same time, we discuss results which we will employ in Volume VI. In Chapter 30 we will return to the calculus of variations.

The theory of partial differential equations of first order is a rather geometric theory as was already observed by J.F. Pfaff and G. Monge and some ideas are best explained first for linear and quasi-linear equations which we do now before turning to fully non-linear equations.

We start with the equation

$$au_x(x,y) + bu_y(x,y) = 0, \quad a,b \in \mathbb{R}, \quad a^2 + b^2 \neq 0, \tag{29.1}$$

which we consider in $\mathbb{R}^2$. Thus we are seeking a C^1-function $u : \mathbb{R}^2 \to \mathbb{R}$ satisfying (29.1). For $(x,y) \in \mathbb{R}^2$, the vector $(u_x(x,y), u_y(x,y))$ is the gradient of u at the point (x,y) and writing (29.1) as

$$\left\langle \begin{pmatrix} a \\ b \end{pmatrix}, \begin{pmatrix} u_x(x,y) \\ u_y(x,y) \end{pmatrix} \right\rangle = 0 \tag{29.2}$$

we find that the geometric meaning of equation (29.1) is that at every point (x,y) the vector $\begin{pmatrix} a \\ b \end{pmatrix}$ and $\operatorname{grad} u(x,y)$ are orthogonal, i.e. $\begin{pmatrix} a \\ b \end{pmatrix} \perp \begin{pmatrix} u_x(x,y) \\ u_y(x,y) \end{pmatrix}$. Let

$$\gamma(t) = \begin{pmatrix} a \\ b \end{pmatrix} t + \begin{pmatrix} x_0 \\ y_0 \end{pmatrix}, \quad t \in \mathbb{R}, \tag{29.3}$$

be the straight line in direction of $\begin{pmatrix} a \\ b \end{pmatrix}$ passing for $t = 0$ through the point $\begin{pmatrix} x_0 \\ y_0 \end{pmatrix}$, and let

$$\sigma(s) = \begin{pmatrix} u_x(x_0,y_0) \\ u_y(x_0,y_0) \end{pmatrix} s + \begin{pmatrix} x_0 \\ y_0 \end{pmatrix} \tag{29.4}$$

be the straight line in direction of $\begin{pmatrix} u_x(x_0, y_0) \\ u_y(x_0, y_0) \end{pmatrix}$ passing for $s = 0$ through $\begin{pmatrix} x_0 \\ y_0 \end{pmatrix}$.

Note that $\begin{pmatrix} b \\ -a \end{pmatrix} \perp \begin{pmatrix} a \\ b \end{pmatrix}$ and therefore with $\begin{pmatrix} x \\ y \end{pmatrix} = \begin{pmatrix} a \\ b \end{pmatrix} t + \begin{pmatrix} x_0 \\ y_0 \end{pmatrix}$ we find that

$$\left\langle \begin{pmatrix} b \\ -a \end{pmatrix}, \begin{pmatrix} x \\ y \end{pmatrix} \right\rangle = \left\langle \begin{pmatrix} b \\ -a \end{pmatrix}, \begin{pmatrix} x_0 \\ y_0 \end{pmatrix} \right\rangle,$$

i.e. γ has the normal form

$$bx - ay = \delta, \qquad \delta = bx_0 - ay_0$$

and for σ we find

$$ax + by = \eta, \qquad \eta = ax_0 + by_0.$$

Under (29.1) it follows that $\dot{\gamma}(0) \perp \dot{\sigma}(0)$ and a direction orthogonal to $\begin{pmatrix} a \\ b \end{pmatrix}$ is given by $\begin{pmatrix} b \\ -a \end{pmatrix}$ implying that for some $\lambda \in \mathbb{R}\backslash\{0\}$

$$u_x(x_0, y_0) = \lambda b \quad \text{and} \quad u_y(x_0, y_0) = -\lambda a. \tag{29.5}$$

Without loss of generality, we may choose $\lambda = 1$ and we note that for $u(x, y) = f(bx - ay)$, $f \in C^1(\mathbb{R})$, we have

$$u_x(x, y) = \frac{\partial}{\partial x} f(bx - ay) = bf'(bx - ay)$$

and

$$u_y(x, y) = \frac{\partial}{\partial y} f(bx - ay) = -af'(bx - ay)$$

which yields

$$au_x + bu_y = abf' - abf' = 0. \tag{29.6}$$

Consider the change of variable

$$\begin{cases} \xi = bx - ay \\ \eta = ax + by, \end{cases} \tag{29.7}$$

which due to the condition $0 \neq a^2 + b^2 = \det \begin{pmatrix} b & a \\ -a & b \end{pmatrix}$ indeed defines a bijective linear mapping from $\mathbb{R}^2$ onto itself. In these new coordinates we find for

$$u(x, y) = v(\xi(x, y), \eta(x, y))$$

498

that

$$u_x = v_\xi \cdot \xi_x + v_\eta \cdot \eta_x = b v_\xi + a v_\eta$$

$$u_y = v_\xi \cdot \xi_y + v_\eta \cdot \eta_y = -a v_\xi + b v_\eta$$

or

$$0 = a u_x + b u_y = a b v_\xi + a^2 v_\eta - a b v_\xi + b^2 v_\eta = (a^2 + b^2) v_\eta$$

which means

$$v_\eta(\xi, \eta) = 0, \tag{29.8}$$

and therefore we have $v(\xi, \eta) = g(\xi)$ for some function $g : \mathbb{R} \to \mathbb{R}$. Reversing the change of variable we arrive again at

$$u(x, y) = g(bx - ay). \tag{29.9}$$

In summary, the geometric interpretation of (29.1) allows us to determine the change of variable (29.7) and after this change we obtain a much simpler equation, namely (29.8).

The function g in (29.9) is arbitrary, but we assume it to be of class C^1. In order to get, if possible, a unique solution we specify $u(x, 0) = h(x)$ for all $x \in \mathbb{R}$ which yields immediately $h(x) = g(bx)$, or for $b \neq 0$ we find for a given function h that

$$u(x, y) := h\left(x - \frac{a}{b}y\right) = g\left(b\left(x - \frac{a}{b}y\right)\right) = g(bx - ay) \tag{29.10}$$

is a solution of (29.1) with the property that $u(x, 0) = h(x)$. This solution has the property that it is constant on the straight lines $bx - ay = c$ (and these are orthogonal to the straight lines $ax + by = d$).

We still assume $b \neq 0$, otherwise we will have $a \neq 0$ and changing the role of the variables x and y will return us to the case $b \neq 0$. When c varies in $\mathbb{R}$, the lines $bx - ay = c$ cover the whole plane $\mathbb{R}^2$ in a unique way, i.e. for every $(x_0, y_0) \in \mathbb{R}^2$ there exists exactly one c such that $bx_0 - ay_0 = c$, the uniqueness is of course a consequence of the fact that for $c_1 \neq c_2$ the two lines $bx - ay = c_1$ and $bx - ay = c_2$ are parallel, see Figure 29.1

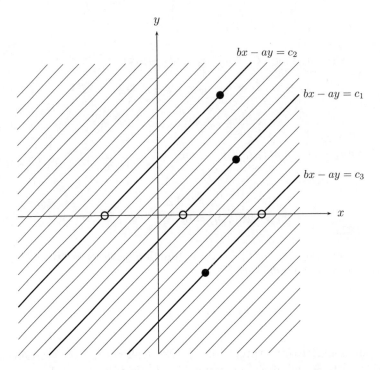

Figure 29.1

Given h as an initial condition $u(x,0) = h(x)$, we can determine the value of $u(x_0, y_0)$, $(x_0, y_0) \in \mathbb{R}^2$, in the following way: first we find $c = c(x_0, y_0)$ such that (x_0, y_0) belongs to the straight line $bx - ay = c$. This line intersects the line $y = 0$ at exactly one point, namely at $\left(\frac{c}{b}, 0\right)$. The value $u(x_0, y_0)$ equals the value $u(x_1, y_1)$ for all (x_1, y_1) on the line $bx - ay = c$, hence it is equal to $u\left(\frac{c}{b}, 0\right) = h\left(\frac{c}{b}\right)$.

Basically, we have covered $\mathbb{R}^2$ by a family of (smooth) curves such that two of these curves do not intersect and on each of the curves a solution of the differential equation must be constant. These curves intersect the line $y = 0$ at exactly one point and given the value of a solution at the point of intersection the value of the solution on the entire curve is determined. However, we need an argument that this construction leads indeed to a C^1-function u of two variables (which in the linear case discussed above is trivial due to the simple form of the curves). We want to study how this idea works for more general equations, namely quasi-linear partial differential equations of first order.

We consider the equation

$$a(x, y, u(x, y))u_x(x, y) + b(x, y, u(x, y))u_y(x, y) - c(x, y, u(x, y)) = 0, \quad (29.11)$$

where $a, b, c : \mathbb{R}^3 \to \mathbb{R}$ are at least continuous functions. In general, the equation

500

(29.11) is not anymore linear in u, but it is linear in the first partial derivatives, i.e. in the highest partial derivatives which enter the equation. Such a partial differential equation is called **quasi-linear**.

We interpret a solution $u : \mathbb{R}^2 \to \mathbb{R}$ as a (parametric) surface S in $\mathbb{R}^3$, compare with Chapter II.8, given by the equation $z = u(x, y)$ or the mapping $(x, y) \mapsto \begin{pmatrix} x \\ y \\ z \end{pmatrix} = \begin{pmatrix} x \\ y \\ u(x,y) \end{pmatrix}$. The vector $\begin{pmatrix} u_x(x,y) \\ u_y(x,y) \\ -1 \end{pmatrix}$ is normal to S at the point $(x, y, u(x, y))$ since the tangent directions to S at $(x, y, u(x, y))$ are given by $\begin{pmatrix} 1 \\ 0 \\ u_x(x,y) \end{pmatrix}$ and $\begin{pmatrix} 0 \\ 1 \\ u_y(x,y) \end{pmatrix}$ and

$$\begin{pmatrix} 1 \\ 0 \\ u_x(x,y) \end{pmatrix} \times \begin{pmatrix} 0 \\ 1 \\ u_y(x,y) \end{pmatrix} = - \begin{pmatrix} u_x(x,y) \\ u_y(x,y) \\ -1 \end{pmatrix},$$

compare with Definition II.8.4.B. Note the freedom we have to choose the orientation, i.e. to decide in which direction the inner and the outer normal at $(x, y, u(x, y))$ points. The geometric interpretation of (29.11) becomes clear when we rewrite the equation as

$$\left\langle \begin{pmatrix} a(x,y,u(x,y)) \\ b(x,y,u(x,y)) \\ c(x,y,u(x,y)) \end{pmatrix} , \begin{pmatrix} u_x(x,y) \\ u_y(x,y) \\ -1 \end{pmatrix} \right\rangle = 0. \tag{29.12}$$

We are looking for a surface S, $z = u(x, y)$, where at each point $(x, y, z) \in S$ a tangent direction is given by $\begin{pmatrix} a(x,y,z) \\ b(x,y,z) \\ c(x,y,z) \end{pmatrix}$. In the following we assume that a, b and c are C^1-functions. We first need some terminology.

Definition 29.1. *Let $a, b, c : \mathbb{R}^3 \to \mathbb{R}$ be C^1-functions and denote by S the surface* $(x, y) \mapsto \begin{pmatrix} x \\ y \\ z \end{pmatrix}$, $z = u(x, y)$, *for $u : \mathbb{R}^2 \to \mathbb{R}$.*

   A. *An **integral surface** of (29.11) is a surface S, $z = u(x, y)$, where u solves (29.11).*

B. The direction $\begin{pmatrix} a(x,y,z) \\ b(x,y,z) \\ c(x,y,z) \end{pmatrix} \in \mathbb{R}^3$ is called the **characteristic direction** associated with the equation (29.11) at the point $(x,y,z) \in \mathbb{R}^3$.

C. By definition a **characteristics** to equation (29.11) is a curve $\gamma : \mathbb{R} \to \mathbb{R}^3$ which is tangential to the characteristic direction at each point.

D. The projection of a characteristics γ onto the (x,y)-plane is called a **characteristic curve** of (29.11).

Let $\gamma : \mathbb{R} \to \mathbb{R}^3$, $\gamma(t) = (x(t), y(t), z(t))$, be a characteristic to (29.11). Then it must be at each point parallel to the characteristic direction, i.e.

$$\dot{x}(t) = a(\gamma(t)), \quad \dot{y}(t) = b(\gamma(t)), \quad \dot{z}(t) = c(\gamma(t)), \tag{29.13}$$

or more explicitly

$$\dot{x}(t) = a(x(t), y(t), z(t)), \quad \dot{y}(t) = b(x(t), y(t), z(t)), \quad \dot{z}(t) = c(x(t), y(t), z(t)) \tag{29.14}$$

which is a system of ordinary differential equations and since by assumption a, b and c are C^1-functions, for every initial value (x_0, y_0, z_0), there exists a unique maximal solution.

If S, $z = u(x,y)$, is a surface being the union of characteristics then S is an integral surface of the equation $au_x + bu_y - c = 0$. Indeed, if $P \in S$ then a characteristics γ passes through P and γ lies completely in S. Thus γ is tangent to S at P since it is orthogonal to the normal $(u_x, u_y, -1)$ at P, which proves the assertion. This observation suggests to find integral surfaces for (29.11) as a union of characteristics. In following up this idea, we first state

Theorem 29.2. Let $P = (x_0, y_0, z_0)$ be a point on an integral surface S, $z = u(x,y)$, of (29.11) and let γ be a characteristics passing through P. Then γ lies completely on S.

Proof. Let $\gamma(t) = (x(t), y(t), z(t))$ be a solution to (29.13) with $(x(t_0), y(t_0), z(t_0)) = (x_0, y_0, z_0)$ for some t_0. We consider the function

$$\varphi(t) := z(t) - u(x(t), y(t)), \quad t \in \mathbb{R}. \tag{29.15}$$

Clearly we have $\varphi(t_0) = z_0 - u(x_0, y_0) = 0$ since $(x_0, y_0, z_0) \in S$. From (29.13) we derive

$$\begin{aligned}
\frac{d\varphi}{dt} &= \frac{dz}{dt} - u_x(x, y)\frac{dx}{dt} - u_y(x, y)\frac{dy}{dt} \\
&= c(x, y, z) - u_x(x, y)a(x, y, z) - u_y(x, y)b(x, y, z) \\
&= c(x, y, \varphi + u(x, y)) - u_x(x, y)a(x, y, \varphi + u(x, y)) \\
&\quad - u_y(x, y)b(x, y, \varphi + u(x, y))
\end{aligned} \tag{29.16}$$

which is an ordinary differential equation for φ. Now $\varphi(t) = 0$ for all t is a solution to (29.16) since u, by assumption, satisfies (29.11). The uniqueness result for ordinary differential equations yields that this is the only solution to (29.16) vanishing at t_0. It follows that $\varphi(t) = 0$ for all t, i.e. $0 = z(t) - u(x(t), y(t))$ for all t, i.e. γ lies entirely in S proving the theorem. □

Corollary 29.3. *Every integral surface of* (29.11) *is the union of characteristics.*

Example 29.4. Let us return to equation (29.1), i.e. to

$$au_x + bu_y = 0, \quad a, b \in \mathbb{R}. \tag{29.17}$$

The characteristic direction at $(x, y, z) \in \mathbb{R}^3$ is given by $\begin{pmatrix} a \\ b \\ 0 \end{pmatrix}$. The equations for a characteristics are

$$\dot{x}(t) = a(x(t), y(t), z(t)) = a, \quad \dot{y}(t) = b(x(t), y(t), z(t)) = b, \quad \dot{z}(t) = 0. \tag{29.18}$$

As a solution of (29.18) we find $z(t) = z_0$, $x(t) = at + x_0$ and $y(t) = bt + y_0$ with $(x_0, y_0, z_0) \in \mathbb{R}^3$. This gives the characteristics $\gamma(t) = \begin{pmatrix} a \\ b \\ 0 \end{pmatrix}t + \begin{pmatrix} x_0 \\ y_0 \\ z_0 \end{pmatrix}$.

The projection of γ onto the (x, y)-plane now determines the characteristic curve $t \mapsto \begin{pmatrix} a \\ b \end{pmatrix}t + \begin{pmatrix} x_0 \\ y_0 \end{pmatrix}$, which is a straight line in direction $\begin{pmatrix} a \\ b \end{pmatrix}$, i.e. we recover our previous result.

The next more lengthy example shows some typical problems. So far we have assumed that all objects, e.g. integral surfaces or characteristics are globally defined, i.e. for all $(x, y) \in \mathbb{R}^2$ or all $t \in \mathbb{R}$. However, for a non-linear equation this is not necessarily the case.

We want to solve the initial value problem

$$u_x + uu_y = u_x + \left(\frac{u^2}{2}\right)_y = 0, \qquad x \geq 0, \, y \in \mathbb{R}, \tag{29.19}$$

with initial condition $u(0, y) = g(y)$. As we will see, g now determines the properties of the characteristics and the characteristic curves. The characteristic directions associated with (29.19) are $\begin{pmatrix} 1 \\ z \\ 0 \end{pmatrix}$ which leads to the following equations for a characteristic

$$\dot{x}(t) = 1, \qquad \dot{y}(t) = z(t), \qquad \dot{z}(t) = 0 \tag{29.20}$$

with solutions

$$x(t) = t + x_0, \qquad y(t) = z_0 t + y_0, \qquad z(t) = z_0. \tag{29.21}$$

Thus the characteristic through $\begin{pmatrix} x_0 \\ y_0 \\ z_0 \end{pmatrix}$ is given by

$$t \mapsto \begin{pmatrix} 1 \\ z_0 \\ 0 \end{pmatrix} t + \begin{pmatrix} x_0 \\ y_0 \\ z_0 \end{pmatrix}, \qquad z_0 = u(x_0, y_0),$$

which gives for the characteristic curve

$$t \mapsto \begin{pmatrix} 1 \\ z_0 \end{pmatrix} t + \begin{pmatrix} x_0 \\ y_0 \end{pmatrix}$$

and therefore $x - x_0 = t = \frac{y - y_0}{z_0}$ or

$$x = \frac{1}{z_0} y + x_0 - \frac{y_0}{z_0}. \tag{29.22}$$

We note that now the characteristics and the characteristic curve depends on z_0, i.e. the initial value. For all values of a sought solution u on the curve $x = \frac{1}{z_0} y + x_0 - \frac{y_0}{z_0}$ we have $u(x, y) = z_0 = g(y_0)$. Our initial curve on the integral surface S, $z = u(x, y)$, is given by

$$s \mapsto \begin{pmatrix} 0 \\ s + y_0 \\ g(s + y_0) \end{pmatrix}$$

504

which yields for the corresponding characteristic curve through $(0, y_0)$ under the initial conditions $(0, y_0, z_0) = (0, y_0, g(y_0))$ that

$$x = \frac{1}{z_0} y - \frac{y_0}{z_0} = \frac{1}{g(y_0)} y - \frac{y_0}{g(y_0)}. \tag{29.23}$$

For each fixed y_0 a straight line in the (x, y)-plane (or half-plane $x \geq 0$) is given by (29.23) being a characteristic curve of (29.19) passing through $(0, y_0)$ and corresponding to the characteristics passing through $(0, y_0, z_0) = (0, y_0, g(y_0))$. We want to discuss the following initial values given by g

i) $g(y) = 1$ for all y;

ii) $g(y) = \begin{cases} 1, & y < 0 \\ y + 1, & 0 \leq y \leq 1 \\ 2, & y > 1; \end{cases}$

iii) $g(y) = \begin{cases} 2, & y < 0 \\ -y + 2, & 0 \leq y \leq 1 \\ 1, & y > 1; \end{cases}$

iv) $g(y) = \begin{cases} 1, & y < 0 \\ 2, & y \geq 0. \end{cases}$

i) The condition $g(y) = 1$ for all y leads to the equation $x = y - y_0$ and the characteristics $t \mapsto \begin{pmatrix} t \\ t + y_0 \\ 1 \end{pmatrix}$ with characteristic curves as in Figure 29.2

505

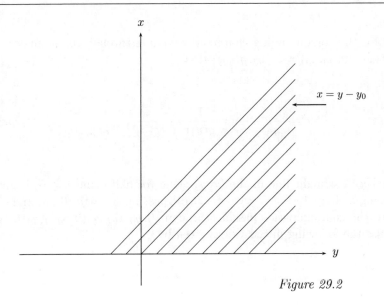

Figure 29.2

We find that each point in the half plane $x \geq 0$ lies on one and only one characteristic curve. The characteristics from the surface $u(x, y) = 1$ for all $x \geq 0$, $y \in \mathbb{R}$, i.e. the function $(x, y) \mapsto 1$ is the (unique) solution to $u_x + u u_y = 0$, $u(0, y) = 1$, $x \geq 0$, $y \in \mathbb{R}$.

ii) The function $g(y) = \begin{cases} 1, & y < 0 \\ y + 1, & 0 \leq y \leq 1 \\ 2, & y > 1 \end{cases}$ has the graph shown in Figure 29.3

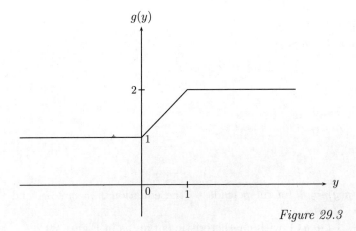

Figure 29.3

and for the characteristic curves we find $x = \frac{1}{g(y_0)}y - \frac{1}{g(y_0)}y_0$ which gives

$$y_0 < 0 : g(y_0) = 1 \quad \text{and} \quad x = y - y_0;$$

$$y_0 \in [0,1] : g(y_0) = y_0 + 1 \quad \text{and} \quad x = \frac{1}{y_0 + 1}y - \frac{y_0}{y_0 + 1};$$

$$y_0 > 1 : g(y_0) = 2 \quad \text{and} \quad x = \frac{y}{2} - \frac{y_0}{2}.$$

This leads to the following Figure 29.4

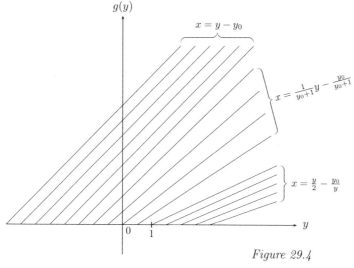

Figure 29.4

We note again that every point in the half-plane $x \geq 0$ lies exactly on one characteristic curve.

iii) The graph of g is given in Figure 29.5

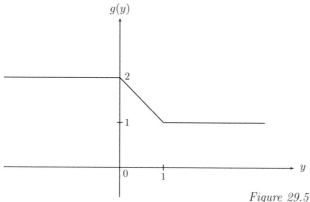

Figure 29.5

The equation $x = \frac{1}{g(y_0)}y - \frac{y_0}{g(y_0)}$ now leads to the following cases

$$y_0 \leq 0 : g(y_0) = 2 \quad \text{and} \quad x = \frac{y}{2} - \frac{y_0}{2};$$

$$y_0 \in [0,1] : g(y_0) = -y_0 + 2 \quad \text{and} \quad x = \frac{y}{2 - y_0} - \frac{y_0}{2 - y_0};$$

$$y_0 \geq 1 : g(y_0) = 1 \quad \text{and} \quad x = y - y_0.$$

For the characteristic curves, we find the following figure

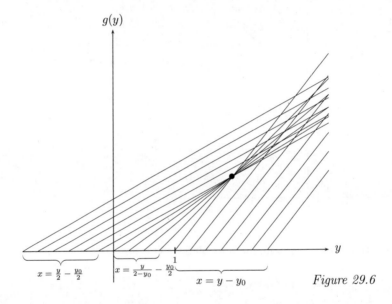

Figure 29.6

Thus each point (x, y) with $0 < x < 1$ and $y \in \mathbb{R}$ lies on only one characteristic curve and in this region we may construct $u(x, y)$ without any difficulty just as before. The same holds for points in the region $x \geq 0$, $y < x + 1$ as well as $x > 0$, $y > 2x$. However, all characteristic curves starting at some $y_0 \in [0, 1]$ pass through the point $(1, 2)$, and for points (x, y) with $x + 1 < y < 2x$, $x > 1$ we find that three characteristic curves pass through them. In this region we cannot (yet) define a solution: on a characteristic a solution must be constant but we have (in general) three different possibilities to choose this value. Thus we need a criterion to decide which value we shall take.

iv) The graph of g is given in Figure 29.7 below

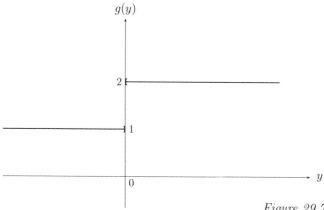

Figure 29.7

For the equation $x = \frac{1}{g(y_0)}y - \frac{1}{g(y_0)}y_0$ we now find

$$y_0 < 0 : g(y_0) = 1 \quad \text{and} \quad x = y - y_0,$$
$$y_0 \geq 0 : g(y_0) = 2 \quad \text{and} \quad x = \frac{y}{2} - \frac{y_0}{2},$$

which leads to characteristic curves as shown in Figure 29.8

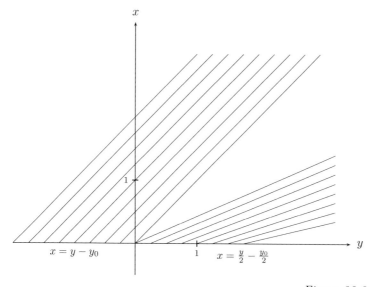

Figure 29.8

Now we find that whenever a point in the half-plane $x \geq 0$ lies on a characteristic curve it lies only on one, but there are points which do not lie on any characteristic curve at all. So how can we define a solution in the entire half-plane $x \geq 0$? Note that the fact that g is discontinuous is not the problem, replacing g by $\tilde{g}$ with a graph as in Figure 29.9 will not change the type of problem.

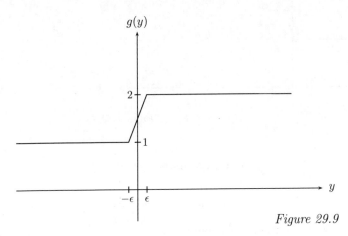

Figure 29.9

In Volume VI, we will pick up these problems when discussing discontinuous solutions of hyperbolic conservation laws. Here, we take the message that quasi-linear first order equations may already lead to severe difficulties. With this in mind, we now turn to general (non-linear) first order equations for functions in n variables. We follow [68] closely, which makes much use of [124].

For a given function $H : \mathbb{R}^n \times \mathbb{R} \times \mathbb{R}^n \to \mathbb{R}$ we are seeking a function $\varphi : \mathbb{R}^n \to \mathbb{R}$ such that φ solves the equation

$$H\left(x_1, \ldots, x_n, \varphi(x_1, \ldots, x_n), \frac{\partial \varphi}{\partial x_1}(x_1, \ldots, x_n), \ldots, \frac{\partial \varphi}{\partial x_n}(x_1, \ldots, x_n)\right)$$

$$= H(x, \varphi(x), \operatorname{grad} \varphi(x)) = 0. \tag{29.24}$$

In order to have no problems with domains and ranges, we assume for the moment that all functions are defined on the largest set possible, i.e. $\mathbb{R}^{2n+1}$ or $\mathbb{R}^n$, respectively. It is convenient to introduce already here some notations that we will use in the next chapter and to emphasize the geometric interpretation of (29.24) (as well as some interpretations related to mechanics). We often set $q := x$, i.e. $q_j = x_j$, $u := \varphi$ or $u = \varphi(x)$, and $p := \operatorname{grad} \varphi$, i.e. $p_j = \frac{\partial \varphi}{\partial x_j}$, or $p_j = \frac{\partial \varphi}{\partial x_j}(x)$. Then (29.24) reads as

$$H(q, u, p) = 0. \tag{29.25}$$

The function φ we assume to be at least of class C^2, while from H we require to be of class C^3 and to satisfy

$$\sum_{j=1}^{n} \left(\frac{\partial H}{\partial p_j}(q, u, p) \right)^2 \neq 0. \tag{29.26}$$

This implies that we always have some dependence on p in (29.25), i.e. (29.24) is always a first order partial differential equation and cannot degenerate to an equation for φ only.

For $(q, u) \in \mathbb{R}^{n+1}$ fixed, the equation (29.25) may have several solutions p. For example if $n = 1$ and if for (q, u) fixed $p \mapsto H(q, u, p)$ is a polynomial of degree $k \in \mathbb{N}$ we may have up to k distinct real solutions for p. As discussed in the linear and quasi-linear case, we are now trying to construct a solution to (29.24) as an integral surface $S \subset \mathbb{R}^{n+1}$, $u = \varphi(x)$, with S being the union of certain curves $\tilde{\gamma} : I \to \mathbb{R}^{n+1}$, i.e. $\tilde{\gamma}(I) \subset S$. Instead of a surface directly we are just seeking for a family $(\tilde{\gamma}_j)_{j \in J}$ of curves $\tilde{\gamma}_j : I_j \to \mathbb{R}^{n+1}$, with the property that $\bigcup_{j \in J} \tilde{\gamma}_j(I_j)$ gives a smooth hypersurface $S = \{(x, \varphi(x))\} \subset \mathbb{R}^{n+1}$. However, looking more closely to the quasi-linear case we see that we also need to determine p. Thus we try to find a family of curves $\gamma : I \to \mathbb{R}^{2n+1}$ such that

$$\gamma(t) = (q(t), u(t), p(t)) = (x(t), \varphi(x(t)), \operatorname{grad} \varphi(x(t))) \tag{29.27}$$

and

$$H(q(t), u(t), p(t)) = 0, \quad t \in I. \tag{29.28}$$

With the help of the projections $(x(t), \varphi(x(t))) = (q(t), u(t))$ we want to construct an integral surface.

Suppose that $\varphi : \mathbb{R}^n \to \mathbb{R}$ is a C^2-solution of (29.24). Taking in (29.24) the partial derivatives with respecto to x_j we obtain

$$H_{q_j} + H_u \varphi_{x_j} + \sum_{i=1}^{n} H_{p_i} \varphi_{x_i x_j} = 0 \tag{29.29}$$

and (29.27) yields

$$\dot{u} = \sum_{i=1}^{n} \varphi_{x_i} \dot{q}_i, \tag{29.30}$$

further for $1 \leq j \leq n$ we get

$$\dot{p}_j = \sum_{i=1}^{n} \dot{q}_i \varphi_{x_i x_j}. \tag{29.31}$$

With the help of (29.29) we get

$$\dot{p}_j + H_{q_j} + H_u \varphi_{x_j} = \sum_{i=1}^{n} (\dot{q}_i - H_{p_i}) \varphi_{x_i x_j}. \tag{29.32}$$

We restrict ourselves now to curves satisfying in addition

$$\dot{q} = H_p, \quad H_p(q, u, p) = \left(\frac{\partial H}{\partial p_1}(q, u, p), \dots, \frac{\partial H}{\partial p_n}(q, u, p) \right). \tag{29.33}$$

(Instead of H_p we might sometimes write $\mathrm{grad}_p H(q, u, p)$). If we determine along an integral curve of (29.33) the terms $p_i = \frac{\partial \varphi}{\partial x_i}$ and $u = \varphi$ we find with the help of (29.32) that

$$\dot{p}_j = -H_{q_j} - H_u \varphi_{x_j} \tag{29.34}$$

and (29.30) implies

$$\dot{u} = \sum_{i=1}^{n} \varphi_{x_i} H_{p_i}(q, u, p). \tag{29.35}$$

Since $\varphi_{x_k}(x(t)) = p_k(t)$ and $\varphi(x(t)) = u(t)$, we eventually arrive at the system

$$\begin{cases} \dot{q}_i = H_{p_i}(q, u, p), \quad 1 \le i \le n, & (29.36\mathrm{a}) \\ \dot{u} = \sum_{i=1}^{n} p_i H_{p_i}(q, u, p) \quad (= \langle p, H_p(q, u, p) \rangle), & (29.36\mathrm{b}) \\ \dot{p}_i = -H_{q_i}(q, u, p) - p_i H_u(q, u, p), \quad 1 \le i \le n. & (29.36\mathrm{c}) \end{cases}$$

Clearly, if $\gamma(t) = (q(t), u(t), p(t))$ is a curve satisfying (29.33) and belonging to the surface $\check{S} = \{(x, \varphi(x), \mathrm{grad}(\varphi(x))) \mid x \in \mathbb{R}^n\}$ then it is a solution of (29.36a) - (29.36c).

Before we continue our investigations we want to fix some terminology.

Definition 29.5. A. *The system* (29.36a) - (29.36c) *is called the system of* ***characteristic differential equations*** *associated with* (29.24).

B. *A solution of* (29.36a) - (29.36c) *which satisfies in addition* $H(q(t), u(t), p(t)) = 0$ *is called a* ***characteristic strip*** *of* (29.36a) - (29.36c) *and is denoted by* Σ_H.

C. *The projection of a characteristic strip onto the* (q, u)-*space is called a* ***characteristics*** *of* (29.24) *and the projection of a characteristics of* (29.24) *onto the* q-*space is called a* ***characteristic curve***.

In analogy to Definition 27.3, we give

Definition 29.6. *Let $\psi : I \to \mathbb{R}^k$ be a solution to the differential equation $G(t, \psi(t), \dot{\psi}(t)) = 0$ and $f : \mathbb{R}^k \to \mathbb{R}$ a function. We call f a **first integral** for $G(t, \psi(t), \dot{\psi}(t)) = 0$ if for every integral curve ψ we have*

$$\frac{\mathrm{d}}{\mathrm{d}t} f(\psi(t)) = 0. \tag{29.37}$$

Lemma 29.7. *The function $H : \mathbb{R}^{2n+1} \to \mathbb{R}$ is a first integral of the characteristic differential equations (29.36a) - (29.36c).*

Proof. Let $\gamma(t) = (q(t), u(t), p(t))$ be an integral curve of (29.37). It follows that

$$\frac{\mathrm{d}H}{\mathrm{d}t} = \sum_{j=1}^{n} H_{q_j} \dot{q}_j + H_u \dot{u} + \sum_{j=1}^{n} H_{p_j} \dot{p}_j$$

and using (29.36a) - (29.36c) we obtain

$$\frac{\mathrm{d}H}{\mathrm{d}t} = \sum_{j=1}^{n} H_{q_j} H_{p_j} + \sum_{j=1}^{n} p_j H_{p_j} H_u - \sum_{j=1}^{n} H_{p_j} H_{q_j} - \sum_{j=1}^{n} p_j H_u H_{p_j} = 0.$$

$\square$

Under our smoothness conditions on H for every point $(q_0, u_0, p_0) \in \mathbb{R}^{2n+1}$ there exists a uniquely determined maximal solution $\gamma : I_{(q_0, u_0, p_0)} \to \mathbb{R}^{2n+1}$ of (29.36a) - (29.36c) passing through (q_0, u_0, p_0). However, for a fixed point (q_0, u_0), but different values p_0^1 and p_0^2, in general the maximal solutions to (29.36a) - (29.36c) passing through (q_0, u_0, p_0^1) and (q_0, u_0, p_0^2) will be different and so will be in general their projections on the (q, u)-space. In such a case there will be several characteristics passing through (q_0, u_0).

Let $\varphi : \mathbb{R}^n \to \mathbb{R}$ be a C^2-function. By Γ_φ we denote the set

$$\Gamma_\varphi = \left\{ (x, \varphi(x), \operatorname{grad} \varphi(x)) \mid x \in \mathbb{R}^n \right\} \subset \mathbb{R}^{2n+1}. \tag{29.38}$$

Lemma 29.8. *Let $\sum_H$, $\sum_H(t) = (q(t), u(t), p(t))$, be a characteristic strip of (29.24). If $\varphi : \mathbb{R}^n \to \mathbb{R}$ is a C^2-solution of (29.24) such that $\sum_H \cap \Gamma_\varphi \neq \emptyset$ then $\sum_H \subset \Gamma_\varphi$ and the corresponding characteristic is a subset of $S = \{(x, \varphi(x)) \mid x \in \mathbb{R}^n\}$, the graph of φ.*

Proof. Let $(q_0, u_0, p_0) \in \sum_H \cap \Gamma_\varphi$. Since $\sum_H$ is an integral curve to (29.36a) - (29.36c) there exists t_0 such that $(q(t_0), u(t_0), p(t_0)) = (q_0, u_0, p_0)$. Denote by $\tilde{\gamma} = \tilde{x}(t)$ the integral curve of

$$\frac{\mathrm{d}\tilde{x}_j(t)}{\mathrm{d}t} = H_{p_j}(\tilde{x}(t), \varphi(\tilde{x}(t)), \operatorname{grad} \varphi(\tilde{x}(t))), \quad 1 \le j \le n, \quad \tilde{x}(t_0) = q_0. \tag{29.39}$$

513

Further let $\sum' \subset \Gamma_\varphi$ be the curve

$$\Sigma' = \left\{ (\tilde{q}(t), \tilde{u}(t), \tilde{p}(t)) \mid \tilde{q}(t) = \tilde{x}(t), \tilde{u}(t) = \varphi\left(\tilde{x}(t)\right), \tilde{p}(t) = \operatorname{grad} \varphi\left(\tilde{x}(t)\right) \right\}. \tag{29.40}$$

The lemma is proved if we can show that by (29.40) a maximal solution to (29.36a) - (29.36c) is given. Indeed, since $\sum' \cap \sum_H \neq \emptyset$ the Theorem of Picard-Lindelöf will imply $\sum_H \subset \sum'$ and therefore $\sum_H \subset \Gamma_\varphi$. From (29.39) and (29.40) we deduce immediately (29.36a). For $\tilde{u}(t) = \varphi\left(\tilde{x}(t)\right)$ and $\tilde{p}(t) = \operatorname{grad} \varphi\left(\tilde{x}(t)\right)$ we find

$$\frac{\mathrm{d}\tilde{u}(t)}{\mathrm{d}t} = \left\langle \dot{\tilde{x}}(t), \operatorname{grad} \varphi\left(\tilde{x}(t)\right) \right\rangle = \left\langle \dot{\tilde{q}}(t), \tilde{p}(t) \right\rangle$$

and since we know that (29.36a) already holds we conclude

$$\dot{\tilde{u}} = \sum_{i=1}^{n} H_{\tilde{p}_i}\left(\tilde{q}, \tilde{u}, \tilde{p}\right) \tilde{p}_i.$$

i.e. (29.36b) is proved. Since φ is a solution to (29.24) and hence (29.29) holds, we have

$$\sum_{i=1}^{n} H_{p_i} \varphi_{x_i x_j} = -H_{q_j} - H_u \varphi_{x_j}$$

and on $\sum'$ we obtain by using once more (29.36a) that

$$\sum_{i=1}^{n} \tilde{x}_i \varphi_{x_i x_j}\left(\tilde{x}\right) = -H_{\tilde{q}_j}\left(\tilde{q}, \tilde{u}, \tilde{p}\right) - H_{\tilde{u}}\left(\tilde{q}, \tilde{u}, \tilde{p}\right) \tilde{p}_j$$

and taking

$$\sum_{i=1}^{n} \tilde{x}_i \varphi_{x_i x_j}\left(\tilde{x}\right) = \frac{\mathrm{d}}{\mathrm{d}t} \varphi_{x_j}\left(\tilde{x}(t)\right) = \dot{\tilde{p}}_j$$

into account we eventually arrive at

$$\dot{\tilde{p}}_j = -H_{\tilde{q}_j} - H_{\tilde{u}} \tilde{p}_j$$

and the lemma is proved. $\qquad\square$

Definition 29.9. *Let* $\varphi : \mathbb{R}^n \to \mathbb{R}$ *be a* C^2*-function and* Γ_φ *be defined as in* (29.38). *We call* Γ_φ ***decomposable into characteristic strips*** *if for every point in* Γ_φ *there exists a characteristic strip of* (29.24) *belonging entirely to* Γ_φ.

Theorem 29.10. *A C^2-function $\varphi : \mathbb{R}^n \to \mathbb{R}$ is a solution of (29.24) if and only if Γ_φ is decomposable into characteristic strips.*

Proof. Suppose that φ is a solution of (29.24). Then Lemma 29.8 implies that Γ_φ is decomposable into characteristic strips. Now let $\varphi : \mathbb{R}^n \to \mathbb{R}$ be a C^2-function and assume that Γ_φ is decomposable into characteristic strips of (29.24). In this case, the surface $S = \{(x, \varphi(x)) \,|\, x \in \mathbb{R}\}$ is decomposable into the corresponding characteristics and for $(q, u) \in S$, $u = \varphi(x)$, $\operatorname{grad} \varphi(x)$ coincides with p-component of a characteristic strip. Along a characteristic strip we have however $H(q, u, p) = 0$ and the theorem follows. □

We want to use these results to solve (29.24) and for this we need some more preparations. Let $F : \mathbb{R}^{n-1} \to \mathbb{R}^n$ be a C^2-function and $W := \{x \in \mathbb{R}^n \,|\, x = F(\xi), \xi \in \mathbb{R}^{n-1}\}$ be a hyper-surface in $\mathbb{R}^n$ without any self-intersection. Moreover, we assume that the Jacobi matrix $J_F(\xi)$ of F has for all $\xi \in \mathbb{R}^{n-1}$ maximal rank. If $v_0 \in C^2(\mathbb{R}^{n-1})$ we can define a function $v : W \to \mathbb{R}$ by $v(x) = v(F(\xi)) = v_0(\xi)$, $x = F(\xi)$. On the other hand, if v is defined in a neighbourhood of W we can find $v_0 : \mathbb{R}^{n-1} \to \mathbb{R}$ such that $v(x) = v(F(\xi)) = v_0(\xi)$.

Now let $u = \varphi(x)$ be a C^2-solution to (29.24) and on W we may consider $x = q = q(\xi)$, $u = u_0(\xi)$ and $p = p(\xi)$, $p = \operatorname{grad} u$. It follows first that

$$\frac{\partial u_0}{\partial \xi_j} = \sum_{i=1}^{n} \frac{\partial u}{\partial x_i} \frac{\partial q_i}{\partial \xi_j}$$

and therefore on W

$$\left\langle p(\xi), \frac{\partial q}{\partial \xi_j} \right\rangle = \frac{\partial u}{\partial \xi_j}, \quad 1 \le j \le n - 1, \tag{29.41}$$

and

$$H(q(\xi), u_0(\xi), p(\xi)) = 0. \tag{29.42}$$

These are n equations we may use to determine $p = p(\xi)$ on W. We assume that there exists an open domain $U \subset \mathbb{R}^{n-1}$ with compact closure $\overline{U}$ such that for every $\xi \in \overline{U}$ the system (29.41), (29.42) has exactly M solutions $p^{(l)}(\xi)$, $1 \le l \le M$, such that

$$\det \begin{pmatrix} \frac{\partial q_1(\xi)}{\partial \xi_1} & \cdots & \frac{\partial q_n(\xi)}{\partial \xi_1} \\ \vdots & & \vdots \\ \frac{\partial q_1(\xi)}{\partial \xi_{n-1}} & \cdots & \frac{\partial q_n(\xi)}{\partial \xi_{n-1}} \\ \frac{\partial H(q(\xi), u_0(\xi), p^{(l)}(\xi))}{\partial p_1} & \cdots & \frac{\partial H(q(\xi), u_0(\xi), p^{(l)}(\xi))}{\partial p_n} \end{pmatrix} \ne 0 \tag{29.43}$$

for all $\xi \in \overline{U}$. By the implicit function theorem, Theorem II.10.5, it follows that $p^{(l)}$, $1 \leq l \leq M$, is, in a neighbourhood U_ϵ of $\overline{U}$, continuously differentiable. We denote the part of W which is parametrized by U_ϵ by W'.

Definition 29.11. *The **Cauchy problem** for equation (29.24) with respect to a hypersurface $W \subset \mathbb{R}^n$ which is assumed to have no self-intersection and with respect to the initial function $\omega : W \to \mathbb{R}$ consists in finding an open neighbourhood $\Omega \subset \mathbb{R}^n$ of W and a C^1-function $u : \Omega \to \mathbb{R}$ such that*

$$H(x, u(x), \operatorname{grad} u(x)) = 0 \quad \text{for } x \in \Omega \tag{29.44}$$

and

$$u|_W = \omega. \tag{29.45}$$

Now we can prove the local solvability of the Cauchy problem for (29.24).

Theorem 29.12. *Suppose that the system (29.41), (29.42) has for all $\xi \in \overline{U}$ exactly M solutions $p^{(l)} = p^{(l)}(\xi)$ and that for these solutions condition (29.43) holds. Let U_ϵ be a neighbourhood of $\overline{U}$ in which we can represent the $p^{(l)}$ as continuously differentiable functions and denote by W' the part of W which we can parametrise with U_ϵ. Further, let $\omega_0 : W \to \mathbb{R}$ be a given C^2-function. Then there exists a neighbourhood Ω of W' such that in Ω the Cauchy problem $H(x, u(x), \operatorname{grad} u(x)) = 0$ and $u|_{W'} = \omega_0$ has exactly M solutions. The set Γ_u is decomposable into characterstic strips starting at the points*

$$\left(x, u, p^{(l)} \right)(0) = \left(x(\xi), \omega_0(\xi), p^{(l)}(\xi) \right), \quad \xi \in U_\xi, \quad 1 \leq l \leq M. \tag{29.46}$$

Proof. As we know, each of the functions $p^{(l)}$, $1 \leq l \leq M$ is in U_ϵ a C^1-function. In the following we fix l and we write p instead $p^{(l)}$. For each $\xi \in U_\epsilon$ we can solve the characteristic differential equations (29.36a) - (29.36c) with initial condition (29.46). Thus we obtain curves $\gamma : (-\epsilon', \epsilon') \to \mathbb{R}^{2n+1}$

$$\gamma(t) = (x(t), u(t), p(t)) = (x(t, \xi), u(t, \xi), p(t, \xi)) \tag{29.47}$$

and $(x(0, \xi), u(0, \xi), p(0, \xi)) = (x(\xi), \omega_0(\xi), p(\xi))$. Without loss of generality, we may take $\epsilon = \epsilon'$. From Lemma 29.7 together with (29.42) (with u_0 being replaced by ω_0) we deduce that the curves γ are characteristic strips. A more sophisticated result on the smooth dependence of initial data for ordinary differential equations, [3], implies that the mapping

$$(t, \xi) \mapsto (x(t, \xi), u(t, \xi), p(t, \xi))$$

is for $t \in (-\epsilon, \epsilon)$ and $\xi \in U_\epsilon$ continuously differentiable. We want to prove that

$$x = x(\tau, \xi) \tag{29.48}$$

defines a local change of coordinates in a neighbourhood $\Omega \subset \mathbb{R}^n$ of W'. For this we note that

$$\det J_x(0, \xi) = \det \begin{pmatrix} \frac{\partial x_1}{\partial \tau} & \cdots & \frac{\partial x_n}{\partial \tau} \\ \vdots & & \vdots \\ \frac{\partial x_1}{\partial \xi_{n-1}} & \cdots & \frac{\partial x_n}{\partial \xi_{n-1}} \end{pmatrix} (0, \xi)$$

$$= \det \begin{pmatrix} \frac{\partial H}{\partial p_1} & \cdots & \frac{\partial H}{\partial p_n} \\ \vdots & & \vdots \\ \frac{\partial x_1}{\partial \xi_{n-1}} & \cdots & \frac{\partial x_n}{\partial \xi_{n-1}} \end{pmatrix} (0, \xi) \neq 0$$

where we have used (29.36a) and (29.43). Hence the existence of Ω on which x is used is a local change of coordinates follows. Thus we can solve (29.48) for (τ, ξ) and with the help of (29.47) it follows that we can now consider u and p as continuously differentiable functions of x. The graph of the function u consists, by construction, of characteristics of (29.24). We want to prove that for $u = u(x)$ we have

$$\operatorname{grad} u(x) = p(x). \tag{29.49}$$

Once (29.49) is proved, it follows that u is a C^2-function and that Γ_u is decomposable into characteristic strips, hence by Theorem 29.10 it follows that u is a solution of the Cauchy problem in Ω. We consider the functions

$$V_j := \frac{\partial u}{\partial \xi_j} - \left\langle p, \frac{\partial x}{\partial \xi_j} \right\rangle, \quad 1 \leq j \leq n-1,$$

$$V_0 := \frac{\partial u}{\partial t} - \left\langle p, \frac{\partial x}{\partial t} \right\rangle.$$

Since (29.48) gives a change of coordinates and

$$\begin{cases} \dfrac{\partial u}{\partial \xi_j} = \left\langle \operatorname{grad} u, \dfrac{\partial x}{\partial \xi_j} \right\rangle, & 1 \leq j \leq n-1, \tag{29.50a} \\[2mm] \dfrac{\partial u}{\partial t} = \left\langle \operatorname{grad} u, \dfrac{\partial x}{\partial t} \right\rangle, \tag{29.50b} \end{cases}$$

in order to prove (29.49) it is sufficient to show that $V_j = 0$ for $1 \leq j \leq n-1$. Since $t \mapsto (x(t), u(t), p(t))$ is a characteristic strip, we have immediately $V_0 = 0$ and (29.41) implies $V_j|_{t=0} = 0$ for $1 \leq j \leq n-1$. We know that the solution (29.47) depends continuously differentiable on t and ξ, therefore we may differentiate the equations (29.36a) - (29.36c) to obtain

$$\frac{\partial V_j}{\partial t} = \frac{\partial^2 p}{\partial \xi_j \partial t} - \left\langle \frac{\partial \varphi}{\partial t}, \frac{\partial x}{\partial \xi_j} \right\rangle - \left\langle p, \frac{\partial^2 x}{\partial t \partial \xi_j} \right\rangle, \quad 1 \le j \le n-1,$$

and

$$\frac{\partial V_0}{\partial \xi_j} = \frac{\partial^2 u}{\partial \xi_j \partial t} - \left\langle \frac{\partial p}{\partial \xi_j}, \frac{\partial x}{\partial t} \right\rangle - \left\langle p, \frac{\partial^2 x}{\partial t \partial \xi_j} \right\rangle,$$

and since $V_0 = 0$ we obtain with (29.36a) - (29.36c) for $1 \le j \le n-1$

$$\frac{\partial V_j}{\partial t} = \frac{\partial V_j}{\partial t} - \frac{\partial V_0}{\partial \xi_j}$$

$$= \left\langle \operatorname{grad}_x H, \frac{\partial x}{\partial \xi_j} \right\rangle + H_u \left\langle p, \frac{\partial x}{\partial \xi_j} \right\rangle + \left\langle \frac{\partial p}{\partial \xi_j}, \operatorname{grad}_p H \right\rangle.$$

On the other hand, we deduce from $H(x, u, p) = 0$ that

$$0 = \frac{\partial}{\partial \xi_j} H(x, u, p) = \left\langle \operatorname{grad}_x H, \frac{\partial x}{\partial \xi_j} \right\rangle + H_u \frac{\partial u}{\partial \xi_j} + \left\langle \frac{\partial p}{\partial \xi_j}, \operatorname{grad}_p H \right\rangle$$

which implies for $1 \le j \le n-1$

$$\frac{\partial V_j}{\partial t} = -H_u \left(\frac{\partial u}{\partial \xi_j} - \left\langle p, \frac{\partial x}{\partial \xi_j} \right\rangle \right) = -H_u V_j.$$

For ξ fixed this is an ordinary differential equation for $V_j = V_j(t)$ and since $V_j(0) = 0$ it follows that $V_j(t) = 0$ for all t. This implies the local solvability of the Cauchy problem. The uniqueness follows from the uniqueness results for achieving ordinary differential equations. $\qquad \square$

Often the function H does not depend explicitly on u, i.e. we have $H = H(q, p)$ and we consider

$$H(q, p) = 0, \quad p = \operatorname{grad} u. \tag{29.51}$$

The system of characteristic differential equations (29.36a) - (29.36c) becomes now

$$\begin{cases} \dot{q}_i = H_{p_i}(q, p), & 1 \le i \le n, & (29.52a) \\ \dot{p}_i = -H_{q_i}(q, p), & 1 \le i \le n, & (29.52b) \end{cases}$$

and it follows

$$\dot{u} = \sum_{i=1}^{n} p_i H_{p_i}(q, p). \tag{29.53}$$

Thus the system (29.36a) - (29.36c) decouples to (29.52a), (29.52b) and (29.53). The function H is usually called the **Hamilton function**, (29.52a), (29.52b) is called the associated **Hamilton system** or the **Hamilton differential equations**. The projections of the solutions of (29.52a), (29.52b), (29.53) onto the (q, p)-space are called the **bicharacteristics**. We will pick up many of these ideas in Volume VI when dealing with the propagation of singularities of partial differential equations, and moreover in the next chapter when we return to the calculus of variations.

Problems

1.   a) For C^1-functions $h : \mathbb{R} \to \mathbb{R}$ and $g : \mathbb{R}^2 \to \mathbb{R}$ verify that $u(x, y) = h(g(x, y))$ satisfies the first order partial differential equation $u_x g_y - u_y g_x = 0$. Give an interpretation of this equation.

    b) For a C^1-function $h : \mathbb{R} \to \mathbb{R}$ consider the function $u : \mathbb{R}^2 \to \mathbb{R}$, $u(x, y) = h(x^2 + y^2)$. What can we say about the graph $\Gamma(u) \subset \mathbb{R}^3$? Prove that u solves the first order partial differential equation $y u_x - x u_y = 0$.

2. Solve the differential equation $\frac{\partial u}{\partial x} + 2 \frac{\partial u}{\partial y} + u - 1 = 0$ subject to the initial condition $u(x, x) = \cos^2 x$, $x \in \mathbb{R}$.

3. In each of the following cases draw the characteristic curves and discuss the solvability of the initial value problem $u_x + \frac{1}{4} u u_y = 0$, $u(0, y) = g(y)$, in $\mathbb{R}_+ \times \mathbb{R}$:

    a) $g(y) = 5$ for all $y \in \mathbb{R}$;

    b)

$$g(y) = \begin{cases} 3, & y \leq 0 \\ 2y + 3, & y \in [0, 1] \\ 5, & y \geq 1. \end{cases}$$

4. Discuss the differential equation $\frac{\partial u}{\partial x} + 5 y^{\frac{4}{5}} \frac{\partial u}{\partial y} = 0$ in a neighbourhood of the point $(0, 0) \in \mathbb{R}^2$.

5. We consider a first order partial differential equation

$$(*) \quad F(x, u(x), \operatorname{grad} u(x)) = 0, \quad x \in \mathbb{R}^n.$$

Suppose that a solution φ depends on n parameters $a = (a_1, \ldots, a_n) \in \mathbb{R}^n$. These parameters may serve to prescribe initial conditions. We call such a solution $\varphi = \varphi(x; a)$ a **complete integral** of $(*)$ if the matrix

$$\begin{pmatrix} \varphi_{a_1} & \varphi_{x_1 a_1} & \cdots & \varphi_{x_n a_1} \\ \vdots & \vdots & & \vdots \\ \varphi_{a_n} & \varphi_{x_1 a_n} & \cdots & \varphi_{x_n a_n} \end{pmatrix}$$

has full rank.

a) Prove that $\varphi(x, y; a, b) := ax + by + f(a, b)$ is a complete integral of the **Clairaut differential equation** $x u_x + y u_y + f(u_x u_y) = u$.

b) Consider the differential equation

$$\frac{\partial u}{\partial x_0} + H(u_{x_1}, \ldots, u_{x_n}) = 0,$$

where $u : \mathbb{R}^{n+1} \to \mathbb{R}$. Show that for $a \in \mathbb{R}^n$ and $b \in \mathbb{R}$ a complete integral to this differential equation is given by

$$\varphi(x_0, x_1, \ldots, x_n; b, a_1, \ldots, a_n) := \langle a, x \rangle - x_0 H(a) + b,$$

where $x = (x_1, \ldots, x_n)$ and $\langle a, x \rangle$ is the scalar product in $\mathbb{R}^n$.

6. Let $u : \mathbb{R}^2 \times (a, b) \to \mathbb{R}$ be a C^1-function and assume $\frac{\partial u(x,y,\xi)}{\partial \xi} = 0$ for all $(x, y, \xi) \in \mathbb{R}^2 \times (a, b)$. Suppose further that we can solve this equation for ξ, i.e. $\xi = \varphi(x, y)$, where we assume that φ is also a C^1-function. We call the function $v(x, y) := u(x, y, \varphi(x, y))$ the **envelope** of the family $(u(\cdot, \xi))_{\xi \in (a,b)}$. (We refer to Chapter II.11 for a discussion of envelopes.)

a) Consider the partial differential equation

$$(*) \quad F(x, y, u(x, y), u_x(x, y), u_y(x, y)) = 0$$

for a C^1-function F. Assume that a family of solutions $w(x, y; \xi)$, $\xi \in (a, b)$, is given and that this family has the envelope $v(x, y) := w(x, y, \varphi(x, y))$ with a C^1-function φ. Prove that v is a further solution to $(*)$.

b) Prove that $w(x, y, \alpha) = x \cos \alpha + y \sin \alpha$ is for every $\alpha \in \mathbb{R}$ a solution to the **eikonal equation** in two dimensions, i.e.

$$\left(\left(\frac{\partial u}{\partial x} \right)^2 + \left(\frac{\partial u}{\partial y} \right)^2 \right)^{\frac{1}{2}} = 1.$$

Find the envelope of the family $(w\cdot, \alpha)_{\alpha \in \mathbb{R}}$ for $x > 0$ and $y \in \mathbb{R}$ and verify that this envelope is a further solution of the eikonal equation in $\mathbb{R}_+ \times \mathbb{R}$.

7. Consider the equation $u_x^2 + u_y^2 = 1$ for $u = u(x, y)$.

    a) Find the corresponding characteristic differential equations.

    b) Let the hypersurface W in Definition 29.11 be given as the trace of the curve $s \mapsto \begin{pmatrix} x(s) \\ y(s) \end{pmatrix} = \begin{pmatrix} f(s) \\ g(s) \end{pmatrix}$ with functions $f, g : I \to \mathbb{R}$ and let $\tilde{h} :$ $W \to \mathbb{R}$ be a function defined on W for which we set $h(s) := \tilde{h}(x(s), y(s)) = \tilde{h}(f(s), g(s))$. Find (29.41)-(29.43) when $u_0 = h$ and $p(s) = \begin{pmatrix} a(s) \\ b(s) \end{pmatrix}$ with functions a and b to be determined.

    c) Prove that the solution to the characteristic differential equations subject to the initial data is given by

$$x(\tau, s) = f(s) + \tau a(s)$$
$$y(\tau, s) = g(s) + \tau b(s)$$
$$u(\tau, s) = \tau + h(s)$$
$$p_1(\tau, s) = a(s)$$
$$p_2(\tau, s) = b(s).$$

8. Let $v \in C^2(\mathbb{R}^2)$ be a solution to $\frac{\partial v(x,y)}{\partial x} + \frac{\partial v(x,y)}{\partial y} = w(x, y)$, $w \in C^1(\mathbb{R}^2)$, and assume that w satisfies $\frac{\partial w(x,y)}{\partial x} - \frac{\partial w(x,y)}{\partial y} = 0$. Prove that v is a solution to the wave equation $\frac{\partial^2 u(x,y)}{\partial x^2} - \frac{\partial^2 u(x,y)}{\partial y^2} = 0$.

30 Aspects of Hamilton-Jacobi Theory

D'Alembert's principle of virtual work led, in the hands of Lagrange, to the introduction of the function nowadays called the Lagrange function and the Lagrange equations (Euler-Lagrange equations) for a system of N particles. Using the Lagrange function and guided by his own work on optics, Hamilton stated his basic principle for mechanics and made this a starting point for investigations which culminated in the Hamilton-Jacobi theory which we first of all shall look at as a theory in analytical mechanics. However, for a mathematician it is also a part of the calculus of variations and the theory of partial differential equations of first order.

Arguably the most authoritative account of the history of these mechanical principles is due to I. Szabó [119], followed by C. Truesdell [122], and a treatise still worth reading is that of E. T. Whittaker [127]. As a more contemporary reference text in mechanics, we refer to the classics of H. Goldstein in its latest edition [45]. On a more mathematical side, we mention V. I. Arnold [6] and R. Abraham and J. E. Marsden [1]. However, this chapter concentrates only on some mathematical aspects and we leave arguments justifying mechanical principles aside.

According to Lagrange and Hamilton we assume that a mechanical system, a system of interacting point masses for example, is described by its **Lagrange function** L. If $x(t)$ denotes the vector of the positions of the particles at time t, i.e. $x(t) \in \mathbb{R}^{3N}$ if we are dealing with N particles, and if $\dot{x}(t)$ denotes the velocity vector of these particles, i.e. the derivative of x with respect to time, then the assumption is that L is a function of x and $\dot{x}$ and maybe (depending on the type of interaction) on t. More specific, in many cases the assumption is that L is of the type

$$L = T - V \tag{30.1}$$

where T is the kinetic energy and V is the potential energy of the system. For a particle with mass $m > 0$, one typically assumes

$$T = \frac{\|\dot{x}(t)\|^2}{2m}, \tag{30.2}$$

thus for N particles with masses $m_1, \ldots, m_N$, positions $x^{(1)}(t), \ldots, x^{(N)}(t) \in \mathbb{R}^3$ and velocities $\dot{x}^{(1)}(t), \ldots, \dot{x}^{(N)}(t) \in \mathbb{R}^3$ we have

$$T = \sum_{j=1}^{N} \frac{\|\dot{x}^{(j)}(t)\|^2}{2m_j}. \tag{30.3}$$

The potential we assume to depend only on $x(t)$, in some cases maybe on t, but to be independent of $\dot{x}(t)$. Suppose now that a mechanical system is described by a Lagrange function $L = L(t, x(t), \dot{x}(t))$. If $t \mapsto x(t)$, $x(t) \in \mathbb{R}^n$, is a (possible) trajectory in the x-space (state space) on which the particles subjected to the dynamics generated by L move, we define the **action** or **action integral** with respect to $x : [a, b] \to \mathbb{R}^n$ as

$$S = S(x) := \int_a^b L(s, x(s), \dot{x}(s)) \, \mathrm{d}s. \tag{30.4}$$

Of course we may, and later on will, turn to

$$S(x, t) := \int_a^t L(s, x(s), \dot{x}(s)) \, \mathrm{d}s, \quad a \le t \le b. \tag{30.5}$$

Hamilton's principle states in its original version that if $L = L(x, \dot{x})$ then the paths of the particles moving from $A = x(a)$ to $B = x(b)$ under the dynamics generated by L are those for which the action becomes a minimum under all "admissible" paths. A more precise formulation is that in the class $D_{A,B} = \{u \in C^2([a, b]; \mathbb{R}^n) \mid u(a) = A, u(b) = B\}$ the path of the system is a minimiser of S, i.e. we are looking at the variational problem

$$\inf_{u \in D_{A,B}} \int_a^b L(u(s), \dot{u}(s)) \, \mathrm{d}s. \tag{30.6}$$

Now we are back in the situation considered in Chapters 26 and 27 where we derived the Euler-Lagrange equations as necessary conditions for a local minimiser of (30.6), and we introduce extremals as solutions of the Euler-Lagrange equations as candidates for a minimiser of (30.6).

Hamilton proposed to describe a mechanical system, instead of with the help of its Lagrange function, by a function we now call the Hamilton function H. It is worth introducing H first as it is done in (classical) books on mechanics as well as on the calculus of variations and then we will see that convex analysis and the Legendre transform (as introduced in convex analysis, see Chapter II.13) gives a more modern framework.

Let $L : \mathbb{R}^{2n+1} \to \mathbb{R}$, $(t, q, \eta) \mapsto L(t, q, \eta)$ be a Lagrange function of class C^2. We write $L_q = \left(\frac{\partial L}{\partial q_1}, \dots, \frac{\partial L}{\partial q_n} \right) = (L_{q_1}, \dots, L_{q_n})$ and $L_\eta = \left(\frac{\partial L}{\partial \eta_1}, \dots, \frac{\partial L}{\partial \eta_n} \right) = (L_{\eta_1}, \dots, L_{\eta_n})$. When dealing with a mechanical system we have $q = q(t) = x(t)$, the position at t, and $\eta = \eta(t) = \dot{x}(t)$, the velocity at t. The Euler-Lagrange equations for the extremal problem $\inf_{x \in D_{A,B}} \int_a^b L(t, x(t), \dot{x}(t)) \, \mathrm{d}t$ are given by

$$\frac{\mathrm{d}}{\mathrm{d}t} L_{\eta_i} - L_{q_i} = 0 \tag{30.7}$$

and by Theorem 27.6 we know that if

$$\det \frac{\partial^2 L}{\partial \eta_k \partial \eta_l} \neq 0 \tag{30.8}$$

then extremals, i.e. solutions of (30.7), are C^2-functions $x : [a, b] \to \mathbb{R}^n$. Moreover, under (30.8) we may solve locally the system of equations

$$p = L_\eta(q, \eta) \tag{30.9}$$

for $\eta = \eta(t, q, p)$ $(= \dot{q}(t))$. The new variable p is called the generalised momentum. The **Hamilton function** associated with L is now introduced as

$$H(t, q, p) := \sum_{i=1}^{n} \eta_i p_i - L(t, q, \eta) \tag{30.10}$$

with $\eta_i = \eta_i(t, q, p)$ $(= \dot{q}_i(t))$. This is the typical way a textbook in mechanics introduces the Legendre transform and then the Hamilton function when starting with the Lagrange function. The process is however well understood in the context of convex analysis, see Chapter II.13. We can introduce the partial conjugate convex function $L^{*,\eta}$ of L with respect to η by

$$H(t, q, p) := L^{*,\eta}(t, q, p) := \sup_{\eta \in \mathbb{R}^n} \left\{ \langle p, \eta \rangle - L(t, q, \eta) \right\}. \tag{30.11}$$

The critical points of

$$\eta \mapsto \langle p, \eta \rangle - L(t, q, \eta), \quad q \in \mathbb{R}^n \text{ fixed} \tag{30.12}$$

are obtained by looking at

$$\frac{\partial}{\partial \eta_j} \left(\langle p, \eta \rangle - L(t, q, \eta) \right) = 0 \tag{30.13}$$

or

$$p_j = \frac{\partial L}{\partial \eta_j}(t, q, \eta) \tag{30.14}$$

which is nothing but (30.9). However we need to guarantee that p_j exists and is uniquely determined in order to obtain a well defined function $H = H(t, q, p)$. A strict convexity assumption on $\eta \mapsto L(t, q, \eta)$ together with the coercivity condition

$$\lim_{\|\eta\| \to \infty} \frac{L(t, q, \eta)}{\|\eta\|} = \infty \quad \text{for all } q \in \mathbb{R}^n \tag{30.15}$$

is sufficient for a C^2-function L, in fact only that $\eta \mapsto L(t, q, \eta)$ is a C^2-function is needed. It is an easy exercise in convex analysis to prove that under these conditions we also have

$$H^*(t, q, \eta) = L(t, q, \eta). \tag{30.16}$$

If $L(t, q, \eta) = T - V$ with $T = T(\eta)$ and $V = V(t, q)$ where T is a positive definite quadratic form in η (a typical kinetic energy of a system of finitely many point masses) we find (under suitable normalisations)

$$H(t, q, p) = T(p) + V(t, q). \tag{30.17}$$

So far we have used the Lagrange function and Hamilton's principle to describe the dynamics associated with L and the corresponding differential equations where the second order Euler-Lagrange equations

$$\frac{\mathrm{d}}{\mathrm{d}t} L_{\eta_j}(t, x(t), \dot{x}(t)) - L_{q_j}(t, x(t), \dot{x}(t)) = 0, \quad 1 \leq j \leq n. \tag{30.18}$$

The Hamilton function $H(t, q, p)$, $p_j = \frac{\partial L}{\partial \eta_j}$, allows us to introduce first order equations for $q = q(t) = x(t)$ and $p = p(t)$. Starting with

$$H(t, q, p) = \sum_{k=1}^{n} \dot{q}_k p_k - L(t, q, \dot{q}),$$

we first observe that

$$H_{q_j} = \sum_{k=1}^{n} p_k \frac{\partial \dot{q}_k}{\partial q_j} - L_{q_j} - \sum_{k=1}^{n} L_{\dot{q}_k} \frac{\partial \hat{q}_k}{\partial q_j}$$

or

$$H_{q_j} = -L_{q_j}$$

and using the Euler-Lagrange equations as well as $p_j = \frac{\partial L}{\partial \dot{q}_j}$ we get for solutions of the Euler-Lagrange equations

$$H_{q_j} = -\frac{\mathrm{d}}{\mathrm{d}t} L_{\dot{q}_j} = \dot{p}_j. \tag{30.19}$$

Moreover, we have

$$H_{p_j} = \dot{q}_j + \sum_{k=1}^{n} p_k \frac{\partial \dot{q}_k}{\partial p_j} - \sum_{k=1}^{n} \frac{\partial L}{\partial \dot{q}_k} \frac{\partial \dot{q}_k}{\partial p_j}$$

and again we may use $p_k = \frac{\partial L}{\partial \dot{q}_k}$ to arrive at

$$H_{p_j} = \dot{q}_j. \tag{30.20}$$

526

Thus we have derived the system, called the **Hamilton system** associated with $H(t, q, p)$

$$\begin{cases} \dot{p}_j = -H_{q_j} & \text{(30.21a)} \\ \dot{q}_j = H_{p_j} & \text{(30.21b)} \end{cases}$$

which holds together with

$$\frac{\partial H}{\partial t} = -\frac{\partial L}{\partial t}. \tag{30.22}$$

By the Hamilton principle, in classical mechanics we are interested in the extremals of $\int_a^b L\left(s, x(s), \dot{x}(s)\right) ds$ where $L\left(s, x(s), \dot{x}(s)\right)$ is the Lagrange function of the system under investigation. We may replace $L\left(s, x(s), \dot{x}(s)\right)$ by $\langle p(s), \dot{q}(s)\rangle - H(s, q(s), p(s))$ and look for the extremals of

$$\int_a^b \left(\langle p(s), \dot{q}(s)\rangle - H\left(s, q(s), p(s)\right)\right) ds. \tag{30.23}$$

However, now q and p are independent functions and this we must take into account when deriving the Euler-Lagrange equations corresponding to (30.23). Thus we must vary q and p independently, i.e. for an extremal (q, p) we look at

$$A(\epsilon, \eta) := \int_a^b \left(\left\langle p(t) + \eta\varphi(t), \dot{q}(t) + \epsilon\dot{\psi}(t) \right\rangle - H(t, q(t) + \epsilon\psi(t), p(t) + \eta\varphi(t)) \right) dt$$

where φ and ψ are elements of $C_0^1\left([a, b]; \mathbb{R}^n\right)$. This yields for $\left.\frac{\partial A}{\partial \epsilon}\right|_{\epsilon=0, \eta=0}$ and $\left.\frac{\partial A}{\partial \eta}\right|_{\epsilon=0, \eta=0}$ now

$$\frac{\partial}{\partial \epsilon} \int_a^b \left(\left\langle p(t) + \eta\varphi(t), \dot{q}(t) + \epsilon\dot{\psi}(t) \right\rangle - H(t, q(t) + \epsilon(t)\psi(t), p(t) + \eta\varphi(t)) \right) dt \bigg|_{\epsilon=0, \eta=0}$$

$$= \int_a^b \left(\left\langle p(t), \dot{\psi}(t) \right\rangle - \langle H_q(t, q(t), p(t)), \psi(t)\rangle \right) dt$$

$$= \int_a^b \left(\langle -\dot{p}(t) - H_q(t, q(t), p(t)), \psi(t)\rangle \right) dt = 0,$$

or

$$\dot{p}(t) = -H_q(t, q(t), p(t)), \tag{30.24}$$

527

and

$$\frac{\partial}{\partial \eta} \int_a^b \left(\left\langle p(t) + \eta\varphi(t), \dot{q}(t) + \epsilon\dot{\psi}(t) \right\rangle - H(t, q(t) + \epsilon(t)\psi(t), p(t) + \eta\varphi(t)) \right) dt \Big|_{\epsilon=0, \eta=0}$$

$$= \int_a^b \left(\langle \varphi(t), \dot{q}(t) \rangle - \langle H_p(t, q(t), p(t)), \varphi(t) \rangle \right) dt$$

$$= \int_a^b \left(\langle \dot{q}(t) - H_p(t, q(t), p(t)), \varphi(t) \rangle \right) dt = 0,$$

implying

$$\dot{q}(t) = H_p(t, q(t), p(t)). \tag{30.25}$$

Thus we can identify the Hamilton system (30.21a), (30.21b) as Euler-Lagrange equations of (30.23).

Remark 30.1. In the case that $H(t, q, p) = H(q, p)$ is independent of t, we note that (30.24) is just (29.36c) and (30.25) is just (29.36a). Moreover, by Lemma 29.7 it follows that H is a first integral of the system (30.24) and (30.25).

Example 30.2. The Lagrange function of the classical harmonic oscillator in $\mathbb{R}^3$ is

$$L(x(t), \dot{x}(t)) = \frac{m}{2}\|\dot{x}(t)\|^2 - \frac{\gamma}{2}\|x(t)\|^2, \quad \gamma > 0. \tag{30.26}$$

The corresponding Hamilton function is

$$H(q(t), p(t)) = \frac{1}{2m}\|p(t)\|^2 + \frac{\gamma}{2}\|q(t)\|^2. \tag{30.27}$$

The Euler-Lagrange equations corresponding to (30.26) are

$$\frac{d}{dt}L_{\dot{x}_j} - L_{x_j} = m\ddot{x}_j(t) + \gamma x_j(t) = 0, \quad 1 \le j \le 3$$

or

$$\ddot{x}_j(t) + \frac{\gamma}{m}x_j(t) = 0. \tag{30.28}$$

The corresponding Hamilton system is given by

$$\dot{q}_j(t) = H_{p_j} = \frac{1}{m}p_j(t)$$

$$\dot{p}_j(t) = -H_{q_j} = -\gamma q_j(t).$$

We also note that $p_j = L_{\dot{x}_j} = m\dot{x}_j$, and $q_j = x_j$ and hence

$$\dot{p}_j(t) = m\ddot{x}_j(t) = -\gamma x_j(t)$$

which is of course (30.28).

The aim in mechanics must be to solve either the Euler-Lagrange equations, or the Hamilton system. These are the equations of motion and under appropriate regularity assumptions they are equivalent. We will now discuss some strategies to solve these equations (if possible) and this will lead us to certain partial differential equations of first order.

The Hamilton system is a system of $2n$ ordinary differential equations for $2n$ unknown functions q_j and p_j, $1 \leq j \leq n$. There are two observations which are rather helpful when dealing with Hamilton systems. The variables q_j and p_j are in a special relationship, they are called **canonical conjugate variables**, and this special relation is the following: if $H(t, q, p) = H(t, q_1, \ldots, q_n, p_1, \ldots, p_n)$ does not depend on q_j, then by (30.21a) it follows that p_j must be constant. The other observation is that first integrals of (30.21a), (30.21b) may allow us to reduce the number of unknown functions.

Example 30.3. Suppose that $H(q, p) = \frac{1}{2m}\|p\|^2 + V(q)$. We know that, in this case, H is a first integral of the Hamilton system. Given the initial values $q(t_0) = q_0$ and $p(t_0) = p_0$ we have, as a preserved quantity, the initial energy $E_0 = \frac{1}{2m}\|p_0\|^2 + V(q_0)$. Thus for a solution $(q(t), p(t))$ of the corresponding Hamilton system we find

$$H(q(t), p(t)) = \frac{1}{2m}\left(p_1^2(t) + \cdots + p_n^2(t)\right) + V(q) = E_0$$

and we get

$$p_1^2(t) = 2mE_0 - 2mV(q(t)) - p_2^2(t) - \cdots - p_n^2(t), \tag{30.29}$$

which allows us to eliminate $p_1(t)$ in $H(q(t), p(t))$ and we have a reduction of the unknown functions to $2n - 1$, in fact to $2n - 2$ taking (30.21a) and (30.21b) into account.

The last example suggests to find as many first integrals for the Hamilton system and to use them to reduce the number of unknown functions. Ideally, in the case where we have $N \leq 2n$ first integrals, we might reduce the number of unknown functions to $2n - N$. In this context we point to Noether's theorem, Theorem 27.7, which allows us to derive the existence of first integrals from the existence of certain symmetries.

We return to the first observation and our aim is to find a transformation theory which allows us to transform a given Hamilton system into a new one, but the new one shall have a Hamilton function depending on less than $2n$ variables. First we give

Definition 30.4. *A coordinate q_j, $1 \leq j \leq n$, is called a **cyclic coordinate** of the Hamilton function $H(t, q, p)$ if $\frac{\partial H}{\partial q_j} = 0$, i.e. H does not depend on q_j.*

Corollary 30.5. *If q_j is a cyclic coordinate of $H(q, p)$ then p_j is a first integral.*

In the following we restrict our considerations to a Hamilton function H not depending explicitly on t, i.e. $H = H(q, p)$. This implies in particular that H is a first integral of the associated Hamilton system. We are now looking for a transformation or a change of variables of class C^1 (or better)

$$\Phi : \mathbb{R}^{2n} \to \mathbb{R}^{2n}, \quad (q, p) \mapsto (Q, P), \tag{30.30}$$

such that in the new coordinates (Q, P) the transformed Hamilton function $K(Q, P) = H(q(Q, P), p(Q, P))$ leads to the Hamilton system

$$\dot{Q}_j = \frac{\partial K}{\partial P_j}, \quad \dot{P}_j = -\frac{\partial K}{\partial Q_j} \tag{30.31}$$

and solutions of (30.21a), (30.21b) are transformed to solutions of (30.31).
The hope is to find a change of variables such that K has more cyclic coordinates than H. Using the chain rule we find for $q(t), p(t)$ solving (30.21a) and (30.21b)

$$\begin{aligned}
\dot{Q}_j(t) &= \frac{d}{dt} Q_j(q(t), p(t)) \\
&= \sum_{l=1}^{n} \frac{\partial Q_j}{\partial q_l} \dot{q}_l(t) + \sum_{l=1}^{n} \frac{\partial Q_j}{\partial p_l} \dot{p}_l(t) \\
&= \sum_{l=1}^{n} \frac{\partial Q_j}{\partial q_l} H_{p_l} - \sum_{l=1}^{n} \frac{\partial Q_j}{\partial p_l} H_{q_l},
\end{aligned} \tag{30.32}$$

and

$$\begin{aligned}
\dot{P}_j(t) &= \frac{d}{dt} P_j(q(t), p(t)) \\
&= \sum_{l=1}^{n} \frac{\partial P_j}{\partial q_l} \dot{q}_l(t) + \sum_{l=1}^{n} \frac{\partial P_j}{\partial p_l} \dot{p}_l(t) \\
&= \sum_{l=1}^{n} \frac{\partial P_j}{\partial q_l} H_{p_l} - \sum_{l=1}^{n} \frac{\partial P_j}{\partial p_l} H_{q_l}.
\end{aligned} \tag{30.33}$$

On the other hand we have

$$\frac{\partial K}{\partial Q_j} = \sum_{l=1}^{n} H_{q_l} \frac{\partial q_l}{\partial Q_j} + \sum_{l=1}^{n} H_{p_l} \frac{\partial p_l}{\partial Q_j} \tag{30.34}$$

and

$$\frac{\partial K}{\partial P_j} = \sum_{l=1}^{n} H_{q_l} \frac{\partial q_l}{\partial P_j} + \sum_{l=1}^{n} H_{p_l} \frac{\partial p_l}{\partial P_j}. \tag{30.35}$$

Aiming for (30.31) we require

$$
\begin{cases}
\dfrac{\partial p_l}{\partial P_j} = \dfrac{\partial Q_j}{\partial q_l}, & \dfrac{\partial p_l}{\partial Q_j} = -\dfrac{\partial P_j}{\partial q_l} & \text{(30.36a)} \\[3mm]
\dfrac{\partial q_l}{\partial P_j} = -\dfrac{\partial Q_j}{\partial p_l}, & \dfrac{\partial q_l}{\partial Q_j} = \dfrac{\partial P_j}{\partial p_l}. & \text{(30.36b)}
\end{cases}
$$

With $\dfrac{\partial P}{\partial p} = \left(\dfrac{\partial P_j}{\partial p_l}\right)_{j,l=1,\ldots,n}$, $\dfrac{\partial P}{\partial q} = \left(\dfrac{\partial P_j}{\partial q_l}\right)_{j,l=1,\ldots,n}$, $\dfrac{\partial Q}{\partial q} = \left(\dfrac{\partial Q_j}{\partial q_l}\right)_{j,l=1,\ldots,n}$ and $\dfrac{\partial Q}{\partial p} = \left(\dfrac{\partial Q_j}{\partial p_l}\right)_{j,l=1,\ldots,n}$, we now claim the condition, see Problem 6,

$$
\begin{pmatrix} \dfrac{\partial Q}{\partial q} & \dfrac{\partial Q}{\partial p} \\[3mm] \dfrac{\partial P}{\partial q} & \dfrac{\partial P}{\partial p} \end{pmatrix}^{-1} = \begin{pmatrix} \left(\dfrac{\partial P}{\partial p}\right)^T & -\left(\dfrac{\partial Q}{\partial p}\right)^T \\[3mm] -\left(\dfrac{\partial P}{\partial q}\right)^T & \left(\dfrac{\partial Q}{\partial q}\right)^T \end{pmatrix} \tag{30.37}
$$

where (30.37) is an equality of $2n \times 2n$-matrices built up as block matrices, each block being an $n \times n$-matrix. Furthermore, we now denote the transpose of a matrix A by A^T (instead of A^t).

Definition 30.6. *A C^∞-coordinate change $\Phi : \mathbb{R}^{2n} \to \mathbb{R}^{2n}$, $(q,p) \mapsto \Phi(q,p) = (Q,P)$ satisfying (30.37) (or (30.36a), (30.36b)) is called a* **canonical transformation**.

We can now reformulate our goal: given a Hamilton function, find a canonical transformation such that the transformed Hamilton function has more cyclic coordinates.

In order to derive (30.37), we have assumed that under a canonical transformation a Hamilton function H is transformed into a new Hamilton function K such that Hamilton's equations are also valid for the transformed system. In addition, no transformation of the parameter t was used. Then in both cases the first variation of the integrals

$$
\int_a^b \left(\langle p(s), \dot{q}(s) \rangle - H(s, q(s), p(s)) \right) ds
$$

and

$$
\int_a^b \left(\langle P(s), \dot{Q}(s) \rangle - K(s, Q(s), P(s)) \right) ds
$$

must vanish, recall that we have to treat q and p as well as Q and P as independent functions. This observation implies that there must exist a function G, which may depend explicitly on s, but also on some of the components of $q(s)$, $p(s)$, $Q(s)$ and $P(s)$, such that

$$\langle p(s), \dot{q}(s) \rangle - H(s, q(s), p(s)) = \left\langle P(s), \dot{Q}(s) \right\rangle - K(s, Q(s), P(s)) + \frac{\mathrm{d}}{\mathrm{d}s} G. \quad (30.38)$$

We call the function G the **generating function of the canonical transformation**. It turns out that the most interesting generating functions are of the type

$$G_1 = G_1(s, q, Q), \quad G_2 = \tilde{G}_2(s, q, P) - \langle Q, P \rangle,$$
$$G_3 = \tilde{G}_3(s, Q, p) + \langle q, p \rangle, \quad G_4 = \tilde{G}_4(s, Q, P) + \langle q, p \rangle - \langle Q, P \rangle.$$

Let us consider for a moment $G = G(s, r, T)$ where (r, T) is one of the pairs (q, Q), (q, P), (Q, p) or (p, P). It follows that

$$\frac{\mathrm{d}}{\mathrm{d}s} G(s, r(s), T(s)) = \frac{\partial G}{\partial s} + \sum_{j=1}^{n} \frac{\partial G}{\partial r_j} \dot{r}_j + \sum_{j=1}^{n} \frac{\partial G}{\partial T_j} \dot{T}_j. \quad (30.39)$$

Now we can compare (30.38) with (30.39) which we do just for the case $r(s) = q(s)$ and $T(s) = Q(s)$, i.e. G_1. The three remaining cases are discussed briefly in Problem 7. It follows that

$$\sum_{j=1}^{n} p_j(s) \dot{q}_j(s) - H(s, q(s), p(s))$$

$$= \sum_{j=1}^{n} P_j(s) \dot{Q}_j(s) - K(s, Q(s), P(s)) + \frac{\partial G_1}{\partial s} + \sum_{j=1}^{n} \frac{\partial G_1}{\partial q_j} \dot{q}_j + \sum_{j=1}^{n} \frac{\partial G_1}{\partial Q_j} \dot{Q}_j$$

implying

$$\sum_{j=1}^{n} \left(p_j - \frac{\partial G_1}{\partial q_j} \right) \dot{q}_j = \sum_{j=1}^{n} \left(P_j + \frac{\partial G_1}{\partial Q_j} \right) \dot{Q}_j + H(s, q(s), p(s)) - K(s, Q(s), P(s)) + \frac{\partial G_1}{\partial t}.$$
$$(30.40)$$

Taking (30.36a) and (30.36b) into account, we may suggest

$$\begin{cases} p_j = \dfrac{\partial G_1}{\partial q_j}(s, q, Q) & (30.41a) \\[2mm] P_j = -\dfrac{\partial G_1}{\partial Q_j}(s, q, Q) & (30.41b) \\[2mm] K = H + \dfrac{\partial G_1}{\partial s}(s, q, Q) & (30.41c) \end{cases}$$

as a transformation, provided that we can solve (30.41b) for q since then (30.41a) will give p_j as a function of Q and P. The equation (30.41c) then gives the new Hamilton function. Note that if G_1 does not depend on s then we have $K = H$.

Example 30.7. (Compare with A. Budó [15]) We pick up Example 30.2 however now for $n = 1$. The Hamilton function is $H(q, p) = \frac{1}{2m}p^2 + \frac{\gamma}{2}q^2$, but for simplicity we choose $m = \gamma = 1$. As a generating function, we suggest

$$G_1(q, Q) = \frac{1}{2}q^2 \cot Q \tag{30.42}$$

to find

$$p = \frac{\partial G_1}{\partial q} = q \cot Q \tag{30.43}$$

and

$$P = -\frac{\partial G_1}{\partial Q} = \frac{1}{2}\frac{q^2}{\sin^2 Q}. \tag{30.44}$$

These equations yield

$$q = \sqrt{2}\sqrt{P}\sin Q \tag{30.45}$$

and

$$p = \sqrt{2}\sqrt{P}\cos Q, \tag{30.46}$$

hence

$$K(Q, P) = H(q(Q, P), p(Q, P)) + \frac{\partial G_1}{\partial t}$$
$$= H(q(Q, P), p(Q, P))$$
$$= \frac{1}{2} \cdot 2P\cos^2 Q + \frac{1}{2} \cdot 2P\sin^2 Q = P.$$

This implies that Q is a cyclic coordinate of $K(Q, P)$ and therefore we must have $\dot{P}(t) = 0$, or $P(t) = c_0$ and further $\dot{Q} = 1$ which gives $Q(t) = t + c_1$. Now (30.45) and (30.46) entail

$$q(t) = \sqrt{2c_0}\sin(t + c_1)$$
$$p(t) = \sqrt{2c_0}\cos(t + c_1) \; (= \dot{q}(t)).$$

As Example 30.7 shows, it is possible to find canonical transformations leading to Hamilton functions with more cyclic coordinates than the original one has. The question which remains is, how can we find such transformations in a more systematic way. A first integral of the Hamilton system is a function $J : \mathbb{R}^{2n+1} \rightarrow$

$\mathbb{R}$, $(t, q, p) \mapsto J(t, q(t), p(t))$, with the property that for a solution (q, p) of the Hamilton system we have $\frac{d}{dt} J(t, q(t), p(t)) = 0$. Since

$$\frac{d}{dt} J(t, q(t), p(t)) = \frac{\partial J}{\partial t} + \sum_{j=1}^{n} \frac{\partial J}{\partial q_j} \dot{q}_j + \sum_{j=1}^{n} \frac{\partial J}{\partial p_j} \dot{p}_j \tag{30.47}$$

$$= \frac{\partial J}{\partial t} + \sum_{j=1}^{n} \left(\frac{\partial J}{\partial q_j} \frac{\partial H}{\partial p_j} - \frac{\partial J}{\partial p_j} \frac{\partial H}{\partial q_j} \right),$$

where we have used the fact that (q, p) solves the Hamilton equations, thus for a first integral J of the Hamilton system we must have

$$\frac{\partial J}{\partial t} + \sum_{j=1}^{n} \left(\frac{\partial J}{\partial q_j} \frac{\partial H}{\partial p_j} - \frac{\partial J}{\partial p_j} \frac{\partial H}{\partial q_j} \right) = 0. \tag{30.48}$$

In the case where J does not depend explicitly on t, we arrive at

$$\{J, H\} := \sum_{j=1}^{n} \left(\frac{\partial J}{\partial q_j} \frac{\partial H}{\partial p_j} - \frac{\partial J}{\partial p_j} \frac{\partial H}{\partial q_j} \right) = 0. \tag{30.49}$$

For two function $J_1, J_2 : \mathbb{R}^{2n+1} \to \mathbb{R}$, $J_l = J_l(t, q, p)$, $l = 1, 2$, we call

$$\{J_1, J_2\} := \sum_{j=1}^{n} \left(\frac{\partial J_1}{\partial q_j} \frac{\partial J_2}{\partial p_j} - \frac{\partial J_1}{\partial p_j} \frac{\partial J_2}{\partial q_j} \right) \tag{30.50}$$

the **Poisson bracket** of J_1 and J_2. Thus a necessary condition for a function $J = J(q, p)$ to be a first integral of the Hamilton system is that its Poisson bracket with Hamilton function vanishes provided H does not depend explicitly on t. The following rules for the Poisson bracket are easy to check

$$\{J_1, J_1\} = 0, \tag{30.51}$$

$$\{J_1, J_2\} = -\{J_2, J_1\}, \tag{30.52}$$

$$\{J_1, J_2 + J_3\} = \{J_1, J_2\} + \{J_1, J_3\}, \tag{30.53}$$

$$\frac{\partial J_1}{\partial q_k} = \{J_1, p_k\}, \tag{30.54}$$

$$\frac{\partial J_1}{\partial p_k} = -\{J_1, q_k\} \tag{30.55}$$

as well as

$$\{q_k, q_l\} = 0, \quad \{p_k, p_l\} = 0 \tag{30.56}$$

and

$$\{q_k, p_l\} = \delta_{kl}. \tag{30.57}$$

Moreover, we can rewrite the Hamilton system as

$$\dot{q}_j = \{q_j, H\}, \quad \dot{p}_j = \{p_j, H\} \tag{30.58}$$

and (30.47) as

$$\frac{\mathrm{d}J}{\mathrm{d}t} = \frac{\partial J}{\partial t} + \{J, H\}. \tag{30.59}$$

Note that (30.59) implies for $J = H$ that

$$\frac{\mathrm{d}H}{\mathrm{d}t} = \frac{\partial H}{\partial t},$$

i.e. if H does not explicitly depend on t then it is a first integral as we already know.

From (30.57) and (30.58) we deduce that necessary and sufficient for a transformation to be a canonical transformation is that for the new canonical variables $Q = Q(q, p)$ and $P = P(q, p)$ we have

$$\{Q_k, Q_l\} = 0, \quad \{P_k, P_l\} = 0 \quad \text{and} \quad \{Q_k, P_l\} = \delta_{kl}. \tag{30.60}$$

A further useful identify for the Poisson bracket is the **Jacobi identity**

$$\{J_1, \{J_2, J_3\}\} + \{J_2, \{J_3, J_1\}\} + \{J_3, \{J_1, J_2\}\} = 0 \tag{30.61}$$

which we will prove in Problem 8. With the help of the Jacobi identity we immediately obtain

Corollary 30.8. *Suppose that H does not depend explicitly on t. Let $J_l = J_l(q, p)$, $l = 1, 2$, be two first integrals of a given Hamilton system. Then $\{J_1, J_2\}$ is a further first integral of the system.*

Proof. We know that $\{J_l, H\} = 0$ and from Jacobi's identity we derive

$$\{\{J_1, J_2\}, H\} = \{J_1, \{J_2, H\}\} + \{J_2, \{H, J_1\}\} = 0.$$

$\square$

Let $H(t, q, p)$ be a Hamilton function on $\mathbb{R}^{2n+1}$ and suppose that we can find a canonical transformation Φ with generating function W such that the transformed Hamilton function $K(t, Q, P)$ vanishes identically. The corresponding Hamilton system becomes

$$\dot{Q}_j = \frac{\partial K}{\partial P_j} = 0 \quad \text{and} \quad P_j = -\frac{\partial K}{\partial Q_j} = 0, \tag{30.62}$$

implying that both Q_j and P_j are constant. We want to investigate the properties of W and try to answer the question whether such a W may exist. We assume here that W is of the type G_1, i.e. $W = W(t, q, Q)$, which gives

$$0 = K = \frac{\partial W(t, q, Q)}{\partial t} + H(t, q, p), \quad q = q(Q, P), \, p = p(Q, P). \tag{30.63}$$

From (30.41a) we derive

$$p_j = \frac{\partial W}{\partial q_j}(t, q, Q) \tag{30.64}$$

and since $Q_j = \gamma_j$ is a constant we find that p_j depends de facto only on q. Substituting (30.64) into (30.63) we obtain, for W as a function of q and the constants $\gamma = (\gamma_1, \ldots, \gamma_n)$, the **Hamilton-Jacobi differential equation**

$$\frac{\partial W}{\partial t} + H\left(t, q_1, \ldots, q_n, \frac{\partial W}{\partial q_1}, \ldots, \frac{\partial W}{\partial q_n}\right) = 0, \quad W = W(t, q, \gamma). \tag{30.65}$$

The Hamilton-Jacobi differential equation is a first order partial differential equation for $(t, q) \mapsto W(t, q, \gamma)$ where γ_j, $1 \leq j \leq n$, is constant. Note that, in general, a first order partial differential equation for a function of $n + 1$ variables will be dependent on $n + 1$ functions, however we are looking only for special solutions in order to derive a canonical transformation, so we do not seek a general solution, but a solution which only depends on $n + 1$ constants. Moreover, since only partial derivatives of W enter the equation, one constant must be additive, note that if W solves (30.65) then so does $W + \kappa$, and we choose this constant to be equal to 0. Thus we are seeking, for a function $(t, q) \mapsto W(t, q, \gamma)$ depending on constants $\gamma_1, \ldots, \gamma_n$, to solve (30.65). We do not want to discuss domain problems for an existent solution $W(t, q, \gamma)$, but a warning is needed. Using the methods of characteristics as developed in the last chapter will not help a lot since the characteristic differential equations will give the original Hamilton system!

Suppose we can find a solution $W(t, q, \gamma)$ of (30.65) and we use this solution as a generating function of a canonical transformation. We now have $Q_j = \gamma_j$ and further

$$p_j = \frac{\partial W}{\partial q_j} \quad \text{and} \quad P_j = -\frac{\partial W}{\partial \gamma_j}. \tag{30.66}$$

The Hamilton system corresponding to $K = 0$ is

$$\dot{Q}_j = 0 \quad \text{and} \quad \dot{P}_j = 0,$$

implying that Q_j and P_j are constants and therefore the equations in (30.66) are "algebraic" equations for p_j and q_j. It turns out that this observation allows us, in some cases of interest, to solve the equations of motions. In other cases, the

Hamilton function may split as

$$H(t,q,p) = H_1\left(t, q^{(1)}, p^{(1)}\right) + H_2\left(t, q^{(2)}, p^{(2)}\right)$$

with $q = \left(q^{(1)}, q^{(2)}\right)$ and $p = \left(p^{(1)}, p^{(2)}\right)$ which leads to a partial decoupling of the Hamilton system and the method developed above might be applied to one of the sub-systems, compare with Problem 10 and 11. In general we refer to books on classical mechanics such as [1], [6], [15] or [45] for more discussions and concrete examples.

In the case where H does not depend explicitly on t we know that H is a first integral and with the conserved initial energy $E = H(q_0, p_0)$ we find from (30.65) with $W = S - Et$ that

$$H\left(q_1, \ldots, q_n, \frac{\partial S}{\partial q_1}, \ldots, \frac{\partial S}{\partial q_n}\right) = E. \tag{30.67}$$

We want to summarise our considerations in two theorems.

Theorem 30.9. *Let* $W = W(t, q_1(t), \ldots q_n(t), \alpha_1, \ldots, \alpha_m)$, $m \leq n$, *be a* C^2- *solution of the Hamilton-Jacobi equations*

$$\frac{\partial W}{\partial t} + H\left(t, q_1, \ldots, q_n, \frac{\partial W}{\partial q_1}, \ldots, \frac{\partial W}{\partial q_n}\right) = 0$$

depending on m *parameters. Then* $\frac{\partial W}{\partial \alpha_j}$, $j = 1, \ldots, m$, *is a first integral of the Hamilton system*

$$\dot{q}_j = H_{p_j}, \quad \dot{p}_j = -H_{q_j}.$$

Proof. We have to prove that along an extremal, i.e. a solution of the Hamilton system, we have $\frac{d}{dt}\frac{\partial W}{\partial \alpha_j} = 0$, $1 \leq j \leq m$. We note that

$$\frac{d}{dt}\frac{\partial W}{\partial \alpha_j} = \frac{\partial^2 W}{\partial t \partial \alpha_j} + \sum_{k=1}^{n} \frac{\partial^2 W}{\partial q_k \partial \alpha_j} \dot{q}_k. \tag{30.68}$$

Further, we obtain from the Hamilton-Jacobi equation

$$\frac{\partial}{\partial \alpha_j}\left(\frac{\partial}{\partial t}W + H\left(t, q_1, \ldots, q_n, \frac{\partial W}{\partial q_1}, \ldots, \frac{\partial W}{\partial q_n}\right)\right)$$

$$= \frac{\partial^2 W}{\partial t \partial \alpha_j} + \sum_{k=1}^{n} \frac{\partial H}{\partial p_k}\frac{\partial^2 W}{\partial q_k \partial \alpha_j} = 0$$

or

$$\frac{\partial^2 W}{\partial t \partial \alpha_j} = -\sum_{k=1}^{n} \frac{\partial H}{\partial p_k} \frac{\partial^2 W}{\partial q_k \partial \alpha_j},$$

which implies with (30.68)

$$\frac{d}{dt} \frac{\partial W}{\partial \alpha_j} = -\sum_{k=1}^{n} \frac{\partial H}{\partial p_k} \frac{\partial^2 W}{\partial q_k \partial \alpha_j} + \sum_{k=1}^{n} \frac{\partial^2 W}{\partial q_k \partial \alpha_j} \dot{q}_k$$

$$= \sum_{k=1}^{n} \frac{\partial^2 W}{\partial q_k \partial \alpha_j} (\dot{q}_k - H_{p_k}) = 0.$$

$\square$

Theorem 30.10. *Let $W(t, q_1, \ldots, q_n, \alpha_1, \ldots, \alpha_n)$ be a C^2-solution of the Hamilton-Jacobi equation and assume that*

$$\det \left(\frac{\partial^2 W}{\partial q_k \partial \alpha_j} \right)_{k,j=1,\ldots,n} \neq 0. \tag{30.69}$$

Let $\beta_1, \ldots, \beta_n \in \mathbb{R}$ be arbitrary constants. Then the n equations

$$\frac{\partial W}{\partial \alpha_j}(t, q_1, \ldots, q_n, \alpha_1, \ldots \alpha_n) = \beta_j, \quad j = 1, \ldots, n,$$

define n functions

$$q_j = q_j(t, \alpha_1, \ldots, \alpha_n, \beta_1, \ldots, \beta_n), \quad j = 1, \ldots, n,$$

of class C^1 which together with the functions

$$p_j = \frac{\partial}{\partial q_j} W(t, q_1, \ldots, q_n, \alpha_1, \ldots, \alpha_n), \quad j = 1, \ldots, n,$$

are solutions of the Hamilton system

$$\dot{q}_j = H_{p_j} \quad and \quad \dot{p}_j = -H_{q_j}, \quad j = 1, \ldots, n.$$

Proof. We choose the function W as a generating function of a canonical transformation with $p_j = \frac{\partial W}{\partial q_j}$, $\beta_j = \frac{\partial W}{\partial \alpha_j}$ and transformed Hamilton function $K = H + \frac{\partial W}{\partial t}$. Since W is a solution of the Hamilton-Jacobi equation we have

$$K = H + \frac{\partial W}{\partial t} = 0$$

and therefore $\dot{\alpha}_j = 0$ and $\dot{\beta}_j = 0$, i.e. α_j and β_j are constant along extremals, so we have

$$\frac{\partial W}{\partial \alpha_j}(t, q_1, \ldots, q_n, \alpha_1, \ldots, \alpha_n) = \beta_j$$

with constants β_j. The condition (30.69) entails that we can solve this equation for $q_1, \ldots, q_n$ and it follows from the construction that the $2n$ functions q_j and p_j solve the Hamilton system. $\qquad\square$

Finally we want to return to the **action** as defined by (30.5) and relate it to the Hamilton-Jacobi equation. Let $W = W(t, q_1(t), \ldots, q_n(t), \alpha_1, \ldots, \alpha_n)$ be as in Theorem 30.10. It follows that

$$\frac{d}{dt}W = \frac{\partial W}{\partial t} + \sum_{k=1}^{n} \frac{\partial W}{\partial q_k} \dot{q}_k = \frac{\partial W}{\partial t} + \sum_{k=1}^{n} p_k \dot{q}_k$$

$$= -H(t, q, p) + \sum_{k=1}^{n} p_k \dot{q}_k = L(t, q(t), \dot{q}(t)),$$

i.e. $\frac{dW}{dt} = L$, or

$$W(t) = \int_a^t L(s, q(s), \dot{q}(s)) \, ds, \tag{30.70}$$

which means that in this case by (30.5) we find that W is just the action S.

Problems

1.   a) Let $f : \mathbb{R}^n \to \mathbb{R}$ be a coercive convex function which is rotational invariant, i.e. $f(x) = g(\|x\|^2)$. Assume that g is a C^1-function and that the function $r \mapsto H(r) = 4g'(r)^2 r$ admits an inverse. Prove that the conjugate convex function f^* of f is also rotational invariant.

   b) Find the conjugate convex function of $h_\alpha(x) = \frac{1}{\alpha}\|x\|^\alpha$ defined on $\mathbb{R}^n$, $\alpha > 1$.

2.   a) For $\alpha > 2$ and $q, \eta \in \mathbb{R}^n$ consider the Lagrange function $L(q, \eta) := \frac{1}{\alpha}\|\eta\|^\alpha - \frac{1}{\alpha}\|q\|^\alpha$. Find the corresponding Hamilton function.

   b) For $1 < \beta < 2$ find the Lagrange function corresponding to the Hamilton function $H(q, p) = \frac{1}{\beta}\|p\|^\beta + \frac{1}{\|q\|^{n-\beta}}$.
   Hint: use the results of Problem 1.

3. In spherical coordinates $x = r \sin \vartheta \cos \varphi$, $y = r \sin \vartheta \sin \varphi$, $z = r \cos \vartheta$, the Hamilton function $\tilde{H}(q,p) = \frac{1}{2m}\|p\|^2 + V(q)$, $q = q(x,y,z)$, has the form

$$(*) \quad H(r,\vartheta,\varphi,p_r,p_\vartheta,p_\varphi) = \frac{1}{2m}\left(p_r^2 + \frac{p_\vartheta^2}{r^2} + \frac{p_\varphi^2}{r^2 \sin^2 \vartheta} \right) + U(r,\vartheta,\varphi),$$

where p_r, p_ϑ, p_φ are the components of p in spherical coordinates and $U(r,\vartheta,\varphi) = V(q)$. For the three potentials $U_1 = U_1(r), U_2 = U_2(r,\vartheta)$ and $U_3 = U_3(r,\varphi)$, find the cyclic coordinates of $(*)$ with U being replaced by U_j.

4. Denote by $J \in M(2n;\mathbb{R})$ the matrix $J = \begin{pmatrix} 0 & \mathrm{id}_n \\ -\mathrm{id}_n & 0 \end{pmatrix}$. We call a matrix $A \in M(2n;\mathbb{R})$ **symplectic** if $A^T J A = J$, where A^T denotes the transpose of A. Prove that J and id_{2n} are symplectic and for a symmetric matrix A we always have $|\det A| = 1$. Now deduce that the set of all symplectic matrices form a subgroup $Sp(2n;\mathbb{R})$ of $GL(2n;\mathbb{R})$. The group $Sp(2n,\mathbb{R})$ is called the **symplectic group** in $\mathbb{R}^{2n}$.

Remark: it is possible to prove that for $A \in Sp(2n;\mathbb{R})$ we always have $\det A = 1$. We will use this result in Problem 5 and 6, a proof can be found in [36] or [89].

5. Let $H(q,p)$ be a Hamilton function.

a) Rewrite the corresponding Hamilton system as

$$\dot{w}(t) = J H_w(w(t)),$$

where $w = \begin{pmatrix} q \\ p \end{pmatrix}$ and J is as in Problem 4.

b) Prove that if a coordinate change $u : \mathbb{R}^{2n} \to \mathbb{R}^{2n}$ has the property that its differential, i.e. its Jacobi matrix, is at every point a symplectic matrix, then it is a canonical transformation.

6. Change our definition of a canonical transformation to the statement of Problem 5 b), i.e. call a coordinate change $u : \mathbb{R}^{2n} \to \mathbb{R}^{2n}$ a canonical transformation if du is at every point a symplectic matrix. Now verify (30.37).

7. Derive the equations analogous to (30.41a)-(30.41c) for a generating function $G_2 = \tilde{G}_2(s,q,p) - \langle Q,P \rangle$.

8. Prove the Jacobi identity (30.61), i.e. for C^2-functions $J_l = J_l(q,p)$ show that

$$\{J_1, \{J_2, J_3\}\} + \{J_2, \{J_3, J_1\}\} + \{J_3, \{J_1, J_2\}\} = 0.$$

9. Let $J_1(t,q,p)$ and $J_2(t,q,p)$ be two first integrals corresponding to the Hamilton function $H(q,p)$. Prove that

$$\frac{d}{dt}\{J_1, J_2\} = \left\{\frac{dJ_1}{dt}, J_2\right\} + \left\{J_1, \frac{dJ_2}{dt}\right\}$$

and deduce that $\{J_1, J_2\}$ is a further first integral.
Note: this result extends Corollary 30.8 to first integrals depending explicitly on t.

10. Consider the Hamilton function for the **Kepler problem** in spherical coordinates, i.e.

$$H(r, \vartheta, \varphi, p_r, p_\vartheta, p_\varphi) = \frac{1}{2m}\left(p_r^2 + \frac{p_\vartheta^2}{r^2} + \frac{p_\varphi^2}{r^2 \sin^2 \vartheta}\right) - \frac{\gamma mM}{r}.$$

State the Hamilton-Jacobi equation for W and then for S, where $W = S - Et$. Make a separation Ansatz $S(r, \vartheta, \varphi) = R(r) + \Theta(\vartheta) + \Phi(\varphi)$ and find equations for R, Θ and Φ. Show that W has the form

$$W(r, \vartheta, \varphi) = -Et + \alpha\varphi + \int \sqrt{\beta - \frac{\alpha^2}{\sin^2 \vartheta}} \, d\vartheta$$

$$+ \int \sqrt{2m(E + \frac{\gamma mM}{r}) - \frac{\beta^2}{r^2}} \, dr$$

with constants α and β where the indefinite integrals are interpreted as a primitive.
Hint: recall that we are searching for one solution of the Hamilton-Jacobi equation or equivalently the equations for S, which depends on three parameters E, α and β. We do not seek the general solution of these equations.

11. The Hamilton function $H(q_2, p_1, p_2) = \frac{1}{2m}(p_1^2 + p_2^2) + mgq_2$ is used to model a throw of a particle of mass m in a constant gravitational field with gravitation constant g. Since H does not depend on t explicitly, the corresponding (reduced) Hamilton-Jacobi equation takes, with $W = S - Et$, the form

$$(*) \quad \frac{1}{2m}\left(\left(\frac{\partial S}{\partial q_1}\right)^2 + \left(\frac{\partial S}{\partial q_2}\right)^2\right) + mgq_2 = E.$$

Find q_2 by first solving $(*)$ with the help of a separation Ansatz.

Appendices

Appendix I: Harmonic Analysis on Locally Compact Abelian Groups

There are many similarities between the expansion of 2π-periodic functions into a Fourier series and the Fourier transform of an integrable function. In fact certain theorems have identical formulations, e.g. Plancherel's theorem. It is natural to ask for a common underlying background and this leads to (abstract) **commutative harmonic analysis** or Fourier analysis on locally compact Abelian groups. In this appendix we want to outline the basic ideas and results of commutative harmonic analysis, one of the most beautiful subjects in mathematics. We do not give proofs and at some places we might be brief. This appendix should serve more as an appetizer than a part of a textbook.

We start with recollecting some group theory. By definition a group $(G, \circ)$ is a non-empty set G together with a binary operation $\circ : G \times G \to G$ which is associative, i.e.

$$(x \circ y) \circ z = x \circ (y \circ z) \quad \text{for all } x, y, z \in G. \tag{A.I.1}$$

Further there exists a **neutral element** $e \in G$ which has the property that

$$x \circ e = e \circ x = x \quad \text{for all } x \in G, \tag{A.I.2}$$

and for every $x \in G$ exists a unique **inverse element** $x^{-1} \in G$ which satisfies

$$x \circ x^{-1} = x^{-1} \circ x = e. \tag{A.I.3}$$

If for all $x, y \in G$ we have

$$x \circ y = y \circ x, \tag{A.I.4}$$

the group G is called a **commutative** or **Abelian group**. In this appendix we are only interested in Abelian groups. Examples are $(\mathbb{Z}, +)$, $(\mathbb{Z}^n, +)$, $(\mathbb{R}, +)$, $(\mathbb{R}^n, +)$, $(\mathbb{C}, +)$, $(\mathbb{C}^n, +)$, $(\mathbb{R}\backslash\{0\}, \cdot)$, $(\mathbb{C}\backslash\{0\}, \cdot)$ as well as $(\mathbb{Q}^n, +)$ or $(\mathbb{Q}\backslash\{0\}, \cdot)$. On $S^1 = T^1 = \{z \in \mathbb{C} \mid |z| = 1\} \hat{=} \{(x_1, x_2) \in \mathbb{R}^n \mid x_1^2 + x_2^2 = 1\} \hat{=} \{e^{i\varphi} \mid 0 \le \varphi \le 2\pi\}$ we can use complex multiplication to turn $(S^1, \cdot)$ into an Abelian group. Subgroups of Abelian groups are of course Abelian too as are products. For example $T^n := S^1 \times \cdots \times S^1$, the n-torus, is an Abelian group with its natural operation as is $\mathbb{Z}^k \times T^m \times \mathbb{R}^n$. In addition quotients G/H are Abelian, note that a subgroup of an Abelian group is always a normal subgroup.

The group S^1 we may identify with the quotient $\mathbb{R}/2\pi\mathbb{Z}$. Note that for $\alpha \in \mathbb{R}$ the set $\alpha\mathbb{Z} = \{y = \alpha k \mid k \in \mathbb{Z}\}$ is a subgroup of $(\mathbb{R}, +)$. Thus $T^1 = S^1 = \mathbb{R}/2\pi\mathbb{Z}$ is well defined as a quotient group and since $2\pi\mathbb{Z}$ is a closed subgroup of $\mathbb{R}$ we also obtain the correct topology on T^1. In the same way we obtain $T^n = \mathbb{R}^n/(2\pi\mathbb{Z})^n = T^1 \times \cdots \times T^1$. Some authors introduce the torus as $\mathbb{R}/\mathbb{Z}$ and the n-fold product is then $T^n = \mathbb{R}^n/\mathbb{Z}^n$. In our definition we can represent elements of T as $e^{i\varphi}$, $\varphi \in [0, 2\pi]$, whereas in the other case a representation of elements of T is given by $e^{2\pi i\varphi}$, $\varphi \in [0, 1]$. To be consistent with our treatment of Fourier series, see below, we choose $T^1 = S^1 = \mathbb{R}/2\pi\mathbb{Z}$.

A **homomorphism** between two groups $(G_1, \circ_1)$ and $(G_2, \circ_2)$ is a mapping $h : G_1 \to G_2$ with the property that

$$h(x \circ_1 y) = h(x) \circ_2 h(y). \tag{A.I.5}$$

The best known examples of homomorphisms are $\exp : (\mathbb{R}, +) \to (\mathbb{R}_+\backslash\{0\}, \cdot)$ and $\ln : (\mathbb{R}_+\backslash\{0\}, \cdot) \to (\mathbb{R}, +)$ where $\mathbb{R}_+\backslash\{0\} = \{x \in \mathbb{R} \mid x > 0\}$. A bijective homomorphism $h : G_1 \to G_2$ is called an **isomorphism** and two groups are called **isomorphic** if there exists an isomorphism between them. In this case, we write $G_1 \cong G_2$. Thus, the Abelian groups $(\mathbb{R}, +)$ and $(\mathbb{R}_+\backslash\{0\}, \cdot)$ are isomorphic. A homomorphism $h : G \to G$ is called an **endomorphism**, a bijective endomorphism is an **automorphism**. An example of an automorphism of S^1 is given by $h_{z_0}(z) := z_0 z$, $z_0 \in S^1$. In harmonic analysis we are interested in special homomorphism, namely in homomorphism $\chi : G \to S^1$. An example is $e_\xi : \mathbb{R}^n \to S^1$, $e_\xi(x) = e^{i\langle x, \xi \rangle}$, $\xi \in \mathbb{R}^n$.

In a next step we want to add topology, i.e. we want to consider topological (Abelian) groups. As a set an Abelian group may carry many topologies, but of special interest are those topologies which make the group operations continuous.

Definition A.I.1. *A **topological group** $(G, \circ, \mathcal{O})$ is a group $(G, \circ)$ on which a topology $\mathcal{O}$ is given which is Hausdorff and for which the mappings $(x, y) \mapsto x \circ y$ and $x \mapsto x^{-1}$ are continuous.*

Note that on $G \times G$ we use of course the product topology. Recall that a topological space is called a **Hausdorff space** if for any two points $x, y, x \neq y$ exists disjoint (open) neighbourhoods, i.e. we can separate x and y be neighbourhoods. Once a topology is introduced on G we can speak about continuous functions $f : G \to \mathbb{C}$. By $C(G)$ we denote the vector space of all continuous functions on G, and $C_0(G)$ is the space of all continuous functions on G with compact support. Recall that for $f : G \to \mathbb{C}$ the **support** $\mathrm{supp} f$ is defined by

$$\mathrm{supp} f := \overline{\{x \in G \mid f(x) = 0\}}. \tag{A.I.6}$$

If $(G, \circ, \mathcal{O})$ is a topological group which is in addition Abelian and for which the topology is locally compact, then we call $(G, \circ, \mathcal{O})$ a **locally compact Abelian**

group. Recall that a topological space is called **locally compact** if every point has a compact neighbourhood. If $(G, \mathcal{O})$ is a compact space we call $(G, \circ, \mathcal{O})$ a **compact (Abelian)** group.

The groups $(\mathbb{R}^n, +)$, $(\mathbb{C}^n, +), n \geq 1$, or $(\mathbb{R}\backslash\{0\}, \cdot)$ and $(\mathbb{C}\backslash\{0\}, \cdot)$ are locally compact topological groups when choosing the Euclidean topology on $\mathbb{R}^n$ and $\mathbb{C}^n$, respectively, and $(S^1, \cdot)$ as well as T^n are compact groups. The groups $(\mathbb{Z}^n, +)$, $n \geq 0$, is a discrete, i.e. every subset of $\mathbb{Z}^n$ is open, and hence it is locally compact too - the compact sets in $\mathbb{Z}^n$ are the finite sets.

The important observation is that for a topological group the topology is completely determined by the system of (open) neighbourhoods of the neutral element. Indeed, if $\mathcal{U}(x)$ denotes the system of open neighbourhoods of x, then for every $U \in \mathcal{U}(x)$ we can find $V \in \mathcal{U}(e)$ such that $U = xV$.

In [101],[98] or [118] many properties of topological groups are discussed, more advanced treatments are given in [57] and [62]. With these preparations we can give

Definition A.I.2. *A **continuous character** of a locally compact Abelian group $(G, \circ, \theta)$ is a **continuous homomorphism** $\chi : G \to S^1$.*

In a following we will write the group operation in an Abelian group additive, i.e. we write $x + y$ (for $x \circ y$), 0 for e and $(-x)$ for x^{-1}. However in S^1 we always use the multiplicative notation, in particular we write $\chi(x)\chi(y)$ for the product of the images of x and y, $x, y \in G$, under a character χ of G. Moreover, we will just write G instead of $(G, +)$ or $(G, +, \mathcal{O})$, provided no confusion may arise.

Let G be a locally compact Abelian group. The set of all continuous characters $\gamma : G \to S^1$ form an Abelian group Γ when we set for $x \in G$ and $\gamma_1, \gamma_2 \in \Gamma$

$$(\gamma_1 + \gamma_2)(x) := \gamma_1(x)\gamma_2(x). \tag{A.I.7}$$

It is convenient to use also the notation

$$< x, \gamma > := \gamma(x) \tag{A.I.8}$$

and it follows that

$$< x + y, \gamma > \; = \; < x, \gamma >< y, \gamma >, \tag{A.I.9}$$
$$< x, \gamma_1 + \gamma_2 > \; = \; < x, \gamma_1 >< x, \gamma_2 >, \tag{A.I.10}$$

as well as

$$< 0, \gamma > = \gamma(0) = 1 \qquad \text{(A.I.11)}$$

and

$$< -x, \gamma > = < x, -\gamma > = < x, \gamma >^{-1} = \overline{< x, \gamma >}. \qquad \text{(A.I.12)}$$

Note that $< x, \gamma > \in S^1 \subset \mathbb{C}$ and therefore $< x, \gamma >^{-1}$ is the multiplicative inverse of the element $< x, \gamma > \in S^1 \subset \mathbb{C}$, and $\overline{< x, \gamma >}$ is the complex conjugate of $< x, \gamma >$.

On $C(G)$ it is well justified to consider for sequences of functions as mode of convergence that of uniform convergence on compact subsets. One can show, as we will do in Volume V, that this convergence is induced by a topology on $C(G)$ which we call the **topology of compact convergence**. Since for a locally compact Abelian group $\Gamma \subset C(G)$, Γ inherits a topology from $C(G)$. With respect to this topology a base of neighbourhoods of $0 \in \Gamma$ is given by the sets

$$U(K, \epsilon) := \{\gamma \in \Gamma \mid |1 - \gamma(x)| < \epsilon \text{ for all } x \in K\} \qquad \text{(A.I.13)}$$

where ϵ varies in $(0, 1)$ and K varies over all compact subsets of G. With this topology Γ is again a locally compact Abelian group which we call the (topological) **dual group** of G. We will later give a further possibility to introduce and hence to characterize the topology on Γ. Given G, we also write G^* for its topological dual group.

Before we can proceed further we need in addition to the algebraic and the topological structure on G the Haar measure. The topology on G induces the Borel σ-field $\mathcal{B}(G)$ on G and hence we may consider measures on G, or more correctly on $\mathcal{B}(G)$. This was our first approach to introduce the Lebesque measure $\lambda^{(n)}$ on $\mathbb{R}^n$ (or $\mathcal{B}^{(n)}$). The Lebesque measure has an important property: it is translation invariant or in our context we can phrase it as being invariant under the action of the group $\mathbb{R}^n$ in the sense that

$$\lambda^{(n)}(B + x) = \lambda^{(n)}(B) \qquad \text{(A.I.14)}$$

for all $B \in \mathcal{B}^{(n)}$ and $x \in \mathbb{R}^n$. A natural question is whether for a given locally compact Abelian group a measure μ on the Borel sets $\mathcal{B}^{(n)}$ exists which is invariant under G, i.e. for which we have

$$\mu(B + x) = \mu(B) \qquad \text{(A.I.15)}$$

for all $B \in \mathcal{B}(G)$ and $x \in G$. A non-trivial result due to [51], see also [126], is that such a measure always exists, is regular and is unique up to a normalization.

Standard references for a proof are, in addition to some work already mentioned, [86] or [93]. We call such a measure μ_G a **Haar measure** on G and for many concrete groups the term Haar measure always entails a fixed normalization, e.g. if G is compact then we will require $\mu(G) = 1$, if G is discrete we require $\mu(\{x\}) = 1$ for every $x \in G$, we will discuss this in more detail below.

Measurability of functions on G refers always to the Borel σ-field and if not stated otherwise, we will integrate with respect to a fixed Haar measure μ_G. Now we can introduce the spaces $L^p(G, \mu_G)$, $1 \leq p < \infty$, of (equivalence classes) of complex-valued measurable functions which are p-fold integrable with respect to μ_G, compare with Chapter III.8. In cases where no confusion may arise we will just write $L^p(G)$. Since the Haar measure is regular, it follows that $C_0(G)$ is dense in $L^p(G)$, $1 \leq p < \infty$.

Next we want to determine some dual groups. Let $\chi \in \mathbb{R}^*$, i.e. χ is a character of the group $(\mathbb{R}, +)$. Since $\chi(0) = 1$ it follows that for some $r > 0$ we have $\rho := \int_0^r \chi(s) \, ds \neq 0$ and therefore

$$\rho\chi(t) = \int_0^r \chi(s)\chi(t) \, ds = \int_0^r \chi(s+t) \, ds = \int_t^{t+r} \chi(s) \, ds,$$

or

$$\chi(t) = \frac{1}{\rho} \int_t^{t+r} \chi(s) \, ds$$

implying that χ is differentiable on $\mathbb{R}$ and

$$\chi'(t) = \frac{1}{\rho}(\chi(r+t) - \chi(t)) = \left(\frac{\chi(r) - 1}{\rho}\right)\chi(t).$$

With $c := \frac{\chi(r)-1}{\rho}$ we have

$$\chi'(t) = c\chi(t)$$

which yields

$$\chi(t) = e^{ct}.$$

Since $|\chi(t)| = 1$ we find with $\xi = -ic$ that

$$\chi(t) = e^{i\xi t}. \tag{A.I.16}$$

In other words we can establish a correspondence $D : \mathbb{R}^* \to \mathbb{R}$, $\chi \mapsto \xi$, $\chi(t) = e^{i\xi t}$. This mapping is surjective since for a given $\xi \in \mathbb{R}$ we can define a continuous character on $\mathbb{R}$ by $\chi_\xi(t) = e^{i\xi t}$. Moreover, if $\xi_1 \neq \xi_2$ then the equality $e^{e\xi_1 t} = e^{i\xi_2 t}$

for all $t \in \mathbb{R}$ implies $e^{i(\xi_1 - \xi_2)t} = 0$ for all $t \in \mathbb{R}$ which holds only for $\xi_1 = \xi_2$. Thus D is a bijective mapping. Moreover D is a group homomorphism since

$$(\chi_{\xi_1} + \chi_{\xi_2})(t) = \chi_{\xi_1}(t)\chi_{\xi_2}(t) = e^{i\xi_1 t}e^{i\xi_2 t}$$
$$= e^{i(\xi_1 + \xi_2)t} = \chi_{\xi_1 + \xi_2}(t).$$

Hence we have proved that $\mathbb{R}^*$ and $\mathbb{R}$ are isomorphic groups. It is not hard to see, see [35] or [101], that for locally compact Abelian groups $G_1, \ldots, G_N$ we have for the dual group

$$(G_1 \times \cdots \times G_N)^* \cong G_1^* \times \cdots \times G_N^* \tag{A.I.17}$$

from which we can now deduce that the dual group $(\mathbb{R}^n)^*$ can be identified with $\mathbb{R}^n$. In the following we will just use this identification and suppress the corresponding isomorphism. From $(T^1)^* = \mathbb{R}^*/(2\pi\mathbb{Z})^*$ one can now deduce, see [35], that

$$(T^1)^* \cong \mathbb{Z}. \tag{A.I.18}$$

Moreover, for $\chi \in \mathbb{Z}^*$ we find for $a = \chi(1) \in S^1$ that $\chi(k) = \chi(1)^k = a^k$ implying that $\chi(k) = e^{iwk}$ for some $w \in T^1 = S^1$, and therefore

$$\mathbb{Z}^* \cong T^1. \tag{A.I.19}$$

Eventually we arrive at

Corollary A.I.3. *For $n \in \mathbb{N}$ we have*

$$(\mathbb{R}^n)^* \cong \mathbb{R}^n, \tag{A.I.20}$$
$$(T^n)^* \cong \mathbb{Z}^n, \tag{A.I.21}$$
$$(\mathbb{Z}^n)^* \cong T^n. \tag{A.I.22}$$

From Corollary A.I.3 we obtain the interesting relation

$$((\mathbb{R}^n)^*)^* \cong (\mathbb{R}^n)^* = \mathbb{R}^n, \tag{A.I.23}$$
$$((T^n)^*)^* \cong (\mathbb{Z}^n)^* = T^n, \tag{A.I.24}$$

and

$$((\mathbb{Z}^n)^*)^* \cong (T^n)^* \cong \mathbb{Z}^n. \tag{A.I.25}$$

This is in fact a general result and content of **Pontryagin's duality theorem** which we formulate for locally compact Abelian groups:

Theorem A.I.4. *For a locally compact Abelian group G we have*

$$(G^*)^* \cong G. \tag{A.I.26}$$

A further noteworthy result is, see [35] or [101],

Theorem A.I.5. *The dual group of a compact group is discrete and the dual group of a discrete group is compact.*

This theorem allows us to introduce a compactification of every locally compact Abelian group in the following way. Given G and choose on G^* the discrete topology. Denote this group for a moment by G_d^*. Now we can consider $(G_d^*)^*$ which must be a compact group.

Theorem A.I.6. *There exists a continuous isomorphism from G onto a dense subgroup of $(G_d^*)^*$.*

This compactification is called the **Bohr compactification** of G, named after Harald Bohr.

An interesting application of Theorem A.I.5 leads to so called solenoid (groups): Take a subgroup of $(\mathbb{Q}, +)$ and on this subgroup choose the discrete topology. The dual group is always a compact Abelian group.
Now we can introduce the Fourier transform on a locally compact Abelian group.

Definition A.I.7. *Let G be a locally compact Abelian group with dual group Γ and fix a Haar measure μ_G on G. The **Fourier transform** of $u \in L^1(G, \mu_G)$ is the function $\hat{u} : \Gamma \to \mathbb{C}$ defined by*

$$\hat{u}(\chi) := \int_G u(x) <-x, \chi> \mu_G(dx). \tag{A.I.27}$$

We also write $(F_G u)(\chi)$ or just $(Fu)(\chi)$ for $\hat{u}(\chi)$. Moreover, when identifying Γ with some known group $\tilde{G}$ by an isomorphism $\chi \mapsto \xi \in \tilde{G}$ we also write

$$\hat{u}(\xi) = \hat{u}(\chi_\xi),$$

where $\chi_\xi \in \Gamma$ corresponds to $\xi \in \tilde{G}$. Note further that by (A.I.12) we may also write

$$\hat{u}(\chi) = \int_G u(x) \overline{<x, \chi>} \mu_G(dx). \tag{A.I.28}$$

We remark that since $| <-x, \chi> | = 1$, the integral in (A.I.27) is for $u \in L^1(G, \mu_G)$ well defined.

Example A.I.8. A. On $\mathbb{R}^n$ we use as Haar measure $\mu_{\mathbb{R}^n} := (2\pi)^{-\frac{n}{2}} \lambda^{(n)}$ and with the identification of $(\mathbb{R}^n)^*$ with $\mathbb{R}^n$ we find

$$\hat{u}(\chi_\xi) = \hat{u}(\xi) = \int_{\mathbb{R}^n} u(x) e^{-ix \cdot \xi} \mu_{\mathbb{R}^n}(dx) = (2\pi)^{-\frac{n}{2}} \int_{\mathbb{R}^n} u(x) e^{-ix \cdot \xi} dx, \tag{A.I.29}$$

where $x \cdot \xi$ is the scalar product in $\mathbb{R}^n$.

B. Let $u : T^1 \to \mathbb{C}$ be a continuous function. We may identify u with a 2π-periodic function and consider its Fourier series $u \backsim \sum_{k \in \mathbb{Z}} c_k e^{ikx}$. When we choose on $T^1 = S^1$ the Haar measure μ_{T^1} such that $\mu_{T^1}(T^1) = 1$ the corresponding Fourier transform $\hat{u}$ of u becomes the sequence $(c_k)_{k \in \mathbb{Z}}$ which is a function on $\mathbb{Z} \doteq (T^1)^*$, i.e.

$$\hat{u}(k) = c_k = \frac{1}{2\pi} \int_0^{2\pi} u(t) e^{-ikt} \, \mathrm{d}t = \frac{1}{2\pi} \int_{-\pi}^{\pi} u(t) e^{-ikt} \, \mathrm{d}t. \tag{A.I.30}$$

Of course, this extends to $u \in L^1(T^1, \mu_{T^1})$.

The Lemma of Riemann-Lebesgue, Theorem 3.16, states that the Fourier transform of $u \in L^1(\mathbb{R}^n)$ is a continuous function vanishing at infinity, and its discrete version, Corollary 3.17, states for the Fourier coefficients c_k of $u \in L^1(T^1)$ that $\lim_{|k| \to \infty} c_k = 0$. We want to formulate this result for a general locally compact Abelian group. Let X be a locally compact topological space and $u : X \to \mathbb{C}$ a function. We say that **u vanishes at infinity** if for every $\epsilon > 0$ there exists a compact set $K_\epsilon \subset X$ such that $|u_{|K_\epsilon^c}| < \epsilon$. By $C_\infty(X)$ we denote the space of all continuous functions $u : X \to \mathbb{C}$ vanishing at infinity. In general we have $C_0(X) \subset C_\infty(X) \subset C_b(X) \subset C(X)$, however for a compact space X we have $C_0(X) = C_\infty(X) = C_b(X) = C(X)$. Now we can state the **lemma of Riemann-Lebesgue**, see [101] for a proof.

Proposition A.I.9. *Let G be a locally compact Abelian group and $u \in L^1(G)$. Then $\hat{u} \in C_\infty(\Gamma)$ and*

$$\|\hat{u}\|_\infty \leq \|u\|_{L^1}. \tag{A.I.31}$$

The estimate (A.I.31) is of course trivial:

$$|\hat{u}(\chi)| = \left| \int_G u(x) < -x, \chi > \mu_G(\mathrm{d}x) \right| \leq \int_G |u(x)| \mu_G(\mathrm{d}x).$$

The next interesting result we want to establish for locally compact Abelian groups is the convolution theorem. We need first

Definition A.I.10. *For a locally compact Abelian group let $u, v : G \to \mathbb{C}$ be two measurable functions such that for every $x \in G$ the function $y \mapsto u(x - y)v(y)$ is μ_G-integrable. The convolution $u * v$ of u and v is defined by*

$$(u * v)(x) := \int_G u(x - y)v(y)\mu_G(\mathrm{d}y). \tag{A.I.32}$$

We refer to Volume III where we have discussed how to define $u * v$ for elements in some L^p-spaces. Most results known to us for the convolution in $\mathbb{R}^n$ do carry over to locally compact Abelian groups.

Theorem A.I.11. *Let G be a locally compact Abelian group.*
A. *When $(u * v)(x)$ is defined we have $(u * v)(x) = (v * u)(x)$.*
B. *For $u \in L^1(G, \mu_G)$ and $v \in L^p(G, \mu_G)$, $1 \le p < \infty$ we have $u * v \in L^p(G, \mu_G)$ and*

$$\|u * v\|_{L^p} \le \|u\|_{L^1} \|v\|_{L^p}. \tag{A.I.33}$$

C. *For $1 < p < \infty$, $\frac{1}{p} + \frac{1}{q} = 1$, $u \in L^p(G, \mu_G)$ and $v \in L^q(G, \mu_G)$ the estimate*

$$\|u * v\|_\infty \le \|u\|_{L^p} \|v\|_{L^q} \tag{A.I.34}$$

holds and $u * v \in C_\infty(G)$.
D. *In $L^1(G, \mu_G)$ is convolution is associative, i.e. for $u, v, w \in L^1(G, \mu_G)$ we have*

$$(u * v) * w = u * (v * w). \tag{A.I.35}$$

For a proof of this theorem we refer to [101] or [35]. Now we can prove the **convolution theorem**

Theorem A.I.12. *For $u, v \in L^1(G, \mu_G)$ we have*

$$(u * v)^\wedge(\chi) = \hat{u}(\chi) \cdot \hat{v}(\chi). \tag{A.I.36}$$

Proof. Using the properties of characters and Fubini's theorem we find

$$(u * v)^\wedge(\chi) = \int_G \left(\int_G u(x - y) v(y) \mu_G(dy) < -x, \chi > \right) \mu_G(dx)$$

$$= \int_G \int_G u(x - y) v(y) < -(x - y), \chi >< -y, \chi > \mu_G(dy) \mu_G(dx)$$

$$= \int_G \int_G u(x - y) < -(x - y), \chi > \mu_G(dx) v(y) < -y, \chi > \mu_G(dy)$$

$$= \int_G u(z) < -z, \chi > \mu_G(dz) \int_G v(y) < -y, \chi > \mu_G(dy)$$

$$= \hat{u}(\chi) \cdot \hat{v}(\chi),$$

where in the last step we have used that μ_G is invariant under the action of G. $\square$

Defining the Fourier transform on some space $L^p(G, \mu_G)$, $p > 1$, needs the distinction whether or not G is compact. If G is compact the regularity of the Haar measure implies that $\mu_G(G) < \infty$ and therefore we use the normalization $\mu_G(G) = 1$. This implies by (III.7.8) the estimate

$$\|u\|_{L^1(G, \mu_G)} \le \|u\|_{L^p(G, \mu_G)}, \quad p \ge 1, \tag{A.I.37}$$

and therefore $L^p(G, \mu_G) \subset L^1(G, \mu_G)$ and the Fourier transform is defined for every $u \in L^p(G, \mu_G)$. However for non-compact groups which have no bounded Haar measure the statement that $L^p(G, \mu_G)$ is a subspace of $L^1(G, \mu_G)$ does not hold as we have seen already in the case $L^2(\mathbb{R}^n)$. It is the isometry property of the Fourier transform on $L^2(G, \mu_G) \cap L^1(G, \mu_G)$ with respect to the norm $\|\cdot\|_{L^2}$ which allows us an extension of the Fourier transform from $L^2(G, \mu_G) \cap L^1(G < \mu_G)$ to $L^2(G, \mu_G)$, i.e. we have again **Plancherel's theorem**

Theorem A.I.13. *Let* $u \in L^2(G, \mu_G) \cap L^1(G, \mu_G)$. *For the Fourier transform of* u *we find* $\hat{u} \in L^2(\Gamma, \mu_\Gamma)$ *and*

$$\|\hat{u}\|_{L^2(\Gamma)} = \|u\|_{L^2(G)} \qquad (A.I.38)$$

Moreover, the image of $L^2(G, \mu_G) \cap L^1(G, \mu_G)$ *under the Fourier transform is a dense subspace of* $L^2(\Gamma, \mu_\Gamma)$.

From Theorem A.I.13 we may deduce

Corollary A.I.14. *The Fourier transform extends from* $L^2(G, \mu_G) \cap L^1(G, \mu_G)$ *to a bijective isometry from* $L^2(G, \mu_G)$ *to* $L^2(\Gamma, \mu_\Gamma)$. *In the following we will denote this extension again by* F *and for* $u \in L^2(G, \mu_G)$ *we will write* $\hat{u}$ *or* Fu *for its Fourier transform. Applying the polarisation identity we obtain for* $u, v \in L^2(G, \mu_G)$ *Parseval's formula*

$$\int_G u(x)\overline{v(x)}\mu_G(dx) = \int_\Gamma \hat{u}(\chi)\overline{\hat{v}(\chi)}\mu_\Gamma(d\chi). \qquad (A.I.39)$$

We refer the reader to Chapter 5 and to Chapter 11 where Plancherel's theorem was discussed for $L^2(T, \mu_T)$ and $L^2(\mathbb{R}^n, \mu_{\mathbb{R}^n})$, respectively.

In a next step we want to investigate the Fourier transform of bounded measures defined on the Borel σ-field of a locally compact Abelian group G with a fixed Haar measure μ_G. Let μ be a bounded measure on the locally compact Abelian group G. Then we can define a function $\hat{\mu}$ on Γ by

$$\hat{\mu}(\chi) := \int_G <-x, \chi> \mu(dx). \qquad (A.I.40)$$

In some cases we will also write $F\mu$ for $\hat{\mu}$. The function $\hat{\mu} : \Gamma \to \mathbb{C}$ is of course bounded since

$$|\hat{\mu}(\chi)| \leq \mu(G), \qquad (A.I.41)$$

and moreover $\hat{\mu}$ is a uniformly continuous function. In addition many standard results known for the Fourier transform of functions transfer to measures. For example we have

$$F(\alpha\mu + \beta\nu) = \alpha F\mu + \beta F\nu, \quad \alpha, \beta \geq 0, \tag{A.I.42}$$

or

$$(\mu * \nu)^{\wedge}(\chi) = \hat{\mu}(\chi)\hat{\nu}(\chi). \tag{A.I.43}$$

A possibility to define $\mu * \nu$ is by requiring $\mu * \nu$ to be the unique measure satisfying for all bounded measurable functions u the identity

$$\int_G u \, \mathrm{d}(\mu * \nu) = \int_G \left(\int_G u(x + y) \, \mathrm{d}\mu \right) \mathrm{d}\nu. \tag{A.I.44}$$

If μ has the density g with respect to the Haar measure μ_G, i.e. $\mu = g\mu_G$, then $g \in L^1(G, \mu_G)$ and $\hat{\mu}(\chi) = \hat{g}(\chi)$.

Definition A.I.15. *Let G be a locally compact Abelian group. A function $u : G \to \mathbb{C}$ is called a **positive definite function** if for all $N \in \mathbb{N}$ and any choice of group elements $x_1, \ldots, x_n \in G$ the matrix $(u(x_k - x_l))_{k,l=1,\ldots,N}$ is positive Hermitian, i.e.*

$$\sum_{k,l=1}^{N} u(x_k - x_l)\lambda_k\overline{\lambda_l} \geq 0 \quad \text{for all } \lambda_1, \ldots, \lambda_N \in \mathbb{C}.$$

As in the case of $\mathbb{R}^n$, see Lemma 13.7, we have

Lemma A.I.16. *The Fourier transform of μ is a positive definite function on Γ.*

In [12] or [101] many properties of positive definite functions are studied. The most important and far reaching result is again **Bochner's** theorem:

Theorem A.I.17. *A continuous function $u : G \to \mathbb{C}$ is positive definite if and only if it is the Fourier transform of a bounded measure μ on Γ, i.e.*

$$u(x) = \int_\Gamma <\chi, x> \mu(\mathrm{d}\chi). \tag{A.I.45}$$

(To have more convenient interpretations we switched the sign of χ compared with our original definition of the Fourier transform.)

We refer in particular to [12] where many properties of the Fourier transform of measures and their relation to positive definite functions are discussed.

A non-trivial question is that of the existence of the inverse Fourier transform. Only in the case of the extension of the Fourier transform to $L^2(G, \mu_G)$ we have so far an answer since $F : L^2(G) \to L^2(\Gamma)$ is a bijective isometry. Even in this case we do not have a formula for F^{-1}. The reader may recall that in the case $G = \mathbb{R}^n$ the construction of the inverse Fourier transform was a non-trivial result. To proceed further we introduce the function space

$$B(G) := \mathrm{span}\{u : G \to \mathbb{C} \mid u(x) = \int_\Gamma < \chi, x > \mu(\mathrm{d}\chi), \ \mu \in \mathcal{M}_b^+(\Gamma)\}$$

where $\mathcal{M}_b^+(\Gamma)$ denotes the bounded non-negative measures on Γ. Thus $B(G)$ consists of linear combinations of continuous positive definite functions on G. Now we can state, compare [101],

Theorem A.I.18. *For $u \in B(G) \cap L^1(G)$ the function $\hat{u}$ belongs to $L^1(\Gamma)$. Moreover, for a fixed Haar measure μ_G on G there exists a unique Haar measure μ_Γ on Γ such that*

$$u(x) = \int_\Gamma < \chi, x > \hat{u}(\chi)\mu_\Gamma(\mathrm{d}\chi). \tag{A.I.46}$$

Starting with Theorem A.I.18 it is now possible to extend the formula (A.I.46) for the inverse Fourier transform to certain Banach spaces, for example to the Wiener algebra $A(G)$ defined by $A(G) = \{u \in L^1(G, \mu_G) \mid \hat{u} \in L^1(\Gamma, \mu_\Gamma)\}$.

When discussing the inverse Fourier transform on $\mathbb{R}^n$ we have made much use of the fact that we can approximate elements in $C_\infty(\mathbb{R}^n)$ or $L^p(\mathbb{R}^n)$, $1 \le p < \infty$, by smooth functions, say by elements from $\mathcal{S}(\mathbb{R}^n)$ or $C_0^\infty(\mathbb{R}^n)$. Such "nice" subsets do not exist in a general locally compact Abelian group, for more see our discussion below. However on T^1 or T^n we can use a "nice" dense subset in our function spaces the set of all trigonometric polynomials. We define on a (general) compact Abelian group a generalized trigonometric polynomial by

$$\mathrm{trig}(x) = \sum_{k=1}^N c_k < x, \chi_k >, \quad c_k \in \mathbb{C}, \tag{A.I.47}$$

for $\chi_k \in \Gamma$. Now one can prove, see [101] that the set of all trigonometric polynomials on a compact Abelian group is dense in $C(G)$ as well as in $L^p(G, \mu_G)$, $1 \le p < \infty$.

On $\mathbb{R}^n$ or T^n we have a natural differentiable structure and we can investigate differentiable functions. Differentiation is a local operation in the precise meaning that

$$\mathrm{supp} D^\alpha u \subset \mathrm{supp} u.$$

In Chapter 13 we could relate the smoothness of a function $u : \mathbb{R}^n \to \mathbb{C}$ to the decay of its Fourier transform. In fact we have done something similar before for functions $u : T^1 \to \mathbb{C}$. Note that we identify now the sequence of the Fourier coefficients $(\hat{u}_k)_{k \in \mathbb{Z}} = (c_k)_{k \in \mathbb{Z}}$ of u as the Fourier transform $\hat{u} : \mathbb{Z} \to \mathbb{C}$ of u and we have seen that the smoothness of u determines the decay of the sequence $(\hat{u}_k)_{k \in \mathbb{Z}}$ of u as the Fourier transform $\hat{u} : \mathbb{Z} \to \mathbb{C}$ of u and we have seen that the smoothness of u determines the decay of the sequence $(\hat{u}_k)_{k \in \mathbb{N}}$ and vice versa. While on a general locally compact Abelian group we can in general not define differentiation as local operation, we may try to use the decay of the Fourier transform of a given function to discuss its properties. Let us recall the definition of the Bessel potential spaces $H^s(\mathbb{R}^n)$ and $H^s(T)$ or the obvious extension $H^s(T^n)$. For $u \in L^2(\mathbb{R}^n)$ we introduce

$$\|u\|_{H^s(\mathbb{R}^n)} = \|(1 + |\cdot|^2)^{\frac{s}{2}} \hat{u}(\cdot)\|_{L^2(\mathbb{R}^n)} \tag{A.I.48}$$

$$= \left(\int_{\mathbb{R}^n} (1 + |\xi|^2)^s |\hat{u}(\xi)|^2 \, d\xi \right)^{\frac{1}{2}}$$

and for $u \in L^2(T^n)$ we set

$$\|u\|_{H^s(T^n)} = \left(\sum_{k \in \mathbb{Z}^n} (1 + |k|^2)^s |\hat{u}_k|^2 \right)^{\frac{1}{2}}. \tag{A.I.49}$$

The spaces $H^s(\mathbb{R}^n)$ and $H^s(T^n)$ are the subspaces of the respective L^2-spaces whose elements have finite norm $\|\cdot\|_{H^s(\mathbb{R}^n)}$ and $\|\cdot\|_{H^s(T^n)}$, respectively. Note that for $u \in H^s(T^n)$ the series in (A.I.49) is absolutely convergent, hence any definition of partial sums will lead to a convergent series and all limits will be equal.

For a compact Abelian group G with countable dual group Γ we will now introduce and study spaces analogous to $H^s(T^n)$ and explain the idea why these spaces should be considered as a tool to measure the smoothness of functions defined on G. Besides T^n, we may consider the following groups: Choose an infinite subgroup Γ of $(\mathbb{Q}^n, +)$ and equip the subgroup with the discrete topology. Now define G as the dual group of Γ, i.e. $G := \Gamma^*$. We start with the following observation

Lemma A.I.19. *Let G be a compact Abelian group and Γ its dual group. For $\chi \in \Gamma \backslash \{0\}$*

$$\int_G \chi(x) \mu_G(dx) = 0. \tag{A.I.50}$$

Proof. Since χ is not the zero element in Γ there exists $x_0 \in G$ such that $\chi(x_0) \neq 1$

557

and we find

$$\int_G \chi(x)\mu_G(\mathrm{d}x) = \int_G \chi(x_0)\chi(x - x_0)\mu_G(\mathrm{d}x)$$

$$= \chi(x_0)\int_G \chi(x - x_0)\mu_G(\mathrm{d}x)$$

$$= \chi(x_0)\int_G \chi(x)\mu_G(\mathrm{d}x),$$

where we have use that μ_G is invariant under the operation of G. Since $\chi(x_0) \neq 1$ by assumption we deduce $\int_G \chi(x)\mu_G(\mathrm{d}x) = 0$. $\qquad\square$

Theorem A.I.20. *Let G be a compact Abelian group with normalized Haar measure μ_G, i.e. $\mu_G(G) = 1$, and countable dual group $\Gamma = \{\chi_k \mid k \in \mathbb{N}_0\}$. Then $\{\chi_k \mid k \in \mathbb{N}_0\}$ is a complete orthonormal system, i.e. an orthonormal basis, in $L^2(G)$.*

Proof. We may assume that χ_0 is the zero element in Γ, i.e. $\chi_0(x) = 1$ for all $x \in G$ which implies

$$\int_G \chi_0(x)\mu_G(\mathrm{d}x) = 1.$$

Moreover, for $\chi_k, \chi_l \in \Gamma$ we find

$$\int_G \chi_k(x)\overline{\chi_l(x)}\mu_G(\mathrm{d}x) = \int_G \chi_{k-l}(x)\mu_G(\mathrm{d}x)$$

and for $k \neq l$ the orthogonality of χ_k and χ_l follows. Since $\chi_k(x) \in S^1$ for all $x \in G$ it follows that

$$\|\chi_k\|_{L^2(G)}^2 = \int_G |\chi_k(x)|^2\mu_G(\mathrm{d}x) = 1.$$

This proves that $\{\chi_k \mid k \in \mathbb{N}_0\}$ is an orthonormal system in $L^2(G, \mu_G)$. However, by Plancherel's theorem we have for all $u \in L^2(G)$ that

$$\|u\|_{L^2(G)}^2 = \|\hat{u}\|_{L^2(\Gamma)} = \sum_{k \in \mathbb{N}_0} |\hat{u}_k|^2 \tag{A.I.51}$$

where $\hat{u}_k = <-x, \chi_k>$. Thus (A.I.51) is the completeness relation for the system $\{\chi_k \mid k \in \mathbb{N}_0\}$ which is now identified as an orthonormal basis in $L^2(G, \mu_G)$. $\qquad\square$

Theorem A.I.20 implies in particular that $u \in L^2(G, \mu_G)$ has a representation

$$u = \sum_{k=1}^{\infty} \hat{u}_k \chi_k(\cdot) \tag{A.I.52}$$

where the series converges in $L^2(G, \mu_G)$. We aim to introduce spaces $H^s(G)$ analogously to the spaces $H^s(T^n)$. This requires an additional observation. For $\mathbb{Z}^n = (T^n)^*$ we could use a natural partition $\mathbb{Z}^n = \bigcup_{j=0}^\infty \mathbb{Z}_j^n$ where $\mathbb{Z}_j^n := \{k \in \mathbb{Z}^n \mid |k| = j\}$ where $|k| = k_1 + \cdots + k_n$. The decay of a sequence $(a_k)_{k \in \mathbb{Z}^n}$, $a_k \in \mathbb{C}$ was defined with the help of a function $h : \mathbb{N}_0 \to \mathbb{R}_+$ by

$$|a_k| \le h(|k|) \tag{A.I.53}$$

and replacing $|k|$ by a norm on $\mathbb{R}^n$ would not make much of a difference provided h has some reasonable behaviour we expect from a function measuring decay, e.g. some monotonicity or satisfying some Peetre-type estimate. For the discrete countable Abelian group Γ such a natural partition does not exist in general, just think of $\mathbb{Q}$. To overcome these difficulties we introduce the notion of an **order generating partition** of Γ by finite subsets as to be a sequence $(\Gamma_m)_{m \in \mathbb{N}}$ of finite subsets of Γ with the properties that $\Gamma_m \ne \emptyset$, $\Gamma_m \subset \Gamma_{m+1}$ and $\bigcup_{m \in \mathbb{N}} \Gamma_m = \Gamma$. Note that the sets $\Gamma_m \backslash \Gamma_{m-1}$, $\Gamma_0 = \emptyset$, then form a partition of Γ. It is now possible to define the limit α of a sequence $(\alpha_\chi)_{\chi \in \Gamma}$, $\alpha_\chi \in \mathbb{C}$, along such an order generating partition if for every $\epsilon > 0$ exists $N = N(\epsilon) \in \mathbb{N}$ such that $\chi \in \Gamma \backslash \Gamma_N$ implies $|\alpha_\chi - \alpha| < \epsilon$. Of course this limit depends in general on the choice of $(\Gamma_m)_{m \in \mathbb{N}}$. In particular we can study the sequence of partial sums $(\sum_{\chi \in \Gamma_m} w(\chi))_{m \in \mathbb{N}}$ for some functions $w : \Gamma \to \mathbb{C}$, and in the case this sequence converges along $(T_m)_{m \in \mathbb{N}}$ we will write

$$\sum_{\chi \in \Gamma} w(\chi) := \lim_{m \to \infty} \sum_{\chi \in \Gamma_m} w(\chi). \tag{A.I.54}$$

Now it is possible to measure with the help of $(\Gamma_m)_{m \in \mathbb{N}}$ and a weight function $w : \Gamma \to \mathbb{R}_+$ the decay of the Fourier transform $\hat{u} : \Gamma \to \mathbb{C}$ of a function $u \in L^2(G, \mu_G)$. We just have to require

$$|\hat{u}_k| \le w(\chi_k) \tag{A.I.55}$$

for all $k \in \mathbb{N}_0$, or a bit more generally, for all but finitely many $k \in \mathbb{N}$.

In the following we fix on Γ an order generating partition $(\Gamma_m)_{m \in \mathbb{N}}$ of $\Gamma = \{\chi_k \mid k \in \mathbb{N}\}$ and for a non-negative function $h : \Gamma \to \mathbb{C}$ we set $h_* : \Gamma \to \mathbb{R}_+$ by

$$h_*(\chi) := (1 + |h(\chi)|^2)^{\frac{1}{2}}. \tag{A.I.56}$$

We define the norm

$$\|u\|_{H_h^s(G)} := \left(\sum_{k=1}^\infty (1 + |h(\chi_k)|^2)^s |\hat{u}_k|^2 \right)^{\frac{1}{2}} \tag{A.I.57}$$

and the space

$$H_h^s(G) := \{u \in L^2(G, \mu_G) \mid \|u\|_{H_h^s(G)} < \infty\}. \tag{A.I.58}$$

It is not hard to see that $H_h^s(G)$ is a Hilbert space with respect to the scalar product

$$< u, v >_{H_h^s(G)} = \sum_{k=1}^{\infty} (1 + |h(\chi_k)|^2)^s \hat{u}_k \overline{\hat{v}_k}. \tag{A.I.59}$$

Clearly, for a bounded function h the norms $\| \cdot \|_{H_h^s(G)}$ are all equivalent to the norm $\| \cdot \|_{L^2(G)}$. A non-trivial, i.e. a proper subspace of $L^2(G, \mu_G)$ is obtained when we require that for some $t_0 > 0$ the series

$$\sum_{\chi \in \Gamma} h_*^{-2t_0}(\chi) := \lim_{m \to \infty} \sum_{\chi \in \Gamma_m} h_*^{-2t_0}(\chi) \tag{A.I.60}$$

converges. Now we can prove

Theorem A.I.21. *Let G, Γ and $(\Gamma_m)_{m \in \mathbb{N}}$ be as mentioned above and assume that for $h : G \to \mathbb{C}$ we have $\sum_{\chi \in \Gamma} h_*^{-2t_0}(\chi)$ for some $t_0 > 0$. Then for $r \geq t_0$ the space $H_h^r(G)$ is continuously embedded into the space $C(G)$ in the sense that for $u \in H_h^r(G)$ the series $\sum_{k=1}^{\infty} u_k \chi_k(\cdot)$ converges absolutely and uniformly on G to a function $\tilde{u} \in C(G)$, $\tilde{u} = u$ μ_G-almost everywhere, and the estimate*

$$|\tilde{u}(x)| \leq c\|u\|_{H_h^r(G)} \tag{A.I.61}$$

holds for all $x \in G$ implying of course $\|\tilde{u}\|_{\infty, G} \leq c\|u\|_{H_h^r(G)}$.

Proof. Let $u \in H_h^r(G)$ and set $P_N(u) := \sum_{k=1}^{N} \hat{u}_k \chi_k$. For $N > M$ it follows that

$$|P_N(u)(x) - P_M(u)(x)| = \left| \sum_{k=M+1}^{N} \hat{u}_k \chi_k(x) \right|$$

$$= \left| \sum_{k=M+1}^{N} \hat{u}_k h_*^r(\chi_k) h_*^{-r}(\chi_k) \chi_k(x) \right|$$

$$\leq \left(\sum_{k=M+1}^{N} h_r^{-2r}(\chi_k) \right)^{\frac{1}{2}} \left(\sum_{k=M+1}^{N} |\hat{u}_k|^2 h_*^{2r}(\chi_k) \right)^{\frac{1}{2}}$$

$$\leq \left(\sum_{k=M+1}^{N} h_*^{-2t_0}(\chi_k) \right)^{\frac{1}{2}} \|u\|_{H_h^r(G)}.$$

Hence $(P_N(u))_{N \in \mathbb{N}}$ is a Cauchy sequence with respect to the supremum norm on G and therefore it converges to a continuous function $\tilde{u} : G \to \mathbb{C}$ which must be μ_G-about everywhere equal to u. The above calculation applied to $P_N(u)$ implies also the estimate (A.I.61). $\qquad \square$

We leave these considerations here and do not ask whether for larger values of r "better" regularity results are possible, not least due to the fact that we need a clarification of "better" regularity for functions defined on G.

Finally in this appendix we want to investigate convolution semigroups of bounded measures on a locally compact Abelian group G with dual group Γ. We start with

Definition A.I.22. *A family $(\mu_t)_{t\geq 0}$ of bounded measures on the locally compact Abelian group G is called a **convolution semigroup** if*

$$\mu_t(G) \leq 1 \text{ for all } t \geq 0; \tag{A.I.62}$$

$$\mu_t * \mu_s = \mu_{t+s}, \ s,t \geq 0, \text{ and } \mu_0 = \epsilon_0, \tag{A.I.63}$$

$$\mu_t \to \epsilon_0 \text{ vaguely as } t \to 0. \tag{A.I.64}$$

Here ϵ_0 denotes the unit mass or Dirac measure at 0, i.e. $\epsilon_0(A) = 1$ if $0 \in A$, $\epsilon_0(A) = 0$ if $0 \notin A$, and $\mu_t \to \epsilon_0$ vaguely does mean that

$$\lim_{t\to 0} \int_G u(x)\mu_t(\mathrm{d}x) = u(0)$$

for all $u \in C_0(G)$.

Let $(\mu_t)_{t\geq 0}$ be a convolution semigroup on G. Since μ_t is a bounded measure, its Fourier transform $\hat{\mu}_k : \Gamma \to G$ exists and is a bounded continuous function, in fact it is a positive definite function. It turns out that the family $(\hat{\mu}_t)_{t\geq 0}$ of positive definite functions admits a representation

$$\hat{\mu}_t(\chi) = e^{-t\psi(\chi)}, \tag{A.I.65}$$

see [12]. The function $\psi : \Gamma \to \mathbb{C}$ is a **continuous negative definite function**, i.e. it is continuous and for every $N \in \mathbb{N}$ and any choice of characters $\chi_1, \ldots, \chi_N$ the matrix $(\psi(\chi_k) + \overline{\psi(\chi_l)} - \psi(\chi_k - \chi_l))_{k,l=1,\ldots,N}$ is positive Hermitian, i.e. for all $\lambda_1, \ldots, \lambda_N \in \mathbb{N}$ we have

$$\sum_{k,l=1}^{N} (\psi(\chi_k) + \overline{\psi(\chi_l)} - \psi(\chi_k - \chi_l))\lambda_k\overline{\lambda_l} \geq 0. \tag{A.I.66}$$

A theorem due to I. J. Schoenberg states that $\psi : \Gamma \to \mathbb{C}$ is a negative definite function if and only if $\psi(0) \geq 0$ and $\chi \mapsto e^{-t\psi(\chi)}$ is for all $t > 0$ positive definite. Moreover we have, see [12]

Theorem A.I.23. *If $(\mu_t)_{t\geq0}$ is a convolution semigroup on G then there exists a unique continuous negative definite function $\psi : \Gamma \to \mathbb{C}$ such that for all $t > 0$ we have (A.I.65). Conversely, given a continuous negative definite function $\psi : \Gamma \to \mathbb{C}$ there exists a unique convolution semigroup $(\mu_t)_{t\geq0}$ on G such that $\hat{\mu}_t(\chi) = e^{-t\psi(\chi)}$.*

With not too much effort one can show, see again [12], that for two continuous negative definite functions ψ_1 and ψ_2 and $\alpha_1, \alpha_2 \geq 0$ the function $\alpha_1\psi_1 + \alpha_2\psi_2$ is again a continuous negative definite function, and that for a continuous negative definite function the function $\overline{\psi}$ and $\operatorname{Re}\psi$ are continuous negative definite functions too. Moreover, $\operatorname{Re}\psi \geq 0$ and $|\psi|^{\frac{1}{2}}$ is subadditive in the sense that

$$|\psi(\chi_1 + \chi_2)|^{\frac{1}{2}} \leq |\psi(\chi_1)|^{\frac{1}{2}} + |\psi(\chi_2)|^{\frac{1}{2}}. \tag{A.I.67}$$

We also remark that $\{\chi \in \Gamma \mid \psi(\chi) = \psi(0)\}$ is a subgroup of Γ.

Now let $u \in L^2(G, \mu_G)$ and $(\mu_t)_{t\geq0}$ be a convolution semigroup on G with associated continuous negative definite function $\psi : \Gamma \to \mathbb{C}$. Since $|e^{-t\psi(\chi)}| \leq e^{-t\operatorname{Re}\psi(\chi)} \leq 1$ and $\hat{u} \in L^2(\Gamma, \mu_\Gamma)$ it follows that $\hat{u}e^{-t\psi}$ belongs to $L^2(\Gamma, \mu_\Gamma)$, hence its inverse Fourier transform belongs to $L^2(G, \mu_G)$ where we use the fact that the Fourier transform is a bijective isometry from $L^2(G, \mu_G)$ onto $L^2(\Gamma, \mu_\Gamma)$. For the inverse Fourier transform of $\hat{u}e^{-t\psi}$ we write $u * \mu_t$. This is well justified since for $u \geq 0$ and $u \in L^2(G, \mu_G) \cap L^1(G, \mu_G)$ we can interpret u as density of a bounded measure $\nu = u\mu_G$ and then the convolution theorem yields

$$(u * \mu_t)^{\wedge}(\chi) = \hat{u}(\chi)\hat{\mu}_t(\chi) = \hat{u}(\chi)e^{-t\psi(\chi)}. \tag{A.I.68}$$

For a general $u \in L^2(G, \mu_G) \cap L^1(G, \mu_G)$ we may use the decomposition $u = u_t - u_-$ to obtain (A.I.68). In the following it is convenient to write $u \mapsto Fu$ for the Fourier transform from $L^2(G, \mu_G)$ to $L^2(\Gamma, \mu_\Gamma)$ and $v \mapsto F^{-1}v$ for the inverse Fourier transform from $L^2(\Gamma, \mu_\Gamma)$ to $L^2(G, \mu_G)$.

Given $u \in L^2(G, \mu_G) \cap L^1(G, \mu_G)$ we define for $t > 0$ the convolution operator T_t by

$$T_t u = u * \mu_t = F^{-1}(\hat{u}\hat{\mu}_t). \tag{A.I.69}$$

Since we know that

$$\|T_t u\|_{L^2(G)} = \|\hat{u}\hat{\mu}_t\|_{L^2(\Gamma)} \leq \|\hat{u}\|_{L^2(\Gamma)} = \|u\|_{L^2(G)}, \tag{A.I.70}$$

for $t > 0$ these operators are linear contractions from $L^2(G, \mu_G) \cap L^1(G, \mu_G)$ to $L^2(G, \mu_G)$. The density of $L^2(G, \mu_G) \cap L^1(G, \mu_G)$ in $L^2(G, \mu_G)$, recall that $C_0(G) \subset L^p(G)$ for all $p \geq 0$, which allows us to extend each T_t to a linear contraction from

$L^2(G, \mu_G)$ to $L^2(G, \mu_G)$. We denote this extension again by T_t. For $t, s > 0$ we observe now

$$T_{t+s}u = F^{-1}(\hat{u}\hat{\mu}_{t+s}) = F^{-1}(\hat{u}\hat{\mu}_s\hat{\mu}_t)$$
$$= F^{-1}(F(T_s u)\hat{\mu}_t) = T_t(T_s u),$$

i.e. the family $(T_t)_{t>0}$ satisfies on $L^2(G, \mu_G)$ the semigroup relation

$$T_{t+s} = T_t \circ T_s, \quad t, s > 0. \tag{A.I.71}$$

Note that $T_0 = \mathrm{id}_{L^2(G,\mu_G)}$ since $\hat{\mu}_0 = 1$.

The convolution semigroup $(\mu_t)_{t\geq 0}$ we can characterize by a much simpler object, namely the function ψ. We may ask whether we can characterize the family $(T_t)_{t\geq 0}$ by an object, an operator, independent of t. For this we look at

$$\left(\frac{T_t u - u}{t}\right)^{\wedge} = \left(\frac{e^{-t\psi} - 1}{t}\right)\hat{u}. \tag{A.I.72}$$

For a fixed u as t tends to 0 the right hand side of (A.I.72) tends pointwisely to $-\psi(\cdot)\hat{u}$, and a candidate for characterizing $(T_t)_{t\geq 0}$ is the operator $Au = -\psi(D)u$, where $(\psi(D)u)^{\wedge} = \psi(\cdot)\hat{u}$. However some care is needed when looking at the inverse Fourier transform in (A.I.72). Suppose for simplicity that ψ is real-valued, hence $\psi \geq 0$, and that $\psi(\cdot)\hat{u} \in L^2$. Then we find

$$\left\| \left(\frac{e^{-t\psi} - 1}{t} + \psi\right)\hat{u} \right\|_{L^2(\Gamma)}^2 = \int_\Gamma \left|\frac{e^{-t\psi} + t\psi - 1}{t}\right|^2 |\hat{u}|^2 \, d\mu_\Gamma.$$

Since for $a \geq 0$ we have

$$\left|\frac{e^{-ta} + ta - 1}{t}\right| \leq \frac{1}{2}a^2 t,$$

we find

$$\left\| \left(\frac{e^{-t\psi} + t\psi - 1}{t}\right)\hat{u} \right\|_{L^2(\Gamma)}^2 \leq \frac{1}{2}\int_\Gamma \psi^4(\chi)|\hat{u}(\chi)|^2 \mu_\Gamma(d\chi).$$

Now we introduce the spaces $H^{s,\psi}(G)$ as

$$H^{\psi,s}(G) = H^s_{\psi_x^{\frac{1}{2}}}(G) = \{u \in L^2(G, \mu_G) \mid \|u\|_{\psi,s} < \infty\} \tag{A.I.73}$$

where

$$\|u\|_{\psi,s} = \|(1 + \psi(\cdot))^{\frac{s}{2}}\hat{u}\|_{L^2(\Gamma)} \tag{A.I.74}$$

$$= \left(\int_\Gamma (1 + \psi(\chi))^s |\hat{u}(\chi)|^2 \mu_\Gamma(d\chi)\right)^{\frac{1}{2}}.$$

It follows that for $u \in H^{\psi,4}(G)$ we can apply Lebesgue's dominated convergence theorem to pass in

$$\left\| \frac{T_t u - u}{t} + \psi(D)u \right\|_{L^2(G)} = \left\| \left(\frac{e^{-t\psi} - 1}{t} + \psi \right) \hat{u} \right\|_{L^2(\Gamma)}$$

to the limit $t \to 0$ where we define for $u \in H^{\psi,4}(G)$ the operator $\psi(D)$ as $\psi(D)u = F^{-1}(\psi(\cdot)\hat{u})$. A more careful analysis will allow $u \in H^{\psi,2}(G)$. This result immediately points at the importance of the spaces $H^{\psi,s}(G)$. We will leave this topic here, being well aware that some calculations need more arguments. These considerations will be picked up in Volume V and also in Volume VI when linking these ideas to one-parameter operator semigroups, pseudo-differential operators and Markov processes. In particular we will see that $A = -\psi(D)$ indeed "generates" the semigroup $(T_t)_{t \geq 0}$ and that $\frac{\mathrm{d}}{\mathrm{d}t} T_t u = A T_t u$ holds for $u \in H^{\psi,2}(G)$.

Appendix II: Convergence of Measures

In this appendix we want to give some additions to our investigations of Borel measures on a topological space. A more systematic treatment will follow in Volume V when we will discuss dual spaces of spaces of continuous functions. In this appendix the underlying topological space G is always a **Polish space**, i.e. its topology is generated by a complete metric and has a countable base.

We call two measures μ_1 and μ_2 on $\mathcal{B}(G)$ **mutually singular** if there exists a Borel set $A \subset G$ such that $\mu_1(A) = 0$ and $\mu_2(A^{\complement}) = 0$. A **regular measure** μ on a Polish space is a measure which is finite on compact sets and for every $A \in \mathcal{B}(G)$ we have

$$\mu(A) = \sup\{\mu(K) \mid K \subset A, \ K \text{ compact}\}$$
$$= \inf\{\mu(U) \mid A \subset U, \ U \text{ open}\}.$$

Such a measure is also called a **Radon measure**, for a more general notation of a Radon measure we refer to [10] or [107]. A **signed Radon measure** on G (or $\mathcal{B}(G)$) is by definition the difference of two mutually singular regular measures on G. The set of all signed Radon measures form a vector space over $\mathbb{R}$ which we denote by $\mathcal{M}(G)$ and the (non-negative) measures on G we denote by $\mathcal{M}_+(G)$. We can now extend Definition 13.11.

Definition A.II.1. *Let G be a Polish space and $(\mu_k)_{k \in \mathbb{N}}$ be a sequence of Radon measures. We call the sequence $(\mu_k)_{k \in \mathbb{N}}$ **vaguely convergent** to the Radon measure μ if for all $\varphi \in C_0(G)$ we have*

$$\lim_{k \to \infty} \int_G \varphi \, \mathrm{d}\mu_k = \int_G \varphi \, \mathrm{d}\mu, \tag{A.II.1}$$

where $C_0(G)$ denotes the space of all continuous functions on G with compact support.

Note that since φ has a compact support the integrals in (A.II.1) are well defined and finite.

Definition A.II.2. *Let G be a Polish space and $(\mu_k)_{k \in \mathbb{N}}$ be a sequence of bounded Radon measures. We say that $(\mu_k)_{k \in \mathbb{N}}$ **converges weakly** to the bounded Radon measure μ if for all $\varphi \in C_b(G)$*

$$\lim_{k \to \infty} \int_G \varphi \, \mathrm{d}\mu_k = \int_G \varphi \, \mathrm{d}\mu \tag{A.II.2}$$

holds where $C_b(G)$ is the space of all bounded continuous functions on G.

We need in Definition A.II.2 the measures μ_k and μ to the bounded in order to assure that the integrals in (A.II.2) are finite for all bounded continuous functions.

The notions of convergence introduced in Definition A.II.1 and Definition A.II.2 are related to convergence in topological vector spaces. In order to understand this remark we first have to establish that the topological dual space of $C_0(G)$ is the space $\mathcal{M}(G)$ as well as that the topological dual space of $C_b(G)$ is the space of bounded Radon measures $\mathcal{M}_b(G)$. Then we have to consider on these dual spaces the corresponding weak topologies. The weak topology on $\mathcal{M}_b(G)$ is also called the **Bernoulli topology**. We will discuss these concepts and ideas in great detail in Volume V.

Our purpose in this volume is (beside others) to understand the Fourier transforms of measures. For this the following result is of importance.

Theorem A.II.3. *A sequence $(\mu_k)_{k \in \mathbb{N}}$ of bounded Radon measures on G converges weakly to a bounded Radon measure μ on G if and only if it converges vaguely and $\lim_{k \to \infty} \mu_k(G) = \mu(G)$.*

Obviously weak convergence implies vague convergence. A proof of the converse statement in Theorem A.II.3 is given in [10], [12] or [107].

An important consequence of Theorem A.II.3 is that for probability measures weak and vague convergence are equivalent. Moreover for a convolution semigroup $(\mu_t)_{t \geq 0}$ of sub-probability measures μ_t on $\mathbb{R}^n$ (or more generally on a locally compact Abelian group) condition (13.21) can be replaced by weak convergence and (13.22) by

$$\lim_{t \to 0} \int_{\mathbb{R}^n} \varphi \, \mathrm{d}\mu_t = \varphi(0) \tag{A.II.3}$$

for all $\varphi \in C_b(\mathbb{R}^n)$.

Definition A.II.4. *We call a subset M of bounded measures **vaguely bounded** if*

$$\sup_{\mu \in M} \left| \int_G \varphi(x) \mu(\mathrm{d}x) \right| < \infty \tag{A.II.4}$$

for all $\varphi \in C_0(G)$.

Theorem A.II.5. *A subset M of bounded measures is relative compact in the vague topology if and only if it is vaguely bounded.*

In $\mathbb{R}^n$ we have also a nice approximation result

Theorem A.II.6. *For a bounded signed measure μ on $\mathbb{R}^n$ there exists a sequence of (finite) linear combinations of Dirac measures converging weakly, and therefore vaguely, to μ.*

The topics discussed in this appendix are treated in more details in [10], [88], [103] and [107].

Appendix III: Generating Functions, Orthonormal Polynomials

In Chapter 9, in particular Problem 5 and 6, and in Chapter 14, Proposition 14.1, we have encountered the concept of a generating function of a given sequence of functions, e.g. for Legendre polynomials and Bessel functions, respectively. In this appendix we want to outline in more details the underlying ideas and to give further applications to concrete sequences of functions.

Let $(\varphi_k)_{k \in \mathbb{N}_0}$ be a sequence of functions $\varphi_k : I \to \mathbb{R}$ where $I \subset \mathbb{R}$ is an interval, or $\varphi_k : G \to \mathbb{C}$ where $G \subset \mathbb{C}$ is a domain. Suppose that for some function $g : I \times B_r(0) \to \mathbb{C}$, $r > 0$, we have

$$g(x, z) = \sum_{k=0}^{\infty} \varphi_k(x) z^k \qquad \text{(A.III.1)}$$

where the power series converges at least in $B_r(0)$ and r is independent of x. Then we call g the **generating function** of the sequence $(\varphi_k)_{k \in \mathbb{N}_0}$. In some cases it is convenient to allow g to have the expansion $\sum_{k \in \mathbb{Z}} \varphi_k(x) z^k$ which is assumed to converge uniformly in x on compact sets in $B_r(0) \backslash \{0\}$, $r > 0$ is assumed to be independent of x.

Example A.III.1. A. The generating function of the **Legendre polynomials** P_l is given by

$$g_L(x, z) = (1 - 2xz + z^2)^{-\frac{1}{2}} = \sum_{k=0}^{\infty} P_k(x) z^k \qquad \text{(A.III.2)}$$

which converges in $|z| < 1$, see Problem 5 and 6 of Chapter 9.

B. The generating function of the **Bessel functions** J_n, $n \in \mathbb{Z}$, is

$$g_B(x, z) = e^{\frac{x}{2}\left(z - \frac{1}{z}\right)} = \sum_{k \in \mathbb{Z}} J_k(x) z^k, \quad z \neq 0, \qquad \text{(A.III.3)}$$

compare with Proposition 14.1.

Example A.III.2. The generating function of the **Hermite polynomials** is the function

$$\tilde{g}_H(x, z) = e^{2xz - z^2} = \sum_{k=0}^{\infty} \frac{H_k(x)}{k!} z^n, \quad z \in \mathbb{C}. \qquad \text{(A.III.4)}$$

569

This follows easily from the observation that with $y = x - z$ we have

$$\left.\frac{d^n}{dz^n}e^{-(x-z)^2}\right|_{z=0} = (-1)^n \left.\frac{d^n}{dy^n}e^{-y^2}\right|_{y=x}$$

$$= e^{-y^2} H_n(y)|_{y=x} = e^{-x^2} H_n(x)$$

and now we may use the Taylor formula for the exponential function to get

$$e^{-(x-z)^2} = \sum_{k=0}^{\infty} e^{-x^2} H_k(x)\frac{z^k}{k!}$$

or

$$e^{2xz-z^2} = \sum_{k=0}^{\infty} \frac{H_k(x)}{k!}z^k.$$

Note that according to our original definition $\tilde{g}_H$ is the generating function of the sequence $\left(\frac{H_n}{n!}\right)_{n\in\mathbb{N}_0}$. However it is often convenient for polynomials to modify (A.III.1) to

$$\tilde{g}(x, z) = \sum_{k=0}^{\infty} \frac{\varphi_k(x)}{k!}z^k.$$

Example A.III.3. A. The generating function of the **Laguerre polynomials** L_n^α is

$$g_{La}(x, z) = \frac{e^{-\frac{xz}{1-z}}}{(1-z)^{\alpha+1}} = \sum_{k=0}^{\infty} L_k^\alpha(x)z^k, \quad x > 0, \quad |z| < 1. \qquad \text{(A.III.5)}$$

B. For the **Chebyshev polynomials**, see Problem 1 of Chapter 20, we have the generating function

$$g_{Ch}(x, z) = \frac{xz - x^2}{1 - 2xz + z^2}, \quad |x| < 1, \quad |z| < 1. \qquad \text{(A.III.6)}$$

(For details we refer to [34] or [85].)

In certain cases one can obtain generating functions with the help of complex variable theory. This holds for example for orthogonal polynomials being eigenfunctions of the equation

$$pu_n'' + qu_n' + \lambda_n u_n = 0$$

under homogeneous boundary conditions. Here p is given, $q = \left(\frac{pw}{w}\right)'$ and orthogonality refers to the measure $w\lambda^{(n)}|_I$ where I is the interval in which the equation is considered. In cases where w and p have appropriate regularity properties it follows that

$$\tilde{g}(x, z) = \sum_{k=0}^{\infty} \frac{u_n(x)}{n!} z^n \qquad (\text{A.III.7})$$

$$= \frac{1}{2\pi i} \int_{\gamma} \frac{w(\xi)}{w(x)} \frac{d\xi}{\xi - x - zp(\xi)}$$

where trγ encloses $x \in I \setminus \partial I$. We refer to [11] for details. Hence $\tilde{g}(x, z)$ is the generating function of the sequence $\left(\frac{u_n}{n!}\right)_{n \in \mathbb{N}_0}$ in the sense of our original definition.

Once we know the generating function of a given sequence of functions $(u_k)_{k \in \mathbb{N}_0}$ we may use it to derive further properties of these functions u_k. We have seen this in Problem 6 to Chapter 9 where we have derived Legendre's differential equation for the Legendre polynomials. In the similar way we may handle Hermite and Laguerre polynomials.

For the Hermite polynomials we find

$$\frac{\partial \tilde{g}_H}{\partial z} - (2x - 2z)\tilde{g}_H = 0. \qquad (\text{A.III.8})$$

With the help of this equality we may first derive some recurrence formulae such as

$$H_{n+1}(x) - 2x H_n(x) + 2n H_{n-1}(x) = 0, \quad n \in \mathbb{N}, \qquad (\text{A.III.9})$$

$$H_n'(x) = 2n H_{n-1}(x) \qquad (\text{A.III.10})$$

or

$$H_{n+1}(x) - 2x H_n(x) + H_n'(x) = 0, \qquad (\text{A.III.11})$$

which eventually yields the **Hermite differential equation**

$$H_n''(x) - 2x H_n'(x) + 2n H_n(x) = 0, \quad n \in \mathbb{N}_0. \qquad (\text{A.III.12})$$

For details we refer to [85] or [34]. With similar arguments we find for the Laguerre polynomials first

$$(1 - z^2)\frac{\partial g_{La}}{\partial z} + (x - (1 - z)(1 + \alpha))g_{La}, \qquad (\text{A.III.13})$$

and eventually we get **Laguerre's differential equation**

$$x\frac{d^2}{dx^2}L_n^\alpha(x) + (\alpha + 1 - x)\frac{d}{dx}L_n^{(\alpha)}(x) + n L_n^\alpha(x) = 0, \qquad (\text{A.III.14})$$

compare with [85].

We close this appendix by introducing the **Jacobi polynomials** $P_n^{(\alpha,\beta)}$. They are obtained as orthonormal basis in $L^2((-1,1), w\lambda^{(1)})$ where we take $w(x) = (1-x)^\alpha(1+x)^\beta$, $\alpha > -1$, $\beta > -1$. Their generating function is

$$g_J(x,z) = \frac{z^{\alpha+\beta}}{h(x,z)(1-z+h(x,z))(1+z+h(x,z))}, \qquad \text{(A.III.15)}$$

and

$$h(x,z) = (1-2xz+z^2)^{\frac{1}{2}}, \qquad \text{(A.III.16)}$$

and the Jacobi polynomials solve the **Jacobi differential equation**

$$(1-x^2)\frac{\mathrm{d}^2}{\mathrm{d}x^2}P_n^{(\alpha,\beta)}(x) + (\beta - \alpha - (\alpha+\beta+2)x)\frac{\mathrm{d}}{\mathrm{d}x}P_n^{(\alpha,\beta)}(x) \qquad \text{(A.III.17)}$$
$$+ n(n+\alpha+\beta+1)P_n^{(\alpha,\beta)}(x) = 0.$$

For $\alpha = \beta > -\frac{1}{2}$ a simplification is possible by introducing the **Gegenbauer polynomials**

$$C_n^\lambda(x) := \frac{\Gamma(2\lambda+n)\Gamma(\lambda+\frac{1}{2})}{\Gamma(2\lambda)\Gamma(\lambda+n+\frac{1}{2})}P_n^{(\lambda-\frac{1}{2},\lambda-\frac{1}{2})}(x) \qquad \text{(A.III.18)}$$

which have the generating function

$$g_G(x,z) = (1-2xz+z^2)^{-\lambda} = \sum_{k=0}^{\infty} C_k^\lambda(x)z^k. \qquad \text{(A.III.19)}$$

Again we refer to [85], [34] and [11] for further details.

Appendix IV: On Brouwer's Fixed Point Theorem

Brouwer's fixed point theorem, Theorem 16.3, was used to prove Schauder's fixed point theorem, Theorem 16.4, and Schauder's theorem in turn was key to proving the existence result of Peano, Theorem 16.19.

Brouwer's theorem is a result on continuous mappings between topological spaces and it is of topological nature. It is non-trivial in the sense that any of its proof requires a deeper result from topology. There have been attempts to give so called "elementary" proofs when considering differentiable mappings between certain differentiable manifolds, in particular compact subsets of $\mathbb{R}^n$ were studied. The price to avoid non-trivial topological results is always paid by a rather complicated, non-trivial analytic calculation and in the end it gives the theorem not in full generality. For this reason we will discuss here a topological proof taking certain (clearly stated) non-trivial results from topology for granted.

We start with a version of Brouwer's fixed point theorem which differs from Theorem 16.3 but will allow us to deduce that theorem.

Theorem A.IV.1. *Every continuous mapping* $f : \overline{B_R(x_0)} \to \overline{B_R(x_0)}$, $x_0 \in \mathbb{R}^n$, $R > 0$, $\overline{B_R(x_0)} \subset \mathbb{R}^n$, *has a fixed point.*

The case $n = 1$ *is easy to prove. In this case we look at a continuous mapping* $f : [-1, 1] \to [-1, 1]$. *A fixed point of* f *is a zero of* id $- f$. *Since* $(\text{id} - f)(-1) \leq 0$ *and* $(\text{id} - f)(1) \geq 0$ *the intermediate value theorem applied to the continuous function* id $- f$ *implies the existence of some* $x_0 \in [-1, 1]$ *with* $(\text{id} - f)(x_0) = 0$, *or* $x_0 = f(x_0)$.

To proceed further we need

Definition A.IV.2. *Let* X *be a topological space and* $M \subset X$. *A continuous mapping* $r : X \to M$ *is called a* **retraction** *if* $r(x) = x$ *for all* $x \in M$. *The set* M *is called a* **retract** *of* X.

Example A.IV.3. In $\mathbb{R}^n$ we consider the mapping $r : \mathbb{R}^n \to \overline{B_R(0)}$ defined by

$$r(x) := \begin{cases} x, & x \in \overline{B_R(0)} \\ \frac{Rx}{\|x\|}, & x \in \overline{B_R(0)}^\complement. \end{cases} \tag{A.IV.1}$$

Obviously r is a continuous mapping and $r(x) = x$ for $x \in \overline{B_R(0)}$.

The first non-trivial result we need is **Tietze's extension** theorem:

Theorem A.IV.4. *Let (X, d) be a metric space and $(Y, \|\cdot\|)$ be a normed space. Let $\emptyset \neq M \subset X$ be closed and $g : M \to Y$ a continuous mapping. Then g admits a continuous extension $\tilde{g} : X \to \mathrm{conv}(f(M))$, where as usual $\mathrm{conv}\, A$ denotes the convex hull of a subset A of a vector space.*

With the help of Tietze's theorem one can prove

Proposition A.IV.5. *Every closed convex subset $M \subset X$ of a normed space X is a retract of X.*

We just have to consider the identity $\mathrm{id} : M \to M$ and apply the extension result of Tietze.

The deepest result we need is

Theorem A.IV.6. *For a closed ball $\overline{B_R(x_0)} \subset \mathbb{R}^n$, $x_0 \in \mathbb{R}^n$, $R > 0$, the boundary $\partial \overline{B_R(x_0)}$ is not a retract of $\overline{B_R(x_0)}$.*

Now we can give an outline of the

Proof of Theorem A.IV.1. Suppose that $f : \overline{B_R(x_0)} \to \overline{B_R(x_0)}$ has no fixed point, i.e. $f(x) \neq x$ for all $x \in B_R(x_0)$. We now construct a retraction $r : \overline{B_R(x_0)} \to \partial \overline{B_R(x_0)}$ in the following way, see Figure A.IV.1.

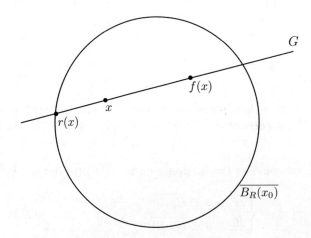

Figure A.IV.1

For $x \in \overline{B_R(x_0)}$ we consider the straight line G passing through x and $f(x)$. We set $r(x)$ to be the intersection point of G and the boundary $\partial B_R(x_0)$ where $r(x)$ is chosen in such a way that x belongs to the line segment connecting $r(x)$ and $f(x)$. This mapping is continuous and hence we have constructed a retraction of $\overline{B_R(x_0)}$ onto $\partial B_R(x_0)$ which is a contradiction. Hence f must have a fixed point. $\quad\square$

Now we can prove Theorem 16.3, i.e.

Theorem A.IV.7. *Every continuous function $f : K \to K$ mapping a compact and convex set $K \subset \mathbb{R}^n$, $K \neq \emptyset$, into itself has at least one fixed point.*

Proof. We choose a closed ball $\overline{B_R(x_0)}$ such that $K \subset \overline{B_R(x_0)}$. By Proposition A.IV.5 there exists a retraction $r : \overline{B_R(x_0)} \to K$, note that if $\tilde{r} : X \to M$ is a retraction and $Y \subset X$ is closed, $M \subset Y$, then $\tilde{r}|_Y : Y \to M$ is a retraction too. For the composition $f \circ r : \overline{B_R(x_0)} \to K$, i.e. $f \circ r : \overline{B_R(x_0)} \to \overline{B_R(x_0)}$, we first observe that it is a continuous mapping and then we deduce from Theorem A.IV.1 that $f \circ r$ must have a fixed point $x_0 = f(r(x_0))$. By construction the points in $\overline{B_R(x_0)}\backslash K$ are fixed under the retraction and for $x \in K$ we have $r(x) = x$, or $f(r(x)) = f(x)$ and the result follows. $\quad\square$

We follows closely the discussion in E. Zeidler [130]. Different approaches to Brouwer's fixed point theorem are discussed in J. Milnor [91] and M. Schechter [105], just to mention two standard reference.

Solutions to Problems of Part 8

Chapter 2

1. We know that $c_0 = \frac{a_0}{2}$, $c_k = \frac{a_k - ib_k}{2}$ and $c_{-k} = \frac{a_{-k} + ib_{-k}}{2}$ for $k \in \mathbb{N}$. This gives

$$a_0 = 2c_0$$

$$a_k = \frac{c_k + c_{-k}}{2}$$

$$b_k = \frac{c_{-k} - c_k}{2i}.$$

However, a_k and b_k must be real-valued which induces $c_{-k} = \overline{c_k}$ for all $k \in \mathbb{Z}$.

2. We need to prove that, for a 2π-periodic function $u : \mathbb{R} \to \mathbb{R}$ which is integrable over any interval of length 2π, we have

$$\int_{a-\pi}^{a+\pi} u(x)\,\mathrm{d}x = \int_{-\pi}^{\pi} u(x)\,\mathrm{d}x$$

for $0 < a < 2\pi$. (In the case a initially does not belong to the interval $(0, 2\pi)$, we can find $m \in \mathbb{Z}$ such that $a + 2m\pi \in (0, 2\pi)$ and the corresponding change of variable yields the reduction.) Thus it remains to show that

$$\int_{-\pi}^{a-\pi} u(x)\,\mathrm{d}x = \int_{\pi}^{a+\pi} u(x)\,\mathrm{d}x.$$

Substituting x by $x + 2\pi$ in the left hand integral gives

$$\int_{-\pi}^{a-\pi} u(x)\,\mathrm{d}x = \int_{\pi}^{a+\pi} u(x + 2\pi)\,\mathrm{d}x = \int_{\pi}^{a+\pi} u(x)\,\mathrm{d}x.$$

3.   a) On $[-\pi, \pi]$, the function $u(x) = |x|$ is even and therefore its 2π-periodic extension is even too. The graph of the extension is given below.

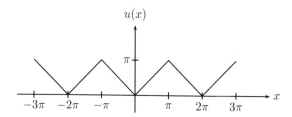

We have only to find the coefficients $a_k, k \in \mathbb{N}_0$, and for these we have

$$a_0 = \frac{1}{\pi} \int_{-\pi}^{\pi} |x|\,\mathrm{d}x = \frac{2}{\pi} \int_0^{\pi} x\,\mathrm{d}x = \pi$$

and

$$a_k = \frac{1}{\pi} \int_{-\pi}^{\pi} |x| \cos kx \, dx = \frac{2}{\pi} \int_0^{\pi} x \cos kx \, dx$$

$$= -\frac{2}{\pi k^2} \int_0^{\pi} \sin kx \, dx = \frac{2}{\pi k^2} \cos kx \Big|_0^{\pi}$$

$$= \frac{2}{\pi k^2} ((-1)^k - 1).$$

Thus we arrive at

$$u(x) = |x| \quad \sim \quad \frac{\pi}{2} + \sum_{k=1}^{\infty} \frac{2}{\pi k^2} ((-1)^k - 1) \cos kx$$

which we can rewrite as

$$|x| \quad \sim \quad \frac{\pi}{2} - \frac{4}{\pi} \sum_{l=0}^{\infty} \frac{\cos (2l + 1)x}{(2l + 1)^2}.$$

b) The function $x \mapsto \cos \gamma x$ is an even function and therefore we need only to calculate the coefficients $a_k, k \in \mathbb{N}_0$. For $k = 0$, we find

$$a_0 = \frac{1}{\pi} \int_{-\pi}^{\pi} \cos \gamma x \, dx = \frac{2}{\pi} \int_0^{\pi} \cos \gamma x \, dx = \frac{2}{\gamma \pi} \sin \gamma \pi.$$

For $k \in \mathbb{N}$, we have

$$\frac{1}{\pi} \int_{-\pi}^{\pi} \cos \gamma x \cos kx \, dx = \frac{2}{\pi} \int_0^{\pi} \cos \gamma x \cos kx \, dx$$

$$= \frac{2}{\pi} \int_0^{\pi} \frac{1}{2} (\cos (\gamma + k)x + \cos (\gamma - k)x) \, dx$$

$$= \frac{1}{\pi} \left(\frac{1}{\gamma + k} \sin (\gamma + k)x \Big|_0^{\pi} + \frac{1}{\gamma - k} \sin (\gamma - k)x \Big|_0^{\pi} \right)$$

$$= \frac{1}{\pi} \left(\frac{1}{\gamma + k} \sin (\gamma + k)\pi + \frac{1}{\gamma - k} \sin (\gamma - k)\pi \right)$$

and using the formula $\sin (\gamma \pm k)\pi = \sin \gamma \pi \cos k\pi \pm \cos \gamma \pi \sin k\pi$, we find for $k \in \mathbb{N}$

$$a_k = \frac{1}{\pi} \left(\frac{1}{\gamma + k} \sin \gamma \pi + \frac{1}{\gamma - k} \sin \gamma \pi \right) \cos k\pi$$

$$= \frac{2\gamma \sin \gamma \pi}{\pi (\gamma^2 - k^2)} \cos k\pi = \frac{2\gamma (-1)^k \sin \gamma \pi}{\pi (\gamma^2 - k^2)}$$

which yields

$$\cos \gamma x \quad \sim \quad \frac{a_0}{2} + \sum_{k=1}^{\infty} a_k \cos kx$$

or

$$\cos \gamma x \quad \sim \quad \frac{\sin \gamma \pi}{\gamma \pi} + \sum_{k=1}^{\infty} \frac{2\gamma(-1)^k \sin \gamma \pi}{\pi(\gamma^2 - k^2)} \cos kx$$

$$= \frac{2}{\pi} \sin \gamma \pi \left(\frac{1}{2\gamma} + \sum_{k=1}^{\infty} (-1)^k \frac{\gamma \cos kx}{\gamma^2 - k^2} \right).$$

4. Before calculating the Fourier coefficients of f, g and h, we first sketch these functions in the interval $[-3\pi, 3\pi)$ (with $\alpha = \frac{1}{2}$).

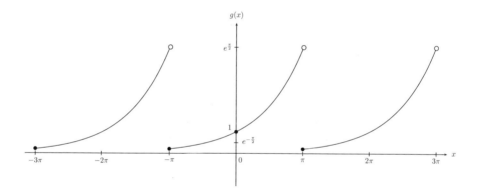

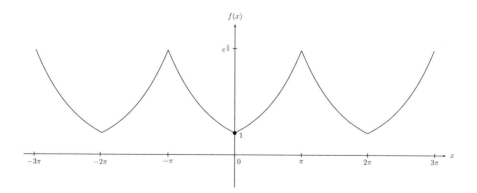

579

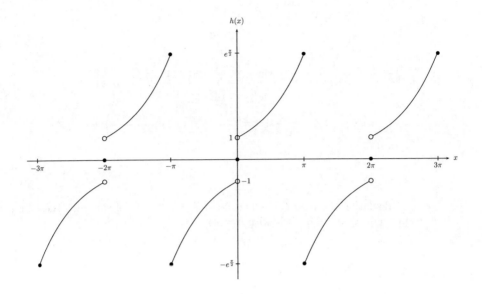

a) The Fourier coefficents of f are given by

$$a_0 = \frac{1}{\pi} \int_{-\pi}^{\pi} e^{\alpha x} \, dx = \frac{1}{\alpha \pi} (e^{\alpha \pi} - e^{-\alpha \pi}) = 2 \frac{\sinh \alpha \pi}{\alpha \pi}$$

$$a_k = \frac{1}{\pi} \int_{-\pi}^{\pi} e^{\alpha x} \cos kx \, dx = \frac{1}{\pi} \frac{\alpha \cos kx + k \sin kx}{\alpha^2 + k^2} e^{\alpha x} \Big|_{-\pi}^{\pi}$$

$$= (-1)^k \frac{2\alpha \sinh \alpha \pi}{\pi(\alpha^2 + k^2)}$$

$$b_k = \frac{1}{\pi} \int_{-\pi}^{\pi} e^{\alpha x} \sin kx \, dx = \frac{1}{\pi} \frac{\alpha \sin kx - k \cos kx}{\alpha^2 + k^2} e^{\alpha x} \Big|_{-\pi}^{\pi}$$

$$= (-1)^{k-1} \frac{2k \sinh \alpha \pi}{\pi(\alpha^2 + k^2)}$$

and we arrive at

$$e^{\alpha x} \Big|_{(-\pi, \pi)} = f(x) \Big|_{(-\pi, \pi)} \sim \frac{2}{\pi} \sinh \alpha \pi \left(\frac{1}{2\alpha} + \sum_{k=1}^{\infty} \frac{(-1)^k}{\alpha^2 + k^2} (\alpha \cos kx - k \sin kx) \right)$$

b) Since g is an even function by construction, all coefficients b_k must be 0. Moreover, we have

$$\frac{a_0}{2} = \frac{1}{2\pi} \int_{-\pi}^{\pi} g(x) \, dx = \frac{1}{\pi} \int_0^{\pi} e^{\alpha x} \, dx = \frac{e^{\alpha x} - 1}{\alpha \pi}$$

and

$$a_k = \frac{1}{\pi} \int_{-\pi}^{\pi} g(x) \cos kx \, dx = \frac{2}{\pi} \int_{0}^{\pi} e^{\alpha x} \cos kx \, dx$$

$$= \frac{2\alpha}{\pi} (-1)^k \frac{e^{\alpha k} - 1}{\alpha^2 + k^2}$$

leading to

$$g(x) \quad \sim \quad \frac{e^{\alpha k} - 1}{\alpha \pi} + \frac{2\alpha}{\pi} \sum_{k=1}^{\infty} (-1)^k \frac{e^{\alpha k} - 1}{\alpha^2 + k^2} \cos kx.$$

c) Since h is odd, we must have $a_k = 0$ for $k \in \mathbb{N}_0$ and furthermore, for $k \in \mathbb{N}$, we have

$$b_k = \frac{1}{\pi} \int_{-\pi}^{\pi} h(x) \sin kx \, dx = \frac{2}{\pi} \int_{0}^{\pi} e^{\alpha x} \sin kx \, dx$$

$$= \frac{2}{\pi} (1 - (-1)^k e^{\alpha \pi}) \frac{k}{\alpha^2 + k^2}$$

and therefore

$$h(x) \quad \sim \quad \frac{2}{\pi} \sum_{k=1}^{\infty} (1 - (-1)^k e^{\alpha \pi}) \frac{k}{\alpha^2 + k^2} \sin kx.$$

5. By assumption we have

$$f(x) = \frac{a_0}{2} + \sum_{k=1}^{\infty} \left(a_k \cos(k\frac{\pi}{L}x) + b_k \sin(k\frac{\pi}{L}x) \right)$$

with a uniformly convergent series. Thus we may multiply f by $\sin k\frac{\pi}{L}\cdot, \cos k\frac{\pi}{L}\cdot$ and integrate term by term. This yields

$$\frac{1}{L} \int_{-L}^{L} f(x) \, dx = a_0 + \sum_{k=1}^{\infty} \left(a_k \frac{1}{L} \int_{-L}^{L} \cos(k\frac{\pi}{L}x) \, dx + b_k \frac{1}{L} \int_{-L}^{L} \sin(k\frac{\pi}{L}x \, dx \right)$$

but

$$\int_{-L}^{L} \cos(k\frac{\pi}{L}x) \, dx = \int_{-L}^{L} \sin(k\frac{\pi}{L}x) \, dx = 0.$$

Furthermore, for $m \in \mathbb{N}$, we get

$$\frac{1}{L} \int_{-L}^{L} \cos(m\frac{\pi}{L}x) f(x) \, dx = \frac{1}{L} \frac{a_0}{2} \int_{-L}^{L} \cos(m\frac{\pi}{L}x) \, dx$$

$$+ \frac{1}{L} \sum_{k=1}^{\infty} (a_k \int_{-L}^{L} \cos(m\frac{\pi}{L}x) \cos(k\frac{\pi}{L}x) \, dx + b_k \int_{-L}^{L} \cos(m\frac{\pi}{L}x) \sin(k\frac{\pi}{L}x)) \, dx.$$

581

We know already that

$$\int_{-L}^{L} \cos\left(m\frac{\pi}{L}x\right) \, dx = 0.$$

Next, we note that using the substitution $\frac{\pi}{L}x = y$, it follows that

$$\int_{-L}^{L} \cos\left(m\frac{\pi}{L}x\right)\sin\left(k\frac{\pi}{L}x\right) \, dx = \frac{L}{\pi}\int_{-\pi}^{\pi} \cos my \sin ky \, dy = 0$$

and

$$\int_{-L}^{L} \cos\left(m\frac{\pi}{L}x\right)\cos\left(k\frac{\pi}{L}x\right) \, dx = \frac{L}{\pi}\int_{-\pi}^{\pi} \cos my \cos ky \, dy = L\delta_{mk}$$

which implies for $k \in \mathbb{N}$

$$\frac{1}{L}\int_{-L}^{L} \cos\left(m\frac{\pi}{L}x\right)f(x) \, dx = a_m.$$

Analogously we get for $k \in \mathbb{N}$

$$\frac{1}{L}\int_{-L}^{L} \sin\left(m\frac{\pi}{L}x\right)f(x) \, dx = b_m.$$

6. The even extension of $\tilde{g}$ has in principle a graph looking as follows.

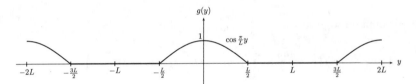

Since g is even, we need to find only the coefficients $a_k, k \in \mathbb{N}_0$, given by

$$a_0 = \frac{1}{L}\int_{-L}^{L} g(y) \, dy$$

and

$$a_k = \frac{1}{L}\int_{-L}^{L} g(y)\cos\left(k\frac{\pi}{L}y\right) \, dy.$$

From the definiton of $\tilde{g}$, we conclude

$$a_0 = \frac{2}{L}\int_{0}^{L} \cos\frac{\pi y}{L} \, dy = \frac{2}{\pi}$$

and

$$a_k = \frac{2}{L}\int_{0}^{L} \tilde{g}(y)\cos\left(k\frac{\pi}{L}y\right) \, dy = \frac{2}{L}\int_{0}^{\frac{L}{2}} \cos\frac{\pi y}{L}\cos\frac{k\pi y}{L} \, dy.$$

The latter integral has the value

$$a_k = \begin{cases} 0 & \text{for k odd} \\ \frac{2(-1)^{\frac{k}{2}}}{\pi(k^2-1)} & \text{for k even} \end{cases}$$

which yields

$$g \quad \sim \quad \frac{1}{\pi} + \frac{L}{2}\cos\frac{\pi y}{L} - \frac{2}{\pi}\sum_{k=1}^{\infty}\frac{(-1)^k}{4k^2-1}\cos\frac{2\pi ky}{L}.$$

7. The series $\sum_{k=1}^{\infty}\frac{\cos kx}{k^2}$ is uniformly and absolutely convergent and therefore we may integrate term by term to find

$$\int_0^x \sum_{k=1}^{\infty}\frac{\cos kt}{k^2}\,dt = \sum_{k=1}^{\infty}\int_0^x\frac{\cos kt}{k^2}\,dt = \sum_{k=1}^{\infty}\frac{\sin kx}{k^3}.$$

On the other hand, we have

$$\int_0^x\left(\left(\frac{t-\pi}{2}\right)^2 - \frac{\pi^2}{12}\right)dt = \frac{x^3-3\pi x^2+2\pi x}{12} = \frac{x(x-\pi)(x-2\pi)}{12}.$$

Thus we have shown

$$\sum_{k=1}^{\infty}\frac{\sin kx}{k^3} = \frac{x(x-\pi)(x-2\pi)}{12}.$$

For $x = \frac{\pi}{2}$, we find

$$\frac{\frac{\pi}{2}\left(\frac{\pi}{2}-\pi\right)\left(\frac{\pi}{2}-2\pi\right)}{12} = \frac{\frac{\pi}{2}\left(-\frac{\pi}{2}\right)\left(-\frac{3\pi}{2}\right)}{12} = \frac{3\pi^3}{96} = \frac{\pi^3}{32}.$$

Whereas on the other hand, we have

$$\sum_{k=1}^{\infty}\frac{\sin\frac{k\pi}{2}}{k^3} = \sum_{l=0}^{\infty}\frac{(-1)^l}{(2l+1)^3},$$

i.e.

$$\sum_{l=0}^{\infty}\frac{(-1)^l}{(2l+1)^3} = \frac{\pi^3}{32}.$$

8. The value of the geometric series $\sum_{k=0}^{\infty}(\frac{e^{it}}{3})^k$ is given by

$$\sum_{k=0}^{\infty}\frac{\cos kt + i\sin kt}{3^k} = \sum_{k=0}^{\infty}\left(\frac{e^{it}}{3}\right)^k = \frac{1}{1-\frac{e^{it}}{3}}$$
$$= \frac{9-3\cos t - 3i\sin t}{10-6\cos t}$$

and therefore

$$\sum_{k=0}^{\infty}\frac{\cos kt}{3^k} = \text{Re}\left(\sum_{k=0}^{\infty}\frac{\cos kt + i\sin kt}{3^k}\right) = \frac{9-3\cos t}{10-6\cos t}$$

9. We may rewrite the series $\sum_{k=0}^{\infty} c_k r^k e^{ik\varphi}$ as $u(z) = \sum_{k=0}^{\infty} c_k z^k$ and the condition $|c_k| \leq M$ yields that

$$\sum_{k=0}^{\infty} |c_k z^k| \leq M \sum_{k=0}^{\infty} |z|^k = \frac{M}{1 - |z|}$$

implying that u is holomorphic in $B_1(0)$. With $c_k = \gamma^k, 0 < \gamma < 1$, we get

$$\sum_{k=0}^{\infty} \gamma^k z^k = \sum_{k=0}^{\infty} (\gamma z)^k$$

which converges uniformly and absolutely for $|\gamma||z| < 1$ i.e. $|z| < \frac{1}{|\gamma|}$ implying that $\sum_{k=0}^{\infty} \gamma^k z^k$ is a holomorphic function in $B_{\frac{1}{\gamma}}(0)$.

Chapter 3

1. We have

$$-i \sum_{k \in \mathbb{Z}} sgn(k) c_k e^{ikx} = -i \sum_{k=1}^{\infty} c_k e^{ikx} - i \sum_{k=1}^{\infty} c_{-k} e^{-ikx}$$

$$= -i \left(\sum_{k=1}^{\infty} c_k e^{ikx} - \overline{c_k e^{ikx}} \right)$$

$$= -2 \sum_{k=1}^{\infty} \text{Im}(c_k e^{ikx})$$

$$= \sum_{k=1}^{\infty} \text{Im}(a_k \cos kx - ib_k \cos kx + ia_k \sin kx - b_k \sin kx)$$

$$= \sum_{k=1}^{\infty} (-b_k \cos kx + a_k \sin kx).$$

2. Let $(d_k)_{k \in \mathbb{Z}}, (e_k)_{k \in \mathbb{Z}} \in l_\infty(\mathbb{Z})$ and note that

$$\|(d_k)_{k \in \mathbb{Z}}\|_\infty = \sup_{k \in \mathbb{Z}} |d_k| < \infty.$$

Obviously, we have $\sup_{k \in \mathbb{Z}} |d_k| \geq 0$ as well as $\sup_{k \in \mathbb{Z}} |d_k| = 0$ if and only if $d_k = 0$. Moreover, for $\lambda \in \mathbb{C}$ we find

$$\|\lambda(d_k)_{k \in \mathbb{Z}}\|_\infty = \|(\lambda d_k)_{k \in \mathbb{Z}}\|_\infty = \sup_{k \in \mathbb{Z}} |\lambda d_k|$$

$$= |\lambda| \sup_{k \in \mathbb{Z}} |d_k| = |\lambda| \|(d_k)_{k \in \mathbb{Z}}\|_\infty$$

and further

$$\|(d_k + e_k)_{k \in \mathbb{Z}}\|_\infty = \sup_{k \in \mathbb{Z}} |d_k + e_k|$$

$$\leq \sup_{k \in \mathbb{Z}} |d_k| + \sup_{k \in \mathbb{Z}} |e_k|$$

$$= \|(d_k)_{k \in \mathbb{Z}}\|_\infty + \|(e_k)_{k \in \mathbb{Z}}\|_\infty$$

proving that $\|.\|_\infty$ is a norm on $l_\infty(\mathbb{Z})$. Now we want to prove that $(l_\infty(\mathbb{Z}), \|.\|_\infty)$ is complete. For this, let $(d^{(l)})_{l\in\mathbb{N}}$ be a Cauchy sequence in $l_\infty(\mathbb{Z})$, i.e. $d^{(l)} = (d_k^{(l)})_{k\in\mathbb{Z}}$ and for every $\epsilon > 0$ there exists $N_0 \in \mathbb{N}$ such that $l, m \geq N_0$ implies

$$\|d^{(l)} - d^{(m)}\|_\infty = \sup_{k\in\mathbb{Z}} |d_k^{(l)} - d_k^{(m)}| < \epsilon. \tag{$*$}$$

Condtion ($*$) implies that for each $k \in \mathbb{Z}$, the sequence $(d_k^{(l)})_{l\in\mathbb{N}}$ is a Cauchy sequence in $\mathbb{C}$, hence it has a limit d_k. We claim $d := (d_k)_{k\in\mathbb{Z}} \in l_\infty(\mathbb{Z})$ and

$$\lim_{l\to\infty} \|d^{(l)} - d\|_\infty = 0.$$

From ($*$) we deduce that for $M \in \mathbb{N}$ and $m, l \geq N_0$, we have

$$\sup_{|k|\leq M} |d_k^{(l)} - d_k^{(m)}| < \epsilon$$

and for $m \to \infty$ we obtain

$$\sup_{|k|\leq M} |d_k^{(l)} - d_k| \leq \epsilon$$

implying that

$$\sup_{k\in\mathbb{Z}} |d_k^{(l)} - d_k| \leq \epsilon.$$

Since $\sup_{k\in\mathbb{Z}} |d_k| \leq \sup_{k\in\mathbb{Z}} |d_k^{(l)}| + \sup_{k\in\mathbb{Z}} |d_k - d_k^{(l)}|$ it follows that $d \in l_\infty(\mathbb{Z})$ and $\lim_{l\to\infty} \|d^{(l)} - d\|_\infty = 0$, i.e. $l_\infty(\mathbb{Z})$ is a complete normed space.

3. Let $(a_k)_{k\in\mathbb{Z}} \in l_p(\mathbb{Z}), p > 1$, and $(b_k)_{k\in\mathbb{Z}} \in l_q(\mathbb{Z}), \frac{1}{p} + \frac{1}{q} = 1$. By Theorem I.23.15, we have

$$\sum_{|k|\leq N} |a_k b_k| \leq \left(\sum_{|k|\leq N} |a_k|^p \right)^{\frac{1}{p}} \left(\sum_{|k|\leq N} |b_k|^q \right)^{\frac{1}{q}}$$

$$\leq \left(\sum_{k\in\mathbb{Z}} |a_k|^p \right)^{\frac{1}{p}} \left(\sum_{k\in\mathbb{Z}} |b_k|^q \right)^{\frac{1}{q}}$$

$$= \|(a_k)_{k\in\mathbb{Z}}\|_p \|(b_k)_{k\in\mathbb{Z}}\|_q$$

Now we may send N to infinity on the left hand side to arrive at

$$\left| \sum_{k\in\mathbb{Z}} a_k b_k \right| \leq \sum_{k\in\mathbb{Z}} |a_k b_k| \leq \|(a_k)_{k\in\mathbb{Z}}\|_p \|(b_k)_{k\in\mathbb{Z}}\|_q.$$

4. We note that

$$\int_{-\pi}^{\pi} \frac{h(t + t_0) \cos mt}{h(t_0)}\, dt = \int_{-\pi}^{\pi} \frac{h(s) \cos m(s - t_0)}{h(t_0)}\, ds$$

$$= \int_{-\pi}^{\pi} \frac{h(s)}{h(t_0)} (\cos ms \cos mt_0 + \sin ms \sin mt_0)\, ds$$

$$= \frac{\cos mt_0}{h(t_0)} \int_{-\pi}^{\pi} h(s) \cos ms\, ds + \frac{\sin mt_0}{h(t_0} \int_{-\pi}^{\pi} h(s) \sin ms\, ds$$

$$= 0$$

585

by (3.36). The result for $\int_{-\pi}^{\pi} \frac{h(t+t_0)\sin mt}{h(t_0)} \, dt$ follows analogously.

5. a) For $k \in \mathbb{N}$, we have

$$\alpha_k = \frac{1}{\pi} \int_{-\pi}^{\pi} f'(t) \cos kt \, dt$$

$$= \frac{1}{\pi} (-f(t)k \sin kt) \Big|_{-\pi}^{\pi} + \frac{1}{\pi} \int_{-\pi}^{\pi} k f(t) \sin kt \, dt$$

$$= k b_k$$

and

$$\beta_k = \frac{1}{\pi} \int_{-\pi}^{\pi} f'(t) \sin kt \, dt$$

$$= \frac{1}{\pi} (f(t)k \cos kt) \Big|_{-\pi}^{\pi} - \frac{1}{\pi} \int_{-\pi}^{\pi} k f(t) \cos kt \, dt$$

$$= -k a_k.$$

b) Since $c_0 = 0$, we find

$$G(t + 2\pi) - G(t) = \int_t^{t+2\pi} g(s) \, ds = \int_{-\pi}^{\pi} g(s) \, ds = 0,$$

i.e. G is 2π-periodic and a C^1-function. For $k \in \mathbb{Z} \setminus \{0\}$, we find

$$\gamma_k = \frac{1}{2\pi} \int_{-\pi}^{\pi} G(t) e^{-ikt} \, dt$$

$$= \frac{1}{2\pi} \int_{-\pi}^{\pi} G(t) \frac{1}{-ik} \frac{d}{dt} (e^{-ikt}) \, dt$$

$$= \frac{1}{2\pi} G(t) \frac{e^{-ikt}}{-ik} \Big|_{-\pi}^{\pi} - \frac{1}{2\pi} \int_{-\pi}^{\pi} G'(t) \frac{e^{-ikt}}{-ik} \, dt$$

$$= \frac{1}{ik} \frac{1}{2\pi} \int_{-\pi}^{\pi} g(t) e^{-ikt} \, dt = \frac{\gamma_k}{ik}.$$

6. a) Since for all $m \in \mathbb{N}$ the estimate

$$\left| \frac{d^m}{dt^m} \sin kt \right| \leq k^m$$

holds and since for all $m \in \mathbb{N}$ the series $\sum_{k=0}^{\infty} \frac{k^m}{4^k}$ converges, it follows from the Weierstrass test together with an iteration of Theorem III.8.4 that

$$\frac{d^m}{dt^m} \sum_{k=0}^{\infty} \frac{\sin kt}{4^k} = \sum_{k=0}^{\infty} \frac{d^m}{dt^m} \frac{\sin kt}{4^k}$$

and the series on the right hand side is representing a continuous function.

b) We have $\frac{d}{dt}\frac{\cos 4^k t}{4^k} = -\sin 4^k t$. Now suppose that $t \to \sum_{k=0}^{\infty}\frac{\cos 4^k t}{4^k}$ was a C^1- function, i.e. an elemental $C^1(\mathbb{R})$. For $m = 4^k$, the Fourier coefficient b_m of its derivative would be -1, thus the sequence of Fourier coefficients of the derivative, a continuous function, would contain an infinite subsequence not converging to 0, which contradicts the lemma of Riemann-Lebesgue.

7. We use the fact that $(Fu')(k) = iku_k$ and find

$$\sum_{|k|\leq N} |u_k| = |u_0| + \sum_{0<|k|\leq N} \frac{1}{|k|}|ku_k|$$

$$\leq |u_0| + \left(\sum_{0<|k|\leq N} \frac{1}{k^2}\right)^{\frac{1}{2}} \left(\sum_{0<|k|\leq N} |ku_k|^2\right)^{\frac{1}{2}}$$

$$\leq |u_0| + \left(2\frac{\pi^2}{6}\right)^{\frac{1}{2}} \left(\int_{-\pi}^{\pi} |u'(x)|^2 \, dx\right)^{\frac{1}{2}}$$

$$= |u_0| + \frac{\pi}{\sqrt{3}}\|u'\|_{L^2}.$$

8. First we note that

$$(f * (g + h))(t) = \frac{1}{2\pi}\int_{-\pi}^{\pi} f(s)(g+h)(t-s) \, ds$$

$$= \frac{1}{2\pi}\int_{-\pi}^{\pi} f(s)g(t-s) \, ds + \frac{1}{2\pi}\int_{-\pi}^{\pi} f(s)h(t-s) \, ds$$

$$= (f * g)(t) + (f * h)(t),$$

and that

$$((\alpha f) * g)(t) = \frac{1}{2\pi}\int_{-\pi}^{\pi} (\alpha f(s))g(t-s) \, ds = \alpha(f * g)(t)$$

$$= \frac{1}{2\pi}\int_{-\pi}^{\pi} f(s)(\alpha g(t-s)) \, ds = (f * (\alpha g))(t).$$

Furthermore, with the substitution $t - s = r$, we find that

$$(f * g)(t) = \frac{1}{2\pi}\int_{-\pi}^{\pi} f(s)g(t-s) \, ds = \frac{1}{2\pi}\int_{-\pi}^{\pi} f(t-r)g(r) \, dr = (g * f)(t).$$

Finally, we observe that

$$((f * g) * h)(t) = \frac{1}{2\pi} \int_{-\pi}^{\pi} (f * g)(t - s)h(s) \ ds$$

$$= \frac{1}{2\pi} \int_{-\pi}^{\pi} (g * f)(t - s)h(s) \ ds$$

$$= \frac{1}{2\pi} \int_{-\pi}^{\pi} \left(\frac{1}{2\pi} \int_{-\pi}^{\pi} f(r)g(t - s - r)h(s) \ dr \right) \ ds$$

$$= \frac{1}{2\pi} \int_{-\pi}^{\pi} \left(\frac{1}{2\pi} \int_{-\pi}^{\pi} f(r)g(t - s - r)h(s) \ ds \right) \ dr$$

$$= \frac{1}{2\pi} \int_{-\pi}^{\pi} f(r)(g * h)(t - r) \ dr$$

$$= (f * (g * h))(t).$$

9. By the convolution theorem, we have

$$|F(g * g * g)(k)| = |(F(g))(k)|^3 = |c_k|^3$$

or

$$|F(g * g * g)(k)| \le \frac{M^3}{|k|^6}, k \ne 0.$$

It follows that the Fourier series associated with $g * g * g$ is uniformly and absolutely convergent and represents $g * g * g$, i.e.

$$(g * g * g)(x) = \sum_{k \in \mathbb{Z}} c_k^3 e^{ikx}$$

and we may differentiate 4-times term by term to get

$$\frac{d^4}{dx^4}(g * g * g)(x) = \sum_{k \in \mathbb{Z}} (ik)^4 c_k^4 e^{ikx}$$

and

$$|(ic_k)^4 c_k^4| \le \frac{M}{k^2}, k \ne 0,$$

hence $g * g * g$ is a C^4-function.

10. The function $g(\cdot) = \chi_{[a,b]}(\cdot)h(\cdot)$ belongs to $L^1(\mathbb{R})$ (extending formally h to $\mathbb{R}\backslash\{[a, b]\}$ by 0) and by Theorem 3.16 we conclude that

$$\lim_{\gamma \to \infty} \int_a^b h(t) \sin \gamma t \ dt = -\text{Im} \lim_{\gamma \to \infty} \int_a^b h(t)e^{-ix\gamma} \ dt = 0$$

as well as

$$\lim_{\gamma \to \infty} \int_a^b h(t) \cos \gamma t \ dt = \text{Re} \lim_{\gamma \to \infty} \int_a^b h(t)e^{-ix\gamma} \ dt = 0.$$

11. We note that

$$\left\| f - \sum_{|k|\leq N} d_k \frac{e^{ik\cdot}}{\sqrt{2\pi}} \right\|_{L^2}^2 = \|f\|_{L^2}^2 - \int_{-\pi}^{\pi} f(t) \sum_{|k|\leq N} d_k \frac{e^{ikt}}{\sqrt{2\pi}} \, dt - \int_{-\pi}^{\pi} \sum_{|k|\leq N} d_k \frac{e^{ikt}}{\sqrt{2\pi}} \overline{f(t)} \, dt$$

$$+ \sum_{|k|,|l|\leq N} d_k \overline{d_l} \int_{-\pi}^{\pi} \frac{e^{i(k-l)t}}{2\pi} \, dt$$

$$= \|f\|_{L^2}^2 - \sum_{|k|\leq N} d_k \overline{c_k} - \sum_{|k|\leq N} \overline{d_k} c_k + \sum_{|k|\leq N} |c_k|^2$$

$$= \|f\|_{L^2}^2 + \sum_{|k|\leq N} |c_k - d_k|^2 - \sum_{|k|\leq N} |c_k|^2.$$

Hence we minimise $\left\| f - \sum_{|k|\leq N} d_k \frac{e^{ik}}{\sqrt{2\pi}} \right\|_{L^2}$ for $d_k = c_k, |k| \leq N$.

Chapter 4

1. For $\alpha > 0$ and $\gamma > 0$, with the change of variable $x \mapsto \gamma y$, we find that

$$\int_0^{\gamma\alpha} \frac{\sin x}{x} \, dx = \int_0^\alpha \frac{\sin \gamma y}{y} \, dy$$

and therefore

$$\int_0^\infty \frac{\sin x}{x} \, dx = \lim_{\gamma\to\infty} \int_0^{\gamma\alpha} \frac{\sin x}{x} \, dx = \lim_{\gamma\to\infty} \int_0^\alpha \frac{\sin \gamma y}{y} \, dy.$$

Next we observe that the function

$$x \mapsto g(x) := \begin{cases} \frac{1}{x} - \frac{1}{2\sin\frac{x}{2}}, & 0 < |x| < 2\pi \\ 0, & x = 0 \end{cases}$$

is continuous on $(-2\pi, 2\pi)$: For $x \neq 0$, this is trivial and for $x = 0$, we observe using $g(x) = \frac{2\sin\frac{x}{2} - x}{2x\sin\frac{x}{2}}, x \neq 0$, and the rules of de l'Hospital, that

$$\lim_{x\to\infty} \left(\frac{1}{x} - \frac{1}{2\sin\frac{x}{2}} \right) = \lim_{x\to\infty} \frac{\cos\frac{x}{2} - 1}{2\sin\frac{x}{2} + x\cos\frac{x}{2}}$$

$$= \lim_{x\to\infty} \frac{-\frac{1}{2}\sin\frac{x}{2}}{2\cos\frac{x}{2} - \frac{x}{2}\sin\frac{x}{2}} = 0$$

By the lemma of Riemann-Lebesgue applied in the version of Problem 10, Chapter 3, we conclude with $\alpha = \pi$, that

$$\lim_{\gamma\to\infty} \int_0^\pi g(x) \sin \gamma x \, dx = 0$$

or

$$\lim_{\gamma\to\infty} \int_0^\pi \left(\frac{\sin \gamma x}{x} - \frac{\sin \gamma x}{2\sin\frac{x}{2}} \right) dx = 0$$

which yields

$$\lim_{\gamma\to\infty}\int_0^\pi \frac{\sin\gamma x}{x}\,dx = \lim_{\gamma\to 0}\int_0^\pi \frac{\sin\gamma x}{2\sin\frac{x}{2}}\,dx.$$

Now we may use the discrete parameter $N\in\mathbb{N}$ instead of γ to find

$$\int_0^\infty \frac{\sin x}{x}\,dx = \lim_{N\to\infty}\int_0^\pi \frac{\sin(N+\frac{1}{2})x}{2\sin\frac{x}{2}}\,dx = \lim_{N\to\infty}\frac{1}{2}\int_0^\pi D_N(x)\,dx = \frac{\pi}{2}.$$

2. We want to apply Theorem 4.5 with $S(t_o) = f(t_0)$. We observe that

$$\int_0^\delta D_N(2s)\frac{f(t_0+2s)+f(t_0-2s)-2f(t_0)}{2}\,ds$$

$$=\int_0^\delta \frac{\sin(2N+1)s}{\sin s}\left(\frac{f(t_0+2s)+f(t_0-2s)-2f(t_0)}{2}\right)\,ds$$

$$=\int_0^\delta \frac{s}{\sin s}\left(\frac{f(t_0+2s)+f(t_0-2s)-2f(t_0)}{2s}\right)\sin(2N+1)s\,ds.$$

Since $\lim_{s\to 0}\frac{s}{\sin s}=1$, the function $s\mapsto\frac{s}{\sin s}$ is continuous on $[0,\delta]$ and we find that

$$\int_0^{\frac{\pi}{2}} D_N(2s)\frac{f(t_0+2s)+f(t_0-2s)-2f(t_0)}{2}\,ds \qquad (*)$$

$$=\int_\delta^\pi \frac{\sin(2N+1)s}{\sin s}\left(\frac{f(t_0+2s)+f(t_0-2s)-2f(t_0)}{2}\right)\,ds$$

$$+\int_0^\delta \sin(2N+1)s\left(\frac{f(t_0+2s)+f(t_0-2s)-2f(t_0)}{2s}\right)\,ds$$

$$+\int_0^\delta \left(\frac{s}{\sin s}-1\right)\sin(2N+1)\left(\frac{f(t_0+2s)+f(t_0-2s)-2f(t_0)}{2s}\right)\,ds.$$

For the first integral, the Riemann-Lebesgue lemma implies that it tends to 0 as N tends to infinity and the second integral tends for N going to infinity to 0 by our assumption. For the third integral, we observe that

$$\limsup_{N\to\infty}\left|\int_0^\delta \left(\frac{s}{\sin s}-1\right)\sin(2N+1)\left(\frac{f(t_0+2s)+f(t_0-2s)-2f(t_0)}{2s}\right)\,ds\right|$$

$$\le \limsup_{N\to\infty}\int_0^\delta \left|\frac{s}{\sin s}-1\right|\left|\frac{f(t_0+2s)+f(t_0-2s)-2f(t_0)}{2s}\right|\,ds$$

$$=\int_0^\delta \left|\frac{s}{\sin s}-1\right|\left|\frac{f(t_0+2s)+f(t_0-2s)-2f(t_0)}{2s}\right|\,ds.$$

We know that $\lim_{s\to 0}\frac{s}{\sin s}=1$ and we note further that

$$|x-\sin x|=\left|\int_0^x 1-\cos t\,dt\right|=\left|\int_0^x\left(\int_0^t \sin s\,ds\right)dt\right|\le\frac{|x|^3}{6}.$$

implying for $-\delta \le s \le \delta$ that

$$\frac{1}{|s|}\left|\frac{s}{\sin s} - 1\right| = \left|\frac{s - \sin s}{s \sin s}\right| \le \frac{|s|^3}{6}\frac{1}{|s||\sin s|} \le \frac{|s|}{6}\left|\frac{s}{\sin s}\right|$$

which implies that the integrand in the third integral is bounded by an integrable function independent of $\delta, 0 < \delta < \frac{\pi}{2}$. This implies that

$$\lim_{\delta \to 0}\int_0^\delta \left|\frac{s}{\sin s} - 1\right|\left|\frac{f(t_0 + 2s) + f(t_0 - 2s) - 2f(t_0)}{2s}\right|ds = 0$$

Thus using $(*)$, first $N \to \infty$ and then $\delta \to 0$ and the result follows.

3.   a) First we note that by Theorem 4.6, the series represents the function $x \mapsto |x|$ for all $|x| < \pi$. In particular for $x = 0$, we find

$$0 = \frac{\pi}{2} - \frac{4}{\pi}\sum_{l=0}^\infty \frac{\cos(2l+1)\cdot 0}{(2l+1)^2} = \frac{\pi}{2} - \frac{4}{\pi}\sum_{k=1}^\infty \frac{1}{(2k-1)^2}$$

implying that

$$\sum_{k=1}^\infty \frac{1}{(2k-1)^2} = \frac{\pi^2}{8}.$$

b) From Example 2.9, we know that

$$x^2 \quad \sim \quad \frac{\pi^2}{3} + \sum_{k=1}^\infty \frac{4(-1)^k}{k^2}\cos kx$$

and by Theorem 4.6 we deduce that the series represents $x \mapsto x^2$ in $(-\pi, \pi)$. With $x = 0$, we get

$$-\frac{\pi^2}{3} = 4\sum_{k=1}^\infty \frac{(-1)^k}{k^2}$$

or

$$\frac{\pi^2}{12} = \sum_{k=1}^\infty \frac{(-1)^{k+1}}{k^2}.$$

4. Since $S_{N+1} = (N+1)\sigma_{N+1} - N\sigma_N$, we find

$$\lim_{N\to\infty}\frac{S_{N+1}}{N} = \lim_{N\to\infty}\left(\frac{N+1}{N}\sigma_{N+1} - \frac{N}{N}\sigma_N\right) = \gamma - \gamma = 0.$$

Now we observe

$$\frac{\gamma_k}{k} = \frac{S_{k+1}}{k} - \frac{k-1}{k}\frac{S_k}{k-1}$$

and using that $\lim_{N\to\infty}\frac{S_{N+1}}{N} = 0$, we deduce $\lim_{k\to\infty}\frac{\gamma_k}{k} = 0$.

591

5. Denote by $S_{N+1} = \sum_{k=0}^{N} \gamma_k$ the partial sum of $\sum_{k=0}^{\infty} \gamma_k$ and observe that

$$\sigma_{N+1} := \frac{1}{N+1} \sum_{l=0}^{N} S_l = \sum_{k=0}^{N} \left(1 - \frac{k}{N+1}\right) \gamma_k$$

or

$$S_{N+1} - \sigma_{N+1} = \frac{1}{N+1} \sum_{k=0}^{N} k\gamma_k.$$

Since $\lim_{k\to\infty} k\gamma_k = 0$, it follows that $\lim_{N\to\infty} \frac{1}{N+1} \sum_{k=0}^{N} k\gamma_k = 0$, implying that $\sum_{k=0}^{\infty} \gamma_k = \lim_{N\to\infty} S_{N+1} = \lim_{N\to\infty} \sigma_N = \gamma$.

6. We start by noting the elementary equalities

$$\sum_{k=1}^{N} \sin kx := \begin{cases} \frac{\cos \frac{x}{2} - \cos(N+\frac{1}{2})x}{2\sin \frac{x}{2}}, & x \neq 0 \\ 0, & x = 0 \end{cases}$$

$$\sum_{k=0}^{N} \sin (k + \frac{1}{2})x := \begin{cases} \frac{\sin^2(N+1)\frac{x}{2}}{\sin \frac{x}{2}}, & x \neq 0 \\ 0, & x = 0 \end{cases}$$

$$\frac{1}{2} + \sum_{k=1}^{N} \cos kx := \begin{cases} \frac{\sin(N+1)x}{2\sin \frac{x}{2}}, & x \neq 0 \\ \frac{1}{2} + N, & x = 0. \end{cases}$$

Therefore we have in the case of the series $\frac{1}{2} + \sum_{k=1}^{\infty} \cos kx$ for $x \neq 0$

$$S_{N+1} = \frac{\sin (N + \frac{1}{2})x}{2\sin \frac{x}{2}}$$

$$\sigma_{N+1} = \frac{1}{N+1} \sum_{k=0}^{N} S_k = \frac{1}{N+1} \frac{1}{\sin \frac{x}{2}} \sum_{k=0}^{N} \sin \left(k + \frac{1}{2}\right)x$$

$$= \frac{1}{N+1} \frac{\sin (N+1)\frac{x}{2}}{(\sin \frac{x}{2})^2}$$

implying that for $x \neq 0$, the series $\frac{1}{2} + \sum_{k=1}^{\infty} \cos kx$ is Cesàro summable to 0. However, for $x = 0$, we have

$$S_{N+1}(0) = \frac{1}{2} + N = \frac{2N+1}{2}$$

and it follows

$$\frac{1}{N+1} \sum_{k=0}^{N} S_k(0) = \frac{1}{N+1} \left(\sum_{k=0}^{N} k + \sum_{k=0}^{N} \frac{1}{2}\right) = \frac{N+1}{2}$$

and hence this series is not Cesàro summable to a finite value.

592

For the series $\sum_{k=1}^{\infty} \sin kx$ we note that

$$S_{N+1} = \frac{\cos \frac{x}{2} - \cos(N + \frac{1}{2})x}{2 \sin \frac{x}{2}}$$

which gives

$$\frac{1}{N+1} \sum_{k=0}^{N} S_k = \frac{1}{N+1} \sum_{k=0}^{N} \frac{\cos \frac{x}{2} - \cos(k + \frac{1}{2})x}{2 \sin \frac{x}{2}}$$

$$= \frac{1}{2} \cot \frac{x}{2} - \frac{1}{N+1} \sum_{k=0}^{N} \frac{\cos(k + \frac{1}{2})x}{2 \sin \frac{x}{2}}.$$

Since $\cos(k + \frac{1}{2})x = \cos kx \cos \frac{x}{2} - \sin kx \sin \frac{x}{2}$, we find

$$\frac{1}{N+1} \sum_{k=0}^{N} \frac{\cos(k + \frac{1}{2})x}{2 \sin \frac{x}{2}} = \frac{1}{N+1} \left(\frac{1}{2} \sum_{k=0}^{N} \cos kx \cot \frac{x}{2} - \frac{1}{2} \sum_{k=0}^{N} \sin kx \right)$$

$$= \frac{1}{2(N+1)} \left(\frac{\sin(N + \frac{1}{2})x}{2 \sin \frac{x}{2}} - \frac{1}{2} \right) \cot \frac{x}{2} - \frac{1}{2(N+1)} \left(\frac{\cos \frac{x}{2} - \cos(N+1)\frac{x}{2}}{2 \sin \frac{x}{2}} \right).$$

Thus we have $\lim_{N \to \infty} \sigma_{N+1} = \frac{1}{2} \cot \frac{x}{2}, x \neq 0$, and hence the series is, for $x \neq 0$, Cesàro summable to $\frac{1}{2} \cot \frac{x}{2}$. For $x = 0$, we have however $\sum_{k=1}^{N} \sin kx = 0$, i.e. the series converges to 0.

7. Since $F_{N+1}(t) = \frac{1}{N+1} \sum_{k=0}^{N} D_k(t)$ and $|D_k(t)| \leq 2k + 1$, we find

$$F_{N+1}(t) \leq \frac{1}{N+1} \sum_{k=0}^{N} (2k + 1) = \frac{1}{N+1} (N+1)^2 = N + 1.$$

Moreover, by (4.18), we have

$$F_{N+1}(t) = \frac{1}{N+1} \frac{\sin^2 \left(\frac{N+1}{2} t \right)}{\sin^2 \left(\frac{t}{2} \right)}.$$

We note that $\sin \frac{t}{2} \geq \frac{t}{\pi}$ for $0 \leq t \leq \pi$ to obtain the second estimate

$$F_N(t) \leq \frac{\pi^2}{(N+1)t^2}.$$

8. The crucial line to see in (4.21) is

$$\sum_{k=m}^{n-1} (u_k - u_{k+1})V_k = \sum_{k=m}^{n-1} u_k V_k - u_{k+1}(V_{k+1} - v_{k+1})$$

$$= \sum_{k=m}^{n-1} u_k V_k - \sum_{k=m}^{n-1} u_{k+1} V_{k+1} + \sum_{k=m}^{n-1} u_{k+1} v_{k+1}.$$

From this, (4.22) is clear, as is (4.23).

9. We start with $\sum_{k=0}^{\infty} r^k e^{ikt} = \sum_{k=0}^{\infty} (re^{it})^k$ and therefore we find

$$\sum_{k=0}^{\infty} r^k e^{ikt} = \frac{1}{1 - re^{it}} = \frac{1 - re^{-it}}{1 - 2r\cos t + r^2}, 0 \leq r < 1.$$

Passing to the real and imaginary parts on both sides, we find

$$\sum_{k=0}^{\infty} r^k \cos kt = \frac{1 - r\cos t}{1 - 2r\cos t + r^2}$$

and

$$\sum_{k=1}^{\infty} r^k \sin kt = \frac{r\sin t}{1 - 2r\cos t + r^2}.$$

This yields that the Abel sum of $\sum_{k=0}^{\infty} \cos kt$ for $t \neq 0$ is equal to $\frac{1-\cos t}{2-2\cos t} = \frac{1}{2}$ whereas the Abel sum of $\sum_{k=1}^{\infty} \sin kt$ is $\frac{\sin t}{2-2\cos t} = \frac{1}{2}\cot\frac{t}{2}$.

10. With $z = re^{it}$, we find

$$\sum_{k=0}^{\infty} \alpha_k r^k (e^{it})^k = \sum_{k=0}^{\infty} \alpha_k r^k \cos kt + i\sum_{k=1}^{\infty} \alpha_k r^k \sin kt.$$

If for $t_0 \in [-\pi, \pi)$, we have $\lim_{r \to 1} h(re^{it_0}) = h(e^{it_0})$ then, by considering the real and imaginary parts we find

$$\mathrm{Re}(h(e^{it_0})) = \lim_{r \to 1} \sum_{k=0}^{\infty} \alpha_k r^k \cos kt$$

and

$$\mathrm{Im}(h(e^{it_0})) = \lim_{r \to 1} \sum_{k=1}^{\infty} \alpha_k r^k \sin kt$$

11. For $|z| < 1$, we have $\log\frac{1}{1-z} = \sum_{k=1}^{\infty} \frac{z^k}{k}$. We determine now $\log\frac{1}{1-z}$ by noting that for $z = e^{it}$

$$\frac{1}{1 - e^{it}} = \frac{1 - e^{-it}}{|1 - e^{it}|^2} = \frac{1}{2} + i\frac{\sin t}{2(1 - \cos t)} = \frac{1}{2|\sin\frac{t}{2}|}\left(\cos\left(\frac{\pi}{2} - \frac{t}{2}\right) + i\sin\left(\frac{\pi}{2} - \frac{t}{2}\right)\right)$$

which yields

$$\log\frac{1}{1 - e^{it}} = -\ln 2\left|\sin\frac{t}{2}\right| + i\frac{\pi - t}{2}$$

and therefore we find that the Abel limit of $\sum_{k=1}^{\infty} \frac{\cos kt}{k}$ is equal to $-\ln 2|\sin\frac{t}{2}|$ for $t \neq 0$.

12. Let f be continuous at t_0. For a given $\epsilon > 0$, choose $\delta > 0$ such that $|s| < \delta$ implies $|f(t_0 - s) - f(t_0)| < \epsilon$. It follows that

$$(K_n * f)(t_0) - f(t_0) = \frac{1}{2\pi} \int_{-\pi}^{\pi} K_n(s)(f(t_0 - s) - f(t_0)) \, ds$$

and therefore

$$\begin{aligned}
|(K_n * f)(t_0) - f(t_0)| &\leq \frac{1}{2\pi} \int_{-\pi}^{\pi} |K_n(s)| |f(t_0 - s) - f(t_0)| \, ds \\
&= \frac{1}{2\pi} \int_{|s| < \delta} |K_n(s)| |f(t_0 - s) - f(t_0)| \, ds \\
&\quad + \frac{1}{2\pi} \int_{\delta \leq |s| \leq \pi} |K_n(s)| |f(t_0 - s) - f(t_0)| \, ds \\
&\leq \frac{\epsilon}{2\pi} \int_{-\pi}^{\pi} |K_n(y)| \, dy + \frac{2\|f\|_\infty}{2\pi} \int_{\delta \leq |s| \leq \pi} |K_n(s)| \, ds \\
&\leq \frac{\epsilon \kappa}{2\pi} + \frac{\|f\|_\infty}{\pi} \int_{\delta \leq |s| \leq \pi} |K_n(s)| \, ds.
\end{aligned}$$

Since for $k \to \infty$ the second term vanishes by assumption, the result follows.

Chapter 5

1.   a) A straightforward calculation yields

$$\begin{aligned}
\|x + y\|^2 &- \|x - y\|^2 + i\|x + iy\|^2 - i\|x - iy\|^2 \\
&= <x,x> + <y,x> + <x,y> + <y,y> - <x,x> - <-y,x> \\
&\quad - <x,-y> - <y,y> + i<x,x> + i<iy,x> + i<x,iy> \\
&\quad + i<iy,iy> - i<x,x> - i<-iy,-iy> - i<x,-iy> \\
&\quad - i<-iy,x> \\
&= 2<x,y> + 2<y,x> + 2<x,y> - 2<y,x> \\
&= 4<x,y>.
\end{aligned}$$

b) With the help of part a), this is now trivial: We need only expand $<Tx, Ty>_{H_2}$ and $<x, y>_{H_1}$ according to (5.37).

2.   a) Let $y_1, y_2 \in M^\perp$ and $\lambda, \mu \in \mathbb{C}$. Then we find for all $x \in M$ that

$$<\lambda y_1 + \mu y_2, x> = \lambda <y_1, x> + \mu <y_2, x> = 0,$$

i.e. $\lambda y_1 + \mu y_2 \in M^\perp$. Moreover, let $(y_k)_{k \in \mathbb{N}}$ be a sequence in $M^\perp$ converging in H to $y \in H$. It follows for $x \in M$ that $<y, x> = \lim_{k \to \infty} <y_k, x> = 0$, thus $y \in M^\perp$ and hence $M^\perp$ is closed.

b) $M_1 \subset M_2$ and $z \in M_2^\perp$. We have to prove that $z \in M_1^\perp$ or $<z, x> = 0$ for all $x \in M_1$, but $<z, x> = 0$ for all $x \in M_2$ and $M_1 \subset M_2$, so $z \in M_1^\perp$.

c) Let $y \in M$ and $x \in M^{\perp}$, thus $< y, x > = 0$, i.e. we have: for all $x \in M^{\perp}$ the equality $< y, x > = 0$ holds which means $y \in (M^{\perp})^{\perp}$ and we have proved that $M \subset M^{\perp\perp}(:= (M^{\perp})^{\perp})$.

d) Since $M \subset \overline{\text{span}M}$ we get from part b) that $(\overline{\text{span}M})^{\perp} \subset M^{\perp}$. We prove now $M^{\perp} \subset (\overline{\text{span}M})^{\perp}$. For $u \in M^{\perp}$ we have $< u, v > = 0$ for all $v \in M$, and therefore $< u, \lambda v_1 + \mu v_2 > = 0$ for all $\lambda, \mu \in \mathbb{C}$ and $v_1, v_2 \in \text{span}M$. For $v \in \overline{\text{span}M}$ exists a sequence $(v_k)_{k \in \mathbb{N}}$, $v_k \in \text{span}M$, converging in H to v and it follows that $< u, v > = \lim_{k \to \infty} < u, v_k > = 0$, hence $< u, v > = 0$ for all $v \in \overline{\text{span}M}$ or $u \in (\overline{\text{span}M})^{\perp}$.

3.   a) Let $x \in \text{span}\{e_k \mid k \in K_1\}$ and $y \in \text{span}\{e_k \mid k \in K_2\}$, i.e. $x = \sum_{j=1}^{N} \lambda_j e_j, j \in K_1, y = \sum_{l=1}^{M} \mu_l e_l, l \in K_2$. Since $< e_j, e_l > = 0$ for $j \in K_1$ and $l \in K_2$, it follows that $< x, y > = 0$. This implies for $x = \lim_{m \to \infty} x_m, x_m \in \text{span}\{e_k \mid k \in K_1\}$ that $< x, y > = 0$, i.e. $\text{span}\{e_k \mid k \in K_1\} \perp \text{span}\{e_k \mid k \in K_2\}$.

Now, for $y = \lim_{n \to \infty} y_n, y_n \in \text{span}\{e_k \mid k \in K_2\}$, we deduce for every $x \in \overline{\text{span}}\{e_k \mid k \in K_1\}$ that $< x, y > = 0$, implying that

$$\overline{\text{span}\{e_k \mid k \in K_1\}} \perp \overline{\text{span}\{e_k \mid k \in K_2\}}.$$

b) The scalar product in $L^2(\mathbb{R})$ is given by $\int_{\mathbb{R}} u(x)\overline{v(x)} \, \mathrm{d}x$. For $u \in H_1$ and $v \in H_2$ it follows that $u \cdot \overline{v} = 0$ a.e. and therefore we find $\int_{\mathbb{R}} u(x)\overline{v(x)} \, \mathrm{d}x$ or $H_1 \perp H_2$.

4.   a) Since $| < u_k - u, \varphi > | \leq \|u_k - u\|\|\varphi\|$ for all $\varphi \in H$, it follows that $\lim_{k \to \infty} \|u_k - u\| = 0$ which implies the weak convergence of $(u_k)_{k \in \mathbb{N}}$ to u.

b) For $\varphi \in l_2(\mathbb{Z}), \varphi = \sum_{k \in \mathbb{Z}} \varphi_k e_k$, we have $\sum_{k \in \mathbb{Z}} |\varphi_k|^2 < \infty$ implying that $\lim_{k \to \infty} \varphi_k = 0$ and $\lim_{k \to -\infty} \varphi_k = 0$. For $\varphi \in l_2(\mathbb{Z})$ and $k \in \mathbb{N}$, we find $< \varphi, e_k > = \varphi_k$ and therefore $\lim_{k \to \infty} < e_k, \varphi > = \lim_{k \to \infty} \overline{< \varphi, e_k >} = \lim_{k \to \infty} \overline{\varphi_k} = 0$, i.e. $(e_k)_{k \in \mathbb{N}}$ converges weakly to 0.

c) For $\varphi \in L^2(\mathbb{T})$ and $k \in \mathbb{N}$ we get

$$< \frac{1}{\sqrt{2\pi}} e^{ik \cdot}, \varphi > = \frac{1}{\sqrt{2\pi}} \int_{-\pi}^{\pi} e^{ikx} \varphi(x) \, \mathrm{d}x$$

$$= \frac{1}{\sqrt{2\pi}} \int_{-\pi}^{\pi} \overline{\varphi(x)} e^{-ikx} \, \mathrm{d}x = \sqrt{2\pi} \, \overline{(F\varphi)(k)}$$

and now the lemma of Riemann-Lebesgue yields

$$\lim_{k \to \infty} < \frac{1}{\sqrt{2\pi}} e^{ik \cdot}, \varphi > = 0$$

i.e. $\left(\frac{1}{\sqrt{2\pi}} e^{ik \cdot} \right)_{k \in \mathbb{N}}$ converges weakly in $L^2(\mathbb{T})$ to 0.

5. We note that

596

$$\frac{1}{2\pi} \int_{-\pi}^{\pi} |h_k(x) - 0|^2 \, dx = \frac{1}{2\pi} \int_{-\pi}^{\pi} \frac{x^{2k}}{\pi^{2k}} \, dx$$

$$= \frac{1}{\pi^{2k+1}} \int_0^{\pi} x^{2k} dx$$

$$= \frac{1}{\pi^{2k+1}} \frac{1}{2k+1} x^{2k+1} \Big|_0^{\pi}$$

$$= \frac{1}{2k+1},$$

implying

$$\lim_{k \to \infty} \frac{1}{2\pi} \int_{-\pi}^{\pi} |h_k(x) - 0|^2 \, dx = 0,$$

i.e. $(h_k)_{k \in \mathbb{N}}$ converges in the quadratic mean to $h_0, h_0(x) = 0$ for all x.

However for $x = \pi$ we find $h_k(\pi) = (\frac{\pi}{\pi})^k = 1$, hence $\lim_{k \to \infty} h_k(\pi) = 1$, and $h_k(-\pi) = (\frac{-\pi}{\pi})^k = (-1)^k$. Thus $(h_k)_{k \in \mathbb{N}}$ does not converge pointwisely on $[-\pi, \pi]$ to h_0. Note however for $|x| < \pi$ we have

$$|h_k(x)| \le \left|\frac{x}{\pi}\right|^k = \rho^k, \rho = \left|\frac{x}{\pi}\right| < 1,$$

and therefore $\lim_{k \to \infty} h_k(x) = h_0(x) = 0$ for $|x| < \pi$.

6. We follow the arguments given in [97]. We consider the dyadic block

$$\sum_{2^m \le |k| < 2^{m+1}} |c_k|$$

where c_k is the kth Fourier coefficient of f. The Cauchy-Schwarz inequality yields

$$\left(\sum_{2^m \le |k| < 2^{m+1}} |c_k| \right)^2 \le \left(\sum_{2^m \le |k| < 2^{m+1}} 1 \right) \left(\sum_{2^m \le |k| < 2^{m+1}} |c_k|^2 \right)$$

and since the sum has $2 \cdot 2^m$ terms, we arrive at

$$\left(\sum_{2^m \le |k| < 2^{m+1}} |c_k| \right)^2 \le 2^{m+1} \left(\sum_{2^m \le |k| < 2^{m+1}} |c_k|^2 \right).$$

Note that

$$\|f_h - f\|_{L^2}^2 = \sum_{k \in \mathbb{Z}} |e^{ikh} - 1|^2 |c_k|^2 \ge \sum_{2^m \le |k| < 2^{m+1}} |e^{ikh} - 1|^2 |c_k|^2.$$

Since $|e^{ikh} - 1|^2 = (\cos kh - 1)^2 + \sin^2 kh = 4\sin^2 \frac{kh}{2}$ it follows with $h = \frac{\pi}{3 \cdot 2^m}$ that $|e^{ikh} - 1| \ge 1$ for $2^m \le |k| < 2^{m+1}$ which implies that $\|f_{\frac{\pi}{3 \cdot 2^m}} - f\|_{L^2}^2 \ge$

597

$\sum_{2^m \leq |k| < 2^{m+1}} |c_k|^2$. Now we find

$$\sum_{0 \neq k \in \mathbb{Z}} |c_k| = \sum_{m=0}^{\infty} \sum_{2^m \leq |k| < 2^{m+1}} |c_k|$$

$$\leq \sum_{m=0}^{\infty} \left(\sum_{2^m \leq |k| < 2^{m+1}} 1 \right)^{\frac{1}{2}} \left(\sum_{2^m \leq |k| < 2^{m+1}} |c_k|^2 \right)^{\frac{1}{2}}$$

$$\leq \sum_{m=0}^{\infty} 2^{\frac{m+1}{2}} \|f_{\frac{\pi}{3 \cdot 2^m}} - f\|_{L^2}$$

$$\leq \kappa \sqrt{2} \frac{\pi^\alpha}{3^\alpha} \sum_{m=0}^{\infty} 2^{\frac{m}{2}} 2^{-m\alpha}$$

where in the last step we used the L^2-Hölder continuity of f. Thus we have the estimate

$$\sum_{0 \neq k \in \mathbb{Z}} |c_k| \leq \kappa \sqrt{2} \frac{\pi^\alpha}{3^\alpha} \sum_{m=0}^{\infty} (2^{\frac{m}{2}})^{1-2\alpha}$$

and the right hand side is finite for $\alpha > \frac{1}{2}$.

7. Since $|c_k|^2 = |\frac{a_k - ib_k}{2}|^2$ for $k > 0$ and $|c_k|^2 = |\frac{a_k - ib_k}{2}|^2$ for $k < 0$ as well as $c_0 = \frac{a_0}{2}$, we find

$$\sum_{k \in \mathbb{Z}} |c_k|^2 = \frac{a_0^2}{4} + \sum_{k=1}^{\infty} \frac{a_k^2 + b_k^2}{2}$$

or

$$\frac{a_0^2}{2} + \sum_{k=1}^{\infty} (a_k^2 + b_k^2) = \frac{1}{\pi} \int_{-\pi}^{\pi} (f(t))^2 \, dt.$$

Now according to Example 2.9, we have

$$x^2 \sim \frac{\pi^2}{3} + \sum_{k=1}^{\infty} \frac{4(-1)^k}{k^2} \cos kx$$

which means $a_0 = \frac{2\pi^2}{3}, a_k = \frac{4(-1)^k}{k^2}, k \in \mathbb{N}, b_k = 0$, and therefore

$$\frac{4\pi^4}{18} + 16 \sum_{k=1}^{\infty} \frac{1}{k^4} = \frac{1}{\pi} \int_{-\pi}^{\pi} t^4 \, dt = \frac{2}{5} \pi^4$$

or

$$\sum_{k=1}^{\infty} \frac{1}{k^4} = \frac{1}{16} \left(\frac{2}{5} \pi^4 - \frac{2}{9} \pi^4 \right) = \frac{\pi^4}{90}.$$

8. Let $(u^{(l)})_{l \in \mathbb{N}}$ be a Cauchy sequence in $H^m(\mathbb{T})$. For every $\epsilon > 0$ there exists $N_0 \in \mathbb{N}$ such that $n, l \geq N_0$ implies

$$\|u^{(n)} - u^{(l)}\|_{H^m} = \left(\sum_{k \in \mathbb{Z}} (1 + |k|^2)^m |u_k^{(n)} - u_k^{(l)}|^2 \right)^{\frac{1}{2}} < \epsilon.$$

For every $k \in \mathbb{Z}$ fixed, $\left((1 + |k|^2)^{\frac{m}{2}} u_k^{(n)} \right)_{n \in \mathbb{N}}$ is a Cauchy sequence in $\mathbb{C}$ with a limit $v_k \in \mathbb{C}$ and we may write $v_k = (1 + |k|^2)^{\frac{m}{2}} u_k$. Thus we obtain a sequence $(u_k)_{k \in \mathbb{Z}}$ to which, by the Riesz-Fischer theorem, corresponds to an element in $L^2(\mathbb{T})$ which we denote by u. Since $(v_k)_{k \in \mathbb{Z}}$ belongs to $L^2(\mathbb{T})$, it follows that u belongs to $H^m(\mathbb{T})$. For $M \in \mathbb{N}$ and $n, l \geq N_0$ we have

$$\left(\sum_{k \leq M} (1 + |k|^2)^m |u_k^{(n)} - u_k^{(l)}|^2 \right)^{\frac{1}{2}} < \epsilon$$

which implies for $n \to \infty$ that

$$\left(\sum_{k \leq M} (1 + |k|^2)^m |u_k - u^{(l)}|^2 \right)^{\frac{1}{2}} \leq \epsilon$$

and therefore

$$\left(\sum_{k \in \mathbb{Z}} (1 + |k|^2)^m |u_k - u_k^{(l)}|^2 \right)^{\frac{1}{2}} \leq \epsilon$$

or

$$\| u - u^{(l)} \|_{H^m} \leq \epsilon.$$

Since $\| u \|_{H^m} \leq \| u - u^{(l)} \|_{H^m} + \| u^{(l)} \|_{H^m}$ it now follows that $u \in H^m(\mathbb{T})$ and further for $l \geq N_0$ that $\| u - u^{(l)} \|_{H^m} < \epsilon$, i.e. $(u^{(l)})_{l \in \mathbb{N}}$ converges in $H^m(\mathbb{T})$ to u, which means that $H^m(\mathbb{T})$ is complete.

9. With $\gamma_k = F(f)(k)$ we define $u \in L^2(\mathbb{T})$ by $u_k = \frac{\gamma_k}{1 + |k|^2}$. This yields

$$\sum_{k \in \mathbb{Z}} |u_k|^2 (1 + |k|^2)^2 = \sum_{k \in \mathbb{Z}} |\gamma_k|^2$$

and therefore u is an element in $H^2(\mathbb{T})$. Moreover, we find

$$F(-\Delta u + u)(k) = (|k|^2 + 1) u_k = \gamma_k = F(f)(k)$$

or

$$F(-\Delta u + u - f)(k) = 0 \text{ for all } k \in \mathbb{Z}.$$

The uniqueness theorem for the Fourier coefficients implies now $-\Delta u + u = f$ a.e.

Chapter 6

1.   a) We introduce $A_n := \int_{n\pi}^{(n+1)\pi} \frac{\sin t}{t} \, dt, n \in \mathbb{N}_0$, and note that

$$\int_0^\infty \frac{\sin t}{t} \, dt = \sum_{n=0}^\infty A_n = \lim_{N \to \infty} \sum_{n=0}^N A_n.$$

The aim is to prove the convergence of the series $\sum_{n=0}^\infty A_n$. A change of variable yields

$$A_n = \int_0^\pi \frac{\sin(t + n\pi)}{t + n\pi} \, dt = (-1)^n \int_0^\pi \frac{\sin t}{t + n\pi} \, dt.$$

Moreover, the integral $B_n := \int_0^\pi \frac{\sin t}{t+n\pi}\, dt$ is non-negative and we have the estimate

$$0 < B_n = \int_0^\pi \frac{\sin t}{t+n\pi}\, dt \leq \int_0^\pi \frac{1}{n\pi}\, dt = \frac{1}{n}.$$

Thus $\sum_{n=0}^\infty A_n = \sum_{n=0}^\infty (-1)^n B_n$ and $(B_n)_{n\in\mathbb{N}}$ is a sequence of positive numbers. Since for $t \in [0, \pi]$ we have $\frac{\sin t}{t+(n+1)\pi} \leq \frac{\sin t}{t+n\pi}$ the sequence $(B_n)_{n\in\mathbb{N}}$ is also decreasing and the Leibniz criterion for alternating series gives the result.

b) If $t \mapsto \frac{\sin t}{t}$ were Lebesgue integrable in $[0, \infty)$ then the integral $\int_0^\infty \left|\frac{\sin t}{t}\right|\, dt$ would be finite. We need to prove that this is not the case. With the help of the calculations of part a), we note that

$$\int_{n\pi}^{(n+1)\pi} \left|\frac{\sin t}{t}\right|\, dt = \int_0^\pi \frac{\sin r}{r+n\pi}\, dr$$

and we consider the integrals

$$C_n := \int_0^\pi \left|\frac{\sin nt}{t}\right|\, dt = \int_0^{n\pi} \left|\frac{\sin s}{s}\right|\, ds.$$

It follows that

$$C_{n+1} - C_n = \int_{n\pi}^{(n+1)\pi} \left|\frac{\sin t}{t}\right|\, dt = \int_0^\pi \frac{\sin r}{r+n\pi}\, dr.$$

Since $\frac{1}{(n+1)\pi} \leq \frac{1}{r+n\pi} \leq \frac{1}{n\pi}$ for $0 \leq r \leq \pi$ we have, using the fact that $\int_0^\pi \sin r\, dr = 2$, that

$$\frac{2}{\pi(n+1)} \leq C_{n+1} - C_n \leq \frac{2}{n\pi}$$

implying that

$$\frac{2}{\pi} \sum_{n=1}^{m-1} \frac{1}{n+1} \leq \sum_{n=1}^{m-1}(C_{n+1} - C_n) \leq \frac{2}{\pi} \sum_{n=1}^{m-1} \frac{1}{n}$$

or

$$\frac{2}{\pi} \sum_{n=1}^{m-1} \frac{1}{n+1} \leq \sum_{n=1}^{m-1} \int_{n\pi}^{(n+1)\pi} \left|\frac{\sin t}{t}\right|\, dt \leq \frac{2}{\pi} \sum_{n=1}^{m-1} \frac{1}{n}.$$

The divergence of the harmonic series yields now the divergence of $\int_0^\infty \left|\frac{\sin t}{t}\right|\, dt$.

2. We note that

$$|x - \sin x| = \left|\int_0^x (1 - \cos t)\, dt\right| = \left|\int_0^x \left(\int_0^t \sin s\, ds\right) dt\right|$$

$$\leq \int_0^x \left(\int_0^t s\, ds\right) dt = \int_0^x \frac{t^2}{2}\, dt = \frac{x^3}{6}.$$

3.   a) Using the substitution $x \mapsto x + \frac{\pi}{k}, k \neq 0$, in the formula

$$c_k = \frac{1}{2\pi} \int_{-\pi}^{\pi} u(x) e^{-ikx} \, dx,$$

we find

$$c_k = \frac{1}{2\pi} \int_{-\pi}^{\pi} u(x) e^{-ikx} \, dx = \frac{1}{2\pi} \int_{-\pi + \frac{\pi}{k}}^{\pi + \frac{\pi}{k}} u\left(x + \frac{\pi}{k}\right) e^{-ik(x + \frac{\pi}{k})} \, dx$$

$$= -\frac{1}{2\pi} \int_{-\pi}^{\pi} u\left(x + \frac{\pi}{k}\right) e^{-ikx} \, dx,$$

and therefore we have for $k \neq 0$

$$c_k = \frac{1}{4\pi} \int_{-\pi}^{\pi} \left(u(x) - u\left(x + \frac{\pi}{k}\right)\right) e^{-ikx} \, dx$$

implying

$$|c_k| \leq \frac{1}{4\pi} \int_{-\pi}^{\pi} \left| u\left(x + \frac{\pi}{k}\right) - u(x) \right| \, dx \leq \frac{1}{4\pi} \omega_1\left(u, \frac{\pi}{k}\right).$$

b) We may use the relations $a_k = c_k + c_{-k}$ and $b_k = i(c_k - c_{-k})$ to arrive at

$$|a_k| \leq 2|c_k| \leq \frac{1}{2\pi} \omega_1\left(u, \frac{\pi}{k}\right)$$

and

$$|b_k| \leq 2|c_k| \leq \frac{1}{2\pi} \omega_1\left(u, \frac{\pi}{k}\right).$$

However, we can repeat the argument directly, i.e.

$$a_k = \frac{1}{\pi} \int_{-\pi}^{\pi} u(x) \cos kx \, dx = \frac{1}{\pi} \int_{-\pi}^{\pi} u\left(x + \frac{\pi}{k}\right) \cos k \left(x + \frac{\pi}{k}\right) \, dx$$

$$= \frac{1}{\pi} \int_{-\pi}^{\pi} u\left(x + \frac{\pi}{k}\right) \cos\left(kx + \pi\right) \, dx = -\frac{1}{\pi} \int_{-\pi}^{\pi} u\left(x + \frac{\pi}{k}\right) \cos kx \, dx$$

where we used that $\cos\left(kx + \pi\right) = \cos kx \cos \pi - \sin kx \sin \pi = -\cos kx$. This now gives

$$a_k = \frac{1}{2\pi} \int_{-\pi}^{\pi} \left(u(x) - u\left(x + \frac{\pi}{k}\right)\right) \cos kx \, dx$$

implying

$$|a_k| \leq \frac{1}{2\pi} \omega_1\left(u, \frac{\pi}{k}\right).$$

Analogously. we can estimate directly $|b_k|$.

601

4. Since u is of bounded variation, the variation sums

$$\sum_{m=1}^{2k} \left| u\left(x + m\frac{\pi}{k}\right) - u\left(x + (m-1)\frac{\pi}{k}\right) \right|$$

are bounded by the total variation $V(u)$ of u. From Problem 3 b) we know that

$$|a_k| \leq \frac{1}{2\pi} \int_{-\pi}^{\pi} \left| u\left(x + \frac{\pi}{k}\right) - u(x) \right| \, dx$$

and analogously

$$|b_k| \leq \frac{1}{2\pi} \int_{-\pi}^{\pi} \left| u\left(x + \frac{\pi}{k}\right) - u(x) \right| \, dx.$$

Taking the periodicity of all functions being involved into account we find

$$\int_{-\pi}^{\pi} \left| u\left(x + \frac{\pi}{m}\right) - u(x) \right| \, dx = \int_{-\pi}^{\pi} \left| u\left(x + m\frac{\pi}{k}\right) - u\left(x + (m-1)\frac{\pi}{k}\right) \right| \, dx$$

implying

$$|a_k| \leq \frac{1}{2\pi} \int_{-\pi}^{\pi} \left| u\left(x + m\frac{\pi}{k}\right) - u\left(x + (m-1)\frac{\pi}{k}\right) \right| \, dx.$$

This yields

$$|a_k| \leq \frac{1}{2\pi k} \int_{0}^{2\pi} \sum_{m=1}^{2k} \left| u\left(x + m\frac{\pi}{k}\right) - u\left(x + (m-1)\frac{\pi}{k}\right) \right| \, dx$$

$$\leq \frac{1}{2\pi k} \int_{0}^{2\pi} V(u) \, dx = \frac{V}{2k}.$$

The estimate for $|b_k|$ goes analogously.

5. We will use the fact that the series $\sum_{k=2}^{\infty} \frac{1}{k \ln k}$ diverges, compare with Problem 4 b) in Chapter I.18. Suppose that $\sum_{k=2}^{\infty} \frac{|\sin kx|}{k \ln k}$ converges. Since $\sin^2 kx \leq |\sin kx|$ this implies the convergence of $\sum_{k=2}^{\infty} \frac{\sin^2 kx}{k \ln k}$ and hence the series $\sum_{k=2}^{\infty} \frac{1 - \cos 2kx}{k \ln k}$ must converge.

We know that the series $\sum_{k=2}^{\infty} \frac{\cos 2kx}{k \ln k}$ converges for $x \neq 0, \pm\pi, \pm 2\pi, \ldots$ and hence the convergence of $\sum_{k=2}^{\infty} \frac{|\sin kx|}{k \ln k}$ will imply the convergence of the harmonic series $\sum_{k=1}^{\infty} \frac{1}{k}$ which is a contradiction.

6. We recall that

$$D_N(t) = \sum_{|k| \leq N} e^{ikt} = 1 + \sum_{0 < |k| \leq N} e^{ikt} = 1 + 2 \sum_{0 < |k| \leq N} \cos kt$$

or

$$D_N^{\cos}(t) = \frac{1}{2} + \sum_{0 < |k| \leq N} \cos kt = \frac{1}{2} D_N(t) = \frac{\sin\left(N + \frac{1}{2}\right)t}{2 \sin \frac{t}{2}}.$$

Further, we note that

$$\sum_{k=1}^{N} \sin kt = \frac{1}{2i} \sum_{k=1}^{N} (e^{ikt} - e^{-ikt})$$

$$= \frac{1}{2i} \left(\sum_{k=0}^{N} e^{ikt} - \sum_{k=0}^{N} e^{-ikt} \right)$$

$$= \frac{1}{2i} \left(\frac{1 - e^{i(N+1)t}}{1 - e^{it}} - \frac{1 - e^{-i(N+1)t}}{1 - e^{-it}} \right)$$

$$= \frac{1}{2i} \frac{1}{e^{\frac{it}{2}} - e^{-\frac{it}{2}}} \left(-(e^{\frac{it}{2}} + e^{-\frac{it}{2}}) + e^{i(N+\frac{1}{2})t} + e^{-i(N+\frac{1}{2})t} \right)$$

$$= \frac{1}{2i} \frac{1}{2i \sin \frac{t}{2}} \left(-2 \cos \frac{t}{2} + 2 \cos (N + \frac{1}{2})t \right)$$

$$= \frac{\cos \frac{t}{2} - \cos (N + \frac{1}{2})t}{2 \sin \frac{t}{2}}.$$

7. We know that

$$|\sigma_N(f)(x)| \leq \frac{1}{2\pi} \int_{-\pi}^{\pi} |f(x - t)| F_N(t) \, dt$$

$$\leq \frac{1}{2\pi} \int_{-\pi}^{\pi} M F_N(t) \, dt = M,$$

where we have used that $F_N(t) \geq 0$ and $\frac{1}{2\pi} \int_{-\pi}^{\pi} F_N(t) \, dt = 1$.

Chapter 7

1. We write $f = u + iv$ and note that u and v are harmonic and satisfy the Cauchy-Riemann differential equation $u_x = v_y$ and $u_y = -v_x$. Further we have for $z \in \tilde{G}$, i.e. $f(z) \neq 0$, that $\ln |f(z)| = \ln(u^2 + v^2)^{\frac{1}{2}}$ and therefore

$$\frac{\partial}{\partial x} \ln(u^2 + v^2)^{\frac{1}{2}} = \frac{u_x u + v_x v}{(u^2 + v^2)},$$

$$\frac{\partial^2}{\partial x^2} \ln(u^2 + v^2)^{\frac{1}{2}} = \frac{u_{xx} u^3 + u_{xx} u v^2 + v_{xx} u^2 v + v_{xx} v^3}{(u^2 + v^2)^2}$$

$$+ \frac{-u_x^2 u^2 - v_x^2 v^2 + u_x^2 v^2 + v_x^2 u^2 - 4 u_x v_x u v}{(u^2 + v^2)^2}$$

and

$$\frac{\partial^2}{\partial y^2} \ln(u^2 + v^2)^{\frac{1}{2}} = \frac{u_{yy} u^3 + u_{yy} u v^2 + v_{yy} u^2 v + v_{yy} v^3}{(u^2 + v^2)^2}$$

$$+ \frac{-u_y^2 u^2 - v_y^2 v^2 + u_y^2 v^2 + v_y^2 u^2 - 4 u_y v_y u v}{(u^2 + v^2)^2}$$

603

which yields

$$(u^2 + v^2)^2 \Delta \ln(u^2 + v^2)^{\frac{1}{2}} = (u_{xx} + u_{yy})(u^3 + uv^2) + (v_{xx} + v_{yy})(u^2 v + v^3)$$
$$- u^2(u_x^2 - v_x^2 + u_y^2 - v_y^2) - v^2(v_x^2 - u_x^2 + v_y^2 - u_y^2)$$
$$- 4uv(u_x v_x + u_y v_y).$$

The fact that u and v are harmonic implies that the first two terms on the right hand side vanish, while the Cauchy-Riemann differential equation yields $u_x^2 = v_y^2$ and $v_x^2 = u_y^2$ as well as $u_x v_x + u_y v_y = -u_x u_y + u_y v_x = 0$. Thus it follows that $(u^2 + v^2)^2 \Delta \ln(u^2 + v^2)^{\frac{1}{2}} = 0$, i.e. $\ln(u^2 + v^2)^{\frac{1}{2}}$ is harmonic in $\tilde{G}$.

2.   a) It is clear that $u(r, \varphi) = \sum_{k=1}^{\infty} a_k r^k \cos \varphi$ is harmonic in $B_r(0)$, compare with (7.8). Our assumptions imply that $\lim_{r \to 1} u(r, \varphi) = \lim_{k \to 1} \sum_{k=1}^{\infty} a_k r^k \cos \varphi = \sum_{k=1}^{\infty} a_k \cos \varphi = u_0(\varphi)$.

    b) The function $u(r, \varphi) := \sum_{k=1}^{\infty} \frac{b_k}{k} r^k \sin k\varphi$ is harmonic in $B_1(0)$, see again (7.8). Furthermore we have

$$\lim_{r \to 1} \frac{\partial u(r, \varphi)}{\partial r} = \lim_{r \to 1} \frac{\partial}{\partial r} \sum_{k=1}^{\infty} \frac{b_k}{k} r^k \sin k\varphi$$

$$= \lim_{r \to 1} \sum_{k=1}^{\infty} b_r r^{k-1} \sin k\varphi$$

$$= \lim_{r \to 1} \frac{1}{r} \sum_{k=1}^{\infty} b_r r^k \sin k\varphi = u_0(\varphi).$$

3. The only norm property which is non-trivial to prove for $\|.\|_{\mathcal{H}^p}$ is the triangle inequality. For

$$M_p(f, r) = \left(\frac{1}{2\pi} \int_0^{2\pi} |f(re^{i\varphi})|^p \, d\varphi \right)^{\frac{1}{p}}$$

we have however

$$M_p(f + g, r) \leq M_p(f, r) + M_p(g, r)$$

which leads, for $r \to 1$, to $\|f + g\|_{\mathcal{H}^p} \leq \|f\|_{\mathcal{H}^p} + \|g\|_{\mathcal{H}^p}$.

The completeness of $(\mathcal{H}^p, \|.\|_{\mathcal{H}^p})$ we can prove by using parts of the proof Theorem 7.4. For a Cauchy sequence $(f_k)_{k \in \mathbb{N}}$ with respect to the norm $\|.\|_{\mathcal{H}^p}$ we can derive, as in Theorem 7.4, that for $|z| \leq r < R$ it holds

$$(R - r)|f_k(z) - f_l(z)| \leq \frac{1}{2\pi} \int_0^{2\pi} |f_k(Re^{i\varphi}) - f_l(Re^{i\varphi})| \, d\varphi$$

and now we use the estimate

$$\frac{1}{2\pi} \int_0^{2\pi} |f_k(Re^{i\varphi}) - f_l(Re^{i\varphi})| \, d\varphi \leq \left(\frac{1}{2\pi} \int_0^{2\pi} |f_k(Re^{i\varphi}) - f_l(Re^{i\varphi})|^p \, d\varphi \right)^{\frac{1}{p}}$$

and the same arguments employed in the proof of Theorem 7.4 we can use now when we replace M_2 by M_p.

4. We can find an integer $m \geq 0$ such that $f(z) = z^m g(z)$ and $g : D \to \mathbb{C}$ is a bounded holomorphic function with $g(0) \neq 0$. We apply the Poisson-Jensen formula to g, i.e.

$$\ln |g(0)| + \ln \sum_{j=1}^{M} \ln \frac{r}{|\zeta_j|} = \frac{1}{2\pi} \int_0^{2\pi} \ln |g(re^{i\varphi})| \, d\varphi$$

where $\zeta_1, \ldots, \zeta_M$ are the zeroes of g in $\overline{B_r}(0)$. But the left hand side in the equality for g cannot decrease, so we have $\mu_r(g) \leq \mu_s(g)$ for $0 < r < s < 1$. Since $\mu_r(f) = \mu_r(g) + M \ln r$ we conclude that

$$\mu_r(f) \leq \mu_s(f) \text{ for } 0 < r < s < 1.$$

(This problem is essentially taken from [103].)

5. Bounded holomorphic functions $f : D \to \mathbb{C}$ clearly belong to the Nevanlinna class $\mathcal{N}$. According to Proposition 7.9 we must have for their zeroes $\zeta_l \in D$ (as always, counted according to their multiplicity) that $\sum_{k=1}^{\infty} (1 - |\zeta_k|) < \infty$. For our concrete case, this would mean that

$$\sum_{k=1}^{\infty} \left(1 - \frac{k-1}{k}\right) = \sum_{k=1}^{\infty} \frac{1}{k} < \infty$$

which is of course not possible. Hence there is no bounded holomorphic function in D with zeroes at $\frac{k-1}{k}, k \in \mathbb{N}$.

6. (We follow [73]). Denote by $G(t)$ a primitive of $f(e^{it}) \in L^1(\mathbb{T})$. The function G is continuous, hence its Fourier series is Abel summable to G. The Abel means of this Fourier series are given by $\sum_{k=1}^{\infty} \frac{a_k}{k} r^k$. If $a_k \geq 0$, this implies already the result. In the general case, we factorise $f = f_1 f_2$ with $f_j \in \mathcal{H}^2$, say $f_j(z) = \sum_{k=1}^{\infty} A_k^{(j)} z^k$. By $g_j(z) = \sum_{k=1}^{\infty} |A_k^{(j)}| z^k$ we obtain again functions $g^{(j)} \in \mathcal{H}^2$ and further $g = g^{(1)} \cdot g^{(2)}$ belongs to $\mathcal{H}^1$, say $g(z) = \sum_{k=1}^{\infty} A_k z^k$. Since $A_k \geq 0$, it follows that $\sum_{k=1}^{\infty} \frac{A_k}{k} < \infty$. Using the formula to calculate the coefficients of the product of two power series, we find

$$|a_k| = \left| \sum_{l=0}^{k} A_l^{(1)} A_{k-l}^{(2)} \right| \leq \sum_{l=0}^{k} |A_l^{(1)}| |A_{k-l}^{(2)}| = A_k$$

implying $\sum_{k=1}^{\infty} \frac{|a_k|}{k} < \infty$.

Chapter 8

1. Suppose that $(c_k)_{k \in \mathbb{Z}}$ is a positive definite sequence in the sense of Definition 8.1 but replace in (8.4) the number N by M, given $N \in \mathbb{N}$. Choose in (8.4) the value $M := 2N + 1$ and $\xi_1 = -N, \xi_2 = -N + 1, \ldots, \xi_{-N} = -1, \xi_{N+1} = 0, \xi_{N+2} = 1, \ldots, \xi_{2N+1} = N$ and (8.6) follows.

On the other hand, assume (8.6) with N replaced by M and consider

$$\sum_{k,l=1}^{N} c_{\xi_k - \xi_l} \lambda_k \overline{\lambda_l}.$$

605

We now choose in (8.6), as a value for M, an integer larger than $\max_{1 \leq k \leq N} |\xi_k|$ and set in $\sum_{k,l=-M}^{M} c_{k-l} \lambda_k \overline{\lambda_l}$ all λ_j for an index j not entering in (8.4) equal to 0.

2. We have

$$\hat{\nu}_k = \frac{1}{2\pi} \int_{-\pi}^{\pi} e^{-ikt} \nu(\mathrm{d}t) = \frac{1}{2\pi} \int_{-\pi}^{\pi} e^{-ikt} (1 - \cos t) \mu(\mathrm{d}t)$$

$$= \frac{1}{2\pi} \int_{-\pi}^{\pi} e^{-ikt} \mu(\mathrm{d}t) - \frac{1}{2\pi} \int_{-\pi}^{\pi} \frac{e^{-i(k+1)t} + e^{-i(k-1)t}}{2} \mu(\mathrm{d}t)$$

$$= \hat{\mu}_k - \frac{\hat{\mu}_{k+1} + \hat{\mu}_{k-1}}{2}.$$

3. Since μ is a measure with L^1-density, the Fourier-Stieltjes coefficients $\hat{\mu}_k$ of μ are just the Fourier coefficients of the density. It follows that

$$\hat{\mu}_k = \frac{1}{2\pi} \int_{-\pi}^{\pi} e^{-ikx} \mu(\mathrm{d}x) = \frac{1}{2\pi} \int_{\frac{\pi}{6}}^{\frac{\pi}{4}} e^{-ikx} \, \mathrm{d}x$$

$$= \frac{-1}{2\pi ik} e^{-ikx} \Big|_{\frac{\pi}{6}}^{\frac{\pi}{4}}$$

$$= \frac{1}{\pi ik} \left(\sin \frac{5\pi k}{24} \sin \frac{\pi k}{24} + i \sin \frac{\pi k}{24} \cos \frac{5\pi k}{24} \right).$$

4. The matrix A_u must be positive Hermitian, hence for all $c \in \mathbb{C}^3$ we must have $< A_u(x,y)c, c > \geq 0$. With c_1, c_2, c_3 as given this gives for all $\lambda \in \mathbb{R}$

$$u(0)(1 + 2\lambda^2) + 2\lambda|u(x) - u(y)| - 2\lambda^2 \mathrm{Re}(u(x-y)) \geq 0$$

implying for $u(x) \neq u(y)$

$$4|u(x) - u(y)|^2 \leq 4(2u(0) - 2\mathrm{Re}(u(x-y))),$$

or, in the case that $u(x) \neq u(y)$, we arrive at

$$|u(x) - u(y)|^2 \leq 2u(0)(u(0) - 2\mathrm{Re}(u(x-y))).$$

In the case that $u(x) = u(y)$ we need only to observe that for a positive definite function we must have $\mathrm{Re}(u(\xi)) \leq u(0)$.

5. The Fourier-Stieltjes coefficients of $\mu^{(n)}$ are given by

$$\hat{\mu}_k^{(n)} = \frac{1}{2\pi} \int_{-\pi}^{\pi} e^{-ix \cdot k} \mu^{(n)}(\mathrm{d}x)$$

and since $e^{ik \cdot} \in C(\mathbb{T})$ we find

$$\lim_{n \to \infty} \frac{1}{2\pi} \int_{-\pi}^{\pi} e^{-ikx} \mu^{(n)}(\mathrm{d}x) = \frac{1}{2\pi} e^{-ik \cdot 0} = \frac{1}{2\pi}.$$

Note that the Fourier-Stieltjes coefficients $c_k = \frac{1}{2\pi}$ for all $k \in \mathbb{Z}$ correspond to the measure ϵ_0.

6. Since $\sin\left(N + \frac{1}{2}\right)t = \sin Nt \cos\frac{t}{2} + \cos Nt \sin\frac{t}{2}$ we get

$$\frac{1}{2}D_N(t) = \frac{1}{2}\frac{\sin\left(N + \frac{1}{2}\right)t}{\sin\frac{t}{2}} = \frac{1}{2}\sin Nt\frac{\cos\frac{t}{2}}{\sin\frac{t}{2}} + \frac{1}{2}\cos Nt\frac{\sin\frac{t}{2}}{\sin\frac{t}{2}}$$

$$= \frac{1}{2}\sin Nt\cot\frac{t}{2} + \frac{1}{2}\cos Nt.$$

7.   a) By the convolution theorem we have

$$F\left(f\Big|_{[-\pi,\pi]} * g\right)(k) = F\left(f\Big|_{[-\pi,\pi]}\right)(k)(Fg)(k).$$

Since $g \in L^1(\mathbb{T})$ we have $|(Fg)(k)| \leq \frac{1}{2\pi}\|g\|_{L^1}$ and since f satisfies the assumptions of Theorem 8.11 it follows with $\alpha, \beta > 0$ that $|F(f|_{[-\pi,\pi]})(k)| \leq \beta e^{-\alpha|k|}$. This implies

$$\left|F\left(f\Big|_{[-\pi,\pi]} * g\right)(k)\right| \leq \frac{\beta\|g\|_{L^1}}{2\pi}e^{-\alpha|k|}$$

and a further application of Theorem 8.11 yields that $f|_{[-\pi,\pi]} * g$ must be the restriction of a function holomorphic in a neighbourhood of $[-\pi,\pi]$.

b) Take h as in the limit, i.e. $h \in C_0^\infty(\mathbb{R})$, $\operatorname{supp} h \subset [-\frac{\pi}{2}, \frac{\pi}{2}]$ and consider the 2π-periodic extension g of $h|_{[-\pi,\pi]}$. The function g is arbitrarily often differentiable and for every $m \in \mathbb{N}_0$ we have

$$|k|^m\left|F\left(h\Big|_{[-\pi,\pi]}\right)\right| = |k|^m|(Fg)(k)| \leq c_0\|g^{(m)}\|_\infty$$

implying that for every $m \in \mathbb{N}_0$ we can find a constant $\kappa(m)$ such that $|(Fg)(k)| \leq \kappa(m)|k|^{-m}$. The coefficients $(Fg)(k)$ cannot satisfy (8.35) since otherwise, by Theorem 8.11, g must be the restriction of a function $\tilde{g}$ holomorphic in an open neighbourhood of $[-\pi,\pi]$. Since $g|_{(-\pi,-\frac{\pi}{2})}$ is identically equal to 0, $\tilde{g}$ must be identically equal to 0 on $(-\pi, -\frac{\pi}{2})$ which would imply for the holomorphic function $\tilde{g}$ that itself must be identically 0.

Thus we find that decaying faster than every power does not imply exponential decay.

8. Since f satisfies the assumption of Theorem 8.11 we have for the Fourier coefficients of $f|_{[-\pi,\pi]}$ the estimate (8.35), i.e. $|(Ff|_{[-\pi,\pi]})(k)| \leq \beta e^{-\alpha|k|}$, which implies for $m \in \mathbb{N}$ the existence of a constant γ_m such that

$$(1 + |k|^2)^{\frac{m+1}{2}}|F(f|_{[-\pi,\pi]})(k)| \leq \gamma_m$$

or

$$(1 + |k|^2)^m|F(f|_{[-\pi,\pi]})(k)|^2 \leq \frac{\gamma_m^2}{1 + |k|^2},$$

hence $f|_{[-\pi,\pi]} \in H^m(\mathbb{T})$ for every $m \in \mathbb{N}_0$, i.e. $f|_{[-\pi,\pi]} \in \cap_{m \in \mathbb{N}}H^m(\mathbb{T})$.

Chapter 9

1. We know that for $f \in L^2(\mathbb{T})$ the Fourier series converges in the norm of $L^2(\mathbb{T})$ to f which we can phrase as: the trigonometric polynomials are dense in $L^2(\mathbb{T})$. However the set of all trigonometric polynomials is not countable. We claim that every trigonometric polynomial can be approximated in $L^2(\mathbb{T})$ by a trigonometric polynomial with rational coefficients. Let $g(t) = \sum_{|k| \leq N} c_k e^{ikt}$ be a trigonometric polynomial. For $\epsilon > 0$ and $|k| \leq N$ we can find rational numbers $\alpha_k \in \mathbb{Q}$ such that $|c_k - \alpha_k| < \frac{\epsilon}{\sqrt{2\pi}(2N+1)}$ implying

$$\left\| g(\cdot) - \sum_{|k| \leq N} \alpha_k e^{ik\cdot} \right\|_{L^2} = \left\| \sum_{|k| \leq N} (c_k - \alpha_k) e^{ik\cdot} \right\|_{L^2}$$

$$\leq \max_{|k| \leq N} |c_k - \alpha_k| \left(\int_{-\pi}^{\pi} \left| \sum_{|k| \leq N} 1 \right|^2 dt \right)^{\frac{1}{2}}$$

$$< \frac{\epsilon}{\sqrt{2\pi}(2N+1)} \cdot (2N+1)\sqrt{2\pi} = \epsilon.$$

Now, given $f \in L^2(\mathbb{T})$ and $\epsilon > 0$, we can find $N \in \mathbb{N}$ such that

$$\left\| f - \sum_{|k| \leq N} c_k e^{ik\cdot} \right\|_{L^2} < \frac{\epsilon}{2}$$

where c_k is the kth Fourier coefficient of f. Furthermore, we can find rational numbers $\alpha_k \in \mathbb{Q}, |\alpha| \leq N$, such that

$$\left\| \sum_{|k| \leq N} c_k e^{ik\cdot} - \sum_{|k| \leq N} \alpha_k e^{ik\cdot} \right\|_{L^2} < \frac{\epsilon}{2}$$

implying that

$$\left\| f - \sum_{|k| \leq N} \alpha_k e^{ik\cdot} \right\|_{L^2} < \epsilon,$$

i.e. we can approximate every f in $L^2(\mathbb{T})$ by a trigonometric polynomial with rational coefficients and the set of all trigonometric polynomials with rational coefficients is countable, hence $L^2(\mathbb{T})$ is separable.

2. Let $u \in L^2_{\text{even}}(\mathbb{T})$ and $v \in L^2_{\text{odd}}(\mathbb{T})$. Their corresponding Fourier series converge in $L^2(\mathbb{T})$ to u and v, respectively. Further, we must have

$$u(t) = \frac{a_0}{2} + \sum_{k=1}^{\infty} a_k \cos kt$$

and

$$v(t) = \sum_{l=1}^{\infty} b_l \sin lt.$$

608

This implies

$$\int_{-\pi}^{\pi} u(t)\overline{v(t)}\, dt = \int_{-\pi}^{\pi} u(t)v(t)\, dt$$

$$= \frac{a_0}{2} \sum_{l=1}^{\infty} \int_{-\pi}^{\pi} b_l \sin lt\, dt + \lim_{M\to\infty} \lim_{N\to\infty} \sum_{|k|\leq M} \sum_{|l|\leq N} \int_{-\pi}^{\pi} a_k b_l \cos kt \sin lt\, dt$$

$$= 0$$

and therefore $L^2_{\text{even}}(\mathbb{T}) \perp L^2_{\text{odd}}(\mathbb{T})$. Now, if $f \in L^2(\mathbb{T})$ we can decompose f into its even and odd parts, $f = f_{\text{even}} + f_{\text{odd}}$ and it follows that $L^2(\mathbb{T}) = L^2_{\text{even}}(\mathbb{T}) \oplus L^2_{\text{odd}}(\mathbb{T})$.

3. Suppose that M is countable. The Gram-Schmidt procedure allows us to replace M by a countable orthonormal system $\{e_k \mid k \in \mathbb{N}\}$ with $\text{span}M = \text{span}\{e_k \mid k \in \mathbb{N}\}$. The element $u := \sum_{k=1}^{\infty} \frac{1}{k^2} e_k$ belongs to H since by Bessel's inequality $(\sum_{k=1}^{N} \frac{1}{k^2} e_k)_{N\in\mathbb{N}}$ is a Cauchy sequence in H. However $u \notin \text{span}\{e_k \mid k \in \mathbb{N}\}$ since it is not a (finite) linear combination of elements of $\{e_k \mid k \in \mathbb{N}\}$, hence it cannot be an element of $\text{span}M$. Thus M must be finite, implying that H is finite dimensional.

4. Choose an orthonormal basis $\{e_k \mid k \in \mathbb{N}\}$ in H and represent $u \in H$ by the convergent series $\sum_{k\in\mathbb{N}} u_k e_k$. By Parseval's equation we have

$$\|u\|_H = \left(\sum_{k=1}^{\infty} |u_k|^2\right)^{\frac{1}{2}},$$

i.e. $(u_k)_{k\in\mathbb{N}} \in l_2(\mathbb{N})$ and the mapping $j : H \to l_2(\mathbb{N}), j(u) = (u_k)_{k\in\mathbb{N}}$, is an isometry. Clearly j is a linear mapping.

5. Our solution is based on [34] and [94]. Cauchy's integral formula for derivatives, Theorem III.21.12, reads for a function f holomorphic in a neighbourhood of $B_r(z_0)$

$$f^{(n)}(z) = \frac{n!}{2\pi i} \int_{|\zeta-z_0|=r} \frac{f(z)}{(\zeta-z)^{n+1}}\, d\zeta$$

for $z \in B_r(z_0)$, and we know that we can replace the circle $|\zeta - z_0| = 1$ by any piecewise continuously differentiable curve γ such that we can deform its trace into the circle $|\zeta - z_0| = r$ without hitting any singularities of f. Since P_n is holomorphic, in fact an entire function, the defining equation for P_n yields

$$P_n(x) = \frac{1}{2^n n!} \frac{d^n}{dx^n}(x^2-1)^n = \frac{1}{2\pi i} \int_{|\zeta-x|=1} \frac{(\zeta^2-1)^n}{2^n(\zeta-x)^{n+1}}\, d\zeta.$$

Now we look at the power series

$$\sum_{n=0}^{\infty} P_n(x) z^n = \frac{1}{2\pi i} \int_{|\zeta-x|=1} \sum_{n=0}^{\infty} \frac{z^n(\zeta^2-1)^n}{2^n(\zeta-x)^{n+1}}\, d\zeta$$

and we rewrite

$$\sum_{n=0}^{\infty} \left(\frac{z}{2}\right)^n \left(\frac{(\zeta^2-1)^n}{(\zeta-1)^{n+1}}\right) = \frac{1}{\zeta-x} \sum_{n=0}^{\infty} \left(\frac{z(\zeta^2-1)}{2(\zeta-x)}\right)^n.$$

We want this geometric series to converge for $|\zeta - x| \leq 1$, i.e. we need

$$\left|\frac{z(\zeta^2-1)}{2(\zeta-x)}\right| < 1$$

or $|\zeta^2 - 1| \leq \frac{2}{|z|}$. Since for $\zeta \in \partial B_1(x)$ we have $|\zeta| \leq 1 + |x|$ and therefore $|\zeta^2 - 1| \leq (1+|x|)^2 + 1 \leq 5$, it follows that for $|z| < \frac{2}{5}$ we have $\left|\frac{z(\zeta^2-1)}{z(\zeta-x)}\right| < 1$. Thus for these values of z we obtain

$$\sum_{n=0}^{\infty} \left(\frac{z(\zeta^2-1)}{2(\zeta-x)}\right)^n = \frac{1}{1 - \frac{z(\zeta^2-1)}{2(\zeta-x)}}$$

implying

$$\sum_{n=0}^{\infty} P_n(x)z^n = \frac{1}{2\pi i} \int_{|\zeta-x|=1} \frac{1}{\zeta-x} \left(\frac{1}{1-\frac{z(\zeta^2-1)}{2(\zeta-x)}}\right) d\zeta$$

$$= \frac{1}{2\pi i} \int_{|\zeta-x|=1} \frac{2}{z - 2x + 2\zeta - 2\zeta^2} d\zeta$$

$$= -\frac{1}{2\pi i} \int_{|\zeta-x|=1} \frac{2}{z(\zeta-\zeta_1)(\zeta-\zeta_2)} d\zeta$$

where $\zeta_1 = \frac{1-(1-2xz+z^2)^{\frac{1}{2}}}{z}$ and $\zeta_2 = \frac{1+(1-2xz+z^2)^{\frac{1}{2}}}{z}$. Thus we have

$$\sum_{n=0}^{\infty} P_n(x)z^n = -\frac{1}{2z\pi i} \int_{|\zeta-x|=1} \frac{2}{(\zeta-\zeta_1)(\zeta-\zeta_2)} d\zeta.$$

If necessary we reduce the bound for $|z|$ and we note that for $|z|$ sufficiently small only ζ_1 lies in the interior of the circle $|\zeta - x| = 1$, while ζ_2 is in the complement of $\overline{B_1(x)}$. The residue theorem, Theorem III.25.20 yields

$$-\frac{1}{2z\pi i} \int_{|\zeta-x|=1} \frac{2}{(\zeta-\zeta_1)(\zeta-\zeta_2)} d\zeta = -\frac{2}{z}\mathrm{res}(g,\zeta_1)$$

where $g(\zeta) = \frac{1}{(\zeta-\zeta_1)(\zeta-\zeta_2)}$ and by Proposition III.25.22 we find

$$\mathrm{res}(g,\zeta_1) = \frac{1}{\zeta_1-\zeta_2} = \frac{-z}{2(1-2xz+z^2)^{\frac{1}{2}}}$$

which eventually gives for $|z|$ sufficiently small

$$\sum_{n=0}^{\infty} P_n(x)z^n = \frac{1}{(1-2xz+z^2)^{\frac{1}{2}}}. \qquad (*)$$

We can interpret $\sum_{n=0}^{\infty} P_n(x)z^n$ as Taylor series of the function $z \mapsto (1 - 2xz + z^2)^{-\frac{1}{2}}$ about $z_0 = 0$ and for $|z| < 1$ this function is holomorphic, hence the series $\sum_{n=0}^{\infty} P_n(x)z^n$ converges in $|z| < 1$.

The function $z \mapsto (1 - 2xz + z^2)^{-\frac{1}{2}}$ is the **generating function of the Legendre polynomials**. In Appendix III we will discuss generating functions in more details.

6.   a) We note that

$$\frac{\partial g(x,t)}{\partial t} = \frac{\partial}{\partial t}(1 - 2xt + t^2)^{-\frac{1}{2}}$$
$$= (x - t)(1 - 2xt + t^2)^{-\frac{3}{2}}$$

and

$$\frac{\partial g(x,t)}{\partial t} = \frac{\partial}{\partial x}(1 - 2xt + t^2)^{-\frac{1}{2}}$$
$$= t(1 - 2xt + t^2)^{-\frac{3}{2}}$$

and the two differential equations follow.

b) Substituting in $(1 - 2xt + t^2)\frac{\partial g(x,t)}{\partial t} + (t - x)g(x,t) = 0$ for $g(x,t)$ the power series $g(x,t) = \sum_{n=0}^{\infty} P_n(x)t^n$ we obtain

$$(1 - xt + t^2)\sum_{n=0}^{\infty} nP_n(x)t^{n-1} + (t - x)\sum_{n=0}^{\infty} P_n(x)t^n = 0.$$

Rearranging the identity according to the powers of t and setting the corresponding coefficients equal to 0 we get

$$(n + 1)P_{n+1}(x) - 2nxP_n(x) + (n - 1)P_{n-1}(x) + P_{n-1}(x) - xP_n(x) = 0$$

or

$$(n + 1)P_{n+1}(x) - (2n + 1)P_n(x) + nP_{n-1}(x) = 0, n \in \mathbb{N}. \qquad (1)$$

Similarly, inserting in the differential equation $(1 - 2xt + t^2)\frac{\partial g(x,t)}{\partial t} - tg(x,t) = 0$ the power series $\sum_{n=0}^{\infty} P_n(x)t^n$ for $g(x,t)$ and proceeding as before we find

$$(1 - 2xt + t^2)\sum_{n=0}^{\infty} P_n'(x)t^n - \sum_{n=0}^{\infty} P_n(x)t^{n+1} = 0$$

or

$$P_{n+1}'(x) - 2xP_n'(x) + P_{n-1}'(x) - P_n(x) = 0, n \in \mathbb{N}. \qquad (2)$$

c) Differentiating (1) once with respect to x and eliminating from this new equation with the help of (2) first $P_{n-1}'(x)$ and then $P_{n+1}'(x)$ we arrive at

$$P_{n+1}'(x) - xP_n'(x) = (n + 1)P_n(x), n \in \mathbb{N}_0, \qquad (3)$$

and

$$x P_n'(x) - P_{n-1}'(x) = n P_n(x), n \in \mathbb{N}_0. \tag{4}$$

Adding these equations gives

$$P_{n+1}'(x) - P_{n-1}'(x) = (2n + 1) P_n(x), n \in \mathbb{N}.$$

Rewriting (3) for $n - 1$, i.e. looking at

$$P_n'(x) - x P_{n-1}'(x) = n P_{n-1}(x),$$

we may use (4) to eliminate $P_{n-1}'(x)$ and we get

$$(1 - x^2) P_n'(x) = n P_{n-1}(x) - n x P_n(x).$$

We differentiate this equality with respect to x and again we eliminate $P_{n-1}'(x)$ to arrive eventually at

$$((1 - x^2) P_n'(x))' + n(n + 1) P_n(x) = 0.$$

In our solution of Problem 6, we closely followed [85] where much more results on orthonormal polynomials can be found.

7.   a) The first four Legendre polynomials are

$$P_0(x) = 1, P_1(x) = x, P_2(x) = \frac{1}{2}(3x^2 - 1), P_3(x) = \frac{1}{2}(5x^3 - 3x).$$

We are seeking a representation of x^3, i.e. we are dealing with the equality

$$x^3 = \lambda_0 P_0(x) + \lambda_1 P_1(x) + \lambda_2 P_2(x) + \lambda_3 P_3(x)$$
$$= \left(\lambda_0 - \frac{\lambda_2}{2}\right) 1 + \left(\lambda_1 - \frac{3}{2}\lambda_3\right) x + \frac{3}{2}\lambda_2 x^2 + \frac{5}{2}\lambda_3 x^3$$

which yields

$$\lambda_0 = 0, \lambda_1 = \frac{4}{15}, \lambda_2 = 0, \lambda_3 = \frac{2}{5}$$

and we find

$$x^3 = \frac{4}{15} P_1(x) + \frac{2}{5} P_3(x).$$

However, in order to use the normalized Legendre polynomials we need an adjustment since $\|P_n\|_{L^2} = \sqrt{\frac{2}{2n+1}}$ and we get

$$x^3 = \frac{4}{15}\sqrt{\frac{2}{3}}\left(\sqrt{\frac{3}{2}} P_1(x)\right) + \frac{2}{5}\sqrt{\frac{2}{5}}\left(\sqrt{\frac{5}{2}} P_3(x)\right).$$

b) We need to find the term

$$\int_{-1}^{1} \chi_{[-1,0]}(t) \sqrt{\frac{2n + 1}{2}} P_n(t) \, dt = \int_{-1}^{0} \sqrt{\frac{2n + 1}{2}} P_n(t) \, dt.$$

From Problem 6 we use the relation

$$P_n(x) = \frac{1}{2n+1}\left(P'_{n+1}(x) - P'_{n-1}(x)\right)$$

and we find

$$\int_{-1}^{1} \chi_{[-1,0]}(t)\sqrt{\frac{2n+1}{2}}P_n(t)\,dt = \frac{\sqrt{2n+1}}{2}\frac{1}{2n+1}\int_{-1}^{0}(P'_{n+1}(t) - P'_{n-1}(t))\,dt$$

$$= \frac{1}{\sqrt{4n+2}}\left(P_{n+1}(t) - P_{n-1}(t)\right)\Big|_{-1}^{0}.$$

Now we may use the generating function $g(x,t)$ to find the special values $P_n(-1)$, $P_n(0)$ and $P_n(1)$. Since $\sum_{n=0}^{\infty} P_n(x)t^n = (1 - 2xt + t^2)^{-\frac{1}{2}}$ we find for $x = 1$ that $\sum_{n=0}^{\infty} P_n(1)t^n = (1 - 2t + t^2)^{-\frac{1}{2}} = \frac{1}{1-t}$ implying that $P_n(1) = 1$ for all $n \in \mathbb{N}$. Furthermore we find that $\sum_{n=0}^{\infty} P_n(-1)t^n = \frac{1}{(1+t)}$ which yields $P_n(-1) = (-1)^n$. For $x = 0$ we find $\sum_{n=0}^{\infty} P_n(0)t^n = (1 + t^2)^{-\frac{1}{2}}$ and the Taylor expansion of $t \mapsto (1 + t^2)^{-\frac{1}{2}}$ about $t_0 = 0$ yields

$$P_{2k}(0) = (-1)^k \frac{1 \cdot 3 \cdot \ldots \cdot (2k-1)}{2 \cdot 4 \cdot \ldots \cdot 2k}, \quad P_{2k+1}(0) = 0.$$

Thus we find for n odd i.e. $n+1$ and $n-1$ even, $n \in \mathbb{N}$, that

$$\frac{1}{\sqrt{4n+2}}\left(P_{n+1}(0) - P_{n-1}(0) - P_{n+1}(-1) + P_{n-1}(-1)\right)$$

$$= \frac{1}{\sqrt{4n+2}}\left(P_{n+1}(0) - P_{n-1}\right),$$

and for n even we note that $n+1$ and $n-1$ are off and therefore

$$\frac{1}{\sqrt{4n+2}}\left(P_{n+1}(0) - P_{n-1}(0) - P_{n+1}(-1) + P_{n-1}(-1)\right) = 0.$$

8. The general formula is given, compare with (9.28), by

$$c_{2k}(g_n) = \,< g_n, \tilde{H}_{2k} >$$

$$= \frac{1}{(2^{2k}(2k)!\sqrt{\pi})^{\frac{1}{2}}}\int_{-\infty}^{\infty} x^{2n} H_{2k}(x)e^{-x^2}\,dx,$$

however it is more convenient to calculate

$$\gamma_{2k}(g_n) = \frac{1}{(2^{2k}(2k)!\sqrt{\pi})^{\frac{1}{2}}}c_{2k}(g_n) = \frac{1}{2^{2k}(2k)!\sqrt{\pi}}\int_{-\infty}^{\infty} x^{2n} H_{2k}(x)e^{-x^2}\,dx.$$

Using the definition of H_{2k} we obtain after $2k$-times integration by parts

$$\gamma_{2k}(g_n) = \frac{1}{2^{2k}(2k)!\sqrt{\pi}} \int_{-\infty}^{\infty} x^{2n} \frac{d^{2k}}{dx^{2k}} e^{-x^2} dx$$

$$= \frac{1}{2^{2k}(2k)!\sqrt{\pi}} \frac{(2n)!}{(2n-2k)!} \int_{-\infty}^{\infty} e^{-x^2} x^{2n-2k} dx$$

$$= \frac{1}{2^{2k}(2k)!\sqrt{\pi}} \frac{(2n)!}{(2n-2k)!} \Gamma\left(n - k + \frac{1}{2}\right).$$

Now the duplication formula of Legendre yields

$$2^{2n-2k} \Gamma\left(n - k + \frac{1}{2}\right)(n-k)! = \sqrt{\pi}(2n-2k)!$$

implying

$$\gamma_{2k}(g_n) = \frac{(2n)!}{2^n(2k)!(k-n)!}.$$

Note that for $n < k$ we see from the integration by parts calculation that $c_{2k}(g_n) = 0$.

9. Applying Leibniz' rule in the definition of $L_n^\alpha(t)$, i.e. in

$$L_n^\alpha(t) = \frac{t^{-\alpha} e^t}{n!} \frac{d^n}{dt^n}(t^{\alpha+n} e^{-t})$$

we get

$$L_n^\alpha(t) = \frac{t^{-\alpha} e^t}{n!} \sum_{k=0}^{n} \binom{n}{k} \frac{d^{n-k} t^{\alpha+n}}{dt^{n-k}} \frac{d^k e^{-t}}{dt^k}$$

$$= t^{-\alpha} e^t \sum_{k=0}^{n} \frac{1}{k!(n-k)!}(n+\alpha)(n-1+\alpha)\ldots(k+1+\alpha)(-1)^k t^{k+\alpha} e^{-t}$$

$$= \sum_{k=0}^{n}(-1)^k \frac{(n+\alpha)(n-1+\alpha)\ldots(k+1+\alpha)}{k!(n-k)!} t^k.$$

Note that using the Pochhammer symbol, see (III.16.29) we can rewrite $L_n^\alpha(t)$ as

$$L_n^\alpha(t) = \sum_{k=0}^{n} \frac{(-1)^k}{k!(n-k)!} \frac{(\alpha+1)_n}{(\alpha+1)_k} t^k.$$

In particular we find that $L_n^\alpha(t)$ is a polynomial of order n. The last remark makes it also easy to prove the orthogonality of the Laguerre polynomials. For $m < n$ we have to look at

$$\int_0^\infty L_m^\alpha(x) L_n^\alpha(x) x^\alpha e^{-x} dx = \frac{1}{n!} \int_0^\infty L_m^\alpha(x) \frac{d^n}{dx^n}(x^{\alpha+n} e^{-x}) dx$$

and integration by parts yields

$$\int_0^\infty L_m^\alpha(x) L_n^\alpha(x) x^\alpha e^{-x} \, dx = \frac{(-1)^n}{n!} \int_0^\infty \left(\frac{d^n}{dx^n} L_m^\alpha(x) \right) x^{\alpha+n} e^{-x} \, dx,$$

but $\frac{d^n}{dx^n} L_m^\alpha(x) = 0$ for all x since by assumption $m < n$. This proves the orthogonality of the system $(L_n^\alpha)_{n \in \mathbb{N}}$ in $L^2((0, \infty), \mu_\alpha)$.

Note that for $m = n$ we find

$$\|L_n^\alpha\|_{L^2((0,1),\mu_\alpha)}^2 = \frac{1}{n!} \int_0^\infty x^{\alpha+n} e^{-x} \, dx = \frac{\Gamma(n + \alpha + 1)}{n!}.$$

(We used [34] for solving this problem.)

Chapter 10

1. We recall how we have derivated the Taylor formula in Chapter II.9. We have considered u as a function on the line segment L connecting x with x_0, which is given by $t \mapsto tx + (1 - t)x_0$. The derivatives of $u|_L$ with respect to t are higher order direction derivatives and now (10.14) follows along the lines as (I.29.13) was proved.

2. We need to prove: given $\epsilon > 0$ there exists $\delta > 0$ such that $\|x - y\| < \delta$ implies $|u(x) - u(y)| < \epsilon$. Since u vanishes at infinity, given $\epsilon > 0$ we can find $R \geq 0$ such that in $\overline{B_R(0)}^c$ we have $|u(x)| < \frac{\epsilon}{4}$. Further we know that on the compact set $\overline{B_R(0)}$, u is uniformly continuous, thus given $\epsilon > 0$ we can find $\delta > 0$ such that $x, y \in \overline{B_R(0)}$ and $\|x - y\| < \delta$ implies $|u(x) - u(y)| < \frac{\epsilon}{2}$.

Now let $x, y \in \mathbb{R}^n$ and $\|x - y\| < \delta$. If $x, y \in \overline{B_R(0)}$ we know that $|u(x) - u(y)| < \frac{\epsilon}{2}$. If $x, y \in \overline{B_R(0)}^c$ we know that $|u(x) - u(y)| < |u(x)| + |u(y)| < \frac{\epsilon}{4} + \frac{\epsilon}{4} = \frac{\epsilon}{2}$.

If $x \in B_R(0)$ and $y \in \overline{B_R(0)}^c$ such that $\|x - y\| < \delta$ we choose a unique point $z \in \partial B_R(0)$ on the line segment connecting x with y and we find

$$|u(x) - u(y)| \leq |u(x) - u(x_1)| + |u(x_1)| + |u(y)| < \frac{\epsilon}{2} + \frac{\epsilon}{4} + \frac{\epsilon}{4} = \epsilon.$$

3.   a) Using the Cauchy-Schwarz inequality we find

$$\left| \int_{\mathbb{R}^n} \int_{\mathbb{R}^n} h(x - y) u(x) v(y) \, dy \, dx \right| \leq \|(h * u) \cdot v\|_{L^1}$$

$$\leq \|h * u\|_{L^1} \|v\|_{L^2}$$

$$\leq \|h\|_{L^1} \|u\|_{L^2} \|v\|_{L^2}.$$

b) By Peetre's inequality we have for $M \geq 0$

$$(1 + \|x - y\|^2)^M \leq 2^M (1 + \|x\|^2)^M (1 + \|y\|^2)^M$$

615

which yields

$$\left| \int_{\mathbb{R}^n} \left(\int_{\mathbb{R}^n} (1 + \|x - y\|^2)^M h(x - y) u(x) v(y) \, dy \right) dx \right|$$

$$\leq 2^M \int_{\mathbb{R}^n} \left(\int_{\mathbb{R}^n} |h(x - y)| (1 + \|x\|^2)^M |u(x)| (1 + \|y\|^2)^M |v(y)| \, dy \right) dx.$$

Since for $u \in \mathcal{S}(\mathbb{R}^n)$ it follows that $(1 + \| \cdot \|^2)^M u \in L^2(\mathbb{R}^n)$ we arrive by part a) at

$$\left(\int_{\mathbb{R}^n} \left(\int_{\mathbb{R}^n} (1 + \|x - y\|^2)^M h(x - y) u(x) v(y) \, dy \right) dx \right)$$

$$\leq 2^M \|h\|_{L^1} \|(1 + \| \cdot \|^2)^M u\|_{L^2} \|(1 + \| \cdot \|^2)^M v\|_{L^2}$$

and therefore the integral is finite.

4.   a) First we note that in the case where $T : \mathbb{R}^n \to \mathbb{R}^n$ is not bijective $u \circ T$ need not belong to $\mathcal{S}(\mathbb{R}^n)$. Indeed if $u(0) \neq 0$ then for $x \in \{y \in \mathbb{R}^n \mid Ty = 0\}$ it follows that $u(x) = u(0)$ and therefore u does not vanish at infinity.

Now we look at $v := u \circ T$ under the assumption that T is bijective. Clearly v is an arbitrarily often differentiable function and since with u also $\partial^\alpha u, \alpha \in \mathbb{N}_0^n$, belongs to $\mathcal{S}(\mathbb{R}^n)$, the chain rule implies that it remains to prove that if $(1 + |y|^2)^m |u(y)| \leq C_m$ for all $m \in \mathbb{N}_0$, then $(1 + |x|^2)^m |v(x)| \leq \tilde{C}_m$ for all $m \in \mathbb{N}_0$. (We urge the reader to have a look at the Faà di Bruno formula in dimension 1, see (I.21.13), in order to understand our argument in more detail.) Now with $Tx = y$ we find $v(x) = u(Tx)$ and

$$(1 + |x|^2)^m |v(x)| = (1 + |T^{-1}y|^2)^m |v(T^{-1}y)|$$
$$= (1 + |T^{-1}y|^2)^m |u(y)| \leq c(1 + |y|^2)^m |u(y)| \leq c_0$$

since $u \in \mathcal{S}(\mathbb{R}^n)$.

   b) We can find $U \in O(n)$ such that

$$U^{-1} A U = \begin{pmatrix} \lambda_1 & & 0 \\ & \ddots & \\ 0 & & \lambda_n \end{pmatrix}$$

with $\lambda_j > 0$, and the multiplicities of the eigenvalues of A are taken into account. The change of coordinate $x = Uy$ yields

$$f(Uy) = e^{-<AUy, Uy>} = e^{-<U^t AUy, y>}$$
$$= e^{-<U^{-1} AUy, y>} = \prod_{j=1}^n e^{-\lambda_j y_j^2}.$$

Since $y^\alpha \partial_y^\alpha \prod_{j=1}^n e^{-\lambda_j y_j^2} = \prod_{j=1}^n y_j^{\alpha_j} \partial_{y_j}^{\alpha_j} e^{-\lambda_j y_j^2}$ it remains to prove that for $\lambda > 0$ the function $z \mapsto e^{-\lambda z^2}, z \in \mathbb{R}$, belongs to $\mathcal{S}(\mathbb{R})$. In light of Problem 5 in Chapter

I.9 where we proved that $\frac{d^n}{dx^n}e^{-x^2} = p_n(x)e^{-x^2}$ with a suitable polynomial p_n the result follows.

c) Clearly, as product and composition of C^∞-functions, the function u, $u(x) = e^{-x^2}\sin(e^{x^2})$ is a C^∞-function. Further, since $|\sin e^{x^2}| \leq 1$ and e^{-x^2} decays faster than any power, the function u also decays faster than any power. However, for u' we find

$$u'(x) = \frac{d}{dx}(e^{-x^2}\sin(e^{x^2}))$$
$$= -2xe^{-x^2}\sin(e^{x^2}) + 2x\cos e^{x^2}.$$

For $x_k = \sqrt{\ln 2\pi k}$, $k \in \mathbb{N}$, we find now

$$|u'(\sqrt{\ln 2\pi k})| = 2\sqrt{\ln 2\pi k}$$

which tends to infinity for $k \to \infty$.

5. We consider

$$x^\beta \partial^\alpha(g(x)u(x)) = x^\beta \sum_{\gamma \leq \alpha}\binom{\alpha}{\gamma}\partial^{\alpha-\gamma}g(x)\partial^\gamma u(x)$$

and we find

$$|x^\beta \partial^\alpha(g(x)u(x))| \leq \sum_{\gamma \leq \alpha}\binom{\alpha}{\gamma}c_{\alpha-\gamma}|x^\beta(1+\|x\|^2)^{k_\alpha-\gamma}\partial^\gamma u(x)|$$
$$\leq c(1+\|x\|^2)^{m_1}\sum_{|\gamma|\leq m_2}|\partial^\gamma u(x)| = cq_{2m_1,m_2}(u),$$

where $m_2 = |\alpha|$ and $m_1 \geq |\beta| + \max_{\gamma \leq \alpha}k_{\alpha-\gamma}$. Hence $g \cdot u \in \mathcal{S}(\mathbb{R}^n)$.

6. Let $(\varphi_k)_{k\in\mathbb{N}}, \varphi_k \in \mathcal{S}(\mathbb{R}^n)$, be a sequence converging in $\mathcal{S}(\mathbb{R}^n)$ to $\varphi \in \mathcal{S}(\mathbb{R}^n)$. By assumption the sequence $(T_1\varphi_k)_{k\in\mathbb{N}}$ converges in $\mathcal{S}(\mathbb{R}^n)$ to $T_1\varphi$ and therefore $(T_2(T_1\varphi_k))_{k\in\mathbb{N}}$ converges in $\mathcal{S}(\mathbb{R}^n)$ to $T_2(T_1\varphi)$ i.e. $T_2 \circ T_1$ is continuous from $\mathcal{S}(\mathbb{R}^n)$ to $\mathcal{S}(\mathbb{R}^n)$.

7. The key tool is Lemma 10.1 which we can reformulate as follows: for every pair $\alpha, \beta \in \mathbb{N}_0^n$ exists $m_{1,1}, m_{1,2}\ldots m_{M,1}, m_{M,2} \in \mathbb{N}_0$ such that

$$p_{\alpha,\beta}(u) \leq C \max_{1\leq j\leq M} q_{m_{j,1},m_{j,2}}(u);$$

and for every pair m_1, m_2 exists $\alpha_1, \beta_1, \ldots, \alpha_N, \beta_N \in \mathbb{N}_0^n$ such that

$$q_{m_1,m_2}(u) \leq C' \max_{1\leq l\leq N} p_{\alpha_l,\beta_l}(u).$$

This implies that whenever $\lim_{k\to\infty} q_{m_1,m_2}(\varphi_k - \varphi) = 0$ for all $m_1, m_2 \in \mathbb{N}_0$, then $\lim_{k\to\infty} p_{\alpha,\beta}(\varphi_k - \varphi) = 0$ for all $\alpha, \beta \in \mathbb{N}_0^n$ and now the equivalence of $i) - iv)$ is obvious.

8. We have taken this problem from [99]. First we note for $u \in \mathcal{S}(\mathbb{R})$ that $u(x) = \int_{-\infty}^{x} u'(t) \, dt$ and therefore

$$\|u\|_\infty \leq \|u'\|_{L^1} = \int_{\mathbb{R}} \left(\frac{1}{1+|x|^2}\right)^{\frac{1}{2}} (1+|x|^2)^{\frac{1}{2}} |u'(x)| \, dx$$

$$\leq \left(\int_{\mathbb{R}} \frac{1}{1+x^2} \, dx\right)^{\frac{1}{2}} \|(1+|\cdot|^2)^{\frac{1}{2}} u'\|_{L^2}$$

$$= c \left(\int_{\mathbb{R}} |1+x^2| |u'(x)|^2 \, dx\right)^{\frac{1}{2}}$$

Since $(x^\beta \partial^\alpha u(x))' = \beta x^{\beta-1} \partial^\alpha u(x) + x^\beta \partial^{\alpha+1} u(x)$ we conclude

$$|x^\beta \partial^\alpha u(x)| = \left|\int_{-\infty}^{x} (\beta t^{\beta-1} \partial^\alpha u(t) + t^\beta \partial^{\alpha+1} u(t)) \, dt\right|$$

$$< |\beta| \|t^{\beta-1} \partial^\alpha u(\cdot)\|_{L^1} + \|t^\beta \partial^{\alpha+1} u\|_{L^1}.$$

With the help of the estimate made above we find

$$\|t^{\beta-1} \partial^\alpha u(\cdot)\|_{L^1} \leq c' \left(\int_{\mathbb{R}} (1+t^2) |t^{\beta-1} \partial^\alpha u(t)|^2 \, dt\right)^{\frac{1}{2}}$$

$$\leq c' (\nu_{\alpha,\beta-1,L^2}(u) + \nu_{\alpha,\beta,L^2}(u))$$

and

$$\|t^\beta \partial^{\alpha+1} u\|_{L^1} \leq c'' (\nu_{\alpha+1,\beta,L^2}(u) + \nu_{\alpha+1,\beta+1,L^2}(u)),$$

thus we have proved

$$p_{\alpha,\beta}(u) \leq \tilde{c} (\nu_{\alpha,\beta-1,L^2}(u) + \nu_{\alpha,\beta,L^2}(u) + \nu_{\alpha+1,\beta,L^2}(u) + \nu_{\alpha+1,\beta+1,L^2}(u)).$$

Conversely, we note that

$$\nu_{\alpha,\beta,L^2}(u) = \left(\int_{\mathbb{R}} t^{2\beta} |\partial^\alpha u(t)|^2 \, dt\right)^{\frac{1}{2}}$$

$$\leq \left(\int_{\mathbb{R}} \frac{1}{1+t^2} \, dt\right)^{\frac{1}{2}} \|(1+|t|^2) t^\beta \partial^\alpha u\|_\infty$$

$$\leq c^* (p_{\alpha,\beta}(u) + p_{\alpha,\beta+2}(u)).$$

9.   a) Since $\varphi \in \mathcal{S}(\mathbb{R}^n)$ implies $\varphi \in L^{p'}(\mathbb{R}^n)$, $1 < p < \infty$, $\frac{1}{p} + \frac{1}{p'} = 1$, Hölder's inequality yields for $1 < p < \infty$ that $|l_h(\varphi)| \leq \|h\|_{L^p} \|\varphi\|_{L^{p'}}$ and for $p = 1$ we find $|l_h(\varphi)| \leq \|h\|_{L^1} \|\varphi\|_\infty$. Then l_h is well-defined and for all $\varphi \in \mathcal{S}(\mathbb{R}^n)$ it is finite. Moreover we note that for $1 < p' < \infty$ it follows that

$$\left|\int_{\mathbb{R}^n} h(x)(\varphi_k(x) - \varphi(x)) \, dx\right| \leq \|h\|_{L^p} \|\varphi_k - \varphi\|_{L^p}$$

and

$$\|\varphi_k - \varphi\|_{L^p}^{p'} = \int_{\mathbb{R}^n} |\varphi_k(x) - \varphi(x)|^{p'} \, dx$$

$$= \int_{\mathbb{R}^n} \frac{1}{(1 + \|x\|^2)^{\frac{(N+1)p'}{2}}} (1 + \|x\|^2)^{\frac{(N+1)p'}{2}} |\varphi_k(x) - \varphi(x)|^{p'} \, dx$$

$$\leq \int_{\mathbb{R}^n} \frac{1}{(1 + \|x\|^2)^{\frac{(N+1)p'}{2}}} \, dx \, \|(1 + \|\cdot\|^2)^{\frac{N+1}{2}} |\varphi_k - \varphi\|_{\infty}^{p'},$$

or

$$|l_h(\varphi_k - \varphi)| \leq \gamma q_{N+1,0}(\varphi_k - \varphi)$$

implying that l_h is sequentially continuous from $\mathcal{S}(\mathbb{R}^n)$ to $\mathbb{C}$.

b) Since μ is a bounded measure and $\|\varphi\|_{\infty} < \infty$ for $\varphi \in \mathcal{S}(\mathbb{R}^n)$ we note that $|l_\mu(\varphi)| \leq \mu(\mathbb{R}^n)\|\varphi\|_{\infty}$, i.e. l_μ is well defined on $\mathcal{S}(\mathbb{R}^n)$ amd finite. Furthermore we observe

$$|l_\mu(\varphi_k - \varphi)| \leq \mu(\mathbb{R}^n)q_{0,0}(\varphi_k - \varphi)$$

implying the sequential continuity of l_μ.

10. For $\alpha, \beta \in \mathbb{N}_0^n$ and $u \in \mathcal{S}(\mathbb{R}^n)$ we have

$$|x^\beta \partial^\alpha (P(x, D)u(x))| = \left| x^\beta \partial^\alpha \sum_{|\sigma| \leq m} a_\sigma(x) \partial^\sigma u(x) \right|$$

$$\leq \sum_{|\sigma| \leq m} \left| x^\beta \sum_{\gamma \leq \alpha} \binom{\alpha}{\gamma} \partial^\gamma a_\sigma(x) \partial^{\alpha + \sigma - \gamma} u(x) \right|$$

$$\leq \sum_{|\sigma| \leq m} \sum_{\gamma \leq \alpha} c_{\gamma,\sigma} |x^\beta (1 + \|x\|^2)^{k_{\gamma,\sigma}} \partial^{\alpha + \sigma - \gamma} u(x)|$$

$$\leq c(1 + \|x\|^2)^{m_1} \sum_{|\sigma| \leq m} \sum_{\gamma \leq \alpha} |\partial^{\alpha + \sigma - \gamma} u(x)|$$

$$\leq c(1 + \|x\|^2)^{m_1} \sum_{|\gamma| \leq m + |\alpha|} |\partial^\gamma u(x)|$$

$$\leq c q_{2m_1, m + |\alpha|}(u),$$

where $m_1 \geq |\beta| + \max_{\substack{\gamma \leq \alpha \\ |\sigma| \leq m}} k_{\gamma,\sigma}$. Thus we have proved

$$p_{\alpha,\beta}(P(x, D)u) \leq c q_{2m_1, m + |\alpha|}(u)$$

which implies the continuity of $P(x, D) : \mathcal{S}(\mathbb{R}^n) \to \mathcal{S}(\mathbb{R}^n)$.

Chapter 11

1.   a) Recall that $\frac{1}{\cosh x} = \frac{2}{e^x + e^{-x}}$ and since $\frac{1}{e^x + e^{-x}} \leq e^{-|x|}$ it follows that

$$p_{0,k}\left(\frac{1}{\cosh x}\right) = \sup_{x \in \mathbb{R}} \left| x^k \frac{1}{\cosh x} \right| < \infty.$$

Next we observe that

$$\frac{d}{dx}\frac{1}{\cosh x} = -2\frac{e^x - e^{-x}}{(e^x + e^{-x})^2} = -2\tanh x\frac{1}{\cosh x}$$

and we claim that for $m \geq 1$ we have

$$\frac{d^m}{dx^m}\left(\tanh x\frac{1}{\cosh x}\right) = \sum_{j=1}^{m+2} c_j^{(m)}\frac{1}{\cosh^j x}P_j^{(m)}(\tanh x)$$

where $P_j^{(m)}(t)$ is a polynomial of degree less than or equal to $m + 1$. For $m = 1$ this was just proved. Now we note

$$\frac{d^{m+1}}{dx^{m+1}}\left(\tanh x\frac{1}{\cosh x}\right) = \frac{d}{dx}\sum_{j=1}^{m+2} c_j^{(m)}\frac{1}{\cosh^j x}P_j^{(m)}(\tanh x)$$

$$= \sum_{j=1}^{m+2}(-jc_j^{(m)})\left(\frac{\cosh' x}{\cosh^{j+1} x}\right)P_j^{(m)}(\tanh x)$$

$$+ \sum_{j=1}^{m+2} c_j^{(m)}\frac{1}{\cosh^j x}\frac{d}{dx}P_j^{(m)}(\tanh x)$$

$$= \sum_{j=1}^{m+2}(-jc_j^{(m)})\frac{1}{\cosh^j x}(\tanh x)P_j^{(m)}(\tanh x)$$

$$+ \sum_{j=1}^{m+2} c_j^{(m)}\frac{1}{\cosh^j x}\tilde{P}_j^{(m)}(\tanh x)\frac{1}{\cosh^2 x}$$

where in the last step we used that $P_j^{(m)}(t)$ has zero-order term 0 and $\tilde{P}_j^{(m)}(t)$ has at most degree m.

Since $|\tanh x| \leq 1$ and $\frac{1}{\cosh^j x} \leq \frac{1}{\cosh x}$ we obtain for all $k \in \mathbb{N}_0$ and $m \in \mathbb{N}_0$ that

$$p_{k,m}\left(\frac{1}{\cosh x}\right) = \sup_{x \in \mathbb{R}}\left|x^k\frac{d^m}{dx^m}\frac{1}{\cosh x}\right| < \infty,$$

i.e. $\frac{1}{\cosh x} \in \mathcal{S}(\mathbb{R})$.

b) We have

$$\int_{\mathbb{R}}\frac{1}{\cosh\sqrt{\frac{\pi}{2}}x}\,dx = \sqrt{2\pi}\left(F\left(\frac{1}{\cosh\sqrt{\frac{\pi}{2}}x}\right)\right)(0)$$

$$= \sqrt{2\pi}\frac{1}{\cosh 0} = \sqrt{2\pi},$$

where we have used that $x \mapsto \frac{1}{\cosh\sqrt{\frac{\pi}{2}}x}$ is a fixed point of the Fourier transform.

2.   a) We find

$$\frac{1}{\sqrt{2\pi}} \int_{\mathbb{R}} e^{-ix\xi} g(x) \; \mathrm{d}x = \frac{1}{\sqrt{2\pi}} \int_{\mathbb{R}} e^{-ix\xi} (xe^{-\frac{x^2}{2}}) \; \mathrm{d}x$$

$$= -\frac{1}{\sqrt{2\pi}} \int_{\mathbb{R}} \left(\frac{\mathrm{d}}{\mathrm{d}x} e^{-\frac{x^2}{2}} \right) e^{-ix\xi} \; \mathrm{d}x$$

$$= \frac{-i\xi}{\sqrt{2\pi}} \int_{\mathbb{R}} e^{-\frac{x^2}{2}} e^{-ix\xi} \; \mathrm{d}x = -i\xi e^{-\frac{\xi^2}{2}}.$$

b) By Plancherel's theorem we know that

$$\int_{\mathbb{R}} |\hat{u}(\xi)|^2 \; \mathrm{d}\xi = \int_{\mathbb{R}} |u(x)|^2 \; \mathrm{d}x$$

and with $u(x) = \frac{\mathrm{d}v}{\mathrm{d}x}(x)$ we get

$$\int \xi^2 |\hat{v}(\xi)|^2 \; \mathrm{d}\xi = \int |v'(\xi)|^2 \; \mathrm{d}\xi.$$

3.   a) With $H_0' = 0$, $H_n' = 2nH_{n-1}$, $H_n(x) = 2xH_{n-1}(x) - 2(n-1)H_{n-2}(x)$ and $H_{n+1}(x) = 2xH_n(x) - 2nH_{n-1}(x)$ which we take from Appendix III, we find for $h_n(x) = e^{-\frac{x^2}{2}}$

$$h_n'(x) = -xe^{-\frac{x^2}{2}} H_n(x) + e^{-\frac{x^2}{2}} H_n'(x) = -xh_n(x) + 2nh_{n-1}(x)$$
$$h_{n+1}(x) = 2xh_n(x) - 2nh_{n-1}(x) = 2xh_n(x) - (xh_n(x) + h_n'(x)) = xh_n(x) - h_n'(x)$$

and starting with $xh_n(x) + h_n'(x) = 2nh_{n-1}(x)$ we find using $h_{n-1}'(x) = xh_{n-1}(x) - h_n(x)$ that

$$h_n(x) + xh_n'(x) + h_n''(x) = 2nh_{n-1}'(x)$$
$$= 2n(xh_{n-1}(x) - h_n(x))$$
$$= x(xh_n(x) + h_n'(x)) - 2nh_n(x)$$
$$= x^2 h_n(x) + xh_n'(x) - 2nh_n(x)$$

implying $h_n''(x) - x^2 h_n(x) + (2n + 1)h_n(x) = 0$.
     b) For $n \neq m$ we find

$$\int_{\mathbb{R}} h_n(x) h_m(x) \; \mathrm{d}x = \int_{\mathbb{R}} H_n(x) e^{-\frac{x^2}{2}} H_m(x) e^{-\frac{x^2}{2}} \; \mathrm{d}x$$

$$= \int_{\mathbb{R}} H_n(x) H_m(x) e^{-x^2} \; \mathrm{d}x = 0$$

where we use the orthogonality of the Hermite polynomails in $L^2(\mathbb{R}, \mu), \mu = m\lambda^{(1)}$, $m(x) = e^{-x^2}$.

621

c) Obviously we have $h_n \in S(\mathbb{R})$ since $x \mapsto e^{-\frac{x^2}{2}}$ belongs to $S(\mathbb{R})$. Since $h_0(x) = e^{-\frac{x^2}{2}}$ we have $h_0 = (-i)^0 \hat{h}_0$ as we know that h_0 is a fixed point of the Fourier transform. Now suppose that $\hat{h}_n = (-i)^n h_n$. We take the Fourier transform in the equality $x h_n(x) - h'_n(x) = h_{n+1}(x)$, see part a), and we find

$$F(x h_n) - F(h'_n) = F h_{n+1}$$

or

$$i(\hat{h}_n)' + i\xi \hat{h}_n = \hat{h}_{n+1},$$

and by our assumption it follows that

$$i(-i)^n h'_n + i(-i)^n \xi h_n = \hat{h}_{n+1},$$

i.e. $-i(-i)^n(\xi h_n - h'_n) = \hat{h}_{n+1}$

which yields

$$\hat{h}_{n+1} = (-i)^{n+1} h_{n+1}.$$

In the next Chapter we will see that the equality $\hat{h}_n = (-i)^n h_n$ extends $L^2(\mathbb{R})$ - the non-trivial part is to extend the Fourier transform to $L^2(\mathbb{R}^n)$ - and in Volume V we will interpret this result as an eigenvalue equation for the unitary operator F in $L^2(\mathbb{R})$.

4. Since $\Lambda^s : \mathbb{R}^n \to \mathbb{R}, \Lambda^s(\xi) = (1 + |\xi|^2)^{\frac{s}{2}}$, belongs to $O_M(\mathbb{R}^n)$ it follows that for every $s \in \mathbb{R}$ and $u \in S(\mathbb{R}^n)$ the function $\Lambda^s(\cdot)\hat{u}(\cdot)$ belongs to $S(\mathbb{R}^n)$ and therefore $\Lambda^s(D)u = F^{-1}(\Lambda^s(\cdot)\hat{u}(\cdot)) \in S(\mathbb{R}^n)$. Moreover, in light of Problem 5 in Chapter 9, it is clear that $\Lambda^s(D)$ is continuous. In particular we have $(\Lambda^s(D)u)\hat{} = \Lambda^s(\cdot)\hat{u}(\cdot)$ and therefore we find

$$\Lambda^{-s}(D)(\Lambda^s(D)u) = F^{-1}(\Lambda^{-s}(\cdot)F(\Lambda^s(D)u))$$
$$= F^{-1}(\Lambda^{-s}(\cdot)\Lambda^s(\cdot)\hat{u}) = u$$

and since $s \in \mathbb{R}$ was arbitrary we arrive on $S(\mathbb{R}^n)$ to the equalities $\Lambda^{-s}(D) \circ \Lambda^s(D) = id = \Lambda^s(D) \circ \Lambda^{-s}(D)$ where id is the identity on $S(\mathbb{R}^n)$. For $s, t \in \mathbb{R}$ we get on $S(\mathbb{R}^n)$

$$(\Lambda^s(D) \circ \Lambda^t(D))u = \Lambda^s(D)(\Lambda^t(\cdot)\hat{u})$$
$$= F^{-1}(\Lambda^s(\cdot)\Lambda^t(\cdot)\hat{u})$$
$$= F^{-1}(\Lambda^{s+t}(\cdot)\hat{u}) = \Lambda^{s+t}(D)u.$$

and since $s + t = t + s$ it follows that $\Lambda^s(D) \circ \Lambda^t(D) = \Lambda^t(D) \circ \Lambda^s(D)$. Thus with the composition of operators on $S(\mathbb{R}^n)$ we see that $\{\Lambda^s(D) : S(\mathbb{R}^n) \to S(\mathbb{R}^n) \mid s \in \mathbb{R}\}$ is an Abelian group.

5. First we note that $\frac{\hat{f}}{(1+\|\cdot\|^2)^k}$ belongs to $\mathcal{S}(\mathbb{R}^n)$ since f does. Now we observe that

$$
\begin{aligned}
(1 - \Delta_n)^k u &= F^{-1}(F((1 - \Delta_n)^k u)) \\
&= F^{-1}((1 + \|\cdot\|^2)^k \hat{u}) \\
&= F^{-1}\left((1 + \|\cdot\|^2)^k \frac{\hat{f}}{(1 + \|\cdot\|^2)^k}\right) \\
&= F^{-1}(\hat{f}) = f.
\end{aligned}
$$

6. First we note that $u * v, P(D)u, P(D)v$ and therefore $(P(D)u) * v$ and $u * P(D)v$ all belong to $\mathcal{S}(\mathbb{R}^n)$. For the Fourier transforms we find

$$
\begin{aligned}
(P(D)(u * v))\hat{}(\xi) &= (2\pi)^{\frac{n}{2}} P(\xi)\hat{u}(\xi)\hat{v}(\xi) \\
&= (2\pi)^{\frac{n}{2}} (P(\xi)\hat{u}(\xi))\hat{v}(\xi) \\
&= (2\pi)^{\frac{n}{2}} (P(D)u)\hat{} \cdot \hat{v}(\xi) \\
&= ((P(D)u) * v)\hat{}(\xi)
\end{aligned}
$$

implying

$$
P(D)(u * v) = (P(D)u) * v
$$

and analogously we find

$$
\begin{aligned}
P(D)(u * v)\hat{}(\xi) &= (2\pi)^{\frac{n}{2}} \hat{u}(P(D)v)\hat{}(\xi) \\
&= (u * P(D)v)\hat{}(\xi)
\end{aligned}
$$

which yields

$$
P(D)(u * v) = u * P(D)v.
$$

7. For $p \geq 2$ we have

$$
\begin{aligned}
\int_{\mathbb{R}^n} |\hat{u}(\xi)|^p \, d\xi &\leq \left(\sup_{\xi \in \mathbb{R}^n} |\hat{u}(\xi)|\right)^{p-2} \int_{\mathbb{R}^n} |\hat{u}(\xi)|^2 \, d\xi \\
&= \|\hat{u}\|_\infty^{p-2} \|\hat{u}\|_{L^2}^2 \\
&\leq (2\pi)^{\frac{(2-p)n}{2}} \|u\|_{L^1}^{p-2} \|u\|_{L^2}^2
\end{aligned}
$$

implying

$$
\|\hat{u}\|_{L^p} \leq (2\pi)^{\frac{(2-p)n}{2p}} \|u\|_{L^1}^{1-\frac{2}{p}} \|u\|_{L^2}^{\frac{2}{p}}.
$$

8. Taking the Fourier transform and noting that $x \mapsto e^{-\frac{x^2}{2}}$ is a fixed point of the Fourier transform whereas the Fourier tranform of $x \mapsto e^{-c|x|^2}$ is given by $y \mapsto \frac{1}{\sqrt{4c\pi}} e^{-\frac{|y|^2}{4c}}$, the convolution theorem yields

$$
F(e^{-\frac{|\cdot|^2}{2}} * u)(\xi) = \sqrt{2\pi} e^{-\frac{|\xi|^2}{2}} \hat{u}(\xi) = \frac{1}{\sqrt{4c\pi}} e^{-\frac{|\xi|^2}{4c}}
$$

which gives

$$\hat{u}(\xi) = \frac{1}{2\pi\sqrt{2c}} e^{-\frac{|\xi|^2}{2}(\frac{1}{2c}-1)}.$$

In order to assure $\hat{u} \in \mathcal{S}(\mathbb{R}^n)$ we need $\frac{1}{2c} - 1 > 0$ or $0 < c < \frac{1}{2}$. In this case, u is a solution of the given integral equation.

9.   a) For f even we obtain

$$(Ff)(\xi) = \frac{1}{\sqrt{2\pi}} \int_{\mathbb{R}} f(x)(\cos x\xi - i\sin x\xi) \, dx$$

$$= \frac{1}{\sqrt{2\pi}} \int_{\mathbb{R}} f(x)\cos x\xi \, dx - \frac{i}{\sqrt{2\pi}} \int_{\mathbb{R}} f(x)\sin x\xi \, dx$$

and since the integrand in the second integral is an odd function it follows that $(Ff)(\xi) = (F_{\cos}f)(\xi)$. Analogously, for g odd we find

$$(Fg)(\xi) = \frac{1}{\sqrt{2\pi}} \int_{\mathbb{R}} g(x)\cos x\xi \, dx - \frac{i}{\sqrt{2\pi}} \int_{\mathbb{R}} g(x)\sin x\xi \, dx$$

$$= -\frac{i}{\sqrt{2\pi}} \int_{\mathbb{R}} g(x)\sin x\xi \, d\xi = -i(F_{\sin}g)(\xi)$$

where we used now that $x \mapsto g(x)\cos x\xi$ is odd.

b) Integration by parts gives

$$F_{\cos}(f')(\xi) = \sqrt{\frac{2}{\pi}} \int_0^\infty f'(x)\cos x\xi \, dx$$

$$= \sqrt{\frac{2}{\pi}} f(x)\cos x\xi \Big|_0^\infty - \sqrt{\frac{2}{\pi}} \int_0^\infty f(x)\left(\frac{d}{dx}\cos x\xi\right) dx$$

$$= -\sqrt{\frac{2}{\pi}} \int_0^\infty f(x)(-\xi\sin x\xi) \, dx - \sqrt{\frac{2}{\pi}} f(0)$$

$$= \xi F_{\sin}(f)(\xi) - \sqrt{\frac{2}{\pi}} f(0)$$

as well as

$$(F_{\sin}f')(\xi) = \sqrt{\frac{2}{\pi}} \int_0^\infty f'(x)\sin x\xi \, dx$$

$$= \sqrt{\frac{2}{\pi}} f(x)\sin x\xi \Big|_0^\infty - \sqrt{\frac{2}{\pi}} \int_0^\infty f(x)\xi\cos x\xi \, dx$$

$$= -\xi(F_{\cos}f)(\xi).$$

10. We follow [102], Lemma 7.21. The complex one-dimensional case is well-known, see Theorem III.22.12. We denote by $A(k), 1 \le k \le n$, the following statement: If $z \in \mathbb{C}^n$ has at least k real coordinates then $f(z) = 0$. The assumption f vanishes on $\mathbb{R}^n$ is just the statement $A(n)$. We want to prove $A(0)$, i.e. $f(z) = 0$ for all $z \in \mathbb{C}^n$.

Suppose that $A(l)$, $1 \leq l \leq n$, is true and consider g_l in (11.28). Choose $a_1, \ldots, a_l$ to be real. This implies that $z_l \mapsto g_l(z_l)$ vanishes on the real line, hence by Theorem III.22.12 it vanishes identically. Therefore $A(l-1)$ is true and the result follows.

Chapter 12

1.   a) For $\xi \neq 0$ we find

$$\hat{f}_M(\xi) = \frac{1}{\sqrt{2\pi}} \int_{\mathbb{R}} e^{-ix\xi} \chi_{[-M,M]}(x) \, dx$$

$$= \frac{1}{\sqrt{2\pi}} \int_{-M}^{M} e^{-ix\xi} \, dx = \frac{2}{\sqrt{2\pi}} \int_{0}^{M} \cos x\xi \, dx$$

$$= \sqrt{\frac{2}{\pi}} \frac{\sin M\xi}{\xi},$$

whereas for $\xi = 0$ it follows that

$$\hat{f}_M(0) = \frac{1}{\sqrt{2\pi}} \int_{-M}^{M} 1 \, dx = \frac{2M}{\sqrt{2\pi}} = \sqrt{\frac{2}{\pi}} M,$$

and we note that $\lim_{\xi \to 0} \hat{f}_M(\xi) = \hat{f}_M(0)$.

b) Let $\xi \neq 0$. Then we have

$$\hat{g}_R(\xi) = \frac{1}{\sqrt{2\pi}} \int_{\mathbb{R}} e^{-ix\xi} \chi_{[-R,R]}(x) \left(1 - \frac{|x|}{R}\right) \, dx$$

$$= \frac{1}{\sqrt{2\pi}} \int_{-R}^{R} e^{-ix\xi} \left(1 - \frac{|x|}{R}\right) \, dx$$

$$= \sqrt{\frac{2}{\pi}} \int_{0}^{R} \left(1 - \frac{x}{R}\right) \cos x\xi \, dx = \sqrt{\frac{2}{\pi}} \frac{1 - \cos R\xi}{R\xi^2}$$

$$= \frac{R}{\sqrt{2\pi}} \left(\frac{\sin \frac{R\xi}{2}}{\frac{R\xi}{2}}\right)^2.$$

For $\xi = 0$ it follows that

$$\hat{g}_R(0) = \frac{R}{\sqrt{2\pi}}$$

and we have again $\lim_{\xi \to 0} \hat{g}_R(\xi) = \hat{g}_R(0)$.

c) We have

$$\hat{h}_a(\xi) = \frac{1}{\sqrt{2\pi}} \int_{\mathbb{R}} e^{-ix\xi} e^{-a|x|} \, dx$$

$$= \sqrt{\frac{2}{\pi}} \int_{0}^{\infty} \cos x\xi e^{-ax} \, dx = \sqrt{\frac{2}{\pi}} \frac{a}{a^2 + \xi^2}.$$

d) A detailed calculation of $\hat{k}_t$ was given in Example III.25.29 using the residue theorem. The result is

$$\hat{k}_t(\xi) = \sqrt{\frac{\pi}{2}} \frac{e^{-t|\xi|}}{t}.$$

Switching from k_t to $\sqrt{\frac{2}{\pi}} t k_t$ we find $\left(\sqrt{\frac{2}{\pi}} t k_t\right)^{\widehat{}}(\xi) = e^{-t|\xi|}$ as we do expect in light of part c).

2. The function $x \mapsto \frac{1}{a^2+x^2}$, $a > 0$, belongs to L^2 and therefore, by Plancherel's theorem we have by Problem 1 d),

$$\int_{\mathbb{R}} \left(\frac{1}{a^2+x^2} \cdot \frac{1}{b^2+x^2}\right) dx = \int_{\mathbb{R}} F\left(\frac{1}{a^2+x^2}\right)(\xi)\overline{F\left(\frac{1}{b^2+x^2}\right)(\xi)} \, d\xi$$

$$= \int_{\mathbb{R}} \sqrt{\frac{\pi}{2}} \frac{e^{-a|\xi|}}{a} \sqrt{\frac{\pi}{2}} \frac{e^{-b|\xi|}}{b} \, d\xi$$

$$= \frac{1}{2ab} \int_{\mathbb{R}} e^{-(a+b)|\xi|} \, d\xi = \frac{\pi}{ab} \int_0^\infty e^{-(a+b)\xi} \, d\xi$$

$$= \frac{\pi}{ab(a+b)}.$$

In particular we find that $\int_{\mathbb{R}} \frac{1}{(a^2+x^2)^2} \, dx = \frac{\pi}{2a^3}$.

3. By the convolution theorem we have

$$(K_t * K_s)^{\widehat{}}(\xi) = \sqrt{2\pi} \hat{K}_t(\xi) \hat{K}_s(\xi)$$

and by Problem 1 d) we know that $\hat{K}_t(\xi) = e^{-t|\xi|}$ which yields $(K_t * K_s)^{\widehat{}}(\xi) = \sqrt{2\pi} e^{-(t+s)|\xi|}$. Taking now the inverse Fourier transform and noting that $\xi \mapsto e^{-r|\xi|}$ is an even function, we find with the help of Problem 1 c) that

$$(K_t * K_s)(x) = \sqrt{2\pi} \cdot \sqrt{\frac{2}{\pi}} \frac{t+s}{(t+s)^2+x^2} = 2\frac{t+s}{(t+s)^2+x^2}$$

$$= 2\sqrt{\frac{\pi}{2}}(t+s)k_{t+s}(x) = \sqrt{2\pi} K_{t+s}(x).$$

4.   a) From Problem 1 d) (or Example III.25.29) we know that

$$e^{-\alpha} = \frac{2}{\pi} \int_0^\infty \frac{\cos \alpha x}{1+x^2} \, dx$$

and it follows further that

$$e^{-\alpha} = \frac{2}{\pi} \int_0^\infty \cos\alpha x \left(\int_0^\infty e^{-(1+x^2)s} \, ds \right) dx$$

$$= \frac{2}{\pi} \int_0^\infty e^{-s} \left(\int_0^\infty e^{-sx^2} \cos\alpha x \, dx \right) ds$$

$$= \frac{2}{\pi} \int_0^\infty e^{-s} \left(\frac{1}{2} \int_{\mathbb{R}} e^{-sx^2} e^{i\alpha x} \, dx \right) ds$$

$$= \int_0^\infty e^{-s} \sqrt{\frac{\pi}{s}} e^{-\frac{\alpha^2}{4s}} \, ds = \frac{1}{\sqrt{\pi}} \int_0^\infty \frac{e^{-s}}{\sqrt{s}} e^{-\frac{\alpha^2}{4s}} \, ds.$$

b) We have with the help of part a)

$$\frac{1}{(2\pi)^{\frac{n}{2}}} \int_{\mathbb{R}^n} e^{-ix\cdot\xi} e^{-\|x\|} \, dx = (2\pi)^{-\frac{n}{2}} \left(\int_{\mathbb{R}^n} \frac{1}{\sqrt{\pi}} \int_0^\infty \frac{e^{-s}}{\sqrt{s}} e^{-\frac{\|x\|^2}{4s}} \, ds \right) e^{-ix\cdot\xi} \, dx$$

$$= (2\pi)^{-\frac{n}{2}} \frac{1}{\sqrt{\pi}} \int_0^\infty \frac{e^{-s}}{\sqrt{s}} \left(\int_{\mathbb{R}^n} e^{-\frac{\|x\|^2}{4s}} e^{-ix\cdot\xi} \, dx \right) ds$$

$$= (2\pi)^{-\frac{n}{2}} \frac{1}{\sqrt{\pi}} \int_0^\infty \frac{e^{-s}}{\sqrt{s}} (2\pi)^{\frac{n}{2}} s^{\frac{n}{2}} e^{-s\|\xi\|^2} \, ds$$

$$= \frac{1}{\sqrt{\pi}} \int_0^\infty e^{-s} s^{\frac{n-1}{2}} e^{-s\|\xi\|^2} \, ds$$

$$= \frac{1}{\sqrt{\pi}} \int_0^\infty e^{-(1+\|\xi\|^2)s} s^{\frac{n-1}{2}} \, ds$$

$$= \frac{1}{\sqrt{\pi}} \frac{1}{(1+\|\xi\|^2)^{\frac{n+1}{2}}} \int_0^\infty e^{-r} r^{\frac{n-1}{2}} \, dr$$

$$= \frac{\Gamma(\frac{n+1}{2})}{\sqrt{\pi}} \frac{1}{(1+\|\xi\|^2)^{\frac{n+1}{2}}}$$

or, helpful for later purposes,

$$\int_{\mathbb{R}^n} e^{-ix\cdot\xi} e^{-\|x\|} \, dx = \frac{(2\pi)^{\frac{n}{2}} \Gamma(\frac{n+1}{2})}{\sqrt{\pi}} \frac{1}{(1+\|\xi\|^2)^{\frac{n+1}{2}}}.$$

c) The substitution $tx = y$ yields

$$\frac{1}{(2\pi)^{\frac{n}{2}}} \int_{\mathbb{R}^n} e^{-ix\cdot\xi} e^{-t\|x\|} \, dx = \frac{1}{(2\pi)^{\frac{n}{2}}} \frac{1}{t^n} \int_{\mathbb{R}^n} e^{-i\frac{y}{t}\cdot\xi} e^{-\|y\|} \, dy$$

$$= \frac{1}{t^n} \frac{\Gamma(\frac{n+1}{2})}{\sqrt{\pi}} \frac{1}{(1+\|\frac{\xi}{t}\|^2)^{\frac{n+1}{2}}}$$

$$= \frac{\Gamma(\frac{n+1}{2})}{\sqrt{\pi}} \frac{t}{(t^2+\|\xi\|^2)^{\frac{n+1}{2}}}.$$

5.　　a) By the transformation theorem we find $\int_{\mathbb{R}} h_\lambda(t)\, dt = 1$. For $\epsilon > 0$ there exists $\delta > 0$ such that $|t| < \delta$ implies that $|f(t_0 - t) - f(t_0)| < \frac{\epsilon}{2}$ and it follows that

$$
(f * h_\lambda)(t_0) - f(t_0) = \int_{\mathbb{R}} (f(t_0 - t) - f(t_0)) h_\lambda(t)\, dt
$$

$$
= \int_{-\delta}^{\delta} (f(t_0 - t) - f(t_0)) h_\lambda(t)\, dt + \int_{(-\delta,\delta)^c} (f(t_0 - t) - f(t_0)) h_\lambda(t)\, dt
$$

$$
\leq \frac{\epsilon}{2} \int_{-\delta}^{\delta} |h_\lambda(t)|\, dt + 2\|f\|_\infty \int_{(-\delta,\delta)^c} |h_\lambda(t)|\, dt
$$

$$
\leq \frac{\epsilon}{2} \|h\|_{L^1} + 2\|f\|_\infty \int_{(-\delta,\delta)^c} |h_\lambda(t)|\, dt.
$$

We note that

$$
\int_{(-\delta,\delta)^c} |h_\lambda(t)|\, dt = \int_{(-\delta\lambda,\delta\lambda)^c} |h_\lambda(t)|\, dt,
$$

and therefore, given $\epsilon > 0$ we can find $\lambda > 0$ such that $\delta\lambda > 0$ implies

$$
\int_{(-\delta\lambda,\delta\lambda)^c} |h_\lambda(t)|\, dt < \frac{\epsilon}{4\|f\|_\infty}.
$$

Thus, for $\epsilon > 0$ given and δ as well as λ being chosen as above it follows that

$$
|(f * h_\lambda)(t_0) - f(t_0)| < \epsilon.
$$

In the case that f is uniformly continuous we obtain the last estimate uniformly in t_0, i.e. $\|f * h_\lambda - f\|_\infty \leq \epsilon$ and the uniform convergence of $f * h_\lambda$ to f on $\mathbb{R}$ follows.

　　b) It follows that

$$
\frac{1}{\sqrt{2\pi}} \int_{-R}^{R} e^{ixt_0} \left(1 - \frac{|x|}{R}\right) \hat{f}(x)\, dx = \frac{1}{2\pi} \int_{-R}^{R} e^{ixt_0} \left(1 - \frac{|x|}{R}\right) \left(\int_{\mathbb{R}} e^{-ixr} f(r)\, dr\right) dx
$$

$$
= \frac{1}{2\pi} \int_{\mathbb{R}} f(r) \left(\int_{-R}^{R} e^{-ix(r-t_0)} \left(1 - \frac{|x|}{R}\right) dx\right) dr
$$

$$
= \frac{1}{2\pi} \int_{\mathbb{R}} f(r) \left(\left(1 - \frac{|\cdot|}{R}\right) \chi_{[-R,R]}(\cdot)\right)^{\widehat{}}(r - t_0)\, dr.
$$

From Problem 1 b) we know that

$$
\left(\left(1 - \frac{|\cdot|}{R}\right) \chi_{[-R,R]}(\cdot)\right)^{\widehat{}}(s) = \frac{2}{\sqrt{2\pi}} \frac{(1 - \cos Rs)}{Rs^2}
$$

implying, with $h(s) = \frac{2}{\sqrt{2\pi} s^2}(1 - \cos s)$ and as before $h_\lambda(s) = \lambda h(\lambda s)$, that

$$
\frac{1}{\sqrt{2\pi}} \int_{-R}^{R} e^{-ixt_0} \left(1 - \frac{|x|}{R}\right) \hat{f}(x)\, dx = \frac{1}{\sqrt{2\pi}} \int_{\mathbb{R}} f(r) h_R(t_0 - r)\, dr.
$$

Now we want to apply part a) for which we need that $\int_{\mathbb{R}} h(s)\,ds = \sqrt{2\pi}$. However

$$\int_{\mathbb{R}} h(s)\,ds = \int_{\mathbb{R}} \frac{2}{\sqrt{2\pi}} \frac{1 - \cos s}{s^2}\,ds$$

$$= \frac{4}{\sqrt{2\pi}} \int_0^\infty \frac{1 - \cos s}{s^2}\,ds$$

$$= \frac{4}{\sqrt{2\pi}} \int_0^\infty \frac{\sin s}{s}\,ds = \frac{4}{\sqrt{2\pi}} \frac{\pi}{2} = \sqrt{2\pi}.$$

Again, in the case that f is uniformly continuous, in light of part a) we may deduce that the convergence is uniform in t_0.

c) If $f \in A(\mathbb{R})$ then $f \in C_\infty(\mathbb{R}) \cap L^1(\mathbb{R})$, in particular f is uniformly continuous and belongs to $L^1(\mathbb{R}) \cap L^\infty(\mathbb{R})$. Thus the statement of part b) holds for all $f \in A(\mathbb{R})$ and it gives a way to calculate $F^{-1}u$, $u \in A(\mathbb{R})$.

6. The missing detail is the passing from $\mathcal{S}(\mathbb{R}^n)$ to $L^1(\mathbb{R}^n)$, but we can rely on $\lim_{\epsilon \to 0} \|(2\pi)^{-\frac{n}{2}} u * \varphi - u\|_{L^1}$ for all $u \in \mathcal{S}(\mathbb{R}^n)$. We note that for $(u_k)_{k \in \mathbb{N}}$, $u_k \in \mathcal{S}(\mathbb{R}^n)$, such that $\lim_{k \to \infty} \|u_k - v\|_{L^1} = 0$ it follows that

$$\|u_k * \varphi - v * \varphi\|_{L^1} = \|(u_k - v) * \varphi\|_{L^1} \leq \|u_k - v\|_{L^1} \|\varphi\|_1,$$

so

$$\lim_{k \to \infty} \|u_k * \varphi - v * \varphi\|_{L^1} = 0.$$

Now we observe

$$\|M_\epsilon(v) - v\|_{L^1} = \|(2\pi)^{-\frac{n}{2}} v * \varphi - v\|_{L^1}$$

$$\leq \|(2\pi)^{-\frac{n}{2}} (v - u_k) * \varphi\|_{L^1} + \|(2\pi)^{-\frac{n}{2}} u_k * \varphi_k - u_k\|_{L^1} + \|u_k - v\|_{L^1},$$

implying that $\lim_{\epsilon \to 0} \|M_\epsilon(v) - v\|_{L^1} = 0$ for all $v \in L^1(\mathbb{R}^n)$.

7. The proof is again a density argument. We know for $u \in \mathcal{S}(\mathbb{R}^n)$ that $T_t \circ T_s = T_t(T_s u)$ or

$$h_{t+s} * u = h_t * (h_s * u).$$

Let $v \in L^2(\mathbb{R}^n)$ and $(u_k)_{k \in \mathbb{N}}$, $u_k \in \mathcal{S}(\mathbb{R}^n)$, be a sequence converging in $L^2(\mathbb{R})$ to v. It follows that for $r > 0$ we have $\lim_{r \to \infty} \|h_r * u_k - h_r v\|_{L^2} = 0$, for which we need only to note that $\|h_r * (u_k - v)\|_{L^2} \leq \|h_r\|_{L^1} \|u_k - v\|_{L^2}$. Now we find in the sense of L^2-limits that

$$h_{t+s} * v = h_{t+s} * \lim_{k \to \infty} u_k$$

$$= \lim_{k \to \infty} h_{t+s} * u_k = \lim_{k \to \infty} h_t * (h_s * u_k)$$

$$= h_t * \lim_{k \to \infty} (h_s * u_k) = h_t * h_s * \left(\lim_{k \to \infty} u_k \right)$$

$$= h_t * h_s * v,$$

and we have proved for all $t, s > 0$ and $v \in L^2(\mathbb{R}^n)$ that

$$T_{t+s} v = T_t(T_s v).$$

8. Since $(1 + \|\xi\|^2)^s \sim 1 + \|\xi\|^{2s}$ we have

$$\|u\|^2_{H^s} \sim \int_{\mathbb{R}^n} (1 + \|\xi\|^{2s}) |\hat{u}(\xi)|^2 \, d\xi$$

$$= \|u\|^2_{L^2} + \int_{\mathbb{R}^n} \|\xi\|^{2s} |\hat{u}(\xi)|^2 \, d\xi$$

and therefore it is sufficient to prove that

$$\int_{\mathbb{R}^n} \|\xi\|^{2s} |\hat{u}(\xi)|^2 \, d\xi = c_{s,n} \int_{\mathbb{R}^n} \int_{\mathbb{R}^n} \frac{|u(x) - u(y)|^2}{\|x - y\|^{n+2s}} \, dx \, dy$$

holds for all $u \in \mathcal{S}(\mathbb{R}^n)$. We have

$$u(x + z) - u(x) = (2\pi)^{-n} \int_{\mathbb{R}^n} e^{ix \cdot \xi} (e^{iz \cdot \xi} - 1) \hat{u}(\xi) \, d\xi$$

and by Plancherel's theorem it follows that

$$\int_{\mathbb{R}^n} |u(x + z) - u(x)|^2 \, dx = \int_{\mathbb{R}^n} (e^{iz \cdot \xi} - 1) |\hat{u}(\xi)|^2 \, d\xi$$

implying

$$\int_{\mathbb{R}^n} \int_{\mathbb{R}^n} \frac{|u(x + z) - u(x)|^2}{\|z\|^{n+2s}} \, dx \, dz = \int_{\mathbb{R}^n} \|\xi\|^{2s} |\hat{u}(\xi)|^2 \left(\int_{\mathbb{R}^n} \frac{|e^{iz \cdot \xi} - 1|^2}{\|\xi\|^{2s} \|z\|^{n+2s}} \, dz \right) d\xi.$$

We claim that the integral

$$I := \int_{\mathbb{R}^n} \frac{|e^{iz \cdot \xi} - 1|^2}{\|\xi\|^{2s} \|z\|^{n+2s}} \, dz$$

is independent of ξ. Once this is proved, the substitution $z \mapsto y - x$ yields

$$\int_{\mathbb{R}^n} \int_{\mathbb{R}^n} \frac{|u(y) - u(x)|^2}{\|y - x\|^{n+2s}} \, dx \, dy = \tilde{c}_{n,s} \int_{\mathbb{R}^n} \|\xi\|^{2s} |\hat{u}(\xi)|^2 \, d\xi.$$

Now we prove that I is independent of ξ. We write $\xi = \|\xi\| \eta$ and the substitiution $w := z\|\xi\|$ gives

$$\int_{\mathbb{R}^n} \frac{|e^{iz \cdot \xi} - 1|^2}{\|\xi\|^{2s} \|z\|^{n+2s}} \, dz = \int_{\mathbb{R}^n} \frac{|e^{iw \cdot \eta} - 1|^2}{\|w\|^{n+2s}} \, dw.$$

Next we choose a rotation such that $R\eta = e_1$, note that η and R depend on ξ, but the unit vector e_1 does not. We find

$$\int_{\mathbb{R}^n} \frac{|e^{iw\eta} - 1|^2}{\|w\|^{n+2s}} \, dw = \int_{\mathbb{R}^n} \frac{|e^{i(R^{-1}w) \cdot e_1} - 1|^2}{\|R^{-1}w\|^{n+2s}} \, d(R^{-1}w),$$

where we have used that $|\det R| = 1$. Thus we find

$$I = \int_{\mathbb{R}^n} \frac{|e^{iy \cdot e_1} - 1|^2}{\|y\|^{n+2s}} \, dy$$

which is indeed independent of ξ. Note that by $0 < s < 1$ the integral is well defined.

9. We write $1 - \cos \xi \cdot y = \frac{(e^{iy \cdot \xi} - 1)(e^{-iy \cdot \xi} - 1)}{2}$ and we find for $u \in \mathcal{S}(\mathbb{R}^n)$ that

$$\int_{\mathbb{R}^n} |\hat{u}(\xi)|^2 \psi(\xi) \, d\xi = \int_{\mathbb{R}^n} \left(\int_{\mathbb{R}^n} (1 - \cos \xi \cdot y) \hat{u}(\xi) \overline{\hat{u}(\xi)} \nu(y) \, dy \right) d\xi$$

$$= \frac{1}{2} \int_{\mathbb{R}^n} \int_{\mathbb{R}^n} (e^{iy \cdot \xi} - 1)(e^{-iy \cdot \xi} - 1) \hat{u}(\xi) \hat{u}(-\xi) \, d\xi \, \nu(y) \, dy$$

$$= \frac{(2\pi)^{-n}}{2} \int_{\mathbb{R}^n} \int_{\mathbb{R}^n} \left(\int_{\mathbb{R}^n} \int_{\mathbb{R}^n} e^{i\sigma \cdot x}(e^{iy \cdot \sigma} - 1)\hat{u}(\sigma) \right.$$

$$\left. e^{i\tau \cdot x}(e^{iy \cdot \tau} - 1)\hat{u}(\tau) \, d\sigma \, d\tau \right) \nu(y) \, dy \, dx$$

$$= \frac{1}{2} \int_{\mathbb{R}^n} \int_{\mathbb{R}^n} (u(x + y) - u(x))(u(x + y) - u(x)) \nu(y) \, dy \, dx$$

$$= \frac{1}{2} \int_{\mathbb{R}^n} \int_{\mathbb{R}^n} |u(x) - u(x + y)|^2 \nu(y) \, dy \, dx$$

and the substitution $x + y \mapsto y$ yields

$$\int_{\mathbb{R}^n} |\hat{u}(\xi)|^2 \psi(\xi) \, d\xi = \frac{1}{2} \int_{\mathbb{R}^n} \int_{\mathbb{R}^n} |u(x) - u(y)|^2 \nu(x - y) \, dy \, dx.$$

10.   a) We note that

$$|\Lambda^s(\xi) - \Lambda^s(\eta)| = |(1 + \|\xi\|^2)^{\frac{s}{2}} - (1 + \|\eta\|^2)^{\frac{s}{2}}|$$

$$= s \left| \int_{(1+\|\eta\|^2)^{\frac{1}{2}}}^{(1+\|\xi\|^2)^{\frac{1}{2}}} t^{s-1} \, dt \right|$$

$$\leq s((1 + \|\xi\|^2)^{\frac{s-1}{2}} + (1 + \|\eta\|^2)^{\frac{s-1}{2}})|(1 + \|\xi\|^2)^{\frac{1}{2}} - (1 + \|\eta\|^2)^{\frac{1}{2}}|$$

$$= s \left(\frac{(1 + \|\xi\|^2)^{\frac{s-1}{2}}}{(1 + \|\eta\|^2)^{\frac{s-1}{2}}} + 1 \right) (1 + \|\eta\|^2)^{\frac{s-1}{2}} |(1 + \|\xi\|^2)^{\frac{1}{2}} - (1 + \|\eta\|^2)^{\frac{1}{2}}|$$

$$\leq s 2^{\frac{s-1}{2}} ((1 + \|\xi - \eta\|^2)^{\frac{s-1}{2}} + 1)(1 + \|\eta\|^2)^{\frac{s-1}{2}} (1 + \|\zeta\|^2)^{-\frac{1}{2}} |\xi - \eta|$$

where ζ is a point on the line segment connecting ξ with η. Thus we obtain

$$|\Lambda^s(\xi) - \Lambda^s(\eta)| \leq c_s (1 + \|\xi - \eta\|^2)^{\frac{s+1}{2}} (1 + \|\eta\|^2)^{\frac{s-1}{2}}$$

$$= c_s (1 + \|\xi - \eta\|^2)^{\frac{s+1}{2}} \Lambda^{s-1}(\eta).$$

b) Taking the Fourier transform we find

$$([\Lambda^s(D), \varphi]u)\hat{}(\xi) = (\Lambda^s(D)(\varphi u))\hat{}(\xi) - (\varphi \Lambda^s(D)u)\hat{}(\xi)$$

$$= \Lambda^s(\xi)(\varphi u)\hat{}(\xi) - (\varphi \Lambda^s(D)u)\hat{}(\xi)$$

$$= \Lambda^s(\xi)(2\pi)^{-\frac{n}{2}} \int_{\mathbb{R}^n} \hat{\varphi}(\xi - \eta)\hat{u}(\eta) \, d\eta - (2\pi)^{-\frac{n}{2}} \int_{\mathbb{R}^n} \hat{\varphi}(\eta - \xi)\Lambda^s(\eta)\hat{u}(\eta) \, d\eta$$

$$= (2\pi)^{-\frac{n}{2}} \int_{\mathbb{R}^n} \hat{\varphi}(\eta - \xi)(\Lambda^s(\xi) - \Lambda^s(\eta))\hat{u}(\eta) d\eta.$$

631

Now we find with part a)

$$|(\Lambda^t(\xi)[\Lambda^s(D),\varphi]u)\hat{}(\xi)| \le (2\pi)^{-\frac{n}{2}} \int_{\mathbb{R}^n} \Lambda^t(\xi)|\hat{\varphi}(\eta-\xi)||\Lambda^s(\xi)-\Lambda^s(\eta)||\hat{u}(\eta)| \, d\eta$$

$$\le \tilde{c} \int_{\mathbb{R}^n} \Lambda^t(\xi)|\hat{\varphi}(\eta-\xi)|(1+\|\xi-\eta\|^2)^{\frac{s+1}{2}}\Lambda^{s-1}(\eta)|\hat{u}(\eta)| \, d\eta$$

$$\le c' \int_{\mathbb{R}^n} \Lambda^t(\xi-\eta)|\hat{\varphi}(\eta-\xi)|(1+\|\xi-\eta\|^2)^{\frac{s+1}{2}}\Lambda^{t+s-1}(\eta)|\hat{u}(\eta)| \, d\eta$$

$$\le c \int_{\mathbb{R}^n} (1+\|\xi-\eta\|^2)^{\frac{t+s+1}{2}}|\hat{\varphi}(\xi-\eta)|\Lambda^{t+s-1}(\eta)|\hat{u}(\eta)| \, d\eta.$$

Since $\hat{\varphi} \in \mathcal{S}(\mathbb{R}^n)$ we conclude that $\tau \mapsto (1+\|\tau\|^2)^{\frac{t+s-1}{2}}|\hat{\varphi}(\tau)|$ belongs to $L^1(\mathbb{R}^n)$ and for $u \in \mathcal{S}(\mathbb{R}^n)$ it follows that $\eta \mapsto \Lambda^{s+t-1}(\eta)|\hat{u}(\eta)|$ belongs to $L^2(\mathbb{R}^n)$. Hence by Young's inequality we find that $\Lambda^t(\cdot)([\Lambda^s(D),\varphi]u)\hat{}(\cdot)$ is an element in $L^2(\mathbb{R}^n)$ and we get

$$\|[\Lambda^s(D),\varphi]u\|_{H^t} \le c\|(1+\|\cdot\|^2)^{\frac{t+s-1}{2}}\hat{\varphi}(\cdot)\|_{L^1}\|\Lambda^{s+t-1}(\cdot)\hat{u}\|_{L^2}$$

$$\le \kappa\|u\|_{H^{s+t-1}}.$$

The result has an important interpretation and consequence. If we measure the order of a linear operator $A : \mathcal{S}(\mathbb{R}^n) \to \mathcal{S}(\mathbb{R}^n)$ with the infimum of all s such that $\|Au\|_{t+s} \le c_t\|u\|_t$ holds for all $t \ge 0$, then $\Lambda^s(D)$ had order s and the operator $M_\varphi(u) := \varphi u$ has for $\varphi \in \mathcal{S}(\mathbb{R}^n)$ the order 0. Thus $M_\varphi \circ \Lambda^s(D)$ and $\Lambda^s(D) \circ M_\varphi$ are operators of order s. Their commutator $[\Lambda^s(D),\varphi]$ however is an operator of order $s-1$, i.e. commutators tend to decrease the order of an operator.

11. Since

$$\left|\frac{e^{-t\|\xi\|^2}-1+t\|\xi\|^2}{t}\right| \le t\|\xi\|^4$$

with the help of Plancherel's theorem we find

$$\left\|\frac{T_tu-u}{t}-\Delta u\right\|_{L^2}^2 = \int_{\mathbb{R}^n} \left|\frac{e^{-t\|\xi\|^2}-1}{t}+\|\xi\|^2\right|^2 |\hat{u}(\xi)|^2 \, d\xi$$

$$\le t^2 \int_{\mathbb{R}^n} (1+\|\xi\|^2)^4|\hat{u}(\xi)|^2 \, d\xi = t^2\|u\|_{H^4}^2.$$

Passing with t to zero we find

$$\lim_{t \to 0} \left\|\frac{T_t-u}{t}-\Delta u\right\|_{L^2} = 0.$$

Thus in the sense of a limit in $L^2(\mathbb{R}^n)$ we find for all $u \in H^4(\mathbb{R}^n)$ that

$$\lim_{t \to 0} \frac{T_tu-u}{t} = -\Delta u \text{ or } \left.\frac{d}{dt}T_t\right|_{t=0} = -\Delta.$$

In Volume V we will see that the result holds already for all $u \in H^2(\mathbb{R}^n)$.

Chapter 13

1. a) We have

$$\hat{\mu}(\xi) = \frac{1}{\sqrt{2\pi}} \int_{\mathbb{R}} e^{-ix\xi} \mu(d\xi)$$

$$= \frac{1}{\sqrt{2\pi}} \int_{\mathbb{R}} e^{-ix\xi} \sum_{l=0}^{m} \binom{m}{l} p^l (1-p)^{m-l} \epsilon_l(dx)$$

$$= \frac{1}{\sqrt{2\pi}} \sum_{l=0}^{m} \binom{m}{l} p^l (1-p)^{m-l} e^{-i\xi l}$$

$$= \frac{1}{\sqrt{2\pi}} \sum_{l=0}^{m} \binom{m}{l} (pe^{-i\xi})^l (1-p)^{m-l}$$

$$= \frac{1}{\sqrt{2\pi}} ((1-p) + pe^{-i\xi})^m.$$

b) We find

$$\hat{\mu}(\xi) = \frac{1}{\sqrt{2\pi}} \int_{\mathbb{R}} e^{-ix\xi} \sum_{l=0}^{\infty} e^{-\sigma} \frac{\sigma^l}{l!} \epsilon_l(dx)$$

$$= \frac{1}{\sqrt{2\pi}} \sum_{l=0}^{\infty} e^{-\sigma} \frac{\sigma^l}{l!} e^{-il\xi}$$

$$= \frac{1}{\sqrt{2\pi}} e^{-\sigma} \sum_{l=0}^{\infty} \frac{(\sigma e^{-i\xi})^l}{l!} = \frac{1}{\sqrt{2\pi}} e^{-\sigma} e^{(\sigma e^{-i\xi})}$$

$$= \frac{1}{\sqrt{2\pi}} e^{\sigma(e^{-i\xi}-1)}.$$

2. If $M_k(\mu) < \infty$ then $M_l(\mu) < \infty$ for all $0 \le l \le k$, recall that μ is a bounded measure, i.e. $M_0(\mu) < \infty$, and note that for $1 \le l \le k-1$ we have $|x|^l \le c_l(1 + |x|^k)$. Since

$$\int_{\mathbb{R}} \frac{d^l}{d\xi^l} e^{-ix\xi} \mu(dx) = (-i)^l \int_{\mathbb{R}} x^l e^{-ix\xi} \mu(dx)$$

is finite we can apply Theorem III.8.4 and differentiate under the integral sign to obtain for $0 \le l \le k$ that

$$\frac{d^l}{d\xi^l} \hat{\mu}(\xi) = \frac{1}{\sqrt{2\pi}} \int_{\mathbb{R}} \frac{d^l}{d\xi^l} e^{-ix\xi} \mu(dx)$$

$$= \frac{(-i)^l}{\sqrt{2\pi}} \int_{\mathbb{R}} \xi^l e^{-ix\xi} \mu(dx).$$

It follows that $\hat{\mu} \in C^k(\mathbb{R})$. In fact, $\hat{\mu} \in C_b^k(\mathbb{R})$ since our calculation implies $|\frac{d^l}{d\xi^l} \hat{\mu}(\xi)| \le \frac{1}{\sqrt{2\pi}} M_l(\mu) < \infty$.

633

3. By Bochner's theorem there exists $\mu \in M_b^+(\mathbb{R})$ such that $\hat{\mu} = \varphi$. For the image measure μ_S of μ under the reflection $Sx = -x$ we find

$$(\mu * \mu_S)\hat{}(\xi) = \sqrt{2\pi}\varphi(\xi)\varphi(-\xi)$$
$$= \sqrt{2\pi}\varphi(\xi)\overline{\varphi(\xi)} = \sqrt{2\pi}|\varphi(\xi)|^2.$$

Thus $|\varphi|^2$ is the Fourier transform of the measure $\frac{1}{\sqrt{2\pi}}\mu * \mu_S$ and therefore it is a positive definite function.

4. (Compare with [12]). First we note that by Problem 16 in Chapter 12 we can conclude that the function $\varphi_a(x) = \chi_{[-a,a]}(x)(1 - \frac{|x|}{a}), a > 0$ is positive definite since it is the Fourier transform of $\xi \mapsto \sqrt{\frac{2}{\pi}}\frac{1-\cos a\xi}{a\xi^2}$. Next we observe that $f|_{(0,\infty)}$ must be strictly decreasing from $f(0) > 0$ to 0. If we can approximate f uniformly by a linear combination of functions φ_{a_k} with non-negative coefficients the result will follow.

For $\epsilon \in (0, f(0))$ there exists $x_0 > 0$ such that $f(x_0) = \epsilon$ and $f|_{[0,x_0]} > \epsilon$ whereas $f|_{(x_0,\infty)} < \epsilon$. The uniform continuity of f on $[0, x_0]$ yields the existence of a_k, $0 = a_0 < a_1 < \cdots < a_{N-1} = x_0$ such that $x, y \in [a_k, a_{k+1}]$, $k = 0, \ldots, N-2$, implies $|f(x) - f(y)| < \epsilon$. With $b_k := \varphi(a_k)$, $k = 0, \ldots, N-1$, we introduce

$$c_k := \frac{b_{k+1} - b_k}{a_{k+1} - a_k}, \quad k = 0, \ldots, N-2,$$

and we clearly have $c_0 < c_1 < \cdots < c_{N-2} < 0$. We define now $g : \mathbb{R} \to [0,\infty)$ in the following way. We require that g is even, continuous and piecewise linear. On $[a_k, a_{k+1}], k = 0, \ldots, N-1$, the graph of g is the line segment connecting (a_k, b_k) with (a_{k+1}, b_{k+1}). Next we set $a_N := a_{N-1} - \frac{b_{N-1}}{c_{N-1}}$ where $c_{N-1} \in (c_{N-2}, 0)$, and we define $g|_{[a_{N-1}, a_N]}$ as the function whose graph is the line segment connecting (a_{N-1}, b_{N-1}) with $[a_N, 0]$. Finally, on $[a_n, \infty)$ we set $g|_{[a_N, \infty)} = 0$. The function g is continuous and satisfies $|f(x) - g(x)| < \epsilon$ for all $x \in \mathbb{R}$. We note that $g(x) = \sum_{k=1}^{N} \frac{a_k}{c_k - c_{k-1}}\varphi_{a_k}(x)$, $\frac{a_k}{c_k - c_{k-1}} > 0$, and hence g is a positive definite function implying that f is a continuous positive definite function.

5.   a) By Problem 4 in Chapter 8 we have for every continuous positive definite function u the estimate

$$|u(\xi) - u(\eta)|^2 \le 2u(0)(u(0) - \text{Re}(u(\xi - \eta))).$$

This gives for $\xi \in \mathbb{R}$ and $h \in \mathbb{R}$ that

$$\frac{|u(\xi + h) - u(\xi)|^2}{h^2} \le \frac{1}{h^2}2|u(0)|(u(0) - \text{Re}(u(h))$$
$$= 2u(0)\text{Re}\left(\frac{u(0) - u(h)}{h^2}\right)$$

implying that $\lim_{h \to 0} \frac{u(\xi+h)-u(\xi)}{h} = 0$ for all $\xi \in \mathbb{R}$, i.e. u is a C^1-function with derivative identically equal to 0, i.e. u must be constant.

634

b) For $\alpha > 2$ it follows that in the case $\xi \neq 0$

$$\frac{1 - e^{-|\xi|^\alpha}}{|\xi|^2} \leq \frac{|\xi|^\alpha}{|\xi|^2} = |\xi|^{\alpha-2},$$

implying that $\lim_{\xi \to 0} \frac{1-e^{-|\xi|^\alpha}}{|\xi|^2} = 0$ and the result is a consequence of part a).

6. The double integral is an improper Riemann integral. So we look first at the integral

$$\int_{Q_R(0)} \int_{Q_R(0)} u(\xi - \eta)\overline{\varphi(\eta)}\varphi(\xi) \, d\xi \, d\eta \qquad (*)$$

where $Q_R(0) = \{x \in \mathbb{R}^n | -R \leq x_j \leq R \text{ for } 1 \leq j \leq n\}$. This is a Riemann integral of a bounded continuous function which we can approximate by Riemann sums

$$\sum_{j,l=1}^{k} u(\xi^j - \xi^l)\overline{\varphi(\xi^j)}\varphi(\xi^l)\Delta\xi_1^j \ldots \Delta\xi_n^j \Delta\xi_1^l \ldots \Delta\xi_n^l.$$

Since u is positive definite these Riemann sums are all non-negative, hence in the limit we conclude that $(*)$ is non-negative. Finally we pass in $(*)$ to the limit $R \to \infty$ to obtain

$$\int_{\mathbb{R}^n} \int_{\mathbb{R}^n} \eta(\xi - \eta)\overline{\varphi(\eta)}\varphi(\xi) \, d\xi \, d\eta \qquad (**)$$

and again as the limit of non-negative terms this integral must be non-negative.

Note that the converse statement holds too: If $(**)$ is for all $\varphi \in \mathcal{S}(\mathbb{R}^n)$ non-negative, then u must be positive definite. A proof is provided in [69].

7. First we remark that $\hat{u} \in C_\infty(\mathbb{R}^n)$ and therefore $\hat{u}$ is bounded, and $\hat{\mu} \in C_b(\mathbb{R}^n)$. Thus, both integrals $\int_{\mathbb{R}^n} \hat{u}(\xi)\mu(d\xi)$ and $\int_{\mathbb{R}^n} u(x)\hat{\mu}(x)dx$ are defined and finite. Fubini's theorem yields

$$\int_{\mathbb{R}^n} \hat{u}(\xi)\mu(d\xi) = (2\pi)^{-\frac{n}{2}} \int_{\mathbb{R}^n} \int_{\mathbb{R}^n} u(x)e^{-ix\cdot\xi} \, dx \, \mu(d\xi)$$

$$= (2\pi)^{-\frac{n}{2}} \int_{\mathbb{R}^n} \int_{\mathbb{R}^n} e^{-ix\cdot\xi}\mu(d\xi) \, u(x) \, dx$$

$$= \int_{\mathbb{R}^n} u(x)\hat{\mu}(x) \, dx.$$

8.   a) By the convolution theorem we find

$$\hat{\mu}(\xi) = (\mu_k * \cdots * \mu_k)\hat{}(\xi)$$

$$= ((2\pi)^{\frac{n}{2}})^{k-1}\hat{\mu}_k(\xi)$$

or

$$(2\pi)^{\frac{n}{2}}\hat{\mu}(\xi) = ((2\pi)^{\frac{n}{2}}\hat{\mu}_k(\xi))^k.$$

635

b) First we note by our assumptions $\hat{\mu}(0) = (2\pi)^{-\frac{n}{2}}$, i.e. μ is a probability measure. For $k \in \mathbb{N}$ we define μ_k by $\hat{\mu}_k(\xi) = (2\pi)^{-\frac{n}{2}} e^{-\frac{1}{k}\psi(\xi)}$. Again μ_k is a probability measure since $\hat{\mu}_k(0) = (2\pi)^{-\frac{n}{2}}$. Furthermore, the construction yields

$$(\mu_k * \cdots * \mu_k)\hat{}(\xi) = ((2\pi)^{\frac{n}{2}})^{k-1}((2\pi)^{-\frac{n}{2}} e^{-\frac{1}{k}\psi(\xi)})^k$$
$$= (2\pi)^{-\frac{n}{2}} e^{-\psi(\xi)} = \hat{\mu}(\xi)$$

or $\mu = \mu_k * \cdots * \mu_k$.

c) In light of part b) we consider on $\mathbb{R}$ the measure μ with Fourier transform $\hat{\mu}(\xi) = \frac{1}{\sqrt{2\pi}} e^{-|\xi|}$. By Problem 1 c) in Chapter 12, we have $\mu = \sqrt{\frac{2}{\pi}} \frac{1}{1+x^2} \lambda^{(1)}$.

Chapter 14

1. We use the power series $J_l(x) = \sum_{k=0}^{\infty} \frac{(-1)^k x^{2k+l}}{2^{2k+l} k!(k+l)!}$ to find

$$\frac{\mathrm{d}}{\mathrm{d}x}(x^l J_l(x)) = \frac{\mathrm{d}}{\mathrm{d}x} \sum_{k=0}^{\infty} \frac{(-1)^k x^{2k+l}}{2^{2k+l} k!(k+l)!}$$
$$= \sum_{k=0}^{\infty} \frac{(-1)^k (2k+2l) x^{2k+2l-1}}{2^{2k+l} k!(k+l)!}$$
$$= \sum_{k=0}^{\infty} \frac{(-1)^k x^{2k+2l-1}}{2^{2k+l-1} k!(k+l-1)!} = x^l J_{l-1}(x).$$

2. Recall that for $l \in \mathbb{N}$ we have

$$x^2 J_l''(x) + x J_l'(x) + (x^2 - l^2) J_l(x) = 0$$

and for $-l, l \in \mathbb{N}$, we have $J_{-l}(x) = (-1)^l J_l(x)$ which yields

$$x^2 J_{-l}''(x) + x J_{-l}'(x) + (x^2 - l^2) J_{-l}(x) = 0.$$

Multiplying the equation for J_l by J_{-l} and that for J_{-l} by J_l and subtracting the results gives

$$x^2 (J_l''(x) J_{-l}(x) - J_{-l}''(x) J_l(x)) + x(J_l'(x) J_{-l}(x) - J_{-l}'(x) J_l(x)) = 0$$

or

$$\frac{\mathrm{d}}{\mathrm{d}x}(x(J_l'(x) J_{-l}(x) - J_{-l}'(x) J_l(x))) = 0$$

implying that $x(J_l'(x) J_{-l}(x) - J_{-l}'(x) J_l(x)) = c$ for some $c \in \mathbb{R}$. But $J_{-l}(x) = (-1)^l J_l(x)$ and hence $J_{-l}'(x) = (-1)^l J_l'(x)$ which gives

$$(-1)^l x(J_l'(x) J_l(x) - J_l'(x) J_l(x)) = 0,$$

thus $J_l'(x) J_{-l}(x) - J_{-l}'(x) J_l(x) = 0$ for all $x \in \mathbb{R}$.

Of course, our solution is too complicated, we could give the result immediately by just looking at the last two lines. However, the calculation leading to $x(J_l'(x)J_{-l}(x) - J_{-l}'(x)J_l(x)) = c$ holds also for the Bernel functions of non-integer order and in this case we can determine c to be $c = \frac{2\sin l\pi}{\pi}$. We will give an interpretation of $J_l'J_{-l} - J_{-l}'J_l$ in Chapter 19.

3. From (14.7) we derive

$$\sum_{k\in\mathbb{Z}} J_k(x+y)e^{ik\varphi} = e^{i(x+y)\sin\varphi}$$

$$= e^{ix\sin\varphi}e^{iy\sin\varphi}$$

$$= \left(\sum_{l\in\mathbb{Z}} J_l(x)e^{il\varphi}\right)\left(\sum_{m\in\mathbb{Z}} J_m(y)e^{im\varphi}\right)$$

$$= \sum_{l,m\in\mathbb{Z}} J_l(x)J_m(y)e^{i(l+m)\varphi}$$

$$= \sum_{k\in\mathbb{Z}}\left(\sum_{m\in\mathbb{Z}} J_m(x)J_{k-m}(y)\right)e^{ik\varphi}.$$

Since $\sum_{k\in\mathbb{Z}} J_k(x+y)e^{ik\varphi} = \sum_{k\in\mathbb{Z}}\left(\sum_{m\in\mathbb{Z}} J_m(x)J_{k-m}(y)\right)e^{ik\varphi}$ has to hold for all φ we arrive at

$$J_k(x+y) = \sum_{m\in\mathbb{Z}} J_m(x)J_{k-m}(y).$$

4. By Theorem 14.6 we have for $[\hat{\chi}_{B_1(0)}]_0$

$$[\hat{\chi}_{B_1(0)}]_0(\rho) = \frac{1}{\rho^{\frac{n}{2}-1}}\int_0^\infty (\chi_{B_1(0)})_0(\gamma)\gamma^{\frac{n}{2}} J_{\frac{n}{2}-1}(\rho\gamma)\,\mathrm{d}\gamma$$

with $n = 2$, i.e.

$$[\hat{\chi}_{B_1(0)}]_0(\rho) = \int_0^\infty (\chi_{B_1(0)})_0(\gamma)\gamma J_0(\rho\gamma)\,\mathrm{d}\gamma.$$

Since $(\chi_{B_1(0)})_0(\gamma) = \chi_{[0,1)}(\gamma)$ we obtain

$$[\hat{\chi}_{B_1(0)}]_0(\rho) = \int_0^1 \gamma J_0(\rho\gamma)\,\mathrm{d}\gamma$$

$$= \int_0^\rho \frac{\eta}{\rho^2} J_0(\eta)\,\mathrm{d}\eta.$$

By Problem 1 we have $\eta J_0(\eta) = \frac{\mathrm{d}}{\mathrm{d}\eta}(\eta J_1(\eta))$ and therefore

$$[\hat{\chi}_{B_1(0)}]_0(\rho) = \int_0^\rho \frac{1}{\rho^2}\frac{\mathrm{d}}{\mathrm{d}\eta}(\eta J_1(\eta))\,\mathrm{d}\eta = \frac{1}{\rho}J_1(\rho).$$

5. We put $a := \|\xi\|\hat{u}_0(\xi), b := \hat{u}_1(\xi)$ and $c := \|\xi\|t$. We have to show $|-a\sin ct + b\cos ct|^2 + |a\cos ct + b\sin ct|^2 = |a|^2 + |b|^2$. A straightforward calculation gives

$$|-a\sin ct + b\cos ct|^2 = |a|^2\sin^2 ct - (a\bar{b} + \bar{a}b)\cos ct\sin ct + |b|^2\cos^2 ct$$

and

$$|a\cos ct + b\sin ct|^2 = |a|^2\cos^2 ct + (a\bar{b} + \bar{a}b)\cos ct\sin ct + |b|^2\sin^2 ct.$$

Adding up these terms yields

$$|-a\sin ct + b\cos ct|^2 + |a\cos ct + b\sin ct|^2$$
$$= |a|^2(\cos^2 ct + \sin^2 ct) + |b|^2(\cos^2 ct + \sin^2 ct)$$
$$= |a|^2 + |b|^2.$$

6.   a) For $u_0, u_1 \in S(\mathbb{R})$ we can deduce that the solution to the initial value problem $\left(\frac{\partial^2}{\partial t^2} - \frac{\partial^2}{\partial x^2}\right)w = 0$, $w(x,0) = u_0(x)$, $\frac{\partial w}{\partial t}(x,0) = u_1(x)$ is given by

$$w(x,t) = \frac{1}{\sqrt{2\pi}}\int_{\mathbb{R}} e^{ix\xi}\left(\hat{u}_0(\xi)\cos\xi t + \hat{u}_1(\xi)\frac{\sin\xi t}{\xi}\right)d\xi$$

where we used that $\xi \mapsto \cos\xi t$ and $\xi \mapsto \frac{\sin\xi t}{\xi}$ are even functions. We write now $\cos\xi t = \frac{e^{i\xi t} + e^{-i\xi t}}{2}$ and $\sin\xi t = \frac{e^{i\xi t} - e^{-i\xi t}}{2i}$ and find

$$w(x,t) = \frac{1}{\sqrt{2\pi}}\int_{\mathbb{R}} e^{ix\xi}\left(\frac{e^{i\xi t} + e^{-i\xi t}}{2}\right)\hat{u}_0(\xi)\,d\xi$$
$$+ \frac{1}{\sqrt{2\pi}}\int_{\mathbb{R}} e^{ix\xi}\left(\frac{e^{i\xi t} - e^{-i\xi t}}{2i}\right)\hat{u}_1(\xi)\,d\xi.$$

For the first integral we get

$$\frac{1}{\sqrt{2\pi}}\int_{\mathbb{R}} \frac{e^{i\xi(x+t)} + e^{-i\xi(x-t)}}{2}\hat{u}_0(\xi)\,d\xi = \frac{1}{2}(u_0(x+t) - u_0(x-t)).$$

For the second integral it follows

$$\frac{1}{\sqrt{2\pi}}\int_{\mathbb{R}} \frac{e^{i\xi(x+t)} - e^{-i\xi(x-t)}}{2i\xi}\hat{u}_1(\xi)\,d\xi = \frac{1}{\sqrt{2\pi}}\int_{\mathbb{R}}\frac{1}{2}\int_{x-t}^{x+t} e^{i\xi y}\,dy\,\hat{u}_1(\xi)\,d\xi$$
$$= \frac{1}{2}\int_{x-t}^{x+t}\frac{1}{\sqrt{2\pi}}\int_{\mathbb{R}} e^{i\xi y}\hat{u}_1(\xi)\,d\xi\,dy$$
$$= \frac{1}{2}\int_{x-t}^{x+t} u_1(y)\,dy.$$

Further we note now $g = u_0$ and $h = u_1$.

b) For $g \in C^2(\mathbb{R})$ and $h \in C^1(\mathbb{R})$ we may differentiate

$$(x,t) \mapsto \frac{g(x+t) + g(x-t)}{2} + \frac{1}{2}\int_{x-t}^{x+t} h(y)\,dy = \alpha(x,t) + \beta(x,t)$$

twice with respect to t and x and we get

$$\frac{\partial\alpha}{\partial x}(x,t) = \frac{g'(x+t) + g'(x-t)}{2}, \quad \frac{\partial^2\alpha}{\partial x^2}(x,t) = \frac{g''(x+t) + g''(x-t)}{2}$$

638

and

$$\frac{\partial \alpha}{\partial t}(x,t) = \frac{g'(x+t) - g'(x-t)}{2}, \quad \frac{\partial^2 \alpha}{\partial t^2}(x,t) = \frac{g''(x+t) + g''(x-t)}{2}$$

implying that $\left(\frac{\partial^2}{\partial t^2} - \frac{\partial^2}{\partial x^2}\right)\alpha(x,t) = \left(\frac{\partial^2}{\partial t^2} - \frac{\partial^2}{\partial x^2}\right)\left(\frac{g(x+t)+g(x-t)}{2}\right) = 0$. Furthermore we have

$$\frac{\partial}{\partial t}\beta(x,t) = h(x+t) + h(x-t), \quad \frac{\partial^2}{\partial t^2}\beta(x,t) = h'(x+t) - h'(x-t)$$

and

$$\frac{\partial}{\partial x}\beta(x,t) = h(x+t) - h(x-t), \quad \frac{\partial^2}{\partial x^2}\beta(x,t) = h'(x+t) - h'(x-t),$$

which yields $\left(\frac{\partial^2}{\partial t^2} - \frac{\partial^2}{\partial x^2}\right)\beta(x,t) = \left(\frac{\partial^2}{\partial t^2} - \frac{\partial^2}{\partial x^2}\right)\left(\frac{1}{2}\int_{x-t}^{x+t} h(y)\, dy\right) = 0$.

Finally we observe that

$$w(x,0) = \frac{g(x) + g(x)}{2} = g(x)$$

and

$$\frac{\partial}{\partial t}w(x,0) = \frac{g'(x) - g'(x)}{2} + \frac{h(x) + h(x)}{2} = h(x),$$

thus g and h are the initial data.

7. Since our initial data belong to $\mathcal{S}(\mathbb{R})$ we may apply the result of Problem 6 a) to find

$$w(x,t) = \frac{1}{2}e^{-\frac{(x+t)^2}{2}} + \frac{1}{2}e^{-\frac{(x-t)^2}{2}}. \tag{$*$}$$

When discussing in Volume VI the wave equation in more detail, we will understand the interpretation of $(*)$ as a superposition of a forward and backward travelling initial "perturbation".

8. We take in $\frac{\partial^2 u}{\partial t^2} = \frac{\partial^2 u}{\partial x^2} - u$ the Fourier transform with respect to x and arrive at

$$\frac{d^2 \hat{u}(\xi,t)}{dt^2} = -\xi^2 \hat{u}(\xi,t) - \hat{u}(\xi,t) = -(|\xi|^2 + 1)\hat{u}(\xi,t)$$

and for the initial values we find $\hat{u}(\xi,0) = \hat{u}_0(\xi)$, $\frac{\partial}{\partial t}\hat{u}(\xi,0) = \hat{u}_1(\xi)$. This initial value problem has the solution

$$\hat{u}(\xi,t) = a_1(\xi)\cos(\sqrt{|\xi|^2 + 1}\,t) + a_2(\xi)\sin(\sqrt{|\xi|^2 + 1}\,t)$$

with

$$a_1(\xi) = \hat{u}_0(\xi) \text{ and } a_2(\xi) = \frac{\hat{u}_1(\xi)}{\sqrt{\xi^2 + 1}}.$$

Thus we come up with

$$u(x,t) = \frac{1}{\sqrt{2\pi}} \int_{\mathbb{R}} e^{ix\xi}\left(\hat{u}_0(\xi)\cos(\sqrt{|\xi|^2+1}t) + \hat{u}_1(\xi)\frac{\sin(\sqrt{|\xi|^2+1}t)}{\sqrt{|\xi|^2+t}}\right) d\xi \qquad (*)$$

The verification that $(*)$ is indeed a solution requires to differentiate under the integral sign and this is justified by the condition $u_0, u_1 \in \mathcal{S}(\mathbb{R})$ implying $\hat{u}_0, \hat{u}_1 \in \mathcal{S}(\mathbb{R})$. We find

$$\frac{\partial^2}{\partial x^2}u(x,t) = \frac{1}{\sqrt{2\pi}} \int_{\mathbb{R}} e^{ix\xi}\left(-\xi^2\left(\hat{u}_0(\xi)\cos(\sqrt{|\xi|^2+1}t)\right.\right.$$
$$\left.\left. + \hat{u}_1(\xi)\frac{\sin(\sqrt{|\xi|^2+1}t)}{\sqrt{|\xi|^2+1}}\right)\right) d\xi$$

$$\frac{\partial^2}{\partial t^2}u(x,t) = \frac{1}{\sqrt{2\pi}} \int_{\mathbb{R}} e^{ix\xi}\left(-(\xi^2+1)\left(\hat{u}_0(\xi)\cos(\sqrt{|\xi|^2+1}t)\right.\right.$$
$$\left.\left. + \hat{u}_1(\xi)\frac{\sin(\sqrt{|\xi|^2+1}t)}{\sqrt{|\xi|^2+1}}\right)\right) d\xi$$

implying that $\left(\frac{\partial^2}{\partial t^2} - \frac{\partial^2}{\partial x^2}\right)u(x,t) = -u(x,t)$ or $\frac{\partial^2}{\partial t^2}u(x,t) = \frac{\partial^2}{\partial x^2}u(x,t) - u(x,t)$.

9. We note that by Problem 16 in Chapter 12 we have

$$\left(\left(\frac{\sin\frac{x}{2}}{\frac{x}{2}}\right)^2\right)^{\hat{}}(\xi) = \sqrt{2\pi}\chi_{[-1,1]}(\xi)(1-|\xi|).$$

Applying the Poisson summation formula to the function $u(t) = \left(\frac{\sin\frac{t}{2}}{\frac{t}{2}}\right)^2$ we find from

$$\sqrt{2\pi}\sum_{k\in\mathbb{Z}} u(x+2\pi k) = \sum_{k\in\mathbb{Z}} \hat{u}(k)e^{ixk}$$

first that the sum of the left hand side is equal to $\sqrt{2\pi}$ since

$$\hat{u}(k) = \sqrt{2\pi}\chi_{[-1,1]}(k)(1-|k|) = \begin{cases} \sqrt{2\pi}, & k = 0 \\ 0, & k \neq 0. \end{cases}$$

Thus we arrive with $x = 2\pi t$ at

$$\sum_{k\in\mathbb{Z}} \left(\frac{\sin\pi(t+k)}{\pi(t+k)}\right)^2 = 1.$$

Noting that $(\sin\pi(t+k))^2 = (\sin\pi t)^2$ for all $k \in \mathbb{Z}$ and $t \in \mathbb{R}$ we get

$$\left(\frac{\sin\pi t}{\pi}\right)^2 \sum_{k\in\mathbb{Z}} \frac{1}{(t+k)^2} = 1$$

or

$$\sum_{k\in\mathbb{Z}} \frac{1}{(t+k)^2} = \frac{\pi^2}{(\sin\pi t)^2}, \quad t\notin\mathbb{Z}.$$

For $t=\frac{1}{2}$ we find further that

$$\sum_{k\in\mathbb{Z}} \frac{1}{(\frac{1}{2}+k)^2} = 4\sum_{k\in\mathbb{Z}} \frac{1}{(1+2k)^2} = 4\sum_{k=0}^{\infty} \frac{1}{(1+2k)^2} + 4\sum_{k=1}^{\infty} \frac{1}{(2k-1)^2}$$

$$= 8\sum_{k=0}^{\infty} \frac{1}{(2k+1)^2}$$

which yields

$$\sum_{k=0}^{\infty} \frac{1}{(2k+1)^2} = \frac{\pi^2}{8}.$$

For $t=\frac{1}{4}$ we find $\frac{\pi^2}{(\sin\frac{\pi}{4})^2} = 2\pi^2$ and

$$\sum_{k\in\mathbb{Z}} \frac{1}{(\frac{1}{4}+k)^2} = 16\sum_{k\in\mathbb{Z}} \frac{1}{(1+4k)^2} = 16\left(\sum_{k=0}^{\infty} \frac{1}{(1+4k)^2} + \sum_{k=1}^{\infty} \frac{1}{(4k-1)^2}\right).$$

However $\sum_{k=0}^{\infty}\frac{1}{(4k+1)^2} + \sum_{k=1}^{\infty}\frac{1}{(4k-1)^2} = \sum_{k=0}^{\infty}\frac{1}{(2k+1)^2}$ and we get a confirmation of the above result.

10. The heat kernel in one dimension is given by $h_t(x) = \frac{1}{\sqrt{4\pi t}}e^{-\frac{|x|^2}{4t}}$ with Fourier transform $\hat{h}_t(\xi) = \frac{1}{\sqrt{2\pi}}e^{-t|\xi|^2}$. Therefore the Poisson summation formula yields

$$\sqrt{2\pi}\sum_{k\in\mathbb{Z}} \frac{1}{\sqrt{4\pi t}}e^{-\frac{|x+2\pi k|^2}{4t}} = \frac{1}{\sqrt{2\pi}}\sum_{k\in\mathbb{Z}} e^{-tk^2}e^{ixk}.$$

We can extend the function on the right hand side to a 2π-periodic function (or we may identify this function as a function on $S^1 = \mathbb{T}^1$) and a direct calculation shows $\left(\frac{\partial}{\partial t} - \frac{\partial^2}{\partial x^2}\right)\left(\frac{1}{\sqrt{2\pi}}\sum_{k\in\mathbb{Z}} e^{-tk^2}e^{ikx}\right) = 0$, i.e. we obtain a 2π-periodic solution to the heat equation (or a solution to the heat equation in $\mathbb{T}^1$).

Solutions to Problems of Part 9

Chapter 15

1. Suppose that for some t_1 we have $w(t_1) \neq 0$. Then $w \neq 0$ in some interval $(-\delta + t_1, t_1 + \delta)$, $\delta > 0$, and in this interval we must have $w(t) = ce^{-\alpha t}$. Moreover it follows that $w(t_1) = ce^{-\alpha t_1}$ or $c = w(t_1)e^{\alpha t_1} \neq 0$. Thus $w(t) = w(t_1)e^{-\alpha(t-t_1)} \neq 0$ for all $t \in \mathbb{R}$. Next we prove that if $w'(t) + \alpha w(t) = 0$ and $w(t_0) = 0$ then w is identically 0. We multiply $w'(t) + \alpha w(t) = 0$ with $w(t)$ and integrate the result from t_0 to t (or t to t_0 if $t < t_0$) and find

$$0 = \int_{t_0}^{t} (w(s)w'(s) + \alpha w^2(s)) \, ds = \int_{t_0}^{t} \left(\frac{1}{2}\frac{d}{ds}w^2(s) + \alpha w^2(s)\right) \, ds$$

or

$$\int_{t_0}^{t} w^2(s) \, ds = -\frac{1}{2\alpha}w^2(t) \leq 0.$$

The continuity of w implies now that $w(s) = 0$ in $[t_0, t]$, but $t \in \mathbb{R}$ was arbitrary, hence $w(s) = 0$ for all $s \in \mathbb{R}$.

2. For $t > a$ we know already that $u'_{a,b}(t) = \frac{t-a}{2}$ and for $t < -b$ we find $u'_{a,b}(t) = -\frac{(t+b)}{2}$. Since for $-b < t < a$ we have $u'_{a,b}(t) = 0$, the fact that the two one-sided derivatives $u'_{a,b,+}(a)$ and $u'_{a,b,-}(b)$ are zero implies that $u_{a,b}$ is a C^1-function on $\mathbb{R}$. The differential equation $u'(t) = |u(t)|^{\frac{1}{2}}$ is satisfied for $t > a$, $t < b$ and in (a, b), and it continues to hold in the two limiting cases $t \to a$ and $t \to b$, thus it holds in $\mathbb{R}$. Obviously we have $u_{a,b}(0) = 0$.

3. For u solving $\dot{u}(t) = |u(t)|^{\frac{1}{4}}$ we find with $z(t) := -u(-t)$ that

$$\dot{z}(t) = \frac{d}{dt}(-u(-t)) = -\frac{d}{dt}(u(-t)) = \dot{u}(-t) = |u(-t)|^{\frac{1}{4}} = |z(t)|^{\frac{1}{4}}.$$

Using the method of separation of variables we find for $c \in \mathbb{R}$

$$\int \frac{du}{|u(t)|^{\frac{1}{4}}} = t + c$$

where we read the "integral" as a symbol for the primitive of the function under the integral sign. This gives

$$u(t) = \left(\frac{3}{4}|t + c|\right)^{\frac{4}{3}}.$$

For $t \geq -c$ a solution is given by $u(t) = (\frac{3}{4}(t + c))^{\frac{4}{3}}$, as we can easily check by calculating $\dot{u}(t) = (\frac{3}{4}(t + c))^{\frac{1}{3}}$. By the first part we know that $-u(-t) = -(\frac{3}{4}(-t+c))^{\frac{4}{3}}$ solves the equation for $-t+c \geq 0$, i.e. $t \leq c$. Moreover, $t \mapsto u_0(t) = 0$ for all t is a solution and hence for every $b < 0 < a$ the function $u_{a,b}$ solves the initial value problem $\dot{u}(t) = |u(t)|^{\frac{1}{4}}$, $u(0) = 0$.

4. Since by assumption $y(t) \neq 0$ for all t we can rewrite the differential equation $y'(t) + \frac{t}{y(t)} = 0$ as $y'(t)y(t) = -t$ or $\frac{1}{2}\frac{d}{dt}y^2(t) = -t$ which we may integrate to obtain

$$\frac{1}{2}\int_{t_0}^{t} \frac{d}{ds}y^2(s)\, ds = -\int_{t_0}^{t} s\, ds$$

or

$$\frac{1}{2}y^2(t) - \frac{1}{2}y^2(t_0) = -\frac{1}{2}t^2 + \frac{1}{2}t_0^2,$$

i.e.

$$y^2(t) + t^2 = t_0^2 + y^2(t_0). \tag{$*$}$$

The equation $(*)$ gives a circle in the $x - y$-plane, $x = t$, with centre in the origin and radius $r_0 = (t_0^2 + y^2(t_0))^{\frac{1}{2}}$, provided that $t_0 \neq 0$ or $y(0) \neq 0$. If $y(t_0) = 0$ then our assumptions are violated and $(*)$ characterizes the point $(0,0)$, i.e. the origin.

5. The system we have to discuss is

$$\dot{q}_j(t) = p_j(t), \quad \dot{p}_j(t) = -q_j(t), \quad 1 \leq j \leq 3. \tag{$*$}$$

We note that this system for six unknown functions consists of three decoupled systems each for two unknown functions, and each of these three systems is of type $\dot{y}(t) = x(t)$, $\dot{x}(t) = -y(t)$. This system has the general solution $x(t) = \alpha \cos(t + \varphi)$ and $y(t) = \alpha \sin(t + \varphi)$ where α and φ will be determined by initial conditions. Thus the general solution to $(*)$ is

$$q(t) = \begin{pmatrix} q_1(t) \\ q_2(t) \\ q_3(t) \end{pmatrix} = \begin{pmatrix} \alpha_1 \cos(t + \varphi_1) \\ \alpha_2 \cos(t + \varphi_2) \\ \alpha_3 \cos(t + \varphi_3) \end{pmatrix}$$

and

$$p(t) = \begin{pmatrix} p_1(t) \\ p_2(t) \\ p_3(t) \end{pmatrix} = \begin{pmatrix} \alpha_1 \sin(t + \varphi_1) \\ \alpha_2 \sin(t + \varphi_2) \\ \alpha_3 \sin(t + \varphi_3) \end{pmatrix}$$

where $\alpha = (\alpha_1, \alpha_2, \alpha_3)$ and φ_1, φ_2, φ_3 are determined by initial conditions. Note that

$$H(q(t), p(t)) = \frac{1}{2}\|q(t)\|^2 + \frac{1}{2}\|p(t)\|^2$$

$$= \frac{1}{2}(\alpha_1^2 \cos^2(t + \varphi_1) + \alpha_2^2 \cos^2(t + \varphi_2) + \alpha_3^2 \cos^2(t + \varphi_3))$$

$$+ \frac{1}{2}(\alpha_1^2 \sin^2(t + \varphi_1) + \alpha_2^2 \sin^2(t + \varphi_2) + \alpha_3^2 \sin^2(t + \varphi_3))$$

$$= \alpha_1^2 + \alpha_2^2 + \alpha_3^2 = \|\alpha\|^2 = \text{const.}$$

This observation will become quite helpful in a more general context in Chapter 30.

6. The equation is of homogeneous type since we can write for $h(t,y) = \frac{10t+8y}{2t-y}$ also $h(t,y) = f(\frac{y}{t})$ where $f(s) = \frac{10+8s}{2-s}$. We introduce the new function $u(t) = \frac{y(t)}{t}$ and arrive at $\dot{u}(t) = \frac{f(u)-u}{t} = \frac{10+6u+u^2}{(2-u)t}$ which gives (as equality for primitive)

$$\int \frac{(2-u)\,du}{u^2+6u+10} = \int \frac{1}{t}\,dt.$$

This yields the relation

$$-\frac{1}{2}\ln(u^2+6u+10) + 5\arctan(u+3) = \ln t + c.$$

Of course, it is now a separate question for which initial values $(t_0, u(t_0))$ we can solve this equation locally, i.e. in a neighbourhood of $(t_0, u(t_0))$, to obtain in some interval $I(t_0)$ a function $u : I(t_0) \to \mathbb{R}$. The tool to tackle this problem is the important function theorem.

7. We may use (15.31) and we find

$$y(t) = e^{\int_0^t \frac{1}{1+s}\,ds} + e^{\int_0^t \frac{1}{1+s}\,ds}\int_0^t (1+s)^2 e^{-\int_0^s \frac{1}{1+r}\,dr}\,ds$$

$$= e^{\ln(1+t)} + e^{\ln(1+t)}\int_0^t (1+s)^2 e^{-\ln(1+s)}\,ds$$

$$= \frac{1}{2}(t+1) + \frac{(t+1)^3}{2}.$$

Now we can easily check that this is indeed the solution:

$$y(0) = \frac{1}{2} + \frac{1}{2} = 1$$

and

$$y'(t) = \frac{1}{2} + \frac{3(t+1)^2}{2} = \frac{1}{1+t}\left(\frac{1}{2}(1+t) + \frac{(1+t)^3}{2}\right) + (1+t)^2$$

$$= \frac{1}{1+t}y(t) + (1+t)^2.$$

8. The transformation $z(t) = y^{-2}(t)$ yields the new initial value problem $\dot{z}(t) = \frac{2}{t}z(t) - 4$, $z(1) = 1$. Using formula (15.31) we find for this problem the solution

$$z(t) = \left(1 + \int_1^t \left(-4e^{-\int_1^r \frac{2}{s}\,ds}\right)dr\right) e^{\int_1^t \frac{2}{s}\,ds} = 4t - 3t^2.$$

Indeed we have $z(1) = 4 - 3 = 1$ and

$$\dot{z}(t) = 4 - 6t = \frac{2}{t}(4t - 3t^3) = \frac{2}{t}z(t) - 4.$$

645

Now it follows that
$$y(t) = z(t)^{-\frac{1}{2}} = \frac{1}{\sqrt{4t - 3t^2}}.$$

We need the conditions that $4t - 3t^3 > 0$ as well as $t > 0$, which holds for $t \in \left(0, \sqrt{\frac{4}{3}}\right)$, note that $1 < \sqrt{\frac{4}{3}}$. Thus in $\left(0, \sqrt{\frac{4}{3}}\right)$ a solution to the original problem is given by $y(t) = \frac{1}{\sqrt{4t - 3t^2}}$. Here comes a further verification: $y(1) = \frac{1}{\sqrt{4-3}} = 1$ and in addition we have
$$y'(t) = \frac{3t - 2}{(4t - 3t^2)^{\frac{3}{2}}}$$

and

$$-\frac{1}{t}y(t) + 2y^3(t) = -\frac{1}{t}\frac{1}{(4t - 3t^2)^{\frac{1}{2}}} + 2\frac{1}{(4t - 3t^2)^{\frac{3}{2}}}$$
$$= \frac{1}{(4t - 3t^2)^{\frac{3}{2}}}\left(-\frac{1}{t}(4t - 3t^2) + 2\right)$$
$$= \frac{3t - 2}{(4t - 3t^2)^{\frac{3}{2}}}.$$

9. a) With $h(t, y) = at^n + y\cos t$ and $g(t, y) = \sin t - by^m$ we have

$$(at^n + y\cos t) + (\sin t - by^m)\dot{y}(t) = h(t, y) + g(t, y)\dot{y}(t) = 0.$$

Since $\frac{\partial h(t,y)}{\partial y} = \cos t$ and $\frac{\partial g(t,y)}{\partial t} = \cos t$ the integrability conditions are fulfilled. A primitive of $h(t, y)$ with respect to t is given by

$$\int (at^n + y\cos t)\, dt = \frac{a}{n+1}t^{n+1} + y\sin t + c(y).$$

In order to find $c(y)$ and a potential function F we try to achieve

$$\frac{\partial}{\partial y}\left(\frac{a}{n+1}t^{n+1} + y\sin t + c(y)\right) = \sin t - by^m$$

or

$$\sin t + c'(y) = \sin t - by^m,$$

which yields as possible value for c, $c(y) = \frac{b}{m+1}y^{m+1}$, and in turn we arrive as candidate for a potential function at

$$F(t, y) = \frac{at^{n+1}}{n+1} + y\sin t - \frac{b}{m+1}y^{m+1}.$$

Indeed we have

$$\frac{\partial F}{\partial t}(t, y) = at^n + y\cos t = h(t, y)$$

and

$$\frac{\partial F}{\partial y}(t, y) = \sin t - by^m = g(t, y).$$

For $m = 0$ we can find y as solution of

$$\frac{at^{n+1}}{n+1} + y \sin t - b = 0$$

or

$$y(t) = \frac{1}{\sin t}\left(b - \frac{at^{n+1}}{n+1}\right).$$

Of course we need to take care on the domain of y since the sine function has zeroes.

b) With $h(t, y) = t^3 + ty^2 - y = t(t^2 + y^2) - y$ and $g(t, y) = t$ we find $\frac{\partial h}{\partial y}(t, y) = 2ty - 1$ and $\frac{\partial g}{\partial t}(t, y) = 1$ and therefore the equation is not exact. The structure of h and g suggests as candidate for an integrating factor $M(t, y) = \frac{1}{t^2+y^2}$. Multiplying the differential equation with $M(t, y)$ we arrive at

$$M(t, y)h(t, y) + M(t, y)g(t, y)\dot{y}(t) = \left(t - \frac{y}{t^2 + y^2}\right) + \frac{t}{t^2 + y^2}\dot{y}(t). \qquad (*)$$

Since

$$\frac{\partial}{\partial y}\left(t - \frac{y}{t^2 + y^2}\right) = \frac{y^2 - t^2}{t^2 + y^2}$$

and

$$\frac{\partial}{\partial t}\left(\frac{t}{t^2 + y^2}\right) = \frac{y^2 - t^2}{t^2 + y^2}$$

it follows that for $(*)$ the integrability condition is satisfied. A primitive of $t \mapsto \left(t - \frac{y}{t^2+y^2}\right)$ is given by

$$\int\left(t - \frac{y}{t^2 + y^2}\right)dt = \frac{t^2}{2} - \arctan\frac{t}{y} + c(y).$$

The condition

$$\frac{\partial}{\partial y}\left(\frac{t^2}{2} - \arctan\frac{t}{y} + c(y)\right) = \frac{t}{t^2 + y^2}$$

yields

$$c(y) = \int\frac{t}{t^2 + y^2}\,dy + \int\frac{\partial}{\partial y}\arctan\frac{t}{y}\,dy$$

$$= \arctan\frac{y}{t} + \arctan\frac{t}{y}$$

which suggests as potential function for $(*)$

$$F(t, y) = \frac{t^2}{2} + \arctan\frac{y}{t}.$$

Indeed we have

$$\frac{\partial F}{\partial t}(t, y) = \frac{\partial}{\partial t}\left(\frac{t^2}{2} + \arctan\frac{y}{t}\right)$$

$$= t - \frac{y}{t^2}\frac{1}{1 + (\frac{y}{t})^2} = t - \frac{y}{t^2 + y^2}$$

647

and

$$\frac{\partial F}{\partial y}(t, y) = \frac{\partial}{\partial y}\left(\frac{t^2}{2} + \arctan\frac{y}{t}\right) = \frac{t}{t^2 + y^2}.$$

Again we may now study the equation $F(t, y) = 0$ to obtain (locally) a solution y of the original differential equation.

10.   a) We have to find the integral

$$\int_0^\infty e^{-st} e^{at}\, dt = \int_0^\infty e^{-(s-a)t}\, dt = \frac{1}{s-a} \qquad \text{for } s > a.$$

b) We consider the integral $\int_0^\infty e^{-st} t^p\, dt$ and the substitution $st = r$ gives

$$\int_0^\infty e^{-st} t^p\, dt = \frac{1}{s^{p+1}} \int_0^\infty r^p e^{-r}\, dr = \frac{\Gamma(p+1)}{s^{p+1}}, \qquad p > -1,\ s > 0,$$

where we have taken advantage from the fact that $t \mapsto e^{-st} t^p$, $s > 0$, $p > -1$ vanishes at infinity and is integrable at 0.

c) Since $t \mapsto e^{-st} \sin\alpha t$, $\alpha \neq 0$, $s > 0$, vanishes at ∞ and is integrable at 0 we obtain

$$\int_0^\infty e^{-st} \sin\alpha t\, dt = \frac{-e^{-st}(s\sin\alpha t + \alpha\cos\alpha t)}{s^2 + \alpha^2}\bigg|_0^\infty = \frac{\alpha}{s^2 + \alpha^2}.$$

11. Using the decomposition $\cosh(\alpha t) = \frac{1}{2}(e^{\alpha t} + e^{-\alpha t})$ we find

$$\mathcal{L}(\cosh a\cdot)(s) = \frac{1}{2}(\mathcal{L}(e^{a\cdot})(s) + \mathcal{L}(e^{-a\cdot})(s)).$$

For $s > a$ we have $\mathcal{L}(e^{a\cdot})(s) = \frac{1}{s-a}$ and for $s > -a$ we get $\mathcal{L}(e^{-a\cdot})(s) = \frac{1}{s+a}$ which yields for $s > |a|$ that

$$\mathcal{L}(\cosh a\cdot)(s) = \frac{1}{2}\left(\frac{s}{s-a} + \frac{1}{s+a}\right) = \frac{s}{s^2 - a^2}.$$

Moreover, since $\sinh(at) = \frac{1}{2}(e^{at} - e^{-at})$, it follows for $s > |a|$ that

$$\mathcal{L}(\sinh a\cdot)(s) = \frac{1}{2}\left(\frac{s}{s-a} - \frac{1}{s+a}\right) = \frac{a}{s^2 - a^2}.$$

12. First we note that

$$\mathcal{L}(u)(s) = \int_0^\infty e^{-st} u(t)\, dt = \sum_{k=0}^\infty \int_{ka}^{(k+1)a} e^{-st} u(t)\, dt.$$

Next we observe that due to the periodicty of u we have

$$\int_{ka}^{(k+1)a} e^{-st} u(t)\, dt = \int_{ka}^{(k+1)a} e^{-st} u(t - ka)\, dt = e^{-kas} \int_0^a e^{-sr} u(r)\, dr$$

which yields

$$\mathcal{L}(u)(s) = \sum_{k=0}^{\infty} e^{-kas} \int_0^a e^{-sr} u(r) \, dr$$

$$= \left(\sum_{k=0}^{\infty} (e^{-as})^k \right) \int_0^a e^{-sr} u(r) \, dr = \frac{1}{1 - e^{as}} \int_0^a e^{-sr} u(r) \, dr,$$

where we used that $a > 0$ and $s > 0$, hence $e^{-as} < 1$.

13. a) For $s > a$ we have

$$\mathcal{L}\left(\frac{t^{k-1}}{(k-1)!} e^{at} \right) = \int_0^\infty \frac{t^{k-1}}{(k-1)!} e^{at} e^{-st} \, dt = \frac{1}{(s-a)^k},$$

or

$$\mathcal{L}^{-1}\left(\frac{1}{(s-a)^k} \right) = \frac{t^{k-1}}{(k-1)!} e^{at}.$$

Now it follows

$$\mathcal{L}^{-1}\left(\frac{2s-3}{(3s-4)^5} \right) = \frac{1}{3^5} \mathcal{L}^{-1}\left(\frac{2s-3}{(s-\frac{4}{3})^5} \right)$$

$$= \frac{1}{3^5} \mathcal{L}^{-1}\left(\frac{2(s-\frac{4}{3}) - \frac{1}{3}}{(s-\frac{4}{3})} \right)$$

$$= \frac{2}{3^5} \mathcal{L}^{-1}\left(\frac{1}{(s-\frac{4}{3})^4} \right) - \frac{1}{3^6} \mathcal{L}^{-1}\left(\frac{1}{(s-\frac{4}{3})^5} \right),$$

which gives

$$\mathcal{L}^{-1}(h)(t) = \frac{2}{3^5} \frac{t^3}{3!} e^{\frac{4a}{3}t} - \frac{1}{3^6} \frac{t^4}{4!} e^{\frac{4a}{3}t}$$

$$= \frac{1}{729} e^{\frac{4a}{3}t} \left(t^3 - \frac{t^4}{24} \right) e^{\frac{4a}{3}t}.$$

b) Using Corollary 15.18 we find for $h(s)$ that $h'(s) = \frac{1}{s+1} - \frac{1}{s}$ and therefore

$$\mathcal{L}^{-1}(h)(t) = -\frac{1}{t} \mathcal{L}^{-1}(h')(t) = -\frac{1}{t} \mathcal{L}^{-1}\left(\frac{1}{s+1} - \frac{1}{s} \right)(t)$$

$$= -\frac{1}{t}(e^{-t} - 1) = \frac{1 - e^{-t}}{t}.$$

14. Taking the Laplace transform we find

$$(s^2 \mathcal{L}(y)(s) - sy(0) - y'(0)) - 3(s\mathcal{L}(y)(s) - y(0)) + 2\mathcal{L}(y)(s) = \frac{2}{s+1}$$

649

and using the initial values we obtain

$$s^2 \mathcal{L}(y)(s) - 2s + 1 - 3s\mathcal{L}(y)(s) + 6 + 2\mathcal{L}(y)(s) = \frac{2}{s+1}$$

which yields

$$\mathcal{L}(y)(s) = \frac{2s^2 - 5s - 5}{(s+1)(s^2 - 3s + 2)} = \frac{1}{3}\frac{1}{s+1} + \frac{4}{s-1} - \frac{7}{3}\frac{1}{s-2}$$

and we eventually arrive at

$$y(t) = \frac{1}{3}e^{-t} + 4e^t - \frac{7}{3}e^{2t}.$$

(The calculation in this example is taken from [112]).

Chapter 16

1. The iteration scheme is given by

$$y_n(t) = 1 + \int_0^t s y_{n-1}(s)\, \mathrm{d}s.$$

The first four iterations give

$$y_0(t) = 1$$

$$y_1(t) = 1 + \int_0^t s\, \mathrm{d}s = 1 + \frac{t^2}{2}$$

$$y_2(t) = 1 + \int_0^t s\left(1 + \frac{s^2}{2}\right) \mathrm{d}s = 1 + \frac{t^2}{2} + \frac{t^4}{2 \cdot 4}$$

$$y_3(t) = 1 + \int_0^t s\left(1 + \frac{s^2}{2} + \frac{s^4}{2 \cdot 4}\right) \mathrm{d}s = 1 + \frac{t^2}{2} + \frac{t^4}{2 \cdot 4} + \frac{t^6}{2 \cdot 4 \cdot 6}.$$

We conjecture

$$y_n(t) = \sum_{k=0}^{n} \frac{t^{2k}}{2^k k!}. \qquad (*)$$

For $n = 0, 1, 2, 3$ this conjecture is true. Moreover we have

$$y_{n+1}(t) = 1 + \int_0^t s\left(\sum_{k=0}^{n} \frac{s^{2k}}{2^k k!}\right) \mathrm{d}s$$

$$= 1 + \int_0^t \sum_{k=0}^{n} \frac{s^{2k+1}}{2^k k!}\, \mathrm{d}s = 1 + \sum_{k=0}^{n} \frac{t^{2(k+2)}}{2^k k!(2k+2)}$$

$$= \sum_{k=0}^{n+1} \frac{t^{2k}}{2^k k!}.$$

Thus by the principle of mathematical induction the function y_n is indeed given by (∗) and for n going to infinity we obtain the limit

$$y(t) = \sum_{k=0}^{\infty} \frac{t^{2k}}{2^k k!} = \sum_{k=0}^{\infty} \frac{\left(\frac{t^2}{2}\right)^k}{k!} = e^{\frac{t^2}{2}}.$$

2. The conditions on k and u ensure by Theorem II.14.32 that $Tu \in C([0, a])$. For $u, v \in C([0, a])$ we find

$$|Tu(x) - Tv(x)| = \left| g(x) + \int_0^x k(x, t, u(t))\, dt - g(x) - \int_0^x k(x, t, v(t))\, dt \right|$$

$$= \left| \int_0^x (k(x, t, u(t)) - k(x, t, v(t)))\, dt \right|$$

$$\leq \int_0^x |k(x, t, u(t)) - k(x, t, v(t))|\, dt$$

$$\leq xL\|u - v\|_\infty \leq aL\|u - v\|_\infty$$

implying that

$$\|Tu - Tv\|_\infty \leq aL\|u - v\|_\infty.$$

If $L < \frac{1}{a}$ then $aL = K < 1$ and the Banach fixed point theorem implies that T has a unique fixed point w, i.e. $Tw = w$, or

$$w(x) = g(x) + \int_0^x k(x, t, w(t))\, dt.$$

3.   a) We know that $(C([a, b]), \|\cdot\|_\infty)$ is a Banach space and it is also an algebra with multiplication defined by $(u \cdot v)(t) := u(t)v(t)$. Therefore the function $t \mapsto 1$ is the unit element. Of course we have $\|1\|_\infty = 1$ and $\|u \cdot v\|_\infty \leq \|u\|_\infty \|v\|_\infty$. Thus $(C([a, b]), \|\cdot\|_\infty)$ is a Banach algebra.

Our aim is to prove that for $x \in X$ the sequence $\left(\sum_{k=0}^N c_k x^k\right)_{k \in \mathbb{N}}$ is a Cauchy sequence in X provided $\|x\|_X < R$. First we observe that

$$\|x^k\|_X = \|x \cdot \ldots \cdot x\|_X \leq \|x\|_X^k$$

and therefore $\sum_{k=0}^{\infty} |c_k| \|x\|_X^k$ converges for $\|x\|_X < R$, hence for $\epsilon > 0$ we can find $N = N(\epsilon) \in \mathbb{N}$ such that $n > m \geq N$ implies that $\sum_{k=m}^n |c_k| \|x\|^k < \epsilon$. Now it follows with $\epsilon > 0$ and $N(\epsilon)$ as above that for $n > m \geq N$ we have

$$\left\| \sum_{k=0}^n c_k x^k - \sum_{k=0}^m c_k x^k \right\|_X = \left\| \sum_{k=m+1}^n c_k x^k \right\|_X$$

$$\leq \sum_{k=m+1}^n \|c_k x^k\|_X \leq \sum_{k=m+1}^n |c_k| \|x\|_X^k < \epsilon.$$

Thus $(\sum_{k=0}^N c_k x^k)_{N \in \mathbb{N}}$ is a Cauchy sequence in X and therefore convergent.

651

4. Using either **Minkowski's integral inequality** or the solution of Problem 11 to Chapter III.9 we find the estimate

$$\|Ku\|_{L^2} = \left(\int_0^1 \left| \int_0^1 k(x,y)u(y)\,dy \right|^2 dx \right)^{\frac{1}{2}} \leq \|k\|_{L^2}\|u\|_{L^2}.$$

With $k(x,y) = (\frac{x+y}{2})^{\frac{1}{2}}$ it follows that

$$\|k\|_{L^2}^2 = \int_0^1 \left(\int_0^1 \frac{x+y}{2}\,dy \right) dx$$

$$= \int_0^1 \frac{x}{2}\,dx + \int_0^1 \frac{y}{2}\,dy = \frac{1}{2}$$

implying that

$$\|Ku\|_{L^2} \leq \frac{1}{\sqrt{2}}\|u\|_{L^2},$$

i.e. K is a contradiction.

For convenience, we give the proof of $\|Ku\|_{L^2} \leq \|k\|_{L^2}\|u\|_{L^2}$ once more here:

$$\|Ku\|_{L^2} = \left(\int_0^1 \left| \int_0^1 k(x,y)u(y)\,dy \right|^2 dx \right)^{\frac{1}{2}}$$

$$\leq \left(\int_0^1 \left(\int_0^1 |k(x,y)u(y)|\,dy \right)^2 dx \right)^{\frac{1}{2}}$$

$$\leq \int_0^1 \left(\int_0^1 |k(x,y)|^2 |u(y)|^2\,dx \right)^{\frac{1}{2}} dy \qquad (*)$$

$$= \int_0^1 \left(\int_0^1 |k(x,y)|^2\,dx \right) |u(y)|\,dy$$

$$\leq \left(\int_0^1 \left(\int_0^1 |k(x,y)|^2\,dx \right) dy \right)^{\frac{1}{2}} \left(\int_0^1 |u(y)|^2\,dy \right)^{\frac{1}{2}} \qquad (**)$$

$$= \|k\|_{L^2}\|u\|_{L^2}.$$

Here we used in $(*)$ Minkowski's integral inequality and in $(**)$ the Cauchy-Schwarz inequality.

5. Since $\mathcal{S}(\mathbb{R}^n)$ is dense in $L^2(\mathbb{R}^n)$ we may first assume $u \in \mathcal{S}(\mathbb{R}^n)$. By Plancherel's theorem and the convolution theorem, i.e. Theorem 12.17 and Theorem 12.13, we find

$$\|Tu\|_{L^2} = \|p * u\|_{L^2} = \|(p * u)^\wedge\|_{L^2}$$

$$= (2\pi)^{\frac{n}{2}}\|\hat{p}\hat{u}\|_{L^2} \leq (2\pi)^{\frac{n}{2}}\|\hat{p}\|_\infty\|\hat{u}\|_{L^2}$$

$$= (2\pi)^{\frac{n}{2}}\|\hat{p}\|_\infty\|u\|_{L^2},$$

and by our assumption we have $(2\pi)^{\frac{n}{2}}\|\hat{p}\|_\infty < 1$.

6. It is sufficient to prove that for every sequence $(\xi_k)_{k\in\mathbb{N}}$, $\xi_k \in \text{conv}(\{y_1,\ldots,y_m\})$, we can find a convergence subsequence $(\xi_{k_n})_{n\in\mathbb{N}}$ with a limit ξ belonging to $\text{conv}(\{y_1,\ldots,y_m\})$. Since $\xi_k \in \text{conv}(\{y_1,\ldots,y_m\})$ we can find numbers $\lambda_1^{(k)},\ldots,$ $\lambda_m^{(k)} \in [0,1]$, $\lambda_1^{(k)} + \cdots + \lambda_m^{(k)} = 1$, such that $\xi_k = \sum_{j=1}^m \lambda_j^{(k)} y_j$. For every j, $1 \le j \le m$, the sequence $(\lambda_j^{(k)})_{k\in\mathbb{N}}$ is bounded. By the Bolzano-Weierstrass theorem $(\lambda_j^{(k)})_{k\in\mathbb{N}}$ has a convergent subsequence, hence we can find a sequence $(k_\nu)_{\nu\in\mathbb{N}}$, $k_\nu \in \mathbb{N}$, such that each of the sequence $(\lambda_j^{k_\nu})_{\nu\in\mathbb{N}}$, $j = 1,\ldots,m$, converges. This implies that $(\xi_{k_\nu})_{\nu\in\mathbb{N}}$ converges. Since $0 \le \lambda_j^{(k_\nu)} \le 1$ and $\sum_{j=1}^m \lambda_j^{k_\nu} = 1$, it follows that the limits for ν tending to infinity must satisfy the same relations, i.e. $\xi := \lim_{\nu\to\infty} \xi_{k_\nu} \in \text{conv}(\{y_1,\ldots,y_m\})$.

7. We just have to apply Proposition 16.8 to obtain a solution in some interval $[t_0, t_0 + \eta_1)$ and in some interval $(t_0 - \delta_1, t_0]$. These two solutions we can merge to a solution in $(t_0 - \delta_1, t_0 + \eta_1)$. The uniqueness follows again from the Lipschitz condition.

Chapter 17

1. With the help of (17.8) we find

$$\dot{v}_1(t) = v_2(t)$$
$$\dot{v}_2(t) = v_3(t)$$

$$\dot{v}_3(t) = t - (1 + t^4)v_1(t) + \frac{e^{-t}}{1 + t^2} v_2(t) + (\cos t)v_3(t)$$

and $v_1(0) = 1$, $v_2(0) = 0$, $v_3(0) = 1$. We can rewrite this system also as

$$\begin{pmatrix} v_1 \\ v_2 \\ v_3 \end{pmatrix}^{\cdot}(t) = \begin{pmatrix} 0 & 1 & 0 \\ 0 & 0 & 1 \\ -(1+t^4) & \frac{e^{-t}}{1+t^2} & \cos t \end{pmatrix} \begin{pmatrix} v_1 \\ v_2 \\ v_3 \end{pmatrix}(t) + \begin{pmatrix} 0 \\ 0 \\ t \end{pmatrix}$$

and $\begin{pmatrix} v_1 \\ v_2 \\ v_3 \end{pmatrix}(0) = \begin{pmatrix} 1 \\ 0 \\ 1 \end{pmatrix}$.

2. We transform the first equation to

$$\dot{u}(t) = v(t)$$
$$\dot{v}(t) = -a(t)w(t) - v(t)$$

and add the second equation, i.e.

$$\dot{w}(t) = -b(t)u(t)$$

with initial conditions $u(0) = 0$, $v(0) = 1$, $w(0) = 1$. Again we can rewrite this system in matrix form as

$$\begin{pmatrix} u \\ v \\ w \end{pmatrix}^{\cdot}(t) = \begin{pmatrix} 0 & 1 & 0 \\ 0 & -1 & -a(t) \\ -b(t) & 0 & 0 \end{pmatrix} \begin{pmatrix} u \\ v \\ w \end{pmatrix}(t)$$

with initial condition $\begin{pmatrix} u \\ v \\ w \end{pmatrix}(0) = \begin{pmatrix} 0 \\ 1 \\ 1 \end{pmatrix}$.

653

3. By the mean-value theorem we have

$$f_j(x) - f_j(y) = \; <(x-y), \mathrm{grad} f_j(\vartheta)>$$

where ϑ is a point on the line segment connecting x with y. Therefore we find

$$|f_j(x) - f_j(y)| \le n\|x-y\|_\infty \|\mathrm{grad} f_j(\vartheta)\|_\infty$$
$$\le Mn\|x-y\|_\infty.$$

Note that in this calculation we used

$$| <(x-y), \mathrm{grad} f_j(\vartheta)> | = \left| \sum_{l=1}^{n} (x_l - y_l) \frac{\partial f_j}{\partial x_l}(\vartheta) \right|$$

$$\le \left\| \frac{\partial f_j}{\partial x_l} \right\|_\infty \sum_{l=1}^{n} |x_l - y_l|.$$

Consequently we find

$$\|f(x) - f(y)\|_\infty \le L\|x-y\|_\infty, \quad L = Mn.$$

4. We have to prove that if $(v_\nu)_{\nu \in \mathbb{N}}$, $v_\nu \in C^k([a,b]; \mathbb{R}^n)$, is a Cauchy sequence with respect to the norm $\|u\|_{k,\infty} := \max_{0 \le l \le k} \|u^{(l)}\|_\infty$ then $(v_\nu)_{\nu \in \mathbb{N}}$ has a limit in $C^k([a,b]; \mathbb{R}^n)$ with respect to the norm $\| \cdot \|_{k,\infty}$. Suppose that for every $\epsilon > 0$ exists $N = N(\epsilon) \in \mathbb{N}$ such that $\nu, \mu \ge N(\epsilon)$ implies that $\|v_\mu - v_\nu\|_{k,\infty} < \epsilon$. This implies that for $m, n \ge N(\epsilon)$ we have for $0 \le l \le k$ and all $1 \le j \le n$ that $\|v_{\mu,j}^{(l)} - v_{\mu,j}^{(l)}\|_\infty < \epsilon$ where $v_{\mu,j}$ is the jth component of v_μ and consequently $v_{\mu,j}^{(l)}$ is the lth derivative of $v_{\mu,j}$. Thus we find that for $0 \le l \le k$ and $1 \le j \le n$ the sequence $(v_{\nu,j}^{(l)})_{\nu \in \mathbb{N}}$ is a Cauchy sequence in $C([a,b]; \mathbb{R})$ with respect to the norm $\| \cdot \|_\infty$. Since $(C([a,b]; \mathbb{R}), \| \cdot \|_\infty)$ is a Banach space each sequence $(v_{\nu,j}^{(l)})_{\nu \in \mathbb{N}}$ has a limit $v_{\infty,j}^{(l)}$ in this Banach space. In particular we have a limit $v_{\infty,j} := v_{\infty,j}^{(0)}$ for $1 \le j \le n$ and it remains to prove that $\frac{d^l}{dx^l} v_{\infty,j} = v_{\infty,j}^{(l)}$. This follows however for $\nu \to \infty$ from the identity

$$v_{\nu,j}^{(l)}(t) = v_{\nu,j}^{(l)}(t_0) + \int_{t_0}^{t} v_{\nu,j}^{(l+1)}(s) \, ds$$

which holds for $t, t_0 \in \mathbb{R}$, $1 \le j \le n$.

5. First we note that for any $k \in \mathbb{N}$ and any real numbers $a_0, \ldots, a_k$ we have the estimate

$$\max_{0 \le j \le k} |a_j| \le \sum_{j=0}^{k} |a_j| \le (k+1) \max_{0 \le j \le k} |a_j|$$

which implies

$$\|u\|_{k,\infty} = \max_{0 \le l \le k} \|u^{(l)}\|_\infty \le \sum_{l=0}^{k} \|u^{(l)}\|_\infty$$

$$\le (k+1) \max_{0 \le l \le k} \|u^{(l)}\|_\infty = \|u\|_{k,\infty},$$

654

i.e. the equivalence of the norms (17.13) and (17.14) follows.

In order to deduce the equivalence of (17.18) and (17.13) it is sufficient to prove that on $C([a, b], \mathbb{R})$ the norms $\|v\|_\infty := \sup_{t \in [a,b]} |v(t)|$ and $\|v\|_{\infty, \lambda} := \sup_{t \in [a,b]} |v(t)| e^{-\lambda t}$ are equivalent. We know that the continuous function $t \mapsto e^{-\lambda t}$ attains on the compact interval its infimum and its supremum, i.e. we have $0 < A \le e^{-\lambda t} \le B$ for all $t \in [a, b]$, which implies

$$A|v(t)| \le |v(t)| e^{-\lambda t} \le B|v(t)|$$

or $A\|v\|_\infty \le \|v\|_{\infty, \lambda} \le B\|v\|_\infty$.

6. The important observation is that since the system consists of a system for two functions and a single equation, the system is partly decoupled and the linearity allows us to write it in matrix form

$$\begin{pmatrix} u \\ v \\ w \end{pmatrix}^{\cdot}(t) = \begin{pmatrix} 0 & 1 & 0 \\ -1 & 0 & 0 \\ 0 & 0 & 1 \end{pmatrix} \begin{pmatrix} u \\ v \\ w \end{pmatrix}(t).$$

Thus the system matrix is a block matrix with two diagonal blocks, namely $A_1 = \begin{pmatrix} 0 & 1 \\ -1 & 0 \end{pmatrix}$ and $A_2 = 1$, all other entries are 0. Thus we find for $k \in \mathbb{N}$

$$\begin{pmatrix} 0 & 1 & 0 \\ -1 & 0 & 0 \\ 0 & 0 & 1 \end{pmatrix}^k = \begin{pmatrix} A_1 & 0 \\ 0 & A_2 \end{pmatrix}^k = \begin{pmatrix} A_1^k & 0 \\ 0 & A_2^k \end{pmatrix}.$$

Now we can use our results for the iteration corresponding to the systems

$$\begin{aligned} \dot{u} &= v \\ \dot{v} &= -u \end{aligned} \qquad \text{and} \qquad \dot{w} = w$$

to find

$$\begin{pmatrix} u \\ v \\ w \end{pmatrix}(t) = e^{t\begin{pmatrix} 0 & 1 & 0 \\ -1 & 0 & 0 \\ 0 & 0 & 1 \end{pmatrix}} \begin{pmatrix} 0 \\ 1 \\ 1 \end{pmatrix} = \begin{pmatrix} \sin t \\ \cos t \\ e^t \end{pmatrix}.$$

Chapter 18

1. a) We have

$$\frac{d}{dt} \begin{pmatrix} \cos t & t^3 & e^t \\ \frac{1}{1+t^2} & 2t & \sin 5t \\ e^{t^2} & \cosh t & 3 \end{pmatrix} = \begin{pmatrix} -\sin t & 3t^2 & e^t \\ \frac{-2t}{(1+t^2)^2} & 2 & 5\cos 5t \\ 2te^{t^2} & \sinh t & 0 \end{pmatrix}.$$

b) We have

$$\int_0^t \begin{pmatrix} \cos s & s^2 \\ e^s & 4s \end{pmatrix} ds = \begin{pmatrix} \sin s & \frac{1}{3}s^3 \\ e^s & 2s^2 \end{pmatrix} \Big|_0^t$$

$$= \begin{pmatrix} \sin t & \frac{1}{3}t^3 \\ e^t - 1 & 2t^2 \end{pmatrix}.$$

655

c) We have

$$\frac{\mathrm{d}}{\mathrm{d}t} \sum_{j=1}^{n} a_{kj}(t)b_{jl}(t) = \sum_{j=1}^{n} a'_{kj}(t)b_{jl}(t) + \sum_{j=1}^{n} a_{kj}(t)b'_{jl}(t)$$

implying

$$\frac{\mathrm{d}}{\mathrm{d}t} A(t)B(t) = A'(t)B(t) + A(t)B'(t)$$

where $A'(t) = \frac{\mathrm{d}}{\mathrm{d}t}A(t)$.

2.   a) With $\|A\|_{\alpha,\beta} := \inf\{\kappa \geq 0 \mid \|Ax\|_\beta \leq \kappa\|x\|_\alpha\}$ we find $\|Ax\|_\beta \leq \|A\|_{\alpha,\beta}\|x\|_\alpha$ and therefore it follows for $x \neq 0$ that $\frac{\|Ax\|_\beta}{\|x\|_\alpha} \leq \|A\|_{\alpha,\beta}$ which in turn implies that

$$\tau := \sup\left\{ \frac{\|Ax\|_\beta}{\|x\|_\alpha} \;\middle|\; x \in \mathbb{R}^n\backslash\{0\} \right\} \leq \|A\|_{\alpha,\beta}.$$

However it also holds $\|Ax\|_\beta \leq \tau\|x\|_\alpha$ for $x \in \mathbb{R}^n\backslash\{0\}$ implying that $\tau \leq \|A\|_{\alpha,\beta}$, i.e. $\tau = \|A\|_{\alpha,\beta}$.

b) Clearly $\|A\|_{\alpha,\beta} \geq 0$ and $\|A\|_{\alpha,\beta} = 0$ implies that $\|Ax\|_\beta = 0$ for all $x \in \mathbb{R}^n$, i.e. $Ax = 0$ for $x \in \mathbb{R}^n$ which means that $A = 0$. For $\lambda \in \mathbb{R}$ we find by part a)

$$\|\lambda A\|_{\alpha,\beta} = \sup\left\{ \frac{\|\lambda Ax\|_\beta}{\|x\|_\alpha} \;\middle|\; x \in \mathbb{R}^n\backslash\{0\} \right\}$$

$$= |\lambda| \sup\left\{ \frac{\|Ax\|_\beta}{\|x\|_\alpha} \;\middle|\; x \in \mathbb{R}^n\backslash\{0\} \right\} = |\lambda|\|A\|_{\alpha,\beta}.$$

Finally we find for $A, B \in M(n; \mathbb{R})$ that

$$\|A + B\|_{\alpha,\beta} = \sup\left\{ \frac{\|Ax + Bx\|_\beta}{\|x\|_\alpha} \;\middle|\; x \in \mathbb{R}^n\backslash\{0\} \right\}$$

$$\leq \sup\left\{ \frac{\|Ax\|_\beta}{\|x\|_\alpha} + \frac{\|Bx\|_\beta}{\|x\|_\alpha} \;\middle|\; x \in \mathbb{R}^n\backslash\{0\} \right\}$$

$$\leq \sup\left\{ \frac{\|Ax\|_\beta}{\|x\|_\alpha} \;\middle|\; x \in \mathbb{R}^n\backslash\{0\} \right\} + \sup\left\{ \frac{\|Bx\|_\beta}{\|x\|_\alpha} \;\middle|\; x \in \mathbb{R}^n\backslash\{0\} \right\}$$

$$= \|A\|_{\alpha,\beta} + \|B\|_{\alpha,\beta}.$$

Hence $\|\cdot\|_{\alpha,\beta}$ is a norm on $M(n; \mathbb{R})$. Clearly we have $\|Ax\|_\beta \leq \|A\|_{\alpha,\beta}\|x\|_\alpha$.

3.   a) Since $\left(\begin{smallmatrix} 0 & 1 \\ 0 & 0 \end{smallmatrix}\right)$ and $\left(\begin{smallmatrix} 1 & 0 \\ 0 & 0 \end{smallmatrix}\right)$ are not the zero element in $M(n, \mathbb{R})$ they will have a strictly positive norm. However $\left(\begin{smallmatrix} 0 & 1 \\ 0 & 0 \end{smallmatrix}\right)\left(\begin{smallmatrix} 1 & 0 \\ 0 & 0 \end{smallmatrix}\right) = \left(\begin{smallmatrix} 0 & 0 \\ 0 & 0 \end{smallmatrix}\right)$ and therefore the product of these two matrices has norm 0.

b) Since $\|A\| < 1$ it follows from

$$\left\| \sum_{k=0}^{\infty} A^k \right\| \leq \sum_{k=0}^{\infty} \|A^k\| \leq \sum_{k=0}^{\infty} \|A\|^k = \frac{1}{1 - \|A\|}$$

656

that the series $\sum_{k=0}^{\infty} A^k$ converges in the norm $\| \cdot \|$. Now we observe that

$$(\mathrm{id}_n - A) \sum_{k=0}^{\infty} A^k = \sum_{k=0}^{\infty} A^k - \sum_{k=0}^{\infty} A^{k+1} = \mathrm{id}_n.$$

4. Since $A^N = 0$ for $k > N$ it follows that $A^k = 0$ and therefore we find

$$\sum_{k=0}^{\infty} c_k (sA)^k = \sum_{k=0}^{N-1} c_k A^k s^k$$
$$= (c_0 \mathrm{id}_n) + (c_1 A)s + \cdots + (c_{N-1} A^{N-1}) s^{N-1}$$

which is a matrix-valued polynomial in s of degree at most $N - 1$.

5. We note first that

$$\cos At = \sum_{k=0}^{\infty} \frac{(-1)^k}{(2k!)} A^{2k} t^{2k}$$

and

$$\sin At = \sum_{k=1}^{\infty} \frac{(-1)^{k-1}}{(2k-1)!} A^{2k-1} t^{2k-1}.$$

Now we find using the uniform convergence of the original series as well as their formal derivatives

$$\frac{\mathrm{d}}{\mathrm{d}t} \cos At = \frac{\mathrm{d}}{\mathrm{d}t} \sum_{k=0}^{\infty} \frac{(-1)^k}{(2k)!} A^{2k} t^{2k}$$
$$= \sum_{k=0}^{\infty} \frac{(-1)^k}{(2k)!} A^{2k} (2k) t^{2k-1}$$
$$= A \sum_{k=0}^{\infty} \frac{(-1)^k}{(2k-1)!} A^{2k-1} t^{2k-1}$$
$$= -A \sum_{k=0}^{\infty} \frac{(-1)^{k-1}}{(2k-1)!} A^{2k-1} t^{2k-1} = -A \sin At$$

and

$$\frac{\mathrm{d}}{\mathrm{d}t} \sin At = \frac{\mathrm{d}}{\mathrm{d}t} \sum_{k=1}^{\infty} \frac{(-1)^{k-1}}{(2k-1)!} A^{2k-1} t^{2k-1}$$
$$= \sum_{k=1}^{\infty} \frac{(-1)^{k-1}}{(2k-1)!} A^{2k-1} (2k-1) t^{2k-2}$$
$$= A \sum_{k=1}^{\infty} \frac{(-1)^{k-1}}{(2(k-1))!} A^{2(k-1)} t^{2(k-1)}$$
$$= A \sum_{k=0}^{\infty} \frac{(-1)^k}{(2k)!} A^{2k} t^{2k} = A \cos At.$$

6.   a) This follows from the convergence of the series $\sum_{k=0}^{\infty} \frac{(iA)^k}{k!}$ and the relation

$$\sum_{k=0}^{\infty} \frac{(iA)^k}{k!} = \sum_{k=0}^{\infty} \frac{(-1)^k}{2k} A^{2k} + i \sum_{k=1}^{\infty} \frac{(-1)^{k-1}}{(2k-1)!} A^{2k-1}.$$

b) We note that by part a) we have

$$\cos A = \frac{e^{iA} + e^{-iA}}{2}$$

and therefore

$$\cos(A + B) = \frac{e^{i(A+B)} + e^{-i(A+B)}}{2}$$
$$= \frac{e^{iA} e^{iB} + e^{-iA} e^{-iB}}{2}$$

where we used in the last step that $[A, B] = 0$. Next we observe that

$$e^{iA} e^{iB} = (\cos A + i \sin A)(\cos A + i \sin B)$$

and

$$e^{-iA} e^{-iB} = (\cos A - i \sin A)(\cos B - i \sin B).$$

Multiplying these two right hand sides gives the result when taking into account that $[A, B] = 0$ implies that commutators such as $[\cos A, \sin B]$ etc. all vanish.

7. We have to find

$$e^{\left(\begin{smallmatrix} 0 & 1 \\ 1 & 0 \end{smallmatrix} \right)t} = \sum_{k=0}^{\infty} \frac{1}{k!} \begin{pmatrix} 0 & 1 \\ 1 & 0 \end{pmatrix}^k t^k.$$

Noting that

$$\begin{pmatrix} 0 & 1 \\ 1 & 0 \end{pmatrix}^0 = \mathrm{id}_2$$

$$\begin{pmatrix} 0 & 1 \\ 1 & 0 \end{pmatrix}^1 = \begin{pmatrix} 0 & 1 \\ 1 & 0 \end{pmatrix}$$

$$\begin{pmatrix} 0 & 1 \\ 1 & 0 \end{pmatrix}^2 = \mathrm{id}_2$$

we find that $\left(\begin{smallmatrix} 0 & 1 \\ 1 & 0 \end{smallmatrix} \right)^{2k} = \mathrm{id}_2$, $k \in \mathbb{N}_0$, and $\left(\begin{smallmatrix} 0 & 1 \\ 1 & 0 \end{smallmatrix} \right)^{2k-1} = \left(\begin{smallmatrix} 0 & 1 \\ 1 & 0 \end{smallmatrix} \right)$, $k \in \mathbb{N}$. This yields

$$e^{\left(\begin{smallmatrix} 0 & 1 \\ 1 & 0 \end{smallmatrix} \right)t} = \sum_{l=0}^{\infty} \frac{1}{(2l)!} \begin{pmatrix} 0 & 1 \\ 1 & 0 \end{pmatrix}^{2l} t^{2l} + \sum_{l=1}^{\infty} \frac{1}{(2l-1)!} \begin{pmatrix} 0 & 1 \\ 1 & 0 \end{pmatrix}^{2l-1} t^{2l-1}$$

$$= \sum_{l=0}^{\infty} \frac{1}{(2l)!} \begin{pmatrix} t^{2l} & 0 \\ 0 & t^{2l} \end{pmatrix} + \sum_{l=1}^{\infty} \frac{1}{(2l-1)!} \begin{pmatrix} 0 & t^{2l-1} \\ t^{2l-1} & 0 \end{pmatrix}$$

$$= \begin{pmatrix} \cosh t & 0 \\ 0 & \cosh t \end{pmatrix} + \begin{pmatrix} 0 & \sinh t \\ \sinh t & 0 \end{pmatrix} = \begin{pmatrix} \cosh t & \sinh t \\ \sinh t & \cosh t \end{pmatrix}.$$

Finally we obtain

$$\begin{pmatrix} u(t) \\ v(t) \end{pmatrix} = \begin{pmatrix} \cosh t & \sinh t \\ \sinh t & \cosh t \end{pmatrix} \begin{pmatrix} 1 \\ 0 \end{pmatrix} = \begin{pmatrix} \cosh t \\ \sinh t \end{pmatrix}.$$

8. We have to discuss the matrix $A = \begin{pmatrix} 1 & -3 & 3 \\ 3 & -5 & 3 \\ 6 & -6 & 4 \end{pmatrix}$. A longer, but straight forward calculation gives that A is diagonalizable with the help of the matrix $U = \begin{pmatrix} 1 & 1 & 1 \\ 1 & 0 & 1 \\ 0 & -1 & 2 \end{pmatrix}$, $U^{-1} = -\frac{1}{2}\begin{pmatrix} 1 & -3 & 1 \\ -2 & 2 & 0 \\ -1 & 1 & -1 \end{pmatrix}$ and we have

$$U^{-1}AU = \begin{pmatrix} -2 & 0 & 0 \\ 0 & -2 & 0 \\ 0 & 0 & 4 \end{pmatrix}.$$

(Details of this calculation may be found in S. Lipschutz, Theory and Problems of Linear Algebra, McGraw-Hill, 1968.) We transform the initial value $\begin{pmatrix} 1 \\ 2 \\ 3 \end{pmatrix}$ with the help of U^{-1} to find

$$w(0) = U^{-1}\begin{pmatrix} 1 \\ 2 \\ 3 \end{pmatrix} = -\frac{1}{2}\begin{pmatrix} 1 & -3 & 1 \\ -2 & 2 & 0 \\ 1 & -1 & 1 \end{pmatrix}\begin{pmatrix} 1 \\ 2 \\ 3 \end{pmatrix} = \begin{pmatrix} 1 \\ -1 \\ 1 \end{pmatrix}$$

and the solution to

$$\dot{w}(t) = \begin{pmatrix} -2 & 0 & 0 \\ 0 & -2 & 0 \\ 0 & 0 & 4 \end{pmatrix}w(t), \quad w(0) = \begin{pmatrix} 1 \\ -1 \\ 1 \end{pmatrix}$$

is given by

$$w(t) = \begin{pmatrix} e^{-2t} \\ -e^{-2t} \\ e^{4t} \end{pmatrix}.$$

The solution to the original problem we obtain now as

$$y(t) = Uw(t) = \begin{pmatrix} 1 & 1 & 1 \\ 1 & 0 & 1 \\ 0 & -1 & 2 \end{pmatrix}\begin{pmatrix} e^{-2t} \\ -e^{-2t} \\ e^{4t} \end{pmatrix}$$

$$= \begin{pmatrix} e^{4t} \\ e^{-2t} + e^{4t} \\ e^{-2t} + 2e^{4t} \end{pmatrix}.$$

Indeed we find $y(0) = \begin{pmatrix} 1 \\ 1+1 \\ 1+2 \end{pmatrix} = \begin{pmatrix} 1 \\ 2 \\ 3 \end{pmatrix}$ and

$$y_1'(t) = 4e^{4t} = e^{4t} - 3e^{-2t} - 3e^{4t} + 3e^{-2t} + 6e^{4t}$$
$$y_2'(t) = -2e^{-2t} + 4e^{4t} = 3e^{4t} - 5e^{-2t} - 5e^{4t} + 3e^{-2t} + 6e^{4t}$$
$$y_3'(t) = -2e^{-2t} + 8e^{4t} = 6e^{4t} - 6e^{-2t} - 6e^{4t} + 4e^{-2t} + 8e^{4t}.$$

659

9. We can diagonalize A and therefore e^{tA} by an orthogonal matrix U, i.e. we have

$$U^{-1}e^{tA}U = e^{\begin{pmatrix} \lambda_1 & & \mathbf{0} \\ & \ddots & \\ \mathbf{0} & & \lambda_n \end{pmatrix}t}$$

and we may choose $\lambda_1 \geq \lambda_k > 0$ for all $1 \leq k \leq n$ where $\lambda_1, \ldots, \lambda_n$ are the eigenvalues of A (counted according to their multiplicity). Therefore we find

$$U^{-1}\int_0^N e^{-\lambda t}e^{tA}\, dtU = \int_0^N U^{-1}e^{-\lambda t}e^{tA}U\, dt$$

$$= \int_0^N e^{-\lambda t}e^{\begin{pmatrix} \lambda_1 & & \mathbf{0} \\ & \ddots & \\ \mathbf{0} & & \lambda_n \end{pmatrix}t}\, dt.$$

Since

$$\int_0^N e^{-\lambda t + \lambda_k t}\, dt = \frac{1}{\lambda_k - \lambda}e^{-t(\lambda - \lambda_k)}\Big|_0^N$$

$$= \frac{1}{\lambda - \lambda_k} - \frac{e^{-N(\lambda - \lambda_k)}}{\lambda - \lambda_k} =: c_{kk}(N)$$

it follows that

$$\lim_{N\to\infty}\int_0^N e^{-\lambda t}e^{\begin{pmatrix} \lambda_1 & & \mathbf{0} \\ & \ddots & \\ \mathbf{0} & & \lambda_n \end{pmatrix}t}\, dt = \begin{pmatrix} \frac{1}{\lambda - \lambda_1} & & \mathbf{0} \\ & \ddots & \\ \mathbf{0} & & \frac{1}{\lambda - \lambda_n} \end{pmatrix}.$$

Eventually we get

$$\lim_{N\to\infty}\int_0^N e^{-\lambda t}e^{tA}\, dt = \lim_{N\to\infty}U\begin{pmatrix} c_{11}(N) & & \mathbf{0} \\ & \ddots & \\ \mathbf{0} & & c_{nn}(N) \end{pmatrix}U^{-1}$$

$$= U\begin{pmatrix} \frac{1}{\lambda - \lambda_1} & & \mathbf{0} \\ & \ddots & \\ \mathbf{0} & & \frac{1}{\lambda - \lambda_n} \end{pmatrix}U^{-1} = (\lambda\mathrm{id} - A)^{-1}.$$

10. We get $w := C^{-1}y$, i.e. $y = Cw$ and arrive at the problem

$$\dot{w} = C^{-1}Ay = C^{-1}ACw = Bw$$

with

$$w(0) = C^{-1}y(0) = C^{-1}\begin{pmatrix} 1 \\ -1 \\ 0 \end{pmatrix}.$$

We note that $C^{-1} = \frac{1}{3} \begin{pmatrix} 0 & -1 & 0 \\ -18 & 9 & 3 \\ 3 & -2 & 0 \end{pmatrix}$ and therefore we find

$$w(0) = \frac{1}{3} \begin{pmatrix} 0 & -1 & 0 \\ -18 & 9 & 3 \\ 3 & -2 & 0 \end{pmatrix} \begin{pmatrix} 1 \\ -1 \\ 0 \end{pmatrix} = \begin{pmatrix} \frac{1}{3} \\ -9 \\ \frac{5}{3} \end{pmatrix}.$$

Now it follows that

$$w(t) = e^{Bt} w(0) = e^{\begin{pmatrix} -3 & 1 & 0 \\ 0 & -3 & 0 \\ 0 & 0 & -3 \end{pmatrix} t} \begin{pmatrix} \frac{1}{3} \\ -9 \\ \frac{5}{3} \end{pmatrix}$$

$$= \begin{pmatrix} e^{-3t} & te^{-3t} & 0 \\ 0 & e^{-3t} & 0 \\ 0 & 0 & e^{-3t} \end{pmatrix} \begin{pmatrix} \frac{1}{3} \\ -9 \\ \frac{5}{3} \end{pmatrix} = \begin{pmatrix} \frac{1}{3}e^{-3t} - 9te^{-3t} \\ -9e^{-3t} \\ \frac{5}{3}e^{-3t} \end{pmatrix}.$$

Just let us check the result

$$\dot{w}(t) = \begin{pmatrix} -e^{-3t} - 9e^{-3t} + 27te^{-3t} \\ 27e^{-3t} \\ -5e^{-3t} \end{pmatrix} = \begin{pmatrix} -10 + 27t \\ 27 \\ -5 \end{pmatrix} e^{-3t}$$

and

$$\begin{pmatrix} -3 & 1 & 0 \\ 0 & -3 & 0 \\ 0 & 0 & -3 \end{pmatrix} \begin{pmatrix} \frac{1}{3}e^{-3t} - 9te^{-3t} \\ -9e^{-3t} \\ \frac{5}{3}e^{-3t} \end{pmatrix} = \begin{pmatrix} -10 + 27t \\ 27 \\ -5 \end{pmatrix} e^{-3t}.$$

Now we find y as

$$y = Cw = \begin{pmatrix} -2 & 0 & 1 \\ -3 & 0 & 0 \\ -3 & 1 & 6 \end{pmatrix} \begin{pmatrix} \frac{1}{3} - 9t \\ -9 \\ \frac{5}{3} \end{pmatrix} e^{-3t}$$

$$= \begin{pmatrix} 1 + 18t \\ -1 + 27t \\ 27t \end{pmatrix} e^{-3t}.$$

Again, we briefly check the result

$$\dot{y}(t) = \begin{pmatrix} 18 \\ 27 \\ 27 \end{pmatrix} e^{-3t} - 3 \begin{pmatrix} 1 + 18t \\ -1 + 27t \\ 27t \end{pmatrix} e^{-3t} = \begin{pmatrix} 15 - 54t \\ 30 - 81t \\ 27 - 81t \end{pmatrix} e^{-3t}$$

and on the other side we have

$$Ay = \begin{pmatrix} 9 & -6 & -2 \\ 18 & -12 & -3 \\ 18 & -9 & -6 \end{pmatrix} \begin{pmatrix} 1 + 18t \\ -1 + 27t \\ 27t \end{pmatrix} e^{-3t} = \begin{pmatrix} 15 - 54t \\ 30 - 81t \\ 27 - 81t \end{pmatrix} e^{-3t}.$$

11. We have

$$A^2 = \begin{pmatrix} 0 & 1 & 0 & 1 \\ 0 & 0 & 1 & 1 \\ 0 & 0 & 0 & 0 \\ 0 & 0 & 0 & 0 \end{pmatrix} \begin{pmatrix} 0 & 1 & 0 & 1 \\ 0 & 0 & 1 & 1 \\ 0 & 0 & 0 & 0 \\ 0 & 0 & 0 & 0 \end{pmatrix} = \begin{pmatrix} 0 & 0 & 1 & 1 \\ 0 & 0 & 0 & 0 \\ 0 & 0 & 0 & 0 \\ 0 & 0 & 0 & 0 \end{pmatrix}$$

and

$$A^3 = A^2 A = \begin{pmatrix} 0 & 0 & 1 & 1 \\ 0 & 0 & 0 & 0 \\ 0 & 0 & 0 & 0 \\ 0 & 0 & 0 & 0 \end{pmatrix} \begin{pmatrix} 0 & 1 & 0 & 1 \\ 0 & 0 & 1 & 1 \\ 0 & 0 & 0 & 0 \\ 0 & 0 & 0 & 0 \end{pmatrix} = \begin{pmatrix} 0 & 0 & 0 & 0 \\ 0 & 0 & 0 & 0 \\ 0 & 0 & 0 & 0 \\ 0 & 0 & 0 & 0 \end{pmatrix}$$

and therefore A is nilpotent of degree 3. Since

$$\sin At = \sum_{k=1}^{\infty} \frac{(-1)^{k-1}}{(2k-1)!} A^{2k-1} t^{2k-1}$$

we find

$$\sin At = At = \begin{pmatrix} 0 & t & 0 & t \\ 0 & 0 & t & t \\ 0 & 0 & 0 & 0 \\ 0 & 0 & 0 & 0 \end{pmatrix}.$$

Moreover, from

$$\cosh At = \sum_{k=0}^{\infty} \frac{A^{2k} t^{2k}}{(2k)!}$$

we deduce

$$\cosh At = \operatorname{id}_4 + \frac{A^2}{2} t^2 = \begin{pmatrix} 1 & 0 & \frac{t^2}{2} & \frac{t^2}{2} \\ 0 & 1 & 0 & 0 \\ 0 & 0 & 1 & 0 \\ 0 & 0 & 0 & 1 \end{pmatrix}$$

12. For $N = \begin{pmatrix} 0 & 1 & 0 & \cdots & 0 \\ \vdots & \ddots & & \ddots & \vdots \\ \vdots & & \ddots & & 0 \\ \vdots & & & \ddots & 1 \\ 0 & \cdots & \cdots & \cdots & 0 \end{pmatrix} \in M(m, \mathbb{R})$ we have

$$e^{tN} = \sum_{k=0}^{\infty} \frac{(tN)^k}{k!} = \sum_{k=0}^{m-1} \frac{t^k N^k}{k!}$$

and it remains to find N^k, $0 \le k \le m - 1$. Clearly we have $N^0 = \operatorname{id}_m$ and $N^1 = N$. We claim that for the matrix elements $b_{\mu,\nu}^{(k)}$ of N^k, $0 \le k \le m - 1$, we have $b_{l,l+k} = 1$, $l = 1, \ldots, n$, and $b_{\mu,\nu} = 0$ otherwise. For $k = 0$ and $k = 1$ this is trivial. Suppose the result holds for N^k. For the matrix elements $b_{\nu,\mu}^{(k+1)}$ we have

$$b_{\nu,\mu}^{(k+1)} = \sum_{\alpha=1}^{n} b_{\nu\alpha}^{(k)} b_{\alpha\mu}^{(1)}.$$

For $\nu = l$ the only non-trivial term $b_{\nu\alpha}^{(k)}$ is $b_{l,l+k}^{(k)} = 1$ and only for $b_{l+k,l+k+1}^{(1)}$ we have a contribution to the sum which gives

$$b_{l,l+k+1}^{(k+1)} = 1 \quad \text{and} \quad b_{\nu,\mu}^{(k+1)} = 0 \quad \text{otherwise,}$$

which is our claim. Now (18.44) follows from adding up the terms

$$\frac{t^k}{k!} N^k = \begin{pmatrix} 0 & \cdots & 0 & \frac{t^k}{k!} & 0 & \cdots & 0 \\ & \ddots & & \ddots & & \ddots & \vdots \\ & & \ddots & & \ddots & & 0 \\ & & & \ddots & & \ddots & \frac{t^k}{k!} \\ & & & & \ddots & & 0 \\ & 0 & & & & \ddots & \vdots \\ & & & & & & 0 \end{pmatrix}.$$

Chapter 19

1.   a) Suppose that for c_1, $c_2 \in \mathbb{R}$ we have $c_1 a_1^t + c_2 a_2^t = 0$ for all $t \in \mathbb{R}$. This implies that

$$c_1 + c_2 e^{(\ln a_2 - \ln a_1)t} = 0 \quad \text{for all } t \in \mathbb{R},$$

or

$$c_1 = -c_2 e^{(\ln a_2 - \ln a_1)t} \quad \text{for all } t \in \mathbb{R}.$$

By assumption we have $\ln a_2 - \ln a_1 > 0$ and therefore it follows that $c_1 = c_2 = 0$, i.e. the set $\{f, g\}$ is an independent set in $C(\mathbb{R})$.

b) We note that $\cos 4x = \cos^2 2x - \sin^2 2x$. Now, with $c_1 = c_2 = 1$ and $c_3 = -1$ we find

$$c_1 \cos 4x + c_2 \sin^2 2x + c_3 \cos^2 2x = c_1 \cos^2 2x - c_1 \sin^2 2x + c_2 \sin^2 2x + c_3 \cos^2 2x = 0$$

for all $x \in \mathbb{R}$ and therefore the set $\{f, g, h\}$ is linearly dependent in $C(\mathbb{R})$.

c) Straightforward differentiation yields $\cosh' t = \sinh t$ and $\sinh' t = \cosh t$, hence $t \mapsto \left(\begin{smallmatrix} \cosh t \\ \sinh t \end{smallmatrix}\right)$ and $t \mapsto \left(\begin{smallmatrix} \sinh t \\ \cosh t \end{smallmatrix}\right)$ are solutions to the system $\dot{y}_1 = y_2$, $\dot{y}_2 = y_1$. Suppose that

$$\lambda \begin{pmatrix} \cosh t \\ \sinh t \end{pmatrix} + \mu \begin{pmatrix} \sinh t \\ \cosh t \end{pmatrix} = 0 \quad \text{for all } t \in \mathbb{R}.$$

For $\lambda \neq 0$ this implies $\cosh t = -\frac{\mu}{\lambda} \sinh t$ which does not hold, and therefore we must have $\lambda = 0$ which however implies $\mu = 0$, i.e. the two mappings $t \mapsto \left(\begin{smallmatrix} \cosh t \\ \sinh t \end{smallmatrix}\right)$ and $t \mapsto \left(\begin{smallmatrix} \sinh t \\ \cosh t \end{smallmatrix}\right)$ are independent in $C(\mathbb{R}; \mathbb{R}^2)$.

2. By inspection we find that $t \mapsto \left(\begin{smallmatrix}\cos t\\ \sin t\end{smallmatrix}\right)$ and $t \mapsto \left(\begin{smallmatrix}-\sin t\\ \cos t\end{smallmatrix}\right)$ are solutions, in fact independent solutions of the system $\dot{y}_1 = -y_2$ and $\dot{y}_2 = y_1$. Moreover we have

$$\begin{pmatrix} \cos t & -\sin t \\ \sin t & \cos t \end{pmatrix}^{\cdot} = \begin{pmatrix} -\sin t & -\cos t \\ \cos t & -\sin t \end{pmatrix}$$

$$= \begin{pmatrix} 0 & -1 \\ 1 & 0 \end{pmatrix} \begin{pmatrix} \cos t & -\sin t \\ \sin t & \cos t \end{pmatrix},$$

verifying (19.12).

3. We note that for $t \in (0, \infty)$

$$\begin{pmatrix} u_1 \\ v_1 \end{pmatrix}^{\cdot}(t) = \begin{pmatrix} -2t \sin t^2 \\ -2 \sin t^2 - 4t^2 \cos t^2 \end{pmatrix}$$

$$= \begin{pmatrix} 0 & 1 \\ -4t^2 & \frac{1}{t} \end{pmatrix} \begin{pmatrix} \cos t^2 \\ -2t \sin t^2 \end{pmatrix}$$

and

$$\begin{pmatrix} u_2 \\ v_2 \end{pmatrix}^{\cdot}(t) = \begin{pmatrix} 2t \cos t^2 \\ 2 \cos t^2 - 4t^2 \sin t^2 \end{pmatrix}$$

$$= \begin{pmatrix} 0 & 1 \\ -4t^2 & \frac{1}{t} \end{pmatrix} \begin{pmatrix} \sin t^2 \\ 2t \cos t^2 \end{pmatrix}.$$

The Wronski determinant of $\left(\begin{smallmatrix}u_1\\ v_1\end{smallmatrix}\right)$ and $\left(\begin{smallmatrix}u_2\\ v_2\end{smallmatrix}\right)$ is given by

$$W(t) = \det \begin{pmatrix} \cos t^2 & \sin t^2 \\ -2t \sin t^2 & 2 \cos t^2 \end{pmatrix}$$

$$= 2t(\cos t^2)^2 + 2t(\sin t^2)^2 = 2t \neq 0$$

which is non-zero for $t \in (0, \infty)$ and therefore $\left(\begin{smallmatrix}u_1\\ v_1\end{smallmatrix}\right)$ and $\left(\begin{smallmatrix}u_2\\ v_2\end{smallmatrix}\right)$ are independent. We observe further that $\dot{W}(t) = 2$. Further we find

$$\operatorname{tr} A(t) = \operatorname{tr} \begin{pmatrix} 0 & 1 \\ -4t^2 & \frac{1}{t} \end{pmatrix} = \frac{1}{t}$$

and therefore $\operatorname{tr} A(t) W(t) = 2 = \dot{W}(t)$ and we have verified (19.20). Finally we note

$$\begin{pmatrix} u_1 & u_2 \\ v_1 & v_2 \end{pmatrix}^{\cdot}(t) = \begin{pmatrix} -2t \sin t^2 & 2t \cos t^2 \\ -2 \sin t^2 - 4t^2 \cos t^2 & 2 \cos t^2 - 4t^2 \sin t^2 \end{pmatrix}$$

$$= \begin{pmatrix} 0 & 1 \\ -4t^2 & \frac{1}{t} \end{pmatrix} \begin{pmatrix} \cos t^2 & \sin t^2 \\ -2t \sin t^2 & 2t \cos t^2 \end{pmatrix}.$$

4. First we note that for n differentiable functions $u_1, \ldots, u_n : (a, b) \to \mathbb{R}$ we have

$$(u_1 \cdot \ldots \cdot u_n)^{\cdot}(t) = \sum_{j=1}^{n} u_1(t) \cdot \ldots \cdot u_{j-1}(t) \dot{u}_j(t) u_{j+1}(t) \cdot \ldots \cdot u_n(t), \qquad (*)$$

664

which follows by induction from the observation that

$$(u_1 \cdot \ldots \cdot u_n)^{\cdot}(t) = \dot{u}_1(t)(u_2 \cdot \ldots \cdot u_n)(t) + u_1(t)(u_2 \cdot \ldots \cdot u_n)^{\cdot}(t).$$

Now we consider the Laplace expansion, see (III.A.I.16), of $\det B(t) = \det(b_1(t), \ldots,$
$b_n(t))$ with $b_j(t) = \begin{pmatrix} b_{1,j}(t) \\ \vdots \\ b_{n,j}(t) \end{pmatrix}$, i.e.

$$\det B(t) = \sum_{\sigma \in S_n} (\operatorname{sgn}\sigma) b_{1,\sigma(1)}(t) \ldots b_{n,\sigma(n)}(t) \tag{**}$$

where S_n denotes the symmetric group acting on n elements and $\operatorname{sgn}\sigma$ is the signum of the permutation σ. From $(*)$ and $(**)$ we deduce

$$(\det B(t))^{\cdot} = \sum_{\sigma \in S_n} (\operatorname{sgn}\sigma)(b_{1,\sigma(1)}(t) \ldots b_{n,\sigma(n)}(t))^{\cdot}$$

$$= \sum_{\sigma \in S_n} (\operatorname{sgn}\sigma) \sum_{j=1}^{n} b_{1,\sigma(1)}(t) \ldots b_{j-1,\sigma(j-1)}(t) \dot{b}_{j,\sigma(j)}(t) b_{j+1,\sigma(j+1)}(t) \ldots b_{n,\sigma(n)}(t)$$

$$= \sum_{j=1}^{n} \det(b_1(t), \ldots, b_{j-1}(t), \dot{b}_j(t), b_{j+1}(t), \ldots, b_n(t)).$$

5. Let $\tilde{u}$ be a special solution to $\dot{u}(t) = A(t)u(t) + f(t)$ and v be a solution to $\dot{v}(t) = A(t)v(t)$. Then it follows that $(v + \tilde{u})^{\cdot}(t) = A(t)v(t) + A(t)\tilde{u}(t) + f(t) = A(t)(v + \tilde{u})(t) + f(t)$, so $v + \tilde{u}$ is a solution to $\dot{u}(t) = A(t)u(t) + f(t)$. Conversely, let w be a solution to $\dot{u}(t) = A(t)w(t) + f(t)$ and $\tilde{u}$ a special solution. Then we have $(w - \tilde{u})^{\cdot}(t) = A(t)w(t) - f(t) - A(t)\tilde{u}(t) + f(t) = A(t)(w - \tilde{u})(t)$, i.e. $w - \tilde{u}$ solves $\dot{v}(t) = A(t)v(t)$.

6. With $A(t) = \begin{pmatrix} \frac{3}{t} & -1 \\ \frac{1}{2t^2} & \frac{1}{2t} \end{pmatrix}$ and $U(t) = \begin{pmatrix} t^2 & -2t^{\frac{5}{2}} \\ t & t^{\frac{3}{2}} \end{pmatrix}$ we find $\det U(t) = t^{\frac{7}{2}}$ and $U(t)^{-1} = \frac{1}{t^{\frac{7}{2}}} U(t)^{-1} = \begin{pmatrix} -t^{\frac{3}{2}} & 2t^{\frac{5}{2}} \\ -t & t^2 \end{pmatrix}$. This leads to

$$U(t)U^{-1}(s) = \begin{pmatrix} t^2 & -2t^{\frac{5}{2}} \\ t & -t^{\frac{3}{2}} \end{pmatrix} \frac{1}{s^{\frac{7}{2}}} \begin{pmatrix} -s^{\frac{3}{2}} & 2s^{\frac{5}{2}} \\ -s & s^2 \end{pmatrix}$$

$$= \frac{1}{s^{\frac{7}{2}}} \begin{pmatrix} -t^2 s^{\frac{3}{2}} + 2t^{\frac{5}{2}} s & 2t^2 s^{\frac{5}{2}} - 2t^{\frac{5}{2}} s \\ -ts^{\frac{3}{2}} + t^{\frac{3}{2}} s & 2ts^{\frac{5}{2}} - t^{\frac{3}{2}} s \end{pmatrix}$$

and therefore

$$U(t)U^{-1}(s) \begin{pmatrix} s^4 \\ s^{\frac{5}{2}} \end{pmatrix} = \frac{1}{s^{\frac{7}{2}}} \cdot s \cdot s^{\frac{5}{2}} \begin{pmatrix} -t^2 s^{\frac{1}{2}} + 2t^{\frac{5}{2}} & 2t^2 s^{\frac{3}{2}} - 2t^{\frac{5}{2}} s \\ -ts^{\frac{1}{2}} + t^{\frac{3}{2}} & 2ts^{\frac{3}{2}} - t^{\frac{3}{2}} \end{pmatrix} \begin{pmatrix} s^{\frac{3}{2}} \\ 1 \end{pmatrix}$$

$$= \begin{pmatrix} -t^2 s^2 + 2(t^2 + t^{\frac{5}{2}})s^{\frac{3}{2}} - 2t^{\frac{5}{2}} s \\ -ts^2 + (t^{\frac{3}{2}} + 2t)s^{\frac{3}{2}} - 2t^{\frac{5}{2}} s \end{pmatrix}.$$

It follows that

$$\int_1^t U(t)U^{-1}(s)\begin{pmatrix}s^4\\s^{\frac{5}{2}}\end{pmatrix}\,ds = \int_1^t \begin{pmatrix}-t^2s^2+2(t^2+t^{\frac{5}{2}})s^{\frac{3}{2}}-2t^{\frac{5}{2}}s\\-ts^2+(t^{\frac{3}{2}}+2t)s^{\frac{3}{2}}-2t^{\frac{5}{2}}s\end{pmatrix}\,ds$$

$$=\begin{pmatrix}\frac{7}{15}t^5-\frac{1}{5}t^{\frac{9}{2}}+\frac{1}{5}t^{\frac{5}{2}}+\frac{7}{15}t^2\\-\frac{1}{3}t^5-2t^{\frac{9}{2}}+\frac{2}{5}t^4+\frac{4}{5}t^{\frac{7}{2}}+t^{\frac{5}{2}}+\frac{1}{3}t^2-\frac{2}{5}t^{\frac{3}{2}}-\frac{4}{5}t\end{pmatrix}.$$

Furthermore we find

$$U(t)U^{-1}(1)\begin{pmatrix}1\\1\end{pmatrix}=U(t)\begin{pmatrix}-1&2\\-1&1\end{pmatrix}\begin{pmatrix}1\\1\end{pmatrix}$$

$$=U(t)\begin{pmatrix}1\\0\end{pmatrix}=\begin{pmatrix}t^2&-2t^{\frac{5}{2}}\\t&t^{\frac{3}{2}}\end{pmatrix}\begin{pmatrix}1\\0\end{pmatrix}=\begin{pmatrix}t^2\\t\end{pmatrix}$$

which yields by (19.32) that

$$\begin{pmatrix}u\\v\end{pmatrix}(t)=U(t)U^{-1}(1)+\int_1^t U(t)U^{-1}(s)\begin{pmatrix}s^4\\s^{\frac{5}{2}}\end{pmatrix}\,ds$$

$$=\begin{pmatrix}t^2\\t\end{pmatrix}+\begin{pmatrix}\frac{7}{15}t^5-\frac{1}{5}t^{\frac{9}{2}}+\frac{1}{5}t^{\frac{5}{2}}+\frac{7}{15}t^2\\-\frac{1}{3}t^5-2t^{\frac{9}{2}}+\frac{2}{5}t^4+\frac{4}{5}t^{\frac{7}{2}}+t^{\frac{5}{2}}+\frac{1}{3}t^2-\frac{2}{5}t^{\frac{3}{2}}-\frac{4}{5}t\end{pmatrix}$$

$$=\frac{1}{15}\begin{pmatrix}7t^5-3t^{\frac{9}{2}}+3t^{\frac{5}{2}}+22t^2\\-5t^5-15t^{\frac{9}{2}}+6t^4+12t^{\frac{7}{2}}+15t^{\frac{5}{2}}+5t^2-6t^{\frac{3}{2}}+3t\end{pmatrix}.$$

Please note, once $A(t)$ and $U(t)$ are known, the rest of the calculation can be done by computational packages.

7.   a) Suppose $\lambda_1 u_1(t)+\cdots+\lambda_N u_N(t)=0$ for all $t\in[0,1]$ and one $\lambda_{j_0}\neq 0$. Changing the enumeration of the functions if necessary we may assume that $\lambda_1\neq 0$. Therefore we find

$$u_1(t)=-\frac{\lambda_2}{\lambda_1}u_2(t)-\cdots-\frac{\lambda_N}{\lambda_1}u_N(t)\quad\text{for all }t\in[0,1].$$

Multiplying this identity with u_1 and integrating over $[0,1]$ yields

$$1=\int_0^1 |u_1(t)|^2\,dt=-\sum_{j=2}^N\frac{\lambda_j}{\lambda_1}\int_0^1 u_j(t)u_1(t)\,dt=0$$

which is a contradiction.

    b) We find

$$\lambda\int_0^1 u(x)v(x)\,dx=\int_0^1 v(x)\frac{d}{dx}\left(a(x)\frac{du(x)}{dx}\right)\,dx=-\int_0^1 a(x)v'(x)u'(x)\,dx$$

and

$$\mu\int_0^1 u(x)v(x)\,dx=\int_0^1 u(x)\frac{d}{dx}\left(a(x)\frac{dv(x)}{dx}\right)\,dx=-\int_0^1 a(x)u'(x)v'(x)\,dx$$

which implies

$$(\lambda - \mu) \int_0^1 u(x)v(x) \, dx = 0.$$

Since $\lambda \neq \mu$ we deduce that $\int_0^1 u(x)v(x) \, dx = 0$ and now part a) implies the independence of the set $\{u, v\}$.

8. We first suppose that $W(u_1, \ldots, u_N)(t) \neq 0$ for all $t \in [a, b]$. Suppose now that the functions $u_1, \ldots, u_N$ are linearly dependent on $[a, b]$. Then we can find $\lambda_1, \ldots, \lambda_N \in \mathbb{R}$ not all equal to 0 such that

$$\lambda_1 u_1 + \cdots + \lambda_N u_N = 0.$$

Taking derivatives we arrive at the system

$$\lambda_1 u_1 + \cdots + \lambda_N u_N = 0$$
$$\lambda_1 \dot{u}_1 + \cdots + \lambda_N \dot{u}_N = 0$$
$$\vdots$$
$$\lambda_1 u_1^{(N-1)} + \cdots + \lambda_N u_N^{(N-1)} = 0$$

in $[a, b]$ which admits for every $t \in [a, b]$ the non-trivial solution $(\lambda_1, \ldots, \lambda_N)$. Hence the determinant of the system must be equal to zero on $[a, b]$, but this is just $W(u_1, \ldots, u_N)(t)$. Thus we have constructed a contradiction.

Now we suppose that the functions $u_1, \ldots, u_N$ are linearly independent on $[a, b]$ but for some t_0 we assume $W(u_1, \ldots, u_N)(t_0) = 0$. It follows that the system

$$\lambda_1 u_1(t_0) + \cdots + \lambda_N u_N(t_0) = 0$$
$$\vdots$$
$$\lambda_1 u_1^{(N-1)}(t_0) + \cdots + \lambda_N u_N^{(N-1)}(t_0) = 0$$

admits a non-trivial solution $(\lambda_1^0, \ldots, \lambda_N^0)$. With this non-trivial solution $(\lambda_1^0, \ldots, \lambda_N^0)$ we consider the function

$$u(t) = \lambda_1^0 u_1(t) + \cdots + \lambda_N^0 u_N(t)$$

defined on $[a, b]$. Since for $1 \leq j \leq N$ we have assumed that $P(D)u_j = 0$ in $[a, b]$ it follows that $P(D)u = 0$ in $[a, b]$ and $u^{(k)}(t_0) = 0$. But the function $t \mapsto z(t) = 0$ for all $t \in [a, b]$ is also a solution to $P(D)w = 0$, $w(t_0) = 0$. Now the uniqueness result implies that u must be identically zero which is a contradiction. (Note that we have applied the uniqueness result to a linear equation of higher order with constant coefficients which is of course equivalent to a linear system of first order with constant coefficients.)

667

9. The Wronski determinant of $u_1, \ldots u_N$ at t_0 is

$$\det \begin{pmatrix} u_1(t_0) & \cdots & u_N(t_0) \\ \vdots & & \vdots \\ u_1^{(k)}(t_0) & \cdots & u_N^{(k)}(t_0) \\ \vdots & & \vdots \\ u_1^{(N-1)}(t_0) & \cdots & u_N^{(N-1)}(t_0) \end{pmatrix} = 0$$

since one of the rows is identically zero. From Corollary 19.11, we deduce now that the functions $u_1, \ldots, u_N$ are not independent.

10. The system can be written as

$$\begin{pmatrix} u \\ v \end{pmatrix}^{\cdot}(t) = \begin{pmatrix} t & t \\ 0 & t \end{pmatrix} \begin{pmatrix} u \\ v \end{pmatrix}(t) = A(t) \begin{pmatrix} u \\ v \end{pmatrix}(t)$$

and a primitive $\tilde{A}(t)$ of $A(t)$ is the matrix $\begin{pmatrix} \frac{t^2}{2} & \frac{t^2}{2} \\ 0 & \frac{t^2}{2} \end{pmatrix}$. Since $A(t) = t \left(\begin{smallmatrix} 1 & 1 \\ 0 & 1 \end{smallmatrix} \right)$ and the given primitive is given by $\frac{t^2}{2} \left(\begin{smallmatrix} 1 & 1 \\ 0 & 1 \end{smallmatrix} \right)$ it follows that $[A(t), \tilde{A}(t)] = 0$ for all t. Now Proposition 19.23 yields

$$\begin{pmatrix} u \\ v \end{pmatrix}(t) = e^{\int_0^t A(s)\, ds} \begin{pmatrix} 1 \\ 1 \end{pmatrix} = \sum_{k=0}^{\infty} \frac{1}{k!} \left(\frac{t^2}{2} \right)^k \begin{pmatrix} 1 & 1 \\ 0 & 1 \end{pmatrix}^k \begin{pmatrix} 1 \\ 1 \end{pmatrix}.$$

Since $\left(\begin{smallmatrix} 1 & 1 \\ 0 & 1 \end{smallmatrix} \right)^k = \left(\begin{smallmatrix} 1 & k \\ 0 & 1 \end{smallmatrix} \right)$ we find further

$$\begin{pmatrix} u \\ v \end{pmatrix}(t) = \sum_{k=0}^{\infty} \frac{1}{k!} \left(\frac{t^2}{2} \right)^k \begin{pmatrix} 1 & k \\ 0 & 1 \end{pmatrix} \begin{pmatrix} 1 \\ 1 \end{pmatrix}$$

$$= \begin{pmatrix} e^{\frac{t^2}{2}} & \frac{t^2}{2} e^{\frac{t^2}{2}} \\ 0 & e^{\frac{t^2}{2}} \end{pmatrix} \begin{pmatrix} 1 \\ 1 \end{pmatrix} = \begin{pmatrix} e^{\frac{t^2}{2}} + \frac{t^2}{2} e^{\frac{t^2}{2}} \\ e^{\frac{t^2}{2}} \end{pmatrix}.$$

We note that $\left(\begin{smallmatrix} u \\ v \end{smallmatrix} \right)(0) = \left(\begin{smallmatrix} 1 \\ 1 \end{smallmatrix} \right)$ and

$$\dot{u}(t) = t e^{\frac{t^2}{2}} + t e^{\frac{t^2}{2}} + \frac{t^2}{2} t e^{\frac{t^2}{2}} = t u(t) + t v(t)$$

$$\dot{v}(t) = t e^{\frac{t^2}{2}} = t v(t).$$

11. We have

$$y_1'' + p y_1' + q y_1 = 0,$$
$$y_2'' + p y_2' + q y_2 = 0$$

and therefore

$$y_2 y_1'' + p y_2 y_1' + q y_2 y_1 = 0$$
$$y_1 y_2'' + p y_1 y_2' + q y_1 y_2 = 0$$

which yields

$$y_1 y_2'' = y_2 y_1'' + p(y_1 y_2' - y_2 y_1') = 0. \qquad (*)$$

Now

$$\frac{\mathrm{d}}{\mathrm{d}x} W(x) = \frac{\mathrm{d}}{\mathrm{d}x}(y_1 y_2' - y_2 y_1') = y_1' y_2' + y_1 y_2'' - y_2' y_1' - y_2 y_1'' = y_1 y_2'' - y_2 y_1''$$

and therefore $(*)$ reads
$$W'(x) + pW(x) = 0$$

and the solution of this equation is obtained by separating variables

$$\int \frac{\mathrm{d}W}{W} = -\int p(x)\,\mathrm{d}x + c$$

or $W(x) = ce^{-P(x)}$, P being a primitive of p.

Chapter 20

1. With $T_\lambda(x) = \sum_{k=0}^{\infty} c_k x^k$ we find

$$T_\lambda'(x) = \sum_{k=0}^{\infty} k c_k x^{k-1}, \quad T_\lambda''(x) = \sum_{k=0}^{\infty} k(k-1) c_k x^{k-2}$$

which yields

$$(1-x^2)T_\lambda''(x) - xT_\lambda'(x) + \lambda^2 T_\lambda(x) = \sum_{k=0}^{\infty}(k+2)(k+1)c_{k+2}x^k - \sum_{k=0}^{\infty}k(k-1)c_k x^k$$

$$- \sum_{k=0}^{\infty} k c_k x^k + \sum_{k=0}^{\infty} \lambda^2 c_k x^k = 0$$

from which we deduce that

$$(k+2)(k+1)c_{k+2} = (k^2 - \lambda^2)c_k$$

or

$$c_{k+2} = \frac{k^2 - \lambda_2}{(k+1)(k+2)} c_k.$$

669

This leads to

$$c_2 = -\frac{\lambda^2}{2!}c_0$$

$$c_{2k} = -\frac{\lambda^2(2-\lambda^2)(4^2-\lambda^2)\ldots((2k-2)^2-\lambda^2)}{(2k)!}c_0, \ k \in \mathbb{N}\setminus\{1\}$$

$$c_{2k+1} = \frac{(1-\lambda^2)(3^2-\lambda^2)\ldots((2k-1)^2-\lambda^2)}{(2k+1)!}c_1.$$

Thus we find for $c_0, \ c_1 \in \mathbb{R}$

$$T_\lambda(x) = c_0\left(1 - \frac{\lambda^2}{2!} - \sum_{k=2}^{\infty}\frac{\lambda^2(2^2-\lambda^2)(4^2-\lambda^2)\ldots((2k-2)^2-\lambda^2)}{(2k)!}x^{2k}\right)$$

$$+ c_1\left(x + \sum_{k=1}^{\infty}\frac{(1-\lambda^2)(3^2-\lambda^2)\ldots((2k-1)^2-\lambda^2)}{(2k+1)!}x^{2k+1}\right).$$

Now, if $\lambda = 2n$ the first series is a finite sum, while for $\lambda = 2n+1$ the second series is a finite sum. Choosing in the first case $c_1 = 0$ and in the second case $c_0 = 0$, then T_{2n} and T_{2n+1}, respectively, become polynomials. It is possible to prove that up to a normalization these are the **Chebyshev polynomials** known from approximation theory.

The convergence of the two series defining T_λ follows for $x \in (-1,1)$ easily by the ratio test. For the first series we find

$$\frac{|x|^{2k+1}(\lambda^2(2^2-\lambda^2)\ldots((2k)^2-\lambda^2))(2k)!}{|x|^{2k}(2k+2)!(\lambda^2(2^2-\lambda^2)\ldots((2k-2)^2-\lambda^2)} = |x|^2\frac{4k^2-\lambda^2}{(2k+1)(2k+2)}.$$

Since $\lim_{k\to\infty}\frac{4k^2-\lambda^2}{(2k+1)(2k+1)} = 1$ we have convergence for $|x| < 1$. The other case goes analogously.

2. For the equation $(1+t^2)\ddot{u}(t) + 2t\dot{u}(t) - 2u(t) = 0$ it is easy to guess that $v(t) = \gamma t, \ \gamma \neq 0$, is a non-trivial solution. In order to obtain a second solution we start with $u(t) = \sum_{k=0}^{\infty}c_k t^k$ which gives

$$\dot{u}(t) = \sum_{k=0}^{\infty}kc_k t^{k-1} \quad \text{and} \quad \ddot{u}(t) = \sum_{k=0}^{\infty}k(k-1)c_k t^{k-2},$$

hence we find

$$(1+t^2)\ddot{u}(t) + 2t\dot{u}(t) - 2u(t) = \sum_{k=0}^{\infty}(k+2)(k+1)c_{k+2}t^k$$

$$+ \sum_{k=0}^{\infty}k(k-1)c_k t^k + \sum_{k=0}^{\infty}2kc_k t^k - \sum_{k=0}^{\infty}2c_k t^k.$$

From this identity we get

$$(k+2)(k+1)c_{k+2} + k(k-1)c_k + 2kc_k - 2c_k = 0$$

670

or

$$c_{k+2} = \frac{2 - k - k^2}{(k+2)(k+1)} c_k = -\frac{(k-1)}{k+1} c_k.$$

Since v is an odd function we are looking for a second solution being even, so we choose $c_1 = 0$ implying $c_{2k+1} = 0$ for all $k \in \mathbb{N}_0$ and

$$c_{2k} = (-1)^k \frac{1}{2k+1} c_0.$$

Eventually we arrive at

$$u(t) = \left(1 + \sum_{k=0}^{\infty} (-1)^k \frac{t^{2k+2}}{2k+1}\right) c_0$$

$$= c_0(1 + t \arctan t), \quad |t| < 1.$$

While the power series converges only for $|t| < 1$, the function $t \mapsto \arctan t$ is defined on $\mathbb{R}$ and it follows that the general solution to the original differential equation is given by

$$w(t) = \gamma t + c_0(1 + t \arctan t).$$

3. We set $u_2(t) := v(t) + c u_1(t) \ln t$ where u_1 is a solution to (20.6) and we require that u_2 solves (20.6) too. Now we note that

$$0 = \ddot{u}_2(t) + \alpha(t)\dot{u}_2(t) + \beta(t)u_2(t)$$
$$= \ddot{v}(t) + \alpha(t)\dot{v}(t) + \beta(t)v(t) + c((u_1(t)\ln t)^{\cdot\cdot} + \alpha(t)(u_1(t)\ln t)^{\cdot} + \beta(t)u_1(t)\ln t)$$
$$= \ddot{v}(t) + \alpha(t)\dot{v}(t) + \beta t v(t) + \ddot{u}_1(t)\ln t + 2\dot{u}_1(t)\frac{1}{t} - u_1(t)\frac{1}{t^2}$$
$$\quad + \alpha(t)\dot{u}(t)\ln t + \alpha(t)u_1(t)\frac{1}{t} + \beta(t)u_1(t)\ln t$$
$$= \ddot{v}(t) + \alpha(t)\dot{v}(t) + \beta(t)v(t) + \ln t(\ddot{v}_1(t) + \alpha(t)\dot{u}_1(t) + \beta(t)u_1(t))$$
$$\quad + \frac{2}{t}\dot{u}_1(t) + \left(\frac{\alpha(t)}{t} - \frac{1}{t^2}\right)u_1(t)$$
$$= \ddot{v}(t) + \alpha(t)\dot{v}(t) + \beta(t)v(t) + \frac{2}{t}\dot{u}_1(t) + \left(\frac{\alpha(t)}{t} - \frac{1}{t^2}\right)u_1(t),$$

and (20.23) is proved.

4. We study the two series separately and apply in each case the ratio test. We have

$$\frac{t^{3(l+1)}}{2 \cdot 5 \cdot 8 \cdot \ldots \cdot (3(l+1) - 1)3^{l+1}(l+1)!} \cdot \frac{2 \cdot 5 \cdot 8 \cdot \ldots \cdot (3l-1)3^l l!}{t^{3l}}$$
$$= \frac{t^3}{3(3l+1)(l+1)}$$

671

as well as

$$\frac{t^{3(l+1)+1}}{4\cdot 7\cdot\ldots\cdot(3(l+1)+1)3^{l+1}(l+1)!}\cdot\frac{4\cdot 7\cdot\ldots\cdot(3l+1)3^l l!}{t^{3l+1}}$$

$$=\frac{t^3}{3(3(l+1)+1)(l+1)}.$$

In each case, for t fixed the term tends to zero implying that both series converge (uniformly and absolutely) on every compact interval, hence compact disc in $\mathbb{C}$. Thus the function is an entire function.

5. We note that

$$\dot{y}_0(t)=\frac{1}{t}J_0(t)+\dot{J}_0(t)\ln t+\sum_{k=1}^{\infty}kc_k t^{k-1},$$

and

$$\ddot{y}(t)=-\frac{1}{t^2}J_0(t)+\frac{2}{t}\dot{J}_0(t)+\ddot{J}_0(t)\ln t+\sum_{k=1}^{\infty}k(k-1)c_k t^{k-2}$$

which yields now

$$0=(t^2\ddot{y}_0(t)+t\dot{y}_0(t)+t^2 y_0(t))$$

$$=t^2\ddot{J}_0(t)\ln t+2t\dot{J}_0(t)-J_0(t)+t\dot{J}_0(t)\ln t+J_0(t)+t^2 J_0(t)\ln t$$

$$+c_1+4c_2 t^2+\sum_{k=3}^{\infty}(k^2 c_k+c_{k-2})t^k$$

$$=2t\dot{J}_0(t)+c_1 t+4c_2 t^2+\sum_{k=3}^{\infty}(k^2 c_k+c_{k-2})t^k.$$

This gives

$$c_1 t+4c_2 t^2+\sum_{k=3}^{\infty}(k^2 c_k+c_{k-2})t^k=\sum_{k=1}^{\infty}\frac{(-1)^{k+1}4k}{2^{2k}(k!)^2}t^{2k}$$

where we have used the power series representation $J_0(t)=\sum_{k=0}^{\infty}\frac{(-1)^k t^{2k}}{2^{2k}(k!)^2}$. Comparing coefficients we arrive at

$$c_{2k-1}=0,\ k\in\mathbb{N},$$

$$c_2=\frac{1}{4},$$

and

$$(2k)^2 c_{2k}+c_{2k-2}=\frac{(-1)^{k+1}4k}{2^{2k}(k!)^2},\quad k\in\mathbb{N}\setminus\{1\},$$

from which we obtain

$$c_{2k}=\frac{(-1)^{k+1}\sum_{l=1}^{k}\frac{1}{l}}{2^{2k}(k!)^2},\quad k\in\mathbb{N}.$$

672

(The reader is invited to give a proof of this formula by induction.) It follows that

$$y_0(t) = J_0(t) \ln t + \sum_{k=1}^{\infty} (-1)^{k+1} \frac{\sum_{l=1}^{k} \frac{1}{l}}{2^{2k}(k!)^2} t^{2k}, \quad t > 0$$

is the sought second solution. The convergence of the series is obvious since $\sum_{l=1}^{k} \frac{1}{l} < k$, the independence we deduce either from the general theory or by a direct analysis.

Instead of y_0 it is more common to use as second and independent solution of the equation $t^2 \ddot{u}(t) + t\dot{u}(t) + t^2 u(t) = 0$ the function

$$Y_0(t) := \frac{2}{\pi}(y_0(t) + (\gamma - \ln 2)J_0(t))$$

where γ is the Euler constant, see Theorem I.18.24, defined by $\gamma := \lim_{n \to \infty} (\sum_{l=1}^{n} \frac{1}{l} - \ln n)$.

6. We start with the Ansatz $y(x) = \sum_{k=0}^{\infty} c_k x^{\rho+k}$, $c_0 \neq 0$, $x > 0$, and we find

$$\left(\rho^2 - \frac{1}{4}\right)c_0 x^\rho + \left((\rho+1)^2 - \frac{1}{4}\right)c_1 x^{\rho+1} + \sum_{k=2}^{\infty} \left(\left((\rho+k)^2 - \frac{1}{4}\right)c_k - c_{k-2}\right)x^{\rho+k} = 0.$$

We deduce that we must have $\rho^2 - \frac{1}{4} = 0$, i.e. $\rho_1 = \frac{1}{2}$ and $\rho_2 = -\frac{1}{2}$. This implies

$$\left((\rho+1)^2 - \frac{1}{4}\right)c_1 = 0$$

as well as

$$\left((\rho+k)^2 - \frac{1}{4}\right)c_k = -c_{k-2}, \quad k \geq 2.$$

For $\rho_1 = \frac{1}{2}$ we find

$$c_{2k-1} = 0, \quad k \in \mathbb{N}$$

and

$$c_{2k} = -\frac{c_{2k-2}}{2k(2k+1)}, \quad k \in \mathbb{N}.$$

For $c_0 = 1$ we obtain

$$c_{2k} = \frac{(-1)^k}{(2k+1)!}$$

which gives a first solution

$$y_1(x) = x^{\frac{1}{2}} \sum_{k=0}^{\infty} \frac{(-1)^k}{(2k+1)!} x^{2k} = \frac{1}{x^{\frac{1}{2}}} \sum_{k=0}^{\infty} \frac{(-1)^k x^{2k+1}}{(2k+1)!}$$

$$= \frac{1}{\sqrt{x}} \sin x.$$

673

Instead y_1 it is common to consider the solution

$$J_{\frac{1}{2}}(x) = \sqrt{\frac{2}{\pi x}} \sin x, \quad x > 0.$$

Since $\rho_1 - \rho_2 = 1$ we now look for a second solution with the help of the Ansatz

$$y_2(x) = x^{-\frac{1}{2}} \sum_{k=0}^{\infty} d_k x^k = \sum_{k=0}^{\infty} d_k x^{k-\frac{1}{2}}.$$

As recurrence formula we obtain now for $d_0 = 1$

$$d_{2k} = \frac{(-1)^k}{(2k)!}, \quad d_{2k+1} = 0, \quad k \in \mathbb{N},$$

and the function y_2 becomes

$$y_2(x) = x^{-\frac{1}{2}} \sum_{k=0}^{\infty} \frac{(-1)^k}{(2k)!} x^{2k} = \frac{1}{\sqrt{x}} \cos x.$$

As standard solution one takes $J_{-\frac{1}{2}}(x) := \sqrt{\frac{2}{\pi x}} \cos x$, $x > 0$. (In solving this problem we benefited much from the calculations made in [64].)

7.   a) Applying the ratio test to the given power series we find

$$\frac{|t|^{2(k+1)}}{2^{2k+2+\nu}(k+1)!\Gamma(k+\nu+2)} \cdot \frac{2^{2k+\nu}k!\Gamma(k+\nu+1)}{|t|^{2k}}$$

$$= \frac{|t|^2}{4(k+1)|k+1+\nu|},$$

implying that for fixed $t \in \overline{D_R(0)}\backslash(-\infty,0)$ the series gives an entire function in ν.

b) The term $Y_\nu(z) = \frac{J_\nu(z)\cos\nu\pi - J_{-\nu}(z)}{\sin\nu\pi}$ is undetermined for $\nu \in \mathbb{Z}$ since $\sin\nu\pi = 0$ for $\nu \in \mathbb{Z}$ and $J_{-\nu}(z) = (-1)^\nu J_\nu(z)$ for $\nu \in \mathbb{Z}$. It follows for $\nu \in \mathbb{R}\backslash\mathbb{Z}$ that

$$\frac{\partial}{\partial\nu}(J_\nu(z)\cos\nu\pi - J_{-\nu}(z)) = \frac{\partial J_\nu}{\partial\nu}(z)\cos\nu\pi - J_\nu(z)\pi\sin\nu\pi - \frac{\partial J_{-\nu}}{\partial\nu}(z)$$

$$= \frac{\partial J_\nu}{\partial\nu}(z)\cos\nu\pi - \frac{\partial J_{-\nu}}{\partial\nu}(z).$$

Since $\frac{\partial}{\partial\nu}\sin\nu\pi = \pi\cos\nu\pi$ we find by the de l'Hospital's rule

$$Y_n(z) = \lim_{\nu\to n} \frac{\frac{\partial}{\partial\nu}(J_\nu(z)\cos\nu\pi - J_{-\nu}(z))}{\frac{\partial}{\partial\nu}\sin\nu\pi}$$

$$= \frac{1}{\pi\cos n\pi}\left(\frac{\partial J_\nu}{\partial\nu}(z)\Big|_{\nu=n}\cos n\pi - \frac{\partial J_{-\nu}}{\partial\nu}(z)\right)$$

$$= \frac{1}{\pi}\left(\frac{\partial J_\nu}{\partial\nu}(z)\Big|_{\nu=n} - (-1)^n\frac{\partial J_{-\nu}}{\partial\nu}(z)\Big|_{\nu=n}\right).$$

c) We know already that for $\nu \in \mathbb{R}\backslash\mathbb{Z}$ the functions J_ν and $J_{-\nu}$ solve the differential equation, hence for these values of ν it follows that Y_ν satisfies the differential equation too. The convergence properties of $J_\nu(z)$ in ν and z allow now to consider the analytic extension with respect to ν of

$$\frac{\mathrm{d}^2}{\mathrm{d}z^2}\left(\frac{J_\nu(z)\cos\nu\pi - J_{-\nu}(z)}{\sin\nu\pi}\right) + \frac{1}{z}\frac{\mathrm{d}}{\mathrm{d}z}\left(\frac{J_\nu(z)\cos\nu\pi - J_{-\nu}(z)}{\sin\nu\pi}\right)$$
$$+ \left(1 - \frac{\nu^2}{z^2}\right)\left(\frac{J_\nu(z)\cos\nu\pi - J_{-\nu}(z)}{\sin\nu\pi}\right)$$

to $n \in \mathbb{Z}$, and the result follows.

8. For $\mu = i$ and $\nu \notin \mathbb{Z}$ the equation becomes

$$t^2 u''(t) + t u'(t) + (-t^2 - \nu^2)u(t) = 0.$$

First we find

$$\tilde{I}_\nu(t) := J_\nu(it) = \sum_{k=0}^\infty \frac{(-1)^k}{k!\Gamma(k+\nu+1)}\left(\frac{it}{2}\right)^{2k+\nu}$$
$$= i^\nu \sum_{k=0}^\infty \frac{(-1)^k i^{2k}}{k!\Gamma(k+\nu+1)}\left(\frac{t}{2}\right)^{2k+\nu},$$

and therefore we find

$$I_\nu(t) = i^{-\nu}J_\nu(it) = \sum_{k=0}^\infty \frac{1}{k!\Gamma(k+\nu+1)}\left(\frac{t}{2}\right)^{2k+\nu}.$$

On the other hand we have with $s = it$, $v(s) = u(t(s))$, $v'(s) = \frac{1}{i}u'(t)$ and $v''(s) = \frac{1}{i^2}u''(t)$ which yields that

$$s^2 v''(s) + s v'(s) + (-s^2 - \nu^2)v(s)$$
$$= (it)^2\left(\frac{1}{i^2}u''(t)\right) + it\left(\frac{1}{i}u'(t)\right) + (-(it)^2 - \nu^2)u(t)$$
$$= t^2 u''(t) + t u''(t) + (t^2 - \nu^2)u(t)$$

implying that $I_\nu(t)$ solves the given differential equation. Analogously we can prove that $I_{-\nu}(t) = i^{-\nu}J_{-\nu}(it)$ is a solution to this equation and now follows that $K_\nu(t) = \frac{\pi}{2}\frac{I_\nu(t)-I_{-\nu}(t)}{\sin\nu\pi}$ is for $\nu \notin \mathbb{Z}$ a further solution.

9. We note that

$$y'(x) = \frac{\mathrm{d}s}{\mathrm{d}x}g'(s(x))$$

and

$$y''(x) = \left(\frac{\mathrm{d}s}{\mathrm{d}x}\right)^2 g''(s(x)) + \frac{\mathrm{d}^2 s}{\mathrm{d}x^2}g'(s(x)).$$

Further we deduce from $s^2 = 4ax$ that $s = \sqrt{4ax}$ and consequently

$$\frac{\mathrm{d}s}{\mathrm{d}x} = \frac{2a}{\sqrt{4ax}} = \frac{2a}{s}$$

and

$$\frac{\mathrm{d}^2 s}{\mathrm{d}x^2} = \frac{-4a^2}{(4ax)^{-\frac{3}{2}}} = \frac{-4a^2}{s^3}.$$

Now we find

$$0 = xy''(x) + y'(x) + ay(x)$$

$$= \frac{s^2}{4a}\left(\left(\frac{\mathrm{d}s}{\mathrm{d}x}\right)^2 g''(s) + \left(\frac{\mathrm{d}^2 s}{\mathrm{d}x^2}\right)g'(s)\right) + \left(\frac{\mathrm{d}s}{\mathrm{d}x}\right)g'(s) + ag(s)$$

$$= ag''(s) - \frac{a}{s}g'(s) + \frac{2a}{s}g'(s) + ag(s)$$

$$= a\left(g''(s) + \frac{1}{s}g'(s) + g(s)\right)$$

which is Bessel's differential equation for g, $s > 0$. It follows that $g(s) = \lambda_1 J_0(s) + \lambda_2 Y_0(s)$, $s > 0$, and therefore

$$y(x) = \lambda_1 J_0(\sqrt{4ax}) + \lambda_0 Y_0(\sqrt{4ax}).$$

10. The important fact to note is that the function Q must be 2π-periodic. From

$$P(\vartheta)Q(\vartheta)\frac{\mathrm{d}^2 U(r)}{\mathrm{d}r^2} + \frac{U(r)Q(\varphi)}{r^2 \sin\vartheta}\frac{\mathrm{d}}{\mathrm{d}\vartheta}\left(\sin\vartheta\frac{\mathrm{d}P(\vartheta)}{\mathrm{d}\vartheta}\right)$$

$$+ \frac{U(r)P(\vartheta)}{r^2 \sin^2\vartheta}\frac{\mathrm{d}^2 Q(\varphi)}{\mathrm{d}\varphi^2} = 0,$$

we can deduce first by dividing by $U(r) \cdot P(\vartheta) \cdot Q(\varphi)$ and multiplying by $r^2 \sin^2\vartheta$ that

$$0 = r^2 \sin^2\vartheta\left(\frac{1}{U(r)}\frac{\mathrm{d}^2 U(r)}{\mathrm{d}r^2} + \frac{1}{r^2 \sin\vartheta P(\vartheta)}\frac{\mathrm{d}}{\mathrm{d}\vartheta}\left(\sin\vartheta\frac{\mathrm{d}P(\vartheta)}{\mathrm{d}\vartheta}\right)\right) + \frac{1}{Q(\varphi)}\frac{\mathrm{d}^2 Q(\varphi)}{\mathrm{d}\varphi^2}$$

which implies with some constant c

$$\frac{1}{Q(\varphi)}\frac{\mathrm{d}^2 Q(\varphi)}{\mathrm{d}\varphi^2} = -c$$

and

$$r^2 \sin^2\vartheta\left(\frac{1}{U(r)}\frac{\mathrm{d}^2 U(r)}{\mathrm{d}r^2} + \frac{1}{r^2 \sin\vartheta P(\vartheta)}\frac{\mathrm{d}}{\mathrm{d}\vartheta}\left(\sin\vartheta\frac{\mathrm{d}P(\vartheta)}{\mathrm{d}\vartheta}\right)\right) = c. \qquad (*)$$

However Q must be 2π-periodic and this requires $-c = -m^2$, $m \in \mathbb{Z}$. Now we turn to $(*)$ with $c = m^2$ and find the equality

$$\frac{r^2}{U(r)}\frac{\mathrm{d}^2 U(r)}{\mathrm{d}r^2} = \left(\frac{m^2}{\sin^2\vartheta} - \frac{1}{\sin\vartheta P(\vartheta)}\frac{\mathrm{d}}{\mathrm{d}\vartheta}\left(\sin\vartheta\frac{\mathrm{d}P(\vartheta)}{\mathrm{d}\vartheta}\right)\right).$$

676

Since both side are dependent on different variables we deduce that they must be constant and we choose this constant to be $l(l+1)$, $l \in \mathbb{N}_0$, see below. It follows that

$$\frac{r^2}{U(r)} \frac{d^2 U(r)}{dr^2} = l(l+1)$$

or

$$\frac{d^2 U(r)}{dr^2} - \frac{l(l+1)}{r^2} U(r) = 0$$

and

$$l(l+1) = \frac{m^2}{\sin^2 \vartheta} - \frac{1}{\sin \vartheta P(\vartheta)} \frac{d}{d\vartheta} \left(\sin \vartheta \frac{dP(\vartheta)}{d\vartheta} \right)$$

or

$$\frac{1}{\sin \vartheta} \frac{d}{d\vartheta} \left(\sin \vartheta \frac{dP(\vartheta)}{d\vartheta} \right) + \left(l(l+1) - \frac{\ln 2}{\sin^2 \vartheta} \right) P(\vartheta) = 0. \qquad (**)$$

The choice of the separation constant $l(l+1)$ takes already the solvability of $(**)$ into account when considered as a certain eigenvalue problem.

Chapter 21

1. The general solution to $\ddot{u} + w^2 u = 0$ in $[a, b]$ is given by $u(t) = \mu \cos wt + \nu \sin wt$.

   a) The boundary conditions imply

   $$\mu \cos wa + \nu \sin wa = \alpha$$
   $$-\mu w \sin wb + \nu w \cos wb = \beta$$

   or

   $$\begin{pmatrix} \cos wa & \sin wa \\ -w \sin wb & w \cos wb \end{pmatrix} \begin{pmatrix} \mu \\ \nu \end{pmatrix} = \begin{pmatrix} \alpha \\ \beta \end{pmatrix}.$$

   We have

   $$\det \begin{pmatrix} \cos wa & \sin wa \\ -w \sin wb & w \cos wb \end{pmatrix} = w \cos w(a - b),$$

   which yields the condition $w \neq 0$ and $w(b - a) \neq \frac{\pi}{2} + \pi k$, $k \in \mathbb{Z}$. With the normalization $w > 0$ the latter condition is $w(b - a) = (\frac{2k+1}{2})\pi$, $k \in \mathbb{N}_0$. Under this condition we find

   $$\mu = \frac{\det \begin{pmatrix} \alpha & \sin wa \\ \beta & w \cos wb \end{pmatrix}}{w \cos w(b - a)} = \frac{\alpha \sin wa - \beta w \cos wb}{w \cos w(b - a)}$$

   and

   $$\nu = \frac{\det \begin{pmatrix} \cos wa & \alpha \\ -w \sin wb & \beta \end{pmatrix}}{w \cos w(b - a)} = \frac{\beta \cos wa + \alpha \sin wb}{w \cos w(b - a)}$$

   and with these values for μ and ν the problem has the unique solution $t \mapsto \mu \cos wt + \nu \sin wt$.

b) We argue as in part a) and note that the boundary conditions imply now the linear system

$$-\mu w \sin wa + \nu w \cos wa = \alpha$$
$$-\mu w \sin wb + \nu w \cos wb = \beta$$

or

$$\begin{pmatrix} -w\sin wa & w\cos wa \\ -w\sin wb & w\cos wb \end{pmatrix} \begin{pmatrix} \mu \\ \nu \end{pmatrix} = \begin{pmatrix} \alpha \\ \beta \end{pmatrix}.$$

The determinant is now given by

$$\det \begin{pmatrix} -w\sin wa & w\cos wa \\ -w\sin wb & w\cos wb \end{pmatrix} = -w^2 \sin w(b-a)$$

and for a unique solution we have to require $w \neq 0$, so we may choose $w > 0$, and $w(b-a) \neq k\pi$, $k \in \mathbb{N}_0$. For μ and ν we find under these assumptions

$$\mu = \frac{\beta w \cos wa - \alpha w \cos wb}{w^2 \sin w(b-a)} = \frac{\beta \cos wa - \alpha \cos wb}{w \sin w(b-a)}$$

and

$$\nu = \frac{\beta \sin wa - \alpha w \sin wb}{w^2 \sin w(b-a)} = \frac{\beta \sin wa - \alpha \sin wb}{w \sin w(b-a)}.$$

2. We search u in the form $u(x,t) = w(x)v(t)$ and from

$$\frac{1}{w(x)v(t)} \frac{\partial^2 w(x)v(t)}{\partial t^2} - \frac{1}{w(x)v(t)} \frac{\partial^2 w(x)v(t)}{\partial^2 x^2} = 0$$

we deduce again

$$\frac{d^2 v(t)}{dt^2} = \lambda v(t), \quad \frac{d^2 w(x)}{d\lambda^2} = \lambda w(x).$$

The boundary conditions become $w(0)v(t) = 0$, $\dot{w}(0)v(t) = 0$ for all $t \geq 0$ and therefore we assume $w(0) = 0$ and $\dot{w}(0) = 0$. This implies $\lambda = -k^2$, $k \in \mathbb{N}$, and $\frac{d^2 v(t)}{dt^2} = -k^2 v(t)$ which has the general solution $v_k(t) = \alpha_k \cos kt + \beta_k \sin kt$. As in the main text we deduce that

$$u(x,t) = \sum_{k=1}^{N} \gamma_k \cos kx \cos kt + \delta_k \cos kx \sin kt, \quad N \in \mathbb{N},$$

solves the wave equation and the boundary condition $\frac{\partial u}{\partial x}(0,t) = 0$ and $\frac{\partial u}{\partial x}(2\pi, t) = 0$.

Indeed we have

$$\frac{\partial}{\partial x} u(x, t) = \sum_{k=1}^{N} (-\gamma_k k \sin kx \cos kt - \delta_k k \sin kx \sin kt),$$

$$\frac{\partial^2}{\partial x^2} u(x, t) = \sum_{k=1}^{N} (-\gamma_k k^2 \cos kx \cos kt - \delta_k k^2 \cos kx \sin kt),$$

$$\frac{\partial}{\partial t} u(x, t) = \sum_{k=1}^{N} (-\gamma_k k \cos kx \sin kt + \delta_k k \cos kx \cos kt),$$

$$\frac{\partial^2}{\partial t^2} u(x, t) = \sum_{k=1}^{N} (-\gamma_k k^2 \cos kx \cos kt - \delta_k k^2 \cos kx \cos kt).$$

In order to satisfy the initial condition we need to require

$$u(x, 0) = \sum_{k=1}^{N} \gamma_k \cos kx = u_0(x)$$

and

$$\frac{\partial}{\partial t} u(x, 0) = \sum_{k=1}^{N} \delta_k k \cos kx = 0.$$

The latter condition implies $\delta_k = 0$ for all $k = 1, \ldots, N$. However, for a given $u_0 \in C^4(\mathbb{R})$ which is 2π-periodic and even we know that it has a convergent Fourier series representation

$$u_0(x) = \sum_{k=1}^{\infty} c_k \cos kx, \quad |c_k| \le \frac{\kappa}{k^4}.$$

This implies that in

$$u(x, t) = \sum_{k=1}^{\infty} c_k \cos kx \cos kt \tag{$*$}$$

we can differentiate with respect to x and with respect to t twice under the sum, hence $\frac{\partial^2 u(x,t)}{\partial t^2} - \frac{\partial^2 u(x,t)}{\partial t^2} = 0$, $\frac{\partial u}{\partial x}(0, t) = \frac{\partial u}{\partial x}(2\pi, t) = 0$ for all $t > 0$ as well as $u(x, 0) = u_0$ and $\frac{\partial u}{\partial t}(x, 0) = 0$, i.e. $(*)$ solves the problem.

3. Recall that $w(x, y) = u_1(x)u_2(y) - u_1(y)u_2(x)$. Moreover, the assertion is equivalent to

$$w(x, a)w(y, b) + w(y, x)w(a, b) = w(x, b)w(y, a).$$

It follows that

$$(u_1(x)u_2(a) - u_1(a)u_2(x))(u_1(y)u_2(b) - u_1(b)u_2(y))$$
$$+ (u_1(y)u_2(x) - u_1(x)u_2(y))(u_1(a)u_2(b) - u_1(b)u_2(a))$$
$$= u_1(x)u_2(a)u_1(y)u_2(b) - u_1(x)u_2(a)u_1(b)u_2(y)$$
$$- u_1(y)u_2(b)u_1(a)u_2(b) + u_1(a)u_2(x)u_1(b)u_2(y)$$
$$+ u_1(y)u_2(x)u_1(a)u_2(b) - u_1(y)u_2(x)u_1(b)u_2(a)$$
$$- u_1(x)u_2(y)u_1(a)u_2(b) + u_1(x)u_2(y)u_1(b)u_2(a)$$
$$= u_1(x)u_1(y)u_2(a)u_2(b)$$
$$+ u_2(x)u_2(y)u_1(a)u_1(b)$$
$$- u_1(y)u_2(x)u_1(b)u_2(a)$$
$$- u_1(x)u_2(y)u_1(a)u_2(b)$$
$$= u_1(x)u_2(b)(u_1(y)u_2(a) - u_1(a)u_2(y))$$
$$+ u_2(x)u_1(b)(u_2(y)u_1(a) - u_1(y)u_2(a))$$
$$= (u_1(x)u_2(b) - u_2(x)u_1(b))w(y,a)$$
$$= w_1(x,b)w(y,a).$$

4. We start with the Ansatz $u(t) = e^{\mu t}$ and find the equation $\mu^2 + \mu b - \lambda = 0$ with the solutions

$$\mu_{1,2} = -\frac{b}{2} \pm \sqrt{\frac{b^2}{4} + \lambda}.$$

For $\frac{b^2}{4} + \lambda > 0$ we have the two independent solutions $t \mapsto e^{\left(-\frac{b}{2}+\sqrt{\frac{b^2}{4}+\lambda}\right)t}$ and $t \mapsto e^{\left(-\frac{b}{2}-\sqrt{\frac{b^2}{4}+\lambda}\right)t}$ and hence the general solution

$$u(t) = Ae^{\left(-\frac{b}{2}+\sqrt{\frac{b^2}{4}+\lambda}\right)t} + Be^{\left(-\frac{b}{2}-\sqrt{\frac{b^2}{4}+\lambda}\right)t}.$$

The boundary conditions $u(0) = u(1)$ imply now

$$u(0) = A + B = 0$$
$$u(1) = Ae^{-\frac{b}{2}+\sqrt{\frac{b^2}{4}+\lambda}} + Be^{-\frac{b}{2}-\sqrt{\frac{b^2}{4}+\lambda}} = 0,$$

which gives first $B = -A$ and then

$$A\left(e^{-\frac{b}{2}+\sqrt{\frac{b^2}{4}+\lambda}} - e^{-\frac{b}{2}-\sqrt{\frac{b^2}{4}+\lambda}}\right) = 0$$

or

$$e^{-\frac{b}{2}+\sqrt{\frac{b^2}{4}+\lambda}} = e^{-\frac{b}{2}-\sqrt{\frac{b^2}{4}+\lambda}},$$

i.e. $\sqrt{\frac{b^2}{4}+\lambda} = -\sqrt{\frac{b^2}{4}+\lambda}$ which means $\frac{b^2}{4} + \lambda = 0$. Thus for $\frac{b^2}{4} + \lambda > 0$ no non-trivial solution exists. For $\frac{b^2}{4} + \lambda = 0$ two independent solutions are given by

$t \mapsto e^{-\frac{b}{2}t}$ and $t \mapsto te^{-\frac{b}{2}t}$. The boundary conditions yield for $v(x) = Ae^{-\frac{b}{2}t} + Bte^{-\frac{b}{2}t}$ that

$$\dot{v}(0) = A = 0$$
$$v(1) = Be^{-\frac{b}{2}} = 0,$$

and again we obtain only the trivial solution. In the case that $\frac{b^2}{4} + \lambda < 0$ we obtain with $w = \sqrt{-\left(\frac{b^2}{4} + \lambda\right)}$ as general solution of the equation

$$u(t) = Ae^{-\frac{b}{2}t}\cos wt + Be^{-\frac{b}{2}t}\sin wt.$$

The boundary conditions give now

$$u(0) = A = 0$$

and

$$u(1) = Be^{-\frac{b}{2}}\sin wt = 0$$

implying for a non-trivial solution $\sqrt{-\frac{b^2}{4} + \lambda} = k\pi$ or $\lambda = k^2\pi^2 - \frac{b^2}{4}$. These are the eigenvalues for our problem for which a non-trivial solution exists.

5. A fundamental system of $(xy'(x))' = 0$ is given by $y_1(x) = 1$ and $y_2(x) = \ln x$. The corresponding Wronski determinant is

$$W(x) = y_1(x)y_2'(x) - y_1'(x)y_2(x)$$
$$= 1 \cdot \frac{1}{x} - 0 \cdot \ln x = \frac{1}{x} \neq 0 \text{ in } [1, e].$$

Further we have

$$w(x, y) = y_1(x)y_2(y) - y_1(y)y_2(x)$$
$$= 1 \cdot \ln y - 1 \cdot \ln x = \ln y - \ln x.$$

Therefore we find with $p_0(y) = y$

$$\frac{w(x, b)w(y, a)}{w(a, b)} \frac{1}{p_0(y)W(y)} = \frac{(\ln b - \ln x)(\ln a - \ln y)}{\ln b - \ln a}$$

and with $a = 1$ and $b = e$ we obtain

$$\frac{w(x, e)w(y, 1)}{w(1, e)} \frac{1}{p_0(y)W(y)} = \frac{-(1 - \ln x)\ln y}{\ln e} = (\ln x - 1)\ln y.$$

Analogously we get

$$\frac{w(x, a)w(y, b)}{w(a, b)} \frac{1}{p_0(y)W(y)} = (1 - \ln y)\ln x$$

681

which yields as Green function for the Dirichlet problem for $(xy'(x))'$

$$g(x,y) = \begin{cases} \ln(x-1)\ln y, & 1 \le y \le x \le e \\ (1-\ln y)\ln x, & 1 \le x \le y \le e. \end{cases}$$

The problem $(xy'(x))' = f(x)$, $y(1) = y(e) = 0$, is solved by

$$u(x) = \int_1^e g(x,y)f(y)\,dy$$

$$= \int_1^x (\ln(x-1)\ln y)f(y)\,dy + \int_x^e ((1-\ln y)\ln x)f(y)\,dy$$

$$= \ln(x-1)\int_1^x (\ln y)f(y)\,dy + \ln x \int_x^e (1-\ln y)f(y)\,dy.$$

With $f(x) = x^m$ we obtain further

$$u(x) = \ln(x-1)\int_1^x y^m \ln y\,dy + \ln x \int_x^e y^m\,dy$$

$$- \ln x \int_x^e y^m \ln y\,dy$$

$$= \ln(x-1)\left(y^{m+1}\left(\frac{\ln y}{m+1} - \frac{1}{(m+1)^2}\right)\Big|_1^x\right)$$

$$+ \ln x \left(\frac{1}{m+1}y^{m+1}\Big|_x^e\right) - \ln x \left(y^{m+1}\left(\frac{\ln y}{m+1} - \frac{1}{(m+1)^2}\right)\Big|_x^e\right).$$

6. A fundamental system is given by $u_1(x) = e^x - e^{-x}$ and $u_2(x) = e^x - e^2 e^{-x}$. The corresponding Wronski determinant is

$$W(x) = u_1(x)u_2'(x) - u_1'(x)u_2(x)$$
$$= 2(e^2 - 1) \ne 0.$$

Moreover we have

$$w(x,y) = u_1(x)u_2(y) - u_1(y)u_2(x)$$
$$= e^{x-y} - e^{-(x-y)} + e^{2-x+y} - e^{2+x-y}$$
$$= 2\sinh(x-y) - 2e^2\sinh(x-y) = 2(1-e^2)\sinh(x-y).$$

It follows with $p_0(y) = 1$ that

$$\frac{w(x,b)w(y,a)}{w(a,b)}\frac{1}{p_0(y)W(y)} = \frac{w(x,b)w(y,a)}{w(a,b)W(y)}$$

$$= \frac{\sinh(x-b)\sinh(y-a)}{\sinh(a-b)}.$$

and with $a = 0$ and $b = 1$ we arrive at

$$\frac{w(x,b)w(y,a)}{w(a,b)W(y)p_0(y)} = \frac{\sinh(x-1)\sinh y}{\sinh 1}$$

as well as

$$\frac{w(x,a)w(y,b)}{w(a,b)W(y)p_0(y)} = \frac{\sinh x \sinh(y-1)}{\sinh 1}.$$

This gives the Green function

$$g(x,y) = \begin{cases} \frac{\sinh(x-1)\sinh y}{\sinh 1}, & 1 \le y \le x \le e \\ \frac{\sinh x \sinh(y-1)}{\sinh 1}, & 1 \le x \le y \le e. \end{cases}$$

7. From Problem 6 to Chapter 9 we know that for $x \in (-1,1)$ we have

$$L_1(x,D)P_n(x) = \frac{d}{dx}\left((1-x^2)\frac{dP_n}{dx}(x)\right) = -n(n+1)P_n(x).$$

By Problem 1 to Chapter 20 we find

$$L_2(x,D)T_n(x) = (1-x^2)\frac{d^2}{dx^2}T_n(x) - x\frac{d}{dx}T_n(x) = n^2 T_n(x).$$

Finally, Bessel's differential equation of order n reads as $x^2 u''(x) + x u'(x) + (x^2 - n^2)u(x) = 0$, or

$$L_3(x,D)J_n(x) = x^2 J_n''(x) + x J_n'(x) + x^2 J_n(x) = n^2 J_n(x).$$

Chapter 22

1.   a) We note that $(\alpha)_k = \frac{\Gamma(\alpha+k)}{\Gamma(\alpha)}$ and using $\Gamma(z+1) = z\Gamma(z)$ we find

$$(\alpha+1)_k = \frac{\Gamma(\alpha+k+1)}{\Gamma(\alpha+1)} = \frac{(\alpha+k)\Gamma(\alpha+k)}{\alpha\Gamma(\alpha)} = \frac{\alpha+k}{\alpha}(\alpha)_k$$

as well as

$$(\alpha-1)_k = \frac{\Gamma(\alpha-1+k)}{\Gamma(\alpha-1)} = \frac{(\alpha-1+k)(\alpha-1)\Gamma(\alpha-1+k)}{(\alpha-1+k)(\alpha-1)\Gamma(\alpha-1)}$$
$$= \frac{(\alpha-1)}{(\alpha-1+k)}\frac{\Gamma(\alpha+k)}{\Gamma(\alpha)} = \frac{(\alpha-1)}{(\alpha-1+k)}(\alpha)_k.$$

b) For $\beta > 0$ and $\gamma - \alpha - \beta > 0$ the improper integral

$$\int_0^1 t^{\beta-1}(1-t)^{\gamma-\alpha-\beta-1}\, dt$$

683

exists and therefore using the limiting case of (22.13) as $z \to 1$ we get

$$_2F_1(\alpha, \beta; \gamma; 1) = \frac{\Gamma(\gamma)}{\Gamma(\beta)\Gamma(\gamma - \beta)} \int_0^1 t^{\beta-1}(1-t)^{\gamma-\alpha-\beta-1} \, dt$$

$$= \frac{\Gamma(\gamma)}{\Gamma(\beta)\Gamma(\gamma - \beta)} B(\beta, \gamma - \alpha - \beta)$$

$$= \frac{\Gamma(\gamma)}{\Gamma(\beta)\Gamma(\gamma - \beta)} \frac{\Gamma(\beta)\Gamma(\gamma - \alpha - \beta)}{\Gamma(\beta + \gamma - \alpha - \beta)}$$

$$= \frac{\Gamma(\gamma)\Gamma(\gamma - \alpha - \beta)}{\Gamma(\gamma - \beta)\Gamma(\gamma - \alpha)}.$$

2. a) We note that

$$(\gamma - \alpha - \beta)_2F_1(\alpha, \beta; \gamma; z) + \alpha(1 - z)_2F_1(\alpha + 1, \beta; \gamma; z)$$
$$- (\gamma - \beta)F(\alpha, \beta - 1; \gamma; z)$$

$$= \sum_{k=0}^{\infty} \left((\gamma - \alpha - \beta)\frac{(\alpha)_k(\beta)_k}{(\gamma)_k k!} + \frac{(\alpha + 1)_k(\beta)_k}{(\gamma)_k k!} \right.$$

$$\left. - (\gamma - \beta)\frac{(\alpha)_k(\beta - 1)_k}{(\gamma)_k k!} - \frac{\alpha(\alpha + 1)_{k-1}(\beta)_{k-1}}{(\gamma)_{k-1}(k - 1)!} \right) z^k$$

$$= \sum_{k=0}^{\infty} \frac{(\alpha)_k(\beta)_{k-1}}{(\gamma)_k k!}((\gamma - \alpha - \beta)(\beta + k - 1) + (\alpha + k)(\beta + k - 1)$$

$$- (\gamma - \beta)(\beta - 1) - (\gamma + k - 1)k)z^k$$

$$= 0$$

where we used that

$$(\gamma - \alpha - \beta)(\beta + k - 1) + (\alpha + k)(\beta + k - 1) - (\gamma - \beta)(\beta - 1) - (\gamma + k - 1)k = 0.$$

b) We have

$$(\gamma - \alpha - 1)_2F_1(\alpha, \beta; \gamma; z) + \alpha_2F_1(\alpha + 1, \beta; \gamma; z) - (\gamma - 1)_2F_1(\alpha, \beta; \gamma - 1; z)$$

$$= \sum_{k=0}^{\infty} \left((\gamma - \alpha - 1)\frac{(\alpha)_k(\beta)_k}{(\gamma)_k k!} + \alpha\frac{(\alpha + 1)_k(\beta)_k}{(\gamma)_k k!} - (\gamma - 1)\frac{(\alpha)_k(\beta)_k}{(\gamma - 1)_k k!} \right) z^k$$

$$= \sum_{k=0}^{\infty} \frac{(\alpha)_k(\beta)_k}{(\gamma)_k k!} \left(\gamma - \alpha - 1 + \alpha\frac{\alpha + k}{\alpha} - (\gamma - 1)\frac{1}{\frac{\gamma - 1}{\gamma - 1 + k}} \right) z^k$$

$$= \sum_{k=0}^{\infty} \frac{(\alpha)_k(\beta)_k}{(\gamma)_k k!}(\gamma - \alpha - 1 + \alpha + k - \gamma + 1 - k)z^k = 0.$$

684

3. We note that $(1)_k = k!$ and $(2)_k = (k+1)!$ and therefore we find

$$z_2F_1(1,1;2;-z) = \sum_{k=0}^{\infty} \frac{(1)_k(1)_k}{(2)_k k!}(-z)^k$$

$$= \sum_{k=0}^{\infty} (-1)^k \frac{k!k!}{k!(k+1)!} z^{k+1}$$

$$= \sum_{k=1}^{\infty} (-1)^{k+1} \frac{z^k}{k!} = \ln(1+z).$$

Furthermore we have

$$z_2F_1\left(\frac{1}{2},1;\frac{3}{2};-z\right) = z\sum_{k=0}^{\infty} \frac{(\frac{1}{2})_k(1)_k}{(\frac{3}{2})_k k!}(-z^2)^k.$$

Since

$$\frac{(\frac{1}{2})_k(1)_k}{(\frac{3}{2})_k k!} = \frac{(\frac{1}{2})_k k!}{(2k+1)(\frac{1}{2})_k k!} = \frac{1}{2k+1}$$

we arrive at

$$z_2F_1\left(\frac{1}{2},1;\frac{3}{2};-z^2\right) = \sum_{k=0}^{\infty} (-1)^k \frac{z^{2k+1}}{2k+1} = \arctan z.$$

4. From the definition follows that

$$(-n)_k = (-n)(-n+1)\cdot\ldots\cdot(-n+k-1)$$

implying that $(-n)_k = 0$ for $k > n$. This implies

$$_2F_1(-n,\beta;\gamma;z) = \sum_{k=0}^{\infty} \frac{(-n)_k(\beta)_k}{k!(\gamma)_k} z^k$$

$$= \sum_{k=0}^{n} \frac{(-n)_k(\beta)_k}{k!(\gamma)_k} z^k.$$

Since $(-n)_k = (-1)^k \frac{(n+1-k)!}{n!}$ we can rewrite the result as

$$_2F_1(-n,\beta;\gamma;z) = \sum_{k=0}^{n} (-1)^k \frac{(n+1-k)!}{k!n!} \frac{(\beta)_k}{(\gamma)_k} z^k.$$

5. We know (22.13), i.e.

$$_2F_1(\alpha,\beta;\gamma;z) = \frac{\Gamma(\gamma)}{\Gamma(\beta)\Gamma(\gamma-\beta)} \int_0^1 t^{\beta-1}(1-t)^{\gamma-\beta-1}(1-tz)^{-\alpha} \, dt.$$

The substitution $t = 1 - s$ yields

$$\int_0^1 t^{\beta-1}(1-t)^{\gamma-\beta-1}(1-tz)^{-\alpha} \, dt$$

$$= \int_0^1 (1-s)^{\beta-1} s^{\gamma-\beta-1}(1-(1-s)z)^{-\alpha} \, ds$$

$$= \int_0^1 (1-s)^{\beta-1} s^{\gamma-\beta-1}\left(1 - \frac{zs}{z-1}\right)^{-\alpha}(1-z)^{-\alpha} \, ds$$

$$= (1-z)^{-\alpha} {}_2F_1\left(\alpha, \gamma-\beta; \gamma; \frac{z}{z-1}\right).$$

6. By (22.23) we have with $s = \frac{z}{z-1}$

$$_2F_1(\alpha, \beta; \gamma; z) = (1-z)^{-\beta} {}_2F_1\left(\gamma-\alpha, \beta; \gamma; \frac{z}{z-1}\right)$$

$$= (1-z)^{-\beta} {}_2F_1(\gamma-\alpha, \beta; \gamma; s).$$

From (22.22) we deduce

$$_2F_1(\gamma-\alpha, \beta; \gamma; s) = (1-s)^{-\gamma+\alpha} {}_2F_1\left(\gamma-\alpha, \gamma-\beta; \gamma; \frac{s}{s-1}\right).$$

We note that

$$1 - s = 1 - \frac{z}{z-1} = -\frac{1}{z-1} = (1-z)^{-1},$$

hence $(1-s)^{-\gamma+\alpha} = (1-z)^{\alpha-\gamma}$, and

$$\frac{s}{s-1} = \frac{\frac{z}{z-1}}{\frac{z}{z-1} - 1} = z,$$

which yields

$$_2F_1(\gamma-\alpha, \beta; \gamma; s) = (1-z)^{\alpha-\gamma} {}_2F_1(\gamma-\alpha, \gamma-\beta; \gamma; z)$$

and eventually we get

$$_2F_1(\alpha, \beta; \gamma; z) = (1-z)^{-\beta}(1-z)^{\alpha-\gamma} {}_2F_1(\gamma-\alpha, \gamma-\beta; \gamma; z).$$

Chapter 23

1. We define a function g by

$$g(t) := \int_{t_0}^t b(s)f(s) \, ds$$

and note that $g(t_0) = 0$ and $f(t) \le a(t) + c(t)g(t)$ as well as

$$g'(t) = b(t)f(t) \le b(t)a(t) + b(t)c(t)g(t). \tag{$*$}$$

Since

$$\frac{d}{dt}\left(g(t)e^{-\int_{t_0}^t b(r)c(r)\,dr}\right) = g'(t)e^{-\int_{t_0}^t b(r)c(r)\,dr}$$
$$- b(t)c(t)g(t)e^{-\int_{t_0}^t b(r)c(r)\,dr}$$

we find

$$g'(t)e^{-\int_{t_0}^t b(r)c(r)\,dr} = \frac{d}{dt}\left(g(t)e^{-\int_{t_0}^t b(r)c(r)\,dr}\right)$$
$$+ b(t)c(t)g(t)e^{-\int_{t_0}^t b(r)c(r)\,dr}.$$

Thus multiplying $(*)$ with $e^{-\int_{t_0}^t b(r)c(r)\,dr}$ yields

$$\frac{d}{dt}\left(g(t)e^{-\int_{t_0}^t b(r)c(r)\,dr}\right) + b(t)c(t)g(t)e^{-\int_{t_0}^t b(r)c(r)\,dr}$$
$$\leq b(t)a(t)e^{-\int_{t_0}^t b(r)c(r)\,dr} + b(t)c(t)g(t)e^{-\int_{t_0}^t b(r)c(r)\,dr},$$

or

$$\frac{d}{ds}\left(g(s)e^{-\int_{t_0}^s b(r)c(r)\,dr}\right) \leq b(s)a(s)e^{-\int_{t_0}^s b(r)c(r)\,dr}. \qquad (**)$$

Integrating $(**)$ from t_0 to t we arrive at

$$g(t)e^{-\int_{t_0}^t b(r)c(r)\,dr} \leq \int_{t_0}^t b(s)a(s)e^{-\int_{t_0}^s b(r)c(r)\,dr}\,ds$$

where we have used that $g(t_0) = 0$. Since $f(t) \leq a(t) + c(t)g(t)$ we eventually arrive at

$$f(t) \leq a(t) + c(t)e^{\int_{t_0}^t b(r)c(r)\,dr}\int_{t_0}^t b(s)a(s)e^{-\int_{t_0}^s b(r)c(r)\,dr}\,ds$$
$$= a(t) + c(t)\int_{t_0}^t b(s)a(s)e^{\int_s^t b(r)c(r)\,dr}\,ds.$$

2. The general solution of the first equation is

$$u(t) = A\cos wt + B\sin wt$$

and the initial conditions $u(0) = 1$ and $\dot{u}(0) = 0$ imply $A = 1$ and $B = 0$. Thus $u(t) = \cos wt$ is the solution. The general solution of the second equation is

$$v_\epsilon(t) = \alpha\cos(w+\epsilon)t + \beta\sin(w+\epsilon)t$$

and for $v(0) = 1$ and $\dot{v}_\epsilon(0) = \eta$ we get $\alpha = 1$ and $\beta = \frac{\eta}{w+\epsilon}$, i.e. $v_{\epsilon,\eta}(t) = \cos(w+\epsilon)t + \frac{\eta}{w+\epsilon}\sin(w+\epsilon)t$ solves the second problem. It follows that

$$|u(t) - v_{\epsilon,\eta}(t)| \leq |\cos wt - \cos(w+\epsilon)t| + \frac{\eta}{w+\epsilon}|\sin(w+\epsilon)t|.$$

By the mean-value theorem we have for all $x, y \in \mathbb{R}$ that $|\cos x - \cos y| \leq |x - y|$ and $|\sin z| \leq |z|$ which implies

$$|u(t) - v_{\epsilon,\eta}(t)| \leq |wt - wt - \epsilon t| + \frac{\eta}{w + \epsilon}(w + \epsilon)|t|$$

$$= (\epsilon + \eta)T \quad \text{for } |t| < T.$$

3. We follow closely the considerations made in Chapter 17 and find

$$\|Tu(t, \lambda) - Tv(t, \lambda)\|_\infty$$

$$= \left\| \int_{t_0}^{t} (R(s, \lambda, u(s, \lambda)) - R(s, \lambda, v(s, \lambda)))e^{-2L(s-t_0)}e^{2L(s-t_0)} \, ds \right\|_\infty$$

$$\leq \int_{t_0}^{t} \|(R(s, \lambda, u(s, \lambda)) - R(s, \lambda, v(s, \lambda)))e^{-2L(s-t_0)}\|_\infty e^{2L(s-t_0)} \, ds$$

$$\leq L\|u - v\|_{\infty,L,K} \int_{t_0}^{t} e^{2L(s-t_0)} \, ds$$

$$\leq \frac{1}{2}e^{2L(t-t_0)}\|u - v\|_{\infty,L,K},$$

implying

$$\|Tu - Tv\|_{\infty,L,K} \leq \frac{1}{2}\|u - v\|_{\infty,L,K}.$$

4. The eigenvalues of A_1 and A_2 are all real and strictly positive and since $[A_1, A_2] = 0$ we can diagonalize A_1 and A_2 simultaneously. Hence $-A_1 A_2$ has strictly negative eigenvalues and the result follows.

5. With $R(u, v) = \left(\begin{smallmatrix} v \\ -\sin u \end{smallmatrix} \right)$ we have equilibrium points at $(k\pi, 0)$. Further we note that

$$(dR)(u, v) = \begin{pmatrix} \frac{\partial v}{\partial u} & \frac{\partial v}{\partial v} \\ -\frac{\partial \sin u}{\partial u} & -\frac{\partial \sin u}{\partial v} \end{pmatrix} = \begin{pmatrix} 0 & 1 \\ -\cos u & 0 \end{pmatrix}$$

and for $(u, v) = (\pi, 0)$ we find $(dR)(\pi, 0) = \left(\begin{smallmatrix} 0 & 1 \\ 1 & 0 \end{smallmatrix} \right)$. The eigenvalues of $\left(\begin{smallmatrix} 0 & 1 \\ 1 & 0 \end{smallmatrix} \right)$ are the solutions of the equation $\lambda^2 - 1 = 0$, i.e. $\lambda_1 = 1$ and $\lambda_2 = -1$, which yields that the system is unstable at $(\pi, 0)$.

6. We have $R(x, y) = \left(\begin{smallmatrix} -2x^3 + y \\ -x - 2y^5 \end{smallmatrix} \right)$ and for an equilibrium point we must have $-2x^3 + y = 0$ and $-x - 2y^5 = 0$, or $y = 2x^3$ and hence $-x = 64x^{15}$. Clearly $(0, 0)$ is an equilibrium point. If $x \neq 0$ then we must have $-1 = 64x^{14}$ which is impossible. Thus we must have $x = 0$ implying $y = 0$, and it follows that $(0, 0)$ is the only equilibrium point.

We try to find a Lyapunov function at $(0, 0)$ by the Ansatz $V(x, y) = \alpha x^2 + \beta y^2$ which gives $\mathrm{grad}V(x, y) = (2\alpha x, 2\beta y)$ and the condition $< \mathrm{grad}V(x, y), R(x, y) > < 0$ is to be satisfied. However

$$< \mathrm{grad}V(x, y), R(x, y) > \; = \; < \begin{pmatrix} 2\alpha x \\ 2\beta y \end{pmatrix}, \begin{pmatrix} -2x^3 + y \\ -x - 2y^5 \end{pmatrix} >$$

$$= -4\alpha x^4 + (2\alpha - 2\beta)xy - 4\beta y^6.$$

With $\alpha = \beta = 1$ it follows that $V(x, y) = x^2 + y^2$ is a strict Lyapunov function for the system at the equilibrium point $(0, 0)$.

7. Note that $\mathrm{grad}h(x, y) = (4x \sin(x^2 + y^2) \cos(x^2 + y^2), 4y \sin(x^2 + y^2) \cos(x^2 + y^2))$. Thus we are dealing with the system

$$\dot{u}_1(t) = -4u_1(t) \sin(u_1^2(t) + u_2^2(t)) \cos(u_1^2(t) + u_2^2(t))$$

and

$$\dot{u}_2(t) = -4u_2(t) \sin(u_1^2(t) + u_2^2(t)) \cos(u_1^2(t) + u_2^2(t)).$$

The point $(0, 0)$ is an equilibrium point which is indeed an isolated equilibrium point. The function $h(x, y) = \sin^2(x^2 + y^2)$ has an isolated zero at $(0, 0)$ and we have since now $R = -\mathrm{grad}h$

$$< \mathrm{grad}h, R > = -\|\mathrm{grad}h\|^2 = -16(x^2 + y^2) \sin^2(x^2 + y^2) \cos^2(x^2 + y^2) < 0$$

for all (x, y) in a neighbourhood of $(0, 0)$. Hence h is a strict Lyapunov function at $(0, 0)$.

Chapter 24

1. Let φ_1, $\varphi_2 \in [f]$ and γ_1, $\gamma_2 \in [g]$ then we have in some neighbourhood U of q that $\varphi_1|_U = \varphi_2|_U$ and in some neighbourhood V of q we have $\gamma_1|_V = \gamma_2|_V$. In $W := U \cap V$ we find $\varphi_1 = \varphi_2$ and $\gamma_1 = \gamma_2$ and therefore

$$\varphi_1|_W + \gamma_1|_W = (\varphi_1 + \gamma_1)|_W = (\varphi_2 + \gamma_2)|_W = \varphi_2|_W + \gamma_2|_W$$

implying

$$[\varphi_1] + [\gamma_1] = [\varphi_1 + \gamma_1] = [\varphi_2 + \gamma_2] = [\varphi_2] + [\gamma_2],$$

but $[\varphi_1] = [\varphi_2] = [f]$ and $[\gamma_1] = [\gamma_2] = [g]$. The second statement is now trivial since for $\lambda \in \mathbb{R}$ and φ_1, $\varphi_2 \in [f]$ we have for a suitable neighbourhood U of q that $(\lambda\varphi_1)|_U = \lambda(\varphi_1|_U) = \lambda(\varphi_2|_U) = (\lambda\varphi_2)|_U$.

2. For $g, h \in [f]$ there exists a neighbourhood U of $0 \in \mathbb{R}$ such that $g|_U = h|_U$. We may conclude that for some open interval $(-\rho, \rho) \subset (-r, r)$ we have $(-\rho, \rho) \subset U$ and $g|_{(-\rho,\rho)} = h|_{(-\rho,\rho)}$, which implies $h(x) = \sum_{k=0}^{\infty} a_k x^k$ in $(-\rho, \rho)$. Thus if $[f]$ admits one element having a convergent power series representation (about 0 and a positive radius of convergence) then all elements of $[f]$ have this property and the power series coincide.

3. Clearly $\mathcal{P}$ is an algebra and D_j is linear. For $p = \sum_{|\alpha| \leq k} a_\alpha x^\alpha$ and $q = \sum_{|\beta| \leq l} b_\beta x^\beta$,

689

$\alpha, \beta \in \mathbb{N}_0^n$, we find

$$D_j(p \cdot q) = D_j \left(\sum_{|\alpha| \le k} \sum_{|\beta| \le l} a_\alpha b_\beta x^{\alpha + \beta} \right)$$

$$= \sum_{|\alpha| \le k} \sum_{|\beta| \le l} a_\alpha b_\beta (\alpha + \beta)_j x^{\alpha + \beta - \epsilon_j}$$

$$= \sum_{|\alpha| \le k} \sum_{|\beta| \le l} a_\alpha b_\beta \alpha_j x^{\alpha + \beta - \epsilon_j}$$

$$+ \sum_{|\alpha| \le k} \sum_{|\beta| \le l} a_\alpha b_\alpha \beta_j x^{\alpha + \beta - \epsilon_j}$$

$$= (D_j p)q + p(D_j q),$$

recall $\alpha - \epsilon_j = 0$ for $\alpha_j = 0$.

4. We have to go back to the definition, i.e. (24.10), and to note that

$$(g \circ f)^*[\phi] = [\phi \circ g \circ f]$$

and

$$(f^* \circ g^*)(\phi) = f^*(g^*[\phi])$$
$$= f^*[\phi \circ g]$$
$$= [\phi \circ f \circ g].$$

5. We know that $M(n; \mathbb{K})$ is a vector space and since $AB \in M(n; \mathbb{K})$ for $A, B \in M(n; \mathbb{K})$ it follows that $[A, B] \in M(n; \mathbb{K})$. Further we note that

$$[A, B] = AB - BA = -(BA - AB) = -[B, A].$$

For $A, B, C \in M(n; \mathbb{K})$ we observe

$$[A, [B, C]] = A(BC - CB) - (BC - CB)A$$
$$= ABC - ACB - BCA + CBA,$$
$$[[A, B], C] = (AB - BA)C - C(AB - BA)$$
$$= ABC - BAC - CAB + CBA,$$
$$[B, [A, C]] = B(AC - CA) - (AC - CA)B$$
$$= BAC - BCA - ACB + CAB,$$

and therefore

$$[[A, B], C] + [B, [A, C]] = ABC - BAC - CAB + CBA$$
$$+ BAC - BCA - ACB + CAB$$
$$= ABC - ACB - BCA + CBA$$
$$= [A, [B, C]].$$

6. Suppose that for $\lambda, \mu, \nu \in \mathbb{R}$ we have $\lambda X + \mu Y + \nu Z = 0$, i.e.

$$\begin{pmatrix} 0 & \lambda & 0 \\ -\lambda & 0 & 0 \\ 0 & 0 & 0 \end{pmatrix} + \begin{pmatrix} 0 & 0 & \mu \\ 0 & 0 & 0 \\ -\mu & 0 & 0 \end{pmatrix} + \begin{pmatrix} 0 & 0 & 0 \\ 0 & 0 & -\nu \\ 0 & \nu & 0 \end{pmatrix} = \begin{pmatrix} 0 & \lambda & \mu \\ -\lambda & 0 & -\nu \\ -\mu & \nu & 0 \end{pmatrix} \overset{!}{=} \begin{pmatrix} 0 & 0 & 0 \\ 0 & 0 & 0 \\ 0 & 0 & 0 \end{pmatrix}.$$

It follows that $\lambda = \mu = \nu = 0$ and therefore $\{X, Y, Z\}$ is an independent set in $M(3; \mathbb{R})$. We want to prove that $\text{span}\{X, Y, Z\}$ is a Lie algebra. For this we have to prove that $A, B \in \text{span}\{X, Y, Z\}$ implies $[A, B] \in \text{span}\{X, Y, Z\}$. Once we have shown that $[X, Y] = Z$, $[Z, X] = Y$ and $[Y, Z] = X$ the result follows since for $A = \alpha_1 X + \alpha_2 Y + \alpha_3 Z$ and $B = \beta_1 X + \beta_2 Y + \beta_3 Z$ it holds:

$$\begin{aligned} [A, B] &= [\alpha_1 X + \alpha_2 Y + \alpha_3 Z, \ \beta_1 X + \beta_2 Y + \beta_3 Z] \\ &= \alpha_1 \beta_2 [X, Y] + \alpha_1 \beta_3 [X, Z] \\ &\quad + \alpha_2 \beta_1 [Y, X] + \alpha_2 \beta_3 [Y, Z] \\ &\quad + \alpha_3 \beta_1 [Z, X] + \alpha_3 \beta_2 [Z, Y] \\ &= (\alpha_1 \beta_2 - \alpha_2 \beta_1)[X, Y] + (\alpha_3 \beta_1 - \alpha_1 \beta_3)[Z, X] + (\alpha_2 \beta_3 - \alpha_3 \beta_2)[Y, Z] \\ &= (\alpha_1 \beta_2 - \alpha_2 \beta_1) Z + (\alpha_3 \beta_1 - \alpha_1 \beta_3) Y + (\alpha_2 \beta_3 - \alpha_3 \beta_2) X, \end{aligned}$$

which is an element of $\text{span}\{X, Y, Z\}$. Next we note

$$\begin{aligned} [X, Y] &= \begin{pmatrix} 0 & 1 & 0 \\ -1 & 0 & 0 \\ 0 & 0 & 0 \end{pmatrix} \begin{pmatrix} 0 & 0 & 1 \\ 0 & 0 & 0 \\ -1 & 0 & 0 \end{pmatrix} - \begin{pmatrix} 0 & 0 & 1 \\ 0 & 0 & 0 \\ -1 & 0 & 0 \end{pmatrix} \begin{pmatrix} 0 & 1 & 0 \\ -1 & 0 & 0 \\ 0 & 0 & 0 \end{pmatrix} \\ &= \begin{pmatrix} 0 & 0 & 0 \\ 0 & 0 & -1 \\ 0 & 1 & 0 \end{pmatrix} = Z, \end{aligned}$$

$$\begin{aligned} [Z, X] &= \begin{pmatrix} 0 & 0 & 0 \\ 0 & 0 & -1 \\ 0 & 1 & 0 \end{pmatrix} \begin{pmatrix} 0 & 1 & 0 \\ -1 & 0 & 0 \\ 0 & 0 & 0 \end{pmatrix} = \begin{pmatrix} 0 & 1 & 0 \\ -1 & 0 & 0 \\ 0 & 0 & 0 \end{pmatrix} \begin{pmatrix} 0 & 0 & 0 \\ 0 & 0 & -1 \\ 0 & 1 & 0 \end{pmatrix} \\ &= \begin{pmatrix} 0 & 0 & 1 \\ 0 & 0 & 0 \\ -1 & 0 & 0 \end{pmatrix} = Y, \end{aligned}$$

$$\begin{aligned} [Y, Z] &= \begin{pmatrix} 0 & 0 & 1 \\ 0 & 0 & 0 \\ -1 & 0 & 0 \end{pmatrix} \begin{pmatrix} 0 & 0 & 0 \\ 0 & 0 & -1 \\ 0 & 1 & 0 \end{pmatrix} - \begin{pmatrix} 0 & 0 & 0 \\ 0 & 0 & -1 \\ 0 & 1 & 0 \end{pmatrix} \begin{pmatrix} 0 & 0 & 1 \\ 0 & 0 & 0 \\ -1 & 0 & 0 \end{pmatrix} \\ &= \begin{pmatrix} 0 & 1 & 0 \\ -1 & 0 & 0 \\ 0 & 0 & 0 \end{pmatrix} = X, \end{aligned}$$

and eventually the statement is proved.

7.   a) We prove that $(H, \circ)$ is a group. Obviously we have $X \circ Y = X + Y +$

$\frac{1}{2}[X,Y] \in H$ for $X, Y \in H$. Next we note that

$$(X \circ Y) \circ Z = \left(X + Y + \frac{1}{2}[X,Y] \right) \circ Z$$

$$= X + Y + \frac{1}{2}[X,Y] + Z$$

$$+ \frac{1}{2}[X + Y + \frac{1}{2}[X,Y], Z]$$

$$= X + Y + Z + \frac{1}{2}[X,Y] + \frac{1}{2}[X+Y, Z]$$

$$= X + Y + Z + \frac{1}{2}[X,Y] + \frac{1}{2}[X, Z] + \frac{1}{2}[Y, Z]$$

and

$$X \circ (Y \circ Z) = X \circ \left(Y + Z + \frac{1}{2}[Y, Z] \right)$$

$$= X + Y + Z + \frac{1}{2}[Y, Z]$$

$$+ \frac{1}{2}[X, Y + Z + \frac{1}{2}[Y, Z]]$$

$$= X + Y + Z + \frac{1}{2}[Y, Z] + \frac{1}{2}[X,Y] + \frac{1}{2}[X, Z],$$

where we have used in both calculations that $[X, [Y, Z]] = 0$. Thus we have proved that $\circ$ is an associative operation on H. For $0 \in H$ we find

$$X \circ 0 = X + 0 + \frac{1}{2}[0, X] = X$$

and with $-X \in H$ it follows that

$$X \circ (-X) = X + (-X) + \frac{1}{2}[X, -X] = 0.$$

Hence we have proved that $(H, \circ)$ with neutral element 0 and $-X$ being the inverse to $X \in H$ is a group. Since in general $[X,Y] \neq [Y, X]$ we do not expect this group to be commutative.

b) For $X = \begin{pmatrix} 0 & x_1 & x_2 \\ 0 & 0 & x_3 \\ 0 & 0 & 0 \end{pmatrix}$, $Y = \begin{pmatrix} 0 & y_1 & y_2 \\ 0 & 0 & y_3 \\ 0 & 0 & 0 \end{pmatrix}$ and $Z = \begin{pmatrix} 0 & z_1 & z_2 \\ 0 & 0 & z_3 \\ 0 & 0 & 0 \end{pmatrix}$ we find

$$[Y, Z] = \begin{pmatrix} 0 & y_1 & y_2 \\ 0 & 0 & y_3 \\ 0 & 0 & 0 \end{pmatrix} \begin{pmatrix} 0 & z_1 & z_2 \\ 0 & 0 & z_3 \\ 0 & 0 & 0 \end{pmatrix} - \begin{pmatrix} 0 & z_1 & z_2 \\ 0 & 0 & z_3 \\ 0 & 0 & 0 \end{pmatrix} \begin{pmatrix} 0 & y_1 & y_2 \\ 0 & 0 & y_3 \\ 0 & 0 & 0 \end{pmatrix}$$

$$= \begin{pmatrix} 0 & 0 & y_1 z_3 - y_3 z_1 \\ 0 & 0 & 0 \\ 0 & 0 & 0 \end{pmatrix}$$

and consequently

$$[X,[Y,Z]] = \begin{pmatrix} 0 & x_1 & x_2 \\ 0 & 0 & x_3 \\ 0 & 0 & 0 \end{pmatrix} \begin{pmatrix} 0 & 0 & y_1z_3 - y_3z_1 \\ 0 & 0 & 0 \\ 0 & 0 & 0 \end{pmatrix}$$

$$- \begin{pmatrix} 0 & 0 & y_1z_3 - y_3z_1 \\ 0 & 0 & 0 \\ 0 & 0 & 0 \end{pmatrix} \begin{pmatrix} 0 & x_1 & x_2 \\ 0 & 0 & x_3 \\ 0 & 0 & 0 \end{pmatrix} = 0.$$

Thus all the assumptions of part a) are satisfied.

8. For $u \in C^\infty(G)$ we have

$$\sum_{j=1}^{n} a_j(x) \frac{\partial}{\partial x_j} \left(\sum_{l=1}^{n} b_l(x) \frac{\partial u(x)}{\partial x_l} \right)$$

$$= \sum_{j=1}^{n} \sum_{l=1}^{n} \left(a_j(x) \frac{\partial b_l(x)}{\partial x_j} \frac{\partial u(x)}{\partial x_j} + a_j(x) b_l(x) \frac{\partial^2 u(x)}{\partial x_j \partial x_l} \right)$$

and

$$\sum_{l=1}^{n} b_l(x) \frac{\partial}{\partial x_l} \left(\sum_{j=1}^{n} a_j(x) \frac{\partial u}{\partial x_j} \right)$$

$$= \sum_{l=1}^{n} \sum_{j=1}^{n} \left(b_l(x) \frac{\partial a_j(x)}{\partial x_l} \frac{\partial u(x)}{\partial x_j} + b_l(x) a_j(x) \frac{\partial^2 u(x)}{\partial x_l \partial x_j} \right)$$

which implies

$$[X,Y]_q = \sum_{j=1}^{n} \left(\sum_{l=1}^{n} a_l(q) \frac{\partial b_j}{\partial x_l}(q) - b_l(q) \frac{\partial a_j}{\partial x_l}(q) \right) \left(\frac{\partial}{\partial x_l} \right)_q.$$

9. Let $Z = \nu \frac{\partial}{\partial x} + \mu \frac{\partial}{\partial y}$ with C^∞-functions ν and μ. Suppose that $Z = \alpha X + \beta Y$ with C^∞-functions α and β. It follows that

$$\alpha \frac{\partial}{\partial x} + x\beta \frac{\partial}{\partial y} \overset{!}{=} \nu \frac{\partial}{\partial x} + \mu \frac{\partial}{\partial y}.$$

However for $(0,y)$ we have $\alpha \frac{\partial}{\partial x} + x\beta \frac{\partial}{\partial y} = \alpha \frac{\partial}{\partial x}$, and so the vector field $Z = \frac{\partial}{\partial y}$ cannot be represented as a linear combination (with C^∞-functions as coefficients) of the two vector fields X and Y.

Now we observe that

$$[X,Y]u = \frac{\partial}{\partial x} \left(x \frac{\partial u}{\partial y} \right) - x \frac{\partial}{\partial y} \left(\frac{\partial u}{\partial x} \right) = \frac{\partial u}{\partial y},$$

or

$$[X,Y] = \frac{\partial}{\partial y}$$

693

and therefore, if $Z = \nu\frac{\partial}{\partial x} + \mu\frac{\partial}{\partial y}$ then we have with $a = \nu$, $b = 0$ and $c = \mu$ the representation

$$Z = aX + bY + c[X, Y] = \nu\frac{\partial}{\partial x} + 0\left(x\frac{\partial}{\partial y}\right) + \mu\frac{\partial}{\partial y}.$$

10. An integral curve γ has to satisfy $\dot{\gamma}(t) = X_\gamma(t)$ where

$$X_p = y\left(\frac{\partial}{\partial x}\right)_p - x\left(\frac{\partial}{\partial y}\right)_p, \quad p = (x, y),$$

thus we have to solve the differential equations

$$\dot{\gamma}_1(t) = \gamma_2(t) \quad \text{and} \quad \dot{\gamma}_2(t) = -\gamma_1(t)$$

or

$$\dot{\gamma}(t) = \begin{pmatrix} 0 & 1 \\ -1 & 0 \end{pmatrix}\gamma(t).$$

A fundamental system for this equation is $\{\left(\begin{smallmatrix} \sin t \\ \cos t \end{smallmatrix}\right), \left(\begin{smallmatrix} \cos t \\ -\sin t \end{smallmatrix}\right)\}$. Thus, once initial conditions are specified, i.e. $\gamma(t_0) = p_0$ for some t_0, we can determine a unique integral curve passing through p_0.

Chapter 25

1. See the following figure.

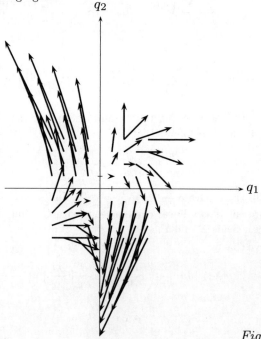

Figure I

2. The system has the general solution $u(t) + A\sin(2t + \varphi_0)$ and $v(t) = 2A\cos(2t + \varphi_0)$. It follows that

$$\frac{u^2}{A^2} + \frac{v^2}{4A^2} = 1$$

which is an ellipse in the (u, v)-axis symmetric to the coordinate axes and half-axis of length A and $2A$ respectively. Therefore we obtain the phase diagram

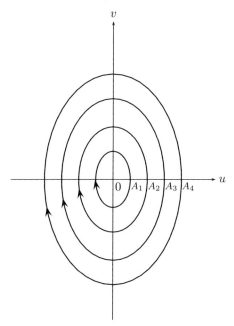

Figure II

3. We note that

$$\dot{r}(t) = \frac{1}{\sqrt{x^2(t) + y^2(t)}}(\dot{x}(t)x(t) + \dot{y}(t)y(t))$$

$$= \frac{1}{r(t)}\left(\left(-y(t) + \frac{x(t)}{r(t)}(1 - r^2(t))\right)x(t) + \left(x(t) + \frac{y(t)}{r(t)}(1 - r^2(t))\right)y(t)\right)$$

$$= \frac{1}{r(t)}\left(\frac{x^2(t) + y^2(t)}{r(t)}(1 - r^2(t))\right) = (1 - r^2(t))$$

and the equation $\dot{r}(t) = 1 - r^2(t)$ admits the constant solution $r(t) = 1$. Such a solution reduces the original system to $\dot{x} = -y$ and $\dot{y} = x$ which has of course the solution $u(t) = \binom{\cos t}{\sin t}$ and we note that indeed $\cos^2(t) + \sin^2(t) = 1$ holds.

Now let v be a solution to the system starting at t_0 at $v(t_0)$, $\|v(t_0)\| < 1$. Suppose that for some t_1 and t_2 we have $u(t_1) = v(t_2)$. This would imply for all $t \in \mathbb{R}$ that

695

$u(t_1 + t) = v(t_2 + t)$ and therefore $\|u(t_1 + t)\| = \|v(t_2 + t)\|$. Since $\|u(t_1 + t)\| = 1$ we arrive at $\|v(t_2 + t)\| = 1$ for all t but by assumption we have $\|v(t_0)\| < 1$ which is a contradiction.

4. The key observation is that if φ is harmonic then $\operatorname{div}(\operatorname{grad}\varphi) = \Delta_n\varphi = 0$. Hence we may apply Liouville's theorem to find

$$\lambda^{(n)}(H_t) = \lambda^{(n)}(B_\rho(p_0)) = w_n\rho^n.$$

Since ϕ_t is a diffeomorphism it maps compact sets onto compact sets and connected sets onto connected sets.

5. The solution to $\dot{u} = Au$ is given by $u(t) = e^{At}q_0$ with $q_0 = u(0)$ which exists for all $t \in \mathbb{R}$ and $q_0 \in \mathbb{R}^n$. Clearly, the flow mapping is $H \mapsto e^{tA}(H)$. Furthermore e^{At} is a linear mapping with inverse e^{-At}, i.e. $e^{At} \in GL(n; \mathbb{R})$. From Theorem III.3.11 we deduce

$$\lambda^{(n)}(H_t) = \lambda^{(n)}(e^{At}H) = |\det e^{At}|\lambda^{(n)}(H)$$

for every Borel set H.

If A is symmetric then $A = U^{-1}\begin{pmatrix} \lambda_1 & & \mathbf{0} \\ & \ddots & \\ \mathbf{0} & & \lambda_n \end{pmatrix}U$ where U is a suitable orthogonal matrix and we have $e^{At} = U^{-1}e^{\begin{pmatrix} \lambda_1 t & & \mathbf{0} \\ & \ddots & \\ \mathbf{0} & & \lambda_n t \end{pmatrix}}U$ and consequently it follows that

$$|\det e^{At}| = \left|\det e^{\begin{pmatrix} \lambda_1 t & & \mathbf{0} \\ & \ddots & \\ \mathbf{0} & & \lambda_n t \end{pmatrix}}\right| = e^{(\lambda_1 + \cdots + \lambda_n)t} = e^{\operatorname{tr}(A)t}.$$ Thus if $\operatorname{tr}(A) = 0$ then $|\det e^{At}| = 1$ and we find $\lambda^{(n)}(H_t) = \lambda^{(n)}(H)$ for all Borel sets H.

6. We set $H_t^x := \phi_t(\overline{B_1(x)})$, $H_t^y := \phi_t(\overline{B_1(y)})$ and $H_t^{x,y} = \phi_t(\overline{B_1(x)} \cup \overline{B_2(y)})$. For general mappings we have $f(A \cup B) = f(A) \cup f(B)$, see (I.A.II.25) and therefore it follows that $H_t^{x,y} = H_t^x \cup H_t^y$. By Liouville's theorem we have $\lambda^{(n)}(H_t^{x,y}) = 2w_n$ and $\lambda^{(n)}(H_t^x) = \lambda^{(n)}(H_t^y) = w_n$. Now we find by (III.1.41)

$$\lambda^{(n)}(H_t^x \cup H_t^y) + \lambda^{(n)}(H_t^x \cap H_t^y) = \lambda^{(n)}(H_t^x) + \lambda^{(n)}(H_t^y)$$

or

$$2w_n + \lambda^{(n)}(H_t^x \cap H_t^y) = 2w_n$$

implying $\lambda^{(n)}(H_t^x \cap H_t^y) = 0$.

Solutions to Problems of Part 10

Chapter 26

1. We consider the figure below.

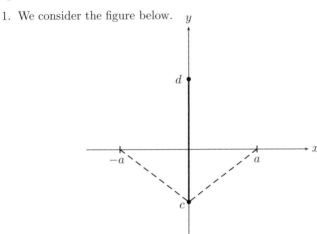

We cannot connect $(-a, 0)$ with $(a, 0)$ by a line segment since the point $(0, 0)$ does not belong to $\mathbb{R}^2 \backslash (c, d)$. The problem is symmetric with respect to the y-axis and therefore the shortest path connected $(-a, 0)$ with $(a, 0)$ is obtained by reflecting the shortest path connecting $(-a, 0)$ with the y-axis at the y-axis. For $|c| < d$ the shortest path connecting $(-a, 0)$ with the y-axis is the line segment connecting $(-a, 0)$ with $(0, c)$, in the case $d < |c|$ the shortest path connecting $(-a, 0)$ with the y-axis is the line segment connecting $(-a, 0)$ with $(0, d)$. In each case the solution is unique. However, if $|c| = d$ then the line segment connecting $(-a, 0)$ with $(0, c) = (0, -d)$ has the same length as does the line segment connecting $(-a, 0)$ with $(0, d)$. Hence in the last case we have not a unique solution.

2. For $n = 1$ the fundamental lemma of the calculus of variation says: If for $f \in C([a, b])$ we have for all $\varphi \in C_0^\infty((a, b)) \subset C([a, b])$ that $\int_a^b f(x)\varphi(x)\, \mathrm{d}x = 0$ then $f = 0$. We know that $C([a, b]) \subset L^2([a, b])$ and further that $C_0^\infty((a, b))$ is dense in $L^2([a, b])$. Therefore the relation $\int_a^b f(x)\varphi(x)\, \mathrm{d}x = 0$ for all $\varphi \in C_0^\infty((a, b))$ means that the element $f \in L^2([a, b])$ is orthogonal to a dense set of $L^2([a, b])$ and hence must be equal to the zero element in $L^2([a, b])$. Since f is by assumption continuous it must be as function identically zero.

3. With $F(t, x, y) = \frac{\sqrt{1+y^2}}{\sqrt{2g}\sqrt{x}}$ we find

$$F_y(t, x, y) = \frac{y}{\sqrt{2g}\sqrt{x}\sqrt{1 + y^2}}$$

697

and

$$F_x(t, x, y) = -\frac{\sqrt{1+y^2}}{2\sqrt{2g}\sqrt{x^3}}$$

which yields

$$\frac{\mathrm{d}}{\mathrm{d}t} F_y(t, u(t), \dot{u}(t)) - F_x(t, u(t), \dot{u}(t))$$

$$= \frac{\mathrm{d}}{\mathrm{d}t}\left(\frac{\dot{u}(t)}{\sqrt{2g}\sqrt{u(t)}\sqrt{1+\dot{u}^2(t)}} + \frac{\sqrt{1+\dot{u}^2(t)}}{2\sqrt{2g}\sqrt{u^3(t)}}\right) = 0.$$

It is more advantageous not to perform the remaining differentiation.

4. With $F(t, x_1, x_2, y_1, y_2) = \frac{\sqrt{1+y_1^2+y_2^2}}{\rho(t, x_1, x_2)}$ we find

$$F_{y_1}(t, x_1, x_2, y_1, y_2) = \frac{y_1}{\rho(t, x_1, x_2)\sqrt{1+y_1^2+y_2^2}},$$

$$F_{y_2}(t, x_1, x_2, y_1, y_2) = \frac{y_2}{\rho(t, x_1, x_2)},$$

$$F_{x_1}(t, x_1, x_2, y_1, y_2) = -\frac{\partial\rho(t, x_1, x_2)}{\partial x_1}\frac{\sqrt{1+y_1^2+y_2^2}}{\rho^2(t, x_1, x_2)},$$

$$F_{x_2}(t, x_1, x_2, y_1, y_2) = -\frac{\partial\rho(t, x_1, x_2)}{\partial x_2}\frac{\sqrt{1+y_1^2+y_2^2}}{\rho^2(t, x_1, x_2)}.$$

From here we find the two Euler-Lagrange equations

$$\frac{\mathrm{d}}{\mathrm{d}t}\frac{\dot{u}(t)}{\rho(t, u(t), v(t))\sqrt{1+\dot{u}^2(t)+\dot{v}^2(t)}} + \frac{\partial\rho(t, u(t), v(t))}{\partial x_1}\frac{\sqrt{1+\dot{u}^2(t)+\dot{v}^2(t)}}{\rho(t, u(t), v(t))} = 0$$

and

$$\frac{\mathrm{d}}{\mathrm{d}t}\frac{\dot{v}(t)}{\rho(t, u(t), v(t))\sqrt{1+\dot{u}^2(t)+\dot{v}^2(t)}} + \frac{\partial\rho(t, u(t), v(t))}{\partial x_2}\frac{\sqrt{1+\dot{u}^2(t)+\dot{v}^2(t)}}{\rho(t, u(t), v(t))} = 0.$$

Again, it is reasonable not to carry out the final differentiation.

5. For $\mathcal{F}_f(q) = \frac{1}{2}\int_a^b \dot{q}(t)^2 \, \mathrm{d}t + \int_a^b f(t)q(t) \, \mathrm{d}t$ we find with $\epsilon > 0$ and $\varphi \in C_0^\infty((a, b))$ that

$$\frac{\mathrm{d}}{\mathrm{d}\epsilon}\mathcal{F}(q + \epsilon\varphi) = \int_a^b (\dot{q}(t) + \epsilon\dot{\varphi}(t))\dot{\varphi}(t) \, \mathrm{d}t + \int_a^b f(t)\varphi(t) \, \mathrm{d}t$$

$$= -\int_a^b \ddot{q}(t)\varphi(t) \, \mathrm{d}t + \int_a^b f(t)\varphi(t) \, \mathrm{d}t + \epsilon\int_a^b \dot{\varphi}^2(t) \, \mathrm{d}t$$

which yields

$$\frac{\mathrm{d}}{\mathrm{d}\epsilon}\mathcal{F}(q + \epsilon\varphi)\bigg|_{\epsilon=0} = \int_a^b (-\ddot{q}(t) + f(t))\varphi(t) \, \mathrm{d}t = 0$$

which must hold for all $\varphi \in C_0^\infty((a,b))$ and the fundamental lemma of the calculus of variation implies $-\ddot{q}(t) + f(t) = 0$ or $\ddot{q}(t) = f(t)$ and $q(a) = q(b)$ follows from the assumptions.

6. The situation has some similarity with the Lagrange multiplier rule, compare with Theorem II.11.3. Since the constraint is $G(u,t) = 0$ for all $t \in [t_1, t_2]$ it follows for an extremal of $\mathcal{F}$ satisfying the constraints that it is also an extremal of

$$\int_{t_1}^{t_2} \{F(t, u(t), \dot{u}(t)) + \lambda(t)G(u(t))\} \, dt.$$

Thus we are looking for the Euler-Lagrange equations for the new Lagrange function $F(t, u(t), \dot{u}(t)) + \lambda(t)G(u(t))$ or $H(t, x, y) = F(t, x, y) + \lambda(t)G(x)$. It follows that

$$H_y(t, x, y) = F_y(t, x, y)$$

and

$$H_x(t, x, y) = F_x(t, x, y) + \lambda(t)G_x(x).$$

From this we deduce as Euler-Lagrange equations for H

$$\frac{d}{dt}F_y(t, x, y) - F_x(t, x, y) - \lambda(t)G_x(x) = 0$$

or

$$\frac{d}{dt}F_y(t, u(t), \dot{u}(t)) - F_x(t, u(t), \dot{u}(t)) - \lambda(t)G_x(u(t)) = 0.$$

7. The Lagrange function under consideration is

$$F(t, x, y) = \frac{1}{2}ty - \frac{1}{2}x + \lambda\sqrt{1 + y^2}$$

which yields

$$F_y(t, x, y) = \frac{1}{2}t + \frac{\lambda y}{\sqrt{1 + y^2}}$$

and

$$F_x = -\frac{1}{2},$$

from which we deduce the Euler-Lagrange equation

$$\frac{d}{dt}\left(\frac{1}{2}t + \frac{\lambda \dot{u}(t)}{\sqrt{1 + \dot{u}^2(t)}}\right) + \frac{1}{2} = 0$$

or

$$\frac{d}{dt}\frac{\lambda \dot{u}(t)}{\sqrt{1 + \dot{u}^2(t)}} + 1 = 0. \qquad (*)$$

Integrating $(*)$ yields with some constant α

$$\frac{\lambda \dot{u}(t)}{\sqrt{1 + \dot{u}^2(t)}} = -t + \alpha.$$

We read this equation as an algebraic equation for $\dot{u}(t)$ which has the two solutions

$$\dot{u}(t) = \pm \frac{t-\alpha}{\lambda^2 - (t-\alpha)^2}.$$

Now we integrate this differential equation to obtain with a further constant β

$$u(t) - \beta = \pm\sqrt{\lambda^2 - (t-\alpha)^2}$$

or with $y = u(t)$ and $x = t$ as coordinates in the $x-y$–plane

$$(x-\alpha)^2 + (y-\beta)^2 = \lambda^2$$

which means that the trace of the curve is a circle. In principle we have obtained a solution of the **isoperimetric problem** in the plane, however the justification for considering the functional we started with needs a few more arguments.

8. Part a) follows from Hölder's inequality

$$\int_0^1 \|\dot{\gamma}(t)\| \, dt \le \left(\int_0^1 \|\dot{\gamma}(t)\|^2 \, dt\right)^{\frac{1}{2}} \left(\int_0^1 1^2 \, dt\right)^{\frac{1}{2}}$$

or

$$l^2(\gamma) \le E(\gamma).$$

Inspecting the calculation leading to the invariance of the arc length with respect to parameter transforms or just doing the change of variables we find for a parametric curve $\gamma : I_1 \to \mathbb{R}^n$ and a change of parameter $\varphi : I_2 \to I_1$, for $\tilde{\gamma} := \gamma \circ \varphi : I_2 \to \mathbb{R}^n$

$$\int_{I_2} \|\dot{\tilde{\gamma}}(s)\|^2 \, ds = \int_{I_2} \|(\gamma \circ \varphi)^\cdot(s)\|^2 \, ds$$

$$= \int_{I_2} \|\dot{\gamma}(\varphi(s))\dot{\varphi}(s)\|^2 \, ds$$

$$= \int_{I_2} \|(\dot{\gamma})(\varphi(s))\|^2 |\dot{\varphi}(s)|^2 \, ds$$

which is in general not equal to $\int_{I_2} \|(\dot{\gamma})(\varphi(s))\| |\dot{\varphi}(s)| \, ds = E(\gamma)$. A concrete counter example for a curve $\gamma : [0,1] \to \mathbb{R}^4$ would be the parameter transformation $\varphi(s) = \sqrt{2}\sin\frac{\pi}{4}s$. This shows part b. Finally assume that $\gamma_0 : [0,1] \to \mathbb{R}^n$ is a minimiser of the energy parameterized proportional to the arc length and suppose for some C^1-curve $\gamma : [0,1] \to \mathbb{R}^n$ parameterized with respect to the arc length and satisfying $\gamma(0) = A$ and $\gamma(1) = B$ we have $l(\gamma) < l(\gamma_0)$. First we observe that $E(\gamma) = l^2(\gamma)$ if $\|\dot{\gamma}(t)\| = c_0$. Now we find by our assumption

$$E(\gamma) = l^2(\gamma) < l^2(\gamma_0) = E(\gamma_0)$$

where in the last step we used that γ_0 is again parameterized proportional to the arc length. This is however a contradiction to the assumption that γ_0 is a minimiser of the energy.

9. We consider

$$< u, v >_E := \int_G < \text{grad} u(x), \text{grad} v(x) > \, dx$$

on $C_0^1(G)$. Clearly $< \cdot, \cdot >_E$ is a bilinear form and $< u, u >_E \geq 0$. Moreover $< u, v >_E = < v, u >_E$, i.e. this bilinear form is symmetric. It remains to prove that $< u, u >_E = 0$ implies $u = 0$. For this we write with $x = (x_1, \ldots, x_n)$ now $x = (x_1, y)$, $x_1 \in \mathbb{R}$, $y \in \mathbb{R}^{n-1}$ and we find

$$\int_G |u(x)|^2 \, dx = \int_G u^2(x_1, \ldots, x_n) \, dx$$

$$= \int_{\mathbb{R}^{n-1}} \left(\int_a^b u^2(x_1, y) \, dx_1 \right) dy$$

$$= \int_{\mathbb{R}^{n-1}} \left(\int_a^b \frac{\partial x_1}{\partial x_1} u^2(x_1, y) \, dx_1 \right) dy$$

$$= - \int_{\mathbb{R}^{n-1}} \left(\int_a^b x_1 \frac{\partial u^2(x_1, y)}{\partial x_1} \, dx_1 \right) dy$$

$$= -2 \int_{\mathbb{R}^{n-1}} \left(\int_a^b x_1 u(x_1, y) \frac{\partial u(x_1, y)}{\partial x_1} \, dx_1 \right) dy$$

$$\leq 2 \max(|a|, |b|) \int_{\mathbb{R}^{n-1}} \left(\int_a^b |u(x_1, y)| \left| \frac{\partial u(x_1, y)}{\partial x_1} \right| \, dx_1 \right) dy$$

$$= 2 \max(|a|, |b|) \int_G |u(x)| \left| \frac{\partial u(x)}{\partial x_1} \right| \, dx$$

$$\leq 2 \max(|a|, |b|) \left(\int_G |u(x)|^2 \, dx \right)^{\frac{1}{2}} \left(\int_G \left| \frac{\partial u(x)}{\partial x_1} \right|^2 \, dx \right)^{\frac{1}{2}}$$

$$\leq 2 \max(|a|, |b|) \left(\int_G |u(x)|^2 \, dx \right)^{\frac{1}{2}} \left(\int_G \|\text{grad} u(x)\|^2 \, dx \right)^{\frac{1}{2}}$$

implying

$$\left(\int_G |u(x)|^2 \, dx \right)^{\frac{1}{2}} \leq 2 \max(|a|, |b|) \left(\int_G \|\text{grad} u(x)\|^2 \, dx \right)^{\frac{1}{2}}$$

and therefore it follows that $|||u|||^2 = < u, u >_E = 0$ implies that u is identically zero.

Chapter 27

1. Let $m := \inf \left\{ \int_0^1 e^{-u'^2(t)} \, dt \mid u \in D \right\}$. For every $u \in D$ we must have $\int_0^1 e^{-u'^2(t)} \, dt > 0$. When we can show that $m = 0$ then it is proved that our problem does not admit a solution, i.e. a minimiser in D. For $k \in \mathbb{N}$ we define

$$u_k(t) = k \left(t - \frac{1}{2} \right)^2 - \frac{k}{4}.$$

701

Clearly $u_k \in C^1([0,1])$ and $u_k(0) = \frac{k}{4} - \frac{k}{4} = 0$ as we find $u_k(1) = \frac{k}{4} - \frac{k}{4} = 0$. Moreover we have

$$\int_0^1 e^{-u_k'^2(t)} \, dt = \int_0^1 e^{-4k^2\left(1-\frac{1}{2}\right)^2} \, dt = \frac{1}{2k} \int_{-k}^{k} e^{-s^2} \, ds.$$

Since $\lim_{k\to\infty} \frac{1}{2k} \int_{-k}^{k} e^{-s^2} \, ds = 0$ it follows that $m = 0$. (We borrowed this example from [xx].)

2. With $F(x,y) = x^2 + y^2$ we find $F_y = 2y$, $F_x = 2x$ and therefore as Euler-Lagrange equation

$$\left(\frac{d}{dt} 2\dot{v}(t)\right) - 2v(t) = 0$$

or $\ddot{v}(t) - v(t) = 0$. Two independent solutions to this equation are given by $v_1(t) = \cosh t$ and $v_2(t) = \sinh t$. The general solution is of the form $\lambda \cosh t + \mu \sinh t$. The boundary conditions imply now

$$0 = \lambda \cosh 0 + \mu \sinh 0 = \lambda$$

and therefore as second condition

$$1 = \mu \sinh 1$$

or $\mu = \frac{1}{\sinh 1}$. Hence $u(t) = \frac{\sinh t}{\sinh 1}$ is the unique solution of the corresponding Euler-Lagrange equations.

3. For a general surface of the revolution S given according to Example III.24.8 as $f(v,w) = \begin{pmatrix} h(v)\cos w \\ h(v)\sin w \\ k(v) \end{pmatrix}$ we have proved that

$$A(S) = 2\pi \int_a^b |h(v)| \sqrt{k'(v)^2 + h'(v)^2} \, dv.$$

In our case we have with $t := v$ that $k(v) = t$ and $h(v) = u(t) > 0$ and therefore

$$A(S) = 2\pi \int_a^b u(t) \sqrt{1 + u'^2(t)} \, dt.$$

4. For $F = F(y)$ we have of course $F_x = 0$ and $F_y = F'$ and therefore the Euler-Lagrange equation is

$$\frac{d}{dt} F_y(\dot{u}(t)) = \frac{d}{dt} F'(\dot{u}(t)) = 0.$$

Thus the function $F' : \mathbb{R} \to \mathbb{R}$ is constant (with respect to t) meaning that $F'(\dot{u}(t)) = c_0$. This implies in turn that $\dot{u}(t) = c_1$ and therefore $u(t) = c_1 t + c_2$, i.e. the extremals are line segments. (Note that we mixed a bit the languages as usual in differential geometry and related field, u is strictly spoken a function, the line segment is its graph, or when we look at $t \mapsto \begin{pmatrix} t \\ u(t) \end{pmatrix}$ as a parametric curve it is the trace of this curve.)

702

5. The Lagrange function under consideration is

$$F(t, x, y) = \sqrt{1 + y^2 \sin^2 t}$$

which is independent of x. Hence by Example 27.4.A we have that F_y is a first integral of the variational problem. Since

$$F_y = \frac{y \sin^2 t}{\sqrt{1 + y^2 \sin^2 t}}$$

we find for the concrete problem

$$\frac{\dot\varphi(\vartheta) \sin^2 \vartheta}{\sqrt{1 + \dot\varphi^2(\vartheta) \sin^2 \vartheta}} = \text{const.}$$

When choosing this constant equal to be zero, it follows that $\dot\varphi(\vartheta) = 0$, or $\varphi(\vartheta) = \varphi_0$ for all $\vartheta \in [\vartheta_1, \vartheta_2]$ which implies that

$$\gamma(\vartheta) = \begin{pmatrix} \sin\vartheta \cos\varphi_0 \\ \sin\vartheta \sin\varphi_0 \\ \cos\vartheta \end{pmatrix},$$

i.e. γ is a meridian segment. In this case we also find

$$l(\gamma) = \int_{\vartheta_1}^{\vartheta_2} 1 \, d\vartheta = \vartheta_2 - \vartheta_1.$$

In particular we can conclude that the meridian connecting the north with the south pole on S^2 has length π.

6. We first note that since F does not depend on t, the energy is a first integral. By Example 27.4.B the energy is given by

$$E(u(t), u'(t)) = \, < u'(t), F_y(u(t), u'(t)) > -F(u(t), u'(t)).$$

With $F(x, y) = \frac{1}{p}\|y\|^p - \frac{1}{p}\|x\|^p$ we have

$$F_{y_j}(x, y) = \frac{\partial}{\partial y_j}\left(\frac{1}{p}(y_1^2 + \cdots + y_n^2)^{\frac{p}{2}} - \frac{1}{p}\|x\|^p\right)$$

$$= y_j(y_1^2 + \cdots + y_n^2)^{\frac{p}{2}-1} = y_j\|y\|^{p-2}.$$

This implies

$$< y, F_y > \, = \sum_{j=1}^{n}(y_j^2\|y\|^{p-2}) = \|y\|^p.$$

Eventually we find

$$E(u(t), u'(t)) = \|u'(t)\|^p - \frac{1}{p}\|u'(t)\|^p + \frac{1}{p}\|u(t)\|^p$$

$$= \frac{p-1}{p}\|u'(t)\|^p + \frac{1}{p}\|u(t)\|^p.$$

703

In particular we recover the result for $p = 2$:

$$F(u(t), u'(t)) = \frac{1}{2}\|u'(t)\|^2 - \frac{1}{2}\|u(t)\|^2$$

and

$$E(u(t), u'(t)) = \frac{1}{2}\|u'(t)\|^2 + \frac{1}{2}\|u(t)\|^2.$$

For $n = 2$ and $u(t) = \left(\begin{smallmatrix} \cos \alpha t \\ \sin \alpha t \end{smallmatrix}\right)$ we find $u'(t) = \alpha \left(\begin{smallmatrix} -\sin \alpha t \\ \cos \alpha t \end{smallmatrix}\right)$ and $\|u'(t)\| = \alpha$, $\|u(t)\| = 1$. Therefore we obtain

$$\begin{aligned}
E(u(t), u'(t)) &= \frac{p-1}{p}\alpha^p + \frac{1}{p} \\
&= \left(\frac{p-1}{p}\right)\left(\frac{pC-1}{p-1}\right) + \frac{1}{p} \\
&= C - \frac{1}{p} + \frac{1}{p} = C.
\end{aligned}$$

7. We consider $\varphi : \mathbb{R} \times \mathbb{R}^3 \times (-\epsilon_0, \epsilon_0) \to \mathbb{R}^3$ where

$$\varphi(t, x, \epsilon) = \begin{pmatrix} \cos \epsilon & 0 & -\sin \epsilon \\ 0 & 1 & 0 \\ \sin \epsilon & 0 & \cos \epsilon \end{pmatrix} \begin{pmatrix} x_1 \\ x_2 \\ x_3 \end{pmatrix}$$

which is a rotation around the x_2-axis. We have $\varphi(t, x, 0) = x$ and for $\epsilon_0 < \pi$ the mapping $\varphi(t, \cdot, \epsilon) : \mathbb{R}^3 \to \mathbb{R}^3$ is a diffeomorphism. Moreover we find

$$\frac{\partial \varphi}{\partial \epsilon}(t, x, \epsilon) = \begin{pmatrix} -x_1 \sin \epsilon - x_3 \cos \epsilon \\ 0 \\ x_1 \cos \epsilon - x_3 \sin \epsilon \end{pmatrix}$$

and so we define

$$\tau(t, x) := \frac{\partial \varphi}{\partial \epsilon}(t, x, \epsilon)\Big|_{\epsilon=0} = \begin{pmatrix} -x_3 \\ 0 \\ x_1 \end{pmatrix}.$$

Moreover we have

$$\begin{aligned}
\dot{\varphi}(t, u(t), \epsilon) &= \frac{d}{dt} \begin{pmatrix} u_1(t) \cos \epsilon - u_3(t) \sin \epsilon \\ u_2(t) \\ u_1(t) \sin \epsilon + u_3(t) \cos \epsilon \end{pmatrix} \\
&= \begin{pmatrix} \dot{u}_1(t) \cos \epsilon - \dot{u}_3(t) \sin \epsilon \\ \dot{u}_2(t) \\ \dot{u}_1(t) \sin \epsilon + \dot{u}_3(t) \cos \epsilon \end{pmatrix}.
\end{aligned}$$

Now we obtain

$$F(\varphi(t, u(t), \epsilon), \dot{\varphi}(t, u(t), \epsilon)) = \frac{1}{4}\|\dot{\varphi}(t, u(t), \epsilon)\|^4 - \frac{1}{4}\|\varphi(t, u(t), \epsilon)\|^4.$$

704

We note that

$$\|\dot{\varphi}(t, u(t), \epsilon)\|^2 = \dot{u}_1^2(t) \cos^2 \epsilon - 2\dot{u}_1(t)\dot{u}_3(t) \cos \epsilon \sin \epsilon + \dot{u}_3^2(t) \sin^2 \epsilon$$
$$+ \dot{u}_2^2(t) + \dot{u}_1^2(t) \sin^2 \epsilon + 2\dot{u}_1(t)\dot{u}_3(t) \sin \epsilon \cos \epsilon + \dot{u}_3^2(t) \cos^2 \epsilon$$
$$= \|\dot{u}(t)\|^2$$

and

$$\|\varphi(t, u(t), \epsilon)\|^2 = u_1^2(t) \cos^2 \epsilon - 2u_1(t)u_3(t) \cos \epsilon \sin \epsilon + u_3^2(t) \sin^2 \epsilon$$
$$+ u_2^2(t) + u_1^2(t) \sin^2 \epsilon + 2u_1(t)u_3(t) \cos \epsilon \sin \epsilon + u_3^2(t) \cos^2 \epsilon$$
$$= \|u(t)\|^2$$

which yields

$$F(\varphi(t, u(t), \epsilon), \dot{\varphi}(t, u(t), \epsilon)) = \frac{1}{4}\|\dot{u}(t)\|^4 - \frac{1}{4}\|u(t)\|^4$$
$$= F(u(t), \dot{u}(t)).$$

Now we may choose in Noether's theorem $W = 0$ which implies $w(t, x) = 0$ and we get as the corresponding first integral according to (27.28)

$$< \tau(t, u(t)), F_y(u(t), \dot{u}(t)) > -w(t, x)$$
$$= < \begin{pmatrix} -u_3(t) \\ 0 \\ u_1(t) \end{pmatrix}, \begin{pmatrix} \dot{u}_1(t)\|\dot{u}(t)\|^2 \\ \dot{u}_2(t)\|\dot{u}(t)\|^2 \\ \dot{u}_3(t)\|\dot{u}(t)\|^2 \end{pmatrix} >$$
$$= (u_1(t)\dot{u}_3(t) - u_3(t)\dot{u}_1(t))\|\dot{u}(t)\|^2.$$

8. Let $\xi_1, \xi_2 \in (a, b)$. We may suppose that $\xi_1 < \xi_2$ and that for some $\epsilon > 0$ we have $[\xi_1 - \epsilon, \xi_2 + \epsilon] \subset (a, b)$. We define the piecewise linear and continuous function $\varphi : [a, b] \to \mathbb{R}$ by

$$\varphi(t) = 1 \text{ on } [\xi_1, \xi_2]$$
$$\varphi(t) = 0 \text{ on } [a, b]\backslash[\xi_1 - \epsilon, \xi_2 + \epsilon]$$
$$\varphi(t) = \frac{1}{\epsilon}(x - \xi_1 + \epsilon) \text{ on } [\xi_1 - \epsilon, \xi_1]$$
$$\varphi(t) = \frac{1}{\epsilon}(\xi_2 - x + \epsilon) \text{ on } [\xi_2, \xi_2 + \epsilon].$$

The support of this continuous function in $[\xi_1 - \epsilon, \xi_2 + \epsilon]$ and therefore we can approximate φ using standard mollifier techniques by elements from $C_0^\infty([a, b])$. For these elements η we have by assumption that

$$\int_a^b f(t)\eta'(t) \, dt = 0,$$

705

and this relation must hold also for the limit φ, i.e. we have

$$\int_a^b f(t)\varphi'(t)\,\mathrm{d}t = 0.$$

Using the definition of φ this implies

$$\frac{1}{\epsilon}\int_{\xi_1-\epsilon}^{\xi_1} f(t)\,\mathrm{d}t - \frac{1}{\epsilon}\int_{\xi_2}^{\xi_2+\epsilon} f(t)\,\mathrm{d}t = 0.$$

Since f is by assumption continuous we obtain for $\epsilon \to 0$ that $f(\xi_1) = f(\xi_2)$. The choice of ξ_1 and ξ_2 was arbitrary, and therefore f must be constant.

9. By assumption we have

$$\int_a^b u(x)\varphi''(x)\,\mathrm{d}x = 0$$

for all $\varphi \in C^2([a,b])$ with $\varphi(a) = \varphi(b) = \varphi'(a) = \varphi'(b) = 0$. Integrating by parts yields

$$0 = \int_a^b u(x)\varphi''(x)\,\mathrm{d}x = -\int_a^b u'(x)\varphi'(x)\,\mathrm{d}x + (u\varphi')|_a^b,$$

and since $(u\varphi')|_a^b = u(b)\varphi'(b) - u(a)\varphi'(a) = 0$ we arrive at

$$\int_a^b u'(x)\varphi'(x)\,\mathrm{d}x = 0$$

for all φ such that $\varphi(a) = \varphi(b) = 0$. Now the Lemma of du Bois-Reymond implies $u'(x) = $ constant or $u(x) = c_0 x + c$.

Chapter 28

1. Suppose that u and v are two C^2-solutions to the problem. Then both must satisfy the Euler-Lagrange equation, i.e. we have

$$\alpha \ddot{u}(t) - \beta u(t) = 0 \quad u(a) = A \text{ and } u(b) = B$$

and

$$\alpha \ddot{v}(t) - \beta v(t) = 0 \quad v(a) = A \text{ and } v(b) = B.$$

For $w = u - v$ we obtain the equation

$$\alpha \ddot{w}(t) - \beta w(t) = 0 \quad w(a) = 0 \text{ and } w(b) = 0$$

or $\ddot{w} = \frac{\beta}{\alpha} w(t)$ and $w(a) = 0$, $w(b) = 0$. Since by assumption $\frac{\beta}{\alpha}$ is not an eigenvalue of this problem it has only the trivial solution $w = 0$ implying $u = v$.

2. We have $F_{x_k x_l} = g_{x_k x_l} = (\text{Hess}(g))_{kl}$, $F_{y_k y_l} = h_{y_k y_l} = (\text{Hess}(h))_{kl}$ and $F_{xy} = 0$. It follows for every $u \in C^3([a,b])$ and all $w, z \in \mathbb{R}^n$ with some $\kappa_1 > 0$ and $\kappa_2 > 0$ that

$$\frac{1}{2} \sum_{k,l=1}^{n} \{ F_{x_k x_l}(u(t), \dot{u}(t)) w_k w_l + 2 F_{x_k y_l}(u(t), \dot{u}(t)) w_k z_l + F_{y_k y_l}(u(t), \dot{u}(t)) z_k z_l \}$$

$$= \frac{1}{2} < \text{Hess}(g)(u(t), \dot{u}(t)) w_k , w_l > + \frac{1}{2} < \text{Hess}(h)(u(t), \dot{u}(t)) z_k , z_l >$$

$$\geq \kappa_1 \|w\|^2 + \kappa_2 \|z\|^2 \geq \min(\kappa_1, \kappa_2)(\|w\|^2 + \|z\|^2).$$

3. The Lagrange function of the problem is $F(x,y) = \frac{1}{2} y^2 + 2 x^2$. We have $F_x = 4x$, $F_{xx} = 4$, $F_y = y$, $F_{yy} = 1$, $F_{xy} = 0$. This gives the Euler-Lagrange equation

$$0 = \frac{d}{dt} F_y - F_x = \ddot{u}(t) - 4u(x), \quad u(0) = u\left(\frac{1}{2}\ln 2\right) = 1.$$

The Ansatz $u(t) = e^{\alpha t}$ yields

$$(\alpha^2 - 4)e^{\alpha t} = 0$$

or $\alpha_{1,2} = \pm 2$. Thus the general solution to the Euler-Lagrange equation is given by $u(t) = ae^{2t} + be^{-2t}$. The boundary conditions lead to the equations

$$a + b = 1 \quad \text{and} \quad 2a + \frac{b}{2} = 1$$

which have the solution $a = \frac{1}{3}$, $b = \frac{2}{3}$. Thus the unique solution of the Euler-Lagrange equation under the given boundary values is $u_0(t) = \frac{1}{3}e^{2t} + \frac{2}{3}e^{-2t}$.

Next we check (28.9) which reads in our concrete case as

$$\frac{1}{2}(F_{xx}(u_0, \dot{u}_0)w^2 + 2 F_{xy}(u_0, \dot{u}_0)wz + F_{yy}(u_0, \dot{u}_0)z^2)$$

$$= \frac{1}{2}(4w^2 + z^2) \geq \frac{1}{2}(w^2 + z^2)$$

for all $w, z \in \mathbb{R}$ implying that u_0 is indeed a local minimum.

4. The Lagrange function of the problem is $F : \mathbb{R} \times \mathbb{R} \to \mathbb{R}$, $F(x,y) = \frac{1}{2} y^2 + \frac{\lambda}{2} x^2$ which leads to $F_x = \lambda x$, $F_{xx} = \lambda$, $F_y = y$, $F_{yy} = 1$ and $F_{xy} = 0$. For the Euler-Lagrange equation we find

$$\frac{d}{dt} F_y(u, \dot{u}) - F_x(u, \dot{u}) = \ddot{u} - \lambda u = 0, \quad u(0) = u(1) = 1.$$

The Ansatz $u(t) = e^{\alpha t}$ leads to the equation $\alpha^2 - \lambda = 0$. If $\lambda > 0$ then the general solution to $\ddot{u} - \lambda u$ is given by $u(t) = ae^{\sqrt{\lambda}t} + be^{-\sqrt{\lambda}t}$, whereas for $\lambda < 0$ the general solution is given by $u(t) = a \cos \sqrt{|\lambda|}t + b \sin \sqrt{|\lambda|}t$. The boundary conditions imply in the first case the equations

$$a + b = 1 \quad \text{and} \quad ae^{-\sqrt{\lambda}t} + be^{-\sqrt{\lambda}t} = 1$$

with the solution

$$a = \frac{e^{-\sqrt{\lambda}} - 1}{e^{-\sqrt{\lambda}} - e^{\sqrt{\lambda}}}, \quad b = \frac{1 - e^{\sqrt{\lambda}}}{e^{-\sqrt{\lambda}} - e^{\sqrt{\lambda}}}.$$

In the second case the boundary conditions become

$$a \cos \sqrt{|\lambda|} 0 + b \sin \sqrt{|\lambda|} 0 = 1, \text{ i.e. } a = 1$$

and

$$a \cos \sqrt{|\lambda|} + b \sin \sqrt{|\lambda|} = 1$$

and we have to discuss the equation

$$b \sin \sqrt{|\lambda|} = 1 - \cos \sqrt{|\lambda|}.$$

If $\sin \sqrt{|\lambda|} \neq 0$, i.e. $\lambda^2 \neq k^2 \pi^2$, $k \in \mathbb{Z}$, then we obtain a unique solution $a = 1$ and $b = \frac{1 - \cos \sqrt{|\lambda|}}{\sin \sqrt{|\lambda|}}$. For $\lambda^2 = k^2 \pi^2$ we must have $1 = \cos \sqrt{|\lambda|}$ which holds only for $\sqrt{|\lambda|} = 2m\pi$, $m \in \mathbb{N}_0$, i.e. $\lambda^2 = 4m^2\pi^2$ or $k = 2m$. In this case the choice of b is arbitrary and the solution to the boundary value is not unique. For every $b \in \mathbb{R}$ the function $u_b(t) = \cos(2m\pi t) + b \sin(2m\pi t)$ solves the boundary value problem. In the case $\lambda = (2m+1)^2 \pi$, $m \in \mathbb{Z}$, no solution to the Euler-Lagrange equation which also satisfies the boundary conditions exists.

Now we check (28.9) for the cases where we have a solution to the Euler-Lagrange equations satisfying the boundary conditions. We have

$$\frac{1}{2}(F_{xx}(u, \dot{u})w^2 + 2F_{xy}(u, \dot{u})wz + F_{yy}(u, \dot{u})z^2)$$
$$= \frac{1}{2}(\lambda w^2 + z^2).$$

It follows that only for $\lambda > 0$ we have the estimate

$$\frac{1}{2}(\lambda w^2 + z^2) \geq \kappa(w^2 + z^2), \quad \kappa > 0,$$

with $\kappa = \min(\lambda, 1)$. Thus in the cases where we have an extremal for $\lambda < 0$ we cannot use Theorem 28.3 to decide whether this extremal is a minimum, whereas for all $\lambda > 0$ this theorem implies that the extremals satisfying the boundary conditions are minimiser.

5. The condition $F_{yy} \geq 0$ means that F is with respect to y convex. (Note that convexity plays in many advanced parts of the calculus of variation a crucial role.)

6. In general we have $R(t, u_0(t), \dot{u}_0(t)) = F_{yy}(t, u_0(t), \dot{u}_0(t))$ and $S(t, u_0(t), \dot{u}_0(t)) = F_{xx}(t, u_0(t), \dot{u}_0(t)) - \frac{d}{dt}F_{xy}(t, u_0(t), \dot{u}_0(t))$ which in our case gives $R = 1$ and $S = -(2m\pi)^2$. Thus we arrive at

$$-\frac{d}{dt}(R(t, u_0(t), \dot{u}_0(t))v(t)) + S(t, u_0(t), \dot{u}_0(t))v(t)$$
$$= -\ddot{v}(t) - (2m\pi)^2 v(t) = 0,$$

or

$$\ddot{v}(t) + (2m\pi)^2 v(t) = 0$$

which is subjected to the condition $v(0) = v(1) = 0$. A solution of this problem is $v(t) = \sin(2m\pi)t$ since $v''(t) = -(2m\pi)^2 \sin(2m\pi)t$ and $v(0) = v(1) = 0$. For $m \in \mathbb{N}$ the function $v(t) = \sin(2m\pi)t$ has however always a further zero for $t_0 = \frac{1}{2}$ since $\sin(m\pi) = 0$ for $m \in \mathbb{N}$. Hence $t_0 = \frac{1}{2}$ is a conjugate point to v.

7. By Problem 4 to Chapter 27 we know that an extremal for the functional $\int_0^1 \dot{u}^3(t)\, dt$ must be a straight line segment $u_0(t) = \alpha t + \beta$ and the boundary conditions imply that $u_0(t) = t$. With $F(t, x, y) = y^3$ we find $F_{xx} = F_{xy} = 0$ and $F_{yy} = 6y$. For the extremal $u_0(t) = t$ we have $\dot{u}_0(t) = 1$ and therefore $R(t, u_0(t), \dot{u}_0(t)) = 6 > 0$ and the Jacobi equation is $\ddot{v}(t) = 0$ which is solved by the line segments $v(t) = \alpha t + \beta$ and hence we can find non-vanishing solution. By Theorem 28.13 we conclude that $u_0(t) = t$ is a minimiser of our problem.

8. We consider the function $g : (-\eta_0, \eta_0) \to \mathbb{R}$,

$$g(\eta) = e^{\int_0^1 F(u+\eta v, \dot{u}+\eta\dot{v})\, dt}$$

and note that $(\delta^2 \mathcal{H})(u, v) = \frac{d^2 g(0)}{d\eta^2}$.

Now we find

$$\frac{dg(\eta)}{d\eta} = \frac{d}{d\eta} e^{\int_0^1 F(u+\eta v, \dot{u}+\eta\dot{v})\, dt}$$

$$= \left(\frac{d}{d\eta} \int_0^1 F(u + \eta v, \dot{u} + \eta\dot{v})\, dt\right) e^{\int_0^1 F(u+\eta v, \dot{u}+\eta\dot{v})\, dt}$$

and further

$$\frac{d^2 g(\eta)}{d\eta^2} = \frac{d}{d\eta}\left(\left(\frac{d}{d\eta} \int_0^1 F(u + \eta v, \dot{u} + \eta\dot{v})\, dt\right) e^{\int_0^1 F(u+\eta v, \dot{u}+\eta\dot{v})\, dt}\right)$$

$$= \left(\frac{d^2}{d\eta^2} \int_0^1 F(u + \eta v, \dot{u} + \eta\dot{v})\, dt\right) e^{\int_0^1 F(u+\eta v, \dot{u}+\eta\dot{v})\, dt}$$

$$+ \left(\frac{d}{d\eta} \int_0^1 F(u + \eta v, \dot{u} + \eta\dot{v})\, dt\right)^2 e^{\int_0^1 F(u+\eta v, \dot{u}+\eta\dot{v})\, dt}$$

which yields

$$(\delta^2 \mathcal{H})(u, v) = ((\delta^2 \mathcal{F})(u, v) + (\delta\mathcal{F})^2(u, v))e^{\int_0^1 F(u+\eta v, \dot{u}+\eta\dot{v})\, dt}.$$

In particular, since $(\delta\mathcal{F})^2(u, v) \geq 0$ and $e^{\int_0^1 F(u+\eta v, \dot{u}+\eta\dot{v})\, dt} > 0$ from $(\delta^2 \mathcal{F})(u, v) > 0$ it follows that $(\delta^2 \mathcal{H})(u, v) > 0$.

Chapter 29

1.   a) A straightforward calculation shows that $u_x = h'g_x$ and $u_y = h'g_Y$ implying

$$u_x g_y - u_y g_x = h' g_x g_y - h' g_y g_x = 0.$$

When we consider the mapping $(x, y) \mapsto \left(\begin{smallmatrix} u(x,y) \\ g(x,y) \end{smallmatrix} \right)$ from $\mathbb{R}^2$ to $\mathbb{R}^2$ then its Jacobi matrix is $\left(\begin{smallmatrix} u_x & u_y \\ g_x & g_y \end{smallmatrix} \right)$ and its Jacobi determinant is $u_x g_y - u_y g_x$. Hence the equation $u_x g_y - u_y g_x = 0$ states that this Jacobi determinant is zero and therefore we shall expect a functional dependence of u and g, i.e. $u = h(g)$.

     b) The graph $\Gamma(U) \subset \mathbb{R}^3$, $\Gamma(u) = \{(x, y, z) \mid (x, y) \in \mathbb{R}^2,\ z = u(x, y)\}$ is rotational invariant with respect to the z-axis and in principle we can consider Γ as the trace of a surface of revolution, compare with Example II.8.7. By a direct calculation (or by applying part a)) we find $u_x(x, y) = 2xh'(x, y)$, $u_y(x, y) = 2yh'(x, y)$ and therefore $yu_x - xu_y = 0$.

2. We parameterize the line on which the initial condition is given by $s \mapsto (s, s)$, $s \in \mathbb{R}$. The characteristic differential equations are

$$\frac{dx(t)}{dt} = 1, \quad \frac{dy(t)}{dt} = 2, \quad \frac{dz(t)}{dt} + z(t) = 1$$

implying

$$x(t) = t + x_0, \quad y(t) = 2t + y_0, \quad z(t) = 1 + (z_0 - 1)e^{-t}.$$

From the initial condition we find

$$x_0 = s, \quad y_0 = s, \quad z_0 = \cos^2 s$$

which yields

$$x(t) = t + s, \quad y(t) = 2t + s, \quad z(t) = 1 + (\cos^2 s - 1)e^{-t} = 1 - e^{-t} \sin^2 s.$$

Now we find

$$2x - y = s$$

and

$$t = x - s = x - 2x + y = -x + y,$$

which gives as candidate for a solution

$$u(x, y) = 1 - e^{x-y} \sin^2(2x - y). \tag{$*$}$$

We now verify that u given by $(*)$ is indeed a solution to our problem. First we note that

$$u(x, x) = 1 - \sin^2 x = \cos^2 x.$$

Moreover we have

$$\frac{\partial}{\partial x} u(x, y) = -\frac{\partial}{\partial x}(e^{x-y} \sin^2(2x - y))$$
$$= -e^{x-y}(\sin^2(2x - y) + 4\sin(2x - y)\cos(2x - y))$$

710

and

$$\frac{\partial}{\partial y} u(x, y) = -2 \frac{\partial}{\partial y} (e^{x-y} \sin^2(2x - y))$$
$$= 2e^{x-y} (\sin^2(2x - y) + 2 \sin(2x - y) \cos(2x - y)).$$

Therefore we find

$$\frac{\partial u(x, y)}{\partial x} + 2 \frac{\partial u(x, y)}{\partial y} + u(x, y) - 1$$
$$= -e^{x-y} \sin^2(2x - y) - 4e^{x-y} \sin(2x - y) \cos(2x - y)$$
$$+ 2e^{x-y} \sin^2(2x - y) + 4e^{x-y} \sin(2x - y) \cos(2x - y)$$
$$+ 1 - e^{x-y} \sin^2(2x - y) - 1 = 0.$$

3. In both cases we have to solve the system

$$\dot{x}(t) = 1, \quad \dot{y}(t) = \frac{1}{4} z(t), \quad \dot{z}(t) = 0$$

which gives for a characteristic passing through $\begin{pmatrix} x_0 \\ y_0 \\ z_0 \end{pmatrix}$ the equation $\gamma(t) = \begin{pmatrix} 1 \\ \frac{1}{4} z_0 \\ 0 \end{pmatrix} t + \begin{pmatrix} x_0 \\ y_0 \\ z_0 \end{pmatrix}$ where $z_0 = u(x_0, y_0)$. The characteristic curves are $t \mapsto \begin{pmatrix} 1 \\ \frac{1}{4} z_0 \\ 0 \end{pmatrix} t + \begin{pmatrix} x_0 \\ y_0 \end{pmatrix}$ which yields the equation $x = \frac{4}{z_0} y_0 + x_0 - \frac{4y_0}{z_0}$. For the curves passing through $(0, y_0)$ we must have $(0, y_0, z_0) = (0, y_0, g(y_0))$ with $z_0 = g(y_0)$ and therefore we have to discuss the lines $x = \frac{4}{g(y_0)} y - \frac{4y_0}{g(y_0)}$.

a) If $g(y) = 5$ for all y we arrive at $x = \frac{4}{5} y - \frac{4y_0}{5}$ and it follows that we can solve the initial value problem, see Figure A below.

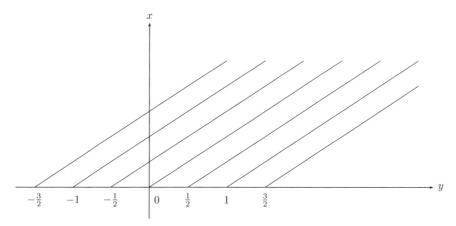

Figure A

b) Now we have $g(y) = \begin{cases} 3, & y \le 0 \\ 2y + 3, & y \in [0, 1] \\ 5, & y \ge 1 \end{cases}$ which leads to the following three

families of curves

$$x = \frac{4}{3}y - \frac{4y_0}{3}, \quad y < 0$$

$$x = \frac{4}{2y_0 + 3}y - \frac{4y_0}{2y_0 + 3}, \quad y \in [0, 1]$$

$$x = \frac{4}{5}y - \frac{4y_0}{5}, \quad y > 1,$$

and again we deduce that the initial value problem is solvable, see Figure B below

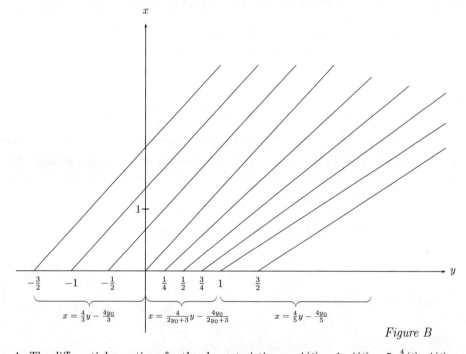

Figure B

4. The differential equations for the characteristics are $\dot{x}(t) = 1$, $\dot{y}(t) = 5y^{\frac{4}{5}}(t)$, $\dot{z}(t) = 0$. It is worth to note that this is a completely decoupled system with $z(t) = z_0$ and $x(t) = t - c_0$. The last equation allows us to use x as parameter and we find as solution of $y'(x) = 5y^{\frac{4}{5}}(x)$ the family of curves given by $y(x) = (x - c)^5$, $c \in \mathbb{R}$. If characteristic curves are the graphs of the functions $y(x) = (x - c)^5$, $c \in \mathbb{R}$, and the line $y(x) = 0$ for all x. For $y = 0$ however it follows that $u_x(x, 0) = 0$ implying that u must be constant, hence u must be constant in a neighbourhood of $(0, 0)$. Thus, when describing for u on the line $y = 0$ a non-constant initial value, we will not have a solution.

5. a) For $\varphi(x, y; a, b) = ax + by + f(a, b)$ we find

$$\varphi_x(x, y; a, b) = a, \qquad \varphi_y(x, y; a, b) = b;$$
$$\varphi_a(x, y; a, b) = x + f_a(a, b), \qquad \varphi_b(x, y; a, b) = y + f_b(a, b);$$
$$\varphi_{xa}(x, y; a, b) = 1, \qquad \varphi_{ya}(x, y; a, b) = 0;$$
$$\varphi_{xb}(x, y; a, b) = 0, \qquad \varphi_{yb}(x, y; a, b) = 1.$$

First it follows that

$$x\varphi_x + y\varphi_y + f(\varphi_x, \varphi_y) = xa + yb + f(a, b) = \varphi,$$

i.e. φ is indeed a solution. For the matrix of interest we find

$$\begin{pmatrix} \varphi_a & \varphi_{xa} & \varphi_{ya} \\ \varphi_b & \varphi_{xb} & \varphi_{yb} \end{pmatrix} = \begin{pmatrix} x + f_a(a, b) & 1 & 0 \\ y + f_b(a, b) & 0 & 1 \end{pmatrix}$$

which has clearly rank 2, i.e. full rank.

b) With $\varphi = \varphi(x_0, x_1, \ldots, x_n; b, a_1, \ldots, a_n) = <a, x> - x_0 H(a) + b$ we have

$$\varphi_{x_0} = -H(a), \qquad \varphi_{x_j} = a_j, \quad 1 \le j \le n;$$
$$\varphi_b = 1, \qquad \varphi_{a_j} = x_j - x_0 H_{a_j}(a);$$
$$\varphi_{x_0 b} = 0, \qquad \varphi_{x_j b} = 0;$$
$$\varphi_{x_0 a_j} = -H_{a_j}, \qquad \varphi_{x_j a_j} = 1.$$

First we observe that

$$\varphi_{x_0} + H(\varphi_{x_1}, \ldots, \varphi_{x_n}) = -H(a) + H(a) = 0,$$

which implies that φ solves the equation. Now we consider the matrix

$$\begin{pmatrix} \varphi_{a_1} & \varphi_{x_0 a_1} & \varphi_{x_1 a_1} & \varphi_{x_2 a_1} & \cdots & \varphi_{x_n a_1} \\ \varphi_{a_2} & \varphi_{x_0 a_2} & \varphi_{x_1 a_2} & \varphi_{x_2 a_2} & \cdots & \varphi_{x_n a_2} \\ \vdots & \vdots & \vdots & \vdots & & \vdots \\ \varphi_{a_n} & \varphi_{x_0 a_n} & \varphi_{x_1 a_n} & \varphi_{x_2 a_n} & \cdots & \varphi_{x_n a_n} \\ \varphi_b & \varphi_{x_0 b} & \varphi_{x_1 b} & \varphi_{x_2 b} & \cdots & \varphi_{x_n b} \end{pmatrix} =$$

$$\begin{pmatrix} x_1 - x_0 H_{a_1}(a) & -H_{a_1}(a) & 1 & 0 & \cdots & 0 \\ x_2 - x_0 H_{a_2}(a) & -H_{a_2}(a) & 0 & 1 & \cdots & 0 \\ \vdots & \vdots & \vdots & \vdots & & \vdots \\ x_n - x_0 H_{a_n}(a) & -H_{a_n}(a) & 0 & 0 & \cdots & 1 \\ 1 & 0 & 0 & 0 & \cdots & 0 \end{pmatrix}.$$

The rank of this matrix is $n + 1$, hence maximal.

6. a) With $v(x, y) = w(x, y, \varphi(x, y))$ we find

$$v_x(x, y) = w_x(x, y, \varphi(x, y)) + w_\xi(x, y, \varphi(x, y))\varphi_x(x, y)$$

713

and
$$v_y(x, y) = w_y(x, y, \varphi(x, y)) + w_\xi(x, y, \varphi(x, y))\varphi_y(x, y).$$
Since $w_\xi(x, y, \varphi(x, y)) = 0$ it follows that
$$v_x(x, y) = w_x(x, y, \varphi(x, y)) \quad \text{and} \quad v_y(x, y) = w_y(x, y, \varphi(x, y))$$
which implies of course
$$\begin{aligned} &F(x, y, v(x, y), v_x(x, y), v_y(x, y)) \\ &= F(x, y, w(x, y, \varphi(x, y)), w_x(x, y, \varphi(x, y)), w_y(x, y, \varphi(x, y))) = 0. \end{aligned}$$

b) First we note that
$$w_x(x, y, \alpha) = \cos\alpha, \quad w_y(x, y, \alpha) = \sin\alpha$$
which implies
$$(w_x^2 + w_y^2)^{\frac{1}{2}} = (\cos^2\alpha + \sin^2\alpha)^{\frac{1}{2}} = 1.$$
Next we observe that
$$\frac{\partial}{\partial\alpha}w(x, y, \alpha) = -x\sin\alpha + y\cos\alpha$$
and the condition $\frac{\partial W}{\partial\alpha}(x, y, \alpha) = 0$ implies $-x\sin\alpha + y\cos\alpha = 0$ or $\alpha = \arctan\frac{y}{x}$. This gives the envelope
$$v(x, y) = x\cos\left(\arctan\frac{y}{x}\right) + y\sin\left(\arctan\frac{y}{x}\right).$$
Since
$$\arctan\frac{y}{x} = \arcsin\frac{y}{\sqrt{x^2 + y^2}}$$
and $\cos\alpha = \sqrt{1 - \sin^2\alpha}$, $-\frac{\pi}{2} \le \alpha \le \frac{\pi}{2}$, we find for $x > 0$ and $y \in \mathbb{R}$ that
$$\begin{aligned} v(x, y) &= x\cos\left(\arctan\frac{y}{\sqrt{x^2 + y^2}}\right) + y\sin\left(\arctan\frac{y}{x}\right) \\ &= x\sqrt{1 - \sin^2\left(\arcsin\frac{y}{\sqrt{x^2 + y^2}}\right)} + y\sin\left(\arcsin\frac{y}{\sqrt{x^2 + y^2}}\right) \\ &= x\sqrt{1 - \frac{y^2}{\sqrt{x^2 + y^2}}} + \frac{y^2}{\sqrt{x^2 + y^2}} \\ &= \frac{x}{\sqrt{x^2 + y^2}}\sqrt{x^2 + y^2 - y^2} + \frac{y^2}{\sqrt{x^2 + y^2}} = \sqrt{x^2 + y^2}. \end{aligned}$$
It follows that $v_x(x, y) = \frac{x}{\sqrt{x^2+y^2}}$ and $v_y(x, y) = \frac{y}{\sqrt{x^2+y^2}}$ and therefore we find
$$(v_x^2 + v_y^2)^{\frac{1}{2}} = \left(\frac{x^2}{x^2 + y^2} + \frac{y^2}{x^2 + y^2}\right)^{\frac{1}{2}} = 1,$$
i.e. v satisfies indeed the eikonal equation.

7.   a) With $H(x, y, u, p_1, p_2) = p_1^2 + p_2^2 - 1$ we find for the characteristic differential equation

$$\dot{x} = p_1, \quad \dot{y} = p_2$$
$$\dot{p}_1 = 0, \quad \dot{p}_2 = 0$$
$$\dot{u} = (p_1^2 + p_2^2) = 1.$$

b) The condition (29.41) becomes

$$\dot{h}(s) = a(s)\dot{f}(s) + b(s)\dot{g}(s) \tag{$*$}$$

and (29.42) takes the form
$$a^2(s) + b^2(s) = 1 \tag{$**$}$$

whereas (29.43) depends on $a(s)$ and $b(s)$ which we now try to determine in such a way that (29.43) holds.

First we note that for $\dot{f}^2 + \dot{g}^2 < \dot{h}^2$ there is no real-valued solution for the system $(*)$ and $(**)$. Indeed, in this case we have

$$\dot{f}^2 + \dot{g}^2 < a^2\dot{f}^2 + 2ab\dot{f}\dot{g} + b^2\dot{g}^2$$

or

$$b^2\dot{f}^2 + a^2\dot{g}^2 < 2ab\dot{f}\dot{g}$$

which is a contradiction. However for $\dot{f}^2 + \dot{g}^2 > \dot{h}^2$ we can always find two distinct solutions which we can derive from the quadratic equation $a^2(\dot{f}^2 + \dot{g}^2) - 2a\dot{f}\dot{h} + \dot{h}^2 - \dot{g}^2 = 0$. In this case, i.e. $\dot{f}^2 + \dot{g}^2 > \dot{h}^2$ we now study (29.43). Suppose that

$$\det \begin{pmatrix} \dot{f} & \dot{g} \\ a & b \end{pmatrix} = \dot{f}b - \dot{g}a = 0$$

then it follows that
$$\dot{f}^2(s)b^2(s) - 2\dot{f}\dot{g}ab + \dot{g}^2a^2 = 0$$

and since
$$\dot{h}^2 = \dot{f}^2a^2 + 2\dot{f}\dot{g}ab + \dot{g}^2b^2 = 0$$

we arrive at
$$\dot{h}^2 = (\dot{f}^2 + \dot{g}^2)a^2 + (\dot{f} + \dot{g}^2)b^2 = (\dot{f}^2 + \dot{g}^2)$$

which is a contradiction to $\dot{h}^2 < \dot{f}^2 + \dot{g}^2$. Thus in this case (29.43) holds too.

c) We fix s and integrate the system given in 6a) where we denote the variable curve parameter by τ. Integrating $\dot{u} = 1$, $u(0) = h(s)$ gives

$$u(\tau, s) = \tau + h(s).$$

The integration of $\dot{p}_1 = 0$ and $\dot{p}_2 = 0$ yields first that $p_1 = \text{const}$ and $p_2 = \text{const}$ and from the initial condition we deduce

$$p_1(\tau, s) = a(s) \quad p_2(\tau, s) = b(s).$$

715

Now we integrate

$$\frac{d}{d\tau}x(\tau, s) = a(s) \quad \text{and} \quad \frac{d}{d\tau}y(\tau, s) = b(s)$$

to obtain

$$x(\tau, s) = f(s) + \tau a(s), \quad y(\tau, s) = g(s) + \tau b(s).$$

8. We have

$$\left(\frac{\partial}{\partial x} - \frac{\partial}{\partial y}\right)\left(\frac{\partial v(x,y)}{\partial x} + \frac{\partial v(x,y)}{\partial y}\right) = \frac{\partial w(x,y)}{\partial x} - \frac{\partial w(x,y)}{\partial y} = 0$$

but

$$\left(\frac{\partial}{\partial x} - \frac{\partial}{\partial y}\right)\left(\frac{\partial v}{\partial x} + \frac{\partial v}{\partial y}\right) = \frac{\partial^2 v}{\partial x^2} - \frac{\partial^2 v}{\partial y \partial x} + \frac{\partial^2 v}{\partial x \partial y} - \frac{\partial^2 v}{\partial y^2} = \frac{\partial^2 v}{\partial x^2} - \frac{\partial^2 v}{\partial y^2}.$$

The interesting observation is that in some sense we can factorize the second order differential operator $\frac{\partial^2}{\partial x^2} - \frac{\partial^2}{\partial y^2}$ according to

$$\frac{\partial^2}{\partial x^2} - \frac{\partial^2}{\partial y^2} = \left(\frac{\partial}{\partial x} - \frac{\partial}{\partial y}\right)\left(\frac{\partial}{\partial x} + \frac{\partial}{\partial y}\right)$$

and may use the system of first order equations

$$\frac{\partial v}{\partial x} + \frac{\partial v}{\partial y} = w \quad \text{and} \quad \frac{\partial w}{\partial x} - \frac{\partial w}{\partial y} = 0$$

in order to study the second order equation.

Chapter 30

1. a) From the definition it follows that

$$f^*(y) = \sup_{x \in \mathbb{R}^n} \{< x, y > -g(\|x\|^2)\}.$$

Since f is coercive and $g \in C^1(\mathbb{R}^n)$, the supremum is attained at a point $\tilde{x}$ which satisfies

$$y_j = 2\tilde{x}_j g'(\|\tilde{x}\|^2).$$

By assumption the function $H(r) = 4g'(r)^2 r$ admits an inverse function and we find

$$\|\tilde{x}\|^2 = H^{-1}(\|y\|^2).$$

Now it follows that

$$f^*(y) = \sup_{x \in \mathbb{R}^n} \{< x, y > -g(\|x\|^2)\}$$
$$= 2 < \tilde{x}, y > -g(\|\tilde{x}\|^2)$$
$$= 2\|\tilde{x}\|^2 g'(\|\tilde{x}\|^2) - g(\|\tilde{x}\|^2)$$
$$= 2H^{-1}(\|y\|^2)g'(H^{-1}(\|y\|^2)) - g(H^{-1}(\|y\|^2))$$

implying that f^* is a rotational invariant function.

b) We have $h_j(x) = \frac{1}{\alpha}\|x\|^\alpha = g_\alpha(\|y\|^2)$ with $g_\alpha(r) = \frac{1}{\alpha}r^{\frac{\alpha}{2}}$. From this we find $H_\alpha(r) = r^{\alpha-1}$ and $H_\alpha^{-1}(s) = s^{\frac{1}{\alpha-1}}$ which yields

$$h_\alpha^*(y) = 2H_\alpha^{-1}(\|y\|^2)g_\alpha'(H^{-1}(\|y\|^2)) - g_\alpha(H^{-1}(\|y\|^2))$$

$$= \frac{1}{\alpha^*}\|y\|^{\alpha^*}$$

2. a) From our general considerations it follows that

$$H(q,p) = L^{*,\eta}(q,p) = \sup_{\eta\in\mathbb{R}^n}\{<p,\eta> - L(q,\eta)\}$$

$$= \sup_{\eta\in\mathbb{R}^n}\{<p,\eta> - \frac{1}{\alpha}\|\eta\|^\alpha\} + \frac{1}{\alpha}\|q\|^\alpha$$

and part b) of Problem 1 yields

$$H(q,p) = \frac{1}{\alpha^*}\|p\|^{\alpha^*} + \frac{1}{\alpha}\|q\|^\alpha, \qquad \frac{1}{\alpha} + \frac{1}{\alpha^*} = 1.$$

b) We know that $L(q,\eta)$ is the partial Legendre transformation of $H(q,p)$ with respect to p, i.e.

$$L(q,\eta) = H^{*,\eta}(q,\eta) = \sup_{p\in\mathbb{R}^n}\{<\eta,p> - H(q,p)\}$$

$$= \sup_{p\in\mathbb{R}^n}\left\{<\eta,p> - \frac{1}{\beta}\|p\|^\beta\right\} - \frac{1}{\|q\|^{n-\beta}}$$

$$= \frac{1}{\beta^*}\|p\|^{\beta^*} - \frac{1}{\|q\|^{n-\beta}}, \qquad \frac{1}{\beta} + \frac{1}{\beta^*} = 1.$$

3. By definition a coordinate q_j is cyclic for H if $\frac{\partial H}{\partial q_j} = 0$. In our case the coordinates are r, ϑ and φ. The kinetic term of H is given by

$$\frac{1}{2m}\left(p_r^2 + \frac{p_\vartheta^2}{r^2} + \frac{p_\varphi^2}{r^2\sin^2\vartheta}\right)$$

and it depends on r and ϑ. Since U is independent of p_r, p_ϑ and p_φ, a coordinate ξ is cyclic for H if and only if $\frac{\partial U}{\partial\xi} = 0$, $\xi \in \{r,\vartheta,\varphi\}$. Since $\frac{\partial U_1}{\partial\varphi} = \frac{\partial U_1(r)}{\partial\varphi} = 0$ as well as $\frac{\partial U_2}{\partial\varphi} = \frac{\partial U_2(r,\vartheta)}{\partial\varphi} = 0$, φ is in these cases a cyclic coordinate. However, since by assumption U_3 depends on φ, in this case φ is not a cyclic coordinate. In each case there is no further cyclic coordinate.

4. First we note that

$$J^2 = \begin{pmatrix} 0 & \mathrm{id}_n \\ -\mathrm{id}_n & 0 \end{pmatrix}\begin{pmatrix} 0 & \mathrm{id}_n \\ -\mathrm{id}_n & 0 \end{pmatrix} = -\mathrm{id}_{2n},$$

$$\mathrm{id}_{2n}^T J\mathrm{id}_{2n} = J,$$

and
$$J^T J J = J^T(-\mathrm{id}_{2n}) = J.$$

Thus J and id_{2n} are symplectic. Furthermore, from $A^T J A$ we deduce that

$$\det(A^T J A) = (\det A)^2 \det J = \det J,$$

and since $\det J \neq 0$ we find $|\det A| = 1$. This result already implies that a symplectic matrix belongs to $GL(2n; \mathbb{R})$. For two symplectic matrices A and B it follows that

$$(AB)^T J (AB) = B^T(A^T J A)B = B^T J B = J,$$

i.e. the product of two symplectic matrices is a symplectic matrix. For A^{-1}, A symplectic, we find

$$(A^{-1})^T J A^{-1} = (A^{-1})^T (A^T J A) A^{-1}$$
$$= (A A^{-1})^T J (A A^{-1}) = J.$$

Thus A^{-1} is symplectic and we have proved that $\mathrm{Sp}(2n; \mathbb{R})$ is a subgroup of $GL(2n; \mathbb{R})$.

Note that for $A \in \mathrm{Sp}(2n; \mathbb{R})$ given in block form $A = \begin{pmatrix} \alpha & \beta \\ \gamma & \delta \end{pmatrix}$, $\alpha, \beta, \gamma, \delta \in M(n; \mathbb{R})$ we derive from the definition of a symplectic matrix the following identities

$$\alpha^T \gamma - \gamma^T \alpha = 0, \quad \beta^T \delta - \delta^T \beta = 0, \quad \alpha^T \delta - \gamma^T \beta = \mathrm{id}_n.$$

5. The Hamilton system corresponding to $H = H(q, p)$ is given by $\dot{q} = H_p(q, p)$ and $\dot{p} = -H_q(q, p)$. Introducing $w = \begin{pmatrix} q \\ p \end{pmatrix}$ and $H_w = \begin{pmatrix} H_q \\ H_p \end{pmatrix}$ we find

$$\dot{w} = \begin{pmatrix} \dot{q} \\ \dot{p} \end{pmatrix} = \begin{pmatrix} H_p(q, p) \\ -H_q(q, p) \end{pmatrix} = \begin{pmatrix} 0 & \mathrm{id}_n \\ -\mathrm{id}_n & 0 \end{pmatrix} \begin{pmatrix} H_q(p, q) \\ H_p(q, p) \end{pmatrix} = J H_w(q, p),$$

or $\dot{w}(t) = J H_w(w(t))$.

Now let $u : \mathbb{R}^{2n} \to \mathbb{R}^{2n}$ be a diffeomorphism, $\zeta \mapsto u(\zeta) = w$ and consider $w(t) = u(\zeta(t))$. It follows that $\dot{w}(t) = (du)(\zeta)\dot{\zeta}(t)$. For $K(\zeta) := (H \circ u)(\zeta)$ we find further

$$K_\zeta(\zeta) = (du)^T(\zeta) H_w(w).$$

We note that $\dot{w} = J H_w(w)$ or using the properties of J we have $-J\dot{w} = H_w(w)$. Together we arrive at

$$K_\zeta(\zeta) = -(du)^T(\zeta) J (du)(\zeta)\dot{\zeta}$$

we aim for $\dot{\zeta} = J K_\zeta(\zeta)$ or $K_\zeta(\zeta) = -J\dot{\zeta}$. Therefore, if we require that $(du)^T(\zeta) J (du)(\zeta) = J$ for every ζ, then we have indeed $\dot{\zeta} = J K_\zeta(\zeta)$ and u is a canonical transformation.

718

6. We assume now that a canonical transformation satisfies the requirements of Problem 5b). It follows that $\begin{pmatrix} \frac{\partial Q}{\partial q} & \frac{\partial Q}{\partial p} \\ \frac{\partial P}{\partial q} & \frac{\partial P}{\partial p} \end{pmatrix}$ is a symplectic matrix, hence by the remark in the end of Problem 4 its determinant is equal to 1. Therefore we find for its inverse

$$\begin{pmatrix} \frac{\partial Q}{\partial q} & \frac{\partial Q}{\partial p} \\ \frac{\partial P}{\partial q} & \frac{\partial P}{\partial p} \end{pmatrix}^{-1} = \frac{1}{\det \begin{pmatrix} \frac{\partial Q}{\partial q} & \frac{\partial Q}{\partial p} \\ \frac{\partial P}{\partial q} & \frac{\partial P}{\partial p} \end{pmatrix}} \begin{pmatrix} \frac{\partial P}{\partial p} & -\frac{\partial P}{\partial q} \\ \frac{\partial Q}{\partial p} & \frac{\partial Q}{\partial q} \end{pmatrix}^{T}$$

$$= \begin{pmatrix} \left(\frac{\partial P}{\partial p}\right)^{T} & -\left(\frac{\partial Q}{\partial p}\right)^{T} \\ -\left(\frac{\partial P}{\partial q}\right)^{T} & \left(\frac{\partial Q}{\partial q}\right)^{T} \end{pmatrix}.$$

7. We have now in (30.39) $r(s) = q(s)$ and $T(s) = P(s)$ and it follows that

$$\sum_{j=1}^{n} p_j(s)\dot{q}_j(s) - H(s, q(s), p(s))$$

$$= \sum_{j=1}^{n} P_j(s)\dot{Q}_j(s) - K(s, Q(s), P(s)) + \frac{\partial \tilde{G}_2(s, q(s), P(s))}{\partial s}$$

$$+ \sum_{j=1}^{n} \frac{\partial \tilde{G}_2(s, q(s), P(s))}{\partial q_j}\dot{q}_j(s) + \sum_{j=1}^{n} \frac{\partial \tilde{G}_2(s, q(s), P(s))}{\partial p_j}\dot{P}_j(s)$$

$$- \sum_{j=1}^{n} P_j(s)\dot{Q}_j(s) - \sum_{j=1}^{n} Q_j(s)\dot{P}_j(s)$$

$$= -K(s, Q(s), P(s)) + \frac{\partial \tilde{G}_2(s, q(s), P(s))}{\partial s} - \sum_{j=1}^{n} Q_j(s)\dot{P}_j(s)$$

$$+ \sum_{j=1}^{n} \frac{\partial \tilde{G}_2(s, q(s), P(s))}{\partial q_j}\dot{q}_j(s) + \sum_{j=1}^{n} \frac{\partial \tilde{G}_2(s, q(s), P(s))}{\partial P_j}\dot{P}_j(s)$$

which yields

$$\sum_{j=1}^{n} \left(p_j - \frac{\partial \tilde{G}_2}{\partial q_j}\right)\dot{q}_j = \sum_{j=1}^{n} \left(\frac{\partial \tilde{G}_2}{\partial P_j} - Q_j\right)\dot{P}_j + H(s, q, p) - K(s, Q, P) + \frac{\partial \tilde{G}_2}{\partial s}.$$

From the last equality we deduce

$$p_j = \frac{\partial \tilde{G}_2}{\partial q_j}, \quad Q_j = -\frac{\partial \tilde{G}_2}{\partial P_j}, \quad K = H + \frac{\partial G_2}{\partial s},$$

note that $\frac{\partial G_2}{\partial s} = \frac{\partial \tilde{G}_2}{\partial s}$.

For the case $G_3 = \tilde{G}_3(s, Q, p) + <q, p>$ we find

$$q_j = -\frac{\partial \tilde{G}_3}{\partial p_j}, \quad P_j = \frac{\partial \tilde{G}_3}{\partial Q_j}$$

and for $G_4 = \tilde{G}_4(s, p, P) + <q, p> - <Q, P>$ we obtain

$$q_j = -\frac{\partial \tilde{G}_4}{\partial p_j}, \quad Q_j = \frac{\partial \tilde{G}_4}{\partial P_j}.$$

8. Since

$$\{J_1, J_2\} = \sum_{j=1}^{n} \left(\frac{\partial J_1}{\partial q_j} \frac{\partial J_2}{\partial p_j} - \frac{\partial J_1}{\partial p_j} \frac{\partial J_2}{\partial q_j} \right)$$

we find

$$\frac{\partial}{\partial q_k}\{J_1, J_2\} = \sum_{j=1}^{n} \left(\frac{\partial^2 J_1}{\partial q_j \partial q_k} \frac{\partial J_2}{\partial p_j} + \frac{\partial J_1}{\partial q_j} \frac{\partial^2 J_2}{\partial q_k \partial p_j} - \frac{\partial^2 J_1}{\partial q_k \partial p_j} \frac{\partial J_2}{\partial q_j} - \frac{\partial J_1}{\partial p_j} \frac{\partial^2 J_2}{\partial q_k \partial q_j} \right)$$

and

$$\frac{\partial}{\partial p_k}\{J_1, J_2\} = \sum_{j=1}^{n} \left(\frac{\partial^2 J_1}{\partial q_j \partial p_k} \frac{\partial J_2}{\partial p_j} + \frac{\partial J_1}{\partial q_j} \frac{\partial^2 J_2}{\partial p_k \partial p_j} - \frac{\partial^2 J_1}{\partial p_k \partial p_j} \frac{\partial J_2}{\partial q_j} - \frac{\partial J_1}{\partial p_j} \frac{\partial^2 J_2}{\partial p_k \partial q_j} \right)$$

which leads to

$$\{J_\alpha, \{J_\beta, J_\gamma\}\}$$
$$= \sum_{k=1}^{n} \sum_{j=1}^{n} \frac{\partial J_\alpha}{\partial q_k} \left(\frac{\partial^2 J_\gamma}{\partial q_j \partial p_k} \frac{\partial J_\beta}{\partial p_j} + \frac{\partial J_\gamma}{\partial q_j} \frac{\partial^2 J_\beta}{\partial p_k \partial p_j} - \frac{\partial^2 J_\gamma}{\partial p_k \partial p_j} \frac{\partial J_\beta}{\partial q_j} - \frac{\partial J_\gamma}{\partial p_j} \frac{\partial^2 J_\beta}{\partial p_k \partial q_j} \right)$$
$$- \sum_{k=1}^{n} \sum_{j=1}^{n} \frac{\partial J_\alpha}{\partial p_k} \left(\frac{\partial^2 J_\gamma}{\partial q_j \partial q_k} \frac{\partial J_\beta}{\partial p_j} + \frac{\partial J_\gamma}{\partial q_j} \frac{\partial^2 J_\beta}{\partial q_k \partial p_j} - \frac{\partial^2 J_\gamma}{\partial q_k \partial p_j} \frac{\partial J_\beta}{\partial q_j} - \frac{\partial J_\gamma}{\partial p_j} \frac{\partial^2 J_\beta}{\partial q_k \partial q_j} \right).$$

It remains to add these terms for $(\alpha, \beta, \gamma) = (1, 2, 3)$, $(\alpha, \beta, \gamma) = (2, 3, 1)$ and $(\alpha, \beta, \gamma) = (3, 1, 2)$ which we leave to the reader. Note that since by assumption all functions are C^2-functions we can always interchange the order when dealing with mixed second order partial derivatives.

9. We note that

$$\frac{\partial}{\partial t}\{J_1(t, q, p), J_2(t, q, p)\}$$
$$= \sum_{j=1}^{n} \frac{\partial}{\partial t} \left(\frac{\partial J_1(t, q, p)}{\partial q_j} \frac{\partial J_2(t, q, p)}{\partial p_j} - \frac{\partial J_1(t, q, p)}{\partial p_j} \frac{\partial J_2(t, q, p)}{\partial q_j} \right)$$
$$= \sum_{j=1}^{n} \left(\frac{\partial}{\partial q_j} \left(\frac{\partial J_1(t, q, p)}{\partial t} \right) \frac{\partial J_2(t, q, p)}{\partial p_j} - \frac{\partial}{\partial p_j} \left(\frac{\partial J_1(t, q, p)}{\partial t} \right) \frac{\partial J_2(t, q, p)}{\partial q_j} \right)$$

$$+ \sum_{j=1}^{n} \left(\frac{\partial J_1(t,q,p)}{\partial q_j} \frac{\partial}{\partial p_j} \left(\frac{\partial J_2(t,q,p)}{\partial t} \right) - \frac{\partial J_1(t,q,p)}{\partial p_j} \frac{\partial}{\partial q_j} \left(\frac{\partial J_2(t,q,p)}{\partial t} \right) \right)$$

$$= \left\{ \frac{\partial J_1}{\partial t}, J_2 \right\} + \left\{ J_1, \frac{\partial J_2}{\partial t} \right\}.$$

Moreover, by (30.59) applied to $\{J_1, J_2\}$ we have

$$\frac{d}{dt}\{J_1, J_2\} = \frac{\partial}{\partial t}\{J_1, J_2\} + \{H, \{J_1, J_2\}\}.$$

With the help of $\frac{\partial}{\partial t}\{J_1, J_2\} = \{\frac{\partial J_1}{\partial t}, J_2\} + \{J_1, \frac{\partial J_2}{\partial t}\}$ and the Jacobi identity we arrive at

$$\frac{d}{dt}\{J_1, J_2\} = \left\{ \frac{\partial J_1}{\partial t}, J_2 \right\} + \left\{ J_1, \frac{\partial J_2}{\partial t} \right\}$$

$$- \{J_1, \{J_2, H\}\} - \{J_2, \{H, J_1\}\}$$

$$= \left\{ \frac{\partial J_1}{\partial t} + \{H, J_1\}, J_2 \right\} + \left\{ J_1, \frac{\partial J_2}{\partial t} + \{H, J_2\} \right\}$$

$$= \left\{ \frac{dJ_1}{dt}, J_2 \right\} + \left\{ J_1, \frac{dJ_2}{dt} \right\} = 0,$$

where the last equality follows from the fact that J_1 and J_2 are first integrals, hence $\{J_1, J_2\}$ is a further first integral.

10. The Hamilton-Jacobi differential equation is now

$$\frac{1}{2m}\left(\left(\frac{\partial W}{\partial r}\right)^2 + \frac{1}{r^2}\left(\frac{\partial W}{\partial \vartheta}\right)^2 + \frac{1}{r^2 \sin^2 \vartheta}\left(\frac{\partial W}{\partial \varphi}\right)^2 \right) - \frac{\gamma m M}{r} + \frac{\partial W}{\partial t} = 0$$

which leads with $W = S - Et$ to

$$\frac{1}{2m}\left(\left(\frac{\partial S}{\partial r}\right)^2 + \frac{1}{r^2}\left(\frac{\partial S}{\partial \vartheta}\right)^2 + \frac{1}{r^2 \sin^2 \vartheta}\left(\frac{\partial S}{\partial \varphi}\right)^2 \right) - \frac{\gamma m M}{r} = E. \qquad (*)$$

Assuming that $S(r, \vartheta, \varphi) = R(r) + \theta(\vartheta) + \Phi(\varphi)$ we find

$$\frac{1}{2m}\left(\left(\frac{dR}{dr}\right)^2 + \frac{1}{r^2}\left(\frac{d\theta}{d\vartheta}\right)^2 + \frac{1}{r^2 \sin^2 \vartheta}\left(\frac{d\Phi}{d\varphi}\right)^2 \right) - \frac{\gamma m M}{r} = E.$$

This yields first

$$-\left(\frac{d\Phi}{d\varphi}\right)^2 = r^2 \sin^2 \vartheta \left(\frac{dR}{dr}\right)^2 + \sin^2 \vartheta \left(\frac{d\theta}{d\vartheta}\right)^2 - rmM \sin^2 \vartheta - 2mEr^2 \sin^2 \vartheta$$

implying that

$$\frac{d\Phi}{d\varphi} = \alpha = \text{constant.}$$

721

Since we are not longing for a general solution of $(*)$ we choose $\Phi(\varphi) = \alpha\varphi$. Now we find

$$\frac{1}{2m}\left(\left(\frac{dR}{dr}\right)^2 + \frac{1}{r^2}\left(\left(\frac{d\theta}{d\vartheta}\right)^2 + \frac{\alpha^2}{\sin^2\vartheta}\right)\right) - \frac{\gamma mM}{r} = E. \tag{$**$}$$

From $(**)$ we get

$$\left(\frac{d\theta}{d\vartheta}\right)^2 + \frac{\alpha^2}{\sin^2\vartheta} = 2mr^2 E + 2\gamma m^2 Mr - r^2\left(\frac{dR}{dr}\right)^2$$

and this implies

$$\left(\frac{d\theta}{d\vartheta}\right)^2 + \frac{\alpha^2}{\sin^2\vartheta} = \beta$$

with some constant β. This differential equation has the solution

$$\theta(\vartheta) = \int \sqrt{\beta - \frac{\alpha^2}{\sin^2\vartheta}}\, d\vartheta.$$

Finally we arrive at

$$\frac{1}{2m}\left(\left(\frac{dR}{dr}\right)^2 + \frac{\beta}{r^2}\right) - \frac{\gamma mM}{r} = E$$

or

$$R(r) = \int \sqrt{2m\left(E + \frac{\gamma mM}{r}\right) - \frac{\beta}{r^2}}\, dr.$$

The function W becomes now

$$W = -Et + \alpha\varphi + \int \sqrt{\beta - \frac{\alpha^2}{\sin^2\vartheta}}\, d\vartheta + \int \sqrt{2m\left(E + \frac{\gamma mM}{r}\right) - \frac{\beta}{r^2}}\, dr$$

which depends on the three parameters E, α and β. As coordinates we now find

$$-Q_1 = \frac{\partial W}{\partial E} = -t + \int \frac{m\, dr}{\sqrt{2m\left(E + \frac{\gamma mM}{r}\right) - \frac{\beta}{r^2}}}\, dr$$

$$Q_2 = \frac{\partial W}{\partial \alpha} = \varphi - \int \frac{\alpha\, d\vartheta}{\sin^2\vartheta\sqrt{\beta - \frac{\alpha^2}{\sin^2\vartheta}}}$$

and

$$Q_3 = \frac{\partial W}{\partial \beta} = \frac{1}{2}\int \frac{d\vartheta}{\sqrt{\beta - \frac{\alpha^2}{\sin^2\vartheta}}} - \frac{1}{2}\int \frac{dr}{r^2\sqrt{2m\left(E + \frac{\gamma mM}{r}\right) - \frac{\beta}{r^2}}}.$$

11. (Compare [15])

With $S(q_1, q_2) = S_1(q_1) + S_2(q_2)$ we find from

$$\frac{1}{2m}\left(\left(\frac{\mathrm{d}S}{\mathrm{d}q_1}\right)^2 + \left(\frac{\mathrm{d}S}{\mathrm{d}q_2}\right)^2\right) + mgq_2 = E$$

the equation

$$\left(\frac{\mathrm{d}S_1(q_1)}{\mathrm{d}q_1}\right)^2 = \gamma, \quad \left(\frac{\mathrm{d}S_2(q_2)}{\mathrm{d}q_2}\right)^2 = (2mE - \gamma) - 2m^2 gq_2$$

which yields

$$S_1(q_1) = \sqrt{\gamma}q_1 + c_1$$

and

$$S_2(q_2) = -\frac{1}{3m^2 g}((2mE - \gamma) - 2m^2 gq_2)^{\frac{3}{2}} + c_2.$$

Choosing $c_1 = c_2 = 0$ we obtain

$$S(q_1, q_2) = \sqrt{\gamma}q_1 - \frac{1}{3m^2 g}((2mE - \gamma) - 2m^2 gq_2)^{\frac{3}{2}}$$

and therefore we have

$$p_1 = \frac{\partial S}{\partial q_1}, \quad p_2 = \frac{\partial S}{\partial q_2}, \quad t - \beta_1 = E, \quad \frac{\partial S}{\partial \gamma} = -\beta_2$$

with constants β_1 and β_2. This leads to

$$-\frac{1}{mg}((2mE - \gamma) - 2m^2 gq_2)^{\frac{1}{2}} = t - \beta_1$$

or

$$q_2(t) = -\frac{1}{2}g(t - \beta_1)^2 + \frac{2mE - \gamma}{2m^2 g} = \frac{1}{2}gt^2 + \gamma_1 t + \gamma_2$$

with γ_1 and γ_2 determined from E, γ and β_1. As we expect, the solution is a parabola.

References

[1] Abraham, R., Marsden, J. E., *Foundations of Mechanics*. 2nd ed. Benjamin/Cumming Publishing Company, Reading MA 1978.

[2] Abramowitz, M., Stegun, J. A., (eds.), *Handbook of Mathematical Functions*. 7th printing. Dover Publications, New York 1970.

[3] Amann, H., *Gewöhnliche Differentialgleichungen*. Walter de Gruyter Verlag, Berlin 1983.
English: *Ordinary Differential Equations. An Introduction to Nonlinear Analysis*. Walter de Gruyter Verlag, Berlin 1990.

[4] Andrews, G.A., Askey, R., Roy, R., *Special Functions*. Cambridge University Press, Cambridge 1999.

[5] Arago, D.F.J., *Eloge Historique de Joseph Fourier*. Mém. Acad. Royale des Sci. 14(1838), LXIX-CXXXVIII.
English: Arago, D.F.J., *Biographies of Distinguished Scientific Men. Vol. I*. Ticknor and Fields, Boston MA 1859, pp. 374-444.

[6] Arnold, V. I., *Mathematical Methods of Classical Mechanics*. Springer-Verlag, New York · Heidelberg · Berlin 1978.

[7] Arnold, V.I., *Gewöhnliche Differentialgleichungen*. Springer Verlag, Berlin 1979.

[8] Bari, N., *A Treatise on Trigonometric Series. Vol I*. Pergamon Press, New York 1964.

[9] Bari, N., *A Treatise on Trigonometric Series. Vol II*. Pergamon Press, New York 1964.

[10] Bauer, H., *Maß-und Integrationstheorie*. Walter de Gruyter Verlag, Berlin 1990.
English: *Measure and Integration Theory*. Walter de Gruyter Verlag, Berlin 2001.

[11] Beals, R., and Wong, R., *Special Functions. A Graduate Text*. Cambridge University Press, 2010.

[12] Berg, C., Forst, G., *Potential Theory on Locally Compact Abelian Groups*. Springer Verlag, Berlin 1975.

[13] Bernoulli, Johannes I, *Problema novum ad cujus solutionen Mathematici invitantur*. Acta Eruditorum (1696), 269.

[14] Bochner, S., *Vorlesungen über Fouriersche Integrale*. Akademische Verlagsgesellschaft, Leipzig 1932.

[15] Budó, A., *Theoretische Mechanik. 7. Aufl*. VEB Deutscher Verlag der Wissenschaften, Berlin 1974.

[16] Burkhardt, H., *Entwicklung nach oscillirenden Functionen und Integration der Differentialgleichungen der mathematischen Physik*. 1. Halbband. Jahresber. D.M.V. 10 (1908), 1-894. 2. Halbband. Jahresber. D.M.V.10 (1909), 895-1904.

[17] Burkhardt, H., *Trigonometrische Reihen und Integrale (bis etwa 1850). Encyclopädie der Mathematischen Wissnschaften. 2. Band: Analysis.* Teil 12, Teubner Verlag, Leipzig 1916, S.819-1354.

[18] Buttazzo, G., Giaquinta, M., Hildebrandt, S., *One-dimensional Variational Problems. An Introduction.* Clarendon Press, Oxford 1998.

[19] Cain, G., Meyer, G.H., *Separation of Variables for Partial Differential Equations.* Chapman & Hall/CRC, Boca Raton FL, 2006.

[20] Cantor, G., *Über die Ausdehnung eines Satzes aus der Theorie der trigonometrischen Reihen.* Math. Ann. 5 (1872), 123-132.

[21] Carleson, L., *On convergence and growth of partial sums and Fourier series.* Acta. Math. 116 (1966), 135-157.

[22] Courant, R., Hilbert, D., *Methoden der mathematischen Physik I. 3. Aufl.* Springer Verlag, Berlin·Heidelberg·New York, 1968.
English: *Methods of Mathematical Physics. I.* Interscience Publishers, New York 1953.

[23] Dacorogna, B., *Introduction to the Calculus of Variations.* Imperial College Press, London 2004.

[24] Dhombres, J., Robert, J.-B., *Fourier 1768-1830. Créateur de la Physique-Mathématique.* Editions Belin, Paris 1998.

[25] Du Bois-Reymond, P., *Über die Fourierschen Reihen.* Nachr. Kgl. Gesellschaft der Wiss. Göttingen, Jahrgang 1873, 571-584.

[26] Du Bois-Reymond, P., *Untersuchungen über die Convergenz und Divergeuz der Fourierschen Darstellungsformeln.* Abhandlungen math. phys. Classe der k. bayerischen Academie der Wissenschaften Bd. XII. 2. Abh. (1876), 1-104.

[27] Duren, P., *Theory of H^p-spaces.* Academic Press, New York 1970.

[28] Duren, P., *Invitation to Classical Analysis.* American Mathematical Society, Providence R.I. 2012.

[29] Fefferman, C.L., Stein, E.M., *H^p-spaces of several variables.* Acta Math. 129 (1972), 137-193.

[30] Fejer, L., *Sur les singularités de la série de Fourier des functions continues.* Ann. Ecole Norm. Sup. 28 (1911), 64-103.

[31] Fischer, G., *Lineare Algebra. 9. Aufl.* Vieweg Verlag, Braunschweig · Wiesbaden 1986.

[32] Fischer, H., Kaul, H., *Mathematik für Physiker.* Band 2. Teubner Verlag, Stuttgart 1998.

[33] Fischer, W., Lieb, I., *Ausgewählte Kapitel aus der Funktionentheorie.* Vieweg Verlag, Braunschweig 1988.

[34] Folland, G., *Fourier Analysis and its Applications.* Wadsworth of Brooks, Pacific Grove CA 1992.

[35] Folland., G., *A Course in Abstract Harmonic Analysis.* CRC Press. Boca Raton FL 1995.

[36] Fomenko, A.T., *Symplectic Geometry.* Gordon and Breach Science Publisher, New York 1988.

[37] Forster, O., *Analysis 1.* Vieweg Verlag, Braunschweig 1978.

[38] Forster, O., *Analysis 2.* Vieweg Verlag, Braunschweig 1979.

[39] Fourier, J-B.J., *Théorie Analytique de la Chaleur.* Firmin Didot, Paris 1822.

[40] Freguglia, P., Giaquinta, M., *The Early Period of the Calculus of Variations.* Birkhäuser Verlag, Basel 2016.

[41] Fritsche, K., Grauert, H., *From Holomorphic Functions to Complex Manifolds.* Springer Verlag, Berlin 2002.

[42] Gasper, G., Rahman, M., *Basic Hypergeometric Series.* 2nd ed. Cambridge University Press, Cambridge 2004.

[43] Gelfand, I.M., Fomin, S.V., *Calculus of Variations.* Prentice Hall Inc., Englewood Cliffs NJ 1963.

[44] Giaquinta, M., Hildebrandt, S., *Calculus of Variations I and II.* Springer Verlag, Berlin·Heidelberg·New York, 1996.

[45] Goldstein, H., Poole, C., Safko, J., *Classical Mechanics.* 3^{rd}ed. Pearson Education, London 2014.

[46] Gradshteyn, J.S., Ryzhik, J.M., *Tables of Integrals, Series and Products.* Corrected and Enlarged Edition. Academic Press, New York 1980.

[47] Grafakos, L., *Classical and Modern Fourier Analysis.* Pearson Education Inc. Upper Saddle River NY, 2004.

[48] Grattan-Guiness, J., *Convolutions in French Mathematics, 1800-1840, 3 Vols.* Birkhäuser Verlag, Basel 1990.

[49] Grattan-Guiness, J., Ravetz, J.R., *Joseph Fourier 1768-1830.* MIT Press, Cambridge MA 1972.

[50] Gröbner, W., *Differentialgleichungen. Erster Teil: Gewöhnliche Differentialgleichungen.* B.I.-Wissenschaftsverlag, Mannheim·Wien·Zürich, 1977.

[51] Haar, A., *Der Massbegriff in der Theorie der kontinuierlichen Gruppen.* Ann. Math. 34 (1933), 147-169.

[52] Halmos, P., *Finite-dimensional Vector Spaces.* D. van Nostrand Comp., Princeton NJ, 1958.

[53] Hardy, G.H., *Orders of Infinity - The 'Infinitärcalcül' of Paul du Bois-Reymond.* Cambridge University Press 1910.

727

[54] Herivel, J., *Jospeh Fourier. The Man and the Physicist.* Clarendon Press, Oxford 1975.

[55] Heuser, H., *Lehrbuch der Analysis. Teil 2.* Teubner Verlag, Stuttgart 1981.

[56] Heuser, H., *Gewöhnliche Differentialgleichungen. 3. Aufl.* Teubner Verlag, Stuttgart 1995.

[57] Hewitt, E., Ross, K.A., *Abstract Harmonic Analysis.* 2nd ed. Vol. I. Springer Verlag, Berlin 1979.

[58] Hildebrandt, S., *Analysis 2.* Springer Verlag, Berlin 2003.

[59] Hildebrandt, S., Tromba, A., *The Parsimonious Universe. Shape and Form in the Natural World.* Copernicus - Springer Verlag, New York 1996.

[60] Hille, E., *Ordinary Differential Equations in the Complex Domain.* Dover Publications, New York 1976.

[61] Hirsch, M., Smale, S., *Differential Equations, Dynamical Systems, and Linear Algebra.* Academic Press, New York, 1974.

[62] Hofmann, K.H., Morris, S.A., *The structure of Compact Groups.* 3rd ed. Walter de Gruyter Verlag, Berlin 2013.

[63] Hörmander, L., *An Introduction to Complex Analysis in Several Variables.* North-Holland Publishing Company, Amsterdam 1973.

[64] Horn, J., Wittich, H., *Gewöhnliche Differentialgleichungen. 6 Aufl.* Walter de Gruyter Verlag, Berlin 1960.

[65] Hugo, V., *Les Misérables. I.* Pagnerre, Paris 1862.
English: *Les Miserables* Alfred Knopf Doubleday Publishing Group, New York 1998.

[66] Hunt, R., *On the convergence of Fourier series. In: Orthogonal Expansions and Their Continuous Analogues.* Proceeding of the Conference at Edwardsville Ill. 1967. (ed. D.T.Haimo), Southern Illinois University Press, Carbondale Ill, 1968, 235-255.

[67] Ince, E.L., *Ordinary Differential Equations.* Dover Publications, New York 1956.

[68] Jacob, N., *Lineare partielle Differentialgleichungen.* Akademie Verlag, Berlin 1995.

[69] Jacob, N., *Pseudo Differential Operators and Markov Processes. Vol I: Fourier Analysis and Semigroups.* Imperial College Press, London 2001.

[70] Jørsboe, O.G., Mejlbro, L., *The Carleson-Hunt Theorem on Fourier Series.* Springer Verlag, Berlin 1982.

[71] Jost, J., Li-Jost, X., *Calculus of Variations.* Cambridge University Press, Cambridge 1998.

[72] Kahane, J.-P., Lemarié-Rieurset, P.G., *Séries de Fourier et Ondelettes.* Cassini, Paris 1998.

[73] Katznelson, Y. *An Introduction to Harmonic Analysis.* 3rd ed. Cambridge University Press, Cambridge 2004.

[74] Klein, F., *Vorlesungen über die hypergeometrische Funktion.* Springer Verlag, Berlin 1933.

[75] Kolmogorov, A.N., *Une série de Fourier-Lebesgue divergente presque partout.* Fund. Math. 4 (1923), 324-328.

[76] Kolmogorov, A.N., *Une série de Fourier-Lebesgue divergente partout.* C.R. Acad. Sci. Paris 183 (1926), 1327-1328.

[77] Koosis, P., *Introduction to H^p-spaces.* 2nd ed. Cambridge University Press, Cambridge 1998.

[78] Körner, T., *Fourier Analysis.* Cambridge University Press, Cambridge 1988.

[79] Körner, T., *Exercises for Fourier Analysis.* Cambridge University Press, Cambridge 1988.

[80] Kot, M., *A First Course in the Calculus of Variations.* American Mathematical Society, Providence RI 2014.

[81] Krantz, S., *Function Theory of Several Complex Variables.* John Wiley & Sons, New York 1982.

[82] Kummer, E.E., *Über die hypergeometrische Reihe $1 + \frac{\alpha \cdot \beta}{1 \cdot \gamma} + \cdots J$ reine angew.* Mathematik 15(1836), 39-83 und 127-182.

[83] Lang, S., *Linear Algebra.* Addison-Wesley Publishing Company, Reading MA 1966.

[84] Landau, L.D., Lifschitz, E.M., *Lehrbuch der Theoretischen Physik. Bd. I. Mechanik. 8. Aufl.* Akademie Verlag, Berlin 1973.
English: *A Course in Theoretical Physics. Vol. I. Mechanics.* 2nd ed. Pergamon Press, Oxford 1969.

[85] Lebedew, N.N., *Spezielle Funktionen und ihre Anwendungen.* B.I.-Wissenschaftsverlag, Mannheim·Wien·Zürich, 1973.
English: *Special Functions and Their Applications.* Prentice-Hall Inc. Englewood Cliffs N.J. 1965.

[86] Loomis, L.H., *An Introduction to Abstract Harmonic Analysis.* D. van Nostrand Company, Princeton NJ 1953.

[87] Lützen, J., *Joseph Liouville 1809-1882. Master of Pure and Applied Mathematics.* Springer Verlag, Berlin 1990.

[88] Malliavin, P., *Integration and Probability.* Springer Verlag, Berlin 1995.

[89] McDuff, D., Salamon, D., *Introduction to Symplectic Topology.* Clarendon Press, Oxford 1995.

[90] Medvedev, F.A., *Scenes from the History of Real Functions.* Birkhäuser Verlag, Basel 1991.

[91] Milnor, J.W., *Topology from the Differential Viewpoint.* The University Press of Virginia, Charlottesville, 5th printing, 1978.

[92] Mozzochi, C.J., *On the Pointwise Convergence of Fourier Series.* Springer Verlag, Berlin 1971.

[93] Nachbin, L., *The Haar Integral.* Van Nostrand Company, Princeton NJ, 1965.

[94] Olver, F.W.J., *Asymptotics and Special Functions.* A.K.Peters, Ltd., Wellesley MA 1997.

[95] Pachpatte, B.G., *Inequalities for Differential and Integral Equations.* Academic Press, New York 1998.

[96] Petersen, B.E., *Introduction to the Fourier Transform and Pseudo-differential Operators.* Pitman Publishing, Boston MA 1983.

[97] Pinsky, M., *Introduction to Fourier Analysis and Wavelets.* Brooks/Cole, Pacific Grove CA 2002.

[98] Pontryagin, L.S., *Topological Groups.* 2nd ed. Gordon and Breach, New York 1954.

[99] Reed, M., Simon, B., *Methods in Modern Mathematical Physics. Vol. I. Functional Analysis.* Academic Press, New York 1972.

[100] Reiter, H., Stegeman, J.D., *Classical Harmonic Analysis on Locally Compact Abelian Groups.* Clarendon Press, Oxford 2000.

[101] Rudin, W., *Fourier Analysis on Groups.* Interscience Publisher, New York 1962.

[102] Rudin, W., *Functional Analysis.* McGraw-Hill Book Company, New York 1973.

[103] Rudin, W., *Real and Complex Analysis.* 2nd ed. McGraw-Hill Book Company, New York 1974.

[104] Schäfke, F.W., *Einführung in die Theorie der speziellen Funktionen der mathematischen Physik.* Springer Verlag, Berlin 1963.

[105] Schechter, M., *An Introduction to Nonlinear Analysis.* Cambridge University Press, Cambridge 2004.

[106] Schiff, J.L., *The Laplace Transform.* Springer Verlag, Berlin 1999.

[107] Schilling, R., *Measures, Integrals and Martingales.* 2nd ed. Cambridge University Press, Cambridge 2017.

[108] Schwartz, L., *Théorie de Distributions. Vol. I.* Hermann, Paris 1950.

[109] Schwartz, L., *Théorie de Distributions. Vol. II.* Hermann, Paris 1951.

[110] Smirnov, W.I., *Lehrgang der höheren Mathematik. Teil III/2. 9. Aufl.* VEB Deutscher Verlag der Wissenschaften, Berlin 1972.

[111] Smirnov, W.I. *Lehrgang der höheren Mathematik. Teil IV. 5. Aufl.* VEB Deutscher Verlag der Wissenschaften, Berlin 1968.

[112] Spiegel, M., *Advanced Mathematics for Engineers and Scientists.* McGraw-Hill Book Company, New York 1971.

[113] Stein, E.M., *Singular Integrals and Differentiability Properties of Functions.* Princeton University Press, Princeton NJ 1970.

[114] Stein, E.M., Shakarchi, R., *Princeton Lectures in Analysis I. Fourier Analysis: An Introduction.* Princeton University Press, Princeton NJ 2003.

[115] Stein, E.M., Weiss, G., *On the theory of harmonic functions of several variables, I: The theory of H^p-spaces.* Acta Math. 103 (1960), 25-62.

[116] Stein, E.M., Weiss, G., *Introduction to Fourier Analysis on Euclidean Spaces.* Princeton University Press, Princeton NJ 1971.

[117] Stromberg, K., *An Introduction to Classical Real Analysis.* Wadsworth Inc. Belmont CA. 1981.

[118] Stroppel, M., *Locally Compact Groups.* EMS Textbooks in Mathematics. European Mathematical Society, Zürich 2006.

[119] Szabo, J., *Geschichte der mechanischen Prinzipien.* 3rd ed. Birkhäuser Verlag, Basel 1987.

[120] Torchinsky, A., *Real-Variable Methods in Harmonic Analysis.* Academic Press, New York 1986.

[121] Triebel, H., *Höhere Analysis.* VEB Deutscher Verlag der Wissenschaften, Berlin 1972.
English: *Higher Analysis.* J. A. Barth Verlag, Leipzig 1992.

[122] Truesdell, C., *Essays in the History of Mechanics.* Springer Verlag, Berlin 1968.

[123] Uchiyama, A., *Hardy Spaces on the Euclidean Space.* Springer Verlag, Berlin 2001.

[124] Vainberg, B.R., *Asymptotic Methods in Equations of Mathematical Physics.* Gordon and Breach Science Publishers, New York 1989.

[125] Walter, W., *Gewöhnliche Differentialgleichungen.* Eine Einführung. Springer Verlag, Berlin 1972.
English: *Ordinary Differential Equations.* Springer-Verlag, New York, 1998.

[126] Weil, A., *L'Intégration dans les Groups Topologiques et ses Applications.* 2^{ieme} ed. Hermann, Paris 1965.

[127] Whittaker, E.T., *A Treatise on the Analytical Dynamics of Particles and Rigid Bodies.* 4th ed., reprinted in "Cambridge Mathematical Library", Cambridge University Press, Cambridge 1988.

[128] Whittaker, E.T., Watson, G.N., *A Course of Modern Analysis.* 4th ed., reprinted in "Cambridge Mathematical Library", Cambridge University Press, Cambridge 1996.

[129] Widder, D.V., *The Laplace Transform.* Princeton University Press, Princeton NJ, 1963 (2nd printing).

[130] Zeidler, E., *Nonlinear Functional Analysis and its Applications I. Fixed-Point Theorems.* Springer Verlag, Berlin 1986.

[131] Zygmund, A., *Trigonometric Series.* 2nd ed. Volume I & II Combined. Cambridge University Press, Cambridge 1988.

Mathematicians Contributing to Analysis (Continued)

Airy, George Biddell (1801-1892).

Bendixson, Ivar Otto (1861 - 1935).

Blaschke, Wilhelm (1885-1962).

Bochner, Salomon (1899-1983).

Brouwer, Luitzen Egbertus Jan (1881-1966).

Cesarò, Ernesto (1859-1906).

Chetaev, Nikolai Gurevich (1902-1959).

Clairaut, Alexis Claude (1713-1765).

Fejer, Leopold (1880-1959).

Fischer, Emil (1875-1954).

Frobenius, Georg Ferdinand (1849-1917).

Fuchs, Immanuel Lazarus (1833-1902).

Gibbs, Josiah Willard (1839-1903).

Gronwall, Thomas Hakon (1877-1932).

Haar, Alfred (1885-1933).

Herglotz, Gustav (1881-1959).

Hildebrandt, Stefan (1936-2015).

Hille, Einar (1894-1980).

Hoèné-Wronski, Jósef-Maria (1778-1853).

Kahane, Jean-Pierre (1926-2017).

Klein, Christian Felix (1849-1925).

Kneser, Adolf (1862-1930).

Kummer, Ernst Eduard (1810-1893).

Lévy, Paul (1886-1971).

Laguerre, Edmund Nicolas (1834-1886).

Ljapunov, Alexander Michailowitsch (1857-1918).

Lotka, Alfred James (1880-1949).

Malliavin, Paul (1925-2010).

Noether, Amalie Emmy (1882-1935).

Painlevé, Paul (1863-1933).
Paley, Raymond (1907-1933).
Parseval des Chénes, Marc-Antoine (1775-1836).
Pfaff, Johann Friedrich (1765-1825).
Plancherel, Michael (1885-1967).
Pontryagin, Lew Semjenowitch (1908-1988).

Riccati, Jacopo Francesco (1676-1754).
Rudin, Walter (1921-2010).

Schauder, Juliusz Pawel (1899-1943).
Schmidt, Erhard (1876-1959).
Schoenberg, Isaac Jacob (1903-1990).
Sobolev, Sergei Lwowich (1908-1989).
Sturm, Jacques Charles François (1803-1855).

Tauber, Alfred (1866-1942).
Tietze, Heinrich Franz Friedrich (1880-1964).

Volterra, Vito (1860-1940).

Watson, George Neville (1886-1965).
Whittaker, Edmund Taylor (1873-1956).

Zeidler, Eberhard (1940-2016).

Subject Index

740

钱钟书在《写在人生边上》一书中解读了伊索寓言故事.在那则《蚂蚁与促织》的故事中写道:"促织饿死了,本身就作蚂蚁的粮食;同样,生前养不活自己的大作家,到了死后偏有一大批人靠他生活.譬如,写回忆怀念文字的亲戚和朋友,写研究论文的批评家和学者."

钱钟书自恃才高,说话刻薄,但细品在理.我们出版者在某种意义上也是一类蚂蚁,靠那些天才人物的营养活着.

本书是英文版系列《分析学教程》中的一部,中文书名或可译为《分析学教程.第4卷,傅里叶分析,常微分方程,变分法》.

本书的作者有两位,一位是尼尔斯·雅各布(Niels Jacob),英国数学家,英国斯旺西大学教授;另一位是克里斯蒂安·P. 埃文斯(Kristian P. Evans),英国数学家,英国斯旺西大学教授.

正如作者在前言中所介绍:

我们这套《分析学教程》的前三卷涵盖的内容必须被视为数学专业本科生进行任何分析学学习的基础.从本卷开始,我们研究的内容对于每个认真学习的分析学专业的学生来说仍然是基础的,但是对于普通的数学专业的学生来说,他们可能会选择模块进行学习.在本卷中,我们将讨论 Fourier 分析、常微分方程和变分法的基础知识(一维情况下的),其中包括一些关于分析动力学的结果,即 Hamilton 力学.

在第 8 部分的 Fourier 分析中,我们理所当然地认为读者对 Lebesgue 积分理论和复分析理论都有很好地理解.我们首先在第 1 章中描述了 Fourier 分析在数学发展中发挥的(并且仍在发挥)非凡的作用,不仅仅是在分析学中,这方面的知识将有助于读者更好地理解某些数学知识的发展. 第 2 章介绍了三角级数和 Fourier 级数. 第 3 章我们讨论了 Fourier 系数的基本性质. 我们给出了级数的实值和复值表示,还讨论了正交性的重要性. 我们研究了 Fourier 系数序列的衰减特性及收敛行为.

第一个主要的章节是第 4 章,我们介绍了逐点收敛结果,并介绍了可和性的概念. 我们还研究了 Dirichlet 核和 Fejer 核,并讨论了符合 Dirichlet 思想的逐点收敛结果以及 Cesàro 和 Abel 可和性. 与第 4 章同样重要的是第 5 章,我们讨论了 Fourier 级数的 L^2 理论,介绍了关于正交级数展开式的更一般的结果,充分探讨了有关正交性和级数展开的思想. 更专业的是第 6 章,在第 6 章我们研究了有关 Fourier 级数收敛的进一步问题,Dini 的结果被证明是某些一致收敛的结果,例如具有 Fourier 系数单调序列的级数. 我们还提供了 Du-Bois Reymond 反例的证明(通过遵循 Fejer 的修正),该反例是在一点具有分歧的 Fourier 级数的连续函数. 最后,我们证明了关于 L^1 函数的 Fourier 级数的 Cesàro 可和性的 Fejer-Lebesgue 定理.

第 7 章重点介绍了单位圆中 Fourier 级数与全纯函数之间的关系,以及它们与调和函数的关系,还介绍了哈代空间 $\mathscr{H}^2$,并研究了它的基本性质. 作为核心工具,我们介绍了 Blaschke 乘积,还讨论了 Nevanlinna 类以及 Poisson-Jensen 公式. 我们给出了空间 $\mathscr{H}^p$,还讨论了 Poisson 积分的边界行为. 在第 8 章中,我们把注意力转向了测度的 Fourier 级数,并介绍了相关的基本概念,比如 Fourier-Stieltjes 系数或正定函数. 假设 Riesz 表示定理(将在第 5 卷中证明),我们证明了 Herglotz-Bochner 定理并讨论了 Poisson 测度积分(圆上). 这导致我们在圆上引入了 Sobolev 空间,并给出了 Sobolev 嵌入定理的第一个版本. 本章以对 Gibbs 现象的讨论结束.

第 9 章延续了第 5 章的思想,我们开始讨论 L^2 函数关于某些完整的正交函数系统的级数展开,例如 Legendre 多项式或 Hermite 多项式. 这是我们将在第 9 部分研究某些二阶线性微分方程 Sturm-Liouville 算子时继续讨论的主题,我们还在附录Ⅲ中收集了一些背景材料.

我们从 Fourier 级数进入到 Fourier 变换的研究. 我们首先介绍了 Schwarz 空间 $S(\mathbb{R}^n)$,在第 5 卷中我们会将它作为一个局部凸拓扑向量空间的例子进行更进一步的研究. 为了之后的目标, 我们对 Friedrichs 光滑算子进行了更详细的研究. 在第 11 章中我们介绍了 $S(\mathbb{R}^n)$ 中的 Fourier 变换, 并对其性质进行了深入的讨论. Fourier 反演定理、Plancherel 定理以及卷积定理(我们在 Fourier 级数中已经研究了后两个定理)都得到了证明, 并强调了它们作为工具的核心作用. 我们将在第 12 章继续讨论 L^p 空间中的 Fourier 变换, 特别是对于 $p=1$ 和 $p=2$ 的情况, 我们再次了解一下 Sobolev 空间和 Bessel 势空间. 第 13 章我们首先了解了子概率度量的卷积半群 $(\mu_t)_t \geqslant 0$, 研究了 $\mathbb{R}^n$ 中 Borel 测度的 Fourier 变换, 还讨论了 Bochner 定理, 我们通过介绍四个选定的主题来结束对 Fourier 变换的研究: 在 Bessel 函数的帮助下旋转不变量函数的 Fourier 变换的表示, 波动方程 Cauchy 问题的第一次讨论, Poisson 求和公式, 以及测不准原理. 在相当广泛的附录 I 中, 我们总结了如何将 Fourier 级数和 Fourier 变换的理论嵌入到交换调和分析中, 即对局部紧 Abel 群的 Fourier 分析.

在我们更详细地讨论第 9 部分常微分方程和第 10 部分变分法导论之前, 有两个一般性说明. 在更高级的分析学的主题中, 常微分方程可能是受高级数学软件的出现影响最大的了. 经典理论已经发展成了某些系统方法, 用来研究常微分方程, 同时给出许多有价值的公式, 然而, 这些公式在具体案例中往往很难评估. 因此, 随着近似方法的发现, 处理公式的分析方法也得到了发展. 如今, 我们可能会使用软件来评估这些公式, 实际上对于某些方程, 一些软件包可以直接找到解决方案. 但这些公式的理论重要性并没有减少, 常微分方程的教学与实践工作将不得不把软件包考虑在内, 并使用它们. 这不适用于诸如我们的"分析学课程", 我们在这里只局限于推导分析理论而避免通过使用分析工具(可以用适当的软件代替)在示例中计算复杂的公式.

应该在几何背景下理解微分方程, 许多书籍都致力于实现这个想法. 然而, 这要求读者对微分几何和可微流形理论有一定的了解, 我们将在第 7 卷中讨论这些主题. 我们在这里尽量减少对几何方面内容的阐释, 将在第 7 卷中讨论常微分方程和动力系统理论中的某些几何内容, 第 25 章和第 10 部分的最后一章除外. 求解一阶偏微分方程需要一种几何方法, 我们将在变分法和分析动力学的背景下研究一阶偏微分方程, 即 Hamilton 系统. 在讨论某些高阶偏微分方程的解的奇异性传播时, Hamilton 系统也非常重要, 例如波动方程. 在正确介绍流形分析之前, 我

745

们必须在第 6 卷中研究此类问题. 但是变分法和常微分方程构成了处理 Hamilton 系统的自然框架. 我们希望我们的方法可以让读者在不掌握高级微分几何知识的情况下也能利用上面这些思想.

第 9 部分以一个介绍性章节开始, 在这一章节中我们讨论了某些最初方法, 比如分离变量, 或者常数的变化, 并修正了某些最初的术语, 比如线性系统、齐次方程、初始值等. 我们还证明了每个高阶方程都可以转化为一阶系统, 使得它们的解相互对应. 在第 16 章中我们提供了初值问题的一般存在定理. Picard-Lindelöf 定理的变式和 Peano 定理被证明了. 我们借助不动点定理证明它们. 虽然 Banach 不动点定理很容易建立, 但证明 Peano 定理所需的 Schander 不动点定理需要 Brouwer 不动点定理来支撑, 附录 Ⅳ 讨论了该定理的证明, 该讨论从代数拓扑学出发, 假设球面作为球的边界是不能回缩的. 我们还讨论了最大解的概念和存在性. 虽然第 16 章只讨论了一个方程, 但在第 17 章中我们将这些结果扩展到了系统中.

由于一阶方程的线性系统的理论与实践的重要性, 我们对其进行了非常详细的讨论. 在第 18 章中我们研究了带常系数系统. 我们证明了系统矩阵与特征值问题的关系, 并推导出解的标准公式. 我们假设读者具有矩阵的 Jordan 范式的背景知识. 而对于常系数系统, 利用系统矩阵的指数来表示结果是一种展示结果的很好的可能性, 并开辟了通往单参数算子半群的道路, 我们将在第 19 章中看到当处理变系数系统时矩阵函数实际上是一个非常有效的工具. 基本结果是通过研究基本系统和 Wronski 行列式的作用获得的. 如上所述, 我们不会过多强调证明具体示例获得的复杂公式, 这里可以使用适合的软件.

具有(实)解析系数的二阶微分方程的 Frobenius 理论不仅导致了 L. Fuchs 和 H. Poincaré 关于微分方程的深刻结果, 现在称为 Fuchsian 微分方程, 还允许我们用系统的方式处理一些通过分离偏微分方程获得的来自数学物理学的最重要的微分方程, 比如, 曲线正交坐标中的 Laplace 方程. 我们全面地证明了 Frobenius 定理, 并详细讨论了几个例子, 例如 Airy 方程, Legendre 方程, Hermite 方程或 Bessel 方程等. 然后我们转而讨论导致 Sturm 振荡定理的解的定性性质. 我们还介绍了(正则)Sturm-Liouville 问题. Sturm-Liouville 算子是第 21 章的主题, 我们第一次遇到边值问题和特征值问题. 我们对这些特征值问题进行了更详细的研究, 特别是当与不同的特征值相关联时,会考虑特征值、特征函数和特征空间以及它们之间的关系. 这也为我们在卷 5 中对(奇异)Sturm-Liouville 算子的谱理论的讨论做了准备. 此外,我们还研究了针对正则 Sturm-Liouville 问题的 Green 函数.

到目前为止,我们研究了许多特殊函数都与超几何微分方程或汇合超几何微分方程有关. 这些方程在常微分方程理论中的核心作用证明,事实上我们需要一些研究方法. 除一般存在性的考虑之外,我们更详细地研究了保持方程不变的变换,以及众所周知的作为超几何函数或汇合超几何函数的表示. 本章的主题是一个当今可能有点被忽视,但仍然很重要且比较经典的主题.

第 23 章讨论了常微分方程的解(或解流形)的定性性质,建立了几个稳定性结果,并引入了 Lyapunov 函数的概念. 我们研究了自主方程和平衡点的解. 最后我们转向梯度系统及其性质的研究.

许多关于常微分方程的教材早就开始讨论相图和流. 我们认为只有在提供适当的几何背景时才应该这样做. 为此,我们详细介绍了 $\mathbb{R}^n$ 子集的切空间和切丛,然后引入作为切丛截面的向量场. 这使我们能够自然而然地扩展到用可微流形的方式来理解微分方程和向量场积分曲线之间的关系. 在第 24 章完成后,在关于微分方程的最后一章中,我们将研究相图以及与微分方程相关的流. 特别地,我们首先讨论了 Liouville 定理.

变分法从开始就与数学物理学紧密相关,特别是与力学和微分几何联系密切. 在研究 Riemann 曲面的全纯函数时,变分法非常重要,比如 Riemann 使用 Dirichlet 原理,在 Weierstrass 批评之后,最终导致引入了 Hilbert 空间和作为求解偏微分方程工具的变分法的直接方法. 对于这些内容,我们会给出一个关于基本观点和结果的简介. 第 26 章建立了一维情形,我们推导出了 Euler-Lagrange 方程作为 C^2 极值的必要条件. 在第 27 章中,我们研究了极值,即 Euler-Lagrange 方程的经典解. 我们讨论了正则性、一阶积分,特别讨论了将对称性与一阶积分联系起来的 Noether 定理的一个版本.

使极值最小化的充分条件更难获得. 在第 28 章中,我们研究了第二种变分并推导出 Legendre 的必要条件. 我们进一步的讨论给出了共轭点 Jacobi 方程的作用.

事实证明,求解一阶偏微分方程是获得进一步进展的关键. 因此,我们增加了一个很长的章节来讨论这些方程. 我们首先介绍了线性方程组的几何概念,然后研究了准线性方程组. 尽管这偏离了我们的主要目标,即一阶偏微分方程和变分法的联系,但准线性偏微分方程可以帮助我们将正则性和存在性问题理解为几何问题(与特征相关). 然后我们给出了一个关于一般非线性一阶偏微分方程的完整的讨论,依赖于"时间"参数 t. 这个讨论为我们提供了第 30 章和第 6 卷中求解双曲型方程时所需要的工具. 我们研究了分析动力学中的一个主要方程 Hamilton-Jacobi 微分方程,我们考虑了循环坐标和运动常数,并讨论了许多力学例子.

与前几卷一样,我们提供了所有(除了第 1 章)章节的问题及其详细的解决方案. 当然,本书针对的是更成熟的读者群,这使得一些计算更简洁.

本书的有些主题是标准的,有些则更高级. 这需要使用不同的引用方法. 对于我们认为是标准的主题,许多书对它们的处理方法都非常相似,我们只给出"全局"参考资料. 但是,无论何时处理更高级或特殊的内容,我们都会非常精确地引用参考资料. 我们希望这种方法能得到读者的赞赏,并得到其他作者的认可.

本书的版权编辑李丹女士为了使读者能快速了解本书的基本内容,特翻译了本书的目录如下:

第 8 部分:Fourier 分析

本书的一大主题是 Fourier 级数. 在数学史上曾对任意形状的初始弦能否用三角级数表示存在争议,后来 Fourier 在研究热传导的过程中解决了这个争议. 例如,在 xOy 平面上随手画的一条曲线,不管我们是否知道它的解析式,还是它不服从任何规律,它都可以展开成 Fourier 级数的表达式

$$f(x) = \frac{a_0}{2} + \sum_{n=1}^{\infty} \left(a_n \cos \frac{n\pi x}{r} + b_n \sin \frac{n\pi x}{r} \right)$$

其中

$$a_n = \frac{1}{r} \int_{-r}^{r} f(x) \cos \frac{n\pi x}{r} \mathrm{d}x, n = 0, 1, 2, \cdots$$

$$b_n = \frac{1}{r} \int_{-r}^{r} f(x) \sin \frac{n\pi x}{r} \mathrm{d}x, n = 1, 2, \cdots$$

所以我们不得不承认它也是函数. 由 Fourier 的结论可知, 函数不在于两个变量是否有依赖关系, 也就是说两个毫不相关的变量所对应的两组数据之间没有规律, 仅仅只是两组数据之间的对应, 它也有表达式, 也是函数, 这打破了我们中学时学习到的变量之间要有依赖关系的束缚, 使函数的本质变成了对应. 这种对应扩大了函数的范围, 不只是满足了现实应用的需要, 也是为了满足数学发展的需要.

并且 Fourier 级数还是解决许多问题的利器, 如:

例1   在环域 (内、外圆半径为 R_1 及 R_2, 中心在原点) 中求解 Laplace 方程的 Dirichlet 问题. 再考虑环域为圆的极限情形.

解   化为极坐标, 可得

$$r^2 \frac{\partial^2 u}{\partial r^2} + r \frac{\partial u}{\partial r} + \frac{\partial^2 u}{\partial \theta^2} = 0$$

边界条件为

$$u(r, \theta + 2\pi) = u(r, \theta), u(R_1, \theta) = f_1(\theta)$$

$$u(R_2, \theta) = f_2(\theta)$$

其中 $f_1(\theta)$ 及 $f_2(\theta)$ 为可展成 Fourier 级数的已知函数. 最后可得

$$u(r, \theta) = \frac{\alpha_0^{(2)} - \alpha_0^{(1)}}{\ln R_2 - \ln R_1} \ln r + \frac{\alpha_0^{(1)} \ln R_2 - \alpha_0^{(2)} \ln R_1}{\ln R_2 - \ln R_1} +$$

$$\sum_{n=1}^{\infty} (R_1^n R_2^{-n} - R_1^{-n} R_2^n)^{-1} \cdot \{ [(\alpha_n^{(1)} R_2^{-n} - \alpha_n^{(2)} R_1^{-n}) r^n - (\alpha_n^{(1)} R_2^n - \alpha_n^{(2)} R_1^n) r^{-n}] \cos n\theta +$$

$$[(\beta_n^{(1)} R_2^{-n} - \beta_n^{(2)} R_1^{-n}) r^n - (\beta_n^{(1)} R_2^n - \beta_n^{(2)} R_1^n) r^{-n}] \sin n\theta \}$$

其中

$$\alpha_0^{(1)} = \frac{1}{2\pi} \int_{-\pi}^{\pi} f_1(\theta) \mathrm{d}\theta$$

$$\alpha_n^{(1)} = \frac{1}{\pi} \int_{-\pi}^{\pi} f_1(\theta) \cos n\theta \mathrm{d}\theta$$

$$\beta_n^{(1)} = \frac{1}{\pi} \int_{-\pi}^{\pi} f_1(\theta) \sin n\theta \mathrm{d}\theta$$

$$\alpha_0^{(2)} = \frac{1}{2\pi}\int_{-\pi}^{\pi} f_2(\theta)\,\mathrm{d}\theta$$

$$\alpha_n^{(2)} = \frac{1}{\pi}\int_{-\pi}^{\pi} f_2(\theta)\cos n\theta\mathrm{d}\theta$$

$$\beta_n^{(2)} = \frac{1}{\pi}\int_{-\pi}^{\pi} f_2(\theta)\sin n\theta\mathrm{d}\theta$$

对圆的极限情形,解为

$$u(r,\theta) = \frac{1}{2\pi}\int_{-\pi}^{\pi} f(t)\frac{R^2 - r^2}{R^2 - 2Rr\cos(t-\theta) + r^2}\mathrm{d}t$$

这个过程中 $f_1(\theta)$ 和 $f_2(\theta)$ 起到了重要作用!

本书还有大量的篇幅来讲 Fourier 变换,这个内容已经进入到复分析中,仅举一例:

例 2 考虑函数

$$f(z) = \frac{\mathrm{e}^{(\pi+\mathrm{i}\omega)z}}{\mathrm{e}^{2\pi z} + 1}$$

这里 ω 是一个实的参变量. 找出 $f(z)$ 的奇点,在图 1 所示的回路上对 $f(z)$ 积分,建立积分

$$A = \int_0^\infty \frac{\cos \omega x}{\cosh \pi x}\mathrm{d}x \quad \text{与} \quad B = \int_0^\infty \frac{\sin \omega x}{\sinh \pi x}\mathrm{d}x$$

之间的关系式,这里 ε 是充分小的数,R 是充分大的数. 在这个关系式中,用 $-\omega$ 代替 ω,计算 A 与 B.

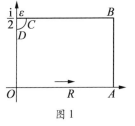

图 1

解 考虑函数

$$f(z) = \frac{\mathrm{e}^{(\pi+\mathrm{i}\omega)z}}{\mathrm{e}^{2\pi z} + 1}$$

这里 ω 是一个实的参变量. 我们在回路 γ 上积分,用半径为 ε 的四分之一的圆代替回路的相应部分,以避开点 $\dfrac{\mathrm{i}}{2}$.

函数 $f(z)$ 的极点是满足

$$2\pi z = \mathrm{i}(\pi + 2k\pi)$$

的 z,也就是

$$z = \frac{\mathrm{i}}{2} + \mathrm{i}k$$

它们都是虚轴上的点. 在 γ 的内部, $f(z)$ 是全纯的, 所以

$$\int_\gamma f(z)\,\mathrm{d}z = 0$$

现在, 这个积分是五个积分(对应于回路的五个部分)的和.

在 OA 上, $z = x$ (实的), 所以

$$\int_{OA} = \int_0^R \frac{\mathrm{e}^{(\pi+\mathrm{i}\omega)x}}{\mathrm{e}^{2\pi x}+1}\mathrm{d}x$$

当 $R \to \infty$ 时, 积分 $\int_{OA}$ 有极限

$$\frac{1}{2}\int_0^\infty \frac{\mathrm{e}^{\mathrm{i}\omega x}}{\cosh \pi x}\mathrm{d}x$$

在 AB 上, 设 $z = R + \mathrm{i}y$. 这时

$$\int_{AB} = \mathrm{i}\int_0^{\frac{1}{2}} \frac{\mathrm{e}^{\pi R-\omega y}\mathrm{e}^{\mathrm{i}(\omega R+\pi y)}}{\mathrm{e}^{2\pi R}\mathrm{e}^{2\mathrm{i}\pi y}+1}\mathrm{d}y$$

我们有

$$\left|\int_{AB}\right| < \int_0^{\frac{1}{2}} \frac{\mathrm{e}^{\pi R-\omega y}}{|\,\mathrm{e}^{2\pi R}\mathrm{e}^{2\mathrm{i}\pi y}+1\,|}\mathrm{d}y$$

y 的值是有界的. 容易验证当 $R \to \infty$ 时, 这个表达式趋向于 0.

在 BC 上, 设 $z = x + \dfrac{\mathrm{i}}{2}$, 可得到

$$\int_{BC} = -\mathrm{i}\mathrm{e}^{-\omega/2}\int_\varepsilon^R \frac{\mathrm{e}^{\pi x+\mathrm{i}\omega x}}{1-\mathrm{e}^{2\pi x}}\mathrm{d}x = +\mathrm{i}\frac{\mathrm{e}^{-\omega/2}}{2}\int_\varepsilon^R \frac{\mathrm{e}^{\mathrm{i}\omega x}}{\sinh \pi x}\mathrm{d}x$$

在 CD 上, 沿着负方向的四分之一小圆绕过简单极点 $\dfrac{\mathrm{i}}{2}$. 我们知道, 这个积分等于这个极点的留数乘以 $-2\mathrm{i}\pi/4$. 该留数等于

$$\left[\frac{\mathrm{e}^{(\pi+\mathrm{i}\omega)z}}{2\pi\mathrm{e}^{2\pi z}}\right]_{z=\mathrm{i}/2} = -\frac{\mathrm{i}}{2\pi}\mathrm{e}^{-\omega/2}$$

因此

$$\int_{CD} = -\frac{1}{4}\mathrm{e}^{-\omega/2}$$

在 DO 上, 设 $z = \mathrm{i}y$. 我们有

$$\int_{DO} = -\mathrm{i}\int_0^{(1/2)-\varepsilon} \frac{\mathrm{e}^{(\pi+\mathrm{i}\omega)\mathrm{i}y}}{\mathrm{e}^{2\mathrm{i}\pi y}+1}\mathrm{d}y = -\frac{\mathrm{i}}{2}\int_0^{(1/2)-\varepsilon} \frac{\mathrm{e}^{-\omega y}}{\cos \pi y}\mathrm{d}y$$

注意, 当 $R \to \infty$ 时, 这些积分的每一个都可取极限. 另外, 当 $\varepsilon = 0$

时,包含 ε 的积分是发散的;当 $\varepsilon \to 0$ 时,只能整体地取极限

$$\frac{1}{2}\int_0^\infty \frac{e^{i\omega x}}{\cosh \pi x}\mathrm{d}x - \frac{1}{4}e^{-\omega/2} + \lim_{\varepsilon \to 0}\left[\frac{i}{2}e^{-\omega/2}\int_\varepsilon^\infty \frac{e^{i\omega x}}{\sinh \pi x}\mathrm{d}x - \right.$$

$$\left.\frac{i}{2}\int_0^{(1/2)-\varepsilon} \frac{e^{-\omega y}}{\cos \pi y}\mathrm{d}y\right] = 0$$

取这个方程的实部,并设

$$A = \int_0^\infty \frac{\cos \omega x}{\cosh \pi x}\mathrm{d}x, B = \int_0^\infty \frac{\sin \omega x}{\sinh \pi x}\mathrm{d}x$$

可得

$$\frac{1}{2}A - \frac{1}{4}e^{-\omega/2} - \lim_{\varepsilon \to 0}\frac{1}{2}e^{-\omega/2}\int_\varepsilon^\infty \frac{\sin \omega x}{\sinh \pi x}\mathrm{d}x = 0$$

由这个公式可以推出,当 $\varepsilon \to 0$ 时,$\int_\varepsilon^\infty (\sin \omega x/\sinh \pi x)\mathrm{d}x$ 有极限,其极限等于 B. 这个积分的收敛性是容易验证的. 特别地,我们有

$$A = \left(B + \frac{1}{2}\right)e^{-\omega/2}$$

用 $-\omega$ 代替 ω,积分 A 保持不变,积分 B 变成 $-B$,有

$$A = \left(-B + \frac{1}{2}\right)e^{\omega/2}$$

解关于 A, B 的线性方程组,得到

$$A = \frac{1}{2\cosh(\omega/2)}$$

$$B = \frac{\sinh(\omega/2)}{2\cosh(\omega/2)}$$

本书的另一个特点是专门介绍了变分法. 有关变分法的简介[①]如下:

变分问题是求解能使某一个量的积分所表示的目标函数达到极小值的未知函数问题. 可以说明求解这样的问题与求解与之对应的 Euler-Lagrange 微分方程是等价的. 变分方法就是把变分问题转化为 Euler-Lagrange 微分方程来求解的一种方法,为此我们先引入几个概念.

[①] 摘自《数学建模 —— 方法导引与案例分析》,方道元,韦明俊编著,浙江大学出版社,2011.

(1) 泛函.

设 S 为一函数集合, 若对于每一个函数 $x(t) \in S$ 有一个实数 J 与之对应, 则称 J 是定义在 S 上的泛函, 记作 $J(x(t))$, 并称 S 为 J 的容许函数集.

例如对于 xOy 平面上过定点 $A(x_1, y_1)$ 和 $B(x_2, y_2)$ 的每一条光滑曲线 $y(x)$, 绕 x 轴旋转得一旋转体, 如图 2 所示.

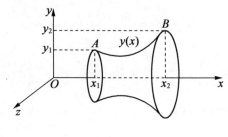

图 2

旋转体的侧面积是曲线 $y(x)$ 的泛函 $J(y(x))$. 由微积分知识不难写出

$$J(y(x)) = \int_{x_1}^{x_2} 2\pi y(x) \sqrt{1 + y'^2(x)} \, \mathrm{d}x \tag{1}$$

容许函数集可表示为

$$\begin{aligned} S = \{ y(x) \mid y(x) &\in C^1[x_1, x_2], \\ y(x_1) &= y_1, y(x_2) = y_2 \} \end{aligned} \tag{2}$$

变分问题的泛函一般可表示为

$$J(x(t)) = \int_{t_1}^{t_2} F(t, x, \dot{x}) \, \mathrm{d}t \tag{3}$$

被积函数 F 包含自变量 t, 未知函数 x 及导数 $\dot{x}$.

(2) 泛函的极值.

泛函 $J(x(t))$ 在 $x_0(t) \in S$ 取得极小值是指, 对于任意一个与 $x_0(t)$ 接近的 $x(t) \in S$, 都有 $J(x(t)) \geqslant J(x_0(t))$. 所谓接近可以用 $\mathrm{d}(x(t), x_0(t)) < \varepsilon$ 来度量, 而距离定义为

$$\mathrm{d}(x(t), x_0(t)) = \max_{t_1 \leqslant t \leqslant t_2} \{ |x(t) - x_0(t)|, |\dot{x}(t) - \dot{x}_0(t)| \}$$

泛函的极大值可以类似地定义. $x_0(t)$ 称为泛函的极值函数或极值曲线.

(3) 泛函的变分.

如同函数的微分是增量的线性主部一样,泛函的变分是泛函增量的线性主部.作为泛函的自变量,函数 $x(t)$ 在 $x_0(t)$ 的增量记作

$$\delta x(t) = x(t) - x_0(t) \tag{4}$$

也称函数的变分.由它引起的泛函的增量记作

$$\Delta J = J(x_0(t) + \delta x(t)) - J(x_0(t))$$

如果 ΔJ 可以表示为

$$\Delta J = L(x_0(t), \delta x(t)) + r(x_0(t), \delta x(t))$$

其中 L 是 δx 的线性项,而 r 是 δx 的高阶项,则 L 称为泛函在 $x_0(t)$ 的变分,记作 $\delta J(x_0(t))$.用变动的 $x(t)$ 代替 $x_0(t)$,就有 $\delta J(x(t))$.

为了求一个变分问题如式(3)的极值,可按如下步骤进行:

对于泛函 $J(x(t))$ 存在变分,在 $x = x_0(t)$ 处为极大或极小时,在此 $x_0(t)$ 处有

$$\delta J = 0 \tag{5}$$

当 $x_0(t), \delta_x$ 不变化时, $J(x_0 + \alpha\delta_x)$ 是 α 的函数.将此写成 $\varphi(\alpha)$ 时,则 $\varphi(\alpha)$ 在 $\alpha = 0$ 处为极大或极小,所以

$$\frac{\partial}{\partial\alpha}\varphi(\alpha)\big|_{\alpha=0} = 0$$

即得

$$\frac{\partial}{\partial\alpha}J(x_0(t) + \alpha\delta x(t))\big|_{\alpha=0} = 0 \tag{6}$$

式(5)是求极值的必要条件.对于含有辅助变量 α 的函数 $x(t,\alpha)$,在 $\alpha = 0$ 时与 $x_0(t)$ 一致,在 $\alpha = 1$ 时与 $x_0(t) + \delta x$ 一致,则 $J(x(t,\alpha))$ 仍是 α 的函数.设它为 $\psi(\alpha)$ 时,则 $\psi(\alpha)$ 在 $\alpha = 0$ 处为极大或极小,所以

$$\frac{\partial}{\partial\alpha}\psi(\alpha)\big|_{\alpha=0} = 0$$

即

$$\frac{\partial}{\partial\alpha}J(x(t,\alpha))\big|_{\alpha=0} = 0 \tag{7}$$

式(7)是比式(6)更为一般的条件.为判断极值我们需要计算相当于函数为二阶可微的泛函的第二变分.

设

$$J(x + \delta x) - J(x) = L(x,\delta x) + \frac{1}{2}B(x,\delta x) + \tau(x,\delta x)(\max|\delta x|)^2 \tag{8}$$

则第二变分按如下定义

$$\delta^2 J = B(x, \delta x) \tag{9}$$

而 B 对于 $\delta x, \delta \dot{x}$ 二次项为线性部分, $\tau(x, \delta x)$ 在 $\delta x \to 0$ 时为收敛于 0 的部分. 为了与第二变分有所区别也称 δJ 为第一变分.

为计算泛函的第二变分, 将 $J(x + \alpha \delta x)$ 对 α 偏微两次, 令 $\alpha = 0$ 即可. 由式(8) 得

$$\Delta J = J(x + \delta \alpha x) - J(x) = \alpha L(x, \delta x) + \frac{1}{2}\alpha^2 B(x, \delta x) +$$
$$\tau(x, \delta \alpha x) \mid \alpha \mid^2 (\max \mid \delta x \mid)^2$$

因此, 得

$$\frac{\partial^2}{\partial \alpha^2} J(x + \delta \alpha x) = B(x, \delta x) + \tau(x, \alpha \delta x)(\max \mid \delta x \mid)^2$$

这里, 设 $\alpha \to 0$ 时第二项收敛于 0, 所以

$$\delta^2 J = \frac{\partial^2}{\partial \alpha^2} J(x + \delta \alpha x) \mid_{\alpha = 0} \tag{10}$$

这样在变分情况下, 对于满足 $\delta J = 0$ 的函数 $x(t)$, $\delta^2 J > 0$ 时泛函为极小, $\delta^2 J < 0$ 时泛函为极大.

其实在《数学分析》教程中渗透进变分法在我国早有传统. 如早在 1978 年 10 月武汉大学数学系编的《数学分析》(下册)(第 1 版) 中就已经有了先例, 比如在最后一节条件极值中就有介绍:

像多元函数的条件极值问题一样, 变分法中也有条件极值问题.
设有泛函

$$J[y(x)] = \int_a^b F(x, y, y') \mathrm{d}x \tag{1}$$

其中 F 和未知函数 $y = y(x)$ 都有二阶连续(偏) 导数, 且限定 $y(a) = \alpha$, $y(b) = \beta$, 并设 $y = y(x)$ 还要满足一个限制条件

$$\int_a^b G(x, y, y') \mathrm{d}x = 0 \tag{2}$$

这里 G 是一个已知函数, 具有二阶连续偏导数, 泛函(1) 在条件(2) 限制下的极值问题就是变分法中的一种条件极值问题.

与解决多元函数的条件极值问题一样, 也可以考虑用 Lagrange 乘数

法,因为有这样的定理:

设泛函(1)在条件(2)下当 $y = y_0(x)$ 时有极值(当然还有 $y(a) = \alpha, y(b) = \beta$ 以及该函数的二阶偏导数连续的假定),则必存在常数 λ,使

$$F^*(y(x)) = \int_a^b (F + \lambda G) \, \mathrm{d}x$$

当 $y = y_0(x)$ 时有极值,即满足 Euler 方程

$$(F_y' + \lambda G_y') - \frac{\mathrm{d}}{\mathrm{d}x}(F_{y'}' + \lambda G_{y'}') = 0 \qquad (3)$$

当然 $y = y_0(x)$ 还要满足条件(2),对 G 也要作这样的假定,即 $G_y' - \frac{\mathrm{d}}{\mathrm{d}x} G_{y'}' \neq 0$,否则式(3)中就不含 G 了.

证明看过本书自会明了,只是过程稍长. 不过这里要感叹的是当年(1978年)《数学分析》可是一次就印了 315 000 册,而今天我们只能印区区 1 000 册,不知道这时代是越变越好了呢,还是相反?

海明威说:"这世界是美好的,值得我们为之奋斗! —— 我相信后半句."

我也是!

刘培杰

2023 年 2 月 26 日

于哈工大

刘培杰数学工作室
已出版(即将出版)图书目录——原版影印

书　名	出版时间	定　价	编号
数学物理大百科全书.第1卷(英文)	2016—01	418.00	508
数学物理大百科全书.第2卷(英文)	2016—01	408.00	509
数学物理大百科全书.第3卷(英文)	2016—01	396.00	510
数学物理大百科全书.第4卷(英文)	2016—01	408.00	511
数学物理大百科全书.第5卷(英文)	2016—01	368.00	512
zeta函数,q-zeta函数,相伴级数与积分(英文)	2015—08	88.00	513
微分形式:理论与练习(英文)	2015—08	58.00	514
离散与微分包含的逼近和优化(英文)	2015—08	58.00	515
艾伦·图灵:他的工作与影响(英文)	2016—01	98.00	560
测度理论概率导论,第2版(英文)	2016—01	88.00	561
带有潜在故障恢复系统的半马尔柯夫模型控制(英文)	2016—01	98.00	562
数学分析原理(英文)	2016—01	88.00	563
随机偏微分方程的有效动力学(英文)	2016—01	88.00	564
图的谱半径(英文)	2016—01	58.00	565
量子机器学习中数据挖掘的量子计算方法(英文)	2016—01	98.00	566
量子物理的非常规方法(英文)	2016—01	118.00	567
运输过程的统一非局部理论:广义波尔兹曼物理动力学,第2版(英文)	2016—01	198.00	568
量子力学与经典力学之间的联系在原子、分子及电动力学系统建模中的应用(英文)	2016—01	58.00	569
算术域(英文)	2018—01	158.00	821
高等数学竞赛:1962—1991年的米洛克斯·史怀哲竞赛(英文)	2018—01	128.00	822
用数学奥林匹克精神解决数论问题(英文)	2018—01	108.00	823
代数几何(德文)	2018—04	68.00	824
丢番图逼近论(英文)	2018—01	78.00	825
代数几何学基础教程(英文)	2018—01	98.00	826
解析数论入门课程(英文)	2018—01	78.00	827
数论中的丢番图问题(英文)	2018—01	78.00	829
数论(梦幻之旅):第五届中日数论研讨会演讲集(英文)	2018—01	68.00	830
数论新应用(英文)	2018—01	68.00	831
数论(英文)	2018—01	78.00	832

刘培杰数学工作室
已出版(即将出版)图书目录——原版影印

书　名	出版时间	定　价	编号
湍流十讲(英文)	2018—04	108.00	886
无穷维李代数:第3版(英文)	2018—04	98.00	887
等值、不变量和对称性(英文)	2018—04	78.00	888
解析数论(英文)	2018—09	78.00	889
《数学原理》的演化:伯特兰·罗素撰写第二版时的手稿与笔记(英文)	2018—04	108.00	890
哈密尔顿数学论文集(第4卷):几何学、分析学、天文学、概率和有限差分等(英文)	2019—05	108.00	891
偏微分方程全局吸引子的特性(英文)	2018—09	108.00	979
整函数与下调和函数(英文)	2018—09	118.00	980
幂等分析(英文)	2018—09	118.00	981
李群,离散子群与不变量理论(英文)	2018—09	108.00	982
动力系统与统计力学(英文)	2018—09	118.00	983
表示论与动力系统(英文)	2018—09	118.00	984
分析学练习.第1部分(英文)	2021—01	88.00	1247
分析学练习.第2部分,非线性分析(英文)	2021—01	88.00	1248
初级统计学:循序渐进的方法:第10版(英文)	2019—05	68.00	1067
工程师与科学家微分方程用书:第4版(英文)	2019—07	58.00	1068
大学代数与三角学(英文)	2019—06	78.00	1069
培养数学能力的途径(英文)	2019—07	38.00	1070
工程师与科学家统计学:第4版(英文)	2019—06	58.00	1071
贸易与经济中的应用统计学:第6版(英文)	2019—06	58.00	1072
傅立叶级数和边值问题:第8版(英文)	2019—05	48.00	1073
通往天文学的途径:第5版(英文)	2019—05	58.00	1074
拉马努金笔记.第1卷(英文)	2019—06	165.00	1078
拉马努金笔记.第2卷(英文)	2019—06	165.00	1079
拉马努金笔记.第3卷(英文)	2019—06	165.00	1080
拉马努金笔记.第4卷(英文)	2019—06	165.00	1081
拉马努金笔记.第5卷(英文)	2019—06	165.00	1082
拉马努金遗失笔记.第1卷(英文)	2019—06	109.00	1083
拉马努金遗失笔记.第2卷(英文)	2019—06	109.00	1084
拉马努金遗失笔记.第3卷(英文)	2019—06	109.00	1085
拉马努金遗失笔记.第4卷(英文)	2019—06	109.00	1086
数论:1976年纽约洛克菲勒大学数论会议记录(英文)	2020—06	68.00	1145
数论:卡本代尔1979:1979年在南伊利诺伊卡本代尔大学举行的数论会议记录(英文)	2020—06	78.00	1146
数论:诺德韦克豪特1983:1983年在诺德韦克豪特举行的Journees Arithmetiques数论大会会议记录(英文)	2020—06	68.00	1147
数论:1985—1988年在纽约城市大学研究生院和大学中心举办的研讨会(英文)	2020—06	68.00	1148

刘培杰数学工作室
已出版（即将出版）图书目录——原版影印

书　名	出版时间	定　价	编号
数论:1987 年在乌尔姆举行的 Journees Arithmetiques 数论大会会议记录(英文)	2020—06	68.00	1149
数论:马德拉斯 1987:1987 年在马德拉斯安娜大学举行的国际拉马努金百年纪念大会会议记录(英文)	2020—06	68.00	1150
解析数论:1988 年在东京举行的日法研讨会会议记录(英文)	2020—06	68.00	1151
解析数论:2002 年在意大利切特拉罗举行的 C. I. M. E. 暑期班演讲集(英文)	2020—06	68.00	1152
量子世界中的蝴蝶:最迷人的量子分形故事(英文)	2020—06	118.00	1157
走进量子力学(英文)	2020—06	118.00	1158
计算物理学概论(英文)	2020—06	48.00	1159
物质,空间和时间的理论:量子理论(英文)	2020—10	48.00	1160
物质,空间和时间的理论:经典理论(英文)	2020—10	48.00	1161
量子场理论:解释世界的神秘背景(英文)	2020—07	38.00	1162
计算物理学概论(英文)	2020—06	48.00	1163
行星状星云	2020—10	38.00	1164
基本宇宙学:从亚里士多德的宇宙到大爆炸(英文)	2020—08	58.00	1165
数学磁流体力学(英文)	2020—07	58.00	1166
计算科学:第 1 卷,计算的科学(日文)	2020—07	88.00	1167
计算科学:第 2 卷,计算与宇宙(日文)	2020—07	88.00	1168
计算科学:第 3 卷,计算与物质(日文)	2020—07	88.00	1169
计算科学:第 4 卷,计算与生命(日文)	2020—07	88.00	1170
计算科学:第 5 卷,计算与地球环境(日文)	2020—07	88.00	1171
计算科学:第 6 卷,计算与社会(日文)	2020—07	88.00	1172
计算科学.别卷,超级计算机(日文)	2020—07	88.00	1173
多复变函数论(日文)	2022—06	78.00	1518
复变函数入门(日文)	2022—06	78.00	1523
代数与数论:综合方法(英文)	2020—10	78.00	1185
复分析:现代函数理论第一课(英文)	2020—07	58.00	1186
斐波那契数列和卡特兰数:导论(英文)	2020—10	68.00	1187
组合推理:计数艺术介绍(英文)	2020—07	88.00	1188
二次互反律的傅里叶分析证明(英文)	2020—07	48.00	1189
旋瓦兹分布的希尔伯特变换与应用(英文)	2020—07	58.00	1190
泛函分析:巴拿赫空间理论入门(英文)	2020—07	48.00	1191
卡塔兰数入门(英文)	2019—05	68.00	1060
测度与积分(英文)	2019—04	68.00	1059
组合学手册.第一卷(英文)	2020—06	128.00	1153
* 一代数、局部紧群和巴拿赫 * 一代数丛的表示.第一卷,群和代数的基本表示理论(英文)	2020—05	148.00	1154
电磁理论(英文)	2020—08	48.00	1193
连续介质力学中的非线性问题(英文)	2020—09	78.00	1195
多变量数学入门(英文)	2021—05	68.00	1317
偏微分方程入门(英文)	2021—05	88.00	1318
若尔当典范性:理论与实践(英文)	2021—07	68.00	1366
伽罗瓦理论.第 4 版(英文)	2021—08	88.00	1408
R 统计学概论	2023—03	88.00	1614
基于不确定静态和动态问题解的仿射算术(英文)	2023—03	38.00	1618

书　名	出版时间	定　价	编号
典型群,错排与素数(英文)	2020—11	58.00	1204
李代数的表示:通过gln进行介绍(英文)	2020—10	38.00	1205
实分析演讲集(英文)	2020—10	38.00	1206
现代分析及其应用的课程(英文)	2020—10	58.00	1207
运动中的抛射物数学(英文)	2020—10	38.00	1208
2—纽结与它们的群(英文)	2020—10	38.00	1209
概率,策略和选择:博弈与选举中的数学(英文)	2020—11	58.00	1210
分析学引论(英文)	2020—11	58.00	1211
量子群:通往流代数的路径(英文)	2020—11	38.00	1212
集合论入门(英文)	2020—10	48.00	1213
酉反射群(英文)	2020—11	58.00	1214
探索数学:吸引人的证明方式(英文)	2020—11	58.00	1215
微分拓扑短期课程(英文)	2020—10	48.00	1216
抽象凸分析(英文)	2020—11	68.00	1222
费马大定理笔记(英文)	2021—03	48.00	1223
高斯与雅可比和(英文)	2021—03	78.00	1224
π与算术几何平均:关于解析数论和计算复杂性的研究(英文)	2021—01	58.00	1225
复分析入门(英文)	2021—03	48.00	1226
爱德华·卢卡斯与素性测定(英文)	2021—03	78.00	1227
通往凸分析及其应用的简单路径(英文)	2021—01	68.00	1229
微分几何的各个方面.第一卷(英文)	2021—01	58.00	1230
微分几何的各个方面.第二卷(英文)	2020—12	58.00	1231
微分几何的各个方面.第三卷(英文)	2020—12	58.00	1232
沃克流形几何学(英文)	2020—11	58.00	1233
彷射和韦尔几何应用(英文)	2020—12	58.00	1234
双曲几何学的旋转向量空间方法(英文)	2021—02	58.00	1235
积分:分析学的关键(英文)	2020—12	48.00	1236
为有天分的新生准备的分析学基础教材(英文)	2020—11	48.00	1237
数学不等式.第一卷.对称多项式不等式(英文)	2021—03	108.00	1273
数学不等式.第二卷.对称有理不等式与对称无理不等式(英文)	2021—03	108.00	1274
数学不等式.第三卷.循环不等式与非循环不等式(英文)	2021—03	108.00	1275
数学不等式.第四卷.Jensen不等式的扩展与加细(英文)	2021—03	108.00	1276
数学不等式.第五卷.创建不等式与解不等式的其他方法(英文)	2021—04	108.00	1277

刘培杰数学工作室
已出版(即将出版)图书目录——原版影印

书　名	出版时间	定　价	编号
冯·诺依曼代数中的谱位移函数:半有限冯·诺依曼代数中的谱位移函数与谱流(英文)	2021—06	98.00	1308
链接结构:关于嵌入完全图的直线中链接单形的组合结构(英文)	2021—05	58.00	1309
代数几何方法.第1卷(英文)	2021—06	68.00	1310
代数几何方法.第2卷(英文)	2021—06	68.00	1311
代数几何方法.第3卷(英文)	2021—06	58.00	1312
代数、生物信息和机器人技术的算法问题.第四卷,独立恒等式系统(俄文)	2020—08	118.00	1199
代数、生物信息和机器人技术的算法问题.第五卷,相对覆盖性和独立可拆分恒等式系统(俄文)	2020—08	118.00	1200
代数、生物信息和机器人技术的算法问题.第六卷,恒等式和准恒等式的相等 问题、可推导性和可实现性(俄文)	2020—08	128.00	1201
分数阶微积分的应用:非局部动态过程,分数阶导热系数(俄文)	2021—01	68.00	1241
泛函分析问题与练习:第2版(俄文)	2021—01	98.00	1242
集合论、数学逻辑和算法论问题:第5版(俄文)	2021—01	98.00	1243
微分几何和拓扑短期课程(俄文)	2021—01	98.00	1244
素数规律(俄文)	2021—01	88.00	1245
无穷边值问题解的递减:无界域中的拟线性椭圆和抛物方程(俄文)	2021—01	48.00	1246
微分几何讲义(俄文)	2020—12	98.00	1253
二次型和矩阵(俄文)	2021—01	98.00	1255
积分和级数.第2卷,特殊函数(俄文)	2021—01	168.00	1258
积分和级数.第3卷,特殊函数补充:第2版(俄文)	2021—01	178.00	1264
几何图上的微分方程(俄文)	2021—01	138.00	1259
数论教程:第2版(俄文)	2021—01	98.00	1260
非阿基米德分析及其应用(俄文)	2021—03	98.00	1261
古典群和量子群的压缩(俄文)	2021—03	98.00	1263
数学分析习题集.第3卷,多元函数:第3版(俄文)	2021—03	98.00	1266
数学习题:乌拉尔国立大学数学力学系大学生奥林匹克(俄文)	2021—03	98.00	1267
柯西定理和微分方程的特解(俄文)	2021—03	98.00	1268
组合极值问题及其应用:第3版(俄文)	2021—03	98.00	1269
数学词典(俄文)	2021—01	98.00	1271
确定性混沌分析模型(俄文)	2021—06	168.00	1307
精选初等数学习题和定理.立体几何.第3版(俄文)	2021—03	68.00	1316
微分几何习题:第3版(俄文)	2021—05	98.00	1336
精选初等数学习题和定理.平面几何.第4版(俄文)	2021—05	68.00	1335
曲面理论在欧氏空间 E_n 中的直接表示(俄文)	2022—01	68.00	1444
维纳—霍普夫离散算子和托普利兹算子:某些可数赋范空间中的诺特性和可逆性(俄文)	2022—03	108.00	1496
Maple中的数论:数论中的计算机计算(俄文)	2022—03	88.00	1497
贝尔曼和克努特问题及其概括:加法运算的复杂性(俄文)	2022—03	138.00	1498

刘培杰数学工作室
已出版(即将出版)图书目录——原版影印

书　名	出版时间	定价	编号
复分析:共形映射(俄文)	2022—07	48.00	1542
微积分代数样条和多项式及其在数值方法中的应用(俄文)	2022—08	128.00	1543
蒙特卡罗方法中的随机过程和场模型:算法和应用(俄文)	2022—08	88.00	1544
线性椭圆型方程组:论二阶椭圆型方程的迪利克雷问题(俄文)	2022—08	98.00	1561
动态系统解的增长特性:估值、稳定性、应用(俄文)	2022—08	118.00	1565
群的自由积分解:建立和应用(俄文)	2022—08	78.00	1570
混合方程和偏差自变数方程问题:解的存在和唯一性(俄文)	2023—01	78.00	1582
拟度量空间分析:存在和逼近定理(俄文)	2023—01	108.00	1583
二维和三维流形上函数的拓扑性质:函数的拓扑分类(俄文)	2023—03	68.00	1584
齐次马尔科夫过程建模的矩阵方法:此类方法能够用于不同目的的的复杂系统研究、设计和完善(俄文)	2023—03	68.00	1594
周期函数的近似方法和特性:特殊课程(俄文)	2023—04	158.00	1622
扩散方程解的矩函数:变分法(俄文)	2023—03	58.00	1623
狭义相对论与广义相对论:时空与引力导论(英文)	2021—07	88.00	1319
束流物理学和粒子加速器的实践介绍:第2版(英文)	2021—07	88.00	1320
凝聚态物理中的拓扑和微分几何简介(英文)	2021—05	88.00	1321
混沌映射:动力学,分形学和快速涨落(英文)	2021—05	128.00	1322
广义相对论:黑洞、引力波和宇宙学介绍(英文)	2021—06	68.00	1323
现代分析电磁均质化(英文)	2021—06	68.00	1324
为科学家提供的基本流体动力学(英文)	2021—06	88.00	1325
视觉天文学:理解夜空的指南(英文)	2021—06	68.00	1326
物理学中的计算方法(英文)	2021—06	68.00	1327
单星的结构与演化:导论(英文)	2021—06	108.00	1328
超越居里:1903年至1963年物理界四位女性及其著名发现(英文)	2021—06	68.00	1329
范德瓦尔斯流体热力学的进展(英文)	2021—06	68.00	1330
先进的托卡马克稳定性理论(英文)	2021—06	88.00	1331
经典场论导论:基本相互作用的过程(英文)	2021—07	88.00	1332
光致电离量子动力学方法原理(英文)	2021—07	108.00	1333
经典域论和应力:能量张量(英文)	2021—05	88.00	1334
非线性太赫兹光谱的概念与应用(英文)	2021—06	68.00	1337
电磁学中的无穷空间并矢格林函数(英文)	2021—06	88.00	1338
物理科学基础数学.第1卷,齐次边值问题、傅里叶方法和特殊函数(英文)	2021—07	108.00	1339
离散量子力学(英文)	2021—07	68.00	1340
核磁共振的物理学和数学(英文)	2021—07	108.00	1341
分子水平的静电学(英文)	2021—08	68.00	1342
非线性波:理论、计算机模拟、实验(英文)	2021—06	108.00	1343
石墨烯光学:经典问题的电解决方案(英文)	2021—06	68.00	1344
超材料多元宇宙(英文)	2021—07	68.00	1345
银河系外的天体物理学(英文)	2021—07	68.00	1346
原子物理学(英文)	2021—07	68.00	1347
将光打结:将拓扑学应用于光学(英文)	2021—07	68.00	1348
电磁学:问题与解法(英文)	2021—07	88.00	1364
海浪的原理:介绍量子力学的技巧与应用(英文)	2021—07	108.00	1365
多孔介质中的流体:输运与相变(英文)	2021—07	68.00	1372
洛伦兹群的物理学(英文)	2021—08	68.00	1373
物理导论的数学方法和解决方法手册(英文)	2021—08	68.00	1374

刘培杰数学工作室
已出版(即将出版)图书目录——原版影印

书　名	出版时间	定价	编号
非线性波数学物理学入门(英文)	2021—08	88.00	1376
波:基本原理和动力学(英文)	2021—07	68.00	1377
光电子量子计量学.第1卷,基础(英文)	2021—07	88.00	1383
光电子量子计量学.第2卷,应用与进展(英文)	2021—07	68.00	1384
复杂流的格子玻尔兹曼建模的工程应用(英文)	2021—08	68.00	1393
电偶极矩挑战(英文)	2021—08	108.00	1394
电动力学:问题与解法(英文)	2021—09	68.00	1395
自由电子激光的经典理论(英文)	2021—08	68.00	1397
曼哈顿计划——核武器物理学简介(英文)	2021—09	68.00	1401
粒子物理学(英文)	2021—09	68.00	1402
引力场中的量子信息(英文)	2021—09	128.00	1403
器件物理学的基本经典力学(英文)	2021—09	68.00	1404
等离子体物理及其空间应用导论.第1卷,基本原理和初步过程(英文)	2021—09	68.00	1405
磁约束聚变等离子体物理:理想MHD理论(英文)	2023—03	68.00	1613
相对论量子场论.第1卷,典范形式体系(英文)	2023—03	38.00	1615
涌现的物理学(英文)	2023—05	58.00	1619
量子化旋涡:一本拓扑激发手册(英文)	2023—04	68.00	1620
非线性动力学:实践的介绍性调查(英文)	2023—05	68.00	1621
拓扑与超弦理论焦点问题(英文)	2021—07	58.00	1349
应用数学:理论、方法与实践(英文)	2021—07	78.00	1350
非线性特征值问题:牛顿型方法与非线性瑞利函数(英文)	2021—07	58.00	1351
广义膨胀和齐性:利用齐性构造齐次系统的李雅普诺夫函数和控制律(英文)	2021—06	48.00	1352
解析数论焦点问题(英文)	2021—07	58.00	1353
随机微分方程:动态系统方法(英文)	2021—07	58.00	1354
经典力学与微分几何(英文)	2021—07	58.00	1355
负定相交形式流形上的瞬子模空间几何(英文)	2021—07	68.00	1356
广义卡塔兰轨道分析:广义卡塔兰轨道计算数字的方法(英文)	2021—07	48.00	1367
洛伦兹方法的变分:二维与三维洛伦兹方法(英文)	2021—08	38.00	1378
几何、分析和数论精编(英文)	2021—08	68.00	1380
从一个新角度看数论:通过遗传方法引入现实的概念(英文)	2021—08	58.00	1387
动力系统:短期课程(英文)	2021—08	68.00	1382
几何路径:理论与实践(英文)	2021—08	48.00	1385
论天体力学中某些问题的不可积性(英文)	2021—07	88.00	1396
广义斐波那契数列及其性质(英文)	2021—08	38.00	1386
对称函数和麦克唐纳多项式:余代数结构与Kawanaka恒等式(英文)	2021—09	38.00	1400
杰弗里·英格拉姆·泰勒科学论文集:第1卷.固体力学(英文)	2021—05	78.00	1360
杰弗里·英格拉姆·泰勒科学论文集:第2卷.气象学、海洋学和湍流(英文)	2021—05	68.00	1361
杰弗里·英格拉姆·泰勒科学论文集:第3卷.空气动力学以及落弹数和爆炸的力学(英文)	2021—05	68.00	1362
杰弗里·英格拉姆·泰勒科学论文集:第4卷.有关流体力学(英文)	2021—05	58.00	1363

刘培杰数学工作室
已出版(即将出版)图书目录——原版影印

书　名	出版时间	定　价	编号
非局域泛函演化方程:积分与分数阶(英文)	2021—08	48.00	1390
理论工作者的高等微分几何:纤维丛、射流流形和拉格朗日理论(英文)	2021—08	68.00	1391
半线性退化椭圆微分方程:局部定理与整体定理(英文)	2021—07	48.00	1392
非交换几何、规范理论和重整化:一般简介与非交换量子场论的重整化(英文)	2021—09	78.00	1406
数论论文集:拉普拉斯变换和带有数论系数的幂级数(俄文)	2021—09	48.00	1407
挠理论专题:相对极大值,单射与扩充模(英文)	2021—09	88.00	1410
强正则图与欧几里得若尔当代数:非通常关系中的启示(英文)	2021—10	48.00	1411
拉格朗日几何和哈密顿几何:力学的应用(英文)	2021—10	48.00	1412

书　名	出版时间	定　价	编号
时滞微分方程与差分方程的振动理论:二阶与三阶(英文)	2021—10	98.00	1417
卷积结构与几何函数理论:用以研究特定几何函数理论方向的分数阶微积分算子与卷积结构(英文)	2021—10	48.00	1418
经典数学物理的历史发展(英文)	2021—10	78.00	1419
扩展线性丢番图问题(英文)	2021—10	38.00	1420
一类混沌动力系统的分歧分析与控制:分歧分析与控制(英文)	2021—11	38.00	1421
伽利略空间和伪伽利略空间中一些特殊曲线的几何性质(英文)	2022—01	68.00	1422
一阶偏微分方程:哈密尔顿—雅可比理论(英文)	2021—11	48.00	1424
各向异性黎曼多面体的反问题:分段光滑的各向异性黎曼多面体反边界谱问题:唯一性(英文)	2021—11	38.00	1425

书　名	出版时间	定　价	编号
项目反应理论手册.第一卷,模型(英文)	2021—11	138.00	1431
项目反应理论手册.第二卷,统计工具(英文)	2021—11	118.00	1432
项目反应理论手册.第三卷,应用(英文)	2021—11	138.00	1433
二次无理数:经典数论入门(英文)	2022—05	138.00	1434
数,形与对称性:数论,几何和群论导论(英文)	2022—05	128.00	1435
有限域手册(英文)	2021—11	178.00	1436
计算数论(英文)	2021—11	148.00	1437
拟群与其表示简介(英文)	2021—11	88.00	1438
数论与密码学导论:第二版(英文)	2022—01	148.00	1423

刘培杰数学工作室

已出版(即将出版)图书目录——原版影印

书　　名	出版时间	定　价	编号
几何分析中的柯西变换与黎兹变换:解析调和容量和李普希兹调和容量、变化和振荡以及一致可求长性(英文)	2021－12	38.00	1465
近似不动点定理及其应用(英文)	2022－05	28.00	1466
局部域的相关内容解析:对局部域的扩展及其伽罗瓦群的研究(英文)	2022－01	38.00	1467
反问题的二进制恢复方法(英文)	2022－03	28.00	1468
对几何函数中某些类的各个方面的研究:复变量理论(英文)	2022－01	38.00	1469
覆盖、对应和非交换几何(英文)	2022－01	28.00	1470
最优控制理论中的随机线性调节器问题:随机最优线性调节器问题(英文)	2022－01	38.00	1473
正交分解法:涡流流体动力学应用的正交分解法(英文)	2022－01	38.00	1475
芬斯勒几何的某些问题(英文)	2022－03	38.00	1476
受限三体问题(英文)	2022－05	38.00	1477
利用马利亚万微积分进行 Greeks 的计算:连续过程、跳跃过程中的马利亚万微积分和金融领域中的 Greeks(英文)	2022－05	48.00	1478
经典分析和泛函分析的应用:分析学的应用(英文)	2022－03	38.00	1479
特殊芬斯勒空间的探究(英文)	2022－03	48.00	1480
某些图形的施泰纳距离的细谷多项式:细谷多项式与图的维纳指数(英文)	2022－05	38.00	1481
图论问题的遗传算法:在新鲜与模糊的环境中(英文)	2022－05	48.00	1482
多项式映射的渐近簇(英文)	2022－05	38.00	1483
一维系统中的混沌:符号动力学,映射序列,一致收敛和沙可夫斯基定理(英文)	2022－05	38.00	1509
多维边界层流动与传热分析:粘性流体流动的数学建模与分析(英文)	2022－05	38.00	1510
演绎理论物理学的原理:一种基于量子力学波函数的逐次置信估计的一般理论的提议(英文)	2022－05	38.00	1511
R^2 和 R^3 中的仿射弹性曲线:概念和方法(英文)	2022－08	38.00	1512
算术数列中除数函数的分布:基本内容、调查、方法、第二矩、新结果(英文)	2022－05	28.00	1513
抛物型狄拉克算子和薛定谔方程:不定常薛定谔方程的抛物型狄拉克算子及其应用(英文)	2022－07	28.00	1514
黎曼-希尔伯特问题与量子场论:可积重正化、戴森-施温格方程(英文)	2022－08	38.00	1515
代数结构和几何结构的形变理论(英文)	2022－08	48.00	1516
概率结构和模糊结构上的不动点:概率结构和直觉模糊度量空间的不动点定理(英文)	2022－08	38.00	1517

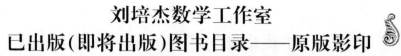

刘培杰数学工作室
已出版（即将出版）图书目录——原版影印

书 名	出版时间	定 价	编号
反若尔当对：简单反若尔当对的自同构(英文)	2022—07	28.00	1533
对某些黎曼－芬斯勒空间变换的研究：芬斯勒几何中的某些变换(英文)	2022—07	38.00	1534
内诣零流形映射的尼尔森数的阿诺索夫关系(英文)	2023—01	38.00	1535
与广义积分变换有关的分数次演算：对分数次演算的研究(英文)	2023—01	48.00	1536
强子的芬斯勒几何和吕拉几何(宇宙学方面)：强子结构的芬斯勒几何和吕拉几何(拓扑缺陷)(英文)	2022—08	38.00	1537
一种基于混沌的非线性最优化问题：作业调度问题(英文)	2023—03	38.00	1538
广义概率论发展前景：关于趣味数学与置信函数实际应用的一些原创观点(英文)	2023—03	48.00	1539
纽结与物理学：第二版(英文)	2022—09	118.00	1547
正交多项式和 q—级数的前沿(英文)	2022—09	98.00	1548
算子理论问题集(英文)	2022—09	108.00	1549
抽象代数：群、环与域的应用导论：第二版(英文)	2023—01	98.00	1550
菲尔兹奖得主演讲集：第三版(英文)	2023—01	138.00	1551
多元实函数教程(英文)	2022—09	118.00	1552
球面空间形式群的几何学：第二版(英文)	2022—09	98.00	1566
对称群的表示论(英文)	2023—01	98.00	1585
纽结理论：第二版(英文)	2023—01	88.00	1586
拟群理论的基础与应用(英文)	2023—01	88.00	1587
组合学：第二版(英文)	2023—01	98.00	1588
加性组合学：研究问题手册(英文)	2023—01	68.00	1589
扭曲、平铺与镶嵌：几何折纸中的数学方法(英文)	2023—01	98.00	1590
离散与计算几何手册：第三版(英文)	2023—01	248.00	1591
离散与组合数学手册：第二版(英文)	2023—01	248.00	1592
分析学教程．第1卷，一元实变量函数的微积分分析学介绍(英文)	2023—01	118.00	1595
分析学教程．第2卷，多元函数的微分和积分，向量微积分(英文)	2023—01	118.00	1596
分析学教程．第3卷，测度与积分理论，复变量的复值函数(英文)	2023—01	118.00	1597
分析学教程．第4卷，傅里叶分析，常微分方程，变分法(英文)	2023—01	118.00	1598

联系地址：哈尔滨市南岗区复华四道街 10 号　哈尔滨工业大学出版社刘培杰数学工作室
网　　址：http://lpj.hit.edu.cn/
邮　　编：150006
联系电话：0451—86281378　　13904613167
E-mail：lpj1378@163.com